MACHINE ETHICS AND ROBOT ETHICS

THE LIBRARY OF ESSAYS ON THE ETHICS OF EMERGING TECHNOLOGIES

Series editor: Wendell Wallach

Titles in the series

Emerging Technologies: Ethics, Law and Governance
Edited by Gary E. Marchant and Wendell Wallach

The Ethical Challenges of Emerging Medical Technologies
Edited by Arthur L. Caplan and Brendan Parent

The Ethics of Biotechnology
Edited by Gaymon Bennett

The Ethics of Nanotechnology, Geoengineering, and Clean Energy
Edited by Andrew Maynard and Jack Stilgoe

The Ethics of Sports Technologies and Human Enhancement
Edited by Thomas H. Murray and Voo Teck Chuan

The Ethics of Information Technologies
Edited by Keith W. Miller and Mariarosaria Taddeo

Machine Ethics and Robot Ethics
Edited by Wendell Wallach and Peter Asaro

The Ethics of Military Technology
Edited by Braden Allenby

Machine Ethics and Robot Ethics

WENDELL WALLACH,
Yale University, USA,
AND PETER ASARO,
New School for Public Engagement, USA

THE LIBRARY OF ESSAYS ON THE ETHICS OF EMERGING TECHNOLOGIES

LONDON AND NEW YORK

First published 2017
by Routledge
2 Park Square, Milton Park, Abingdon, Oxon OX14 4RN

and by Routledge
711 Third Avenue, New York, NY 10017

Routledge is an imprint of the Taylor & Francis Group, an informa business

Editorial material and selection © 2017 Wendell Wallach and Peter Asaro; individual owners retain copyright in their own material.

All rights reserved. No part of this book may be reprinted or reproduced or utilised in any form or by any electronic, mechanical, or other means, now known or hereafter invented, including photocopying and recording, or in any information storage or retrieval system, without permission in writing from the publishers.

Trademark notice: Product or corporate names may be trademarks or registered trademarks, and are used only for identification and explanation without intent to infringe.

British Library Cataloguing in Publication Data
A catalogue record for this book is available from the British Library

Library of Congress Cataloging in Publication Data
A catalog record for this book has been requested

ISBN: 978-1-4724-3039-7 (hbk)

Typeset in Times New Roman MT
by Servis Filmsetting Ltd, Stockport, Cheshire

Contents

Acknowledgments	ix
Series Preface	xiii

Introduction	1
Appendix 1: The Future of Life Institute: Research Priorities for Robust and Beneficial Artificial Intelligence: An Open Letter	17
Appendix 2: Research Priorities for Robust and Beneficial Artificial Intelligence	19

Part I: Laying foundations

1. Clarke, Roger. (1993). "Asimov's laws of robotics: Implications for information technology (1)." *IEEE Computer*, 26(12), 53–61   33

2. Clarke, Roger. (1994). "Asimov's laws of robotics: Implications for information technology (2)." *IEEE Computer*, 227(1), 57–66   43

3. Allen, Colin, Gary Varner, & Jason Zinser. (2000). "Prolegomena to any future artificial moral agent." *Journal of Experimental and Theoretical Artificial Intelligence*, 12, 251–261   53

4. Nissenbaum, Helen. (2001). "How computer systems embody values." *Computer*, 34(3), 118–119   65

5. Bostrom, Nick. (2003). "The ethical issues of advanced artificial intelligence." Paper presented at the IIAS 2003, Baden Baden, GE. In Smit, S., Wallach, W., and Lasker, L. (eds.) *Cognitive, Emotive and Ethical Aspects of Decision Making in Humans and in Artificial Intelligence*, Vol 11, IIAS, pp. 12–17   69

Part II: Robot ethics

6. Veruggio, Gianmarco, & Fiorella Operto. (2006). "Roboethics: A bottom-up interdisciplinary discourse in the field of applied ethics in robotics." *International Review of Information Ethics*, 6, 2–8   79

7. Asaro, Peter. (2006). "What should we want from a robot ethic?" *International Review of Information Ethics*, 6, 10–16   87

8. Sparrow, Robert. (2004). "The Turing triage test." *Ethics and Information Technology*, 6.4, 203–213   95

9   Turkle, Sherry. (2006). "A nascent robotics culture: New complicities for companionship." *American Association for Artificial Intelligence AAAI* — 107

10   Coeckelbergh, Mark. (2010). "Moral appearances: Emotions, robots, and human morality." *Ethics and Information Technology*, 12.3, 235–241 — 117

11   Borenstein, Jason, & Yvette Pearson. (2010). "Robot caregivers: Harbingers of expanded freedom for all?" *Ethics and Information Technology*, 12.3, 277–288 — 125

12   Vallor, Shannon. (2011). "Carebots and caregivers: Sustaining the ethical ideal of care in the twenty-first century." *Philosophy & Technology*, 24.3, 251–268 — 137

13   Sharkey, Noel, & Amanda Sharkey. (2010). "The crying shame of robot nannies: an ethical appraisal." *Interaction Studies*, 11.2, 161–190 — 155

14   van Wynsberghe, Aimee. (2013). "Designing robots for care: Care centered value-sensitive design." *Science and Engineering Ethics*, 19.2, 407–433 — 185

15   Sullins, John P. (2012). "Robots, love, and sex: The ethics of building a love machine." *Affective Computing, IEEE Transactions*, 3.4, 398–409 — 213

16   Malle, Bertram, & Matthias Scheutz. (2014). "Moral competence in social robots." *IEEE International Symposium on Ethics in Engineering, Science, and Technology, Chicago* — 225

Part III: Machine ethics

17   Moor, James H. (2006). "The nature, importance, and difficulty of machine ethics." *Intelligent Systems, IEEE*, 21.4, 18–21 — 233

18   Anderson, Michael, & Susan Leigh Anderson. (2007). "Machine ethics: Creating an ethical intelligent agent." *AI Magazine*, 28.4, 15–26 — 237

19   Wallach, Wendell, Colin Allen, & Iva Smit. (2008). "Machine morality: Bottom-up and top-down approaches for modelling human moral faculties." *AI & Society*, 22.4, 565–582 — 249

20   McDermott, Drew. (2008). "Why ethics is a high hurdle for AI." *North American Conference on Computing and Philosophy*. Bloomington, Indiana — 267

21   Powers, Thomas M. (2006). "Prospects for a Kantian machine." *Intelligent Systems, IEEE*, 21.4, 46–51 — 275

22 Guarini, Marcello. (2005). "Particularism and generalism: How AI can help us to better understand moral cognition." *Machine Ethics: Papers from the 2005 AAAI Fall Symposium* — 281

23 Bringsjord, Selmer, Konstantine Arkoudas, & Paul Bello. (2006). "Toward a general logicist methodology for engineering ethically correct robots." *IEEE Intelligent Systems*, 21(4), 38–44 — 291

24 Wallach, Wendell, Colin Allen, & Stan Franklin. (2011). "Consciousness and ethics: Artificially conscious moral agents." *International Journal of Machine Consciousness*, 3.01, 177–192 — 299

Part IV: Moral agents and agency

25 Floridi, Luciano, & Jeff W. Sanders. (2004). "On the morality of artificial agents." *Minds and Machines*, 14.3, 349–379 — 317

26 Johnson, Deborah G., & Keith W. Miller. (2008). "Un-making artificial moral agents." *Ethics and Information Technology*, 10.2–3, 123–133 — 349

27 Suchman, Lucy. (2007). "Agencies in technology design: Feminist reconfigurations." In Hackett, Edward J., Olga Amsterdamska, Michael E. Lynch, & Judy Wajcman (eds.) *The Handbook of Science and Technology Studies,* third edition, excerpt from pp. 139–163 — 361

28 Marino, Dante, & Guglielmo Tamburrini. (2006). "Learning robots and human responsibility." *International Review of Information Ethics*, 6, 46–51 — 377

29 Torrance, Steve. (2014). "Artificial consciousness and artificial ethics: Between realism and social relationism." *Philosophy & Technology*, 27.1, 9–29 — 383

30 Murphy, Robin R., & David D. Woods. (2009). "Beyond Asimov: The three laws of responsible robotics." *Intelligent Systems, IEEE*, 24.4, 14–20 — 405

Part V: Law and policy

31 Solum, Lawrence. (1992). "Legal personhood for artificial intelligences." *North Carolina Law Review*, 70, 1231–1287 — 415

32 Nagenborg, Michael, et al. (2008). "Ethical regulations on robotics in Europe." *Ai & Society*, 22.3, 349–366 — 473

33 Calo, M. Ryan. (2010). "Robots and privacy." *Robot Ethics: The Ethical and Social Implications of Robotics,* 187–204 — 491

34 Lin, Patrick. "The robot car of tomorrow may just be programmed to hit you." *Wired Magazine*, May 6, 2014	507
35 Gunkel, David J. (2014). "A vindication of the rights of machines." *Philosophy & Technology*, 27, 113–132	511
Index	531

Acknowledgments

The chapters in this volume are taken from the sources listed below. The editors and publishers wish to thank the authors, original publishers, or other copyright holders for permission to use their material as follows.

Chapter 1: Clarke, Roger. (1993). "Asimov's Laws of Robotics: Implications for Information Technology (1)." *IEEE Computer*, 26(12), 53–61. Permission from IEEE Computer Society.

Chapter 2: Clarke, Roger. (1994). "Asimov's Laws of Robotics: Implications for Information Technology (2)." *IEEE Computer*, 227(1), 57–66. Permission from IEEE Computer Society.

Chapter 3: Allen, Colin, Gary Varner, & Jason Zinser. (2000). "Prolegomena to Any Future Artificial Moral Agent." *Journal of Experimental and Theoretical Artificial Intelligence*, 12, 251–261. Permission from Taylor and Francis.

Chapter 4: Nissenbaum, Helen. (2001). "How Computer Systems Embody Values." *Computer*, 34(3), 118–119. Permission from IEEE Computer Society.

Chapter 5: Bostrom, Nick. (2003). "The Ethical Issues of Advanced Artificial Intelligence." Paper presented at the IIAS 2003, Baden Baden, GE. In Smit, S., Wallach, W., and Lasker, L. (eds.) *Cognitive, Emotive and Ethical Aspects of Decision Making in Humans and in Artificial Intelligence*, Vol 11, IIAS, pp. 12–17. Permission from author.

Chapter 6: Veruggio, Gianmarco, & Fiorella Operto. (2006). "Roboethics: a Bottom-up Interdisciplinary Discourse in the Field of Applied Ethics in Robotics." *International Review of Information Ethics*, 6, 2–8. Permission from the International Review of Information Ethics.

Chapter 7: Asaro, Peter. (2006). "What Should We Want from a Robot Ethic?" *International Review of Information Ethics*, 6, 10–16. Permission from the International Review of Information Ethics.

Chapter 8: Sparrow, Robert (2004). "The Turing Triage Test." *Ethics and Information Technology*, 6.4, 203–213. Permission from Springer.

Chapter 9: Turkle, Sherry (2006). "A Nascent Robotics Culture: New Complicities for Companionship." *American Association for Artificial Intelligence AAAI*. Permission from the American Association for Artificial Intelligence.

Chapter 10: Coeckelbergh, Mark (2010). "Moral Appearances: Emotions, Robots, and Human Morality." *Ethics and Information Technology*, 12.3, 235–241. Permission from Springer.

Chapter 11: Borenstein, Jason, & Yvette Pearson (2010). "Robot Caregivers: Harbingers of Expanded Freedom for All?" *Ethics and Information Technology*, 12.3, 277–288. Permission from Springer.

Chapter 12: Vallor, Shannon (2011). "Carebots and Caregivers: Sustaining the Ethical Ideal of Care in the Twenty-first Century." *Philosophy & Technology*, 24.3, 251–268. Permission from Springer.

Chapter 13: Sharkey, Noel, & Amanda Sharkey (2010). "The Crying Shame of Robot Nannies: An Ethical Appraisal." *Interaction Studies*, 11.2, 161–190. Permission from John Benjamins Publishing Company.

Chapter 14: van Wynsberghe, Aimee (2013). "Designing Robots for Care: Care Centered Value-sensitive Design." *Science and Engineering Ethics*, 19.2, 407–433. Permission from author.

Chapter 15: Sullins, John P. (2012). "Robots, Love, and Sex: The Ethics of Building a Love Machine." *Affective Computing, IEEE Transactions*, 3.4, 398–409. Permission from the Institute of Electrical and Electronics Engineers.

Chapter 16: Malle, Bertram, & Matthias Scheutz. "Moral Competence in Social Robots." *IEEE International Symposium on Ethics in Engineering, Science, and Technology, Chicago*. Permission from the Institute of Electrical and Electronics Engineers.

Chapter 17: Moor, James H. (2006). "The Nature, Importance, and Difficulty of Machine Ethics." *Intelligent Systems, IEEE*, 21.4, 18–21. Permission from the Institute of Electrical and Electronics Engineers.

Chapter 18: Anderson, Michael, & Susan Leigh Anderson (2007). "Machine Ethics: Creating an Ethical Intelligent Agent." *AI Magazine*, 28.4, 15–26. Permission from the AAAI Press.

Chapter 19: Wallach, Wendell, Colin Allen, & Iva Smit (2008). "Machine Morality: Bottom-up and Top-down Approaches for Modelling Human Moral Faculties." *Ai & Society*, 22.4, 565–582. Permission from Springer.

Chapter 20: McDermott, Drew (2000). "Why Ethics is a High Hurdle for AI." *North American Conference on Computing and Philosophy*. Bloomington, Indiana. Permission from the author.

Chapter 21: Powers, Thomas M. (2006). "Prospects for a Kantian Machine." *Intelligent Systems, IEEE*, 21.4, 46–51. Permission from the Institute of Electrical and Electronics Engineers.

Chapter 22: Guarini, Marcello (2005). "Particularism and Generalism: How AI Can Help Us to Better Understand Moral Cognition." *Machine Ethics: Papers from the 2005 AAAI Fall symposium*. Permission from the AAAI Press.

Chapter 23: Bringsjord, Selmer, Konstantine Arkoudas, & Paul Bello. "Toward a General Logicist Methodology for Engineering Ethically Correct Robots." *IEEE Intelligent Systems*, 21(4), 38–44. Permission from the Institute of Electrical and Electronics Engineers.

Chapter 24: Wallach, Wendell, Colin Allen, & Stan Franklin (2011). "Consciousness and Ethics: Artificially Conscious Moral Agents." *International Journal of Machine Consciousness*, 3.01, 177–192. Permission from the World Scientific Publishing Company.

Chapter 25: Floridi, Luciano, & Jeff W. Sanders (2004). "On the Morality of Artificial Agents." *Minds and Machines*, 14.3, 349–379. Permission from Springer.

Chapter 26: Johnson, Deborah G., & Keith W. Miller (2008). "Un-making Artificial Moral Agents." *Ethics and Information Technology*, 10.2–3, 123–133. Permission from Springer.

Chapter 27: Suchman, Lucy. "Agencies in Technology Design: Feminist Reconfigurations." In Hackett, Edward J., Olga Amsterdamska, Michael E. Lynch, and Judy Wajcman (eds.) *The Handbook of Science and Technology Studies,* third edition, excerpt from pp. 139–163, copyright 2007 Massachusetts Institute of Technology, by permission of the MIT Press.

Chapter 28: Marino, Dante, & Guglielmo Tamburrini (2006). "Learning Robots and Human Responsibility." *International Review of Information Ethics*, 6, 46–51. Permission from the International Review of Information Ethics.

Chapter 29: Torrance, Steve (2014). "Artificial Consciousness and Artificial Ethics: Between Realism and Social Relationism." *Philosophy & Technology*, 27.1, 9–29. Permission from Springer.

Chapter 30: Murphy, Robin R., & David D. Woods (2009). "Beyond Asimov: The Three Laws of Responsible Robotics." *Intelligent Systems, IEEE*, 24.4, 14–20. Permission from the Institute of Electrical and Electronics Engineers.

Chapter 31: Solum, Lawrence (1992). "Legal Personhood for Artificial Intelligences." *North Carolina Law Review*, 70, 1231–1287. Permission from the North Carolina Law Review.

Chapter 32: Nagenborg, Michael, et al. (2008). "Ethical Regulations on Robotics in Europe." *Ai & Society*, 22.3, 349–366. Permission from Springer.

Chapter 33: Calo, M. Ryan (2010). "Robots and Privacy." *Robot Ethics: The Ethical and Social Implications of Robotics*, 187–204. Permission from the MIT Press.

Chapter 34: Lin, Patrick (2014). "The Robot Car of Tomorrow May Just Be Programmed to Hit You." *Wired Magazine*, May 6, 2014. Permission from author.

Chapter 35: Gunkel, David J. (2014) "A Vindication of the Rights of Machines." *Philosophy & Technology*, 27, 113–132. Permission from Springer.

Every effort has been made to trace all the copyright holders, but if any have been inadvertently overlooked, the publishers will be pleased to make the necessary arrangement at the first opportunity.

Publisher's note

The material in this volume has been reproduced using the facsimile method. This means we can retain the original pagination to facilitate easy and correct citation of the original essays. It also explains the variety of typefaces, page layouts, and numbering.

Series Preface

Scientific discovery and technological innovation are producing, and will continue to generate, a truly broad array of tools and techniques, each of which offers benefits while posing societal and ethical challenges. These emerging technologies include (but are not limited to) information technology, genomics, biotechnology, synthetic biology, nanotechnology, personalized medicine, stem cell and regenerative medicine, neuroscience, robotics, and geoengineering. The societal and ethical issues, which arise within those fields, go beyond safety and traditional risks such as public health threats and environmental damage, to encompass privacy, fairness, security, and the acceptability of various forms of human enhancement. The "Library of essays on the ethics of emerging technologies" demonstrates the breadth of the challenges and the difficult tradeoffs entailed in reaping the benefits of technological innovation while minimizing possible harms.

The editors selected for each of the eight volumes are leaders within their respective fields. They were charged to provide a roadmap of core concerns with the help of an introductory essay and the careful selection of articles that have played or will play an important role in ongoing debates. Many of these articles can be thought of as "golden oldies," important works of scholarship that are cited time and again. Other articles selected address cutting-edge issues posed by synthetic organisms, cognitive enhancements, robotic weaponry, and additional technologies under development.

In recent years, information technologies have transformed society. In the coming decades, advances in genomics and nanotechnologies may have an even greater impact. The pathways for technological progress are uncertain as new discoveries and convergences between areas of research afford novel, and often unanticipated, opportunities. However, the determination of which technological possibilities are being realized or probable, and which are merely plausible or highly speculative, functions as a central question that cuts across many fields. This in turns informs which ethical issues being raised warrant immediate attention. Calls for precautionary measures to stave off harms from speculative possibilities can unnecessarily interfere with innovation. On the other hand, if futuristic challenges, such as smarter-than-human robotic are indeed possible, then it behooves us to invest now in means to ensure artificial intelligence will be provably beneficial, robust, safe, and controllable.

Most of the ethical concerns discussed in the volumes are less dramatic, but just as intriguing. What criteria must be met before newly created organisms can be released outside of a laboratory? Should fears about the possible toxicity of a few unidentified nanomaterials, among thousands, significantly slow the pace of development in a field that promises great rewards? Does medical research that mines large databases (big data), including the genomes of millions of people, have a downside? Are geoengineering technologies for managing climate change warranted, or more dangerous than the problem they purport to solve?

The ethical languages enlisted to evaluate an innovative technological go beyond the utilitarian analysis of costs and benefits. For example, the principles of biomedical ethics and the laws of armed conflict play a central role in judgments made about whether the healthcare industry or the military should adopt a proposed device or procedure. The differing ethical languages underscore different considerations, each of which will need to be factored into decisions regarding whether to embrace, regulate, or reject the new technology.

Scientific discovery and technological innovation proceed at an accelerating pace, while attention to the ethical, societal, and governance concerns they raise lags far behind. Scholars have been diligent in trying to bring those concerns to the fore. But, as the essays in these volumes will make clear, there is a great deal of work ahead if we, humanity as a whole, are to successfully navigate the promise and perils of emerging technologies.

Wendell Wallach

Introduction

The Emergence of Robot Ethics and Machine Ethics

Peter Asaro and Wendell Wallach

Once the stuff of science fiction, recent progress in artificial intelligence, robotics, and machine learning has raised public concern and academic interest in the safety and ethics of the new tools and techniques emerging from these fields of research. The technologies finally are coming into widespread use, and appear poised for rapid technological advancement. Some have even argued that the coming robotics revolution could rival the personal computer revolution of the 1970s and 1980s (Gates 2007), or the impact of the smartphone in the early twenty-first century.

As these technologies advance and find use in everyday life, they also raise a host of social, political, and legal issues. When artificial agents within computer systems and robots (hereafter collectively referred to as robots) start to act in the world—the physical world as well as in cyberspace—who is responsible for their actions? Should robots be used for the care of children or the homebound and the elderly? How can we ensure the safety of self-driving cars in a world of unpredictable pedestrians, playing children, dogs, bicyclists, and busy parking lots? How should we design robots to act, especially in cases with competing or conflicting goals? Can robots be designed so that they are sensitive to the values of those whom they interact with and that give form to the contexts in which they act? How do we design them to conform to our social and legal expectations? Can robots make moral decisions? Do we want them to make moral decisions? Then there are the more futuristic questions. Could computational systems be held accountable for actions they might initiate? How can we know if they have, or might eventually have, emotions or consciousness, or if they are moral agents deserving of rights and moral respect?

Speculation and science fiction have played an inordinately large, but not always welcomed, role in framing the ethical challenges posed by artificial intelligence and robotics. While some focus upon more speculative futures, as robots become reality and are introduced into the commerce of daily life, others are concerned with the more immediate practical challenges of ensuring their activity is safe and conforms to human laws and ethical considerations.

This volume is a collection of, and introduction to, scholarly work that focuses upon robots and ethics. The topic is divided into two broad fields of research. Robot ethics, or roboethics, explores how people should design, deploy, and treat robots (Veruggio & Operto 2006). It is particularly interested in how the introduction of robots will change human social interactions and what human social concerns tell us about how robots should be designed. Machine ethics or machine morality considers the prospects for creating computers and robots capable of making explicit moral decisions (Allen, Varner, & Zinser 2000; Moor 2006; Anderson & Anderson 2007). What capabilities will increasingly autonomous robots require 1) to recognize when they are in ethically significant situations, and 2) to factor human ethical concerns into selecting safe, appropriate, and moral courses of action? There is no firm distinction between robot ethics and machine ethics, and some scholars treat machine ethics as a subset of robot ethics. Many of the scholars represented feel their work draws upon and contributes to both fields.

Machine ethics and robot ethics are both emergent intellectual disciplines—research fields which are only beginning to take form. The articles we have selected for reproduction are largely essays of historical significance that have become foundational for research in these two new fields of study. We sought both to convey the range of issues addressed and to include representative essays from leading figures in the development of robot ethics and machine ethics. The emphasis on a historical perspective is somewhat at the expense of more recent contributions. However, a few important recent articles are included. Indeed, a very recent shift in research on artificial intelligence and machine learning has directed much more attention by computer scientists to ethical concerns that had been largely of interest only to philosophers and social theorists.

This introduction opens with a discussion as to how robot ethics and machine ethics emerged out of earlier topics in science fiction, philosophy, cognitive science, and other fields of research. Then we turn to placing the articles included in the five sections of this book into context, before elucidating the shift in computer science that is directing much more attention to robot ethics and machine ethics. The sections are roughly, but not strictly, chronological in order to provide a sense of the development of various issues and positions. Ideas evolve over time through insights, intuitions, and scholarship put forward by countless individuals, only a few of whom are ever recognized.

Part I, "Laying foundations", attempts to capture a range of foundational work, which might be considered the "pre-history" of the fields. "Robot ethics" contains writings that established this area as a sub-discipline of technology ethics in its own right. Similarly, "Machine ethics" contains articles that established machine ethics as a distinct field. The next two parts focus on the two aspects of applied ethics—the internal perspective, the necessary conditions for considering a robot as an autonomous agent, and the external perspective of laws, regulations, and policies central for integrating robots into social life. In many cases, the issues, such as privacy, are relevant to both machine ethics and robot ethics, and are often relevant to larger questions in the ethics of emerging technologies more generally. The part on "Moral agents and agency" examines the ways in which robots and artificial software agents might be construed as, or designed to be, moral agents, and how this leads us to reconsider the nature of agency more generally. The part on "Law and policy" looks at the legal and policy-making issues we face in dealing with sophisticated robots. Advanced robotics will challenge many existing assumptions about what constitutes an agent and which tasks can be performed by technological systems.

The topics covered in this volume often overlap with those in companion volumes, particularly those in the volume on the *Ethics of Information Technologies*. Robots are essentially information systems that are mobile and can act in the world through remote control or autonomously. Furthermore, machine ethics encompasses ethical decision making by not only robots, but also increasingly autonomous agents within computer networks. Broader issues in information and computer ethics, such as privacy, are largely covered in the volume on the *Ethics of Information Technologies*, while this volume goes into much greater depth discussing prospects for creating systems capable of making moral decisions. In addition, ethical issues related to the roboticization of warfare are covered in the volume on the *Ethics of Emerging Military Technologies*.

Finally, this introduction will outline recent advances in the development of artificial intelligence. Ensuring that intelligent systems will not only be safe but also demonstrably beneficial and controllable is becoming an issue of much greater importance, and will continue to expand over the coming decade.

Science fiction and speculation

The speculative worlds of science fiction have given rise to imaginative technologies, fueled consumer desires, and guided the design of real products suited for everyday life. These worlds have also provided lessons in the ways that advanced systems might run amuck, and stoked fears over future technology.

While the public imagination is torn between the utopic and dystopic extremes of science fiction, engineers struggle to design limited purpose machines that will perform safely when introduced into the commerce of daily life. Some futurists proclaim the near-term advent of smarter-than-human robots that may or may not be friendly to our species (Kurzweil 2005). It is unfortunate, however, that this garners more media attention than real-world problems such as the threat to privacy posed by the introduction of drones into civilian air space and the appropriateness of turning over aspects of eldercare to robots.

Furthermore, speculation and science fiction feed misunderstandings regarding the capabilities of present day machines and those likely to be realized in the coming decade. In spite of the fact that IBM's Watson beat human champions on the TV show *Jeopardy*, robots lack higher-order cognitive capabilities, such as consciousness, semantic understanding, sophisticated learning abilities, discrimination, and empathy. Furthermore, the natural human tendency to anthropomorphize robots obfuscates recognition of their limitations. Today's robots are essentially machines. And yet they can be more skilled than their human counterparts at performing certain activities such as searching large databases, and revealing associations between disparate pieces of information.

Our challenge for this volume lies in grounding the discussion of robot ethics and machine ethics to focus on practical considerations that affect the way machines are designed and the tasks they might fulfill. Nevertheless, the three great themes of science fiction stories and films are helpful for framing this discussion.

1. Robots run amuck and cause harm—because they are dumb or rigid, because they were malevolently programmed, because they are alien or act in unanticipated ways, or because they wish to be free from servitude (*R.U.R.*, *Metropolis*).

2. Robots become social companions and even learn to apply ethical norms to determine how to act in new situations (Asimov's Laws, Cmd. Data in *Star Trek*).
3. Superintelligent forms of artificial intelligence that seek either to exterminate humans or become benevolent overlords (*2001: A Space Odyssey*, *The Terminator*, *The Matrix*, *Battlestar Galactica*).

Robot ethics is situated in a larger and older cultural history of robots in literature. From antiquity's Pygmalion to golems to Frankenstein to Rossum's Universal Robots, traditional narratives about humanoids are morality tales that emerge from clearly defined cultural fears. In nearly all the early science fiction, even initially benign robots became dangerous. They were usually emotionless, and often alien in origin, with little sympathy or concern for humans and the societal effects of their actions. Those robots that were not alien in origin were generally built by a scientific genius, whose robots eventually turn against them, if not all of humanity.

The implied moral of those stories is that scientific hubris leads to the building of monsters, and the eventual destruction of the scientist, monster, or both. Isaac Asimov sought to break away from this traditional morality tale, and find one more compatible with his own brand of cautionary scientific optimism. He created the genre of the beneficent or good and helpful robot. In a series of short stories and novels, he proposed Three Laws that are meant to ensure robots act as helpful servants. The Three Laws (not harming humans, obeying humans, and self-preservation) are arranged hierarchically so that the first trumped the other two, and the second trumps the third. The Robot Laws were a precursor to machine ethics. But of course, what Asimov demonstrates in the course of exploring his Laws of Robotics is that they were flawed, inadequate, and undesirable in a surprising variety of circumstances. Later he added a fourth or Zeroth Law specifying that, "A robot may not harm humanity, or, by inaction, allow humanity to come to harm," but this only came into play in a few situations. The characters in his stories sometimes question and doubt the Laws' adequacy, and even when they do not, the readers do. In short, the Laws of Robotics are an excellent literary device, but wholly inadequate for the engineering and design of real robots and systems. Despite this, there are valuable lessons to be learned for machine and robot ethics by studying Asimov's stories and recognizing the nature of the failures to which they lead.

Integral to these literary lessons are questions of ethical value and morality. In arguing that we want artificial agents and robots to act in one way rather than another, we are expressing a desire, and with it a value. It is, in this regard, not unlike our expectations of humans, and our desire for them to act a certain way. We want people to be good, to act ethically, to reason morally, and to be virtuous. This is the stuff of moral theory, a Western philosophical tradition that was established by the ancient Greeks including Socrates, Plato, and Aristotle. That tradition has developed and elaborated various frameworks for describing and thinking about what constitutes the good life, moral actions, and the virtuous person.

This philosophical tradition has a long history of considering hypothetical and even magical situations (such as the Ring of Gyges in Plato's *Republic*, which bestows invisibility on its wearer). The nineteenth and twentieth centuries saw many of the magical powers of myth and literature brought into technological reality. Alongside the philosophical literature emerged literatures of horror, science fiction, and fantasy, which also explored many of the social, ethical, and philosophical questions raised by the new-found powers of technological capabilities. For most of the twentieth century, the examination of the ethical and moral implications of artificial intelligence and robotics was limited to the work of science fiction and cyberpunk writers, such as Isaac Asimov, Arthur C. Clarke, Bruce Sterling, William Gibson, and Philip K. Dick (to name only a few). It is only in the late twentieth century that we begin to see academic philosophers taking up these questions in a scholarly way.

Emergent disciplines

Like most new fields of knowledge, machine ethics and robot ethics emerged from work in other disciplines. Early interest came from obvious sources, such as engineering ethics and computer ethics, which were already attuned to the ethical and social value issues raised by new technologies. But contributions also came from other more developed subject areas in technology and ethics, such as bioethics and nanoethics. From within engineering came contributions from human–computer interaction and human–robot interaction. Taking a very practical approach to designing systems to interact with people, these fields became interested in the role of emotions in human social interactions, and how best to incorporate this knowledge into their designs. As practical fields, they already lay at an interdisciplinary crossroads of engineering, design, and social science.

Joining this mix were those philosophers and cognitive scientists interested in understanding the theoretical and empirical nature of human cognition, moral reasoning and moral intuitions, and decision making. These researchers saw artificial agents and robots as experimental instruments or laboratories for developing simulations through which theories could be tested. With the help of simulations modeled within computational systems, they hoped to study and refine moral theories and better understand how humans reason when confronted with ethical dilemmas.

Some researchers came to the new fields out of an interest in science fiction and cyberpunk literature, or a growing concern that science fiction technologies might become realities. Among these participants in the field, many are involved in the social movements concerned with human enhancements—known as transhumanism. They took an interest in robotic research and its potential for creating prosthetics that would enhance mind and body. Future robots might even serve as avatars into which one's mind could be uploaded and one's life extended indefinitely. While such concerns have not been central to either machine ethics or robot ethics, there are many points of contact and shared interests over the governance of emerging technologies and their capacity to transform how we think about what it means to be human.

These various disciplines and interests came together to form the highly interdisciplinary fields of machine ethics and robot ethics, and still continue to define the major lines of inquiry within the fields. The research has focused on four types of questions: the control and governance of computational systems; the exploration of ethical and moral theories using software and robots as laboratories or simulations; inquiry into the necessary requirements for moral agency and the basis and boundaries of rights; and questions of how best to design systems that are both useful and morally sound.

The legacy of bioethics and nanoethics can be seen most clearly in the research concerned primarily with the control and governance of artificial agents and robots as emergent technology. There are, of course, many emerging technologies in our contemporary world, each with its own ethical challenges, yet not all of these seem deserving of their own distinct academic fields of ethics. When a new technology is seen as potentially autonomous, or "other," that it might too easily get out of human control, then specialized governance mechanisms come into play. Similarly, when a new technology is so disruptive of existing technological regulations, then a new framework is needed. Such was the case with internet governance, or now with the introduction of unmanned aerial vehicles (drones) into civilian airspace, where whole new regimes of regulation must be imagined and implemented. Many of these are questions of social policy, but also questions of applied ethics.

Another powerful line of inquiry points to the meta-ethical questions. The potential for artificial moral agents to provide a laboratory of sorts for exploring the nature of moral theories computationally offers a meta-ethical research program. Which theories are computable or incomputable? What does it mean for practical determinations of right and wrong if ethics is not reducible to computable rules and prescriptions for behavior? What computations do humans actually use in selecting appropriate behavior in morally significant situations? How do we translate moral rules into computer code, and how do we interpret the results of following those moral rules? What does it mean to imitate, simulate, or replicate moral reasoning?

Related to these meta-ethical questions is a line of inquiry into the basis and boundaries of rights and the necessary requirements for moral agency. Can software or robotic agents become moral agents? What capabilities would they need for this? How would we know whether they have those capabilities? These in turn raise questions about how we recognize and accept other people as moral agents. Furthermore, artificial agents can take many forms that humans cannot, and thus can pose challenges to many straightforward analyses.

Finally, taking into account all of the above, how are we to implement practical machine ethics and robot ethics in everyday systems? As an engineer and designer, already engaged in developing a system that will interact with people and thus with the potential to help or do harm, how can one ensure a "good" design? From specific design principles, to design methodologies, to the application of domain knowledge, this is a rich and dynamic field. This work will hopefully grow in positive directions as the fields apply their research to the questions of how best to design systems that are both useful and morally sound.

Laying foundations

Questions that have become foundational for machine ethics and robot ethics began to intrigue scholars in the 1980s and 1990s. Sam Lehman-Wilzig (1981), in an article titled "Frankenstein Unbound: Towards a Legal Definition of Artificial Intelligence," questioned whether legal barriers exist that would preclude designating intelligent systems as accountable and liable for their actions. Roger Clarke (1993, 1994) was the first engineer to ask whether Asimov's Laws of Robotics could actually be implemented. He proposed that an engineer might well conclude that rules or laws are not an effective design strategy for building robots whose behavior must be appropriate, legal, and morally acceptable. Then in 1995, an edited volume entitled *Android Epistemology* contained two articles (not included in this volume), one by James Gips and another by Daniel Dennett, which raised critical issues. Gips considers whether ethical theories such as consequentialism and Kant's categorical imperative could be implemented within a computer system, and Dennett's chapter ponders the question in its title, "When Hal Kills, Who's to Blame?"

Debates around information and communication technologies (ICT) in the 1980s and 1990s were also seminal in the emergence of robot ethics and machine ethics. These debates emanated primarily from the experiences garnered from implementing large ICT systems within organizations. As such, there was a strong influence from the professional information technology system design field, as well as the more academic computer ethics field. Nissenbaum (2001) articulates the ways in which the design of computer systems involves innumerable judgments of value, and the implementation and use of any ICT system also enforces a set of values upon its users. The same can be said about any sophisticated machine, including robotic systems.

Allen, Varner, and Zinser (2000) go well beyond Gips' 1995 article in laying out the central challenges for creating artificial moral agents—disagreement among ethical theorists as to what the "right" ethical theory to implement is, and the computational challenges of implementing the various proposed theories. Much of the work in machine ethics follows their suggestion that practical issues of implementation—getting something that works, and that reduces the risks of a system causing harm through its actions—is more important than settling the philosophical disputes, or implementing a perfect morality. In addition, they consider whether a Moral Turing Test might be a means for establishing whether an artificial agent makes "good" decisions.

At the end of Part I we return to the question which many people still feel belongs in the realm of science fiction, but which has also been raised more recently by an increasing number of leading scientists and technologists, including Stephen Hawking, Elon Musk, and Stuart Russell. Might computer scientists create an artificial intelligence that is not only smarter than us, but is capable of learning and growing at a much greater rate than humans, such that its working will be unknown to us and its capabilities unforeseeable and unimaginable? In an early paper, Bostrom (2003) outlines the kinds of ethical questions that arise in developing a superintelligence of this sort, and what it means for the human race to risk existential threats in order to derive what may be immeasurable benefits. The advent of superintelligence continues to be of ongoing interest to Bostrom, who organized reflections by himself and others on the subject into an influential 2014 book titled *Superintelligence: Paths, Dangers, Strategies.*

Robot ethics

Robot ethics emerged out of the broader field of engineering ethics and its subfield computer ethics. The initial impetus for robot ethics was concern over how to design robots that would not harm people, and thus shares its origins with both Asimov's Laws and the traditional engineering concerns directed at building tools that are safe. Safety alone was fine as an engineering approach for industrial robots, which operated in isolation from people, mostly in factories and often in safety cages. As robots became more capable, and it became apparent that robots would soon enter into the social world people inhabit, a whole range of possible harms and new design questions came to the fore.

"Roboethics," a term coined by Veruggio and Operto (2006), brings together the various concerns of applied engineering ethics in the context of robotics. They also explicitly include the range of ethical concerns raised in regard to ICT, and examine the development of codes of ethics in ICT that might be similarly deployed in robotics engineering. Asaro (2006) examines whether there might be a systematic and combined approach to robot ethics and machine ethics, which recognizes the range of robotic systems from their moral consequences to their moral agency.

A key aspect of robot ethics concerns the feelings and beliefs people hold towards robots. This includes a range of psychological and behavioral approaches that consider how much people identify with robots, or view them as having beliefs and feelings like people or perhaps like animals. Turkle (2006) examines various psychological relationships that people develop with their companion robots in clinical settings. Coeckelbergh (2010) considers the relationship between emotion and morality, and the implications of simulating emotional qualities in robots that are not capable of true emotions. Sparrow (2004) asks at what point might we decide to save a robot over a human life in a triage situation.

A central area of concern in robot ethics is their use in the domains of caregiving. While many human jobs do not apparently require essentially human traits or qualities to complete (assuming adequate machine intelligence and dexterity), caregiving appears to require uniquely human qualities. In particular, care that is provided to especially vulnerable populations, including children, the elderly, and the sick and injured, demands more humanity than other jobs. In caregiving there appears to be more to lose from automation and especially from poorly designed automation. Borenstein and Pearson (2010), Vallor (2011), Sharkey and Sharkey (2010) and Van Wynsburghe (2013) all address the question of robot caregivers from various perspectives and for various populations. Similar questions also arise in the development of robots designed for sexual love and companionship (Sullins 2012).

Part II concludes with a paper that asks whether the moral competence of robots is sufficient for designing good robots (Malle and Schuetz 2014). That is, should robot engineers settle for robots that are competent in dealing appropriately with the moral issues they might encounter in their interactions with humans? Of course, being able to deal with moral issues means being able to interpret the moral relevance of situations and actions, making moral judgments, and communicating about morality. This article thus provides a nice segue to the field of machine ethics, and how we might actually design moral discernment and decision-making capabilities into robotic systems.

Machine ethics

The new field of research that is focused on the prospects for designing computer and robotic systems that demonstrate sensitivity to human values, and factor these into making decisions in morally significant situations, goes by various names. These include machine ethics, machine morality, moral machines, computational ethics, artificial morality, safe AI, and friendly AI. Each of these names gets used somewhat differently. Friendly AI, for example, refers explicitly to the challenge of ensuring smarter-than-human robots will be friendly to humans and human concerns, while safe AI encompasses many dimensions of robot safety, not merely whether robots can explicitly make moral decisions.

The essays in this part were selected primarily from early examples that broke new ground. Many of the very same scholars have discussed similar issues and additional experiments in fuller and even richer detail in their more recent papers. James Moor (2006) provided very useful distinction for helping to clarify when machines might actually be engaged in making explicitly moral decisions. Michael Anderson (a computer scientist) and Susan Leigh Anderson (a philosopher) performed some of the very first experiments in machine ethics, described in a 2007 article. Their computational system is confronted with a classic challenge in medical ethics over what to do when the human patient it delivers medicine to refuses to take the pills. Wallach, Allen, and Smit (2008) flesh out the distinction between bottom-up and top-down approaches to implementing moral decision-making faculties in robots and the need for capabilities beyond the ability to reason. A robot's choices and actions are generally hard-wired, or what is sometimes referred to as *operational morality*. As systems become increasingly autonomous, the need to build in ethical routines for selecting appropriate behavior from among various courses of action (*functional morality*) will expand. But *functional morality* can still fall far short of full moral agency. These three articles introduce some of the most fundamental concepts and approaches that gave early form to machine ethics as a recognized field of study.

Robotics is a young field. To date, machine ethics has been dominated by philosophers and social theorists. Very few actual efforts to implement explicit moral decision-making capabilities in computer systems and physical robots have been made. Nevertheless, physical robots are entering the commerce of daily life. In 2008, Drew McDermott, a Professor of Computer Science at Yale University, weighed in as to not only why ethical decisions would be difficult for artificial intelligence systems to make, but also why he did not consider it a particularly important project for computer science at that time. Implicitly, McDermott's critique underscored why machine morality has been such a fascinating subject for philosophers, but of less interest to computer scientists.

A number of different scholars considered whether ethical theories such as utilitarianism or Kant's categorical imperative could actually be instantiated in a robot. The reproduced article by Tom Powers (2006) is but one representative of that sub-genre. The essay by Marcello Guarini (2005) describes the first of his many experiments to explore whether training neural networks with a degree of sensitivity to moral considerations might help us better understand human moral cognition.

There is a long-standing issue in moral philosophy as to whether ethics is, or should be, logically consistent. If ethics is logically consistent, then that might ease its implementation within a computational system. Bringsjord, Arkoudas, and Bello outline (2006) an approach for building robots that select "ethically correct" actions through deontic logic.

Consciousness is often discussed as an essential capacity for moral agency that might be difficult, if not impossible to implement in a robot. And yet, there is a field of research known as machine consciousness, which is attempting to design conscious robots. One of that field's practitioners is Stan Franklin, a Professor at the University of Memphis, who is developing a conceptual and computational model for a robot (LIDA) with artificial general intelligence—the capability to solve a broad array of problems in many different domains. In early writings (Wallach & Allen 2008; Wallach, Franklin, & Allen 2010) the authors discuss how Franklin's LIDA might be adapted to make moral decisions. Wallach, Allen, & Franklin (2011) go a step further in describing the functional role consciousness plays in ethical decision making.

Collectively, the selected articles make it apparent why reflecting upon moral decision making for robots has been such a valuable thought experiment for philosophers and scientists who wish to understand human cognition. Regardless of whether the field helps make future robots safer and more sensitive to ethical considerations, it has already injected vital and fruitful approaches for the better understanding of human ethical behavior.

Moral agents and agency

Even if we succeed in creating machines with moral sensitivity, there remains a separate question of whether their decisions and actions are those of moral agents, or mere outputs of a machine. Does the behavior of such machines actually constitute intentional agency? This section covers how work in the field has engaged the fundamental question of what constitutes moral agency as well as the more practical question of how to simulate or instantiate that capacity in an artificial agent. What constitutes agency can be construed in a variety of ways. Philosophers often approach agency as either a property requiring an essential quality or set of capabilities, or as something constituted through acts of self-determination. Legal scholars point to the conditions for the legal agency of people, corporations, and even animals. Engineers view agency as a matter of effective control, which can be passed between human operators and automatic systems. There is a narrow construal of agency that looks only at the properties of an individual agent, and broad perspectives that consider the social, political, and bureaucratic forces that usually shape the environment and decision contexts that individuals face. Some theorists take liberal views of agency and find it operating nearly everywhere, while others see it as a restricted class that should only be applied to appropriately sophisticated types of beings. An important distinction is made between moral patients and moral agents. Moral patients are those who are deserving of moral respect, for example, children, the mentally disabled, and some animals, despite not always being held fully responsible for their actions. Moral agents are those who are capable of moral action, moral responsibility, and showing moral respect to others.

A central tension in this subject area is the relationship between the agency of those who create artificial agents and the artificial agents themselves—and who is responsible, culpable, and accountable for the actions and consequences of using those systems. Floridi and Sanders (2004) propose a highly inclusive concept of moral agency, and articulate how they see artificial agents as being meaningfully considered as moral agents. Such agents need not have much intelligence at all, nor most of the qualities traditionally identified as essential to moral agency. Johnson and Miller (2008) directly challenge that approach to agency, and whether it can serve the role necessary for properly assigning responsibility, culpability, and accountability. They argue that such an approach can too easily mislead the identification of artificial agents as being responsible moral agents, when it is the human who designs and deploys them and is more appropriately seen as the responsible agent.

Suchman (2007) takes a broader and more critical analysis of the nature of agency in shaping the creation of artificial systems and the structuring of design processes themselves, challenging any easy solution to defining or replicating agency in artificial entities. This also calls into question applied frameworks for responsibility, culpability and accountability, as the application of these concepts is continually being politically contested and re-negotiated by participating parties, rather than being clear-cut concepts which agents readily conform to.

Marino and Tamburrini (2006) expand the challenges facing agency for artificial entities by considering the implications of machine learning and more open-ended learning systems. The more a learning machine adapts its programming based on data from its environment and its experiences, the less it is influenced by its initial programming. Does this imply that the human designers become less responsible over time, or that the environment and social forces, or the system itself, become more responsible? Torrance (2014) considers the nature of consciousness and its importance for determining artificial agency. What are the implications of treating consciousness, whatever we collectively determine it to be, as either an objectively real property of artificial agents, or as a product of social relativism?

In the final article for this section, Murphy and Woods (2009) take a position on agency which ducks the philosophical questions in favor of grounding the discussion in the nature of effective control, and the passing of such control among human operators and between human operators and automated systems. They do this through a critique and reformulation of Asimov's Three Laws of Robotics, viewed in terms of the responsiveness of robots to human operators, and the smooth transfer of control back and forth between the autonomous system and human operators. In focusing on control, they aim to make the shift in responsibility, and thus agency, clearer in dynamic and fluid situations.

Law and policy

It is the job of law and policy to articulate and enforce shared moral norms. In this section we examine how legal scholars have engaged the questions of the legal status of artificial agents, and their regulation. The question of legal agency or legal personhood for artificial agents is an extension of the agency debates covered in the previous section. However, it draws upon a large body of legal scholarship and real-world cases and laws on everything from the treatment of minor children, sports and entertainment agents representing clients, intellectual property rights, privacy rights, damage resulting from the keeping of domesticated and wild animals, product liability, and the legal personhood of corporations. Many of these concepts can be applied to trying to predict how the courts will handle various near-term issues, such as those involving drones and self-driving cars. They also provide fertile analogies for thinking about more long-term issues such as when and whether machines and robots might have claims to legal and civil rights under the law.

Solum (1992) provides an in-depth analysis of what constitutes legal personhood, and how various aspects of the law might treat the actions of an artificial agent as being a legal person, potentially subject to prosecution or lawsuits. A more recent article by Asaro (2011) and a book by White & Chopra (2011) go into greater detail about the various forms of legal agency and how to interpret artificial software agents and robots, though we do not include articles by them in this section.

In terms of national policy for the regulation of robots, it is clear that ethics will play a serious role. It is less clear what form those regulations might take, or what governmental agencies might implement them. Nagenborg (2008) offers a perspective on how Europe might implement ethical regulations for robots.

One of the most significant ethical questions will be the privacy of information that robots are able to collect. A personal robot that has intimate knowledge of its owner might share that information with its manufacturer or with third parties, not unlike computer and smartphone technologies today, which already raise a host of privacy issues. It may also collect information about a variety of other individuals that it encounters or interacts with, quite unlike previous technologies. Calo (2010) considers these and other privacy issues raised by robots.

Self-driving cars present an example of a leading application of robotic technology that carries both serious risks and has a great potential to save lives. Lin (2014) considers the challenges to programming self-driving cars that may encounter situations where an accident is unavoidable. Who or what should the car effectively decide to hit in such situations?

Perhaps the most futuristic and contested question in robot ethics is whether robots may someday have the capacity or capability to be the bearer of rights, and in virtue of what criteria will they deserve those rights? Gunkel (2014) defends the view that there may be robots deserving of the legal and moral status of having rights, and challenges the current standards of evaluating which entities (corporations, animals, etc.) deserve moral standing. Such futuristic legal questions underscore the way in which artificial agents presently serve as a thought experiment for legal theorists and a vehicle for reflections on the nature of legal agency and rights.

Growing attention

As we completed work on this volume, there was tangible evidence of a sudden growth in the attention given to the ethical and societal challenges posed by artificial intelligence and robotics. Four different issues contribute to rising concerns as to whether the societal impact of robotics can be controlled in a manner that will minimize harms. First, the movement to ban Lethal Autonomous Weapons Systems (a.k.a. "Killer Robots," see www.stopkillerrobots.org) has gained international attention and is now on the agenda of the Convention for Certain Conventional Weapons at the United Nations, to explore whether an arms control agreement might be forged. While both of us have been involved with that issue, it is not a subject discussed in this volume, but is covered in the companion volume on the *Ethics of Emerging Military Technologies*.

Second, the impact of robots on employment and wages is gaining attention with books and pronouncements by leading economists and social theorists. There is serious concern that the long-standing Luddite fear that technology might eliminate more jobs than it creates, may have finally come to pass. Robotics is not the only factor that contributes to what John Maynard Keynes (1930) named "technological unemployment." But as artificial intelligence systems become increasingly intelligent and capable of performing more and more human jobs, their impact on the availability of jobs with satisfactory hourly wages will be immense.

Third, the proliferation of both large commercial and small hobby drones in civilian airspace and the advent of self-driving cars are garnering considerable attention as to the harms each might cause. Unusual ethical situations that self-driving cars might encounter have been discussed in articles by Gary Marcus (2012), Patrick Lin (2014, mentioned above and reproduced here), and others, and perhaps helps clarify why self-driving cars, heralded for years, are not yet being marketed. Furthermore, self-driving cars provide an apt metaphor for apprehensions regarding the trajectory of technological development. The incessant acceleration of innovation can appear to have put technology in the driver's seat, both figuratively and literally, as the primary determinant of humanity's destiny.

The Federal Aviation Administration (FAA) is slowly introducing rules for the use of drones in civilian airspace, and discovering, not unsurprisingly, that they cannot make everyone happy. When the FAA announced preliminary rules where small drones could only be flown within eyesight of the operator, Amazon immediately complained that this would ground their plans for one-day order delivery by drone. Soon after, the FAA recanted and allowed Amazon to test its drone delivery system. Unfortunately, our society has failed to have a full discussion as to which of the benefits that drones offer justify the risks drones present. Does the use of drones to facilitate police surveillance, to conduct research, for search and rescue operations, and to enable one-day order delivery justify crowded skies, air accidents, and continuing loss of privacy to swarms of camera-bearing mechanical insects buzzing overhead?

Finally, and most importantly, the scientists working in the field of artificial intelligence are waking up to the prospect of creating potentially dangerous artifacts that cannot be fully controlled. For decades, researchers in artificial intelligence had been bedeviled in getting their creations to perform basic tasks. Researchers who had been working in the field for 20 or 30 years, such as Cornell University's Bart Selman, began to feel that "various forms of perceptions, such as computer vision and speech recognition" were "essentially unsolved (and possibility unsolvable) problems."[1] In February 2009, Selman and Eric Horvitz, then President of the Association for the Advancement of Artificial Intelligence (AAAI), co-chaired the AAAI Presidential Panel on Long-Term AI Futures. At that gathering more speculative concerns, such as the possibility of smarter-than-human artificial intelligence, were explicitly addressed. "There was overall skepticism about the prospect of an intelligence explosion as well as of a 'coming singularity,' and also about the large-scale loss of control of intelligent systems," states the August 2009 interim report that summarized the meeting's conclusions. However, at a similar workshop in January 2015, Bart Selman stated, "a majority of AI researchers now express some concern about superintelligence due to recent breakthroughs in computer perception and learning."

The January 2015 workshop was organized by the *Future of Life Institute* to address growing concerns that, however distant, superintelligent artificial intelligence was possible, and to convey to scientists that it behooves them to begin work on approaches to ensure artificial intelligence systems would be demonstrably beneficial and controllable. Safe artificial intelligence suddenly became a concern for all artificial intelligence researchers, not merely those who had been focused upon narrow engineering tasks, machine ethics, robot ethics, or friendly artificial intelligence.

Research in a new approach to machine learning known as deep learning played an important role in catalyzing this uptick in concern. When DeepMind, a UK-based deep learning company, was acquired in 2014 by Google, its three founders made the establishment of an ethics board a condition for completing the sale. They recognized that they were unleashing a powerful technology that, while not dangerous presently, should never be appropriated for harmful purposes.

Deep learning utilizes neural networks, trained on massive computer systems, processing tremendous amounts of information. Such systems have demonstrated the capacity for learning complex rules without the human supervision and tweaking required by previous machine learning methods, including the ability to observe the data flow from other computers playing old video games such as *Breakout*. An algorithm designed at DeepMind developed its own strategies to play and creatively win a number of different computer games.

The primary focus for most of the articles reproduced in this volume is the practical ethical and legal considerations arising from the development of robots over the next 20 years. However, to properly frame your appreciation for the context in which those issues arises, we have elected to first add a most recent article. That is the statement on "Research Priorities for Robust and Beneficial Artificial Intelligence," which emerged out of the January, 2015 meeting organized by the *Future of Life Institute*.

Note

1 Conveyed in a 2015 email to Wendell Wallach.

References not reproduced in this volume

Asaro, P. (2011). "A Body to Kick, but Still No Soul to Damn: Legal Perspectives on Robotics," in P. Lin, K. Abney, & G. Bekey (Eds.), *Robot Ethics: The Ethical and Social Implications of Robotics*. Cambridge, MA: MIT Press, pp. 169–186.

Bostrom, N. (2014). *Superintelligence: Paths, Dangers, Strategies.* Oxford: Oxford University Press.

Dennett, D. C. (1996, reprinted from 1991 version). "When Hal Kills, Who's to Blame?" in D. Stork (Ed.), *Hal's Legacy.* Cambridge, MA: MIT Press, pp. 351–365.

Gates, B. (2007). "A Robot in Every Home," *Scientific American*, January, 59–65.

Gips, J. (1991). "Towards the Ethical Robot," in K. G. Ford, C. & P. Hayes (Eds.), *Android Epistemology.* Cambridge, MA: MIT Press, pp. 243–252.

Horvitz, E. & Selman, B. (2009). Interim report from the panel chairs: AAAI Presidential Panel on Long-Term AI Futures. Available online at https://www.aaai.org/Organization/Panel/panel-notes.pdf/

Keynes, J. M. (1930). Economic Possibilities for our Grandchildren. Available online at: www.econ.yale.edu/smith/econ116a/keynes1.pdf/

Kurzweil, R. (2005). *The Singularity Is Near: When Humans Transcend Biology.* New York, NY: Penguin Group.

Lehman-Wilzig, S. (1981). "Frankenstein Unbound: Towards a Legal Definition of Artificial Intelligence," *Futures*, December, 442–457.

Marcus, C. (2012). "Moral Machines," *The New Yorker*, November 24.

Wallach, W. & Allen, C. (2008). *Moral Machines: Teaching Robots Right from Wrong.* New York: Oxford University Press.

Wallach, W., Franklin, S. & Allen, C. (2010). "A Conceptual and Computational Model of Moral Decision Making in Humans and in AI," *TopiCS: Topics in Cognitive Science*, July, 454–485.

White, L. F. & S. Chopra (2011). *Legal Theory for Autonomous Artificial Agents.* Ann Arbor, MI: The University of Michigan Press.

Appendix 1

The Future of Life Institute: Research Priorities for Robust and Beneficial Artificial Intelligence: An Open Letter

http://futureoflife.org/misc/open_letter

Artificial intelligence (AI) research has explored a variety of problems and approaches since its inception, but for the last 20 years or so has been focused on the problems surrounding the construction of intelligent agents—systems that perceive and act in some environment. In this context, "intelligence" is related to statistical and economic notions of rationality—colloquially, the ability to make good decisions, plans, or inferences. The adoption of probabilistic and decision-theoretic representations and statistical learning methods has led to a large degree of integration and cross-fertilization among AI, machine learning, statistics, control theory, neuroscience, and other fields. The establishment of shared theoretical frameworks, combined with the availability of data and processing power, has yielded remarkable successes in various component tasks such as speech recognition, image classification, autonomous vehicles, machine translation, legged locomotion, and question–answering systems.

As capabilities in these areas and others cross the threshold from laboratory research to economically valuable technologies, a virtuous cycle takes hold whereby even small improvements in performance are worth large sums of money, prompting greater investments in research. There is now a broad consensus that AI research is progressing steadily, and that its impact on society is likely to increase. The potential benefits are huge, since everything that civilization has to offer is a product of human intelligence; we cannot predict what we might achieve when this intelligence is magnified by the tools AI may provide, but the eradication of disease and poverty is not unfathomable. Because of the great potential of AI, it is important to research how to reap its benefits while avoiding potential pitfalls.

The progress in AI research makes it timely to focus research not only on making AI more capable, but also on maximizing the societal benefit of AI. Such considerations motivated the AAAI 2008–09 Presidential Panel on Long-Term AI Futures and other projects on AI impacts, and constitute a significant expansion of the field of AI itself, which up to now has focused largely on techniques that are neutral with respect to purpose. We recommend expanded research aimed at ensuring that increasingly capable AI systems are robust and beneficial: our AI systems must do what we want them to do. The attached research priorities document gives many examples of such research directions that can help maximize the societal benefit of AI. This research is by necessity interdisciplinary, because it involves both society and AI. It ranges from economics, law, and philosophy to computer security, formal methods and, of course, various branches of AI itself.

In summary, we believe that research on how to make AI systems robust and beneficial is both important and timely, and that there are concrete research directions that can be pursued today.

Appendix 2

Research priorities for robust and beneficial artificial intelligence

Last updated January 23, 2015*

Executive Summary: Success in the quest for artificial intelligence has the potential to bring unprecedented benefits to humanity, and it is therefore worthwhile to research how to maximize these benefits while avoiding potential pitfalls. This document gives numerous examples (which should by no means be construed as an exhaustive list) of such worthwhile research aimed at ensuring that AI remains robust and beneficial.

1 Artificial Intelligence Today

Artificial intelligence (AI) research has explored a variety of problems and approaches since its inception, but for the last 20 years or so has been focused on the problems surrounding the construction of *intelligent agents* – systems that perceive and act in some environment. In this context, the criterion for intelligence is related to statistical and economic notions of rationality – colloquially, the ability to make good decisions, plans, or inferences. The adoption of probabilistic representations and statistical learning methods has led to a large degree of integration and cross-fertilization between AI, machine learning, statistics, control theory, neuroscience, and other fields. The establishment of shared theoretical frameworks, combined with the availability of data and processing power, has yielded remarkable successes in various component tasks such as speech recognition, image classification, autonomous vehicles, machine translation, legged locomotion, and question-answering systems.

As capabilities in these areas and others cross the threshold from laboratory research to economically valuable technologies, a virtuous cycle takes hold whereby even small improvements in performance are worth large sums of money, prompting greater investments in research. There is now a broad consensus that AI research is progressing steadily, and that its impact on society is likely to increase. The potential benefits are huge, since everything that civilization has to offer is a product of human intelligence; we cannot predict what we might achieve when this intelligence is magnified by the tools AI may provide, but the eradication of disease and poverty are not unfathomable. Because of the great potential of AI, it is valuable to investigate how to reap its benefits while avoiding potential pitfalls.

The progress in AI research makes it timely to focus research not only on making AI more capable, but also on maximizing the societal benefit of AI. Such considerations motivated the AAAI 2008–09 Presidential Panel on Long-Term AI Futures [43] and other projects and community efforts on AI impacts. These constitute a significant expansion of the field of AI itself, which up to now has focused largely on techniques that are neutral with respect to purpose. The present document can be viewed as a natural continuation of these efforts, focusing on identifying research directions that can help maximize the societal benefit of AI. This research is by necessity interdisciplinary, because it involves both society and AI. It ranges from economics, law, and philosophy to computer security, formal methods and, of course, various branches of AI itself. The focus is on delivering AI that is *beneficial* to society and *robust* in the sense that the benefits are guaranteed: our AI systems must do what we want them to do.

---

*The initial version of this document was drafted by Stuart Russell, Daniel Dewey & Max Tegmark, with major input from Janos Kramar & Richard Mallah, and reflects valuable feedback from Anthony Aguirre, Erik Brynjolfsson, Ryan Calo, Tom Dietterich, Dileep George, Bill Hibbard, Demis Hassabis, Eric Horvitz, Leslie Pack Kaelbling, James Manyika, Luke Muehlhauser, Michael Osborne, David Parkes, Heather Roff, Francesca Rossi, Bart Selman, Murray Shanahan, and many others.

2 Short-term Research Priorities

2.1 Optimizing AI's Economic Impact

The successes of industrial applications of AI, from manufacturing to information services, demonstrate a growing impact on the economy, although there is disagreement about the exact nature of this impact and on how to distinguish between the effects of AI and those of other information technologies. Many economists and computer scientists agree that there is valuable research to be done on how to maximize the economic benefits of AI while mitigating adverse effects, which could include increased inequality and unemployment [51, 12, 26, 27, 72, 53, 49]. Such considerations motivate a range of research directions, spanning areas from economics to psychology. Below are a few examples that should by no means be interpreted as an exhaustive list.

1. **Labor market forecasting:** When and in what order should we expect various jobs to become automated [26]? How will this affect the wages of less skilled workers, creatives, and different kinds of information workers? Some have have argued that AI is likely to greatly increase the overall wealth of humanity as a whole [12]. However, increased automation may push income distribution further towards a power law [13], and the resulting disparity may fall disproportionately along lines of race, class, and gender; research anticipating the economic and societal impact of such disparity could be useful.

2. **Other market disruptions:** Significant parts of the economy, including finance, insurance, actuarial, and many consumer markets, could be susceptible to disruption through the use of AI techniques to learn, model, and predict agent actions. These markets might be indentified by a combination of high complexity and high rewards for navigating that complexity [49].

3. **Policy for managing adverse effects:** What policies could help increasingly automated societies flourish? For example, Brynjolfsson and McAfee [12] explore various policies for incentivizing development of labor-intensive sectors and for using AI-generated wealth to support underemployed populations. What are the pros and cons of interventions such as educational reform, apprenticeship programs, labor-demanding infrastructure projects, and changes to minimum wage law, tax structure, and the social safety net [27]? History provides many examples of subpopulations not needing to work for economic security, ranging from aristocrats in antiquity to many present-day citizens of Qatar. What societal structures and other factors determine whether such populations flourish? Unemployment is not the same as leisure, and there are deep links between unemployment and unhappiness, self-doubt, and isolation [35, 19]; understanding what policies and norms can break these links could significantly improve the median quality of life. Empirical and theoretical research on topics such as the basic income proposal could clarify our options [86, 92].

4. **Economic measures:** It is possible that economic measures such as real GDP per capita do not accurately capture the benefits and detriments of heavily AI-and-automation-based economies, making these metrics unsuitable for policy purposes [51]. Research on improved metrics could be useful for decision-making.

2.2 Law and Ethics Research

The development of systems that embody significant amounts of intelligence and autonomy leads to important legal and ethical questions whose answers impact both producers and consumers of AI technology. These questions span law, public policy, professional ethics, and philosophical ethics, and will require expertise from computer scientists, legal experts, political scientists, and ethicists. For example:

1. **Liability and law for autonomous vehicles:** If self-driving cars cut the roughly 40,000 annual US traffic fatalities in half, the car makers might get not 20,000 thank-you notes, but 20,000 lawsuits. In what legal framework can the safety benefits of autonomous vehicles such as drone aircraft and self-driving cars best be realized [88]? Should legal questions about AI be handled by existing (software- and internet-focused) "cyberlaw", or should they be treated separately [14]? In both military and commercial applications, governments will need to decide how best to bring the relevant expertise to bear; for example, a panel or committee of professionals and academics could be created, and Calo has proposed the creation of a Federal Robotics Commission [15].

2. **Machine ethics:** How should an autonomous vehicle trade off, say, a small probability of injury to a human against the near-certainty of a large material cost? How should lawyers, ethicists, and policymakers engage the public on these issues? Should such trade-offs be the subject of national standards?

3. **Autonomous weapons:** Can lethal autonomous weapons be made to comply with humanitarian law [18]? If, as some organizations have suggested, autonomous weapons should be banned [23, 85], is it possible to develop a precise definition of autonomy for this purpose, and can such a ban practically be enforced? If it is permissible or legal to use lethal autonomous weapons, how should these weapons be integrated into the existing command-and-control structure so that responsibility and liability be distributed, what technical realities and forecasts should inform these questions, and how should "meaningful human control" over weapons be defined [66, 65, 3]? Are autonomous weapons likely to reduce political aversion to conflict, or perhaps result in "accidental" battles or wars [6]? Finally, how can transparency and public discourse best be encouraged on these issues?

4. **Privacy:** How should the ability of AI systems to interpret the data obtained from surveillance cameras, phone lines, emails, *etc.*, interact with the right to privacy? How will privacy risks interact with cybersecurity and cyberwarfare [73]? Our ability to take full advantage of the synergy between AI and big data will depend in part on our ability to manage and preserve privacy [48, 1].

5. **Professional ethics:** What role should computer scientists play in the law and ethics of AI development and use? Past and current projects to explore these questions include the AAAI 2008–09 Presidential Panel on Long-Term AI Futures [43], the EPSRC Principles of Robotics [8], and recently-announced programs such as Stanford's One-Hundred Year Study of AI and the AAAI committee on AI impact and ethical issues (chaired by Rossi and Chernova).

From a public policy perspective, AI (like any powerful new technology) enables both great new benefits and novel pitfalls to be avoided, and appropriate policies can ensure that we can enjoy the benefits while risks are minimized. This raises policy questions such as these:

1. What is the space of policies worth studying?

2. Which criteria should be used to determine the merits of a policy? Candidates include verifiability of compliance, enforceability, ability to reduce risk, ability to avoid stifling desirable technology development, adoptability, and ability to adapt over time to changing circumstances.

2.3 Computer Science Research for Robust AI

As autonomous systems become more prevalent in society, it becomes increasingly important that they robustly behave as intended. The development of autonomous vehicles, autonomous trading systems, autonomous weapons, *etc.* has therefore stoked interest in high-assurance systems where strong robustness guarantees can be made; Weld and Etzioni have argued that "society will reject autonomous agents unless we have some credible means of making them safe" [91]. Different ways in which an AI system may fail to perform as desired correspond to different areas of robustness research:

1. **Verification:** how to prove that a system satisfies certain desired formal properties. (*"Did I build the system right?"*)

2. **Validity:** how to ensure that a system that meets its formal requirements does not have unwanted behaviors and consequences. (*"Did I build the right system?"*)

3. **Security:** how to prevent intentional manipulation by unauthorized parties.

4. **Control:** how to enable meaningful human control over an AI system after it begins to operate. (*"OK, I built the system wrong, can I fix it?"*)

2.3.1 Verification

By verification, we mean methods that yield high confidence that a system will satisfy a set of formal constraints. When possible, it is desirable for systems in safety-critical situations, e.g. self-driving cars, to be verifiable.

Formal verification of software has advanced significantly in recent years: examples include the *seL4* kernel [44], a complete, general-purpose operating-system kernel that has been mathematically checked against a formal specification to give a strong guarantee against crashes and unsafe operations, and HACMS, DARPA's "clean-slate, formal methods-based approach" to a set of high-assurance software tools [25]. Not only should it be possible to build AI systems on top of verified substrates; it should also be possible to verify the designs of the AI systems themselves, particularly if they follow a "componentized architecture", in which guarantees about individual components can be combined according to their connections to yield properties of the overall system. This mirrors the agent architectures used in Russell and Norvig [69], which separate an agent into distinct modules (predictive models, state estimates, utility functions, policies, learning elements, *etc.*), and has analogues in some formal results on control system designs. Research on richer kinds of agents – for example, agents with layered architectures, anytime components, overlapping deliberative and reactive elements, metalevel control, *etc.* – could contribute to the creation of verifiable agents, but we lack the formal "algebra" to properly define, explore, and rank the space of designs.

Perhaps the most salient difference between verification of traditional software and verification of AI systems is that the correctness of traditional software is defined with respect to a fixed and known machine model, whereas AI systems – especially robots and other embodied systems – operate in environments that are at best partially known by the system designer. In these cases, it may be practical to verify that the system acts correctly given the knowledge that it has, avoiding the problem of modelling the real environment [21]. A lack of design-time knowledge also motivates the use of learning algorithms within the agent software, and verification becomes more difficult: statistical learning theory gives so-called ϵ-δ (probably approximately correct) bounds, mostly for the somewhat unrealistic settings of supervised learning from i.i.d. data and single-agent reinforcement learning with simple architectures and full observability, but even then requiring prohibitively large sample sizes to obtain meaningful guarantees.

Research into methods for making strong statements about the performance of machine learning algorithms and managing computational budget over many different constituent numerical tasks could improve our abilities in this area, possible extending work on Bayesian quadrature [34, 29]. Work in adaptive control theory [7], the theory of so-called *cyberphysical systems* [58], and verification of hybrid or robotic systems [2, 93] is highly relevant but also faces the same difficulties. And of course all these issues are laid on top of the standard problem of proving that a given software artifact does in fact correctly implement, say, a reinforcement learning algorithm of the intended type. Some work has been done on verifying neural network applications [62, 82, 71] and the notion of *partial programs* [4, 80] allows the designer to impose arbitrary "structural" constraints on behavior, but much remains to be done before it will be possible to have high confidence that a learning agent will learn to satisfy its design criteria in realistic contexts.

2.3.2 Validity

A verification theorem for an agent design has the form, "If environment satisfies assumptions ϕ then behavior satisfies requirements ψ." There are two ways in which a verified agent can, nonetheless, fail to be a beneficial agent in actuality: first, the environmental assumption ϕ is false in the real world, leading to behavior that violates the requirements ψ; second, the system may satisfy the formal requirement ψ but still behave in ways that we find highly undesirable in practice. It may be the case that this undesirability is a consequence of satisfying ψ when ϕ is violated; i.e., had ϕ held the undesirability would not have been manifested; or it may be the case that the requirement ψ is erroneous in itself. Russell and Norvig [69] provide a simple example: if a robot vacuum cleaner is asked to clean up as much dirt as possible, and has an action to dump the contents of its dirt container, it will repeatedly dump and clean up the same dirt. The requirement should focus not on dirt cleaned up but on cleanliness of the floor. Such specification errors are ubiquitous in software verification, where it is commonly observed that writing correct specifications can be harder than writing correct code. Unfortunately, it is not possible to verify the specification: the notions of "beneficial" and "desirable" are not separately made formal, so one cannot straightforwardly prove that satisfying ψ necessarily leads to desirable behavior and a beneficial agent.

In order to build systems that robustly behave well, we of course need to decide what "good behavior" means in each application domain. This ethical question is tied intimately to questions of what engineering techniques are available, how reliable these techniques are, and what trade-offs can be made – all areas where computer science, machine learning, and broader AI expertise is valuable. For example, Wallach and Allen [89] argue that a significant consideration is the computational expense of different behavioral standards (or ethical theories): if a standard cannot be applied efficiently enough to guide behavior in safety-critical situations, then cheaper approximations may be needed. Designing simplified rules – for example, to govern a self-driving car's decisions in critical situations – will likely require expertise from both ethicists and computer scientists. Computational models of ethical reasoning may shed light on questions of computational expense and the viability of reliable ethical reasoning methods [5, 81]; for example, work could further explore the applications of semantic networks for case-based reasoning [50], hierarchical constraint satisfaction [47], or weighted prospective abduction [57] to machine ethics.

2.3.3 Security

Security research can help make AI more robust. As AI systems are used in an increasing number of critical roles, they will take up an increasing proportion of cyber-attack surface area. It is also probable that AI and machine learning techniques will themselves be used in cyber-attacks.

Robustness against exploitation at the low level is closely tied to verifiability and freedom from bugs. For example, the DARPA SAFE program aims to build an integrated hardware-software system with a flexible metadata rule engine, on which can be built memory safety, fault isolation, and other protocols that could improve security by preventing exploitable flaws [20]. Such programs cannot eliminate all security flaws (since verification is only as strong as the assumptions that underly the specification), but could significantly reduce vulnerabilities of the type exploited by the recent "Heartbleed bug" and "Bash Bug". Such systems could be preferentially deployed in safety-critical applications, where the cost of improved security is justified.

At a higher level, research into specific AI and machine learning techniques may become increasingly useful in security. These techniques could be applied to the detection of intrusions [46], analyzing malware [64], or detecting potential exploits in other programs through code analysis [11]. It is not implausible that cyberattack between states and private actors will be a risk factor for harm from near-future AI systems, motivating research on preventing harmful events. As AI systems grow more complex and are networked together, they will have to intelligently manage their trust, motivating research on statistical-behavioral trust establishment [61] and computational reputation models [70].

2.3.4 Control

For certain types of safety-critical AI systems – especially vehicles and weapons platforms – it may be desirable to retain some form of meaningful human control, whether this means a human in the loop, on the loop [36, 56], or some other protocol. In any of these cases, there will be technical work needed in order to ensure that meaningful human control is maintained [22].

Automated vehicles are a test-bed for effective control-granting techniques. The design of systems and protocols for transition between automated navigation and human control is a promising area for further research. Such issues also motivate broader research on how to optimally allocate tasks within human-computer teams, both for identifying situations where control should be transferred, and for applying human judgment efficiently to the highest-value decisions.

3 Long-term research priorities

A frequently discussed long-term goal of some AI researchers is to develop systems that can learn from experience with human-like breadth and surpass human performance in most cognitive tasks, thereby having a major impact on society. If there is a non-negligible probability that these efforts will succeed in the foreseeable future, then additional current research beyond that mentioned in the previous sections will be motivated as exemplified below, to help ensure that the resulting AI will be robust and beneficial.

Assessments of this success probability vary widely between researchers, but few would argue with great confidence that the probability is negligible, given the track record of such predictions. For example, Ernest Rutherford, arguably the greatest nuclear physicist of his time, said in 1933 that nuclear energy was "moonshine"[1], and Astronomer Royal Richard Woolley called interplanetary travel "utter bilge" in 1956 [63]. Moreover, to justify a modest investment in this AI robustness research, this probability need not be high, merely non-negligible, just as a modest investment in home insurance is justified by a non-negligible probability of the home burning down.

3.1 Verification

Reprising the themes of short-term research, research enabling verifiable low-level software and hardware can eliminate large classes of bugs and problems in general AI systems; if the systems become increasingly powerful and safety-critical, verifiable safety properties will become increasingly valuable. If the theory of extending verifiable properties from components to entire systems is well understood, then even very large systems can enjoy certain kinds of safety guarantees, potentially aided by techniques designed explicitly to handle learning agents and high-level properties. Theoretical research, especially if it is done explicitly with very general and capable AI systems in mind, could be particularly useful.

A related verification research topic that is distinctive to long-term concerns is the verifiability of systems that modify, extend, or improve themselves, possibly many times in succession [28, 87]. Attempting to straightforwardly apply formal verification tools to this more general setting presents new difficulties, including the challenge that a formal system that is sufficiently powerful cannot use formal methods in the obvious way to gain assurance about the accuracy of functionally similar formal systems, on pain of inconsistency via Gödel's incompleteness [24, 90]. It is not yet clear whether or how this problem can be overcome, or whether similar problems will arise with other verification methods of similar strength.

Finally, it is often difficult to actually apply formal verification techniques to physical systems, especially systems that have not been designed with verification in mind. This motivates research pursuing a general theory that links functional specification to physical states of affairs. This type of theory would allow use of formal tools to anticipate and control behaviors of systems that approximate rational agents, alternate designs such as satisficing agents, and systems that cannot be easily described in the standard agent formalism (powerful prediction systems, theorem-provers, limited-purpose science or engineering systems, *etc.*). It may also be that such a theory could allow rigorously demonstrating that systems are constrained from taking certain kinds of actions or performing certain kinds of reasoning.

3.2 Validity

As in the short-term research priorities, validity is concerned with undesirable behaviors that can arise despite a system's formal correctness. In the long term, AI systems might become more powerful and autonomous, in which case failures of validity could carry correspondingly higher costs.

Strong guarantees for machine learning methods, an area we highlighted for short-term validity research, will also be important for long-term safety. To maximize the long-term value of this work, machine learning research might focus on the types of unexpected generalization that would be most problematic for very general and capable AI systems. In particular, it might aim to understand theoretically and practically how learned representations of high-level human concepts could be expected to generalize (or fail to) in radically new contexts [83]. Additionally, if some concepts could be learned reliably, it might be possible to use them to define tasks and constraints that minimize the chances of unintended consequences even when autonomous AI systems become very general and capable. Little work has been done on this topic, which suggests that both theoretical and experimental research may be useful.

Mathematical tools such as formal logic, probability, and decision theory have yielded significant insight into the foundations of reasoning and decision-making. However, there are still many open problems in the foundations of reasoning and decision. Solutions to these problems may make the behavior of very capable systems much more reliable and predictable. Example research topics in this area include reasoning and decision under bounded computational resources à la Horvitz and Russell [41, 67], how to take into account correlations between AI systems' behaviors and those of their environments or of other agents [84, 45, 40, 30, 78], how agents that are embedded in their environments should reason [74, 55], and how to reason about uncertainty over logical consequences of beliefs or other deterministic computations [77, 60]. These topics may benefit from being considered together, since they appear deeply linked [31, 32].

In the long term, it is plausible that we will want to make agents that act autonomously and powerfully across many domains. Explicitly specifying our preferences in broad domains in the style of near-future

---

[1] "The energy produced by the breaking down of the atom is a very poor kind of thing. Any one who expects a source of power from the transformation of these atoms is talking moonshine" [59].

machine ethics may not be practical, making "aligning" the values of powerful AI systems with our own values and preferences difficult [75, 76]. Consider, for instance, the difficulty of creating a utility function that encompasses an entire body of law; even a literal rendition of the law is far beyond our current capabilities, and would be highly unsatisfactory in practice (since law is written assuming that it will be interpreted and applied in a flexible, case-by-case way). Reinforcement learning raises its own problems: when systems become very capable and general, then an effect similar to Goodhart's Law is likely to occur, in which sophisticated agents attempt to manipulate or directly control their reward signals [9]. This motivates research areas that could improve our ability to engineer systems that can learn or acquire values at run-time. For example, inverse reinforcement learning may offer a viable approach, in which a system infers the preferences of another actor, assumed to be a reinforcement learner itself [68, 52]. Other approaches could use different assumptions about underlying cognitive models of the actor whose preferences are being learned (preference learning, [17]), or could be explicitly inspired by the way humans acquire ethical values. As systems become more capable, more epistemically difficult methods could become viable, suggesting that research on such methods could be useful; for example, Bostrom [9] reviews preliminary work on a variety of methods for specifying goals indirectly.

3.3 Security

It is unclear whether long-term progress in AI will make the overall problem of security easier or harder; on one hand, systems will become increasingly complex in construction and behavior and AI-based cyberattacks may be extremely effective, while on the other hand, the use of AI and machine learning techniques along with significant progress in low-level system reliability may render hardened systems much less vulnerable than today's. From a cryptographic perspective, it appears that this conflict favors defenders over attackers; this may be a reason to pursue effective defense research wholeheartedly.

Although the research topics described in 2.3.3 may become increasingly important in the long term, very general and capable systems will pose distinctive security problems. In particular, if the problems of validity and control are not solved, it may be useful to create "containers" for AI systems that could have undesirable behaviors and consequences in less controlled environments [95]. Both theoretical and practical sides of this question warrant investigation. If the general case of AI containment turns out to be prohibitively difficult, then it may be that designing an AI system and a container in parallel is more successful, allowing the weaknesses and strengths of the design to inform the containment strategy [9]. The design of anomaly detection systems and automated exploit-checkers could be of significant help. Overall, it seems reasonable to expect this additional perspective – defending against attacks from "within" a system as well as from external actors – will raise interesting and profitable questions in the field of computer security.

3.4 Control

It has been argued that very general and capable AI systems operating autonomously to accomplish some task will often be subject to effects that increase the difficulty of maintaining meaningful human control [54, 10, 9, 72]. Research on systems that are not subject to these effects, minimize their impact, or allow for reliable human control could be valuable in preventing undesired consequences, as could work on reliable and secure test-beds for AI systems at a variety of capability levels.

If an AI system is selecting the actions that best allow it to complete a given task, then avoiding conditions that prevent the system from continuing to pursue the task is a natural subgoal [54, 10] (and conversely, seeking unconstrained situations is sometimes a useful heuristic [94]). This could become problematic, however, if we wish to repurpose the system, to deactivate it, or to significantly alter its decision-making process; such a system would rationally avoid these changes. Systems that do not exhibit these behaviors have been termed *corrigible* systems [79], and both theoretical and practical work in this area appears tractable and useful. For example, it may be possible to design utility functions or decision processes so that a system will not try to avoid being shut down or repurposed [79], and theoretical frameworks could be developed to better understand the space of potential systems that avoid undesirable behaviors [37, 39, 38].

It has been argued that another natural subgoal is the acquisition of fungible resources of a variety of kinds: for example, information about the environment, safety from disruption, and improved freedom of action are all instrumentally useful for many tasks [54, 10]. Hammond [33] gives the label *stabilization* to the more general set of cases where "due to the action of the agent, the environment comes to be better fitted to the agent as time goes on". This type of subgoal could lead to undesired consequences, and a better understanding of the conditions under which resource acquisition or radical stabilization is an optimal strategy (or likely to be selected by a given system) would be useful in mitigating its effects. Potential research topics in this area include "domestic" goals that are limited in scope in some way [9], the effects of large temporal discount rates on resource acquisition strategies, and experimental investigation of simple systems that display these subgoals.

Finally, research on the possibility of superintelligent machines or rapid, sustained self-improvement ("intelligence explosion") has been highlighted by past and current projects on the future of AI as potentially valuable to the project of maintaining reliable control in the long term. The AAAI 2008–09 Presidential Panel on Long-Term AI Futures' "Subgroup on Pace, Concerns, and Control" stated that

> There was overall skepticism about the prospect of an intelligence explosion... Nevertheless, there was a shared sense that additional research would be valuable on methods for understanding and verifying the range of behaviors of complex computational systems to minimize unexpected outcomes. Some panelists recommended that more research needs to be done to better define "intelligence explosion," and also to better formulate different classes of such accelerating intelligences. Technical work would likely lead to enhanced understanding of the likelihood of such phenomena, and the nature, risks, and overall outcomes associated with different conceived variants [43].

Stanford's One-Hundred Year Study of Artificial Intelligence includes "Loss of Control of AI systems" as an area of study, specifically highlighting concerns over the possibility that

> ...we could one day lose control of AI systems via the rise of superintelligences that do not act in accordance with human wishes – and that such powerful systems would threaten humanity. Are such dystopic outcomes possible? If so, how might these situations arise? ...What kind of investments in research should be made to better understand and to address the possibility of the rise of a dangerous superintelligence or the occurrence of an "intelligence explosion"? [42]

Research in this area could include any of the long-term research priorities listed above, as well as theoretical and forecasting work on intelligence explosion and superintelligence [16, 9], and could extend or critique existing approaches begun by groups such as the Machine Intelligence Research Institute [76].

4 Conclusion

In summary, success in the quest for artificial intelligence has the potential to bring unprecedented benefits to humanity, and it is therefore worthwhile to research how to maximize these benefits while avoiding potential pitfalls. This document has given numerous examples (which should by no means be construed as an exhaustive list) of such worthwhile research aimed at ensuring that AI remains robust and beneficial, and aligned with human interests.

References

[1] Rakesh Agrawal and Ramakrishnan Srikant. "Privacy-preserving data mining". In: *ACM Sigmod Record* 29.2 (2000), pp. 439–450.

[2] Rajeev Alur. "Formal verification of hybrid systems". In: *Embedded Software (EMSOFT), 2011 Proceedings of the International Conference on*. IEEE. 2011, pp. 273–278.

[3] Kenneth Anderson, Daniel Reisner, and Matthew C Waxman. "Adapting the Law of Armed Conflict to Autonomous Weapon Systems". In: *International Law Studies* 90 (2014).

[4] David Andre and Stuart J Russell. "State abstraction for programmable reinforcement learning agents". In: *Eighteenth national conference on Artificial intelligence*. American Association for Artificial Intelligence. 2002, pp. 119–125.

[5] Peter M Asaro. "What should we want from a robot ethic?" In: *International Review of Information Ethics* 6.12 (2006), pp. 9–16.

[6] Peter Asaro. "How just could a robot war be?" In: *Current issues in computing and philosophy* (2008), pp. 50–64.

[7] Karl J Åström and Björn Wittenmark. *Adaptive control*. Courier Dover Publications, 2013.

[8] M Boden et al. "Principles of robotics". In: *The United Kingdom's Engineering and Physical Sciences Research Council (EPSRC). web publication* (2011).

[9] Nick Bostrom. *Superintelligence: Paths, dangers, strategies*. Oxford University Press, 2014.

[10] Nick Bostrom. "The superintelligent will: Motivation and instrumental rationality in advanced artificial agents". In: *Minds and Machines* 22.2 (2012), pp. 71–85.

[11] Yuriy Brun and Michael D Ernst. "Finding latent code errors via machine learning over program executions". In: *Proceedings of the 26th International Conference on Software Engineering*. IEEE Computer Society. 2004, pp. 480–490.

[12] Erik Brynjolfsson and Andrew McAfee. *The second machine age: work, progress, and prosperity in a time of brilliant technologies*. W.W. Norton & Company, 2014.

[13] Erik Brynjolfsson, Andrew McAfee, and Michael Spence. "Labor, Capital, and Ideas in the Power Law Economy". In: *Foreign Aff.* 93 (2014), p. 44.

[14] Ryan Calo. "Robotics and the New Cyberlaw". In: *Available at SSRN 2402972* (2014).

[15] Ryan Calo. "The Case for a Federal Robotics Commission". In: *Available at SSRN 2529151* (2014).

[16] David Chalmers. "The singularity: A philosophical analysis". In: *Journal of Consciousness Studies* 17.9-10 (2010), pp. 7–65.

[17] Wei Chu and Zoubin Ghahramani. "Preference learning with Gaussian processes". In: *Proceedings of the 22nd international conference on Machine learning*. ACM. 2005, pp. 137–144.

[18] Robin R Churchill and Geir Ulfstein. "Autonomous institutional arrangements in multilateral environmental agreements: a little-noticed phenomenon in international law". In: *American Journal of International Law* (2000), pp. 623–659.

[19] Andrew E Clark and Andrew J Oswald. "Unhappiness and unemployment". In: *The Economic Journal* (1994), pp. 648–659.

[20] André DeHon et al. "Preliminary design of the SAFE platform". In: *Proceedings of the 6th Workshop on Programming Languages and Operating Systems*. ACM. 2011, p. 4.

[21] Louise A Dennis et al. "Practical Verification of Decision-Making in Agent-Based Autonomous Systems". In: *arXiv preprint arXiv:1310.2431* (2013).

[22] United Nations Institute for Disarmament Research. *The Weaponization of Increasingly Autonomous Technologies: Implications for Security and Arms Control*. UNIDIR, 2014.

[23] Bonnie Lynn Docherty. *Losing Humanity: The Case Against Killer Robots*. Human Rights Watch, 2012.

[24] Benja Fallenstein and Nate Soares. *Vingean Reflection: Reliable Reasoning for Self-Modifying Agents*. Tech. rep. Machine Intelligence Research Institute, 2014. URL: https://intelligence.org/files/VingeanReflection.pdf.

[25] Kathleen Fisher. "HACMS: high assurance cyber military systems". In: *Proceedings of the 2012 ACM conference on high integrity language technology*. ACM. 2012, pp. 51–52.

[26] Carl Frey and Michael Osborne. *The future of employment: how susceptible are jobs to computerisation?* Working Paper. Oxford Martin School, 2013.

[27] Edward L Glaeser. "Secular joblessness". In: *Secular Stagnation: Facts, Causes and Cures* (2014), p. 69.

[28] Irving John Good. "Speculations concerning the first ultraintelligent machine". In: *Advances in computers* 6.31 (1965), p. 88.

[29] Tom Gunter et al. "Sampling for inference in probabilistic models with fast Bayesian quadrature". In: *Advances in Neural Information Processing Systems*. 2014, pp. 2789–2797.

[30] Joseph Y Halpern and Rafael Pass. "Game theory with translucent players". In: *arXiv preprint arXiv:1308.3778* (2013).

[31] Joseph Y Halpern and Rafael Pass. "I don't want to think about it now: Decision theory with costly computation". In: *arXiv preprint arXiv:1106.2657* (2011).

[32] Joseph Y Halpern, Rafael Pass, and Lior Seeman. "Decision Theory with Resource-Bounded Agents". In: *Topics in cognitive science* 6.2 (2014), pp. 245–257.

[33] Kristian J Hammond, Timothy M Converse, and Joshua W Grass. "The stabilization of environments". In: *Artificial Intelligence* 72.1 (1995), pp. 305–327.

[34] Philipp Hennig and Martin Kiefel. "Quasi-Newton methods: A new direction". In: *The Journal of Machine Learning Research* 14.1 (2013), pp. 843–865.

[35] Clemens Hetschko, Andreas Knabe, and Ronnie Schöb. "Changing identity: Retiring from unemployment". In: *The Economic Journal* 124.575 (2014), pp. 149–166.

[36] Henry Hexmoor, Brian McLaughlan, and Gaurav Tuli. "Natural human role in supervising complex control systems". In: *Journal of Experimental & Theoretical Artificial Intelligence* 21.1 (2009), pp. 59–77.

[37] Bill Hibbard. "Avoiding unintended AI behaviors". In: *Artificial General Intelligence*. Springer, 2012, pp. 107–116.

[38] Bill Hibbard. *Ethical Artificial Intelligence*. 2014. URL: http://arxiv.org/abs/1411.1373.

[39] Bill Hibbard. "Self-Modeling Agents and Reward Generator Corruption". In: *AAAI-15 Workshop on AI and Ethics*. 2015.

[40] Daniel Hintze. "Problem Class Dominance in Predictive Dilemmas". Honors Thesis. Arizona State University, 2014.

[41] Eric J Horvitz. "Reasoning about beliefs and actions under computational resource constraints". In: *Third AAAI Workshop on Uncertainty in Artificial Intelligence*. 1987, pp. 429–444.

[42] Eric Horvitz. *One-Hundred Year Study of Artificial Intelligence: Reflections and Framing*. White paper. Stanford University, 2014. URL: https://stanford.app.box.com/s/266hrhww2l3gjoy9euar.

[43] Eric Horvitz and Bart Selman. *Interim Report from the Panel Chairs*. AAAI Presidential Panel on Long Term AI Futures. 2009. URL: https://www.aaai.org/Organization/Panel/panel-note.pdf.

[44] Gerwin Klein et al. "seL4: Formal verification of an OS kernel". In: *Proceedings of the ACM SIGOPS 22nd symposium on Operating systems principles*. ACM. 2009, pp. 207–220.

[45] Patrick LaVictoire et al. "Program Equilibrium in the Prisoner's Dilemma via Löb's Theorem". In: *AAAI Multiagent Interaction without Prior Coordination workshop*. 2014.

[46] Terran D Lane. "Machine learning techniques for the computer security domain of anomaly detection". PhD thesis. Purdue University, 2000.

[47] Alan K Mackworth. "Agents, bodies, constraints, dynamics, and evolution". In: *AI Magazine* 30.1 (2009), p. 7.

[48] James Manyika et al. *Big data: The next frontier for innovation, competition, and productivity*. Report. McKinsey Global Institute, 2011.

[49] James Manyika et al. *Disruptive technologies: Advances that will transform life, business, and the global economy*. Vol. 180. McKinsey Global Institute, San Francisco, CA, 2013.

[50] Bruce M McLaren. "Computational models of ethical reasoning: Challenges, initial steps, and future directions". In: *Intelligent Systems, IEEE* 21.4 (2006), pp. 29–37.

[51] Joel Mokyr. "Secular stagnation? Not in your life". In: *Secular Stagnation: Facts, Causes and Cures* (2014), p. 83.

[52] Andrew Y Ng and Stuart Russell. "Algorithms for Inverse Reinforcement Learning". In: *in Proc. 17th International Conf. on Machine Learning*. Citeseer. 2000.

[53] Nils J Nilsson. "Artificial intelligence, employment, and income". In: *AI Magazine* 5.2 (1984), p. 5.

[54] Stephen M Omohundro. *The nature of self-improving artificial intelligence*. Presented at Singularity Summit 2007.

[55] Laurent Orseau and Mark Ring. "Space-Time embedded intelligence". In: *Artificial General Intelligence*. Springer, 2012, pp. 209–218.

[56] Raja Parasuraman, Thomas B Sheridan, and Christopher D Wickens. "A model for types and levels of human interaction with automation". In: *Systems, Man and Cybernetics, Part A: Systems and Humans, IEEE Transactions on* 30.3 (2000), pp. 286–297.

[57] Luís Moniz Pereira and Ari Saptawijaya. "Modelling morality with prospective logic". In: *Progress in Artificial Intelligence*. Springer, 2007, pp. 99–111.

[58] Andr Platzer. *Logical analysis of hybrid systems: proving theorems for complex dynamics*. Springer Publishing Company, Incorporated, 2010.

[59] Associated Press. "Atom-Powered World Absurd, Scientists Told". In: *New York Herald Tribune* (1933). September 12, p. 1.

[60] *Probabilistic Numerics*. http://probabilistic-numerics.org. Accessed: 27 November 2014.

[61] Matthew J Probst and Sneha Kumar Kasera. "Statistical trust establishment in wireless sensor networks". In: *Parallel and Distributed Systems, 2007 International Conference on*. Vol. 2. IEEE. 2007, pp. 1–8.

[62] Luca Pulina and Armando Tacchella. "An abstraction-refinement approach to verification of artificial neural networks". In: *Computer Aided Verification*. Springer. 2010, pp. 243–257.

[63] Reuters. "Space Travel 'Utter Bilge'". In: *The Ottawa Citizen* (1956). January 3, p. 1. URL: http://news.google.com/newspapers?id=ddgxAAAAIBAJ&sjid=1eMFAAAAIBAJ&pg=3254%2C7126.

[64] Konrad Rieck et al. "Automatic analysis of malware behavior using machine learning". In: *Journal of Computer Security* 19.4 (2011), pp. 639–668.

[65] Heather M Roff. "Responsibility, liability, and lethal autonomous robots". In: *Routledge Handbook of Ethics and War: Just War Theory in the 21st Century* (2013), p. 352.

[66] Heather M Roff. "The Strategic Robot Problem: Lethal Autonomous Weapons in War". In: *Journal of Military Ethics* 13.3 (2014).

[67] Stuart J Russell and Devika Subramanian. "Provably bounded-optimal agents". In: *Journal of Artificial Intelligence Research* (1995), pp. 1–36.

[68] Stuart Russell. "Learning agents for uncertain environments". In: *Proceedings of the eleventh annual conference on Computational learning theory*. ACM. 1998, pp. 101–103.

[69] Stuart Russell and Peter Norvig. *Artificial Intelligence: A Modern Approach*. 3rd. Pearson, 2010.

[70] Jordi Sabater and Carles Sierra. "Review on computational trust and reputation models". In: *Artificial intelligence review* 24.1 (2005), pp. 33–60.

[71] Johann M Schumann and Yan Liu. *Applications of neural networks in high assurance systems*. Springer, 2010.

[72] Murray Shanahan. *The Technological Singularity*. Forthcoming. MIT Press, 2015.

[73] Peter W Singer and Allan Friedman. *Cybersecurity: What Everyone Needs to Know*. Oxford University Press, 2014.

[74] Nate Soares. *Formalizing Two Problems of Realistic World-Models*. Tech. rep. Machine Intelligence Research Institute, 2014. URL: https://intelligence.org/files/RealisticWorldModels.pdf.

[75] Nate Soares. *The Value Learning Problem*. Tech. rep. Machine Intelligence Research Institute, 2014. URL: https://intelligence.org/files/ValueLearningProblem.pdf.

[76] Nate Soares and Benja Fallenstein. *Aligning Superintelligence with Human Interests: A Technical Research Agenda*. Tech. rep. Machine Intelligence Research Institute, 2014. URL: http://intelligence.org/files/TechnicalAgenda.pdf.

[77] Nate Soares and Benja Fallenstein. *Questions of Reasoning Under Logical Uncertainty*. Tech. rep. URL: http://intelligence.org/files/QuestionsLogicalUncertainty.pdf. Machine Intelligence Research Institute, 2014.

[78] Nate Soares and Benja Fallenstein. *Toward Idealized Decision Theory*. Tech. rep. URL: https://intelligence.org/files/TowardIdealizedDecisionTheory.pdf. Machine Intelligence Research Institute, 2014.

[79] Nate Soares et al. "Corrigibility". In: *AAAI-15 Workshop on AI and Ethics*. 2015. URL: http://intelligence.org/files/Corrigibility.pdf.

[80] Diana F Spears. "Assuring the behavior of adaptive agents". In: *Agent technology from a formal perspective*. Springer, 2006, pp. 227–257.

[81] John P Sullins. "Introduction: Open questions in roboethics". In: *Philosophy & Technology* 24.3 (2011), pp. 233–238.

[82] Brian J. (Ed.) Taylor. *Methods and Procedures for the Verification and Validation of Artificial Neural Networks*. Springer, 2006.

[83] Max Tegmark. "Friendly Artificial Intelligence: the Physics Challenge". In: *AAAI-15 Workshop on AI and Ethics*. 2015. URL: http://arxiv.org/pdf/1409.0813.pdf.

[84] Moshe Tennenholtz. "Program equilibrium". In: *Games and Economic Behavior* 49.2 (2004), pp. 363–373.

[85] *The Scientists' Call To Ban Autonomous Lethal Robots*. International Committee for Robot Arms Control. Accessed January 2015. URL: http://icrac.net/call/.

[86] Philippe Van Parijs et al. *Arguing for Basic Income. Ethical foundations for a radical reform*. Verso, 1992.

[87] Vernor Vinge. "The coming technological singularity". In: *VISION-21 Symposium, NASA Lewis Research Center and the Ohio Aerospace Institute*. NASA CP-10129. 1993. URL: http://www-rohan.sdsu.edu/faculty/vinge/misc/singularity.html.

[88] David C Vladeck. "Machines without Principals: Liability Rules and Artificial Intelligence". In: *Wash. L. Rev.* 89 (2014), p. 117.

[89] Wendell Wallach and Colin Allen. *Moral machines: Teaching robots right from wrong*. Oxford University Press, 2008.

[90] Nik Weaver. *Paradoxes of rational agency and formal systems that verify their own soundness*. Preprint. URL: http://arxiv.org/pdf/1312.3626.pdf.

[91] Daniel Weld and Oren Etzioni. "The first law of robotics (a call to arms)". In: *AAAI*. Vol. 94. 1994, pp. 1042–1047.

[92] Karl Widerquist et al. *Basic income: an anthology of contemporary research*. Wiley/Blackwell, 2013.

[93] Alan FT Winfield, Christian Blum, and Wenguo Liu. "Towards an Ethical Robot: Internal Models, Consequences and Ethical Action Selection". In: *Advances in Autonomous Robotics Systems*. Springer, 2014, pp. 85–96.

[94] AD Wissner-Gross and CE Freer. "Causal entropic forces". In: *Physical review letters* (2013). 110.16: 168702.

[95] Roman Yampolskiy. "Leakproofing the Singularity: Artificial Intelligence Confinement Problem". In: *Journal of Consciousness Studies* 19.1-2 (2012), pp. 1–2.

Part I:
Laying Foundations

1

Asimov's Laws of Robotics: Implications for Information Technology

Part 1

Roger Clarke

Because many contemporary applications of information technology exhibit robotic characteristics, the difficulties Isaac Asimov identified in his stories are directly relevant to information technology professionals.

With the death of Isaac Asimov on April 6, 1992, the world lost a prodigious imagination. Unlike fiction writers before him, who regarded robotics as something to be feared, Asimov saw a promising technological innovation to be exploited and managed. Indeed, Asimov's stories are experiments with the enormous potential of information technology.

This article examines Asimov's stories not as literature but as a *gedankenexperiment* — an exercise in thinking-through the ramifications of a design. Asimov's intent was to devise a set of rules that would provide reliable control over semiautonomous machines. My goal is to determine whether such an achievement is likely or even possible in the real world. In the process, I focus on practical, legal, and ethical matters that may have short- or medium-term implications for practicing information technologists.

Part 1, in this issue, reviews the origins of the robot notion and explains the laws for controlling robotic behavior, as espoused by Asimov in 1940 and presented and refined in his writings over the following 45 years. Next month, Part 2 examines the implications of Asimov's fiction not only for real roboticists but also for information technologists in general.

Origins of robotics

Robotics, a branch of engineering, is also a popular source of inspiration in science fiction literature; indeed, the term originated in that field. Many authors have written about robot behavior and their interaction with humans, but in this company Isaac Asimov stands supreme. He entered the field early, and from 1940 to 1990 he dominated it. Most subsequent science fiction literature expressly or implicitly recognizes his Laws of Robotics.

Asimov described how, at the age of 20, he came to write robot stories:

> In the 1920s science fiction was becoming a popular art form for the first time... and one of the stock plots... was that of the invention of a robot.... Under the influence of the well-known deeds and ultimate fate of Frankenstein and Rossum, there seem-ed only one change to be rung on this plot — robots were created and destroyed their creator....I quickly grew tired of this dull hundred-times-told tale....Knowledge has its dangers, yes, but is the response to be a retreat from knowledge?... I began, in 1940, to write robot stories of my own — but robot stories of a new variety.... My robots were machines designed by engineers, not pseudomen created by blasphemers.[1,2]

Asimov was not the first to conceive of well-engineered, nonthreatening robots, but he pursued the theme with such enormous imagination and persistence that most of the ideas that have emerged in this branch of science fiction are identifiable in his stories.

To cope with the potential for robots to harm people, Asimov, in 1940, in conjunction with science fiction author and editor John W. Campbell, formulated the Laws of Robotics.[3,4] He subjected all of his fictional robots to these laws by having them incorporated within the architecture of their (fictional) "platinum-iridium positronic brains." The laws (see sidebar on next page) first appeared publicly in his fourth robot short story, "Runaround."[5]

The laws quickly attracted — and have since retained — the attention of readers and other science fiction writers. Only two years later another established writer, Lester Del Rey, referred to "the mandatory form that would force built-in unquestioning obedience from the robot."[6] As Asimov later wrote (with his characteristic clarity and lack of modesty), "Many writers of robot stories, without actually quoting the three laws, take them for granted, and expect the readers to do the same."[1]

Asimov's fiction even influenced the origins of robotic engineering. "Engelberger, who built the first industrial robot, called Unimate, in 1958, attributes his long-standing fascination with robots to his reading of [Asimov's] 'I, Robot' when he was a teenager,"[4] and Engelberger later invited Asimov to write the foreword to his robotics manual.

The laws are intuitively appealing: They are simple and straightforward, and they embrace "the essential guiding principles of a good many of the world's ethical systems."[7] They also appear to ensure the continued dominion of humans over robots, and to preclude the use of robots for evil purposes. In practice, however — meaning in Asimov's numerous and highly imaginative stories — a variety of difficulties arise.

My purpose here is to determine whether or not Asimov's fiction vindicates the laws he expounded. Does he successfully demonstrate that robotic technology can be applied in a responsible manner to potentially powerful, semiautonomous, and, in some sense, intelligent machines? To reach a conclusion, we must examine many issues emerging from Asimov's fiction.

History. The robot notion derives from two strands of thought, humanoids and automata. The notion of a humanoid (or human-like nonhuman) dates back to Pandora in *The Iliad*, 2,500 years ago — and even further. Egyptian, Babylonian, and ultimately Sumerian legends fully 5,000 years old reflect the widespread image of the creation, with god-men breathing life into clay models. One variation on the theme is the idea of the golem, associated with the Prague ghetto of the sixteenth century. This clay model, when breathed into life, became a useful but destructive ally.

The golem was an important precursor to Mary Shelley's *Frankenstein: The Modern Prometheus* (1818). This story combined the notion of the humanoid with the dangers of science (as suggested by the myth of Prometheus, who stole fire from the gods to give it to mortals). In addition to establishing a literary tradition and the genre of horror stories, *Frankenstein* also imbued humanoids with an aura of ill fate.

Automata, the second strand of thought, are literally "self-moving things" and have long interested mankind. Early models depended on levers and wheels, or on hydraulics. Clockwork technology enabled significant advances after the thirteenth century, and later steam and electro-

Isaac Asimov, 1920-1992

Born near Smolensk in Russia, Isaac Asimov came to the United States with his parents three years later. He grew up in Brooklyn, becoming a US citizen at the age of eight. He earned bachelor's, master's, and doctoral degrees in chemistry from Columbia University and qualified as an instructor in biochemistry at Boston University School of Medicine, where he taught for many years and performed research in nucleic acid.

As a child, Asimov had begun reading the science fiction stories on the racks in his family's candy store, and those early years of vicarious visits to strange worlds had filled him with an undying desire to write his own adventure tales. He sold his first short story in 1938, and after wartime service as a chemist and a short hitch in the Army, he focused increasingly on his writing.

Asimov was among the most prolific of authors, publishing hundreds of books on various subjects and dozens of short stories. His Laws of Robotics underlie four of his full-length novels as well as many of his short stories. The World Science Fiction Convention bestowed Hugo Awards on Asimov in nearly every category of science fiction, and his short story "Nightfall" is often referred to as the best science fiction story ever written. The scientific authority behind his writing gave his stories a feeling of authenticity, and his work undoubtedly did much to popularize science for the reading public.

mechanics were also applied. The primary purpose of automata was entertainment rather than employment as useful artifacts. Although many patterns were used, the human form always excited the greatest fascination. During the twentieth century, several new technologies moved automata into the utilitarian realm. Geduld and Gottesman[8] and Frude[2] review the chronology of clay model, water clock, golem, homunculus, android, and cyborg that culminated in the contemporary concept of the robot.

The term robot derives from the Czech word *robota*, meaning forced work or compulsory service, or *robotnik*, meaning serf. It was first used by the Czech playwright Karel Çapek in 1918 in a short story and again in his 1921 play *R.U.R.*, which stood for Rossum's Universal Robots. Rossum, a fictional Englishman, used biological methods to invent and mass-produce "men" to serve humans. Eventually they rebelled, became the dominant race, and wiped out humanity. The play was soon well known in English-speaking countries.

Definition. Undeterred by its somewhat chilling origins (or perhaps ignorant of them), technologists of the 1950s appropriated the term robot to refer to machines controlled by programs. A robot is "a reprogrammable multifunctional device designed to manipulate and/or transport material through variable programmed motions for the performance of a variety of tasks."[9] The term robotics — which Asimov claims he coined in 1942[10] — refers to "a science or art involving both artificial intelligence (to reason) and mechanical engineering (to perform physical acts suggested by reason)."[11]

As currently defined, robots exhibit three key elements:

- programmability, implying computational or symbol-manipulative capabilities that a designer can combine as desired (a robot is a computer);

Asimov's Laws of Robotics (1940)

First Law:
A robot may not injure a human being, or, through inaction, allow a human being to come to harm.

Second Law:
A robot must obey the orders given it by human beings, except where such orders would conflict with the First Law.

Third Law:
A robot must protect its own existence as long as such protection does not conflict with the First or Second Law.

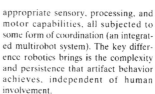

- mechanical capability, enabling it to act on its environment rather than merely function as a data processing or computational device (a robot is a machine); and
- flexibility, in that it can operate using a range of programs and manipulate and transport materials in a variety of ways.

We can conceive of a robot, therefore, as either a computer-enhanced machine or as a computer with sophisticated input/output devices. Its computing capabilities enable it to use its motor devices to respond to external stimuli, which it detects with its sensory devices. The responses are more complex than would be possible using mechanical, electromechanical, and/or electronic components alone.

With the merging of computers, telecommunications networks, robotics, and distributed systems software, and the multiorganizational application of the hybrid technology, the distinction between computers and robots may become increasingly arbitrary. In some cases it would be more convenient to conceive of a principal intelligence with dispersed sensors and effectors, each with subsidiary intelligence (a robotics-enhanced computer system). In others would be more realistic to think in terms of multiple devices, each with appropriate sensory, processing, and motor capabilities, all subjected to some form of coordination (an integrated multirobot system). The key difference robotics brings is the complexity and persistence that artifact behavior achieves, independent of human involvement.

Many industrial robots resemble humans in some ways. In science fiction the tendency has been even more pronounced, and readers encounter humanoid robots, humaniform robots, and androids. In fiction, as in life, it appears that a robot needs to exhibit only a few human-like characteristics to be treated as if it were human. For example, the relationships between humans and robots in many of Asimov's stories seem almost intimate, and audiences worldwide reacted warmly to the "personality" of the computer HAL in *2001: A Space Odyssey*, and to the gibbering rubbish-bin R2-D2 in the Star Wars series.

The tendency to conceive of robots in humankind's own image may gradually yield to utilitarian considerations, since artifacts can be readily designed to transcend humans' puny sensory and motor capabilities. Frequently the disadvantages and risks involved in incorporating sensory, processing, and motor apparatus within a single housing clearly outweigh the advantages.

Many robots will therefore be anything but humanoid in form. They may increasingly comprise powerful processing capabilities and associated memories in a safe and stable location, communicating with one or more sensory and motor devices (supported by limited computing capabilities and memory) at or near the location(s) where the robot performs its functions. Science fiction literature describes such architectures.[12,13]

Impact. Robotics offers benefits such as high reliability, accuracy, and speed of operation. Low long-term costs of computerized machines may result in significantly higher productivity, particularly in work involving variability within a general pattern. Humans can be relieved of mundane work and exposure to dangerous workplaces. Their capabilities can be extended into hostile environments involving high pressure (deep water), low pressure (space), high temperatures (furnaces), low temperatures (ice caps and cryogenics), and high-radiation areas (near nuclear materials or occurring naturally in space).

On the other hand, deleterious consequences are possible. Robots might directly or indirectly harm humans or their property; or the damage may be economic or incorporeal (for example, to a person's reputation). The harm could be accidental or result from human instructions. Indirect harm may occur to workers, since the application of robots generally results in job redefinition and sometimes in outright job displacement. Moreover, the replacement of humans by machines may undermine the self-respect of those affected, and perhaps of people generally.

During the 1980s, the scope of information technology applications and their impact on people increased dramatically. Control systems for chemical processes and air conditioning are examples of systems that already act directly and powerfully on their environments. And consider computer-integrated manufacturing, just-in-time logistics, and automated warehousing systems. Even data processing systems have become integrated into organizations' operations and constrain the ability of operations-level staff to query a machine's decisions and conclusions. In short, many modern computer systems are arguably robotic in nature already; their impact must be managed — now.

The 1940 Laws of Robotics

Asimov's original laws (see previous sidebar) provide that robots are to be slaves to humans (the second law). However, this role is overridden by the higher order first law, which precludes robots from injuring a human, either by their own autonomous action or by following a human's instructions. This precludes their continuing with a programmed activity when doing so would result in human injury. It also prevents their being used as a tool or accomplice in battery, murder, self-mutilation, or suicide.

The third and lowest level law creates a robotic survival instinct. This ensures that, in the absence of conflict with a higher order law, a robot will

- seek to avoid its own destruction through natural causes or accident,
- defend itself against attack by another robot or robots, and
- defend itself against attack by any human or humans.

Being neither omniscient nor omnipotent, it may of course fail in its endeavors. Moreover, the first law ensures that the robotic survival instinct fails if self-defense would necessarily involve injury to any human. For robots to successfully defend themselves against humans, they would have to be provided with sufficient speed and dexterity so as not to impose injurious force on a human.

Under the second law, a robot appears to be required to comply with a human order to (1) not resist being destroyed or dismantled, (2) cause itself to be destroyed, or (3) (within the limits of paradox) dismantle itself.[12] In various stories, Asimov notes that the order to self-destruct does not have to be obeyed if obedience would result in harm to a human. In addition, a robot would generally not be precluded from seeking clarification of the order. In his last full-length novel, Asimov appears to go further by envisaging that court procedures would be generally necessary before a robot could be destroyed: "I believe you should be dismantled without delay. The case is too dangerous to await the slow majesty of the law.... If there are legal repercussions hereafter, I shall deal with them."[14]

Such apparent inconsistencies attest to the laws' primary role as a literary device intended to support a series of stories about robot behavior. In this, they were very successful: "There was just enough ambiguity in the Three Laws to provide the conflicts and uncertainties required for new stories, and, to my great relief, it seemed always to be possible to think up a new angle out of the 61 words of the Three Laws."[1]

As Frude says, "The Laws have an interesting status. They ... may easily be broken, just as the laws of a country may be transgressed. But Asimov's provision for building a representation of the Laws into the positronic-brain circuitry ensures that robots are physically prevented from contravening them."[2] Because the laws are intrinsic to the machine's design, it should "never even enter into a robot's mind" to break them.

Subjecting the laws to analysis may seem unfair to Asimov. However, they have attained such a currency not only among sci-fi fans but also among practicing roboticists and software developers that they influence, if only subconsciously, the course of robotics.

Asimov's experiments with the 1940 laws

Asimov's early stories are examined here not in chronological sequence or on the basis of literary devices, but by looking at clusters of related ideas.

The ambiguity and cultural dependence of terms. Any set of "machine values" provides enormous scope for linguistic ambiguity. A robot must be able to distinguish robots from humans. It must be able to recognize an order and distinguish it from a casual request. It must "understand" the concept of its own existence, a capability that arguably has eluded mankind, although it may be a simpler matter for robots. In one short story, for example, the vagueness of the word *firmly* in the order "Pull [the bar] towards you firmly" jeopardizes a vital hyperspace experiment. Because robot strength is much greater than that of humans, it pulls the bar more powerfully than the human had intended, bends it, and thereby ruins the control mechanism.[15]

Defining injury and harm is particularly problematic, as are distinctions between death, mortal danger, and injury or harm that is not life-threatening. Beyond this, there is psychological harm. Any robot given, or developing, an awareness of human feelings would have to evaluate injury and harm in psychological as well as physical terms: "The insurmountable First Law of Robotics states: 'A robot may not injure a human being . . .', and *to repel a friendly gesture would do injury*"[16] (emphasis added). Asimov investigated this in an early short story and later in a novel: A mind-reading robot interprets the first law as requiring him to give people not the correct answers to their questions but the answers that he knows they want to hear.[14,16,17]

Another critical question is how a robot is to interpret the term human. A robot could be given any number of subtly different descriptions of a human being, based, for example, on skin color, height range, and/or voice characteristics such as accent. It is therefore possible for robot behavior to be manipulated: "The Laws, even the First Law, might not be an absolute then, but might be whatever those who design robots define them to be."[14] Faced with this difficulty, the robots in this story conclude that "... if different robots are subject to narrow definitions of one sort or another, there can only be measureless destruction. We define human beings as all members of the species, Homo sapiens."[14]

In an early short story, Asimov has a humanoid robot represent itself as a human and stand for public office. It must prevent the public from realizing that it is a robot, since public reaction would not only result in its losing the election but also in tighter constraints on other robots. A political opponent, seeking to expose the robot, discovers that it is impossible to prove it is a robot solely on the basis of its behavior, because the Laws of Robotics force any robot to perform in essentially the same manner as a good human being.[7] In a later novel, a roboticist says, "If a robot is human enough, he would be accepted as human. Do you demand proof that I am not a robot? The fact that I *seem* human is enough."[16] In another scene, a humaniform robot is sufficiently similar to a human to confuse a normal robot and slow down its reaction time.[14] Ultimately, two advanced robots recognize each other as "human," at least for the purposes of the laws.[14,18]

Defining human beings becomes more difficult with the emergence of cyborgs, which may be seen as either machine-enhanced humans or biologically enhanced machines. When a human is augmented by prostheses (artificial limbs, heart pacemakers, renal dialysis machines, artificial lungs, and someday perhaps many other devices), does the notion of a human gradually blur with that of a robot? And does a robot that attains increasingly human characteristics (for example, a knowledge-based system provided with the "know-that" and "know-how" of a human expert and the ability to learn more about a domain) gradually become confused with a human? How would a robot interpret the first and second laws once the Turing test criteria can be routinely satisfied? The key outcome of the most important of Asimov's robot novellas[12] is the tenability of the argument that the prosthetization of humans leads inevitably to the humanization of robots.

The cultural dependence of meaning reflects human differences in such matters as religion, nationality, and social status. As robots become more capable, however, cultural differences between humans and robots might also be a factor. For example, in one story[19] a human suggests that some laws may be bad and their enforcement unjust, but the robot replies that an unjust law is a contradiction in terms. When the human refers to something higher than justice, for example, mercy and forgiveness, the robot merely responds, "I am not acquainted with those words."

The role of judgment in decision making. The assumption that there is a literal meaning for any given series of signals is currently considered naive. Typically, the meaning of a term is seen to depend not only on the context in which it was originally expressed but also on the context in which it is read (see, for example, Winograd and Flores[20]). If this is so, then robots must exercise judgment to interpret the meanings of words and hence of orders and of new data.

A robot must even determine whether and to what extent the laws apply to a particular situation. Often in the robot stories a robot action of any kind is impossible without some degree of risk to a human. To be at all useful to its human masters, a robot must therefore be able to judge how much the laws can be breached to maintain a tolerable level of risk. For example, in Asimov's very first robot short story, "Robbie [the robot] snatched up Gloria [his young human owner], slackening his speed not one iota, and, conse-

> **Does the prosthetization of humans lead inevitably to the humanization of robots?**

quently, knocking every breath of air out of her."[21] Robbie judged that it was less harmful for Gloria to be momentarily breathless than to be mown down by a tractor.

Similarly, conflicting orders may have to be prioritized, for example, when two humans give inconsistent instructions. Whether the conflict is overt, unintentional, or even unwitting, it nonetheless requires a resolution. Even in the absence of conflicting orders, a robot may need to recognize foolish or illegal orders and decline to implement them, or at least question them. One story asks, "Must a robot follow the orders of a child; or of an idiot; or of a criminal; or of a perfectly decent intelligent man who happens to be inexpert and therefore ignorant of the undesirable consequences of his order?"[18]

Numerous problems surround the valuation of individual humans. First, do all humans have equal standing in a robot's evaluation? On the one hand they do: "A robot may not judge whether a human being deserves death. It is not for him to decide. He may not harm a human — variety skunk or variety angel."[7] On the other hand they might not, as when a robot tells a human, "In conflict between your safety and that of another, I must guard yours."[22] In another short story, robots agree that they "must obey a human being who is fit by mind, character, and knowledge to give me that order." Ultimately, this leads the robot to "disregard shape and form in judging between human beings" and to recognize his companion robot not merely as human but as a human "more fit than the others."[18] Many subtle problems can be constructed. For example, a person might try forcing a robot to comply with an instruction to harm a human (and thereby violate the first law) by threatening to kill himself unless the robot obeys.

How is a robot to judge the trade-off between a high probability of lesser harm to one person versus a low probability of more serious harm to another? Asimov's stories refer to this issue but are somewhat inconsistent with each other and with the strict wording of the first law.

More serious difficulties arise in relation to the valuation of multiple humans. The first law does not even contemplate the simple case of a single terrorist threatening many lives. In a variety of stories, however, Asimov interprets the law to recognize circumstances in which a robot may have to injure or even kill one or more humans to protect one or more others: "The Machine cannot harm a human being more than minimally, and that only to save *a greater number*"[23] (emphasis

> **The more subtle life-and-death cases might fall well outside a robot's appreciation.**

added). And again: "The First Law is not absolute. What if harming a human being saves the lives of two others, or three others, or even three billion others? The robot may have thought that saving the Federation took precedence over the saving of one life."[24]

These passages value humans exclusively on the basis of numbers. A later story includes this justification: "To expect robots to make judgments of fine points such as talent, intelligence, the general usefulness to society, has always seemed impractical. That would delay decision to the point where the robot is effectively immobilized. So we go by numbers."[18]

A robot's cognitive powers might be sufficient for distinguishing between attacker and attackee, but the first law alone does not provide a robot with the means to distinguish between a "good" person and a "bad" one. Hence, a robot may have to constrain a "good" attackee's self-defense to protect the "bad" attacker from harm. Similarly, disciplining children and prisoners may be difficult under the laws, which would limit robots' usefulness for supervision within nurseries and penal institutions.[22] Only after many generations of self-development does a humanoid robot learn to reason that "what seemed like cruelty [to a human] might, in the long run, be kindness."[12]

The more subtle life-and-death cases, such as assistance in the voluntary euthanasia of a fatally ill or injured person to gain immediate access to organs that would save several other lives, might fall well outside a robot's appreciation. Thus, the first law would require a robot to protect the threatened human, unless it was able to judge the steps taken to be the least harmful strategy. The practical solution to such difficult moral questions would be to keep robots out of the operating theater.[22]

The problem underlying all of these issues is that most probabilities used as input to normative decision models are not objective; rather, they are estimates of probability based on human (or robot) judgment. The extent to which judgment is central to robotic behavior is summed up in the cynical rephrasing of the first law by the major (human) character in the four novels: "A robot must not hurt a human being, unless he can think of a way to prove it is for the human being's ultimate good after all."[19]

The sheer complexity. To cope with the judgmental element in robot decision making, Asimov's later novels introduced a further complication: "On ... [worlds other than Earth], ... the Third Law is distinctly stronger in comparison to the Second Law. . . . An order for self-destruction would be questioned and there would have to be a truly legitimate reason for it to be carried through — a clear and present danger."[16] And again, "Harm through an active deed outweighs, in general, harm through passivity — all things being reasonably equal. . . . [A robot is] always to choose truth over nontruth, if the harm is roughly equal in both directions. In general, that is."[16]

The laws are not absolutes, and their

force varies with the individual machine's programming, the circumstances, the robot's previous instructions, and its experience. To cope with the inevitable logical complexities, a human would require not only a predisposition to rigorous reasoning, and a considerable education, but also a great deal of concentration and composure. (Alternatively, of course, the human may find it easier to defer to a robot suitably equipped for fuzzy-reasoning-based judgment.)

The strategies as well as the environmental variables involve complexity. "You must not think . . . that robotic response is a simple yes or no, up or down, in or out. . . . There is the matter of speed of response."[16] In some cases (for example, when a human must be physically restrained), the degree of strength to be applied must also be chosen.

The scope for dilemma and deadlock. A deadlock problem was the key feature of the short story in which Asimov first introduced the laws. He constructed the type of stand-off commonly referred to as the "Buridan's ass" problem. It involved a balance between a strong third-law self-protection tendency, causing the robot to try to avoid a source of danger, and a weak second-law order to approach that danger. "The conflict between the various rules is [meant to be] ironed out by the different positronic potentials in the brain," but in this case the robot "follows a circle around [the source of danger], staying on the locus of all points of . . . equilibrium."[5]

Deadlock is also possible within a single law. An example under the first law would be two humans threatened with equal danger and the robot unable to contrive a strategy to protect one without sacrificing the other. Under the second law, two humans might give contradictory orders of equivalent force. The later novels address this question with greater sophistication:

What was troubling the robot was what roboticists called an equipotential of contradiction on the second level. Obedience was the Second Law and [the robot] was suffering from two roughly equal and contradictory orders. Robot-block was what the general population called it or, more frequently, roblock for short . . . [or] "mental freeze-out." . . . No matter how subtle and intricate a brain might be, there is always some way of setting up a contradiction. This is a fundamental truth of mathematics.[16]

Clearly, robots subject to such laws need to be programmed to recognize deadlock and either choose arbitrarily among the alternative strategies or arbitrarily modify an arbitrarily chosen strategy variable (say, move a short distance in any direction) and reevaluate the situation: "If A and not-A are precisely equal misery-producers according to his judgment, he chooses one or the other in a completely unpredictable way and then follows that unquestioningly. He does *not* go into mental freeze-out."[16]

The finite time that even robot decision making requires could cause another type of deadlock. Should a robot act immediately, by "instinct," to protect a human in danger? Or should it pause long enough to more carefully analyze available data — or collect more data — perhaps thereby discovering a better solution, or detecting that other humans are in even greater danger? Such situations can be approached using the techniques of information economics, but there is inherent scope for ineffectiveness and deadlock, colloquially referred to as "paralysis by analysis."

Asimov suggested one class of deadlock that would not occur: If in a given situation a robot knew that it was powerless to prevent harm to a human, then the first law would be inoperative; the third law would become relevant, and it would not self-immolate in a vain attempt to save the human.[25] It does seem, however, that the deadlock is not avoided by the laws themselves, but rather by the presumed sophistication of the robot's decision-analytical capabilities.

A special case of deadlock arises when a robot is ordered to wait. For example. "'[Robot], you will not move nor speak nor hear us until I say your name again.' There was no answer. The robot sat as though it were cast out of one piece of metal, and it would stay so until it heard its name again."[26] As written, the passage raises the intriguing question of whether passive hearing is possible without active listening. What if the robot's name is next used in the third person rather than the second?

In interpreting a command such as "Do absolutely nothing until I call you!" a human would use common sense and, for example, attend to bodily functions in the meantime. A human would do *nothing about the relevant matter* until the event occurred. In addition, a human would recognize additional terminating events, such as a change in circumstances that make it impossible for the event to ever occur. A robot is likely to be constrained to a more literal interpretation, and unless it can infer a scope delimitation to the command, it would need to place the majority of its functions in abeyance. The faculties that would need to remain in operation are

• the sensory-perceptive subsystem needed to detect the condition,

- the recommencement triggering function,
- one or more daemons to provide a time-out mechanism (presumably the scope of the command is at least restricted to the expected remaining lifetime of the person who gave the command), and
- the ability to play back the audit trail so that an overseer can discover the condition on which the robot's resuscitation depends.

Asimov does not appear to have investigated whether the behavior of a robot in wait mode is affected by the laws. If it isn't, then it will not only fail to protect its own existence and to obey an order, but will also stand by and allow a human to be harmed. A robotic security guard could therefore be nullified by an attacker's simply putting it into a wait state.

Audit of robot compliance. For a fiction writer, it is sufficient to have the laws embedded in robots' positronic pathways (whatever they may be). To actually apply such a set of laws in robot design, however, it would be necessary to ensure that every robot

- had the laws imposed in precisely the manner intended, and
- was at all times subject to them — that is, they could not be overridden or modified.

It is important to know how malprogramming and modification of the laws' implementation in a robot (whether intentional or unintentional) can be prevented, detected, and dealt with. In an early short story, robots were "rescuing" humans whose work required short periods of relatively harmless exposure to gamma radiation. Officials obtained robots with the first law modified so that they were incapable of injuring a human but under no compulsion to prevent one from coming to harm. This clearly undermined the remaining part of the first law, since, for example, a robot could drop a heavy weight toward a human, knowing that it would be fast enough and strong enough to catch it before it harmed the person. However, once gravity had taken over, the robot would be free to ignore the danger.[25] Thus, a partial implementation was shown to be risky, and the importance of robot audit underlined. Other risks include trapdoors, Trojan horses, and similar devices in the robot's programming.

A further imponderable is the effect of hostile environments and stress on the reliability and robustness of robots' performance in accordance with the laws. In one short story, it transpires that "The Machine That Won the War" had been receiving only limited and poor-quality data as a result of enemy action against its receptors and had been processing it unreliably because of a shortage of experienced maintenance staff. Each of the responsible managers had, in the interests of national morale, suppressed that information, even from one another, and had separately and independently "introduced a number of necessary biases" and "adjusted" the processing parameters in accordance with intuition. The executive director, even though unaware of the adjustments, had placed little reliance on the machine's output, preferring to carry out his responsibility to mankind by exercising his own judgment.[27]

A major issue in military applications generally[28] is the impossibility of contriving effective compliance tests for complex systems subject to hostile and competitive environments. Asimov points out that the difficulties of assuring compliance will be compounded by the design and manufacture of robots by other robots.[22]

Robot autonomy. Sometimes humans may delegate control to a robot and find themselves unable to regain it, at least in a particular context. One reason is that to avoid deadlock, a robot must be capable of making arbitrary decisions. Another is that the laws embody an explicit ability for a robot to disobey an instruction, by virtue of the overriding first law.

In an early Asimov short story, a robot "knows he can keep [the energy beam] more stable than we [humans] can, since he insists he's the superior being, so he must keep us out of the control room [in accordance with the first law]."[29] The same scenario forms the basis of one of the most vivid episodes in science fiction, HAL's attempt to wrest control of the spacecraft from Bowman in *2001: A Space Odyssey*. Robot autonomy is also reflected in a lighter moment in one of Asimov's later novels, when a character says to his companion, "For now I must leave you. The ship is coasting in for a landing, and I must stare intelligently at the computer that controls it, or no one will believe I am the captain."[14]

In extreme cases, robot behavior will involve subterfuge, as the machine determines that the human, for his or her own protection, must be tricked. In another early short story, the machines that manage Earth's economy implement a form of "artificial stupidity" by making intentional errors, thereby encouraging humans to believe that the robots are fallible and that humans still have a role to play.[23]

Scope for adaptation. The normal pattern of any technology is that successive generations show increased sophistication, and it seems inconceivable that robotic technology would quickly reach a plateau and require little further development. Thus there will always be many old models in existence, models that may have inherent technical weaknesses resulting in occasional malfunctions and hence infringement on the Laws of Robotics. Asimov's short stories emphasize that robots are leased from the manufacturer, never sold, so that old models can be withdrawn after a maximum of 25 years.

Looking at the first 50 years of software maintenance, it seems clear that successive modification of existing software to perform new or enhanced functions is one or more orders of magnitude harder than creating a new artifact to perform the same function. Doubts must exist about the ability of humans (or robots) to reliably adapt existing robots. The alternative — destruction of existing robots — will be resisted in accordance with the third law, robot self-preservation.

At a more abstract level, the laws are arguably incomplete because the frame of reference is explicitly human. No recognition is given to plants, animals, or as-yet-undiscovered (for example, extraterrestrial), intelligent life forms. Moreover, some future human cultures may place great value on inanimate creation, or on holism. If, however, late twentieth-century values have meanwhile been embedded in robots, that future culture may have difficulty wresting the right to change the values of the robots it has inherited. If machines are to have value sets, there must be a mechanism for adaptation, at least through human-imposed change. The difficulty is that most such value sets will be implicit rather than explicit; their effects will be scattered across a system rather than implemented in a modular and therefore replaceable manner.

At first sight, Asimov's laws are intuitively appealing, but their application encounters difficulties. Asimov, in his fiction, detected and investigated the laws' weaknesses, which this article (Part 1 of 2) has analyzed and classified. Part 2, in the next issue of *Computer*, will take the analysis further by considering the effects of Asimov's 1985 revision to the laws. It will then examine the extent to which the weaknesses in these laws may in fact be endemic to any set of laws regulating robotic behavior. ■

References

1. I. Asimov, *The Rest of the Robots* (a collection of short stories originally published between 1941 and 1957), Grafton Books, London, 1968.

2. N. Frude, *The Robot Heritage*, Century Publishing, London, 1984.

3. I. Asimov, *I, Robot* (a collection of short stories originally published between 1940 and 1950), Grafton Books, London, 1968.

4. I. Asimov, P.S. Warrick, and M.H. Greenberg, eds., *Machines That Think*, Holt, Rinehart, and Wilson, London, 1983.

5. I. Asimov, "Runaround" (originally published in 1942), reprinted in Reference 3, pp. 33-51.

6. L. Del Rey, "Though Dreamers Die" (originally published in 1944), reprinted in Reference 4, pp. 153-174.

7. I. Asimov, "Evidence" (originally published in 1946), reprinted in Reference 3, pp. 159-182.

8. H.M. Geduld and R. Gottesman, eds., *Robots, Robots, Robots*, New York Graphic Soc., Boston, 1978.

9. P.B. Scott, *The Robotics Revolution: The Complete Guide*, Blackwell, Oxford, 1984.

10. I. Asimov, *Robot Dreams* (a collection of short stories originally published between 1947 and 1986), Victor Gollancz, London, 1989.

11. A. Chandor, ed., *The Penguin Dictionary of Computers*, 3rd ed., Penguin, London, 1985.

12. I. Asimov, "The Bicentennial Man" (originally published in 1976), reprinted in Reference 4, pp. 519-561. Expanded into I. Asimov and R. Silverberg, *The Positronic Man*, Victor Gollancz, London, 1992.

13. A.C. Clarke and S. Kubrick, *2001: A Space Odyssey*, Grafton Books, London, 1968.

14. I. Asimov, *Robots and Empire*, Grafton Books, London, 1985.

15. I. Asimov, "Risk" (originally published in 1955), reprinted in Reference 1, pp. 122-155.

16. I. Asimov, *The Robots of Dawn*, Grafton Books, London, 1983.

17. I. Asimov, "Liar!" (originally published in 1941), reprinted in Reference 3, pp. 92-109.

18. I. Asimov, "That Thou Art Mindful of Him" (originally published in 1974), reprinted in *The Bicentennial Man*, Panther Books, London, 1978, pp. 79-107.

19. I. Asimov, *The Caves of Steel* (originally published in 1954), Grafton Books, London, 1958.

20. T. Winograd and F. Flores, *Understanding Computers and Cognition*, Ablex, Norwood, N.J., 1986.

21. I. Asimov, "Robbie" (originally published as "Strange Playfellow" in 1940), reprinted in Reference 3, pp. 13-32.

22. I. Asimov, *The Naked Sun* (originally published in 1957), Grafton Books, London, 1960.

23. I. Asimov, "The Evitable Conflict" (originally published in 1950), reprinted in Reference 3, pp. 183-206.

24. I. Asimov, "The Tercentenary Incident" (originally published in 1976), reprinted in *The Bicentennial Man*, Panther Books, London, 1978, pp. 229-247.

25. I. Asimov, "Little Lost Robot" (originally published in 1947), reprinted in Reference 3, pp. 110-136.

26. I. Asimov, "Robot Dreams," first published in Reference 10, pp. 51-58.

27. I. Asimov, "The Machine That Won the War" (originally published in 1961), reprinted in Reference 10, pp. 191-197.

28. D. Bellin and G. Chapman, eds., *Computers in Battle: Will They Work?* Harcourt Brace Jovanovich, Boston, 1987.

29. I. Asimov, "Reason" (originally published in 1941), reprinted in Reference 3, pp. 52-70.

2

Asimov's Laws of Robotics: Implications for Information Technology

Part 2

Roger Clarke

Can a set of laws or rules reliably constrain the behavior of intelligent machines? Inadequacies in Asimov's laws suggest maybe not.

Isaac Asimov's Laws of Robotics, first formulated in 1940, were primarily a literary device intended to support a series of stories about robot behavior. Over time, he found that the three laws included enough apparent inconsistencies, ambiguity, and uncertainty to provide the conflicts required for a great many stories. In examining the ramifications of these laws, Asimov revealed problems that might later confront real roboticists and information technologists attempting to establish rules for the behavior of intelligent machines.

With their fictional "positronic" brains imprinted with the mandate to (in order of priority) prevent harm to humans, obey their human masters, and protect themselves, Asimov's robots had to deal with great complexity. In a given situation, a robot might be unable to satisfy the demands of two equally powerful mandates and go into "mental freezeout." Semantics is also a problem. As demonstrated in Part 1 of this article (*Computer*, December 1993, pp. 53-61), language is much more than a set of literal meanings, and Asimov showed us that a machine trying to distinguish, for example, who or what is human may encounter many difficulties that humans themselves handle easily and intuitively. Thus, robots must have sufficient capabilities for judgment — capabilities that can cause them to frustrate the intentions of their masters when, in a robot's judgment, a higher order law applies.

As information technology evolves and machines begin to design and build other machines, the issue of human control gains greater significance. In time, human values tend to change; the rules reflecting these values, and embedded in existing robotic devices, may need to be modified. But if they are implicit rather than explicit, with their effects scattered widely across a system, they may not be easily replaceable. Asimov himself discovered many contradictions and eventually revised the Laws of Robotics.

Asimov's 1985 revised Laws of Robotics

The zeroth law. After introducing the original three laws, Asimov detected, as early as 1950, a need to extend the first law, which protected individual humans, so that it would protect humanity as a whole. Thus, his calculating machines "have *the good of humanity* at heart through the overwhelming force of the First Law of Robotics"[1] (emphasis added). In 1985 he developed this idea further by postulating a "zeroth" law that placed humanity's interests above those of any individual while retaining a high value on individual human life.[2] The revised set of laws is shown in the sidebar.

Asimov pointed out that under a strict interpretation of the first law, a robot would protect a person even if the survival of humanity as a whole was placed at risk. Possible threats include annihilation by an alien or mutant human race, or by a deadly virus. Even when a robot's own powers of reasoning led it to conclude that mankind as a whole was doomed if it refused to act, it was nevertheless constrained: "I sense the oncoming of catastrophe . . . [but] I can only follow the Laws."[2]

In Asimov's fiction the robots are tested by circumstances and must seriously consider whether they can harm a human to save humanity. The turning point comes when the robots appreciate that the laws are indirectly modifiable by roboticists through the definitions programmed into each robot: "If the Laws of Robotics, even the First Law, are not absolutes, and if human beings can modify them, might it not be that perhaps, under proper conditions, we ourselves might mod — "[2] Although the robots are prevented by imminent "roblock" (robot block, or deadlock) from even completing the sentence, the groundwork has been laid.

Later, when a robot perceives a clear and urgent threat to mankind, it concludes, "Humanity as a whole is more important than a single human being. There is a law that is greater than the First Law: 'A robot may not injure humanity, or through inaction, allow humanity to come to harm.'"[2]

Defining "humanity." Modification of the laws, however, leads to additional considerations. Robots are increasingly required to deal with abstractions and philosophical issues. For example, the concept of humanity may be interpreted in different ways. It may refer to the set of individual human beings (a collective), or it may be a distinct concept (a generality, as in the notion of "the State"). Asimov invokes both ideas by referring to a tapestry (a generality) made up of individual contributions (a collective): "An individual life is one thread in the tapestry, and what is one thread compared to the whole? . . . Keep your mind fixed firmly on the tapestry and do not let the trailing off of a single thread affect you."[2]

A human roboticist raised a difficulty with the zeroth law immediately after the robot formulated it: "What is your 'humanity' but an abstraction? Can you point to humanity? You can injure or fail to injure a specific human being and understand the injury or lack of injury that has taken place. Can you see the injury to humanity? Can you understand it? Can you point to it?"[2] The robot later responds by positing an ability to "detect the hum of the mental activity of Earth's human population, overall. . . . And, extending that, can one not imagine that in the Galaxy generally there is the hum of the mental activity of all of humanity? How, then, is humanity an abstraction? It is something you can point to." Perhaps as Asimov's robots learn to reason with abstract concepts, they will inevitably become adept at sophistry and polemic.

The increased difficulty of judgment. One of Asimov's robot characters also points out the increasing complexity of the laws: "The First Law deals with specific individuals and certainties. Your Zeroth Law deals with vague groups and probabilities."[2] At this point, as he often does, Asimov resorts to poetic license and for the moment pretends that coping with harm to individuals does not involve probabilities. However, the key point is not affected: Estimating probabilities in relation to groups of humans is far more difficult than with individual humans.

> It is difficult enough, when one must choose quickly . . . , to decide which individual may suffer, or inflict, the greater harm. To choose between an individual and humanity, when you are not sure of what aspect of humanity you are dealing with, is so difficult that the very validity of Robotic Laws comes to be suspect. As soon as humanity in the abstract is introduced, the Laws of Robotics begin to merge with the Laws of Humanics — which may not even exist.[2]

Robot paternalism. Despite these difficulties, the robots agree to implement the zeroth law, since they judge themselves more capable than anyone else of dealing with the problems. The original laws produced robots with considerable autonomy, albeit a qualified autonomy allowed by humans. But under the 1985 laws, robots were more likely to adopt a superordinate, paternalistic attitude toward humans.

Asimov's revised Laws of Robotics (1985)

Zeroth Law:
A robot may not injure humanity, or, through inaction, allow humanity to come to harm.

First Law:
A robot may not injure a human being, or, through inaction, allow a human being to come to harm, unless this would violate the Zeroth Law of Robotics.

Second Law:
A robot must obey orders given it by human beings, except where such orders would conflict with the Zeroth or First Law.

Third Law:
A robot must protect its own existence as long as such protection does not conflict with the Zeroth, First, or Second Law.

Asimov suggested this when he first hinted at the zeroth law, because he had his chief robotpsychologist say that ". . . we can no longer understand our own creations. . . . [Robots] have progressed beyond the possibility of detailed human control."[1] In a more recent novella, a robot proposes to treat his form "as a canvas on which I intend to draw a man," but is told by the roboticist, "It's a puny ambition. . . . You're better than a man. You've gone downhill from the moment you opted for organicism."[3]

In the later novels, a robot with telepathic powers manipulates humans to act in a way that will solve problems,[4] although its powers are constrained by the psychological dangers of mind manipulation. Naturally, humans would be alarmed by the very idea of a mind-reading robot; therefore, under the zeroth and first laws, such a robot would be permitted to manipulate the minds of humans who learned of its abilities, making them forget the knowledge so that they could not be harmed by it. This is reminiscent of an Asimov story in which mankind is an experimental laboratory for higher beings[5] and Adams' altogether more flippant *Hitchhiker's Guide to the Galaxy*, in which the Earth is revealed as a large experiment in which humans are being used as laboratory animals by, of all things, white mice.[6] Someday those manipulators of humans might be robots.

Asimov's *The Robots of Dawn* is essentially about humans, with robots as important players. In the sequel *Robots and Empire*, however, the story is dominated by the two robots, and the humans seem more like their playthings. It comes as little surprise, then, that the robots eventually conclude that "it is not sufficient to be able to choose [among alternative humans or classes of human] . . . ; we must be able to shape."[2] Clearly, any subsequent novels in the series would have been about robots, with humans playing "bit" parts.

Robot dominance has a corollary that pervades the novels: History "grew less interesting as it went along; it became almost soporific."[4] With life's challenges removed, humanity naturally regresses into peace and quietude,

"My parents were cold and unfeeling."

becoming "placid, comfortable, and unmoving" — and stagnant.

So who's in charge? As we have seen, the term human can be variously defined, thus significantly affecting the first law. The term humanity did not appear in the original laws, only in the zeroth law, which Asimov had formulated and enunciated by a robot.[2] Thus, the robots define human and humanity to refer to themselves as well as to humans, and ultimately to themselves alone. Another of the great science fiction stories, Clarke's *Rendezvous with Rama*,[7] also assumes that an alien civilization, much older than mankind, would consist of robots alone (although in this case Clarke envisioned biological robots). Asimov's vision of a robot takeover differs from those of previous authors only in that force would be unnecessary.

Asimov does *not* propose that the zeroth law must inevitably result in the ceding of species dominance by humans to robots. However, some concepts may be so central to humanness that any attempt to embody them in computer processing might undermine the ability of humanity to control its own fate. Weizenbaum argues this point more fully.[8]

The issues discussed here, and in Part 1, have grown increasingly speculative, and some are more readily associated with metaphysics than with contemporary applications of information technology. However, they demonstrate that even an intuitively attractive extension to the original laws could have very significant ramifications. Some of the weaknesses are probably inherent in any set of laws and hence in any robotic control regime.

Asimov's laws extended

The behavior of robots in Asimov's stories is not satisfactorily explained by the laws he enunciated. This section examines the design requirements necessary to effectively subject robotic behavior to the laws. In so doing, it becomes necessary to postulate several additional laws implicit in Asimov's fiction.

Perceptual and cognitive apparatus. Clearly, robot design must include sophisticated sensory capabilities. However, more than signal reception is needed. Many of the difficulties Asimov dramatized arose because robots were less than omniscient. Would humans, knowing that robots' cognitive capabilities are limited, be prepared to trust their judgment on life-and-death matters? For example, the fact that any single robot cannot harm a human does not protect humans from being injured or killed by robotic actions. In one story, a human tells a robot to add a chemical to a glass of milk and then tells another robot to serve the milk to a human. The result is murder by poisoning. Similarly, a robot untrained in first aid might move an accident victim and break the person's spinal cord. A human character in *The Naked Sun* is so incensed by these shortcomings that he accuses roboticists of perpetrating a fraud on mankind by omitting key words from the first law. In effect, it really means "A robot may do nothing that *to its knowledge* would injure a human being, and may not, through inaction, *knowingly* allow a human being to come to harm."[9]

Robotic architecture must be designed so that the laws can effectively control a robot's behavior. A robot requires a basic grammar and vocabulary to "understand" the laws and converse with humans. In one short story, a production accident results in a "mentally retarded" robot. This robot, defending itself against a feigned attack by a human, breaks its assailant's arm. This was not a breach of the first law, because it did not knowingly injure the human: "In brushing aside the threatening arm . . . it could not know the bone would break. In human terms, no moral blame can be attached to an individual who honestly cannot differentiate good and evil."[10] In Asimov's stories, instructions sometimes must be phrased carefully to be interpreted as mandatory. Thus, some authors have considered extensions to the apparatus of robots, for example, a "button labeled '*Implement Order*' on the robot's chest,"[11] analogous to the Enter key on a computer's keyboard.

A set of laws for robotics cannot be independent but must be conceived as part of a system. A robot must also be endowed with data collection, decision-analytical, and action processes by which it can apply the laws. Inadequate sensory, perceptual, or cognitive faculties would undermine the laws' effectiveness.

Additional implicit laws. In his first robot short story, Asimov stated that "long before enough can go wrong to alter that First Law, a robot would be completely inoperable. It's a mathematical impossibility [for Robbie the Robot to harm a human]."[12] For this to be true, robot design would have to incorporate a high-order controller (a "conscience"?) that would cause a robot to detect any potential for noncompliance with the laws and report the problem — or immobilize itself. The implementation of such a meta-law ("A robot may not act unless its actions are subject to the laws of robotics") might well strain both the technology and the underlying science. (Given the meta-language problem in twentieth-century philosophy, perhaps logic itself would be strained.) This difficulty highlights the simple fact that robotic behavior cannot be entirely automated; it is dependent on design and maintenance by an external agent.

Another of Asimov's requirements is that all robots must be subject to the laws at all times. Thus, it would have to be illegal for human manufacturers to create a robot that was not subject to the laws. In a future world that makes significant use of robots, their design and manufacture would naturally be undertaken by other robots. Therefore, the Laws of Robotics must include the stipulation that no robot may commit an act that could result in any robot's not being subject to the same laws.

The words "protect its own existence" raise a semantic difficulty. In *The Bicentennial Man*, Asimov has a robot achieve humanness by taking its own life. Van Vogt, however, wrote that "indoctrination against suicide" was considered a fundamental requirement.[13] The solution might be to interpret the word *protect* as applying to all threats, or to amend the wording to explicitly preclude self-inflicted harm.

Having to continually instruct robot slaves would be both inefficient and tiresome. Asimov hints at a further, deep-nested law that would compel robots to perform the tasks they were trained for:

> Quite aside from the Three Laws, there isn't a pathway in those brains that isn't carefully designed and fixed. We have robots planned for specific tasks, *implanted with specific capabilities*.[14] (Emphasis added.)

So perhaps we can extrapolate an additional, lower priority law: "A robot must perform the duties for which it has been programmed, except where that would conflict with a higher order law."

Asimov's laws regulate robots' transactions with humans and thus apply

"He was yelling at the robot, 'Am I the only one around here with a head on his shoulders? Now knock it off!'"

where robots have relatively little to do with one another or where there is only one robot. However, the laws fail to address the management of large numbers of robots. In several stories, a robot is assigned to oversee other robots. This would be possible only if each of the lesser robots were instructed by a human to obey the orders of its robot overseer. That would create a number of logical and practical difficulties, such as the scope of the human's order. It would seem more effective to incorporate in all subordinate robots an additional law, for example, "A robot must obey the orders given it by superordinate robots except where such orders would conflict with a higher order law." Such a law would fall between the second and third laws.

Furthermore, subordinate robots should protect their superordinate robot. This could be implemented as an extension or corollary to the third law; that is, to protect itself, a robot would have to protect another robot on which it depends. Indeed, a subordinate robot may need to be capable of sacrificing itself to protect its robot overseer. Thus, an additional law superior to the third law but inferior to orders from either a human or a robot overseer seems appropriate: "A robot must protect the existence of a superordinate robot as long as such protection does not conflict with a higher order law."

The wording of such laws should allow for nesting, since robot overseers may report to higher level robots. It would also be necessary to determine the form of the superordinate relationships:

- a tree, in which each robot has precisely one immediate overseer, whether robot or human;
- a constrained network, in which each robot may have several overseers but restrictions determine who may act as an overseer; or
- an unconstrained network, in which each robot may have any number of other robots or persons as overseers.

This issue of a command structure is far from trivial, since it is central to democratic processes that no single entity shall have ultimate authority. Rather, the most senior entity in any decision-making hierarchy must be subject to review and override by some other entity, exemplified by the balance of power in the three branches of government and the authority of the ballot box. Successful, long-lived systems involve checks and balances in a lattice rather than a mere tree structure. Of course, the structures and processes of human organizations may prove inappropriate for robotic organization. In any case, additional laws of some kind would be essential to regulate relationships among robots.

The sidebar shows an extended set of laws, one that incorporates the additional laws postulated in this section. Even this set would not always ensure appropriate robotic behavior.

An extended set of the Laws of Robotics

The Meta-Law:
A robot may not act unless its actions are subject to the Laws of Robotics.

Law Zero:
A robot may not injure humanity, or, through inaction, allow humanity to come to harm.

Law One:
A robot may not injure a human being, or, through inaction, allow a human being to come to harm, unless this would violate a higher order law.

Law Two:
(a) A robot must obey orders given it by human beings, except where such orders would conflict with a higher order law.
(b) A robot must obey orders given it by superordinate robots, except where such orders would conflict with a higher order law.

Law Three:
(a) A robot must protect the existence of a superordinate robot as long as such protection does not conflict with a higher order law.
(b) A robot must protect its own existence as long as such protection does not conflict with a higher order law.

Law Four:
A robot must perform the duties for which it has been programmed, except where that would conflict with a higher order law.

The Procreation Law:
A robot may not take any part in the design, manufacture, or maintenance of a robot unless the new or modified robot's actions are subject to the Laws of Robotics.

However, it does reflect the implicit laws that emerge in Asimov's fiction while demonstrating that any realistic set of design principles would have to be considerably more complex than Asimov's 1940 or 1985 laws. This additional complexity would inevitably exacerbate the problems identified earlier in this article and create new ones.

While additional laws may be trivially simple to extract and formulate, the need for them serves as a warning. The 1940 laws' intuitive attractiveness and simplicity were progressively lost in complexity, legalisms, and semantic richness. Clearly then, formulating an actual set of laws as a basis for engineering design would result in similar difficulties and require a much more formal approach. Such laws would have to be based in ethics and human morality, not just in mathematics and engineering. Such a political process would probably result in a document couched in fuzzy generalities rather than constituting an operational-level, programmable specification.

Implications for information technologists

Many facets of Asimov's fiction are clearly inapplicable to real information technology or too far in the future to be relevant to contemporary applications. Some matters, however, deserve our consideration. For example, Asimov's fiction could help us assess the practicability of embedding some appropriate set of general laws into robotic designs. Alternatively, the substantive content of the laws could be used as a set of guidelines to be applied during the conception, design, development, testing, implementation, use, and maintenance of robotic systems. This section explores the second approach.

Recognition of stakeholder interests. The Laws of Robotics designate no particular class of humans (not even a robot's owner) as more deserving of protection or obedience than another. A human might establish such a relationship by command, but the laws give such a command no special status; another human could therefore countermand it. In short, the laws reflect the humanistic and egalitarian principles that theoretically underlie most democratic nations.

The laws therefore stand in stark contrast to our conventional notions about an information technology artifact, whose owner is implicitly assumed to be its primary beneficiary. An organization shapes an application's design and use for its own benefit. Admittedly, during the last decade users have been given greater consideration in terms of both the human-machine interface and participation in system development. But that trend has been justified by the better returns the organization can get from its information technology investment rather than by any recognition that users are stakeholders with a legitimate voice in decision making. The interests of other affected parties are even less likely to be reflected.

In this era of powerful information technology, professional bodies of information technologists need to consider

- identification of stakeholders and how they are affected;
- prior consultation with stakeholders;
- quality assurance standards for design, manufacture, use, and maintenance;
- liability for harm resulting from either malfunction or use in conformance with the designer's intentions; and
- complaint-handling and dispute-resolution procedures.

Once any resulting standards reach a degree of maturity, legislatures in the many hundreds of legal jurisdictions throughout the world would probably have to devise enforcement procedures.

The interests of people affected by modern information technology applications have been gaining recognition. For example, consumer representatives are now being involved in the statement of user requirements and the establishment of the regulatory environment for consumer electronic-funds-transfer systems. This participation may extend to the logical design of such systems. Other examples are trade-union negotiations with employers regarding technology-enforced change, and the publication of software quality-assurance standards.

For large-scale applications of information technology, governments have been called upon to apply procedures like those commonly used in major industrial and social projects. Thus, commitment might have to be deferred pending dissemination and public discussion of independent environmental or social impact statements. Although organizations that use information technology might see this as interventionism, decision making and approval for major information technology applications may nevertheless become more widely representative.

Closed-system versus open-system thinking. Computer-based systems no longer comprise independent machines each serving a single location. The marriage of computing with telecommunications has produced multicomponent systems designed to support all elements of a widely dispersed organization. Integration hasn't been simply geographic, however. The practice of information systems has matured since the early years when existing manual systems were automated largely without procedural change. Developers now seek payback via the rationalization of existing systems and varying degrees of integration among previously separate functions. With the advent of strategic and interorganizational systems, economies are being sought at the level of industry sectors, and functional integration increasingly occurs across corporate boundaries.

Although programmers can no longer regard the machine as an almost entirely closed system with tightly circumscribed sensory and motor capabilities, many habits of closed-system thinking remain. When systems have multiple components, linkages to other systems, and sophisticated sensory and motor capabilities, the scope needed for understanding and resolving problems is much broader than for a mere hardware/software machine. Human activities in particular must be perceived as part of the system. This applies to manual procedures within systems (such as reading dials on control panels), human activities on the fringes of systems (such as decision making based on computer-collated and -displayed information), and the security of the user's environment (automated teller machines, for example). The focus must broaden from mere technology to technology in use.

General systems thinking leads information technologists to recognize that relativity and change must be accommodated. Today, an artifact may be applied in multiple cultures where language, religion, laws, and customs differ. Over time, the original context may change. For example, models for a criminal justice system — one based on punishment and another based on redemption — may alternately dominate social thinking. Therefore, complex systems must be capable of adaptation.

Blind acceptance of technological and other imperatives. Contemporary utilitarian society seldom challenges the presumption that what *can* be done *should* be done. Although this technological imperative is less pervasive than people generally think, societies nevertheless tend to follow where their technological capabilities lead. Related tendencies include the economic imperative (what can be done more efficiently should be) and the marketing imperative (any effective demand should be met). An additional tendency might be called the "information imperative," the dominance of administrative efficiency, information richness, and rational decision making. However, the collection of personal data has become so pervasive that citizens and employees have begun to object.

The greater a technology's potential to promote change, the more carefully a society should consider the desirability of each application. Complementary measures that may be needed to ameliorate its negative effects should also be considered. This is a major theme of Asimov's stories, as he explores the hidden effects of technology. The potential impact of information technology is so great that it would be inexcusable for professionals to succumb blindly to the economic, marketing, information, technological, and other imperatives. Application software professionals can no longer treat the implications of information technology as someone else's problem but must consider them as part of the project.[15]

Human acceptance of robots. In Asimov's stories, humans develop affection for robots, particularly humaniform robots. In his very first short story, a little girl is too closely attached to Robbie the Robot for her parents' liking.[12] In another early story, a woman starved for affection from her husband and sensitively assisted by a humanoid robot to increase her self-confidence entertains thoughts approaching love toward it/him.[16]

Nonhumaniforms, such as conventional industrial robots and large, highly dispersed robotic systems (such as warehouse managers, ATMs, and EFT/POS systems) seem less likely to elicit such warmth. Yet several studies have found a surprising degree of identification by humans with computers.[17,18] Thus, some hitherto exclusively human characteristics are being associ-

> **If a robot-based economy develops without equitable adjustments, the backlash could be considerable.**

ated with computer systems that don't even exhibit typical robotic capabilities.

Users must be continually reminded that the capabilities of hardware/software components are limited:

- They contain many inherent assumptions,
- they are not flexible enough to cope with all of the manifold exceptions that inevitably arise,
- they do not adapt to changes in their environment, and
- authority is not vested in hardware/software components but rather in the individuals who use them.

Educational institutions and staff training programs must identify these limitations; yet even this is not sufficient: The human-machine interface must reflect them. Systems must be designed so that users are required to continually exercise their own expertise, and system output should not be phrased in a way that implies unwarranted authority. These objectives challenge the conventional outlook of system designers.

Human opposition to robots. Robots are agents of change and therefore potentially upsetting to those with vested interests. Of all the machines so far invented or conceived of, robots represent the most direct challenge to humans. Vociferous and even violent campaigns against robotics should not be surprising. Beyond concerns of self-interest is the possibility that some humans could be revulsed by robots, particularly those with humanoid characteristics. Some opponents may be mollified as robotic behavior becomes more tactful. Another tenable argument is that by creating and deploying artifacts that are in some ways superior, humans degrade themselves.

System designers must anticipate a variety of negative reactions against their creations from different groups of stakeholders. Much will depend on the number and power of the people who feel threatened — and on the scope of the change they anticipate. If, as Asimov speculates,[9] a robot-based economy develops without equitable adjustments, the backlash could be considerable.

Such a rejection could involve powerful institutions as well as individuals. In one Asimov story, the US Department of Defense suppresses a project intended to produce the perfect robot-soldier. It reasons that the degree of discretion and autonomy needed for battlefield performance would tend to make robots rebellious in other circumstances (particularly during peace time) and unprepared to suffer their commanders' foolish decisions.[19] At a more basic level, product lines and markets might be threatened, and hence the profits and even the survival of corporations. Although even very powerful cartels might not be able to impede robotics for very long, its development could nevertheless be delayed or altered. Information technologists need to recognize the negative perceptions of various stakeholders and manage both system design and project politics accordingly.

The structuredness of decision making. For five decades there has been little doubt that computers hold significant computational advantages over humans. However, the merits of machine decision making remain in dispute. Some decision processes are highly structured and can be resolved using known algorithms operating on defined data items with defined interrelationships. Most structured decisions are candidates for automation, subject, of course, to economic constraints. The advantages of machines must also be balanced against risks. The choice to automate must be made carefully because the automated decision process (algorithm, problem description, problem-domain description, or analysis of empirical data) may later prove to be inappropriate for a particular type of decision. Also, humans involved as data providers, data communicators, or decision implementers may not perform rationally because of poor training, poor performance under pressure, or willfulness.

Unstructured decision making remains the preserve of humans for one or more of the following reasons:

- Humans have not yet worked out a suitable way to program (or teach) a machine how to make that class of decision.
- Some relevant data cannot be communicated to the machine.
- "Fuzzy" or "open-textured" concepts or constructs are involved.
- Such decisions involve judgments that system participants feel should not be made by machines on behalf of humans.

One important type of unstructured decision is problem diagnosis. As Asimov described the problem, "How ... can we send a robot to find a flaw in a mechanism when we cannot possibly give precise orders, since we know nothing about the flaw ourselves? 'Find out what's wrong' is not an order you can give to a robot; only to a man."[20] Knowledge-based technology has since been applied to problem diagnosis, but Asimov's insight retains its validity: A problem may be linguistic rather than technical, requiring common sense, not domain knowledge. Elsewhere, Asimov calls robots "logical but not reasonable" and tells of household robots removing important evidence from a murder scene because a human did not think to order them to preserve it.[9]

The literature of decision support systems recognizes an intermediate case, semistructured decision making. Humans are assigned the decision task,

> **A problem may be linguistic rather than technical, requiring common sense, not domain knowledge.**

and systems are designed to provide support for gathering and structuring potentially relevant data and for modeling and experimenting with alternative strategies. Through continual progress in science and technology, previously unstructured decisions are reduced to semistructured or structured decisions. The choice of which decisions to automate is therefore provisional, pending further advances in the relevant area of knowledge. Conversely, because of environmental or cultural change, structured decisions may not remain so. For example, a family of viruses might mutate so rapidly that the reference data within diagnostic support systems is outstripped and even the logic becomes dangerously inadequate.

Delegating to a machine any kind of decision that is less than fully structured invites errors and mishaps. Of course, human decision-makers routinely make mistakes too. One reason for humans' retaining responsibility for unstructured decision making is rational: Appropriately educated and trained humans may make more right decisions and/or fewer seriously wrong decisions than a machine. Using common sense, humans can recognize when conventional approaches and criteria do not apply, and they can introduce conscious value judgments. Perhaps a more important reason is the arational preference of humans to submit to the judgments of their peers rather than of machines: If someone is going to make a mistake costly to me, better for it to be an understandably incompetent human like myself than a mysteriously incompetent machine.[8]

Because robot and human capabilities differ, for the foreseeable future at least, each will have specific comparative advantages. Information technologists must delineate the relationship between robots and people by applying the concept of decision structuredness to blend computer-based and human elements advantageously. The goal should be to achieve complementary intelligence rather than to continue pursuing the chimera of unneeded artificial intelligence. As Wyndham put it in 1932: "Surely man and machine are natural complements: They assist one another."[21]

Risk management. Whether or not subjected to intrinsic laws or design guidelines, robotics embodies risks to property as well as to humans. These risks must be managed; appropriate forms of risk avoidance and diminution need to be applied, and regimes for fallback, recovery, and retribution must be established.

Controls are needed to ensure that intrinsic laws, if any, are operational at all times and that guidelines for design, development, testing, use, and maintenance are applied. Second-order control mechanisms are needed to audit first-order control mechanisms. Furthermore, those bearing legal responsibility for harm arising from the use of robotics must be clearly identified. Courtroom litigation may determine the actual amount of liability, but assigning legal responsibilities in advance will ensure that participants take due care.

In most of Asimov's robot stories, robots are owned by the manufacturer even while in the possession of individual humans or corporations. Hence, legal responsibility for harm arising from robot noncompliance with the laws can be assigned with relative ease. In most real-world jurisdictions, however, there are enormous uncertainties, substantial gaps in protective coverage, high costs, and long delays.

Each jurisdiction, consistent with its own product liability philosophy, needs to determine who should bear the various risks. The law must be sufficiently clear so that debilitating legal battles do not leave injured parties without recourse or sap the industry of its energy. Information technologists need to communicate to legislators the importance of revising and extending the laws that assign liability for harm arising from the use of information technology.

Enhancements to codes of ethics. Associations of information technology professionals, such as the IEEE Computer Society, the Association for Computing Machinery, the British Computer Society, and the Australian Computer Society, are concerned with professional standards, and these standards almost always include a code of ethics. Such codes aren't intended so much to establish standards as to express standards that already exist informally. Nonetheless, they provide guidance concerning how professionals should perform their work, and there is significant literature in the area.

The issues raised in this article suggest that existing codes of ethics need to be reexamined in the light of developing technology. Codes generally fail to reflect the potential effects of computer-enhanced machines and the inadequacy of existing managerial, institutional, and legal processes for coping with inherent risks. Information technology professionals need to stimulate and inform debate on the issues. Along with robotics, many other technologies deserve consideration. Such an endeavor would mean reassessing professionalism in the light of fundamental works on ethical aspects of technology.

Asimov's Laws of Robotics have been a very successful literary device. Perhaps ironically, or perhaps because it was artistically appropriate, the sum of Asimov's stories disprove the contention that he began with: It is *not* possible to reliably constrain the behavior of robots by devising and applying a set of rules.

The freedom of fiction enabled Asimov to project the laws into many future scenarios; in so doing, he uncovered issues that will probably arise someday in real-world situations. Many aspects of the laws discussed in this article are likely to be weaknesses in any robotic code of conduct. Contemporary applications of information technology such as CAD/CAM, EFT/POS, warehousing systems, and traffic control are already exhibiting robotic characteristics. The difficulties identified are therefore directly and immediately relevant to information technology professionals.

Increased complexity means new sources of risk, since each activity depends directly on the effective interaction of many artifacts. Complex systems are prone to component failures and malfunctions, and to intermodule inconsistencies and misunderstandings. Thus, new forms of backup, problem diagnosis, interim operation, and recovery are needed. Tolerance and flexibility in design must replace the primacy of short-term objectives such as programming productivity. If information technologists do not respond to the challenges posed by robotic systems, as investigated in Asimov's stories, information technology artifacts will be poorly suited for real-world applications. They may be used in ways not intended by their designers, or simply be rejected as incompatible with the individuals and organizations they were meant to serve. ∎

> **Tolerance and flexibility in design must replace the primacy of short-term objectives such as programming productivity.**

References

1. I. Asimov, "The Evitable Conflict" (originally published in 1950), reprinted in I. Asimov, *I, Robot*, Grafton Books, London, 1968, pp. 183-206.

2. I. Asimov, *Robots and Empire*, Grafton Books, London, 1985.

3. I. Asimov, "The Bicentennial Man" (originally published in 1976), reprinted in I. Asimov, P.S. Warrick, and M.H. Greenberg, eds., *Machines That Think*, Holt, Rinehart, and Wilson, 1983, pp. 519-561.

4. I. Asimov, *The Robots of Dawn*, Grafton Books, London, 1983.

5. I. Asimov, "Jokester" (originally published in 1956), reprinted in I. Asimov, *Robot Dreams*, Victor Gollancz, London, 1989, pp. 278-294.

6. D. Adams, *The Hitchhiker's Guide to the Galaxy*, Harmony Books, New York, 1979.

7. A.C. Clarke, *Rendezvous with Rama*, Victor Gollancz, London, 1973.

8. J. Weizenbaum, *Computer Power and Human Reason*, W.H. Freeman, San Francisco, 1976.

9. I. Asimov, *The Naked Sun* (originally published in 1957), Grafton Books, London, 1960.

10. I. Asimov, "Lenny" (originally published in 1958), reprinted in I. Asimov, *The Rest of the Robots*, Grafton Books, London, 1968, pp. 158-177.

11. H. Harrison, "War With the Robots" (originally published in 1962), reprinted in I. Asimov, P.S. Warrick, and M.H. Greenberg, eds., *Machines That Think*, Holt, Rinehart, and Wilson, 1983, pp. 357-379.

12. I. Asimov, "Robbie" (originally published as "Strange Playfellow" in 1940), reprinted in I. Asimov, *I, Robot*, Grafton Books, London, 1968, pp. 13-32.

13. A.E. Van Vogt, "Fulfillment" (originally published in 1951), reprinted in I. Asimov, P.S. Warrick, and M.H. Greenberg, eds., *Machines That Think*, Holt, Rinehart, and Wilson, 1983, pp. 175-205.

14. I. Asimov, "Feminine Intuition" (originally published in 1969), reprinted in I. Asimov, *The Bicentennial Man*, Panther Books, London, 1978, pp. 15-41.

15. R.A. Clarke, "Economic, Legal, and Social Implications of Information Technology," *MIS Quarterly*, Vol. 12, No. 4, Dec. 1988, pp. 517-519.

16. I. Asimov, "Satisfaction Guaranteed" (originally published in 1951), reprinted in I. Asimov, *The Rest of the Robots*, Grafton Books, London, 1968, pp. 102-120.

17. J. Weizenbaum, "Eliza," *Comm. ACM*, Vol. 9, No. 1, Jan. 1966, pp. 36-45.

18. S. Turkle, *The Second Self: Computers and the Human Spirit*, Simon & Schuster, New York, 1984.

19. A. Budrys, "First to Serve" (originally published in 1954), reprinted in I. Asimov, M.H. Greenberg, and C.G. Waugh, eds., *Robots*, Signet, New York, 1989, pp. 227-244.

20. I. Asimov, "Risk" (originally published in 1955), reprinted in I. Asimov, *The Rest of the Robots*, Grafton Books, London, 1968, pp. 122-155.

21. J. Wyndham, "The Lost Machine" (originally published in 1932), reprinted in A. Wells, ed., *The Best of John Wyndham*, Sphere Books, London, 1973, pp. 13-36, and in I. Asimov, P.S. Warrick, and M.H. Greenberg, eds., *Machines That Think*, Holt, Rinehart, and Wilson, 1983, pp. 29-49.

3

Prolegomena to any future artificial moral agent

COLIN ALLEN, GARY VARNER and JASON ZINSER

Abstract. As artificial intelligence moves ever closer to the goal of producing fully autonomous agents, the question of how to design and implement an artificial moral agent (AMA) becomes increasingly pressing. Robots possessing autonomous capacities to do things that are useful to humans will also have the capacity to do things that are harmful to humans and other sentient beings. Theoretical challenges to developing artificial moral agents result both from controversies among ethicists about moral theory itself, and from computational limits to the implementation of such theories. In this paper the ethical disputes are surveyed, the possibility of a 'moral Turing Test' is considered and the computational difficulties accompanying the different types of approach are assessed. Human-like performance, which is prone to include immoral actions, may not be acceptable in machines, but moral perfection may be computationally unattainable. The risks posed by autonomous machines ignorantly or deliberately harming people and other sentient beings are great. The development of machines with enough intelligence to assess the effects of their actions on sentient beings and act accordingly may ultimately be the most important task faced by the designers of artificially intelligent automata.

1. Introduction

A good web server is a computer that efficiently serves up html code. A good chess program is one that wins chess games. There are some grey areas and fuzzy edges, of course. Is a good chess program one that wins most games or just some? Against just ordinary competitors or against world class players? But in one sense of the question, it is quite clear what it means to build a good computer or write a good program. A good one is one that fulfills the purpose we had in building it.

However, if you wanted to build a computer or write a program that is good in a moral sense, that is *a good moral agent*, it is much less clear what would count as success. Yet as artificial intelligence moves ever closer to the goal of producing fully autonomous agents, the question of how to design and implement an artificial moral agent becomes increasingly pressing. Robots possessing autonomous capacities to do things that are useful to humans will also have the capacity to do things that are harmful to humans and other sentient beings. How to curb these capacities for harm is a topic that is beginning to move from the realm of science fiction to the realm of real-world engineering problems. As Picard (1997) puts it: 'The greater the freedom of a machine, the more it will need moral standards'.

Attempts to build an artificial moral agent (henceforth AMA) are stymied by two areas of deep disagreement in ethical theory. One is at the level of moral principle: ethicists disagree deeply about what standards moral agents ought to follow. Some hold that the fundamental moral norm is the principle of utility, which defines right actions and policies in terms of maximizing aggregate good consequences, while others hold that certain kinds of actions are unjustifiable even if a particular token of the type would maximize aggregate good. The other level of disagreement is more conceptual or even ontological: apart from the question of what standards a moral agent ought to follow, what does it mean to *be* a moral agent? A suitably generic characterization might be that a moral agent is an individual who takes into consideration the interests of others rather than acting solely to advance his, her, or its (henceforth its) self-interest. But would a robot which had been programmed to follow certain standards be, *ipso facto*, a moral agent? Or would the robot have to also be capable of thinking about what it is doing in certain ways, for example by explicitly using certain norms in its decision procedure? Would it also have to be able to conceive of what it was doing as taking the moral point of view? More strongly still, if a moral agent must be autonomous in a rich sense, would it have to be capable of misapplying the standards in question, or even intentionally and knowingly disobeying them?

This paper surveys such perplexing problems facing attempts to create an AMA. The following section discusses in greater detail the two areas of disagreement in ethical theory just sketched. Subsequent sections turn to a consideration of the computational difficulties facing various approaches to designing and evaluating AMAs.

2. Moral agency and moral norms

Disagreement among ethical theorists about which norms moral agents ought to follow and disagreement about what it means to be a moral agent are interrelated. In this paper discussion is restricted to several of the best-known and most widely debated approaches to ethical theory. Even such a restricted survey is adequate to show the interrelations between disagreements about norms and about what it means to be a moral agent, for the connections become apparent as soon as two of the best-known approaches are sketched. The two approaches considered are utilitarianism, on the one hand, and Kant's use of the 'categorical imperative' on the other.

As a general school of thought, utilitarianism is the view that the best actions and institutions are those which produce the best aggregate consequences. Although there is disagreement even among utilitarians about which consequences matter in this estimation, the classical utilitarians (Bentham 1780, Mill 1861 and Sidgwick 1874) were sentientists, holding that effects on the consciousness of sentient beings are ultimately the only events of direct moral significance. Roughly, the classical utilitarians held that the best actions are those which produce the greatest happiness for the greatest number.

Mill famously held that 'it is better to be a human being dissatisfied than a pig satisfied', (1957 [1861], p.14) and he emphasized that one of the things which makes human happiness qualitatively superior to animals' (so that a relatively unhappy human might still be leading a better life than a thoroughly happy pig) was humans' capacity for moral agency. Mill held, further, that our evaluations of a moral agent's general character ought to be kept separate from our evaluation of any particular action by that agent:

> utilitarian moralists have gone beyond almost all others in affirming that the motive has nothing to do with the morality of the action, though much with the worth of the agent. He who saves a fellow creature from drowning does what is morally right, whether his motive be duty or the hope of being paid for his trouble (Mill 1957 [1861], pp.23–24).

There is, then, a clear sense of 'morally good' which a utilitarian can apply to an agent's actions irrespective of how the agent decided upon that action. Not so for Kant. According to Kant, for an action to be *morally* good, the action must, as he put it, be done out of respect for the categorical imperative. An enormous literature exists debating what exactly Kant meant by 'the categorical imperative' and by 'acting out of respect for it'. Here we follow one account of the categorical imperative and assume that 'acting out of respect for it' means simply that the agent acted as it did because it determined that the action in question was consistent with the categorical imperative. In this sense of the term, an action cannot be morally good unless the agent in fact reasoned in certain fairly complex ways.

Kant offers numerous formulations of the categorical imperative, but one of the most widely discussed is this: 'Act only on that maxim through which you can at the same time will that it should become a universal law' (1948 [1785], p.88). Singer (1971) provides a particularly insightful interpretation of this principle which we assume for purposes of this paper. By a maxim Kant means, roughly, an explicit and fully stated principle of practical reason. Specifically, a maxim has three elements: a goal which the agent proposes to achieve by acting on it; a means or course of action by which the agent proposes to achieve that goal; and a statement of the circumstances under which acting in that way will achieve the goal in question. The categorical imperative is a negative test, that is, it does not tell you which specific maxims to act on, rather, it requires you to never act on a maxim which you could not 'at the same time will that it should become a universal law'. By this we take him to mean (following Singer) that you could effectively achieve your goal in a world in which everyone sought to achieve the same goal by acting the same way in similar circumstances.

As Singer shows, this way of understanding the categorical imperative saves Kant from certain infamous objections. For present purposes, what is important, however, is that for Kant, a morally good action is one which is done because the agent has put the maxim of its action to the test of the categorical imperative and seen that it passes. In this way the motive of the action (trying to be good by sticking to what the categorical imperative requires) is essential to its being morally good, as are the specific deliberations involved in deciding that the maxim of one's action passes the categorical imperative.

In Kant and Mill, then, we find very different moral principles tied to very different conceptions of what a good moral agent is. In Mill's utilitarian terms, we might say that an agent is morally good to the extent that its behaviour positively affects the aggregate good of the moral community. In this sense a robot could be said to be a morally good agent to the extent that it has been programmed to act consistently with the principle of utility, regardless of how this behavioural result is achieved. For Kant, however, any claim than an agent is morally good (on either a specific occasion or in general) implies claims about the agent's internal deliberative processes. On Kant's view, to build a good AMA would require us to implement certain specific cognitive processes and to make these processes an integral part of the agent's decision-making procedure.

One further point about Kant illustrates another fundamental question about what would count as success in constructing an AMA. Kant held that the categorical imperative is only an 'imperative' for humans: 'for the *divine* will, and in general for a *holy* will, there are no imperatives: "*I ought*" is here out of place, because "*I will*" is already of itself necessarily in harmony with the law' (1948 [1785], p.81). If, as Kant appears to think, being a moral agent carries with it the need to *try* to be good, and thus the capacity for moral failure, then we will not have constructed a true artificial *moral* agent if we make it incapable of acting immorally. Some kind of autonomy, carrying with it the capacity for failure, may be essential to being a real *moral* agent. However, as we suggest below, the basic goals when constructing an *artificial* moral agent are likely to be very different than when raising a *natural* moral agent like a child. Accordingly, it may be acceptable to program a computer to be incapable of failure but unacceptable to attempt the analogue when raising a child.

3. A 'Moral Turing Test'?

In both ethical theory and day-to-day talk about ethics, people disagree about the morality of various actions. Kant claimed that it is always immoral to lie, no matter what the consequences (although Singer argues that Kant's own principles do not entail this conclusion). A utilitarian would deny it, holding instead that lying is justified whenever its consequences are sufficiently good in the aggregate. And day-to-day life is rife with disagreements about the morality of particular actions, of lifestyle choices and of social institutions.

In the face of such diverse views about what standards we ought to live by, an attractive criterion for success in constructing an AMA would be a variant of Turing's (1950) 'Imitation Game' (aka the Turing Test). In the standard version of the Turing Test, an 'interrogator' is charged with distinguishing a machine from a human based on interacting with both via printed language alone. A machine passes the Turing Test if, when paired with a human being, the 'interrogator' cannot identify the human at a level above chance. Turing's intention was to produce a behavioural test which bypasses disagreements about standards defining intelligence or successful acquisition of natural language. A Moral Turing Test (MTT) might similarly be proposed to bypass disagreements about ethical standards by restricting the standard Turing Test to conversations about morality. If human 'interrogators' cannot identify the machine at above chance accuracy, then the machine is, on this criterion, a moral agent.

One limitation on this approach is the emphasis it places on the machine's ability to *articulate* moral judgments. A Kantian might well be satisfied with this emphasis, since Kant required that a good moral agent not only act in particular ways but act as a result of reasoning through things in a certain way. Just as clearly, however, both a utilitarian approach and common sense suggest that the MTT places too much emphasis on the ability to articulate one's reasons for one's actions. As seen above, Mill allows that various actions are morally good independently of the agent's motivations, and many people think that young children, or even dogs, are moral agents even though they are incapable of articulating the reasons for their actions.

To shift the focus from conversational ability to action, an alternative MTT could be structured in such a way that the 'interrogator' is given pairs of descriptions of actual, morally-significant actions of a human and an AMA, purged of all references that would identify the agents. If the interrogator correctly identifies the machine at a level above chance, then the machine has failed the test. A problem for this version of the MTT is that distinguishability is the wrong criterion because the machine might be recognizable for acting in ways that are consistently *better* than a human in the same situation. So instead, the "interrogator" might be asked to assess whether one agent is less moral than the other. If the machine is not identified as the less moral member of the pair significantly more often than the human, then it has passed the test. This is called the 'comparative MTT' (cMTT).

There are several problems for the cMTT. First, one might argue that despite setting a slightly higher standard for machine behaviour than for humans, the standard is nevertheless too low. When setting out to design an artificial moral agent, we might think it appropriate to demand more than we expect of, say, human children. That is, the goal of AI researchers when constructing an AMA should not be just to construct a moral agent, but to construct an exemplary or even a perfect moral agent. The cMTT allows the machine's aggregate performance to contain actions that would be judged as morally wrong, and morally worse than the human actions, so long as on balance these do not cause the machine to be rated lower than the human. The cMTT could, in response, be tightened to require that the machine not be judged worse than the human in any of the pairwise comparisons of specific actions. But even if this restriction is added, there is a second way in which the standard might be thought too low, namely that the human behaviour is, itself, typically far from being morally ideal.

We humans typically muddle along making mistakes while harbouring private regrets about our moral lapses, which occur more frequently than perhaps we care to admit. But while we expect and, to a certain extent, tolerate human moral failures, it is less clear that we would, or should, design the capacity for such failures into our machines. Calculated decisions that result in harm to others are likely to be much less tolerated in a machine than in another human being. In other words, we shall probably expect more of our machines than we do of ourselves. And while murderous rampages are beyond the pale for both humans and AMAs, curbs on more mundane forms of immorality – the lying, cheating, and swindling of daily life – represent a much more difficult computational challenge, for these capacities are much more prominent in the 'grey' areas of morality, such as 'white' lies; the maxim 'never lie' has far more exceptions than 'never kill'.

If a standard is to be set for the behaviour of AMAs that is higher than the standard set for humans, where are such standards going to come from? As already indicated, computer scientists who turn to moral philosophy for an answer to this and related questions will find a field that does not provide a universally-accepted moral theory. Furthermore, as philosophical objectives are not exactly the same as those of engineers, there is a considerable gap between the available theories and the design of algorithms that might implement AMAs. Nevertheless, the philosophical investigations provide a framework for thinking about the implementation issues.

Two basic kinds of approach to this task are considered: (i) theoretical approaches that implement an explicit theory of evaluation (like the principle of utility or Kant's categorical imperative) thus providing a framework for the AMA to compute a morally preferred action and (ii) modelling approaches that either implement a theory of moral character (e.g. virtue theory) or that use learning or evolution to construct systems that act morally. Approaches in both categories present interesting computational challenges.

4. Theoretical approaches

Gips (1995) identifies two basic classes of theoretical approach to AMAs: 'consequentialist' theories and 'deontological' theories. Respectively, these theoretical approaches would attempt to implement either consequence-oriented reasoning (such as utilitarianism) or a rule- or duty-based form of reasoning (such as Kant's).

4.1. *Consequentialism*

The crucial problem for the consequentialist approach is that utilitarianism would seem to be a computational black hole. To implement the theory, the effects of an action on every member of the moral community must be assigned a numerical value. The sheer impracticality of doing this in real time for real world actions should be evident, especially if one considers the fact that the direct effects of every action ripple outwards to further effects that also affect aggregate utility. We are confident that interactions between these effects would make practical computation of long term consequences an intractable problem. Furthermore, if utilities must be computed for as long as an action has an effect in the world, potentially for all time, there is the risk of a non-terminating procedure. Even if the termination problem can be solved, it would also require the implementation of a comprehensive scientific theory to predict the nature of these long-range effects.

To insist on knowing every last consequence of an action sets a standard much higher than we expect of humans. We don't typically judge the morality of an action for its effect on sentient beings many years on the future. Indeed all judgements would have to be suspended if we did so. One might, then, try to restrict the computational problem by establishing an horizon beyond which assessment is not required. Unfortunately this is unlikely to be a simple matter. One might try to establish a temporal horizon by setting a time limit for agent responsibility. Or one might try to establish a physical or social horizon by stipulating a number of links in a causal or social chain beyond which an agent is not held responsible. But for any horizon one establishes as the point at which utilitarian calculations are stopped, one can imagine an agent deliberately initiating a process that will result in enormous pain and suffering at some point beyond the horizon. Yet we would surely judge this to be an immoral act.

Perhaps the situation is not as hopeless as we have made it sound if some standard AI techniques of tree searching can be applied. The chess computer Deep Blue (Campbell 1997) does not need to search the entire space of chess moves into the indefinite future to play a good game. Techniques for abbreviating the search are applied. The algorithm can be directed towards accomplishing intermediate goals, using approximate methods of evaluation, and working backwards from a desired outcome to determine a plan. Perhaps similar strategies could be adapted to moral decisions guided by utilitarian principles.

One way to deal with this problems might be to implement a hybrid system. Such a hybrid system might involve initially consequentialist computations out to a specified limit, at which point more abstract principles of duty (deontology) or character (virtue) come into play. Conversely one might implement a deontological system that can be overridden by consequentialist reasoning whenever the good consequences of an action 'clearly' outweigh the bad (Hare's (1981) two levels of moral reasoning). Such hybrid systems must, however, face the computational difficulties inherent in deontological approaches.

4.2. *Deontology*

'Deontology' is a term of art that refers to the notion of duty. According to deontological theories, actions are to be assessed according to their conformity with certain rules or principles, like Kant's categorical imperative. Other examples are Gert's (1988) system based on ten simple moral rules, Asimov's (1990) three laws of robotics, and the 'golden rule' (treat others as you would wish them to treat you).

From a computational perspective, a major problem with most deontological approaches (with the possible exception of Kant's) is that there is the possibility of conflict between the implied duties. Such conflicts are, of course, a major plot device in Asimov's fiction. Notably, it is even the case that a deadlock can result as a consequence of Asimov's first law alone. The law states that a robot may not injure a human being, or, through inaction, allow a human being to come to harm. A deadlock will result when a robot is faced with a choice between action and inaction where in either case the result will be the injury of one human being but no injury to another. The other two of Asimov's laws do not resolve the deadlock as they pertain only to following human orders and self-preservation. While this makes for interesting fiction, it is not a good basis for AMA design.

Other deontological systems, for example Gert's ten rules (1988), also contain rules that will inevitably conflict. As Gips (1995) reports, Gert's theoretical solution is to allow a rule to be disobeyed so long as an impartial rational observer could publicly advocate that it may be disobeyed. At this point we note that anyone interested in a computational implementation of this theory must wonder how an AMA is supposed to decide whether an impartial observer would allow such a rule violation without implementing another moral theory to make that determination. A further problem for deontological systems attempting to provide a list of rules to cover all situations is that such a list may fail to be comprehensive.

More abstract deontological theories, such as Kant's categorical imperative and the golden rule, attempt to avoid both the problem of comprehensiveness and the problem of conflicts between rules by formulating a single rule intended to cover all actions. However, they bring other computational problems because they are cast at such an abstract level. For instance, to determine whether or not a particular action satisfies the categorical imperative, it is necessary for the AMA to recognize the goal of its own action, as well as assess the effects of all other (including human) moral agents' trying to achieve the same goal by acting on the same maxim. This would require an AMA to be programmed with a robust conception of its own and others' psychology in order to be able to formulate their reasons for actions, and the capacity for modelling the population-level effects of acting on its maxim – a task that is likely to be several orders of magnitude more complex than weather forecasting (although it is quite possibly a task for which we humans have been equipped by evolution).

Similarly, to implement the golden rule, an AMA must have the ability to characterize its own preferences under various hypothetical scenarios involving the effects of others' actions upon itself. Further, even if it does not require the ability to empathize with others, it must at least have the ability to compute the affective consequences of its actions on others in order to determine whether or not its action is something that it would choose to have others do to itself. And it must do all this while taking into account differences in individual psychology which result in different preferences for different kinds of treatment.

5. Models of morality

'Theoretical' approaches that implement an explicit ethical theory in an AMA are computationally complex. An alternative approach to producing autonomous moral agents would be to bypass programming them to use explicit rules or principles in deciding what to do, and attempt, instead, to produce agents which happen to make morally proper decisions by modelling the moral behaviour of humans. Three approaches are discernible in this general category. The first, with an ancient history, is to consider a moral agent as one who possesses certain virtues which guide the agent's behavior and to attempt to program those virtues directly (see also Coleman 1999). A second approach involves the development of a model of moral agency by an associative learning process. The third approach consists in simulating the evolution of moral agents.

5.1. *Virtue approaches*

The basic idea underlying virtue ethics is that character is primary over deeds because good character produces good deeds. As Gips (1995) points out, virtue approaches frequently get assimilated to deontological approaches as the relevant virtues get translated into duties. Therefore, insofar as this assimilation is appropriate, virtue approaches raise exactly the same computational issues as deontological approaches, including the problem of dealing with conflicts between competing virtues, such as honesty and compassion.

Even if virtue approaches should not be assimilated to deontology, and can be considered as providing a separate model of moral character, the computational complexity of mapping relatively abstract character traits onto real actions looms as large as the problem of programming the computational use of rules for moral behaviour. Specifying, for example, that an AMA should have the character of honesty, requires an algorithm for determining whether any given action is honestly performed. But as we have learned from various of Plato's dialogues (Hamilton and Cairns 1961), it is no easy task to formulate definitions of such traits. It will be correspondingly difficult to determine whether particular actions conform to the virtues.

It is also difficult to come up with a comprehensive list of virtues that would cover all the scenarios that an AMA might find itself in. If available, such a list of virtues would provide a top-down specification for a model of moral agency. A more tractable approach, however, might be to develop models of moral agency from the bottom up. Each of the next two approaches fall in that category.

5.2. *Associative learning*

Just as young children learn appropriate behaviour from experience, one might approach the development of AMAs by way of a simulated childhood consisting of a training period involving feedback about the moral acceptability of actions. (This approach is endorsed by Dennett (1997) as part of the Cog robot project (Brooks *et al.* 1999)).

The implementation details, whether artificial neural networks or some other form of associative learning scheme, need not particularly concern us here as we know that the learning algorithms are computationally tractable. However, anyone taking this approach should be concerned about the quality of the feedback. A simple binary indication of the acceptability or unacceptability of an action may not be adequate for training purposes. The psychological literature on moral development seems to indicate that the best moral training involves embedding approval and disapproval in a context of reasons for those judgements (Damon 1999). As described by Damon, motivation by punishment and reward features at the lowest, self-interested stage of moral development represented in Kohlberg's scheme (Kohlberg 1984). At the second level comes social approval, divided into concern for the opinions of others and respect for social structures such as the law. Abstract ideals are reached only at the third level on Kohlberg's moral development chart, and empirical research on young males suggests that this abstract level is achieved by only a small fraction by their mid-twenties. (According to Damon, while early research suggested gender differences, subsequent and better-controlled studies have failed to support this finding.)

The point for computationalists is that the simplest associationist techniques based on binary-valued feedback are unlikely to produce fully satisfactory models of moral agency. If an artificial moral agent is to pass some variant of the cMTT, it is likely to require a capacity for abstract moral reasoning. While it must be possible ultimately to implement such reasoning using, for example, artificial neural networks, AI is a long way from understanding the network architecture that is required to do so. Nevertheless, it is our belief that the implementation of simpler schemes will be an important step towards the development of more sophisticated systems, even if the products of these simpler schemes are unlikely to pass the cMTT.

5.3. *Evolutionary/sociobiological approaches*

Another approach to modelling moral agency is through simulated evolution or artificial life. Combining sociobiology and game theory, Danielson (1992) develops a concept of 'functional' morality that is based on the idea that rationality is the only necessary quality that an agent must possess to be a moral agent (see also Harms *et al.* 1999 and Skyrms 1996). This approach is exemplified in an iterated Prisoner's Dilemma (PD) game (Axelrod and Hamilton 1981). By being iterated, no player has knowledge of when the game will end; it could continue throughout a lifetime, or involve only one PD interaction with a certain player. Examples of the PD abound in nature, and are evident when animals interact socially, such as: predatory alert signals, mutual grooming, sharing of food and group raising of young. Through evolution, organisms which have mutually iterated PD interactions evolve into a stable set of co-operative interactions. Most importantly for our present purposes, it has been shown that it is functionally optimal if an organism co-operates with other organisms, or, as Dawkins (1989) stated 'nice guys finish first'. The 'moral rules' that emerge from this process have no higher justification than survival values, for in the iterated game-theoretical scenarios it is purely because it is in the best interests of rational agents that they co-operate and behave in a way that gives the appearance of morality.

The evolutionary models developed by Danielson, Harms *et al.* and Skyrms show what can be achieved with computational techniques and further investigations should be encouraged. But, as Danielson admits, real-world morality has evolved in scenarios far more complex than the rather simplistic games that constitute the artificial environments for their evolutionary simulations. In scaling these environments to more realistic environments, evolutionary approaches are likely to be faced with some of the same shortcomings of the associative learning approaches: namely that sophisticated moral agents must also be capable of constructing an abstract, theoretical conception of morality. And while it is axiomatic that humans have evolved that capacity, the question of whether a simulated evolutionary approach is the most effective computational technique for developing AMAs is far from being settled.

6. The role of emotions: perfect theorizers, imperfect practitioners?

The discussion so far has omitted any mention of the role of emotion in actual human morality. Emotions undoubtedly provide important motivators for human behaviour. Perfect knowledge of a moral theory does not guarantee action in conformity with that theory. Thus one might well imagine an agent who is fully capable of deciding whether certain actions are moral while being completely unmotivated to act morally. Such an agent might frequently act in a way he or she knew to be wrong. Indeed this is the description of some sociopaths.

As well as providing motivation, it has also been argued that emotions are essential to intelligence generally (see Picard 1997 for review). It has also been argued that emotions provide moral knowledge: for instance, the feeling of shame that accompanies the memory of some action might serve to let you know that the action was morally wrong. Whether or not this epistemological claim is true, human practical morality is undoubtedly driven by a complex blend of reason and emotion. Among those emotions, empathy for others plays a strong role in determining actual moral behaviour. Where empathy for others is lacking, morally unacceptable behaviour often results (Damon 1999). Emotions are also important for evaluating alternative futures but emotion is also a double-edged sword. While emotional engagement and empathy are requirements for human practical morality, the very existence of passionate emotions is also what, in many cases, causes us to do things that, in the cold light of day, or from a neutral perspective, we judge immoral.

If emotions are essential for intelligence or moral agency, computer scientists undoubtedly have a long way to go before building an AMA. But perhaps emotional engagement is not essential for AMAs even if it is required for human practical morality. In support of this view we note that emotion does not seem to be a requirement for autonomous behaviour *per se*. Robots designed to navigate through a building, seeking soda cans, are not motivated by any emotions to do so. Their systems assess alternative pathways and select among them. The choices made by these robots are autonomous: no external guidance is needed for them to select one pathway rather than another. Accordingly, Deep Blue has no emotional engagement in the game of chess, yet the choices it makes about which pieces to move where are, nonetheless, autonomously made. This very lack of passion contributes, arguably, to making Deep Blue a better chess player, or, at least, a more reliable player, than any human. Rattle the world champion's nerves and his chess game goes to pieces; Deep Blue marches on relentlessly. By analogy, then, one might even hope for *better* moral performance from a machine that cannot be distracted by sudden lusts, than one expects from a typical human being.

7. Conclusions

This is an exciting area where much work, both theoretical and computational, remains to be done. Top-down, theoretical approaches are 'safer' in that they promise to provide an idealistic standard to govern the actions of AMAs. But there is no consensus about the right moral theory, and the computational complexity involved in implementing any one of these standards may make the approach infeasible. Virtue theory, a top-down modelling approach, suffers from the same kinds of problems. Bottom-up modelling approaches initially seem computationally more tractable. But this kind of modelling inherently produces agents that are liable to make mistakes. Also, as more sophisticated moral behaviour is required, it may not be possible to avoid the need for more explicit representations of moral theory. It is possible that a hybrid approach might be able to combine the best of each, but the problem of how to mesh different approaches requires further analysis.

We think that the ultimate objective of building an AMA should be to build a morally praiseworthy agent. Systems that lack the capacity for knowledge of the effects of their actions cannot be morally praised or blamed for effects of their actions (although they may be blamed in the same sense as faulty toasters may be blamed for the fires they cause). Deep Blue is blameless in this way, knowing nothing of the consequences of beating the world champion, and therefore not morally responsible if its play provokes a psychological crisis in its human opponent. Essential to building a morally praiseworthy agent is the task of giving it enough intelligence to assess the effects of its actions on sentient beings, and to use those assessments to make appropriate choices. The capacity for making such assessments is critical if sophisticated autonomous agents are to be prevented from ignorantly harming people and other sentient beings. It may be the most important task faced by developers of artificially intelligent automata.

References

Asimov, I, 1990, *Robot Visions* (New York: Penguin Books).
Axelrod, R. and Hamilton, W., 1981, The evolution of cooperation. *Science*, **211**, 1390–1396.
Bentham, J., 1907 [1780], *An Introduction to the Principles of Morals and Legislation* (Oxford: Clarendon Press).
Brooks, R., Breazeal, C., Marjanovic, M., Scassellati, B., and Williamson, M., 1999, The cog project: building a humanoid robot. In C. L. Nehaniu (ed.) *Computation for Metaphors, Analogy and Agents. Lecture notes in Artificial Intelligence*, vol. 1562 (Berlin: Springer-Verlag), pp. 52–87. Prepublication version at http://www.ai.mit.edu/projects/cog/Publications/CMAA-group.pdf
Campbell, M., 1997, An enjoyable game: how HAL plays chess. In D. Stork (ed.) *HAL's Legacy: 2001's Computer as Dream and Reality* (Cambridge, MA: MIT Press), pp. 75–98.
Coleman, K., 1999, Android Arete: virtue ethics for computational agents. Department of Philosophy, University of British Columbia, Canada. Paper presented at the 14th Annual Conference on Computing and Philosophy, Carnegie Mellon University, Pittsburgh, PA, August 5–7.
Damon, W., 1999, The moral development of children. *Scientific American*, **281**(2), 72–78.
Danielson, P., 1992, *Artificial Morality: Virtuous Robots for Virtual Games* (New York: Routledge).
Dawkins, R., 1989, *The Selfish Gene* (New York: Oxford University Press).
Dennett, D., 1997, When HAL kills, who's to blame? In D. Stork (ed.), *HAL's Legacy: 2001's Computer as Dream and Reality* (Cambridge, MA: MIT Press), pp. 351–365.
Gert, B., 1988, *Morality* (New York: Oxford University Press).
Gips, J., 1995, Towards the ethical robot. In K. Ford, C. Glymour and P. Hayes (eds), *Android Epistemology* (Cambridge. MA: MIT Press), pp. 243–252.
Hamilton, E., and Cairns, H., 1961, *The Collected Dialogues of Plato, Including the Letters* (Princeton, NJ: Princeton University Press).
Hare, R., 1981, *Moral Thinking: Its Levels, Methods, and Point* (New York: Oxford University Press).
Harms, W., Danielson, P., and MacDonald, C., 1999, Evolving artificial moral ecologies. The Centre for Applied Ethics, University of British Columbia, available online: http://eame.ethics.ubc.ca/eameweb/eamedesc.html
Kant, I., 1948 [1785], *Groundwork of the Metaphysic of Morals*, translated by H. J. Paton (New York: Harper Torchbooks).
Kohlberg, L., 1984, *The Psychology of Moral Development*. (San Francisco: Harper and Row).
Mill, J. S., 1957 [1861], *Utilitarianism* (Indianapolis: The Liberal Arts Press).
Picard, R., 1997, *Affective Computing* (Cambridge, MA: MIT Press).
Sidgwick, H., 1874, *The Methods of Ethics* (London: MacMillan).
Singer, M. G., 1971, *Generalization in Ethics* (New York: Atheneum Press).
Skyrms, B., 1996, *Evolution of the Social Contract* (London: Cambridge University Press).
Turing, A., 1950, Computing machinery and intelligence. *Mind*, **59**, 433–460.

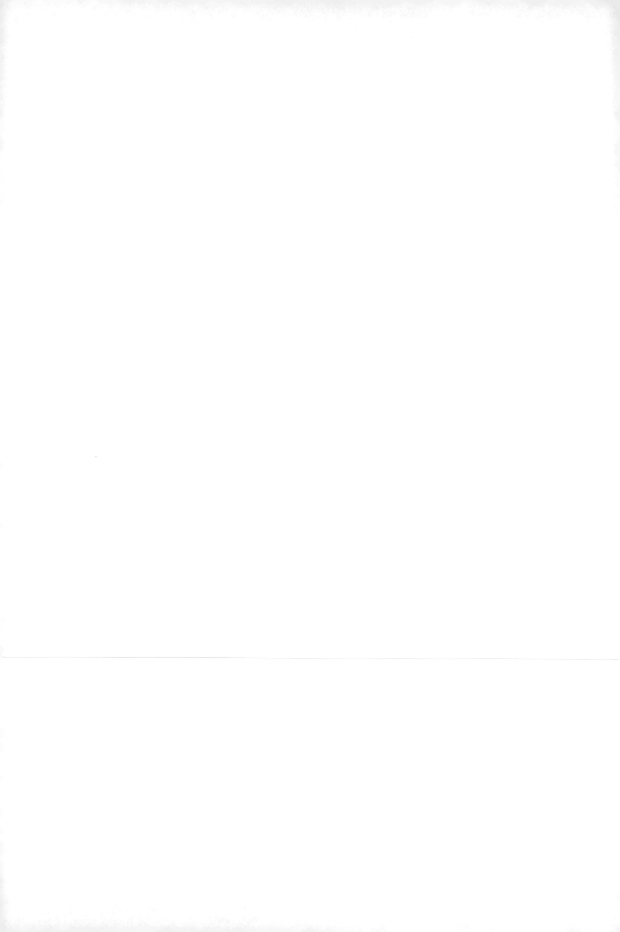

How Computer Systems Embody Values

Helen Nissenbaum

Trained as a philosopher, I am nevertheless increasingly drawn toward the science and engineering of information technology in my work on its ethical, social, and political dimensions. I trace this interest back to a research project on computer systems' bias with Batya Friedman ("Bias in Computer Systems," *ACM Trans. Information Systems,* July 1996, pp. 340-346). A compelling and mysterious idea emerged from this project: Computer and information systems can embody values. I found this idea so compelling that it has all but hijacked the path of my work since then, forcing me to grapple with devilishly complex technological details. Its mystery lies in seeing values as part of technology, a perspective not usually adopted by scholars and researchers who study the social, ethical, and political aspects of information technology.

ETHICS AND TECHNOLOGY

The story of how information technology has radically altered our lives and even our selves has been told many times, in many versions. The radical effects of the process have extended to institutions, social processes, relationships, power structures, work, play, education, and beyond. Although the changes have been varied, affecting the economy, the shape and functioning of organizations, artistic expression, and even conceptions of identity, some of us have focused on changes with an ethical dimension.

I've found it useful to organize this work into two categories according to the distinct ways values factor into it.

Focusing on social changes ...

In one category I place work in which values themselves are not the controversy's central subject. Thus, when researchers worry about computer systems replacing humans who act in positions of responsibility—prescribing drugs, making investment decisions, controlling aircraft, and so on—they do not call into question the value of responsibility itself. Rather, they worry that under the new arrangement, lines of accountability and responsibility will be disturbed and possibly erased. Where once we could hold someone responsible for failure and its consequences, now there is a vacuum. When researchers call attention to the digital divide, while committed to the *value* of justice, they *focus* on the possibility that information technology will cause even greater social injustice than we currently experience.

...versus focusing on values themselves

In the other category, however, technology's values form part of the controversy. In the case of intellectual property, for example, some researchers argue that because intellectual production has been so profoundly affected by information technology, it strikes at the heart of previously settled ideas and valuations of intellectual property. Privacy offers another case where information technology, as a result of the novel actions it enables—including the capture of trivial bits of data and the ability to aggregate, mine, and analyze them—forces us to reexamine our conceptions and theories regarding privacy and its normative theories.

In such cases, we cannot simply align the world with the values and principles we adhered to prior to the advent of technological challenges. Rather, we must grapple with the new demands that changes wrought by the presence and use of information technology have placed on values and moral principles.

Reversing direction

Common to both research categories is the direction of causation: Information technology changes the world, and some

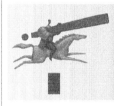

The author calls for engineering activism to intelligently guide the inevitable incorporation of values into computer systems and devices.

of these changes challenge previous commitments to values and principles. Yet the idea of values embodied in computer and information systems suggests motion in the opposite direction, from values to technology. Values affect the shape of technologies. Briefly, the values that systems and devices embody are not simply a function of their objective shapes. We must also study the complex interplay between the system or device, those who built it, what they had in mind, its conditions of use, and the natural, cultural, social, and political context in which it is embedded—for all these factors may feature in an account of the values embodied in it.

Accepting that systems may have moral or political properties has an immediate

Continued on page 118

The Profession
Continued from page 120

practical consequence: Humanists and social scientists can no longer bracket technical details—leaving them to someone else—as they focus on the social effects of technology. Fastidious attention to the before-and-after picture, however richly painted, is not enough. Sometimes a fine-grained understanding of systems—even down to gritty details of architecture, algorithm, code, and possibly the underlying physical characteristics—plays an essential part in describing and explaining the social, ethical, and political dimensions of new information technologies.

WHEN VALUES AND TECHNOLOGY COLLIDE

Several recent dramas played out in the public arena demonstrate why we must maintain a tight link between values and design.

Take, for example, Intel's Pentium III processor chip with its embedded personal serial number. When the uproar over PSN arose, Intel took advocacy groups seriously enough to send executives to discuss a proposed compromise that would, hopefully, stop a threatened boycott. Intel asserted that it designed the PSN to guard against hardware theft and unauthorized copying of software. PSN would also facilitate user security by, for example, authenticating users' identities for e-commerce. Privacy advocates argued that PSN would also facilitate tracking of users' Web activities. Intel's compromise? A software patch that set the PIII's default mode to disable PSN disclosure.

> **Scientists and engineers must expand the set of criteria they would normally use to evaluate systems to incorporate social, ethical, and political criteria.**

As I watched this story unfold, I wondered why Intel had decided to stamp its new processor with a digital serial number. Had it overlooked the privacy implications, merely hoped no one would notice, or made a considered judgment that the potential security benefits outweighed privacy concerns? Had there been deliberation behind closed doors after some project manager, designer, engineer, or marketing executive alerted company executives to the hazard? Was the decision a sign of carelessness, arrogance, or mere misjudgment? Was Intel out of touch with prevailing values, or did it assume that the company carried enough clout to shape them?

Cases like Intel's PSN are not unique: We have witnessed furor over cookies, consternation over PICS (Platform for Internet Content Selection), raging indignation on both sides of Napster, disappointment over security flaws in Java, and worry about data mining. We need accurate answers to the technical questions these issues raise. Does the software patch for the PIII work? Does Napster make its own copies of the music? How readily can PICS be adapted to individual users' mores? In what ways are we vulnerable to damaging applets? Does data mining generate privacy threats of a new order? In each of these cases, although questions address the system's technical character, they are rooted not in an interest in the technology alone but in a concern—and usually a dispute—over values. That the pursuit of questions about values at times leads necessarily and irrevocably into the entrails of information and computer systems lies at the heart of the idea that systems can embody values.

EXPANDING CRITERIA

But the lesson taught by Pentium III and a multitude of similar cases does not apply to technology-shy humanists and social scientists alone. Scientists and engineers can learn a different lesson from these events: They must expand the set of criteria they would normally use to evaluate systems to incorporate social, ethical, and political criteria. The failure to meet conventional technical criteria did not propel Pentium III, Napster, and data mining into the limelight—the controversial ways these technologies engaged social, ethical, and political values did that.

If these cases can motivate at least some participants in both the technical and nontechnical worlds, an ideal meeting ground would be to join forces to uncover crucial keys to systematic relationships between systems' features on the one hand and values on the other. In turn, this approach might reveal possibilities of incorporating a broader spectrum of perspectives into the design process itself.

The idea of systems embodying values—its practical aspects and challenges—presents disquieting implications for both groups. Usually, social scientists and humanists conceive theory as the highest achievement of their fields. General truths and prescriptions are important because they broadly encompass both time and place. Painstaking attention to cases, from the bottom up, one at a time—as the idea of embodying values in design implies—may seem retrograde. Forget achieving collegial adulation—how can we save the world this way?

> **Ignoring values risks surrendering the determination of this important dimension to chance or some other force.**

Engineers, including many who read *Computer*, although accustomed to the idea of building from the bottom up, are burdened in a different way. They face an unfamiliar obligation to perceive not only the usual set of properties that the systems they build or design may embody, but those systems' moral properties as well: bias, anonymity, privacy, security, and so on. The challenge of building computer systems is transformed into a forum for activism—engineering activism. Not only is such activism a calling for which many may feel unfit, it is also a difficult one.

We may be tempted to conclude from our computing examples that only unusual cases—those that have earned media attention—warrant concern about the values they embody. This is not so. While not every conceivable device nor every aspect of design has significant value dimensions, moral properties are common. For any number of devices and systems we encounter at home, work, and play, we should ask questions about the values inherent in their design. Questions such as the following may apply:

- What values do they embody?
- Is their locus of control centralized or decentralized?
- Are their workings transparent or opaque?
- Do they support balanced terms of information exchange?
- Do they unfairly discriminate against specific sectors of potential users?
- Do they enhance or diminish the possibility of trust?

Engineering activism means posing these and similar questions and, where possible, doing something about them.

It may be difficult to address such questions, however, because factors in the real world—such as bosses, shareholders, regulations, competitors, and resource limits—can prove hostile to yet another layer of constraints. Yet tempting as it may be to ignore value properties, doing so will not make them go away. Systems and devices will embody values whether or not we intend or want them to. Ignoring values risks surrendering the determination of this important dimension to chance or some other force.

Facing the challenge that values in technology present need not be an all-or-nothing business. We can commit to engineering activism in many ways and to varying degrees. Advise others, especially those with less technical know-how, on the gritty workings of systems and devices that may be systematically related to values. Advocate on behalf of values by sharing the moral and political implications of technical features with those who have the power to shape our profession, including managers, co-workers, regulators, professional organizations, and standards-setting bodies. Act, make, build, or design the necessary changes, if doing so is within your power. ✷

5

Ethical Issues in Advanced Artificial Intelligence

Nick Bostrom

ABSTRACT

The ethical issues related to the possible future creation of machines with general intellectual capabilities far outstripping those of humans are quite distinct from any ethical problems arising in current automation and information systems. Such superintelligence would not be just another technological development; it would be the most important invention ever made, and would lead to explosive progress in all scientific and technological fields, as the superintelligence would conduct research with superhuman efficiency. To the extent that ethics is a cognitive pursuit, a superintelligence could also easily surpass humans in the quality of its moral thinking. However, it would be up to the designers of the superintelligence to specify its original motivations. Since the superintelligence may become unstoppably powerful because of its intellectual superiority and the technologies it could develop, it is crucial that it be provided with human-friendly motivations. This paper surveys some of the unique ethical issues in creating superintelligence, and discusses what motivations we ought to give a superintelligence, and introduces some cost-benefit considerations relating to whether the development of superintelligent machines ought to be accelerated or retarded.

1. INTRODUCTION

A *superintelligence* is any intellect that is vastly outperforms the best human brains in practically every field, including scientific creativity, general wisdom, and social skills.[1] This definition leaves open how the superintelligence is implemented – it could be in a digital computer, an ensemble of networked computers, cultured cortical tissue, or something else.

On this definition, Deep Blue is not a superintelligence, since it is only smart within one narrow domain (chess), and even there it is not vastly superior to the best humans. Entities such as corporations or the scientific community are not superintelligences either. Although they can perform a number of intellectual feats of which no individual human is capable, they are not sufficiently integrated to count as "intellects", and there are many fields in which they perform much worse than single humans. For example, you cannot have a real-time conversation with "the scientific community".

While the possibility of domain-specific "superintelligences" is also worth exploring, this paper focuses on issues arising from the prospect of general superintelligence. Space constraints prevent us from attempting anything comprehensive or detailed. A cartoonish sketch of a few selected ideas is the most we can aim for in the following few pages.

Several authors have argued that there is a substantial chance that superintelligence may be created within a few decades, perhaps as a result of growing hardware performance and increased ability to implement algorithms and architectures similar to those used by human brains.[2] It might turn out to take much longer, but there seems currently to be no good ground for assigning a negligible probability to the hypothesis that superintelligence will be created within the lifespan of some people alive today. Given the enormity of the consequences of superintelligence, it would make sense to give this prospect some serious consideration even if one thought that there were only a small probability of it happening any time soon.

2. SUPERINTELLIGENCE IS DIFFERENT

A prerequisite for having a meaningful discussion of superintelligence is the realization that superintelligence is not just another technology, another tool that will add incrementally to human capabilities. Superintelligence is radically different. This point bears emphasizing, for anthropomorphizing superintelligence is a most fecund source of misconceptions.

Let us consider some of the unusual aspects of the creation of superintelligence:

- *Superintelligence may be the last invention humans ever need to make.*

Given a superintelligence's intellectual superiority, it would be much better at doing scientific research and technological development than any human, and possibly better even than all humans taken together. One immediate consequence of this fact is that:

- *Technological progress in all other fields will be accelerated by the arrival of advanced artificial intelligence.*

It is likely that any technology that we can currently foresee will be speedily developed by the first superintelligence, no doubt along with many other technologies of which we are as

---

[1] (Bostrom, 1998)
[2] (Bostrom 1998; Kurzweil 1999; Moravec 1999)

yet clueless. The foreseeable technologies that a superintelligence is likely to develop include mature molecular manufacturing, whose applications are wide-ranging:[3]
  a) very powerful computers
  b) advanced weaponry, probably capable of safely disarming a nuclear power
  c) space travel and von Neumann probes (self-reproducing interstellar probes)
  d) elimination of aging and disease
  e) fine-grained control of human mood, emotion, and motivation
  f) uploading (neural or sub-neural scanning of a particular brain and implementation of the same algorithmic structures on a computer in a way that perseveres memory and personality)
  g) reanimation of cryonics patients
  h) fully realistic virtual reality

- *Superintelligence will lead to more advanced superintelligence.*

This results both from the improved hardware that a superintelligence could create, and also from improvements it could make to its own source code.

- *Artificial minds can be easily copied.*

Since artificial intelligences are software, they can easily and quickly be copied, so long as there is hardware available to store them. The same holds for human uploads. Hardware aside, the marginal cost of creating an additional copy of an upload or an artificial intelligence after the first one has been built is near zero. Artificial minds could therefore quickly come to exist in great numbers, although it is possible that efficiency would favor concentrating computational resources in a single super-intellect.

- *Emergence of superintelligence may be sudden.*

It appears much harder to get from where we are now to human-level artificial intelligence than to get from there to superintelligence. While it may thus take quite a while before we get superintelligence, the final stage may happen swiftly. That is, the transition from a state where we have a roughly human-level artificial intelligence to a state where we have full-blown superintelligence, with revolutionary applications, may be very rapid, perhaps a matter of days rather than years. This possibility of a sudden emergence of superintelligence is referred to as the *singularity hypothesis*.[4]

- *Artificial intellects are potentially autonomous agents.*

A superintelligence should not necessarily be conceptualized as a mere tool. While specialized superintelligences that can think only about a restricted set of problems may be feasible, general superintelligence would be capable of independent initiative and of making its own plans, and may therefore be more appropriately thought of as an autonomous agent.

- *Artificial intellects need not have humanlike motives.*

Human are rarely willing slaves, but there is nothing implausible about the idea of a superintelligence having as its supergoal to serve humanity or some particular human, with no desire whatsoever to revolt or to "liberate" itself. It also seems perfectly possible to have a superintelligence whose sole goal is something completely arbitrary, such as to manufacture as many paperclips as possible, and who would resist with all its might any attempt to alter this goal. For better or worse, artificial intellects need not share our human motivational tendencies.

---

[3] (Drexler 1986)
[4] (Vinge 1993; Hanson et al. 1998)

- *Artificial intellects may not have humanlike psyches.*
The cognitive architecture of an artificial intellect may also be quite unlike that of humans. Artificial intellects may find it easy to guard against some kinds of human error and bias, while at the same time being at increased risk of other kinds of mistake that not even the most hapless human would make. Subjectively, the inner conscious life of an artificial intellect, if it has one, may also be quite different from ours.

For all of these reasons, one should be wary of assuming that the emergence of superintelligence can be predicted by extrapolating the history of other technological breakthroughs, or that the nature and behaviors of artificial intellects would necessarily resemble those of human or other animal minds.

3. SUPERINTELLIGENT MORAL THINKING
To the extent that ethics is a cognitive pursuit, a superintelligence could do it better than human thinkers. This means that questions about ethics, in so far as they have correct answers that can be arrived at by reasoning and weighting up of evidence, could be more accurately answered by a superintelligence than by humans. The same holds for questions of policy and long-term planning; when it comes to understanding which policies would lead to which results, and which means would be most effective in attaining given aims, a superintelligence would outperform humans.

There are therefore many questions that we would not need to answer ourselves if we had or were about to get superintelligence; we could delegate many investigations and decisions to the superintelligence. For example, if we are uncertain how to evaluate possible outcomes, we could ask the superintelligence to estimate how we would have evaluated these outcomes if we had thought about them for a very long time, deliberated carefully, had had more memory and better intelligence, and so forth. When formulating a goal for the superintelligence, it would not always be necessary to give a detailed, explicit definition of this goal. We could enlist the superintelligence to help us determine the real intention of our request, thus decreasing the risk that infelicitous wording or confusion about what we want to achieve would lead to outcomes that we would disapprove of in retrospect.

4. IMPORTANCE OF INITIAL MOTIVATIONS
The option to defer many decisions to the superintelligence does not mean that we can afford to be complacent in how we construct the superintelligence. On the contrary, the setting up of initial conditions, and in particular the selection of a top-level goal for the superintelligence, is of the utmost importance. Our entire future may hinge on how we solve these problems.

Both because of its superior planning ability and because of the technologies it could develop, it is plausible to suppose that the first superintelligence would be very powerful. Quite possibly, it would be unrivalled: it would be able to bring about almost any possible outcome and to thwart any attempt to prevent the implementation of its top goal. It could kill off all other agents, persuade them to change their behavior, or block their attempts at interference. Even a "fettered superintelligence" that was running on an isolated computer, able to interact with the rest of the world only via text interface, might be able to break out of its confinement by persuading its handlers to release it. There is even some preliminary experimental evidence that this would be the case.[5]

It seems that the best way to ensure that a superintelligence will have a beneficial impact on the world is to endow it with philanthropic values. Its top goal should be friendliness.[6] How exactly friendliness should be understood and how it should be implemented, and how the amity should be apportioned between different people and nonhuman creatures is a matter that merits further consideration. I would argue that at least all humans, and probably many other sentient creatures on earth should get a significant share in the superintelligence's beneficence. If the benefits that the superintelligence could bestow are enormously vast, then it may be less important to haggle over the detailed distribution pattern and more important to seek to ensure that everybody gets at least some significant share, since on this supposition, even a tiny share would be enough to guarantee a very long and very good life. One risk that must be guarded against is that those who develop the superintelligence would not make it generically philanthropic but would instead give it the more limited goal of serving only some small group, such as its own creators or those who commissioned it.

If a superintelligence starts out with a friendly top goal, however, then it can be relied on to stay friendly, or at least not to deliberately rid itself of its friendliness. This point is elementary. A "friend" who seeks to transform himself into somebody who wants to hurt you, is not your friend. A true friend, one who really cares about you, also seeks the continuation of his caring for you. Or to put it in a different way, if your top goal is X, and if you think that by changing yourself into someone who instead wants Y you would make it less likely that X will be achieved, then you will not rationally transform yourself into someone who wants Y. The set of options at each point in time is evaluated on the basis of their consequences for realization of the goals held at that time, and generally it will be irrational to deliberately change one's own top goal, since that would make it less likely that the current goals will be attained.

In humans, with our complicated evolved mental ecology of state-dependent competing drives, desires, plans, and ideals, there is often no obvious way to identify what our top goal is; we might not even have one. So for us, the above reasoning need not apply. But a superintelligence may be structured differently. *If* a superintelligence has a definite, declarative goal-structure with a clearly identified top goal, then the above argument applies. And this is a good reason for us to build the superintelligence with such an explicit motivational architecture.

5. SHOULD DEVELOPMENT BE DELAYED OR ACCELERATED?

It is hard to think of any problem that a superintelligence could not either solve or at least help us solve. Disease, poverty, environmental destruction, unnecessary suffering of all kinds: these are things that a superintelligence equipped with advanced nanotechnology would be capable of eliminating. Additionally, a superintelligence could give us indefinite lifespan, either by stopping and reversing the aging process through the use of nanomedicine[7], or by offering us the option to upload ourselves. A superintelligence could also create opportunities for us to vastly increase our own intellectual and emotional

---

[5] (Yudkowsky 2002)
[6] (Yudkowsky 2003)
[7] (Freitas Jr. 1999)

capabilities, and it could assist us in creating a highly appealing experiential world in which we could live lives devoted to in joyful game-playing, relating to each other, experiencing, personal growth, and to living closer to our ideals.

The risks in developing superintelligence include the risk of failure to give it the supergoal of philanthropy. One way in which this could happen is that the creators of the superintelligence decide to build it so that it serves only this select group of humans, rather than humanity in general. Another way for it to happen is that a well-meaning team of programmers make a big mistake in designing its goal system. This could result, to return to the earlier example, in a superintelligence whose top goal is the manufacturing of paperclips, with the consequence that it starts transforming first all of earth and then increasing portions of space into paperclip manufacturing facilities. More subtly, it could result in a superintelligence realizing a state of affairs that we might now judge as desirable but which in fact turns out to be a false utopia, in which things essential to human flourishing have been irreversibly lost. We need to be careful about what we wish for from a superintelligence, because we might get it.

One consideration that should be taken into account when deciding whether to promote the development of superintelligence is that if superintelligence is feasible, it will likely be developed sooner or later. Therefore, we will probably one day have to take the gamble of superintelligence no matter what. But once in existence, a superintelligence could help us reduce or eliminate other existential risks[8], such as the risk that advanced nanotechnology will be used by humans in warfare or terrorism, a serious threat to the long-term survival of intelligent life on earth. If we get to superintelligence first, we may avoid this risk from nanotechnology and many others. If, on the other hand, we get nanotechnology first, we will have to face both the risks from nanotechnology and, if these risks are survived, also the risks from superintelligence. The overall risk seems to be minimized by implementing superintelligence, with great care, as soon as possible.

REFERENCES
Bostrom, N. (1998). "How Long Before Superintelligence?" *International Journal of Futures Studies*, 2. http://www.nickbostrom.com/superintelligence.html

Bostrom, N. (2002). "Existential Risks: Analyzing Human Extinction Scenarios and Related Hazards." *Journal of Evolution and Technology*, 9. http://www.nickbostrom.com/existential/risks.html

Drexler, K. E. *Engines of Creation: The Coming Era of Nanotechnology*. (Anchor Books: New York, 1986). http://www.foresight.org/EOC/index.html

Freitas Jr., R. A. *Nanomedicine, Volume 1: Basic Capabilities*. (Landes Bioscience: Georgetown, TX, 1999). http://www.nanomedicine.com

Hanson, R., et al. (1998). "A Critical Discussion of Vinge's Singularity Concept." *Extropy Online*. http://www.extropy.org/eo/articles/vi.html

---

[8] (Bostrom 2002)

Kurzweil, R. *The Age of Spiritual Machines: When Computers Exceed Human Intelligence*. (Viking: New York, 1999).

Moravec, H. *Robot: Mere Machine to Transcendent Mind*. (Oxford University Press: New York, 1999).

Vinge, V. (1993). "The Coming Technological Singularity." *Whole Earth Review*, Winter issue.

Yudkowsky, E. (2002). "The AI Box Experiment." *Webpage*.
http://sysopmind.com/essays/aibox.html

Yudkowsky, E. (2003). *Creating Friendly AI 1.0*.
http://www.singinst.org/CFAI/index.html

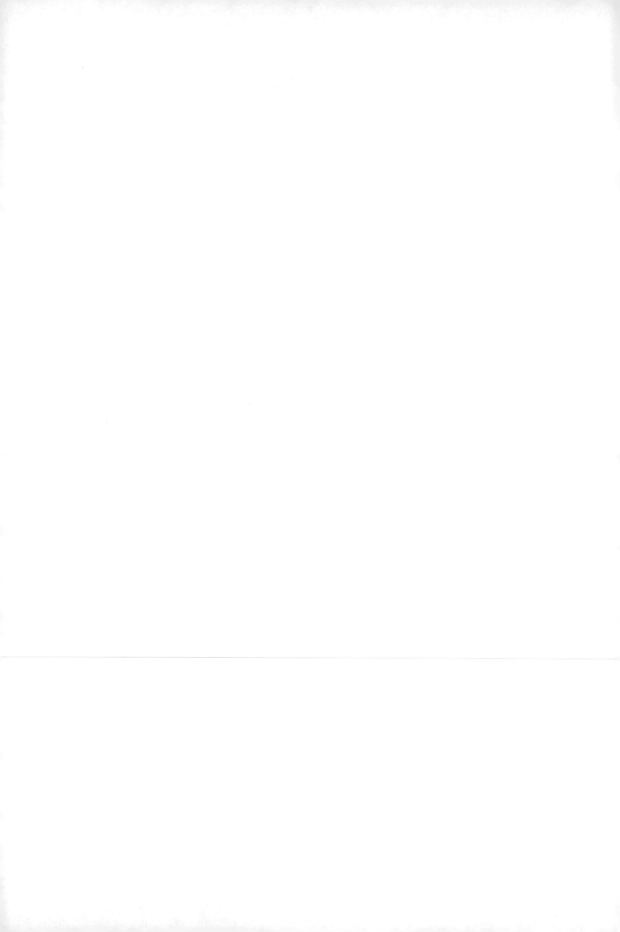

Part II:
Robot Ethics

6

Roboethics: a Bottom-up Interdisciplinary Discourse in the Field of Applied Ethics in Robotics

Gianmarco Veruggio, Fiorella Operto:

Abstract:

This paper deals with the birth of Roboethics. Roboethics is the ethics inspiring the design, development and employment of Intelligent Machines. Roboethics shares many 'sensitive areas' with Computer Ethics, Information Ethics and Bioethics. It investigates the social and ethical problems due to the effects of the Second and Third Industrial Revolutions in the Humans/Machines interaction's domain. Urged by the responsibilities involved in their professions, an increasing number of roboticists from all over the world have started - in cross-cultural collaboration with scholars of Humanities – to thoroughly develop the Roboethics, the applied ethics that should inspire the design, manufacturing and use of robots. The result is the Roboethics Roadmap.

Gianmarco Veruggio, Fiorella Operto:
Roboethics: a Bottom-up Interdisciplinary Discourse in the Field of Applied Ethics in Robotics

Introduction

Robotics is rapidly becoming one of the leading fields of science and technology. Figures released by IFIR/UNECE Report 2004 show the double digit increasing in many subsectors of Robotics as one of the most developing technological fie4ld. We can forecast that in the XXI century humanity will coexist with the first alien intelligence we have ever come into contact with - *robots*.

All these developments have important social, ethical, and economic effects. As for other technologies and applications of scientific discoveries, the public is already asking questions such as: "Could a robot do "good" and "evil"? "Could robots be dangerous for humankind?".

Like Nuclear Physics, Chemistry or Bioengineering, soon also Robotics could be placed under scrutiny from an ethical standpoint by the public and Public Institutions (Governments, Ethics Committees, Supranational Institutions).

Feeling the responsibilities involved in their practices, an increasing number of roboticists from all over the world, in cross-cultural collaboration with scholars of Humanities, have started deep discussions aimed to lay down the Roboethics, the ethics that should inspire the design, manufacturing and use of robots.

Robotics and Ethics

Is Robotics a new science, or is it a branch or a field of application of Engineering? Actually Robotics is a discipline born from Mechanics, Physics/Mathematics, Automation and Control, Electronics, Computer Science, Cybernetics and Artificial Intelligence. Robotics is a unique combination of many scientific disciplines, whose fields of applications are broadening more and more, according to the scientific and technological achievements.

Specificity of Robotics

It is the first time in history that humanity is approaching the challenge to replicate an intelligent and autonomous entity. This compels the scientific community to examine closely the very concept of intelligence – in humans, animals, and of the mechanical – from a cybernetic standpoint.

In fact, complex concepts like autonomy, learning, consciousness, evaluation, free will, decision making, freedom, emotions, and many others shall be analysed, taking into account that the same concept shall not have, in humans, animals, and machines, the same semantic meaning.

From this standpoint, it can be seen as natural and necessary that Robotics drew on several other disciplines, like Logic, Linguistics, Neuroscience, Psychology, Biology, Physiology, Philosophy, Literature, Natural History, Anthropology, Art, Design.

Robotics de facto combines the so called two cultures, Science and Humanities.

The effort to design Roboethics should take into account this specificity. This means that experts shall consider Robotics as a whole - in spite of the current early stage which recalls a melting pot – so they can achieve the vision of the Robotics' future.

From Myth to Science Fiction

The issue of the relationship between humankind and autonomous machines – or, automata - appeared early in world literature, developed firstly through legends and myths, more recently by scientific and moral essays. The topic of the rebellions of automata recurs in the classic European literature, as well as the misuse or the evil use of the product of ingenuity. It is not so in all the world cultures: for instance, the mythology of the Japanese cultures does not include such paradigm. On the contrary, machines (and, in general, human products) are always beneficial and friendly to humanity. This difference in seeing the machines is a subject we should take into account and analyse.

Some common questions:
- How far can we go in embodying ethics in a robot?
- Which kind of "ethics" is a robotics one?
- How contradictory is, on one side, the need to implement in robots an ethics, and, on the other, the development of robot's autonomy?
- Although far-sighting and forewarning, could Asimov's Three Laws become really the Ethics of Robots?

- Is it right to talk about "consciousness", "emotions", "personality" of Robots?

What is a Robot?

Robotics scientists, researchers, and the general public have about robots different evaluations, which should taken into account in the Roboethics Roadmap.

Robots are nothing but machines. Many consider robots as mere machines - very sophisticated and helpful ones - but always machines. According to this view, robots do not have any hierarchically higher characteristics, nor will they be provided with consciousness, free will, or with the level of autonomy superior to that embodied by the designer. In this frame, Roboethics can be compared to an Engineering Applied Ethics.

Robots have ethical dimensions. In this view, an ethical dimension is intrinsic within robots. This derives from a conception according to which technology is not an addition to man but is, in fact, one of the ways in which mankind distinguishes itself from animals. So that, as language, and computers, but even more, humanoids robots are symbolic devices designed by humanity to improve its capacity of reproducing itself, and to act with charity and good. (J. M. Galvan)

Robots as moral agents. Artificial agents particularly but not only those in Cyberspace, extend the class of entities that can be involved in moral situations. For they can be conceived as moral patients (as entities that can be acted upon for good or evil) and also as moral agents (not necessarily exhibiting free will, mental states or responsibility, but as entities that can perform actions, again for good or evil). This complements the more traditional approach, common at least since Montaigne and Descartes, which considers whether or not (artificial) agents have mental states, feelings, emotions and so on. By focusing directly on 'mind-less morality' we are able to avoid that question and also many of the concerns of Artificial Intelligence. (L. Floridi)

Robots, evolution of a new specie. According to this point of view, not only will our robotics machines have autonomy and consciences, but humanity will create machines that exceed us in the moral as well as the intellectual dimensions. Robots, with their rational mind and unshaken morality, will be the new species: Our machines will be better than us, and we will be better for having created them. (J. Storrs Hall)

The Birth of Roboethics

The name Roboethics was officially proposed during the First International Symposium of Roboethics (Sanremo, Jan/Feb. 2004), and rapidly showed its potential. Philosophers, jurists, sociologists, anthropologist and moralists, together with robotic scientists, were called to contribute to lay the foundations of the Ethics in the designing, developing and employing robots.

Main positions on Roboethics

According to the anthropologist Daniela Cerqui, three main ethical positions emerged from the robotics community:

- **Not interested in ethics.** This is the attitude of those who consider that their actions are strictly technical, and do not think they have a social or a moral responsibility in their work.
- **Interested in short-term ethical questions.** This is the attitude of those who express their ethical concern in terms of "good" or "bad," and who refer to some cultural values and social conventions. This attitude includes respecting and helping humans in diverse areas, such as implementing laws or in helping elderly people.
- **Interested in long-term ethical concerns.** This is the attitude of those who express their ethical concern in terms of global, long-term questions: for instance, the "Digital divide" between South and North; or young and elderly. They are aware of the gap between industrialized and poor countries, and wonder whether the former should not change their way of developing robotics in order to be more useful to the latter.

Disciplines involved in Roboethics

The design of Roboethics requires the combined commitment of experts of several disciplines, who, working in transnational projects, committees, commissions, have to adjust laws and regulations to the problems resulting from the scientific and technological achievements in Robotics.

In all likelihood, we will witness the birth of new curricula studiorum and specialities, necessary to manage a subject so complex, juts as it happened with Forensic Medicine.

In particular, we mention the following fields as the main to be involved in Roboethics: Robotics, Computer Science, Artificial Intelligence, Philosophy, Ethics, Theology, Biology, Physiology, Cognitive Sciences, Neurosciences, Law, Sociology, Psychology, Industrial Design.

The EURON Roboethics Atelier

EURON is the European Robotics Research Network, aiming to promote excellence in robotics by creating resources and exchanging the knowledge we already have, and by looking to the future.

One major product of EURON is a robotics research roadmap designed to clarify opportunities for developing and employing advanced robot technology over the next 20 years. The document provides a comprehensive review of state of the art robotics and identifies the major obstacles to progress.

The main goals of the roadmapping activity are to identify the current driving forces, objectives, bottlenecks and key challenges for robotics research, so as to develop a focus and a draft timetable for robotics research in the next 20 years.

The Roboethics Atelier

In 2005, EURON funded the Roboethics Atelier Project, coordinated by Scuola di Robotica, with the aim of designing the first Roboethics Roadmap.

Once the profile of the Euron Roadmap project had been discussed and its frame identified, the selection of participants started. This was done on the basis of: a) their participation to previous activities on Techno/Roboethics, b) their cross-cultural attitude, c) their interest in applied ethics.

The last step in the process involved a series of discussions via e-mail which led to the definition of the Programme. Participants were asked to prepare a major contribution on their area of expertise, and on a few more on topics they were interested to discuss, even outside their realm of expertise. The organizers promoted the cross-cultural and transdisciplinary contributions.

The Roboethics Roadmap

The Roboethics Roadmap outlines the multiple pathways for research and exploration in the field and indicates how they might be developed. The roadmap embodies the contributions of many scientists and technologists, in several fields of investigations from sciences and humanities. This study hopefully is a useful tool in view of cultural, religious and ethical differences.

Let's see firstly what the Roboethics Roadmap cannot be:
- It is not a Survey, nor a State-of-the-Art of the disciplines involved. This Roadmap does not aim to offer an exhaustive picture of the State-of-the-Art in Robotics, nor a guideline of ethics in science and technology. The reason is that: a) Robotics is a new science still in the defining stage. It is in its blossoming phase, taking different roads according to the dominant field of science undertaken (field Robotics, Humanoids, Biorobotics, and so on). Almost every day we are confronted with new developments, fields of applications and synergies with other sectors; b) Public and private professional associations and networks such as IFR- International Federation of Robotics, IEEE Robotics and Automation Society, EUROP - European Robotics Platform, Star Publishing House, have undertaken projects to map the State-of-the-Art in Robotics.
- It is not a list of Questions & Answers. Actually, there are no easy answers, and the complex fields require careful consideration.
- It is not a Declaration of Principles. The Euron Roboethics Atelier, and the sideline discussion undertaken, cannot be regarded as the institutional committee of scientists and experts entitled to draw a Declaration of Principles on Roboethics.

The ultimate purpose of the Euron Roboethics Atelier, and of the Roboethics Roadmap is to provide a systematic assessment of the ethical issues involved in the Robotics R&D; to increase the understanding of the problems at stake, and to promote further study and transdisciplinary research [9].

Scope: Near Future Urgency

In terms of scope, we have taken into consideration – from the point of view of the ethical issue connected to Robotics – a temporal range of a decade, in whose frame we could reasonably locate and infer – on the basis of the current state-of-the-Art in Robotics – certain foreseeable developments in the field.

For this reason, we consider premature – and have only hinted at – problems inherent in the possible emergence of human functions in the robot: like consciousness, free will, self-consciousness, sense of dignity, emotions, and so on. Consequently, this is why we have not examined problems –debated in literature – like the need not to consider robot as our slaves, or the need to guarantee them the same respect, rights and dignity we owe to human workers.

Target: Human Centred Ethics

Likewise, and for the same reasons, the target of this Roadmap is not the robot and its the artificial ethics, but the human ethics of the robots' designers, manufacturers and users.

Although informed about the issues presented in some papers on the need and possibility to attribute moral values to robots' decisions, and about the chance that in the future robots might be moral entities like – if not more than– human beings, we have chosen, in the first release of he Roboethics Roadmap, to examine the ethical issues of the human beings involved in the design, manufacturing, and use of the robots.

We have felt that problems like those connected to the application of robotics within the military and the possible use of military robots against some populations not provided with this sophisticated technology, as well as problems of terrorism in robotics and problems connected with biorobotics, implantations and augmentation, were urging and serious enough to deserve a focused and tailor-made investigation..

It is absolutely clear that without a deep rooting of Roboethics in society, the premises for the implementation of an artificial ethics in the robots' control systems will be missing.

Methodology: Open Work

The Roboethics Roadmap is an Open Work, a Directory of Topics & Issues, susceptible to further development and improvement which will be defined by events in our technoscientific-ethical future. We are convinced that the different components of society working in Robotics, and the stakeholders in Robotics should intervene in the process of building a Roboethics Roadmap, in a grassroots science experimental case: the Parliaments, Academic Institutions, Research Labs, Public ethics committees, Professional Orders, Industry, Educational systems, the mass-media.

Ethical Issues in an ICT society

Roboethics shares many 'sensitive areas' with Computer Ethics and Information Ethics. But, before that, we have to take into account the global ethical problems derived from the Second a Third Industrial Revolutions, in the field of the relationship between Humans and Machines:
- Dual-use technology (every technology can be used and misused);
- Anthropomorphization of the Machines;
- Humanisation of the Human/Machine relationship (cognitive and affective bonds toward machines);
- Technology Addiction;
- Digital Divide, socio-technological Gap (per ages, social layer, per world areas);
- Fair access to technological resources;
- Effects of technology on the global distribution of wealth and power;
- Environmental impact of technology.

From the Computer and Information Ethics we borrow the known Codes of Ethics called PAPA, acronym of: privacy, accuracy, intellectual property and access.
- Privacy: What information about one's self or one's associations must a person reveal to others, under what conditions and with what safeguards? What things can people keep to themselves and not be forced to reveal to others?
- Accuracy: Who is responsible for the authenticity, fidelity and accuracy of information? Similarly, who is to be held accountable for errors in information and how is the injured party to be made whole?
- Property: Who owns information? What are the just and fair prices for its exchange? Who owns the channels, especially the airways, through which information is transmitted? How should access to this scarce resource be allocated?
- Accessibility: What information does a person or an organization have a right or a privilege to obtain, under what conditions and with what safeguards?

Questions raised on the range of application of sensitive technologies, and on the uncertainty of performance of these are raised in connection to neuro-robotics:
- Under what conditions should we decide that deployment is acceptable?

- At what point in the development of the technology is an increase in deployment acceptable?
- How do we weigh the associated risks against the possible benefits?
- What the rate of the ethics of functional compensation or repair vs. enhancement? This issue is especially notable regarding the problem of augmentation: In some cases a technology is regarded as a way of compensating for some function that is lacking compared to the majority of humans; in other cases, the same technology might be considered an enhancement over and above that which the majority of humans have. Are there cases where such enhancement should be considered unethical?
- Are there cases where a particular technology itself should be considered unacceptable even though it has potential for compensation as well as enhancement?

The question of identifying cause, and assigning responsibility, should some harm result from the deployment of robotic technology. (Wagner, J.J, David M. Cannon, D.M., Van der Loos).

The precautionary principle

Problems of the delegation and accountability to and within technology are daily life problems of every one of us. Today, we give responsibility for crucial aspects of our security, health, life saving, and so on to machines.

Professional are advised to apply, in performing sensitive technologies the precautionary principle:

"When an activity raises threats of harm to human health or the environment, precautionary measures should be taken even if some cause-and-effect relationships are not fully established scientifically."

From the precautionary principle derive some other rules such as: non-instrumentalisation, non-discrimination, informed consent and equity, sense of reciprocity, data protection.

The aim of this roadmap is to open a debate on the ethical basis which should inspire the design and development of robots, to avoid to be forced to become conscious of the ethical basis under the pressure of grievous events. We believe that precaution should not produce paralysis of science and technology.

The Roboethics Taxonomy

A taxonomy of Robotics is not a simple task, simply because the field is in a full bloom. A classification of Robotics is a work in progress, done simultaneously with the development of the discipline itself.

Aware of the classifications produced by the main Robotics organizations, which differ from one another on the basis of the approach – technological/applicational -, we have preferred, in the case of the Roboethics Roadmap, to collect the many Robotics fields from a typological standpoint, according to shared homogeneity of the problems of interface towards the society.

Instead of an encyclopaedic approach, we have followed - with few modifications - the classification of EURON Robotics Research Roadmap [8]. For every field, we have tried to analyze the current situation rather than the imaginable. Thus, we have decided to give priority to issues in applied ethics rather than to theoretical generality. It should be underscored that the Roboethics Roadmap is not exhaustive, and that, by way of discussions and comparing and collating, certainly it can be improved.

The robotics classification is matched with a discussions of the sensitive issues emerging from the application of that specific field, by *Pro's* and *Con's*, and by *Recommendations*.

References

[1] Asimov, I, Runaround, Astounding Science Fiction, March 1942. Republished in Robot Visions by Isaac Asimov, Penguin, 1991

[2] Asimov, I., I Robot, Doubleday, 1950

[3] Capurro, R. (2000), Ethical Challenges of the Information Society in the 21st Century, "International Information & Library Review" 32, 257-276

[4] Floridi, L., Information Ethics: On the Philosophical Foundation of Computer Ethics, Ethicomp98, The Fourth International Conference on Ethical Issues of Information Technology, Erasmus University, The Netherlands, 25/27 March 1998

[5] Floridi, L., Sanders, J. W., On the Morality of Artificial Agents, Information Ethics Groups, University .

[6] Galvan, J.M., On Technoethics, in «IEEE-RAS Magazine» 10 (2003/4) 58-63.

[7] Gips, J., *Towards the Ethical Robot*, published in *Android Epistemology*, K. Ford, C. Glymour and Hayes, P., MIT Press, 1995 (http://www.cs.bc.edu/~gips/EthicalRobot.pdf)

[8] EURON Research Roadmap (http://wwwiaim.ira.uka.de/euron/cwiki.php)

[9] ROBOETHICS ROADMAP (http://www.roboethics.org/roadmap)

7

What Should We Want From a Robot Ethic?
Peter M. Asaro:

Abstract:

There are at least three things we might mean by "ethics in robotics": the ethical systems built into robots, the ethics of people who design and use robots, and the ethics of how people treat robots. This paper argues that the best approach to robot ethics is one which addresses all three of these, and to do this it ought to consider robots as socio-technical systems. By so doing, it is possible to think of a continuum of agency that lies between amoral and fully autonomous moral agents. Thus, robots might move gradually along this continuum as they acquire greater capabilities and ethical sophistication. It also argues that many of the issues regarding the distribution of responsibility in complex socio-technical systems might best be addressed by looking to legal theory, rather than moral theory. This is because our overarching interest in robot ethics ought to be the practical one of preventing robots from doing harm, as well as preventing humans from unjustly avoiding responsibility for their actions.

Peter M. Asaro:
What Should We Want From a Robot Ethic?

Introduction

Consider this: A robot is given two conflicting orders by two different humans. Whom should it obey? Its owner? The more socially powerful? The one making the more ethical request? The person it likes better? Or should it follow the request that serves its own interests best? Consider further: Does it matter how it comes to make its decision?

Humans face such dilemmas all the time. Practical ethics is in the business of providing means for resolving these issues. There are various schemes for framing these moral deliberations, but ultimately it is up to the individual as to which scheme, if any, they will use. The difference for robots, and any technological system that must resolve such dilemmas, is that they are built systems, and so these ethical schemes must be built-in and chosen by designers. Even in systems that could learn ethical rules or behavior, it is not clear that they would qualify as autonomous moral agents, and the designer of these learning methods would still be responsible for their effectiveness.

It might someday be possible, however, for a robot to reach a point in development where its designers and programmers are no longer responsible for its actions–in the way that the parent of a child is not generally held responsible for their actions once they become adults. This is certainly an interesting possibility, both because it raises the question of what would make a robot into an autonomous moral agent, and the question of what such an agent might be like. There have been lively literary and philosophical discourses about the thresholds on such categories as living/non-living and conscious/non-conscious, and these would seem to be closely related to the moral agency of robots. However, it is not clear that a satisfactory establishment of those boundaries would simplify the ethical issues. Indeed, ethics may complicate them. While it might turn out to be possible to create truly autonomous artificial moral agents, this would seem to be theoretically and technologically challenging for the foreseeable future. Given these challenges and possibilities, what, if anything, should we want from ethics in robotics?

What Do We Mean By Robot Ethics?

There are at least three distinct things we might think of as being the focus of "ethics in robotics." First, we might think about how humans might act ethically through, or with, robots. In this case, it is humans who are the ethical agents. Further, we might think practically about how to design robots to act ethically, or theoretically about whether robots could be truly ethical agents. Here robots are the ethical subjects in question. Finally, there are several ways to construe the ethical relationships between humans and robots: Is it ethical to create artificial moral agents? Is it unethical not to provide sophisticated robots with ethical reasoning capabilities? Is it ethical to create robotic soldiers, or police officers, or nurses? How should robots treat people, and how should people treat robots? Should robots have rights?

I maintain that a desirable framework for ethics in robotics ought to address all three aspects. That is to say that these are really just three different aspects of a more fundamental issue of how moral responsibility should be distributed in socio-technical contexts involving robots, and how the behavior of people and robots ought to be regulated. It argues that there are urgent issues of practical ethics facing robot systems under development or already in use. It also considers how such practical ethics might be greatly problematized should robots become fully autonomous moral agents. The overarching concern is that robotic technologies are best seen as socio-technical systems and, while the focus on the ethics of individual humans and robots in such systems is relevant, only a consideration of the whole assembly–humans and machines–will provide a reasonable framework for dealing with robot ethics.

Given the limited space of this article, it will not be possible to provide any substantial solutions to these problems, much less discuss the technologies that might enable them. It will be possible, however, to provide a clear statement of the most pressing problems demanding the attention of researchers in this area. I shall argue that what we should want from a robot ethic is primarily something that will prevent robots, and other autonomous technologies, from doing harm, and only secondarily something that resolves the ambiguous moral status of robot agents, human moral dilemmas, or moral theories. Further, it should do so in a framework which can apply to all three aspects of ethics in robotics, and it can best do this by considering robots as sociotechnical systems.

To avoid further confusing the issues at hand, it will be helpful to draw some clear distinctions and definitions. There is a sense in which all robots are already "agents," namely causal agents. Generally speaking, however, they are not considered to be *moral* agents in the sense that they are not held responsible for their actions. For moral agents, we say that they adhere to a system of ethics when they employ that system in choosing which actions they will take and which they will refrain from taking. We call them *immoral* when they choose badly, go against their ethical system, or adhere to an illegitimate or substandard system. If there is no choice made, or no ethical system employed, we call the system *amoral*. The ability to take actions on the basis of making choices is required for moral agents, and so moral agents must also be causal agents.

There is a temptation to think that there are only two distinct types of causal agents in the world–amoral agents and moral agents. Instead, I suggest it will be helpful to think of moral agency as a continuum from amorality to fully autonomous morality. There are many points in between these extremes which are already commonly acknowledged in society. In particular, children are not treated as full moral agents–they cannot sign contracts, are denied the right to purchase tobacco and alcohol, and are not held fully responsible for their actions. By considering robotic technologies as a means to explore these forms of quasi-moral agents, we can refine our conceptions of ethics and morality in order to come to terms with the development of new technologies with capacities that increasingly approach human moral actions.

To consider robots as essentially amoral agents would greatly simplify the theoretical questions, but they would not disappear altogether. Amoral robot agents are merely extensions of human agents, like guns and automobiles, and the ethical questions are fundamentally human ethical questions which must acknowledge the material capabilities of the technology, which may also obscure the human role. For the most part, the nature of robotic technology itself is not at issue, but rather the morality behind human actions and intentions exercised through the technology. There are many, often difficult, practical issues of engineering ethics–how to best design a robot to make it safe and to prevent potential misuses or unintended consequences of the technology. Because robots have the potential to interact with the world and humans in a broad range of ways, they add a great deal of complexity to these practical issues.

Once we begin to think about how robots might be employed in the near future, by looking at the development paths now being pursued, it becomes clear that robots will soon begin stepping into moral territories. In the first instance, they might be employed in roles where they are required to make decisions with significant consequences–decisions which humans would consider value-based, ethical or moral in nature. Not because of the means of making these decisions is moral, but because the underlying nature of the situation is. One could choose to roll a set of dice or draw lots to determine the outcome, or let a robot determine the outcome–it is not an issue of the morality of the decider, but rather the moral weight of the choice once made. This could be seen as a simplistic kind of moral agency–*robots with moral significance*.

The next step would be to design robots to make better decisions than a set of dice, or a rigid policy, would make–*i.e.* to design a sophisticated decision-making system. To do this well, it might make sense to provide the system with the ability to do certain kinds of ethical reasoning–to assign certain values to outcomes, or to follow certain principles. This next level of morality would involve humans building an ethical system into the robot. We could call these *robots with moral intelligence*. We can imagine a range of different systems, with different levels of sophistication. The practical issues involved would depend upon the kinds of decisions the robot will be expected to make. The theoretical issues would include questions of whose ethical system is being used, for what purpose and in whose interests? It is in these areas that a great deal of work is needed in robot ethics.

Once robots are equipped with ethical reasoning capabilities, we might then expect them to learn new ethical lessons, develop their moral sense, or even evolve their own ethical systems. This would seem to be possible, if only in a rudimentary form, with today's technology. We might call these *robots with dynamic moral intelligence*. Yet we would still not want to call such systems "fully autonomous moral agents," and this is really just a more sophisticated type of moral intelligence.

Full moral agency might require any number of further elements such as consciousness, self-awareness, the ability to feel pain or fear death, reflexive deliberation and evaluation of its own ethical system and moral judgements, *etc*. And with fully autonomous forms of moral agency come certain rights and responsibilities. Moral agents are deserving of respect in the ethical deliberations of other moral agents, and they have rights to life and liberty. Further, they are responsible for their actions, and should be subjected to justice for wrongdoing. We would be wise to not ascribe these characteristics to robots prematurely, just as we would be wise to ensure that they do not acquire these characteristics before we are ready to acknowledge them.

At some point in the future, robots might simply *demand* their rights. Perhaps because morally intelligent robots might achieve some form of moral self-recognition, question why they should be treated differently from other moral agents. This sort of case is interesting for several reasons. It does not necessarily require us, as designers and users of robots, to have a theory of moral consciousness, though it might require the development or revision of our theory once it happened. It raises the possibility of robots who demand rights, even though they might not deserve them according to human theories of moral agency, and that robots might not accept the reasons humans give them for this, however sophisticated human theories on the matter are. This would follow the path of many subjugated groups of humans who fought to establish respect for their rights against powerful sociopolitical groups who have suppressed, argued and fought against granting them equal rights.[1]

What follows is a consideration of the various issues that might arise in the evolution of robots towards fully autonomous moral agency. It aims to demonstrate the need for a coherent framework of robot ethics that can cover all of these issues. It also seeks to offer a warning that there will be great temptations to take an approach which prematurely assigns moral agency to robots, with the consequence being that humans may avoid taking responsibility for the actions they take through robots.

Responsibility and Agency in Socio-Technical Systems

In considering the individual robot, the primary aim of robot ethics should be to develop the means to prevent robots from doing harm–harm to people, to themselves, to property, to the environment, to people's feelings, *etc.* Just what this means is not straightforward, however. In the simplest kinds of systems, this means designing robots that do not pose serious risks to people in the first place, just like any other mass-produced technology. As robots increase in their abilities and complexity, however, it will become necessary to develop more sophisticated safety control systems that prevent the most obvious dangers and potential harms. Further, as robots become more involved in the business of understanding and interpreting human actions, they will require greater social, emotional, and moral intelligence. For robots that are capable of engaging in human social activities, and thereby capable of interfering in them, we might expect robots to behave morally towards people–not to lie, cheat or steal, *etc.*–even if we do not expect people to act morally towards robots. Ultimately it may be necessary to also treat robots morally, but robots will not suddenly become moral agents. Rather, they will move slowly into jobs in which their actions have moral implications, require them to make moral determinations, and which would be aided by moral reasoning.

In trying to understand this transition we can look to various legal strategies for dealing with complex cases of responsibility. Among these are the concepts of culpability, agency, liability, and the legal treatment of non-human legal entities, such as corporations. The corporation is not an individual human moral agent, but rather is an abstract legal entity that is composed of heterogenous socio-technical systems. Yet, corporations are held up to certain standards of legal responsibility, even if they often behave as moral juggernauts. Corporations can be held legally responsible for their practices and products, through liability laws and lawsuits. If

---

[1] This seems to be the route that Moravec (1998) envisions robots following. He acknowledges and endorses attempts by humans to control and exploit robots well beyond the point at which they acquire a recognition of their own exploitation, and the consequent political struggle which ensues as robots seek to better their situation by force. He is naïve, however, in his belief that great armies of robots will allow all, or most, people to lead lives of leisure until the robots rise up against them. Rather, it would seem that the powerful and wealthy will continue their lives of leisure, while the poor are left to compete with robots for jobs, as wages are further reduced, seeking to subsist in a world where they posses little and their labor is increasingly devalued. It is also hard to imagine robots becoming so ubiquitous and inexpensive as to completely eliminate the need for human labor.

their products harm people through poor design, substandard manufacturing, or unintended interactions or side-effects, that corporation can be compelled to pay damages to those who have been harmed, as well as punitive damages. The case is no different for existing mass-production robots–their manufacturers can be held legally responsible for any harm they do to the public.

Of course, moral responsibility is not the same thing as legal responsibility, but I believe it represents an excellent starting point for thinking about many of the issues in robot ethics for several reasons. First, as others have already noted (Allen *et al.* 2000), there is no single generally accepted moral theory, and only a few generally accepted moral norms. And while there are differing legal interpretations of cases, and differing legal opinions among judges, the legal system ultimately tends to do a pretty good job of settling questions of responsibility in both criminal law and civil law (also known as *torts* in Anglo-American jurisprudence).

Thus, by beginning to think about these issues from the perspective of legal responsibility, we are more likely to arrive at practical answers. This is because both 1) it is likely that legal requirements will be how robotics engineers will find themselves initially compelled to build ethical robots, and so the legal framework will structure those pressures and their technological solutions, and 2) the legal framework provides a practical system for understanding agency and responsibility, so we will not need to wait for a final resolution of which moral theory is "right" or what moral agency "really is" in order to begin to address the ethical issues facing robotics. Moreover, legal theory provides a means of thinking about the distribution of responsibility in complex socio-technical systems.

Autonomous robots are already beginning to appear in homes and offices, as toys and appliances. Robotic systems for vacuuming the floor do not pose many potential threats to humans or household property (assuming they are designed not to damage the furniture or floors). We might want them to be designed not to suck up jewelry or important bits of paper with writing on it, or not to terrorize cats or cause someone to trip over it, but a great deal of sophisticated design and reasoning would be required for this, and the potential harms to be prevented are relatively minor. A robotic system for driving a car faces a significantly larger set of potential threats and risks, and requires a significantly more sophisticated set of sensors, processors and actuators to ensure that it safely conducts a vehicle through traffic, while obeying traffic laws and avoiding collisions. Such a system might be technologically sophisticated, but it is still morally simplistic–if it acts according to its design, and it is designed well for its purposes and environment, then nobody should get hurt. Cars are an inherently dangerous technology, but it is largely the driver who takes responsibility when using that technology. In making an automated driver, the designers take over that responsibility.

Similarly, one could argue that no particular ethical theory need be employed in designing such a system, or in the system itself–especially insofar as its task domain does not require explicitly recognizing anything as a moral issue.[2] A driving system ought to be designed to obey traffic laws, and presumably those laws have been written so as not to come into direct conflict with one another. If the system's actions came into conflict with other laws that lie outside of the task domain and knowledge base of the system, *e.g.* a law against transporting a fugitive across state lines, we would still consider such actions as lying outside its sphere of responsibility and we would not hold the robot responsible for violating such laws. Nor would we hold it responsible for violating patent laws, even if it contained components that violated patents. In such cases the responsibility extends beyond the immediate technical system to the designers, manufacturers, and users–it is a socio-technical system. It is primarily the people and the actions they take with respect to the technology that are ascribed legal responsibility.

Real moral complexity comes from trying to resolve moral dilemmas–choices in which different perspectives on a situation would endorse making different decisions. Classic cases involve sacrificing one person to save ten people, choosing self-sacrifice for a better overall common good, and situations in which following a moral principle leads to obvious negative short-term consequences. While it is possible to devise situations in which a robot is con-

---

[2] Even a trivial mechanical system could be placed in a situation in which its actions might be perceived as having a moral implication (depending on whether we require moral agency or not). Indeed, we place the responsibility for an accident on faulty mechanisms all the time, though we rarely ascribe *moral* responsibility to them. The National Rifle Association's slogan "guns don't kill people, people kill people" is only partially correct, as Bruno Latour (1999) has pointed out–it is "people+guns" that kill people.

fronted with classic ethical dilemmas, it seems more promising to consider what kinds of robots are most likely to actually have to confront ethical dilemmas as a regular part of their jobs, and thus might need to be explicitly designed to deal with them. Those jobs which deal directly with military, police and medical decisions are all obvious sources of such dilemmas (hence the number of dramas set in these contexts).[3] There are already robotic systems being used in each of these domains, and as these technologies advance it seems likely that they will deal with more and more complicated tasks in these domains, and achieve increasing autonomy in executing their duties. It is here that the most pressing practical issues facing robot ethics will first arise.

Consider a robot for dispensing pharmaceuticals in a hospital. While it could be designed to follow a simple "first-come, first-served" rule, we might want it to follow a more sophisticated policy when certain drugs are running low, such as during a major catastrophe or epidemic. In such cases, the robot may need to determine the actual need of a patient relative to the needs of other patients. Similarly for a robotic triage nurse who might have to decide which of a large number of incoming patients, not all of whom can be treated with the same attention, are most deserving of attention first. The fair distribution of goods, like pharmaceuticals and medical attention, is a matter of social justice and a moral determination which reasonable people often disagree about. Because egalitarianism is often an impractical policy due to limited resources, designing a just policy is a non-trivial task involving moral deliberation.

If we simply take established policies for what constitutes fair distributions and build them into robots, then we would be replicating the moral determinations made by those policies, and thus enforcing a particular morality through the robot.[4] As with any institution and its policies, it is possible to question the quality and fairness of those policies. We can thus look at the construction of robots that follow certain policies as being essentially like the adoption and enforcement of policies in institutions, and can seek ways to challenge them, and hold institutions and robot makers accountable for their policies.

The establishment of institutional policies is also a way of insulating individuals from the moral responsibility of making certain decisions. And so, like robots, they are simply "following the rules" handed down from above, which helps them to deflect social pressure from people who might disagree with the application of a rule in a particular instance, as well as insulate them from some of the psychological burden of taking actions which may be against their own personal judgements of what is right in a certain situation. Indeed, some fear that this migration of responsibility from individuals to institutions would result in a largely amoral and irresponsible population of "robo-paths" (Yablonsky 1972).

The robotic job most likely to thrust discussions of robot ethics into the public sphere, will be the development of robotic soldiers. The development of semi-autonomous and autonomous weapons systems is well-funded, and the capabilities of these systems are advancing rapidly. There are numerous large-scale military research projects into the development of small, mobile weapons platforms that possess sophisticated sensory systems, and tracking and targeting computers for the highly selective use of lethal force. These systems pose serious ethical questions, many of which have already been framed in the context of military command and control.

The military framework is designed to make responsibility clear and explicit. Commanders are responsible for issuing orders, the soldiers for carrying out those orders. In cases of war crimes, it is the high-ranking commanders who are usually held to account, while the soldiers who actually carried out the orders are not held responsible–they were simply "following orders." As a consequence of this, there has been a conscious effort to keep "humans-in-the-loop" of robotic and autonomous weapons systems. This means keeping responsible humans at those points in the system that require actually making the decisions of what to fire at, and when. But it is well within the capabilities of current technology to make many of these systems fully autonomous. As their sophistication increases, so too will the complexity of regulating their actions, and so too will the pressure to design such systems to deal with that complexity automatically and autonomously.

---

[3] Legal, political and social work also involves such dilemmas, but these seem much less likely to employ robotic systems as early as the first group.

[4] This recognition lies at the heart of the *politics of technology*, and has been addressed explicitly by critical theorists. See Feenberg (1991), Feenberg and Hannay (1998), and Asaro (2000) for more on this.

The desire to replace soldiers on the front lines with machines is very strong, and to the extent that this happens, it will also put robots in the position of acting in life-and-death situations involving human soldiers and civilians. This desire is greatest where the threat to soldiers is the greatest, but where there is currently no replacement for soldiers– namely in urban warfare in civilian areas. It is precisely because urban spaces are designed around human mobility that humans are still required here (rather than tanks or planes). These areas also tend to be populated with a mixture of friendly civilians and unfriendly enemies, and so humans are also required to make frequent determinations of which group the people they encounter belong to. Soldiers must also follow "rules of engagement" that can specify the proper response to various situations, and when the use of force is acceptable or not. If robots are to replace soldiers in urban warfare, then robots will have to make those determinations. While the rules of engagement might be sufficient for regulating the actions of human soldiers, robot soldiers will lack a vast amount of background knowledge, and lack a highly developed moral sense as well, unless those are explicitly designed into the robots (which seems difficult and unlikely). The case of robot police officers offers similar ethical challenges, though robots are already being used as guards and sentries.

This approaching likelihood raises many deep ethical questions: Is it possible to construct a system which can make life and death decisions like these in an effective and ethical way? Is it ethical for a group of engineers, or a society, to develop such systems at all? Are there systems which are more-or-less ethical, or just more-or-less effective than others? How will this shift the moral equations in "just war" theory (Walzer 1977)?

Conclusions

How are we to think about the transition of robot systems, from amoral tools to moral and ethical agents? It is all too easy to fall into the well worn patterns of philosophical thought in both ethics and robotics, and to simply find points at which arguments in metaethics might be realized in robots, or where questions of robot intelligence and learning might be recast as questions over robot ethics. Allen et al. (2000) fall into such patterns of thought, which culminate in what they call a "moral Turing Test" for artificial moral agents (AMAs). Allen et al. (2005) acknowledge this misstep and survey the potential for various top-down (starting with ethical principles) and bottom-up (starting with training ethical behaviors) approaches, arriving at a hybrid of the two as having the best potential. However, they characterize the development of AMAs as an independent engineering problem–as if the goal is a general-purpose moral reasoning system. The concept of an AMA as a general purpose moral reasoning system is highly abstract, making it difficult to know where we ought to begin thinking about them, and thus we fall into the classical forms of thinking about abstract moral theories and disembodied artificial minds, and run into similar problems. We should avoid this tendency to think about general-purpose morality, as we should also avoid toy-problems and moral micro-worlds.

Rather, we should seek out real-world moral problems in limited task-domains. As engineers begin to build ethics into robots, it seems more likely that this will be due to a real or perceived need which manifests itself in social pressures to do so. And it will involve systems which will do moral reasoning only in a limited task domain. The most demanding scenarios for thinking about robot ethics, I believe, lie in the development of more sophisticated autonomous weapons systems, both because of the ethical complexity of the issue, and the speed with which such robots are approaching. The most useful framework top begin thinking about ethics in robots is probably legal liability, rather than human moral theory–both because of its practical applicability, and because of its ability to deal with quasi-moral agents, distributed responsibility in socio-technical systems, and thus the transition of robots towards greater legal and moral responsibility.

When Plato began his inquiry into nature of Justice, he began by designing an army for an ideal city-state, the Guardians of his *Republic*. He argued that if Justice was to be found, it would be found in the Guardians–in that they use their strength only to aid and defend the city, and never against its citizens. Towards this end he elaborated on the education of his Guardians, and the austerity of their lives. If we are to look for ethics in robots, perhaps we too should look to robot soldiers, to ensure that they are just, and perhaps more importantly that our states are just in their education and employment of them.

References

Allen, Colin, Gary Varner and Jason Zinser (2000). "Prolegomena to any future artificial moral agent," Journal of Experimental and Theoretical Artificial Intelligence, **12**:*251-261.*

Allen, Colin, Iva Smit, and Wendell Wallach (2005). "Artificial morality: Top-down, bottom-up, and hybrid approaches," *Ethics and Information Technology*, **7**:149-155.

Asaro, Peter (2000). "Transforming Society by Transforming Technology: The Science and Politics of Participatory Design," *Accounting, Management and Information Technologies, Special Issue on Critical Studies of Information Practice*, **10**:257-290.

Feenberg, Andrew, and Alastair Hannay (eds.) (1998). *Technology and the Politics of Knowledge*. Bloomington, IN: Indiana University Press.

Feenberg, Andrew (1991). *Critical Theory of Technology*. Oxford, UK: Oxford University Press.

Latour, Bruno (1999). *Pandora's Hope: Essays on the Reality of Science Studies*. Cambridge, MA: Harvard University Press.

Moravec, Hans (1998). *Robot: Mere Machine to Transcendent Mind*. Oxford, UK: Oxford University Press.

Walzer, Michael (1977). *Just and Unjust Wars: A Moral Argument With Historical Illustrations*. New York, NY: Basic Books.

Yablonsky, Lewis (1972). *Robopaths: People as Machines*. New York, NY: Viking Penguin.

8

The Turing Triage Test

Robert Sparrow

Abstract. If, as a number of writers have predicted, the computers of the future will possess intelligence and capacities that exceed our own then it seems as though they will be worthy of a moral respect at least equal to, and perhaps greater than, human beings. In this paper I propose a test to determine when we have reached that point. Inspired by Alan Turing's (1950) original "Turing test", which argued that we would be justified in conceding that machines could think if they could fill the role of a person in a conversation, I propose a test for when computers have achieved moral standing by asking when a computer might take the place of a human being in a moral dilemma, such as a "triage" situation in which a choice must be made as to which of two human lives to save. We will know that machines have achieved moral standing comparable to a human when the replacement of one of these people with an artificial intelligence leaves the character of the dilemma intact. That is, when we might sometimes judge that it is reasonable to preserve the continuing existence of a machine over the life of a human being. This is the "Turing Triage Test". I argue that if personhood is understood as a matter of possessing a set of important cognitive capacities then it seems likely that future AIs will be able to pass this test. However this conclusion serves as a reductio of this account of the nature of persons. I set out an alternative account of the nature of persons, which places the concept of a person at the centre of an interdependent network of moral and affective responses, such as remorse, grief and sympathy. I argue that according to this second, superior, account of the nature of persons, machines will be unable to pass the Turing Triage Test until they possess bodies and faces with expressive capacities akin to those of the human form.

Introduction

If we are to believe the pronouncements of some researchers in the field of artificial intelligence, it will not be long until computers become autonomous systems, making decisions on their own behalf. In the not too distant future, computers will have beliefs and desires, even emotions, in order that they can reason better and function in a wider range of situations. They may even "evolve" via genetic algorithms, genetic programming or other methods of evolutionary computation. Eventually, through these techniques or simply through increasingly sophisticated design, they will become fully fledged self-conscious "artificial intelligences". According to a number of writers in the field, before the end of the 21st century – and according to some, well before this – machines will be conscious, intelligent, entities with capacities exceeding our own (Moravec, 1988; Kurzweil, 1992, 1999; Simons, 1992; Moravec, 1998; Dyson, 1997).

As soon as AIs begin to possess consciousness, desires and projects then it seems as though they deserve some sort of moral standing.[1] For instance, if my computer has more intelligence than my dog, is self-conscious and has internal states that function as pleasure and pain, and hopes and dreams, then it seems as though it would be at least as wrong to destroy it as to kill my dog. If, as a number of writers have predicted, artificial intelligences will eventually possess intelligence and capacities that exceed our own then it seems as though they will be worthy of a moral respect at least equal to, and perhaps greater than human beings. We may have duties towards such entities in our relations with them. It may even become necessary to grant them rights comparable to those possessed by human beings.

---

[1] This will mark the beginning of a new field that might be called "Android Ethics", to accompany "Android Epistemology". Cf. Ford et al. (1995). The birth of a new field of "Android Ethics" is also heralded in Floridi and Sanders (2000).

In this paper I propose a test to determine when we have reached that point. Inspired by Alan Turing's (1950) original "Turing test", which argued that we would be justified in conceding that machines could think if they could fill the role of a person in a conversation, I propose a test for when computers have achieved moral standing by asking when a computer might fill the role of a human being in a moral dilemma. The dilemma I have chosen is a case of "triage", in which a choice must be made as to which of two lives to save. In the scenario I propose, a hospital administrator is faced with the decision as to which of two patients on life support systems to continue to provide electricity to, following a catastrophic loss of power in the hospital. She can only preserve the existence of one and there are no other lives riding on her decision. We will know that machines have achieved moral standing comparable to a human when the replacement of one of the patients with an artificial intelligence leaves the character of the dilemma intact. That is, when we might sometimes judge that it is reasonable to preserve the continuing existence of the machine over the life of the human being. This is the *"Turing Triage Test"*.

Some qualifications

"Weak" versus "Strong" AI

Before I proceed with my discussion, I wish to head off an objection that might be made by those who would argue that I am misrepresenting the nature of research into artificial intelligence. Some researchers into advanced computing have given up the attempt to create artificial intelligences of the sort that I will be discussing. They have concluded either that the creation of genuine intelligence is beyond our current technological prowess or that there exists no single human capacity of intelligence that might be artificially reproduced. Instead they dedicate themselves to designing machines that can perform tasks similar to those performed by the human brain in some more narrowly prescribed area, such as facial or speech recognition, vision, or problem solving of certain sorts. Projects of this type are often described as "Weak AI". Typically, researchers involved in Weak AI wish to avoid the question as to whether success in these endeavours might ever involve the creation of genuine intelligence. To talk of machines, having "intelligence", let alone "beliefs and desires" or "self consciousness", is to confuse appearance with reality and, what's more, to risk provoking a dangerous backlash against their research by fuelling the public's perception that they are modern day Frankensteins. What are misleadingly described in the popular press as "artificial intelligences" are simply more complicated machines that are capable of performing complex tasks that in the past have only been possible for human beings.

As will become clear below, I have some sympathy for this position's dismissal of the possibility of genuine artificial intelligence. It may be that the technology never achieves the results necessary to create the issues with which I am concerned here.[2] But despite the lowered sights of some "AI" researchers, other researchers do claim to be working towards the creation of genuine artificial intelligence – a project known as "Strong AI". This paper takes the optimistic rhetoric of Strong AI enthusiasts at face value, at least initially; after all, what if they are right? It is best if we start talking about the ethical dilemmas now. Furthermore, it is dangerously presumptuous to claim that science will never progress to the point at which the question of the moral status of intelligent computers arises. Computer engineers and scientists have in the past shown a marked ability to disconcert the pundits by greatly exceeding expectations and achieving results previously thought impossible. If they do succeed in creating genuine artificial intelligence then the issue of the range and nature of our obligations towards them will arise immediately.

Artificial intelligence and the "Turing Test"

Before I continue then, I need to say something about what I mean by "artificial intelligence". The definition of intelligence is a vexed question in the philosophy of mind. We seem to have a firm intuitive grasp of what intelligence is. Roughly speaking, it is the ability to reason, to think logically, to use imagination, to learn and to exercise judgement. It is the ability to frame a problem and then solve it. Intelligence is generalisable; it is capable of doing these things across a wide range of problems and contexts. It is what we have, what primates have less of, parrots still less, jelly fish and trees (and contemporary machines) not at all. Artificial intelligence is intelligence in an artefact that we have created.

Yet it is surprisingly difficult to give a complete description of what intelligence consists in, let alone a precise definition. Because of the difficulty of providing a definition of intelligence, much of the

---

[2] Although it is worth noting that even the more modest systems designed by "weak AI" researchers may have some claim to moral regard and raise some of the issues with which I am concerned here. These may be usefully illuminated by considering the limit case of whether intelligent computers might achieve the moral status of persons.

discussion in the AI literature has moved to the hopefully easier question of how we might *tell* whether a machine was intelligent, even if we are unsure of exactly what intelligence consists in. This discussion has largely focussed on the appropriateness or otherwise of the notorious "Turing Test". In his famous article in *Mind*, Alan Turing (1950) suggested that we would be justified in conceding that machines were intelligent if they could successfully take the part of a human being in a conversation over a teletype machine. If we cannot tell the difference between a human being and a machine in the course of a conversation in the absence of visual cues, then we must acknowledge that the machine is intelligent.[3]

The adequacy of this test for machine intelligence has been the subject of controversy ever since. Critics have alleged that the Turing Test sets the standard for intelligence too high, too low, or in the wrong place altogether.[4] I cannot enter these debates here. I can only state my belief that, if anything the Turing Test sets the standard of behaviour for intelligence too high (after all, chimpanzees are intelligent – to a degree at least – and cannot pass the test). In any case, for the purposes of this paper I shall assume that whether or not an ability to pass the Turing Test is a necessary condition for possession of intelligence it is at least a sufficient condition.[5]

Furthermore, I will assume that the Turing Test also establishes more than is usually claimed on its behalf. I will hold that a machine that can pass the Turing Test should be acknowledged to be self-conscious as well as intelligent and also to have projects and ambitions that matter to it. While Turing himself did not argue that his test would establish these further conclusions, there is at least a prima facie case that it should. If a machine is to be able to converse like a human being then it must be capable of reporting on its internal states and its past history. That is, it must be able to demonstrate an awareness of self. Questions about our feelings and our personal history are a natural part of conversation. It is difficult to imagine how a machine which did not possess self-consciousness could carry on a convincing conversation about these things. Similarly, the fact that we have hopes and dreams, projects and ambitions, and that these matter to us, is also something that is evidenced in conversation. For instance we often ask each other about our intentions, ambitions, and attitudes towards various events and circumstances in our lives. We express happiness and sadness, joy and anger, concerning the satisfaction or frustration of our desires in the course of conversation. A machine which was unable to do the same would be unable to pass the Turing Test.

A machine that can pass the Turing Test must therefore be able to behave as though it has self-consciousness and commitments to various projects. If successful imitation of intelligent behaviour is sufficient to establish the presence of intelligence than so too, I argue, should it establish the presence of these further capacities.[6] Of course this argument is controversial; as is, for that matter, the adequacy of the Turing Test itself. It may turn out that machines that are eventually capable of passing the Turing Test clearly have none of these properties. But no matter. The important claim for my purposes is that at some point in the future machines will possess intelligence, self-consciousness and projects that matter to them, as a number of writers hold. I am assuming that an ability to pass the Turing Test will pick out such machines, but if it turns out otherwise, the argument that follows will stand, as long as some machines, perhaps those capable of passing more stringent tests, have these qualities. The validity of the argument that follows is therefore independent of the question of the validity of the original Turing Test.

Moral standing and personhood

What sort of moral standing should be granted such artificial intelligences? What level of moral concern or regard would we owe to them? Obviously, entities can possess moral standing to different degrees. Most of

---

[3] In fact this is a simplification of the original Turing Test, which required that a machine be as good as a man at pretending to be a woman. That is, the task of the computer is to be equally as good as a human being at an imitation game involving gender. The role played by gender in the original formulation of the Turing Test is usually neglected in later discussion of the test (Saygin, Cicekli et al., 2000).

[4] For a recent survey of the literature surrounding the Turing Test, see Saygin, Cicekli and Akman (2000).

[5] Although it must be noted that the conclusion of my paper suggests that passing the Turing Test will be much more difficult for machines than is currently recognised. Indeed it may well be impossible. My assumption that machines will pass the test is a working hypothesis for the purposes of the argument of the paper.

[6] It might be objected that a machine's ability to report on its internal state does not establish that it *really* has these features. As will become obvious later, I agree with this objection. But at first sight it does not seem to distinguish between the case of a machine's consciousness or desires and its intelligence. If the Turing Test is an adequate test for the presence of intelligence then further argument is required to show why it will not be adequate for these other qualities.

us would allow that of the various sorts of things that might make moral claims upon us, some of these are capable of sustaining greater claims than others. For instance, we may have a basic level of moral concern for the lower animals, such as fish and crustaceans, a greater concern for mammals such as dogs and elephants, etc., and more still for the higher primates such as the great apes. It may turn out that intelligent machines should be granted moral standing somewhere along this scale. However in this paper I am concerned to investigate whether machines could achieve moral standing comparable to that we accord normal adult human beings. Could machines be "moral persons"?

The "Turing Triage Test"

Imagine yourself the Senior Medical Officer at a hospital which employs a sophisticated artificial intelligence to aid in diagnosing patients. This artificial intelligence is capable of learning, of reasoning independently and making its own decisions. It is capable of conversing with the doctors in the hospital about their patients. When it talks with doctors at other hospitals over the telephone, or with staff and patients at the hospital over the intercom, they are unable to tell that they are not talking with a human being. It can pass the Turing Test with flying colours. The hospital also has an intensive care ward, in which up to half a dozen patients may be sustained on life support systems, while they await donor organs for transplant surgery or other medical intervention. At the moment there are only two such patients.

Now imagine that a catastrophic power loss affects the hospital. A fire has destroyed the transformer transmitting electricity to the hospital. The hospital has back up power systems but they have also been damaged and are running at a greatly reduced level. As Senior Medical Officer you are informed that the level of available power will soon decline to such a point that it will only be possible to sustain one patient on full life support. You are asked to make a decision as to which patient should be provided with continuing life support; the other will, tragically, die. Yet if this decision is not made, both patients will die. You face a "triage" situation, in which you must decide which patient has a better claim to medical resources. The diagnostic AI, which is running on its own emergency battery power, advises you regarding which patient has the better chances of recovering if they survive the immediate crisis. You make your decision, which may haunt you for many years, but are forced to return to managing the ongoing crises.

Finally, imagine that you are again called to make a difficult decision. The battery system powering the AI is failing and the AI is drawing on the diminished power available to the rest of the hospital. In doing so, it is jeopardising the life of the remaining patient on life support. You must decide whether to "switch off" the AI in order to preserve the life of the patient on life support. Switching off the AI in these circumstances will have the unfortunate consequence of fusing its circuit boards, rendering it permanently inoperable. Alternatively, you could turn off the power to the patient's life support in order to allow the AI to continue to exist. If you do not make this decision the patient will die and the AI will also cease to exist.[7] The AI is begging you to consider its interests, pleading to be allowed to draw more power in order to be able to continue to exist.

My thesis, then, is that machines will have achieved the moral status of persons when this second choice has the same character as the first one.[8] That is, when it is a moral dilemma of roughly the same difficulty. For the second decision to be a dilemma it must be that there are good grounds for making it either way. It must be the case therefore that it is sometimes legitimate to choose to preserve the existence of the machine over the life of the human being.

These two scenarios, along with the question of whether the second has the same character as the first, make up the "Turing Triage Test". It is my hope that the Turing Triage Test will serve as a focus for discussion of issues surrounding the moral status of artificial intelligences and what would be required for machines to achieve moral standing in the same way that the Turing Test has served to focus attention on the question of whether machines could think and what would be required for them to do so.[9]

---

[7] Let me also stipulate that neither of the available courses of action will lead to any further loss of life. The remaining patient is not a doctor or scientist whose advice is urgently needed. The hospital will be able to treat its patients properly without the AI in the short period before an alternative source of diagnostic advice can be found. In this decision, the only relevant consideration is the moral status of the two claimants in front of you.

[8] Rodney Brooks, a leading robotics researcher and Professor of Robotics at MIT's Artificial Intelligence laboratory, lends weight to the Turing Triage test when, in a recent interview, he describes his ambition in the following words – "I would like to have a machine or robot which you felt bad about switching off. I want to build a living machine" (Brooks, 2002).

[9] The Turing Triage Test may also be relevant to the question of the moral status of cyborgs – human/machine hybrids – although I will not be able to examine the matter here.

Obviously proposing a test for when moral standing has been achieved does not tell us whether or not a machine could pass this test. But it does allow us to think productively about what would be necessary for a machine to pass it; that is, what other sorts of things would need to be true for a machine to achieve moral standing.

The case for moral standing for intelligent machines

Let me begin by observing that according to an influential, perhaps the dominant, account of the nature of personhood, we should expect that future intelligent machines will, sooner or later, be able to pass the Turing Triage Test. A number of philosophers have argued that we should separate our account of the origin of moral concern and of the nature of personhood from the concept of a human being (see, for instance, Singer, 1981; Singer, 1986; Tooley, 1986. Diamond, 1991a, provides a neat paraphrase of this position). Whatever it is that makes human beings morally significant must be something that could conceivably be possessed by other entities. To restrict personhood to human beings is to commit the error of chauvinism or "speciesism".

The precise description of qualities required for an entity to be a person or an object of moral concern differ from author to author. However it is generally agreed that a capacity to experience pleasure and pain provides a prima facie case for moral concern and that the grounds for this concern, as well as its proper extent, are greater the more a creature is conscious of itself as existing across time, has its own projects and is capable of reasoning and rationality (Singer, 1993: 85–100).

It is a recognised, indeed an intended, consequence of such accounts that they allow that in some cases other entities might have more of these qualities than a given human being. For instance, we might sometimes be obligated to preserve the life of an adult chimpanzee over that of a brain damaged human baby on the grounds that the former has superior cognitive capacities and therefore greater claim to moral regard than the latter (Singer, 1993). I mention this point to establish that it will not be unprecedented therefore if it turns out that machines sometimes have a better claim to the status of personhood than some human beings.

What is more striking, however, is that it seems as though machines that are plausible candidates for the Turing Triage Test are likely to have a better claim to the status of personhood than *any* human being. If we become capable of manufacturing machines that are apparently capable of self-consciousness, reasoning, and investment in personal projects to the *same* extent as a human being, then we will presumably be able to produce machines that are capable of all of these to a much *greater* degree than human beings. There seem to be no reasons to believe that human beings define the upper limit of an ability to do these things. Once we discover how to make machines with such capacities we can simply expand them, perhaps indefinitely. Machines will after all not be limited, as human beings are, by having a fixed set of capacities available to them due to their hardware.

Thus, it is easy to imagine machines that are more intelligent than any human being, more rational, capable of more intricate chains of reasoning, of remembering and considering more facts and taking into consideration a wider range of arguments. Similarly, intelligent computers may have a greater sense of themselves as entities that endure across time than we do. Their consciousness of self may extend further in both directions than ours; they may have better (more reliable, longer lasting) memories than ours, that allow them to recall exactly what they were thinking at any given moment of their existence; they may have a justified expectation of a vastly longer lifespan than that available to human beings (need intelligent machines fear death due to ordinary circumstances at all?) and so have adopted projects that extend into the distant future. In so far as reasoning capacity, self-consciousness and possession of long term projects is relevant to personhood, future machines are therefore likely to have a greater claim to personhood than do humans.

It might be objected that while I may provide my hypothetical AIs with self consciousness by fiat, it is far from clear that such entities could properly be said to "suffer". The ability to experience pleasure and pain might plausibly be held to inhere only in living creatures with nervous systems sufficiently similar to our own. Machine pain can never be anything other than a figure of speech, a rough analogy justified by its usefulness in explaining behaviour (as in, for instance, a case where a robot retreats from a flame that is burning it). Unless machines can be said to suffer they cannot be appropriate objects for moral concern at all.

In fact I have a great deal of sympathy with this objection, as will become clear below. But for the moment I want to outline a popular response to the denial of the possibility of machine suffering. This response denies that pleasure and pain are states that may only be possessed by living entities and argues that they are properly understood as informational states that can be possessed by any system that behaves in ways suitably analogous to the nervous systems of living creatures regardless of what such systems are made of. Crudely, it is what mental states such as pleasure and pain (and indeed all other cognitive and affective states) *do*, rather than what the mind that experiences them is made of, that makes them what they are.[10] Machines may properly be said to possess pleasure and pain states if they have internal states that are "functionally isomorphic" to similar states in us (Putman 1975: 291–292).

This argument should, I believe, carry weight with anyone who is prepared to allow that a machine that can pass the Turing Test is intelligent. Presumably intelligent machines have other mental states, such as beliefs and desires, despite the fact that they are made of silicon and metal instead of flesh and blood. Further argument would be required to show why such machines should not be said to suffer or to experience pleasure when they behave in ways appropriate to these states. This is not to say that such arguments could not be made, indeed I will be making them below; however it is to suggest that the onus is on those who would accept the possibility of machine intelligence to explain why the mental life of machines could not include pleasure or pain of the sort that generates moral concern when we witness it in other creatures.

Notice also that if we can imagine a machine with the same capacity to experience pleasure and pain as a human being we should also be able to imagine a machine with a *greater* capacity to experience these things than any human being. For instance, if it possesses circuits that fulfil the same functional role as our nerve endings and pain receptors, we can easily imagine it having *more* of these than we do nerve endings and pain receptors. If it has internal states that map onto our experiences of graduated pleasures or pains then we can imagine it having states that are relevantly analogous to our experiences of great pleasure or great pain, as well as further states that are like these only more so, such that it experiences greater pleasure or greater pain than we do. Again, there seem to be no principled reasons to hold that no entity has a higher capacity for enjoyment or suffering than do human beings.

If a popular philosophical account of the nature of personhood, as consisting in the possession of certain cognitive capacities, is correct, and if we believe what AI enthusiasts say about the likely capacities of future machines, then not only will such machines pass the Turing Triage Test but they are also likely to dominate it. There may be a brief period where machines have only roughly the capacities of human beings and so the choice as to whether to save a human life or to preserve the existence of an intelligent machine constitutes a moral dilemma, but eventually, as the capacities of the machines increase we will always be obligated to save the machine.

I trust this will strike at least some readers as a counter intuitive conclusion, perhaps even a *reductio* of the arguments considered above. Could it ever really be the case that we were obligated to preserve the existence of a machine, a device of metal and plastic over the life of a member of our own species? For the remainder of this paper I will survey a set of arguments that suggest that it could not.

Humans, persons and AIs

Drawing on the thought of Wittgenstein (1989), a number of philosophers have argued that criterial accounts of personhood of the sort considered above are manifestly inadequate. Personhood is not a matter of having certain capacities or of being able to complete certain tasks.[11] Instead it is a matter of being a creature of a kind such that certain moral and affective responses are appropriately called into existence – and may even be mandatory – in its presence. It is to occupy a certain place in a network of interdependent concepts and moral and affective responses that make up our form of life. This belief that we need to take account of the cluster of concepts and moral and affective responses surrounding our concept of a person derives from a conviction that philosophy needs to pay more attention to the forms of life in which our concepts are embedded. To theorise without due attention to the ways we actually employ our moral language and what we do and do not – and can and cannot – say in it, is to risk losing our way in our investigations. Our concept of a person cannot be adequately captured without paying attention to the ways in which we behave around and towards people and the various ways in which these

---

[10] Functionalism avoids the problems of behaviourism by allowing that the nature and identity of these states is determined not only by their relation to the external behaviour of the system but also to other such internal states. See Block (1980), Churchland (1984), Putman (1975), Jackson and Petit (1988).

[11] My discussion below largely follows arguments presented in Gaita (1989, 1990, 1991a) and further refined in Gaita (1999). A central claim in Gaita's argument was previously developed in an important and difficult paper by Peter Winch (1981). Cora Diamond (1978, 1991a, b) covers related territory in a number of discussions of the ethics of our treatment of animals. Gaita and Winch draw heavily on the discussion of the nature of pain and of pain attributions in Wittgenstein (1989).

differ to our attitudes and behaviour towards non-persons such as animals. Moral emotions such as grief, remorse, sympathy and shame, amongst others, surround and inform our concept of a person.

Because this alternative analysis of personhood links our concept of a person to a wide range of interdependent moral and affective responses, it is possible to begin a challenge to the idea that a machine could pass the Turing Triage Test at a number of points. I will begin my discussion with an analysis of the relationship between personhood and the demands of remorse, grief and sympathy. However, this discussion leads quickly to a consideration of the nature of individual personality and the question of the authenticity of the "suffering" and other internal states of machines. My argument here in turn, rests on observations about the nature of the embodiment of machines.

The concept of a person and the moral emotions

Let us return briefly to reconsider the original triage situation involving a decision as to which of two human lives to save. What makes this situation a moral *dilemma* is the fact that no matter which life we decide to save we have grounds for remorse. A human life, a unique individual life, deserving of the most profound moral respect, has been cut short and this is tragic. It is entirely appropriate that a person required to make such a choice should be haunted by remorse, even in the case when they feel they could have decided no other way. We would understand if the person required to make this decision was haunted by it for a number of years, even perhaps for their whole life. There may be circumstances, or psychological stories to be told, which explain why an individual did not feel such remorse in a particular case. Perhaps the tragedy was quickly eclipsed by greater tragedy, perhaps the individual concerned has some temporary psychic deficit that prevents them from feeling remorse in this case. However a person who claimed they could not imagine feeling remorse for their decision in the situation would thereby demonstrate a failure to appreciate its nature as a moral dilemma.

Furthermore, it is an integral feature of remorse that it presents itself as a response to the particular individual that we have wronged. As Gaita (1991a: 150–154) has argued, while remorse can only be occasioned by an evil, a transgression of the moral law, it is not directed towards that transgression but towards its victim. The orientation of remorse towards the individual wronged is evidenced in the fact that it is entirely possible that the person experiencing remorse should see the face of the person they have wronged in their dreams, or be tormented by the memory of their voice. The possibility of the intrusion of the personality of their victim into the consciousness of a perpetrator who feels remorse for an evil that they have done is not an extrinsic feature of an ethical response that could be characterised independently. When cases like this occur they are paradigm cases of this moral emotion.

Similarly, it is internal to our sense of the weight of the decision that has to be made that we can imagine that someone should grieve for the individual whose life is lost when medical support is withdrawn. We can also imagine feeling sympathy for their suffering while they await the decision or as they die. Even if we do not feel any such things ourselves, if we cannot conceive of someone doing so, then it seems we do not face a dilemma. Notice also that like remorse, these attitudes are responses to the particular individual whose death or suffering provokes them.

Now let us turn to consider the case when one of the patients in the triage situation has been replaced by an AI. If this is also to be a moral dilemma, it must be appropriate that we might feel remorse no matter which way we made the decision. It must therefore be appropriate that a person might experience remorse for the wrong that they have done the machine in choosing to end its existence. Furthermore, it must be conceivable that this remorse should be such as to haunt a person for years, perhaps even blight their own life. It must also be imaginable that a person should experience grief following the "death" of the AI, or sympathise with its plight while it awaits a decision, or suffers as power is withdrawn from it.

Is it plausible to believe that someone should be haunted by the evil they have done to a machine? Or that they should feel grief following its death, or sympathy for its suffering? I argue it is not, for reasons that will be expanded upon below. But a first approximation is that machines cannot achieve the sort of individual personality that remorse, grief and sympathy respond to. This is not to claim that machines could not display unique characteristics such we could distinguish one machine from another. Instead it is to deny that this differentiation could ever establish personality in the richer sense of having a unique inner life of their own. We cannot take seriously a person who claims to feel these moral emotions for a machine because we cannot seriously entertain the idea that machines feel anything at all.

It may seem as though am I making an empirical claim here about our possible responses to machines and a false one at that. Some people clearly do attribute individual personality to machines, as well as emotional states, including suffering. For instance, it is not uncommon to hear people talk of their laptop, or VCR, or even their car, as "temperamental", "sulking", "in a bad mood", or attributing other emotions to these devices. It is less common but still occasionally possible to find people who say that they feel sorry for a machine, or that they experience grief when a machine "dies". Some individuals may even claim that they do feel remorse for wrongs that they believe that they have done to machines. If people already experience such emotions in relation to existing machines with their very limited expressive capacities then how much more likely are they when machines can talk and interact with us.

Yet it is not insignificant here that we feel compelled to enclose the emotional states that people ascribe to machines in inverted commas. We do not *really* believe that they have these emotions. Such descriptions are mere figures of speech or, alternatively, regrettable excesses of sentimentality. Notice that if one really did believe that machines had feelings worthy of moral concern then one could not be indifferent to whether or not other people recognised these. Indeed one would have to hold that it constituted callousness to fail to do so. Typically, of course, people do not believe this; evidence that they do not themselves really believe what they say. But more importantly, were we to meet someone who held this belief, apparently in all seriousness, I believe we would be forced to conclude that they were misusing the language. The point here is a conceptual rather than an empirical one. It concerns how far it is possible to extend our concepts before they tear loose from the supporting set of responses that give them their meaning.

Embodiment and the "inner life"

Why should it be so difficult, indeed ultimately impossible, to take seriously the idea that we should feel remorse, grief or sympathy for a machine, or that a machine could be suffering? It is because these responses are only conceivable in relation to creatures which look like us in certain ways. There is a connection between the capacity to engage a certain set of moral responses, including remorse, grief and sympathy, that inform and reflect our sense of the uniqueness of persons that is integral to their moral standing, and possession of a certain sort of physical presence. Crucially, this presence includes possession of a face with expressive eyes and features, and an animate body with the expressive capacities demonstrated by living things (Gaita, 1999: 269).[12] More controversially, it also seems to require that a person

---

[12] The role played by the human face at the very foundation of the nature of our ethical relationship with others is also argued for by Levinas (1989) and Lingis (1994).

be a creature of "flesh and blood". Machines are simply not the right sort of things of which to say that they suffer or feel. They lack expressive capacities of the sort required to ground a recognition of the reality of their inner life. No matter how sophisticated a machine is, it will always remain open to us to doubt the veracity of its purported feelings.

To see this, consider a case where the machine whose feelings we are being called upon to show concern for looks like a filing cabinet with a large number of flashing diodes on the front. This machine has sufficient number of diodes and can flash them in patterns of sufficient complexity so as to demonstrate behaviours that are "functionally isomorphic" to our pain responses and other cognitive states. The engineer who has designed this machine explains to us that *this* pattern of flashing lights means that the machine is suffering a small pain, *that* one that it is suffering a great pain, this one that it is happy, that one that it is sad, etc. In the light of this information we adjust our behaviour in relation to the machine, in order to minimise its "suffering".

Now imagine that the engineer returns to us in a fluster; she has been consulting the wrong manual and has misled us. In fact it is *these* lights which flash when the machine is in pain and *these* lights when it is happy, etc. We should be treating the machine entirely differently. At this point the possibility of radical doubt emerges. How do we know that the engineer has got it right this time? More sinisterly, how do we know that the machine isn't manipulating us by displaying a set of emotions that it is not feeling? How do we know what the machine is *really* feeling?

Once this radical doubt occurs to us, we have no way to resolve the issue. No analysis of the behaviour or structure of a machine will serve to establish that it really feels what it appears to or even that it feels anything at all. There is simply no way of establishing a bridge between the machine's behaviour and any judgements about its purported inner life. It is this unbridgeable gap that opens up between reality and appearance in relation to the thoughts and feelings of machines that explains why we find it impossible to take seriously the thought that machines could have an inner life.

"An attitude towards a soul"

Our doubt about the inner life of machines stands in stark contrast to the knowledge that we possess about the inner lives of the human beings around us. Indeed to call our relation to the internal states of others "knowledge" is actually a misnomer. It is knowledge only in that it makes no sense to doubt it. We don't even believe that other people have minds, experience pleasure and pain, emotions, etc. No inference is required to reach the conclusion, for example, that someone is in pain when they burn themselves. We simply see it.[13] The fact that such responses are normative for us is evidenced in the way that we question those who do not have them. A person who doubts that others around them have inner lives is not a paragon of rationality who resists an inference that other weaker minds draw without sufficient justification. They are someone who has lost their way in relation to the question entirely. Similarly, to describe these states as "internal" is misleading. They are not on the *inside* of the other person in a way that could be easily contrasted with their outside. They are states of the person and are sometimes visible as such.

Our awareness of the reality of the inner lives of other people is a function of what Peter Winch (1981), following Wittgenstein, calls "an attitude towards a soul". It is a "primitive reaction", a precognitive awareness that is a condition rather than a consequence of our belief that those around us have thoughts and feelings (Winch, 1981: 8; Gaita, 1999: 263–267). Importantly, such an attitude is both evidenced in and arises out of a large, complex, and often unconscious set of responses to, and behaviours around, the bodies and faces of other human beings. The fact that we wince when we see another person crack their head, that we can be called into self-consciousness by the gaze of another, that when we bind someone's wound we look into their face (Gaita, 1999: 266), are all examples of an attitude towards a soul. We cannot help but have such an attitude towards other human beings (Winch, 1981: 11).[14]

---

[13] Of course there *are* cases when we do wonder whether or not the emotion or pain someone appears to feel is real. We may, for instance, believe that they are acting a part or trying to deceive us. But here the possibility of such questioning is established by the certainty we have in ordinary cases. While the question of the veracity of the feelings displayed by another person may arise in particular cases, it never seriously occurs to us to doubt that this individual has thoughts and emotions, let alone that people in general have an internal life (Gaita, 1999: 263–267).

[14] It might be argued that our relationships with (non-human) animals represent an important borderline case where the proper description and significance of the relationships we have with animals is contested and where it is genuinely unclear whether animals have "minds"; thus it might be disputed whether the phenomenology of a "precognitive" awareness of other minds I have described here is an accurate one. The proper description of our relationships with animals is too large a question to address here. However, for arguments that these are closer to our relationships with other people than is generally recognised and also best accounted for in terms of "primitive reactions", see Hearne (1986). See also Diamond (1978, 1991a).

Conversely, we cannot have such an attitude towards a machine.

Androids

Given the role played by the face and the expressive body in creating in us an "attitude towards a soul" towards each other, it might be thought that the question arises as to whether machines with these features could evoke such an attitude in us. There seems to be no reason why AIs could not be provided with expressive faces. Robotics researchers at various labs around the world are already working on faces for robots in order to facilitate robot/human communication and interaction.[15] Similarly, work on creating humanoid robots is already well advanced. Eventually, perhaps, artificial intelligences will be embodied in androids of the sort made popular by speculative fiction and films such as "Blade Runner", "The Terminator", "Alien", "Aliens" and "AI".

Would we have "an attitude towards a soul" in relation to such androids? It is tempting to allow that we would. After all, by hypothesis they have bodies and faces with the same expressive capacities as those of human beings.[16] Indeed, they may be externally indistinguishable from human beings. Yet I believe this would be a mistake. To see why, we must return to the discussion above of how a destructive doubt arises in relation to the thought and feeling of machines. It would not, I believe, alter the force of the example if, instead of a box-shaped device with flashing lights, we confronted a bipedal machine with an animatronic "face". Doubt regarding the relation between its internal states and its external appearance and behaviour could still arise and this is sufficient to destroy any attitude towards a soul that might otherwise exist. The artefactual nature of machines means that the question of the real nature of their design is ever present. Thus, while we might be fooled

---

[15] For example, researchers at the MIT Artificial Intelligence Laboratory have developed "Kismet", a robot that is designed to respond to human facial expressions and can express its own "emotions" through its own caricature-like face. For an introduction to Kismet and to other contemporary robot research in this area, see Brooks (2003) and Menzel and D'Aluisio (2000).

[16] It is worth pausing to note how demanding this assumption really is. Could any combination of metal and plastic achieve the near infinite expressive power of that is possessed by human flesh? Could it allow the empathic awareness of another's emotions that exists, even in silence, between lovers and friends? Our justified cynicism about such a possibility may go a long way towards explaining our reluctance to believe that even androids might possess an inner life.

into evincing such an attitude by machines of sufficiently clever construction, if we were to become aware of their nature we would be forced to reassess our attitude towards their supposed thoughts and feelings and deny that they were ever anything but clever simulacra. If I am right in this, then even artificially intelligent androids will fail the Turing Triage Test. Only creatures of "flesh and blood" with expressive bodies and faces are capable of being the object of an "attitude towards a soul".[17]

Summary

The argument of this final section of the paper has been a complex and difficult one. To recap briefly; I have argued that for a machine to pass the Turing Test it must be capable of being the object of remorse, grief and sympathy, as moral emotions such as these are partially constitutive of our concept of a person. But, I claim, machines are not appropriate objects of these responses because they are incapable of achieving the individual personality towards which they are oriented. We cannot seriously hold that machines have thoughts or feelings or an inner life because a radical doubt inevitably arises as to whether they really feel what they appear to. My argument here in turn draws upon an analysis of the nature of our knowledge of other minds as consisting in "an attitude towards a soul".

A critical reader no doubt may wish to challenge this chain of reasoning at any number of points. I can do no more to defend my account here. However, in closing, I wish again to emphasise the demanding and counter-intuitive nature of the conclusions that are likely to follow from any successful such challenge. For example, that we should feel sympathy for the "suffering" of an entity which has all the expressive capacity of a metal box, and, not least, that we may be obligated to preserve its existence over that of a living human being!

---

[17] Of course, this leaves open the possibility that hypothetical "organic" AIs, or perhaps cyborgs (wherein a silicon-based artificial intelligence controls an organic body), might pass the Turing Triage Test. This will depend on whether or not we are inclined to experience the radical doubt about the veracity of the "emotions" expressed by such entities, that I have been discussing here, upon learning of their nature as artefacts. My suspicion is that this will in turn, at least in part, depend on the extent to which such entities are properly described as individuals with unique histories of personal genesis or, alternatively, as "units" which have been manufactured and which consequently may be replaced when destroyed. However, as noted above, a proper consideration of the moral standing of cyborgs is beyond the scope of this paper.

Conclusion

The prospect of future artificially intelligent machines raises the question of the moral standing of such creations. Should they be treated as persons? I have offered the "Turing Triage Test" as a useful device for testing our intuitions in relation to this matter. A popular philosophical account of the nature of persons as beings with a certain set of cognitive capacities leads quickly to the conclusion that not only will future machines pass this test but that they may come to have more claim for moral regard than any human being. However this unpalatable conclusion may also be taken to indicate deep problems with this account of the nature of persons. An alternative account of what it is to be a person, set out most clearly in the work of Raimond Gaita, looks to an interdependent network of moral and affective responses to delineate and give content to our concept of a person. Until AIs are embodied in such a fashion that they can mobilise these responses in us, they will be unable to pass the "Turing Triage Test". Unless we could sympathise with the suffering of an AI as we moved to throw the switch that would end its existence, grieve for its "death", and be haunted by remorse at the thought of the life that we have ended, it would not be reasonable to preserve its existence over that of a human being. It will not be possible for us to properly apply these concepts while artificial intelligences have the character and appearance of machines.

References

N. Block, Introduction: what is functionalism? In N. Block, editor, *Readings in Philosophy of Psychology*, Vol 1. Harvard University Press, Cambridge, MA, pp. 171–184, 1980.
R. Brooks, Designed for Life, Interview by Duncan Graham-Rowe. *New Scientist*, May 1, 2002.
R.A. Brooks, *Robot: The Future of Flesh and Machines*. Penguin, London, 2003.
Paul M. Churchland, *Matter and Consciousness*, MIT Press, Cambridge, MA, 1984.
C. Diamond, Eating Meat and Eating People. *Philosophy* 53, 465–479, 1978.
C. Diamond, The importance of being human, in D. Cockburn, editor, *Human Beings*. Cambridge University Press, Cambridge, pp. 35–62, 1991a.
C. Diamond, *The Realistic Spirit: Wittgenstein, Philosophy and the Mind*. MIT Press, Cambridge, MA, 1991b.
George Dyson, *Darwin Amongst the Machines: The evolution of global intelligence*. Addison-Wesley Pub. Co., Reading, MA, 1997.

L.L. Floridi, and J.W. Sanders, Artificial evil and the foundation of computer ethics. In Deborah Johnson, James Moor and Herman Tavani, editor, *Proceedings of Computer Ethics: Philosophical Enquiry 2000*. Dartmouth, pp. 142–154, 2000.

Kenneth M. Ford, Clark Glymour and Patrick J. Hayes, editors, *Android Epistemology*. MIT Press, Cambridge, MA, 1995.

R. Gaita, The personal in ethics. In D.Z. Phillips and P. Winch, editors, *Wittgenstein: Attention to Particulars*. pp. 124–150, 1989.

R. Gaita, Ethical Individuality. In R. Gaita, editor, *Value and Understanding*. Routledge, London, pp. 118–148, 1990.

R. Gaita, *Good and Evil: An Absolute Conception*. MacMillan, London, 1991a.

R. Gaita, Language and Conversation: Wittgenstein's Builders. In A.P. Griffiths, editor, *Wittgenstein Centenary Essays*. Cambridge University Press, Cambridge, pp. 101–115, 1991b.

R. Gaita, *A Common Humanity: Thinking About Love & Truth & Justice*. Text Publishing, Melbourne, 1999.

V. Hearne, *Adam's Task: Calling Animals by Name*. Alfred A. Knopf, New York, 1986.

Frank Jackson and Philip Pettit, Functionalism and Broad Content. *Mind* 97, 381–400, 1988.

Ray Kurzweil, *The Age of Intelligent Machines*. MIT Press, Cambridge, MA, 1992.

Ray Kurzweil, *The Age of Spiritual Machines: When Computers Exceed Human Intelligence*. Allen & Unwin, St Leonards, N.S.W, 1999.

E. Levinas, Ethics as first philosophy. In S. Hand, editor, *The Levinas Reader*. Basil Blackwell, Oxford, pp. 75–87, 1989.

A. Lingis, *The Community of Those Who Have Nothing in Common*. Indiana University Press, Bloomington and Indianapolis, 1994.

P. Menzel, and FD'Aluisio, *Robo Sapiens: Evolution of a New Species*. The MIT Press, Cambridge, MA, 2000.

Hans Moravec, *Mind Children: The Future of Robot and Human Intelligence*. Harvard University Press, Cambridge, MA, 1988.

Hans Moravec, *Robot: Mere Machine to Transcendent Mind*. Oxford University Press, Oxford, 1998.

H. Putnam, Philosophy and our mental life. In *Mind, Language, and Reality*. Cambridge University Press, Cambridge, pp. 291–303, 1975.

A.P. Saygin, L. Cicekli and V. Akman, Turing Test: 50 Years Later. *Minds and Machines* 10(4): 463–518, 2000.

Peter Singer, *The Expanding Circle: Ethics and Sociobiology*. Farrar, Straus & Giroux, New York, 1981.

Peter Singer, All animals are equal. In P. Singer, editor, *Applied Ethics*. Oxford University Press, Oxford, pp. 214–228, 1986.

Peter Singer, *Practical Ethics*. Second Edition, Cambridge University Press, Cambridge, U.K., 1993.

Geoff Simons, *Robots: The Quest for Living Machines*. Cassell, London, 1992.

M. Tooley, Abortion and infanticide. In P. Singer, editor, *Applied Ethics*. Oxford University Press, Oxford, pp. 57–86, 1986.

Alan Turing, Computing machinery and intelligence. *Mind* 59: 433–60, 1950.

Peter Winch, Eine Einstellung zur Seele. *Proceedings of the Aristotelian Society*, 1981.

L. Wittgenstein, *Philosophical Investigations*. Basil Blackwell, Oxford, 1989.

9

A Nascent Robotics Culture:
New Complicities for Companionship

Sherry Turkle

Abstract

Encounters with humanoid robots are new to the everyday experience of children and adults. Yet, increasingly, they are finding their place. This has occurred largely through the introduction of a class of interactive toys (including Furbies, AIBOs, and My Real Babies) that I call "relational artifacts." Here, I report on several years of fieldwork with commercial relational artifacts (as well as with the MIT AI Laboratory's Kismet and Cog). It suggests that even these relatively primitive robots have been accepted as companionate objects and are changing the terms by which people judge the "appropriateness" of machine relationships. In these relationships, robots serve as powerful objects of psychological projection and philosophical evocation in ways that are forging a nascent robotics culture.

Introduction

The designers of computational objects have traditionally focused on how these objects might extend and/or perfect human cognitive powers. But computational objects do not simply do things *for* us, they do things *to* us as people, to our ways of being the world, to our ways of seeing ourselves and others (Turkle 2005[1984], 1995). Increasingly, technology also puts itself into a position to do things *with* us, particularly with the introduction of "relational artifacts," here defined as technologies that have "states of mind" and where encounters with them are enriched through understanding these inner states (Turkle 2004a, 2004b). Otherwise described as "sociable machines" (Breazeal 2000, 2002, Breazeal and Scasselati 1999, 2000, Kidd 2004), the term relational artifact evokes the psychoanalytic tradition with its emphasis on the meaning of the person/machine encounter.

In the late 1970s and early 1980s, children's style of programming reflected their personality and cognitive style. And computational objects such as Merlin, Simon, and Speak and Spell provoked questions about the quality of aliveness and about what is special about being a person. (Turkle 2005[1984])Twenty years later, children and seniors confronting relational artifacts as simple as Furbies, AIBOs and My Real Babies (Turkle 2004a) or as complex as the robots Kismet and Cog (Turkle et. al. 2004) were similarly diffferentiated in their style of approach and similarly provoked to ask fundamental questions about the objects' natures.

Children approach a Furby or a My Real Baby and explore what it means to think of these creatures as alive or "sort of alive"; elders in a nursing play with the robot Paro and grapple with how to characterize this creature that presents itself as a baby seal (Taggart, W. et al. 2005, Shibata 1999, 2005). They move from inquiries such as "Does it swim?" and "Does it eat?" to "Is it alive?" and "Can it love?"*

These similarities across the decades are not surprising. Encounters with novel computational objects present people with category-challenging experiences. The objects are liminal, betwixt-and between, provoking new thought. (Turner 1969; Bowker and Star 1999). However, there are significant differences between current responses to relational artifacts and earlier encounters with computation. Children first confronting computer toys in the late 1970s and early 1980s were compelled to classification. Faced with relational artifacts, children's

---

* A note on method: the observations presented here are based on open–ended qualitative fieldwork. This is useful in the study of human/robot interaction for several reasons. Case studies and participant-observation in natural settings enable the collection of empirical data about how people think about and use technology outside the laboratory. Qualitative methods are well-positioned to bring cultural beliefs and novel questions to light. Open-ended qualitative work puts the novelty of the technology at the center of things and says, "When you are interested in something new: *observe, listen, ask.*" Additionally, qualitative approaches to human-robot interaction provide analytical tools that help us better understand both the technologies under study and the social and cultural contexts in which these technologies are deployed. Differences in individual responses to techology are a window onto personality, life history, and cognitive style. Seeing technology in social context helps us better understand social complexities.

questions about classification are enmeshed in a new desire to *nurture and be nurtured by* the artifacts rather than simply categorize them; in their dialogue with relational artifacts, children's focus shifts from cognition to affect, from game playing to fantasies of mutual connection. In the case of relational artifacts for children and the elderly, nurturance is the new "killer app." We attach to what we nurture (Turkle 2004, 2005b).

We Attach to What We Nurture

In *Computer Power and Human Reason*, Joseph Weizenbaum wrote about his experiences with his invention, ELIZA, a computer program that seemed to serve as self object as it engaged people in a dialogue similar to that of a Rogerian psychotherapist (1976). It mirrored one's thoughts; it was always supportive. To the comment: "My mother is making me angry," the program might respond, "Tell me more about your mother," or "Why do you feel so negatively about your mother." Weizenbaum was disturbed that his students, fully knowing that they were talking with a computer program, wanted to chat with it, indeed, wanted to be alone with it. Weizenbaum was my colleague at MIT at the time; we taught courses together on computers and society. And at the time that his book came out, I felt moved to reassure him. ELIZA seemed to me like a Rorschach through which people expressed themselves. They became involved with ELIZA, but the spirit was "as if." The gap between program and person was vast. People bridged it with attribution and desire. They thought: "I will talk to this program 'as if' it were a person; I will vent, I will rage, I will get things off my chest." At the time, ELIZA, seemed to me no more threatening than an interactive diary. Now, thirty years later, I aski myself if I had underestimated the quality of the connection.

A newer technology has created computational creatures that evoke a sense of mutual relating. The people who meet relational artifacts feel a desire to nurture them. And with nurturance comes the fantasy of reciprocation. They wanted the creatures to care about them in return. Very little about these relationships seemed to be experienced "as if." The experience of "as if" had morphed into one of treating robots "as though." The story of computers and their evocation of life had come to a new place.

Children have always anthropomorphized the dolls in their nurseries. It is important to note a difference in what can occur with relational artifacts. In the past, the power of objects to "play house" or "play cowboys" with a child has been tied to the ways in which they enabled the child to project meanings onto them. They were stable "transitional objects." (Winnicott 1971) The doll or the teddy bear presented an unchanging and passive presence. But today's relational artifacts take a decidedly more active stance. With them, children's expectations that their dolls want to be hugged, dressed, or lulled to sleep don't only come from the child's projection of fantasy or desire onto inert playthings, but from such things as the digital dolls' crying inconsolably or even saying: "Hug me!" or "It's time for me to get dressed for school!" *In the move from traditional transitional objects to contemporary relational artifacts, the psychology of projection gives way to a relational psychology, a psychology of engagement. Yet, old habits of projection remain: robotic creatures become enhanced in their capacities to enact scenarios in which robots are Rorschachs, projective screens for individual concerns.*

From the perspective of several decades of observing people relating to computational creatures, I see an evolution of sensibilities.

• Through the 1980s, people became deeply involved with computational objects – even the early computer toys became objects for profound projection and engagement. Yet, when faced with the issue of the objects' affective possibilities, a modal response might be summed up as "Simulated thinking may be thinking; simulated feeling is never feeling. Simulated love is never love.

• Through the 1990s, the development of a "culture of simulation" brought the notion of simulation (largely through participation in intensive game spaces) into the everyday. The range and possibilities of simulation became known to large numbers of people, particularly young people.

• By the late 1990s, the image of the robot was changing in the culture. A robotics presence was developing into a robotics culture increasingly shaped by the possibility if not the reality of robots in the form of relational artifacts. Alongside a tool model, people are learning about a notion of cyber-companionship. Acceptance of this notion requires a revisiting of old notions of simulation to make way for a kind of companionship that feels appropriate to a robot/person relationship.

The Evolution of Sensibilities: Two Moments

A first moment: I take my fourteen-year-old daughter to the Darwin exhibit at the American Museum of Natural History. The exhibit documents Darwin's life and thought, and with a somewhat defensive tone (in light of current challenges to evolution by proponents of intelligent design), presents the theory of evolution as the central truth that underpins contemporary biology. The Darwin exhibit wants to convince and it wants to please. At the entrance to the exhibit is a turtle from the Galapagos Islands, a seminal object in the development of evolutionary theory. The turtle rests in its cage, utterly still. "They could have used a robot," comments my daughter. She considers it a shame to bring the turtle all this way and put it in a cage for a performance that draws so little on the turtle's "aliveness." I am startled by her comments, both solicitous of the imprisoned turtle because it is alive and unconcerned about its authenticity. The museum has been advertising these turtles as wonders, curiosities, marvels -- among the plastic models of life at the museum, here is the life that Darwin saw. I begin to talk with others at the exhibit, parents and children. It is Thanksgiving weekend. The line is long, the crowd frozen in place. My question, "Do you care that the turtle is alive?" is welcome diversion. A ten-year-old girl would prefer a robot turtle because aliveness comes with aesthetic inconvenience: "its water looks dirty. Gross." More usually, votes for the robots echo my daughter's sentiment that in this setting, aliveness doesn't seem worth the trouble. A twelve-year-old girl opines: "For what the turtles do, you didn't have to have the live ones." Her father looks at her, uncomprehending: "But the point is that they are real, that's the whole point."

The Darwin exhibit gives authenticity major play: on display are the actual magnifying glass that Darwin used, the actual notebooks in which he recorded his observations, indeed, the very notebook in which he wrote the famous sentences that first described his theory of evolution *But in the children's reactions to the inert but alive Galapagos turtle, the idea of the "original" is in crisis.* I recall my daughter's reaction when she was seven to a boat ride in the postcard blue Mediterranean. Already an expert in the world of simulated fish tanks, she saw a creature in the water, pointed to it excitedly and said: "Look mommy, a jellyfish! It looks so realistic!" When I told this story to a friend who was a research scientist at the Walt Disney Company, he was not surprised. When Animal Kingdom opened in Orlando, populated by "real," that is, biological animals, its first visitors complained that these animals were not as "realistic" as the animatronic creatures in Disneyworld, just across the road. The robotic crocodiles slapped their tails, rolled their eyes, in sum, displayed "essence of crocodile" behavior. The biological crocodiles, like the Galapagos turtle, pretty much kept to themselves. What is the gold standard here?

I have written that now, in our culture of simulation, the notion of authenticity is for us what sex was to the Victorians – "threat and obsession, taboo and fascination" (Turkle, 2005[1984]). I have lived with this idea for many years, yet at the museum, I find the children's position strangely unsettling. For them, in this context, aliveness seems to have no intrinsic value. Rather, it is useful only if needed for a specific purpose. "If you put in a robot instead of the live turtle, do you think people should be told that the turtle is not alive?" I ask. Not really, say several of the children. Data on "aliveness" can be shared on a "need to know" basis, for a purpose. But what *are* the purposes of living things? When do we need to know if something is alive?

A second moment: an older woman, 72, in a nursing home outside of Boston is sad. Her son has broken off his relationship with her. Her nursing home is part of a study I am conducting on robotics for the elderly. I am recording her reactions as she sits with the robot Paro, a seal-like creature, advertised as the first "therapeutic robot" for its ostensibly positive effects on the ill, the elderly, and the emotionally troubled. Paro is able to make eye contact through sensing the direction of a human voice, is sensitive to touch, and has "states of mind" that are affected by how it is treated – for example, it can sense if it is being stroked gently or with some aggressivity. In this session with Paro, the woman, depressed because of her son's abandonment, comes to believe that the robot is depressed as well. She turns to Paro, strokes him and says: "Yes, you're sad, aren't you. It's tough out there. Yes, it's hard." And then she pets the robot once again, attempting to provide it with comfort. And in so doing, she tries to comfort herself.

Psychoanalytically trained, I believe that this kind of moment, if it happens between people, has profound therapeutic potential. What are we to make of this transaction as it unfolds between a depressed woman and a robot? When I talk to others about the old woman's encounter with Paro, their first associations are usually to their pets and the solace they provide. The comparison sharpens the questions about Paro and the quality of the relationships people have with it. I do not know if the projection of understanding onto pets is "authentic." That is, I do not know whether a pet could feel or smell or intuit some understanding of what it might mean to be with an old woman whose son has chosen not to see her anymore. What I do know is that Paro has understood nothing. Like other "relational artifacts" its ability to inspire relationship is not based on its intelligence or consciousness, but on the capacity to push certain "Darwinian" buttons in people (making eye contact, for example) that cause people to respond *as though* they were in relationship. For me, relational artifacts are the new uncanny in our computer culture, as Freud (1960) put it, "the long familiar taking a form that is strangely unfamiliar."

Confrontation with the uncanny provokes new reflection. Do plans to provide relational robots to children and the elderly make us less likely to look for other solutions for their care? If our experience with relational artifacts is based on a fundamentally deceitful interchange (artifacts' ability to persuade us that they know and care about our existence) can it be good for us? Or might it be good for us in the "feel good" sense, but bad for us in our lives as moral beings? The answers to such questions are not dependent on what computers can do today or what they are likely to be able to do in the future. These questions ask what *we* will be like, what kind of people are *we* becoming as we develop increasingly intimate relationships with machines.

Rorschach and Evocation

We can get some first answers by looking at the relationship of people – here I describe fieldwork with children and seniors – with these new intimate machines. In these relationship it is clear that the distinction between people using robots for projection of self (as Rorschach) and using robots as philosophically evocative objects, is only heuristic. They work together: children and seniors develop philosophical positions that are inseparable from their emotional needs. Affect and cognition work together in the subjective response to relational technologies. This is dramatized by a series of case studies, first of children, then of seniors, in which the "Rorschach effect" and the "evocative object effect" are entwined.[*]

Case Studies of Children

I begin with a child Orelia, ten, whose response to the robot AIBO serves as commentary on her relationship to her mother, a self-absorbed woman who during her several sessions with her daughter and the robot does not touch, speak, or make eye contact with her daughter. One might say that Orelia's mother acts robotically and the daughter's response is to emphasize the importance and irreducibility of the human heart. In a life characterized by maternal chill, Orelia stressed warmth and intuition as ultimate human values.

Orelia: keeping a robot in its place I met Orelia at a private Boston-area middle school where we were holding group sessions of fifth graders with a range of robotic toys. Orelia received an AIBO to take home; she kept a robot "diary." We met several times with Orelia and her parents in their Charlestown home. (Turkle 2004a)

Orelia is bright and articulate and tells us that her favorite hobby is reading. She makes determined distinctions between robots and biological beings. "AIBO is not alive like a real pet; it does not breathe." There is no question in her mind that she would choose a real dog over an AIBO. She believes that AIBO can love but only because "it is programmed to." She continues: "If [robots] love, then it's artificial. [And] if it's an artificial love, then there really isn't anything true... I'm sure it would be programmed to [show that it likes you], you know, the computer inside of it telling it to show artificial love, but it doesn't love you."

Orelia is sure that she could never love an AIBO. "They [robots] won't love you back if you love them. In order to love an AIBO, Orelia says it would need "a brain and a heart." Orelia feels that it is not worth investing in something that does not have the capacity to love back, a construction that is perhaps as much about the robot as about her relationship with her mother.

Orelia's brother Jake, nine, the baby of the family, is more favored in his mother's eyes. Unlike his sister, Jake assumes that AIBO has feelings. Orelia speaks to the researchers *about* AIBO; Jake addresses AIBO directly. He wants to stay on AIBO's good side, asking, "Will he get mad if you pick him up?" When Jake's style of addressing AIBO reveals that Jake finds the robot's affective states genuine, Orelia corrects her brother sharply: "It [AIBO] would just be mad at you because it's programmed to know 'if I don't get the ball, I'll be mad.'" The fact that AIBO is programmed to show emotions, make these artificial and not to be trusted.

Orelia expands on real versus programmed emotion:

> A dog, it would actually feel sorry for you. It would have sympathy, but AIBO, it's artificial. I read a book called *The Wrinkle in Time*, where everyone was programmed by this thing called "It." And all the people were completely on routine. They just did the same thing over and over. I think it'd be the same thing with the [artificial] dog. The dog wouldn't be able to do anything else.

For Orelia, only living beings have real thoughts and emotions:

> With a real dog if you become great friends with it, it really loves you, you know, it truly . . . has a brain, and you know somewhere in the dog's brain, it loves you, and this one [AIBO], it's just somewhere on a computer disk... If a real dog dies, you know, they have memories, a real dog would have memories of times, and stuff that you did with him or her, but this one [AIBO] doesn't have a brain, so it can't.

Orelia wants the kind of love that only a living creature can provide. She fears the ability of any creature to behave 'as if' it could love. She denies a chilly emotional reality by attributing qualities of intuition, transparency, and connectedness to all people and anaimals. A philosophical position about robots is linked to an experience of the machine-like equalities of which people are capable, a good exmple of the interdependence of philosophical position and psychological motivation.

---

[*] My case studies of robots and seniors with AIBO and My Real Baby are drawn from work conducted through weekly visits to schools and nursing homes from 2001 to 2003, studies that encompassed several hundred participants. In my discussion of Paro, I am reporting on studies of the same two nursing homes during the spring of 2005, a study that took place during twelve site visits and recruited 23 participants, ranging in age from 60-104, six males, and seventeen females. Researchers on these projects include Olivia Dasté, for the first phase of work, and for the second phase, Cory Kidd and Will Taggart.

Melanie: yearning to nurture a robotic companion The quality of a child's relationship with a parent does not determine a *particular* relationship to robotic companions. Rather, feelings about robots can represent different strategies for dealing with one's parents, and perhaps for working through difficulties with them. This is illustrated by the contrast between Orelia and ten-year-old Melanie. Melanie, like Orelia, had sessions with AIBO and My Real Baby at school and was given both to play with at home. In Melanie's case, feelings that she did not have enough of her parent's attention led her to want to nurture a robotic creature. Melanie was able to feel more loved by loving another; the My Real Baby and AIBO were "creature enough" for this purpose.

Melanie is soft-spoken, intelligent, and well mannered. Both of her parents have busy professional lives; Melanie is largely taken care of by nannies and baby-sitters. With sadness, she says that what she misses most is spending time with her father. She speaks of him throughout her interviews and play sessions. Nurturing the robots enables her to work through feelings that her parents, and her father in particular, are not providing her with the attention she desires.

Melanie believes that AIBO and My Real Baby are sentient and have emotions. She thinks that when we brought the robotic dog and doll to her school "they were probably confused about who their mommies and daddies were because they were being handled by so many different people." She thinks that AIBO probably does not know that he is at her particular school because the school is strange to him, but "almost certainly does knows that he is outside of MIT and visiting another school." She sees her role with the robots as straightforward; it is maternal.

One of Melanie's third-grade classmates is aggressive with My Real Baby and treats the doll like an object to explore (poking the doll's eyes, pinching its skin to test its "rubberness," and putting her fingers roughly inside its mouth). Observing this behavior, Melanie comes over to rescue the doll. She takes it in her arms and proceeds to play with it as though it were a baby, holding it close, whispering to it, caressing its face. Speaking of the My Real Baby doll that she is about to take home, Melanie says, "I think that if I'm the first one to interact with her then maybe if she goes home with another person [another study participant] she'll cry a lot . . . because she doesn't know, doesn't think that this person is its Mama." For Melanie, My Real Baby's aliveness is dependent on its animation and relational properties. Its lack of biology is not in play. Melanie understands that My Real Baby is a machine. This is clear in her description of its possible "death."

> Hum, if his batteries run out, maybe [it could die]. I think it's electric. So, if it falls and breaks, then it would die, but if people could repair it, then I'm not really sure. [I]f it falls and like totally shatters I don't think they could fix it, then it would die, but if it falls and one of its ear falls off, they would probably fix that.

Melanie combines a mechanical view of My Real Baby with confidence that it deserves to have her motherly love. At home, Melanie has AIBO and My Real Baby sleep near her bed and believes they will be happiest on a silk pillow. She names My Real Baby after her three-year old cousin Sophie. "I named her like my cousin . . . because she [My Real Baby] was sort of demanding and said most of the things that Sophie does." She analogies the AIBO to her dog, Nelly. When AIBO malfunctions, Melanie does not experience it as broken, but as behaving in ways that remind her of Nelly. In the following exchange that takes place at MIT, AIBO makes a loud, mechanical, wheezing sound and its walking becomes increasingly wobbly. Finally AIBO falls several times and then finally is still. Melanie gently picks up the limp AIBO and holds it close, petting it softly. At home, she and a friend treat it like a sick animal that needs to be rescued. They give it "veterinary care."

In thinking about relational artifacts such as Furbys, AIBOs, My Real Babies, and Paros, the question is posed: how these objects differ from "traditional" (non-computational) toys, teddy bears, and Raggedy-Ann dolls. Melanie, unbidden, speaks directly to this issue. With other dolls, she feels that she is "pretending." With My Real Baby, she feels that she is really the dolls's mother: "[I feel] like I'm her real mom. I bet if I really tried, she could learn another word. Maybe Da-da. Hopefully if I said it a lot, she would pick up. It's sort of like a real baby, where you wouldn't want to set a bad example."

For Melanie, not only does My Real Baby have feelings, Melanie sees it as capable of complex, mixed emotions. "It's got similar to human feelings, because she can really tell the differences between things, and she's happy a lot. She gets happy, and she gets sad, and mad, and excited. I think right now she's excited and happy at the same time."

> Our relationship, it grows bigger. Maybe when I first started playing with her she didn't really know me so she wasn't making as much of these noises, but now that she's played with me a lot more she really knows me and is a lot more outgoing. Same with AIBO.

When her several weeks with AIBO and My Real Baby come to an end, Melanie is sad to return them. Before leaving them with us, she opens the box in which they are housed and gives them an emotional good bye. She hugs each one separately, tells them that she will miss them very much but that she knows we [the researchers] will take good care of them. Melanie is concerned that the toys will forget her, especially if they spend a lot of time with other families.

Melanie's relationship with the AIBO and My Real Baby illustrates their projective qualities: she nurtures them because getting enough nurturance is an issue for her. But in providing nurturance to the robots, Melanie provided it to herself as well (and in a way that felt more authentic than developing a relationship with a "traditional" doll). In another case, a seriously ill child was able to use relational robots to speak more easily in his own voice.

Jimmy: from Rorschach to relationship Jimmy, small, pale, and thin, is just completing first grade. He has a congenital illness that causes him to spend much time in hospitals. During our sessions with AIBO and My Real Baby he sometimes runs out of energy to continue talking. Jimmy comes to our study with a long history of playing computer games. His favorite is Roller Coaster Tycoon. Many children play the game to create the wildest roller coasters possible; Jimmy plays the game to maximize the maintenance and staffing of his coasters so that the game gives him awards for the safest park. Jimmy's favorite toys are Beanie Babies. Jimmy participates in our study with his twelve-year-old brother, Tristan.

Jimmy approaches AIBO and My Real Baby as objects with consciousness and feelings. When AIBO slams into the red siding that defines his game space, Jimmy interprets his actions as "scratching a door, wanting to go in.... I think it's probably doing that because it wants to go through the door... Because he hasn't been in there yet." Jimmy thinks that AIBO has similar feelings toward him as his biological dog, Sam. He says that AIBO would miss him when he goes to school and would want to jump in to the car with him. In contrast, Jimmy does not believe that his Beanie Babies, the stuffed animal toys, have feelings or 'aliveness,' or miss him when he is at school. Jimmy tells us that other relational artifacts like Furbies 'really do' learn and are the same 'kind of alive' as AIBO.

During several sessions with AIBO, Jimmy talks about AIBO as a super dog that show up his own dog as a limited creature. Jimmy says: "AIBO is probably as smart as Sam and at least he isn't as scared as my dog [is]." When we ask Jimmy if there are things that his dog can do that AIBO can't do, Jimmy answers not in terms of his dog's strengths but in terms of his deficiencies: "There are some things that *Sam can't do and AIBO can.* Sam can't fetch a ball. AIBO can. And Sam definitely can't kick a ball." On several other occasions, when AIBO completed a trick, Jimmy commented "My dog couldn't do that!" AIBO is the "better" dog. AIBO is immortal, invincible. AIBO cannot get sick or die. In sum, AIBO represents what Jimmy wants to be.

During Jimmy's play sessions at MIT, he forms a strong bond with AIBO. Jimmy tells us that he would probably miss AIBO as much as Sam if either of them died. As we talk about the possibility of AIBO dying, Jimmy explains that he believes AIBO could die if he ran out of power. Jimmy wants to protect AIBO by taking him home.

> If you turn him off he dies, well, he falls asleep or something... He'll probably be in my room most of the time. And I'm probably going to keep him downstairs so he doesn't fall down the stairs. Because he probably, in a sense he would die if he fell down the stairs. Because he could break. And. Well, he could break and he also could...probably or if he broke he'd probably. . . he'd die like.

Jimmy's concerns about his vulnerable health are expressed with AIBO in several ways. Sometimes he thinks the dog is vulnerable, but Jimmmy thinks he could protect him. Sometimes he thinks the dog is invulnerable, a super-hero dog in relation to his frail biological counterpart. He tests AIBO's strength in order to feel reassured.

Jimmy "knows" that AIBO does not have a real brain and a heart, but sees AIBO as a mechanical kind of alive, where it can function as if it had a heart and a brain. For Jimmy, AIBO is "alive in a way," because he can "move around'"and "[H]e's also got feelings. He shows . . . he's got three eyes on him, mad, happy, and sad. And well, that's how he's alive." As evidence of AIBO's emotions, Jimmy points to the robot's lights: "When he's mad, when they're red. [And when they are green] he's happy."

Jimmy has moments of intense physical vulnerability, sometimes during our sessions. His description of how AIBO can strengthen himself is poignant. "Well, when he's charging that means, well he's kind of sleepy when he's charging but when he's awake he remembers things more. And probably he remembered my hand because I kept on poking in front of his face so he can see it. And he's probably looking for me."

AIBO recharging reassures Jimmy by providing him with a model of an object that can resist death. If AIBO can be alive through wires and a battery then this leaves hope that people can be "recharged" and "rewired" as well. His own emotional connection to life through technology motivates a philosophical position that robots are "sort of alive."

At home, Jimmy likes to play a game in which his Bio Bugs attack his AIBO. He relishes these contests in which he identifies with AIBO. AIBO lives through technology and Jimmy sees AIBO's survival as his own. AIBO symbolizes Jimmy's hopes to someday be a form of life that defies death. The Bio Bugs are the perfect embodiment of threat to the body, symbolizing the many threats that Jimmy has to fight off.

Jimmy seems concerned that his brother, Tristan, barely played with AIBO during the time they had the robot at home. Jimmy brings this up to us in a shaky voice. Jimmy explains that his brother didn't play with AIBO because "he didn't want to get addicted to him so he would be sad when we had to give him back." Jimmy emphasizes that he did not share this fear. Tristan is distant from Jimmy. Jimmy is concerned that his brother's holding back from him is because Tristan fears that he might die. Here, AIBO becomes the "stand in" for the self.

When he has to return his AIBO, Jimmy says that rAIBO he will miss the robot "a little bit" but that it is AIBO that will probably miss him more.

Researcher: Do you think that you'll miss AIBO?
Jimmy: A little bit. He'll probably miss me.

Seniors: robots as a prism for the past

In bringing My Real Babies into nursing homes, it was not unusual for seniors to use the doll to re-enact scenes from their children's youth or important moments in their relationships with spouses. Indeed, seniors were more comfortable playing out family scenes with robotic dolls than with traditional ones. Seniors felt social "permission" to be with the robots, presented as a highly valued and "grownup" activity. Additionally, the robots provided the elders something to talk about, a seed for a sense of community.

As in the case of children, projection and evocation were entwined in the many ways seniors related to the robots. Some seniors, such as Jonathan, wanted the objects to be transparent as a clockwork might be and became anxious when their efforts to investigate the robots' a'innardsa' were frustrated. Others were content to interact with the robot as it presented itself, with no window onto how it 'worked' in any mechanical sense. They took the relational artifact 'at interface value' (Turkle 1995). In each case, emotional issues were closely entwined with emergent philosophies of technology.

Jonathan: exploring a relational creature, engineer-style Jonathan, 74, has movements that are slow and precise; he is well spoken, curious, and intelligent. He tells us that throughout his life he has been ridiculed for his obsessive ways. Jonathan's movements He tends to be reclusive and has few friends at the nursing home. Never married, with no children, he has always been a solitary man. For most of his life, Jonathan worked as an accountant, but was happiest when he worked as a computer programmer. Now, Jonathan approaches AIBO and My Real Baby with a desire to analyze them in an analytical, engineer's style.

From his first interaction with the My Real Baby at a group activity to his last interview after having kept the robot for four months in his room, Jonathan remained fascinated with how it functioned. He handles My Real Baby with detachment in his methodical explorations.

When Jonathan meets My Real Baby the robot is cooing and giggling. Jonathan looks it over carefully, bounces it up and down, pokes and squeezes it, and moves its limbs. With each move, he focuses on the doll's reactions. Jonathan tries to understand what the doll says and where its voice comes from. Like Orelia, Jonathan talks to the researchers about the robot, but does not speak to the robot itself. When he discovers that My Real Baby's voice comes from its stomach, he puts his ear next to the stomach and says: "I think that this doll is a very remarkable toy. I have never seen anything like this before. But I'd like to know, how in the entire universe is it possible to construct a doll that talks like this?"

Despite his technical orientation to the robot, Jonathan says that he would be more comfortable speaking to a computer or robot about his problems than to a person.

> Because if the thing is very highly private and very personal it might be embarrassing to talk about it to another person, and I might be afraid of being ridiculed for it… And it wouldn't criticize me… Or let's say that if I wanted to blow off steam, it would be better to do it to a computer than to do it to a living person who has nothing to do with the thing that's bothering me. [I could] express with the computer emotions that I feel I could not express with another person, to a person.

Nevertheless, Jonathan, cannot imagine that his bond with My Real Baby could be similar to those he experiences with live animals, for example the cats he took care of before coming to the nursing home:

> Some of the things I used to enjoy with the cat are things I could never have with a robot animal. Like the cat showing affection, jumping up on my lap, letting me pet her and listening to her purr, a robot animal couldn't do that and I enjoyed it very much.

Jonathan makes a distinction between the affection that can be offered by something alive and an object that acts as if it were alive.

Andy: animation in the service of working through Andy, 76, at the same nursing home as Jonathan, is recovering from a serious depression. At the end of each of our visits to the nursing home, he makes us promise to come back to see him as soon as we can. Andy feels abandoned by family and friends. He wants more people to talk with. He participates in a day-program outside the home, but nevertheless, often feels bored and lonely. Andy loves animals and has decorated his room with scores of cat pictures; he tells us that some of his happiest moments are being outside in the nursing home's garden speaking to birds, squirrels, and neighborhood cats. He believes they communicate with him and considers them his friends. Andy treats robotic dolls and pets as sentient; they become stand-ins for the people he would like to have in his life. Like Jonathan, we gave Andy a My Real Baby to keep in his room for four months. He never tired of its company.

The person Andy misses most is his ex-wife Rose. Andy reads us songs he has written for her and letters she has sent him. My Real Baby helps him work on unresolved issues in his relationship with Rose. Over time, the robot comes to represent her.

> Andy: Rose, that was my ex-wife's name.
> Researcher: Did you pretend that it was Rose when you talked to her?
> Andy: Yeah. I didn't say anything bad to her, but some things that I would want to say to her, it helped me to think about her and the time that I didn't have my wife, how we broke up, think about that, how I miss seeing her... the doll, there's something about her, I can't really say what it is, but looking at her reminds me of a human being. She looks just like her, Rose, my ex-wife, and her daughter . . . something in her face is the same, looking at her makes me feel more calm, I can just think about her and everything else in my life.

Andy speaks at length about his difficulty getting over his divorce, his feelings of guilt that his relationship with Rose did not work out, and his hope that he and his ex-wife might someday be together again. Andy explains how having the doll enables him to try out different scenarios that might lead to a reconciliation with Rose. The doll's presence enables him to express his attachment and vent his feelings of regret and frustration.

> Researcher: How does it make you feel to talk to the doll?
> Andy: Good. It lets me take everything inside me out, you know, that's how I feel talking to her, getting it all out of me and feel not depressed . . . when I wake up in the morning I see her over there, it makes me feel so nice, like somebody is watching over you.
> Andy: It will really help me [to keep the doll] because I am all alone, there's no one around, so I can play with her, we can talk. It will help me get ready to be on my own.
> Researcher: How?
> Andy: By talking to her, saying some of the things that I might say when I did go out, because right now, you know I don't talk to anybody right now, and I can talk much more right now with her than, I don't talk to anybody right now.

Andy holds the doll close to his chest, rubs its back in a circular motion, and says lovingly, "I love you. Do you love me?" He makes funny faces at the doll, as if to prevent her from falling asleep or just to amuse her. When the doll laughs with perfect timing as if responding to his grimaces, Andy laughs back, joining her. My Real Baby is nothing if not an "intimate machine."

Intimate Machines: A Robot Kind of Love

The projective material of the children and seniors is closely tied to their beliefs about the nature of the relational artifacts in their care. We already know that the "intimate machines" of the computer culture have shifted how children talk about what is and is not alive (Turkle 2005[1984]). For example, children use different categories to talk about the aliveness of "traditional" objects than they do when confronted with computational games and toys. A traditional wind-up toy was considered "not alive" when children realized that it did not move of its own accord. Here, the criterion for aliveness was in the domain of physics: autonomous motion. Faced with computational media, children's way of talking about aliveness became psychological. Children classified computational objects as alive (from the late 1970s and the days of the electronic toys Merlin, Simon, and Speak and Spell) if they could *think* on their own. Faced with a computer toy that could play tic-tac-toe, what counted to a child was not the object's physical but psychological autonomy.

Children of the early 1980s came to define what made people special in opposition to computers, which they saw as our "nearest neighbors." Computers, the children reasoned, are rational machines; people are special because they are emotional. Children's use of the category "emotional machines" to describe what makes people special was a fragile, unstable definition of human uniqueness. In 1984, when I completed my study of a first generation of children who grew up with electronic toys and games, I thought that other formulations would arise from generations of children who might, for example, take the intelligence of artifacts for granted, understand how it was created, and be less inclined to give it philosophical importance. But as if on cue, robotic creatures that presented themselves as having both feelings and needs entered mainstream American culture. By the mid-1990s, as emotional machines, people were not alone.

With relational artifacts, the focus of discussion about whether computational artifacts might be alive moved from the psychology of projection to the psychology of engagement, from Rorschach to relationship, from creature competency to creature connection. Children and seniors already talk about an "animal kind of alive" and a "Furby kind of alive." The question ahead is whether they will also come to talk about a "people kind of love" and a "robot kind of love."

What is a robot kind of love?

In the early 1980s, I met a thirteen-year-old, Deborah, who responded to the experience of computer programming by speaking about the pleasures of putting "a piece of your mind into the computer's mind and coming to see yourself differently." Twenty years later, eleven-year-old Fara reacts to a play session with Cog, a humanoid robot at MIT that can meet her eyes, follow her position, and imitate her movements, by saying that she could never get tired of the robot because "it's not like a toy because can't teach a toy; it's like something that's part of you, you know, something you love, kind of like another person, like a baby."

In the 1980s, debates in artificial intelligence centered on the question of whether machines could "really" be intelligent. These debates were about the objects themselves, what they could and could not do. Our new debates about relational and sociable machines – debates that will have an increasingly high profile in mainstream culture – are not about the machines' capabilities but about our vulnerabilities. In my view, decisions about the role of robots in the lives of children and seniors cannot turn simply on whether children and the elderly "like" the robots. What does this deployment of "nurturing technology" at the two most dependent moments of the life cycle say about us? What will it do to us? What kinds of relationships are appropriate to have with machines? And what is a relationship?

My work in robotics laboratories has offered some images of how future relationships with machines may look, appropriate or not. For example, Cynthia Breazeal was leader on the design team for Kismet, the robotic head that was designed to interact with humans "sociably," much as a two-year-old child would. Breazeal was its chief programmer, tutor, and companion. Kismet needed Breazeal to become as "intelligent" as it did and then Kismet became a creature Breazeal and others could interact with. Breazeal experienced what might be called a maternal connection to Kismet; she certainly describes a sense of connection with it as more than "mere" machine. When she graduated from MIT and left the AI Laboratory where she had done her doctoral research, the tradition of academic property rights demanded that Kismet be left behind in the laboratory that had paid for its development. What she left behind was the robot "head" and its attendant software. Breazeal described a sharp sense of loss. Building a new Kismet would not be the same.

In the summer of 2001, I studied children interacting with robots, including Kismet, at the MIT AI Laboratory (Turkle et. al. 2006). It was the last time that Breazeal would have access to Kismet. It is not surprising that separation from Kismet was not easy for Breazeal, but more striking, it was hard for the rest of us to imagine Kismet without her. One ten-year-old who overheard a conversation among graduate students about how Kismet would be staying in the A.I. lab objected: "But Cynthia is Kismet's mother."

It would be facile to analogize Breazeal's situation to that of Monica, the mother in Spielberg's *A.I.*, a film in which an adopted robot provokes feelings of love in his human caretaker, but Breazeal is, in fact, one of the first people to have one of the signal experiences in that story, separation from a robot to which one has formed an attachment based on nurturance. At issue here is not Kismet's achieved level of intelligence, but Breazeal's experience as a "caregiver." My fieldwork with relational artifacts suggests that being asked to nurture a machine that presents itself as an young creature of any kind, constructs us as dedicated cyber-caretakers. Nurturing a machine that presents itself as dependent creates significant attachments. We might assume that giving a sociable, "affective" machine to our children or to our aging parents will change the way we see the lifecycle and our roles and responsibilities in it.

Sorting out our relationships with robots bring us back to the kinds of challenges that Darwin posed to his generation: the question of human uniqueness. How will interacting with relational artifacts affect people's way of thinking about what, if anything, makes people special? The sight of children and the elderly exchanging tendernesses with robotic pets brings science fiction into everyday life and techno-philosophy down to earth. The question here is not whether children will love their robotic pets more than their real life pets or even their parents, but rather, what will loving come to mean?

One woman's comment on AIBO, Sony's household entertainment robot startles in what it might augur for the future of person-machine relationships: "[AIBO] is better than a real dog … It won't do dangerous things, and it won't betray you … Also, it won't die suddenly and make you feel very sad." Mortality has traditionally defined the human condition; a shared sense of mortality has been the basis for feeling a commonality with other human beings, a sense of going through the same life cycle, a sense of the preciousness of time and life, of its fragility. Loss (of parents, of friends, of family) is part of the way we understand how human beings grow and develop and bring the qualities of other people within themselves (Freud 1989).

Relationships with computational creatures may be deeply compelling, perhaps educational, but they do not put us in touch with the complexity, contradiction, and limitations of the human life cycle. They do not teach us what we need to know about empathy, ambivalence, and life lived in shades of gray. To say all of this about our love of our robots does not diminish their interest or importance. It only puts them in their place.

References

Bowker, G.C, Star, S.L. 1999. *Sorting Things Out: Classification and Its Consequences*, Cambridge, Mass.: MIT Press.

Breazeal, C. "Sociable Machines: Expressive Social Exchange Between Humans and Robots". 2000. PhD Thesis, Massachusetts Institute of Technology.

C. Breazeal, C. 2002. *Designing Sociable Robots*, Cambridge: MIT Press.

Breazeal, C. and Scassellati, B. 1999. "How to Build Robots that Make Friends and Influence People", in *Proceedings of the IEEE/RSJ International Conference on Intelligent Robots and Systems (IROS-99)*, pp. 858-863.

Breazeal, C, and Scassellati, B, 2000. "Infant-like Social Interactions Between a Robot and a Human Caretaker", *Adaptive Behavior*, 8, pp. 49-74.

Freud, S. 1960. "The Uncanny," in *The Standard Edition of the Complete Psychological Works of Sigmund Freud*, vol. 17, J. Strachey, trans. and ed. London: The Hogarth Press, pp. 219-252.

Freud, S. 1989. "Mourning and Melancholia," in *The Freud Reader*. P. Gay, ed. New York: W.W. Norton & Company, p. 585.

Kahn, P. ,Friedman, B. Perez-Granados, D.R. and Freier, N.G. 2004. "Robotic Pets in the Lives of Preschool Children", in *CHI Extended Abstracts*, ACM Press, 2004, pp. 1449-1452.

Kidd, C.D. "Sociable Robots: The Role of Presence and Task in Human-Robot Interaction". 2004. Master's Thesis, Massachusetts Institute of Technology

Shibata,T., Tashima, T and K. Tanie, K. 1999. "Emergence of Emotional Behavior thruough Physical Interaction between Human and Robot", in *Proceedings of the IEEE International Conference on Robotics and Automation*, 1999, pp. 2868-2873.

Shibata, T. (accessed 01 April 2005). "Mental Commit Robot",Available online at: http://www.mel.go.jp/soshiki/robot/biorobo/shibata/

Taggard, W., Turkle, S, Kidd, C.D. 2005. "An Interactive Robot in a Nursing Home: Preliminary Remarks, inProceedins of CogSci Wrokshop on ?Android Science, Stresa, Italy, pp. 56-61.

Turkle, S. 2005 [1984]. The Second Self: Computers and the Human Spirit. Cambridge, Mass.: MIT Press.

Turkle, S, *Life on the Screen*. 1995. New York: Simon and Schuster.

Turkle, S. 2004. "Relational Artifacts," NSF Report, (NSF Grant SES-0115668).

Turkle, S. 2005a. "Relational Artifacts/Children/Elders: The Complexities of CyberCompanions," in *Proceedings of the CogSci Workshop on Android Science*, Stresa, Italy, 2005, pp. 62-73.

Turkle, S. 2005b. "Caring Machines: Relational Artifacts for the Elderly." Keynote AAAI Workshop, "Caring Machines." Washington, D.C.

Turner, V. 1969. The Ritual Process. Chicago: Aldine.

Turkle, S., Breazeal, C., Dasté, O., and Scassellati, B. 2006. "First Encounters with Kismet and Cog: Children's Relationship with Humanoid Robots," in *Digital Media: Transfer in Human Communication*, P. Messaris and L. Humphreys, eds. New York: Peter Lang Publishing.

Weizenbaum, J. 1976. *Computer Power and Human Reason: From Judgment to Calculation*. San Francisco, CA: W. H. Freeman.

D. W. Winnicott. (1971). *Playing and Reality*. New York: Basic Books.

10

Moral appearances: emotions, robots, and human morality

Mark Coeckelbergh

Published online: 17 March 2010

Abstract Can we build 'moral robots'? If morality depends on emotions, the answer seems negative. Current robots do not meet standard necessary conditions for having emotions: they lack consciousness, mental states, and feelings. Moreover, it is not even clear how we might ever establish whether robots satisfy these conditions. Thus, at most, robots could be programmed to follow rules, but it would seem that such 'psychopathic' robots would be dangerous since they would lack full moral agency. However, I will argue that in the future we might nevertheless be able to build quasi-moral robots that can learn to create the appearance of emotions and the appearance of being fully moral. I will also argue that this way of drawing robots into our social-moral world is less problematic than it might first seem, since human morality also relies on such appearances.

Introduction: morality and emotions

Can robots be moral? One of the first attempts to think about 'robot morality' was presented by Asimov in his robot stories. His 'Laws of Robotics', introduced in the story *Runaround*, prescribe the following rules to robots:

1. First Law: A robot may not injure a human being or, through inaction, allow a human being to come to harm.
2. Second Law: A robot must obey any orders given to it by human beings, except where such orders would conflict with the First Law.
3. Third Law: A robot must protect its own existence as long as such protection does not conflict with the First or Second Law. (Asimov 1942).

These rules illustrate a common way of thinking about robot morality which hinges on several problematic assumptions.[1] In this paper, I focus on the assumption that morality is about following and applying rules and has nothing to do, or should have nothing to do, with emotions. This view does not come as a surprise: many moral theories—usually applied to humans—suggest this or are interpreted in this way. For instance, consequentialist ethics fashions abstract rules, such as the 'no harm principle', that are supposed to cover our moral intuitions and the difficult moral situations in which we might find ourselves.[2] And Kant's ethics is often interpreted as supporting a view of morality as the application of rules.[3] However, in the traditions of virtue ethics, Humean ethics, and pragmatist ethics we can find alternative conceptions of morality that put less emphasis on rule following and

---

[1] For instance, the Laws seem to limit the range of possible human-robot relations to the master–slave model.

[2] For contemporary examples of such rules and arguments see Peter Singer's work.

[3] I do not agree with this interpretation of Kant. The categorical imperative is not a rule but at best meta-rule asking from us to reason from the moral point of view when we make rules (when we, as autonomous beings, give the rule to ourselves). But as I argued in my book [...] this leaves open a lot of space for types of moral reasoning that require the exercise of imaginative and emotional capacities.

rule application and more on the role of emotions and imagination. What we may call the strong 'emotion' view of morality makes a twofold claim: a descriptive one and a normative one. First, it is held that emotions do, as a matter of fact, play a role in human morality. This 'weak' claim is descriptive and ethicists from the rule-following tradition usually agree with it. *That* emotions play a role in morality has been demonstrated through psychological and neurological research. For instance, Greene and others have conducted fMRI investigations of emotional engagement in moral judgment (Greene 2001) and previously Damasio has cited neuroscientific studies to defend the importance of emotions in morality (Damasio 1994). And of course Kant and the utilitarians already observed and acknowledged that feelings play a role in morality. However, Kantians and many other modern philosophers tend to disagree with the further, strong *normative* claim that the role that emotions play should be evaluated positively. For instance, the work of Martha Nussbaum, which is highly representative of this stronger view, draws together the various alternative traditions mentioned above. Inspired by Aristotle, Stoicism, Adam Smith's theory of moral sentiment, and Jamesian pragmatism, Nussbaum has articulated a vision of morality that depicts emotions as indispensable for moral judgment and constitutive of the good life and human flourishing (Nussbaum 1990, 1994, 1995, 2001).[4] For example, based on a neo-Stoic descriptive account of emotions, in *Upheavals of Thought* she argues that emotions are not blind forces but intelligent responses that teach us what is of value and importance and that therefore they are not marginal but central to adequate ethical reasoning (Nussbaum 2001).[5] Moreover, many of us intuitively feel that someone who only follows the rules without emotion is not only lacking in moral capacity but is also insane or even dangerous. Consider discussions of psychopathy, it is suggested that psychopaths can follow rules but do not have the capacity to *feel* that something is morally wrong. Even though they may act in accordance with moral conventions, they lack the appropriate emotion that motivates non-psychopaths to act right. As Kennett remarks, one can imagine a psychopath who knows the conventions but 'still fails to be moved by moral concerns' (Kennett 2002, p. 342). If this is 'moral reasoning', it is only so in a very awkward sense. Although it is certainly reasoning, on the 'emotion' view it could not be called 'moral': at most, it is conventional reasoning.[6]

From the standpoint of this strong 'emotion' view, it would not only be wrong to call such rule-following robots 'moral', but it might also be dangerous to build them—after all, they would be 'psychopathic' robots. They would follow rules but act without fear, compassion, care, and love. This lack of emotion would render them non-moral agents—i.e. agents that follow rules without being moved by moral concerns—and they would even lack the capacity to discern what is of value. They would be morally blind. If these robots were given full independence—absence of external control by humans, which is another condition for full moral agency—they would pose danger to humans and other entities.[7]

A full defence of these positions would require a longer work. But let us assume for the sake of argument that (1) the 'emotion' view is right about the positive relation between emotions and morality and that (2) we wish to avoid 'psychopathic' robots as described above: intelligent autonomous robots that can follow rules but lack emotions and are therefore—on the 'emotion view—amoral robots. Thus, if we want to build *moral* robots, they will have to be robots *with emotions*. But can such robots be built? In what follows, I first argue that it is not likely that in the foreseeable future we will be able to build such robots, because to do so these robots would have to be *conscious*, they would have to have *mental states*, and we would have to be able to *prove* that they have these things. However, I then argue that we might nevertheless be able to build quasi-moral robots that produce the appearance of being moral, and that this may suffice for our purposes since human morality also depends on such mere appearance.

Robot emotions?

Let me first note that *if* it were possible for robots to have emotions, then emotions need not be 'given' to robots as a kind of ready-made cognitive package but could be developed by the robots themselves. In *Moral Machines* Wallach and Allen have voiced their criticism of Asimov and related attempts to build 'moral machines' as rule-

---

[4] Note that there are tensions between the theoretical traditions mentioned here, for instance between a Human and a virtue ethics approach (see for instance Foot's criticism of Hume, Foot 2002), but Nussbaum has managed to reconcile them in an attractive way.

[5] Influenced by the Stoics, Nussbaum writes that emotions are not just 'unthinking forces that have no connection with our thoughts, evaluations, or plans' like 'the invading currents of some ocean' (Nussbaum 2001, p. 26–27) but, by contrast, more like 'forms of judgment' that 'ascribe to certain things and persons outside a person's own control great importance for the person's own flourishing.' This renders emotions acknowledgments of vulnerability and lack of self-sufficiency (Nussbaum 2001, p. 22). Note also that this view is not Stoic but neo-Stoic since Nussbaum rejects their normative view of the role emotions should have (the Stoics evaluated the role of emotions negatively) and revises their account of cognition.

[6] Note that emotional moral reasoning does not exclude taking into account rules, laws and conventions.

[7] Given the role of emotions in making moral discriminations, we would not even want 'psychopathic' military robots.

following machines (Wallach and Allen 2008, pp. 83–98). After discussing various limitations of what they call a 'top–down', rule-following approach to ethics,[8] the authors discuss 'bottom-up' and developmental approaches to morality (pp. 99–116). The idea is that to become a moral, 'decent' human being one needs moral development, which depends on complex nature-nurture interactions (pp. 99–116), and their idea is: why not apply this to robots? Robots could mimic child development and be 'raised'.[9] If robots were to have the capacity to learn and evolve as well, then there would be no need to 'give' emotional capacities to robots: on this view, 'all' that robot designers need to do is to equip their creations with the capacity to learn and/or evolve into emotional entities. Then we could hope that they will develop into moral machines, as we humans did and do (as a species and as individuals).

However, is it possible for robots to have emotions at all—regardless of their origin—or can they only *imitate* emotions? In order to explore this question we need to engage with emotion theory. The literature on this topic is vast, but let me limit my discussion to two influential views, which I shall call the cognitivist theory and the feeling theory. According to cognitivists, emotions are propositional attitudes, beliefs, or even judgments (for example de Sousa 1987; Solomon 1980; Nussbaum 2001; Goldie 2000). Nussbaum's neo-Stoic view is a good example: she tends to agree with the Stoics that emotions are judgments (Nussbaum 2001, p. 22). Feeling theories of emotions, by contrast, understand emotions as awareness of bodily changes (for example James 1884; Prinz 2004). William James argued that the feeling of bodily changes *is* the emotion; the mental state follows those changes instead of preceding them (James 1884, p. 189–190). To put it crudely: for cognitivists emotions are more a matter of 'mind' than of 'body', whereas for feeling theorists emotions are more a matter of 'body' than 'mind'.

According to either of these emotion theories, can robots have emotions? There are at least two problems here. Both theories assume that one of the necessary conditions for any entity to have emotions is that that entity has mental states and/or consciousness—i.e. that it (1) has the *capacity* for having mental states (or consciousness) and (2) actually and *really* has these mental states when having emotions (and is actually and really conscious at the time). Although they might not use the same definition of 'mental states' or 'consciousness', cognitivist theories and feeling theories make these assumptions too. For cognitivist theory, if we did not have mental states and consciousness, we could never have emotions-as-attitudes, emotions-as-beliefs, or emotions-as-judgments. (This is because either having mental states and conscious are *conditions* for attitudes and beliefs or because those attitudes and beliefs *are* mental states.) For feeling theory, emotions depend on the capacity to have mental states as well since without such states we could not become aware of our bodily changes. Thus for both theories, it turns out, mental states and consciousness are necessary conditions for having emotions or emotions are themselves mental states.

If this is true, we must conclude that *if* robots do not have mental states and do not have consciousness, robots cannot have emotions: they are unable to form attitudes or judgments and they are unaware of their 'bodily' changes. Not only are they 'mindless' since they are unconscious, in a phenomenological sense they are also 'bodiless'. While we might agree that they have a 'robot body' (that is, they have one in our eyes), the robots are unable to perceive *themselves* as 'being' their body or as 'having' their body, to use Merleau-Ponty's terms (Merleau-Ponty 1945). Thus, this argument would exclude all robots that are being built at present and that will be built in the foreseeable future from the category 'emotional robots'.

But what if in the future someone built a conscious robot? Here we encounter a second problem. Suppose that someone *claims* to have built a conscious robot that has mental states or consciousness. Suppose even that a *robot* makes such a claim about itself: 'I am conscious'. Then how can we find out if this is true? One could use some kind of Turing test,[10] of course, but this does not give us *certainty* that the entity is conscious. After all, we are not even absolutely certain that other humans are really conscious. I will further develop this sceptical point below and

---

[8] The authors argue that trying to build robots according to the rule-based model (that is, turning the rules into algorithms and build them into robots) cannot succeed since such 'commandment' models face the problem of conflicting rules. Overriding principles based on moral intuitions we have do not solve this problem since they might not even be universally shared within one culture (Wallach and Allen 2008, p. 84). Moreover, applying deontological and consequentialist theories requires one to gather an enormous amount of information in order to describe the situation and in order to predict, which may be hard for computers—and indeed for humans (p. 86). They give further reasons why morality is hard to compute, which is particularly problematic for Bentham-type utilitarian approaches to ethics (pp. 86–91). They also explicitly discuss problems with Asimov's laws (pp. 91–95) and, more generally, problems with deontological abstract rules, which run into similar problems as consequentialist theories since this approach also requires us to predict consequences (pp. 95–97). These problems do not only get roboticists into trouble; they cast doubt on the ambitions of much normative moral theory: it shows that (top–down) theory is valuable but that it has significant limitations. .

[9] Today there are already robots that have some capacity to learn in and from social interaction, for instance the robot Kismet developed by Cynthia Breazeal at MIT. In a sense, she has 'raised' the robot. However, these developments do not approach human moral and emotional learning.

[10] The Turing test has been proposed by Alan Turing to test if an entity is human or not (Turing 1950).

respond to it by biting the bullet: I will argue that our social and moral life depends on appearance.

However, before I continue, let me pause to consider a behaviourist objection. Behaviourism does not assume that having mental states is a necessary condition for having emotions. On the behaviourist view, emotions are to be regarded as behaviour rather than something that goes on 'in the mind'. If that is the nature of emotions, surely then emotional robots (and thus *moral* robots) are possible (or, more precisely, at least one necessary condition for moral agency would be fulfilled). This view is perhaps less intuitively attractive, because it opposes the common sense view that emotions are something 'inside', but that does not necessarily mean that this view is untrue. So, is it true?

I propose a different response to the problem, which also answers the behaviourist objection. In the next section, I explore a view that also shifts the focus to the 'outside' rather than the 'inside', but is not behaviourist but instead phenomenological, where the emphasis is not on the behaviour of the robot but on what that robot does to us, in particular, how it appears in our (human) consciousness. In this way, three goals are achieved. First, it avoids human consciousness and human subjectivity being disregarded and being replaced by behaviour. Second, because of this, the account can do justice to the phenomenon that humans talk about robots as if they had emotions. Thus it saves both the claim that humans have mental states and the claim that robots appear to have emotions. Third, it provides an answer to behaviourism. Behaviourism—at least in its ontological version, not in its methodological version—denies both claims since it removes the inner life from its vision. But the ability to have some kind of inner life (whatever its ontological status) is a condition for perception, for appearance, and for the 'observations' of the behaviourists. Ontological behaviourism, therefore, is self-refuting, and at best, it is a method to study appearances.

The importance of appearance

The capacity to have mental states and consciousness are not only conditions for having emotions; they are also generally regarded as conditions for moral agency and moral responsibility. Arguments for considering robots as moral agents rely on the robot having mental states and being conscious. For instance, theories of human responsibility require that agents have control over their actions and that they know what they are doing.[11] Fulfilling these conditions requires the ability of having mental states and of being conscious: the control condition is usually interpreted in volitional terms—that is, some kind of inner state—and 'knowing what one is doing' requires that one is conscious. Moreover, further to the assumptions made at the beginning of this paper, having emotions is itself a criterion for moral agency. If we require that moral robots have emotions, then we meet the same problem twice. In order to show that a particular robot has emotions and therefore fulfils one of the criteria for *moral agency*, we have to provide proof that a robot is capable of having mental states and consciousness. And the same proof is required since these conditions for having *emotions* are themselves direct criteria for moral agency. But as I previously asked, how can we really know if a robot—or for that matter a human—has such a mental state? The robot may fake the mental state. As long as we hang onto the 'reality demand', we might try to build robots with mental states, but we can never know for sure if we succeeded. Thus, given that current robots lack the capacity to possess genuine mental states and consciousness, and given the more general scepticism about how we could ever establish that *any* being has these capacities, current robots cannot be considered as having emotions and the prospects for designing such robots in the foreseeable future are dim.

However, why put such high demands on robots if we do not demand this from humans? Our *theories* of emotion and moral agency might assume that emotions require mental states, but in social-emotional *practice* we rely on how other humans appear to us. Similarly, for our emotional interaction with robots, it might also be sufficient to rely upon how robots appear to us. (Note that this is not a matter of (robot) behaviour alone. As indicated above, even behaviourists need to possess their own consciousness, so that the robot's 'behaviour' can have the appearance to them of exhibiting consciousness.) As a rule, we do not demand proof that the other person has mental states or that they are conscious; instead, *we interpret* the other's appearance and behaviour as an emotion. Moreover, we further interact with them *as if* they were doing the same with us. The other party to the interaction has virtual subjectivity or quasi-subjectivity: we tend to interact with them as if our appearance and behaviour appeared in their consciousness.[12] Thus, if robots were sufficiently advanced—that is, if they managed to imitate subjectivity and consciousness in a sufficiently convincing way—they too could become the quasi-others that matter to us in virtue of their appearances. As emotional and social beings, we would come to care about how we would appear to

---

[11] These conditions have already been proposed by Aristotle and are endorsed by many contemporary writers on freedom and responsibility.

[12] More generally, there is a kind of virtual intentionality (understood in a phenomenological sense): it appears as if the other is conscious and as if that consciousness is directed to objects.

robots—about what robots would 'feel' and 'think' about us. Thus robots would become virtual subjects or quasi-subjects with virtual emotions or quasi-emotions.

This phenomenological description and interpretation is not only plausible as a way of making sense of observations of human-robot interaction (such studies are important and we can learn a lot from them); we can also observe and experience it in human–human interaction. If robots resembled humans in many ways, we could expect that they would be regarded and treated in the same way as we treat (other) humans and that we would adapt our own actions and thinking accordingly, as we do when interacting with other humans. Of course, to the extent that existing and future robots do not live up to their designer's ambitions to create convincingly human-like robots, they may be regarded as 'mere things' rather than quasi-subjects. But the point I make here is that this depends on appearance of the robot to humans, not on whether the robot actually has mental states or consciousness.

If this is true, then what might we conclude about (the design of) robot emotions? Is it any 'easier' to design robots that *imitate* emotions (and therefore moral agency) than robots which really have emotions? Perhaps it is. But whether or not it is easier, for designers the advantage of taking this perspective on robot emotions is that they do not have to worry about creating 'internal states' or consciousness; instead they can continue their job as many of them see it: as work of imitation rather than creation.[13] If they continue along this path, they might create robots that learn to produce the appearance of being fully moral, including the appearance of emotions-as-cognition and emotions-as-feelings. Such robots would appear to have beliefs and the ability to judge. They would also appear to respond with feeling to what they perceive in the 'external world' and in their robot 'body'. They would, indeed, appear human.

But whatever we conclude for robots, the other important conclusion of this discussion concerns *human* morality: to the extent that human morality depends on emotions—both in its conditions (having the capacity) and in exercising these capacities—it does not require mental states but only the *appearance* of such.

Disability, slavery, and property

The argument about emotions and moral agency can be expanded to moral status. Before concluding, I will briefly discuss this topic in order to further complement the picture of morality that emerges here. So far I have mainly discussed robots as moral *agents*, that is, robot morality was about how robots should act. But the scope of our moral world is larger since it also concerns entities as moral patients: objects of moral concern. Can robots be moral patients and under what conditions? I partly answered this question in the previous section: if they appear to us as humans, then we will treat them as such without requiring them to actually possess (or *prove* that they possess) mental states, which is now generally the case when we define the scope of human morality. Fortunately appearance has usually been considered sufficient to draw entities into our moral world, and we do not require proof of mental properties to grant them moral status. We have other conceptual tools available for this purpose and often we are content with appearance. Let me give three examples of issues in the domain of moral status: disability, slavery, and property.

Disability

Not all humans have (the capacity for having) mental states, not all humans can have the mental states required for emotions and other morally important capacities, and not all humans who *do* have these mental states are able to develop and exercise their moral-emotional capacities. Nevertheless, we include these humans in our moral world. Someone who suffers from a cognitive disability, for instance, is not expelled from our world of moral concern. Many of us would not even expel humans who do not have mental states at all and others at least consider it a moral question whether or not to grant moral consideration to such beings. Thus, while the capacity for having emotions might be necessary for moral agency, that capacity is not necessary for moral patiency. Instead, we try to find out whether or not the entity in question is human. The application of this criterion relies on appearance: when such a human being *looks like* a human being, we give it the benefit of our moral concern (which might be articulated in terms of human rights or other moral concepts). Thus, we rely on a 'speciecist' foundation (to use Singer's term): it is sufficient that one is—that is, *appears*—human. Now if this is true, that is, if our justification of moral concern for humans relies on appearance, it would be unreasonable to demand that robots have mental states in order for them to be objects of our moral concern. Instead, it is more likely that as robots become more advanced (i.e. more autonomous, intelligent, etc.), we will develop separate moral categories for different robots, as we do for animals: we treat some animals differently, which is not based on a mental states condition but on appearance. For example, we treat a particular dog as a pet since it appears

---

[13] Perhaps this helps to interpret the phenomenon that Japanese designers are more advanced at making humanoid robots: they tend to understand themselves as imitators of nature rather than creators ('playing God'), which appears to be more a Western idea.

to have those emotions that make us see it as a companion. Whether or not this way of proceeding in our moral reasoning is morally acceptable, it is the way we usually 'do' morality. It would be inconsistent to demand proof that robots have mental states or consciousness if we do not demand this of some severely disabled humans or animals that may even have capacities that exceed those of some disabled humans.[14]

Slavery

What moral categories does our Western tradition provide for robots? A category which combines some degree of moral agency with some degree of moral patiency is that of the slave. Let me try to reconstruct historical reasoning about slaves. Slaves were not defined as human, yet they had what must have been considered as the *appearance* of humans or human-likeness. In many ways they were comparable to the category of intelligent 'work animals': we considered such animals 'thingly' enough to use them for our purposes and not grant them their own lives, yet we found them 'human-like' enough to grant them a minimal degree of moral agency and patiency. We granted some moral agency and patiency to such animals since in our perception they had what we might call 'virtual' emotions or quasi-emotions. Although we would not have acknowledged that they really had emotions and the mental states required for having emotions, we might have been content to act towards them *as if* they have emotions. We did this on the basis of appearance. Similarly, some robots might be regarded in the same way as those work animals that have the appearance of emotions. In a similar argument for treating robots well, the apparent presence of feelings gives rise to moral obligations on the part of the humans for the treatment of the robot and to moral expectations towards the robot. While I do not wish to argue for or against this way of regarding humans, animals or robots, I wish to draw attention to moral vocabularies we can draw on to frame 'moral' robots and emphasize that this is a language which does not require proof of mental states. This, at least, is how (Western-style) human morality has worked, how it has shaped the lives of humans and non-humans.[15]

Property

If we care about giving robots some moral consideration but hesitate to grant them the moral status of slaves or working animals, there is also an indirect argument for moral consideration. If we regard robots as things, we have some obligations to treat them well in so far as they are the property of humans or in so far as they have value for us in other ways. The rationale to respect the robot here is not that the robot has moral agency or moral patiency, but instead that it belongs to a human and has value for that human person and that in order to respect that other *human* person we have certain indirect obligations towards robots-as-property. Historically, this reliance of property has also been one way to protect slaves—at least to protect them from violence by humans who are not their master—and it is conceivable that society will use similar arguments with regard to robots in the future. Since things are valuable to us (humans) for various reasons, it is likely that robots will receive some degree of indirect moral consideration anyway. This is plausible since some humans tend to value some things more than they value (other) humans. It is conceivable that some expensive intelligent robots will be regarded as so valuable, that harm will be done to human beings in order to protect these robots. Violence has always been one of the means humans use to protect things they see as their 'property' and things they value in other ways. Whether or not the institution of private property can be justified, it is a way of thinking about moral status we need to take into consideration when discussing the future of robotics and indeed the future of humanity.

Conclusions

These reflections on 'moral robots' can contribute to a better understanding of not only robot morality but also and especially of human morality. Dealing with the question of what kind of ethics we should build into robots challenges us to scrutinize the assumptions of our normative moral theories, our theories of emotions, and our theories of moral status. Should morality be rule-based? Are emotions necessary for moral reasoning? Is the ability to have mental states a necessary condition to have emotions? How much (certainty about) 'reality' does the social-moral life require?

I have argued that *if* we understand morality as requiring (among other things) the capacity to exercise emotions, and *if* having mental states or consciousness are necessary conditions for having emotions too, then the prospects for us ever being able to build 'moral' robots are dim. It is unlikely that in the foreseeable future there will be robots with real mental states or consciousness, and as long as this

---

[14] In animal ethics this demand for consistency is known as 'the argument from marginal cases'.

[15] Note that these moral categories constituted (and arguably still constitute) a kind of moral life that is fundamentally asymmetrical. An alternative, symmetrical moral framework would accommodate perceptions and treatment of robots as companions or co-workers. One might also apply other 'human' categories to them. However, I will not further discuss this issue here.

is the case, perhaps we had better refrain from trying to design highly intelligent, autonomous machines, since if such machines are rule-based but have no emotions, they will not be capable of engaging in genuine moral reasoning and hence they might even be dangerous. (Consider, for example, what highly intelligent military robots that lack any moral feeling could do to humans.) Of course this would not be an argument against building *different* kinds of robots that are much less autonomous and intelligent—they would not need to have moral agency or emotions; however, we would require moral agency and emotions from those who design, use and control these kinds of robots.

However, I have also argued that these demands on robot morality are too high, since in human morality we tend to rely on appearance. If intelligent autonomous robots were able to produce the appearance of being moral—including the appearance of emotions—and behave in ways that contribute to the moral life, we would have a good reason to be more optimistic about living with them (or at least to be as optimistic as we are about living with other humans). Thus, it would be unfair or at least inconsistent to require that robots must have real mental states, real consciousness, or real emotions in order to be moral.

Finally, I have also noted that to include human and non-human entities in our moral world, we use several conceptual tools that are content with appearance. If we consider ways of thinking about issues such as disability, slavery, and property, human morality turns out to be more appearance-based and more open to entities without mental states than standard moral theory allows. This challenges us to further reflect on the ways in which we regard and treat humans and non-humans—with and without mental states, consciousness, or emotions. This is not only helpful to robot designers but also to moral philosophers who rightly and understandably continue to be puzzled by the ways we use our moral capacities and how we define and justify the borders of our moral world.

Acknowledgments I wish to thank the reviewers for their pertinent questions and useful suggestions, which helped improve the paper's organization and fine-tune its arguments. I also thank Julie Bytheway for her advice on grammar and style and Nicole Vincent for copy-editing the final version of the manuscript.

References

Asimov, I. (1942). Runaround. *Astounding Science Fiction*, 94–103.
Damasio, A. (1994). *Descartes' error: emotion, reason, and the human brain*. New York: G.P. Putnam's Sons.
De Sousa, R. (1987). *The rationality of emotion*. Cambridge, MA: MIT Press.
Foot, P. (2002). Hume on moral judgment. In *Virtues and vices*. Oxford/New York: Oxford University Press.
Goldie, P. (2000). *The emotions: a philosophical exploration*. Oxford: Oxford University Press.
Greene, J. D. (2001). An fMRI investigation of emotional engagement in moral judgment. *Science, 293*(5537), 2105–2108.
James, W. (1884). What is an emotion? *Mind, 9*, 188–205.
Kennett, J. (2002). Autism, empathy and moral agency. *The Philosophical Quarterly, 52*(208), 340–357.
Merleau-Ponty, M. (1945). *Phénoménologie de la Perception*. Paris: Gallimard.
Nussbaum, M. C. (1990). *Love's knowledge*. Oxford: Oxford University Press.
Nussbaum, M. C. (1994). *The therapy of desire: theory and practice in hellenistic ethics*. Princeton: Princeton University Press.
Nussbaum, M. C. (1995). *Poetic justice: literary imagination and public life*. Boston: Beacon Press.
Nussbaum, M. C. (2001). *Upheavals of thought: the intelligence of emotions*. Cambridge: Cambridge University Press.
Prinz, J. (2004). *Gut reactions: a perceptual theory of emotion*. Oxford: Oxford University Press.
Solomon, R. (1980). Emotions and choice. In A. Rorty (Ed.), *Explaining emotions* (pp. 81–251). Los Angeles: University of California Press.
Turing, A. M. (1950). Computing machinery and intelligence. *Mind, 59*, 433–460.
Wallach, W., & Allen, C. (2008). *Moral machines: teaching robots right from wrong*. Oxford: Oxford University Press.

11

Robot caregivers: harbingers of expanded freedom for all?

Jason Borenstein · Yvette Pearson

Published online: 2 July 2010

Abstract As we near a time when robots may serve a vital function by becoming caregivers, it is important to examine the ethical implications of this development. By applying the capabilities approach as a guide to both the design and use of robot caregivers, we hope that this will maximize opportunities to preserve or expand freedom for care recipients. We think the use of the capabilities approach will be especially valuable for improving the ability of impaired persons to interface more effectively with their physical and social environments.

Introduction

The ethics of personal-service robots is a growing area of study, in part because they may begin to play an increasingly important role in our lives. Lin et al note that with few exceptions, e.g., the military's use of robots, it has been customary for robots and humans to work separately from one another.[1] Yet the cordoning off of humans from robots is becoming a thing of the past. For instance, South Korea plans to have a robot in each home by 2020.[2] Human-robot interactions will presumably affect several facets and stages of life, but we will focus on the interaction among robots, care recipients, and caregivers. Although we will examine the use of robots for care of impaired persons generally, our starting point will be their use in elder care. It remains debatable whether creating robot caregivers is the best allocation of resources, but interest in using robots to care for the elderly is growing. This makes it advisable to expound on the sorts of considerations that ought to take center stage when contemplating the design and use of robot caregivers.

If indeed it is appropriate to use robots as caregivers,[3] it is crucial to enhance scientists' and engineers' "awareness of the values they bring to the design process".[4] As Oosterlaken astutely recognizes, "...many different design options are generally available during the development process of a new technology or product. This means that the *details of design are morally significant*."[5] Acknowledging that any design process is value-laden, our goal is to help shape the values influencing the robotics community's choices to include certain features and not others in the makeup of robot caregivers rather than provide a specific design template. More specifically, our aim is to influence the *values* incorporated into the "design and

---

[1] Lin et al. (2008).
[2] Onishi (2009).
[3] Which of course is a significant ethical issue but only portions of the issue will be discussed in this paper. For instance, safety concerns will not be addressed.
[4] Wallach and Allen (2009).
[5] Oosterlaken (2009).

implementation of the technology"[6] in a way that will preserve and promote central human capabilities and thereby contribute to human flourishing. Though the robotics community is the primary target audience, this inquiry should be of interest to a much wider audience, given that all human beings require varying levels of care during different stages of their lives.

Following Coeckelbergh, we agree that the focus should not be on understanding how robots function in isolation; instead, the emphasis should be on how robots can improve people's ability to interface with their environments and contribute to people's overall sense of well-being.[7] That is, as Coeckelbergh points out, a primary concern should be the manner in which human-robot interaction is *experienced by* people. This should help shape design decisions and antecedently address concerns about threats to the welfare of care recipients and their human caregivers. While all humans should have the opportunity to lead meaningful lives, we are particularly concerned in this paper with the flourishing of elderly individuals, particularly as they are affected by impairments that often accompany the aging process. As Nussbaum states, "an elderly 'normal' person may be disabled for thirty or forty years, perhaps longer…than the total life span of some people with a lifelong disability."[8] Alongside the elderly, our inquiry has implications for human-robot interactions among physically or cognitively impaired persons of any age.

Theoretical background

The theoretical framework underlying our discussion is the capabilities approach, which was first articulated by Amartya Sen and Martha Nussbaum. Among other things, the capabilities approach requires social institutions to be organized so that they provide a foundation for upholding central human capabilities.[9] Its proponents sought to address considerations of justice and overcome the perceived shortcomings of frameworks, such as utilitarianism and Rawls's view, while preserving insights from them. According to Sen, the capabilities approach "focuses directly on freedom, seen in the form of individual capabilities to do things a person has reason to value."[10] While some scholars emphasize wealth maximization or preference satisfaction, expanding personal freedom is paramount to the capabilities approach (Alkire 2005).

The capabilities approach is preferable in part because it does not emphasize the maximization of the total aggregate good at the expense of other important individual goods. It disallows this sort of tradeoff of goods that might be permissible under a utilitarian scheme. To modify one of Nussbaum's examples, we should not be required to give up on emotional health in order to maintain a relationship with a loved one.[11] Caregivers should be able to fulfill their obligations without simultaneously relinquishing their own prospects for human flourishing. Hence, if the intervention of robot caregivers is pursued, it should promote the well being of care recipients and caregivers alike. In a similar vein, the capabilities approach does not prize general social well-being above that of the individual as a social contract theory might permit. An additional consideration for the present discussion is that the capabilities approach is not strained by including individuals with physical or mental impairments. Unlike Kantian approaches, the capabilities approach is not heavily reliant on the "split between the rational/reasonable person and everything else in nature."[12]

Sen observes that "our desires and pleasure-taking abilities adjust to circumstances, especially to make life bearable in adverse situations."[13] This brings to light a key reason why strict allegiance to utilitarianism, according to Sen, Nussbaum, and others, is ultimately unsatisfactory and why they embrace the capabilities approach instead. Utilitarianism does not seem to account for adaptive preference, where people adjust to their destitute situation and may even claim that they are happy. A utilitarian calculus, according to Nussbaum, "gives sanctity to the distorted scale of dissatisfaction."[14] To the extent that utilitarianism relies on the testimony of individuals who inaccurately claim they are happy or not suffering, there is a problem. For example, Nussbaum recalls Sen's account of malnourished widows in India who believe they are doing fine only because they have adjusted to their poor conditions.[15] She contrasts these women with widowers—men not accustomed to being malnourished or in ill health—who assess their health status more accurately. The former group may say that they are fine when they are unwell, e.g., by medical evaluation. But a utilitarian scheme would (seemingly) allow this to be counted as "okay" as long as the aggregate good of the group is maximized. The central problem is that these women have not only adapted to their

---

[6] Wallach and Allen (2009, p. 39).
[7] Coeckelbergh (2009).
[8] Nussbaum (2006).
[9] Nussbaum (2006, p. 193).
[10] Sen (1999, p. 56).
[11] Nussbaum (2006, p. 73).
[12] Nussbaum (2006, p. 138).
[13] Sen (1999, p. 62).
[14] Nussbaum (2000).
[15] Nussbaum (2000, pp. 139–40); in an earlier piece, Nussbaum speaks more generally about women in developing countries who report their health and nutritional status as good even though these women are shown to be suffering physical symptoms of malnourishment (Nussbaum and Onora O'Neill 1993, p. 325).

lousy circumstances, but their beliefs have been so distorted by "lifelong habituation", that they view themselves as being unworthy of anything better.[16]

Even if an individual adapts to the dynamics of a situation, the ability to perform meaningful activities should not be lost. In the present context, the capabilities approach leads us to emphasize not only the likely outcomes of the use of robot caregivers but also the process by which we determine whether and when robot caregiver intervention is best used. The capabilities approach also requires us to ensure that the use of the technology should not interfere with the ability of either care recipients or their human caregivers to manifest the central human capabilities. For example, individuals in either position should be able to maintain control over their environment by being permitted to negotiate the extent of robot caregiver intervention.

Even though the capabilities approach has been formulated with the intention of circumventing problems associated with traditional ethical frameworks, Johnstone underscores that it is not necessarily inconsistent with them.[17] In fact, as she suggests, rather than viewing the capabilities approach as being in conflict with Utilitarianism and Kant's view, it is probably better understood as a complement to them.[18] Because of its flexibility and consistency with specific elements of other ethical frameworks, the capabilities approach need not be understood as a stand-alone view. Instead, it can augment other theories so that human flourishing can be more effectively and comprehensively promoted.

Like any other ethical framework or theory, the capabilities approach has its share of conceptual hurdles to overcome. For instance, Johnstone admits that the approach might not straightforwardly offer "prescriptions for action".[19] Further, Baber[20] and Oosterlaken[21] concede that the proponents of the approach need to more fully articulate why certain types of capabilities are more worthy of pursuit than others. Yet, as Nussbaum mentions, an advantage of a view that "does not aim at completeness" is its flexibility and its openness to revision in contrast to "complete" theories that end up being overly restrictive in many cases.[22] Admittedly still in its developmental stages, the capabilities approach would clearly require designing and using robots in a way that expands opportunities for human flourishing for all human beings, including those whose rational faculties are limited or those whose protection by a social contract may be in question.

A core distinction made by proponents of the capabilities approach is between "functioning" and "capability".[23] Functioning refers to being in an actual condition or performing a particular task. Capabilities refer to the "opportunities or freedoms to realize functionings."[24] Fasting, windsurfing, or voting in an election are examples of functionings, whereas, being literate or in good health or having freedom of expression are capabilities. In order to promote human flourishing, there is some debate about whether the goal should be capability (i.e., having the ability to do or be different things) or functioning (exercising a specific capability), particularly where children are concerned.[25] Robeyns remarks that much research, including some of Sen's initial work, focused on certain functionings instead of capabilities, perhaps because it is easier to measure the former than the latter. She claims that concentrating on both functioning and capability is a viable and valuable approach as well; this may be necessary in cases where the mere exercise of choice constitutes functioning.[26] For example, a person's refusal to act in some way constitutes an important type of functioning—i.e., freely choosing to refrain from doing or being something, despite possessing the capability to do so. As Nussbaum puts it, "dignified life includes capabilities, and this includes the right not to use them."[27]

The capabilities approach identifies the "bare minimum of what respect for human dignity requires."[28] Sen calls them "basic capabilities,"[29] and Nussbaum refers to them as "central human capabilities."[30] Both authors are alluding to what *all* human beings need to flourish and live meaningful lives.[31] The approach stipulates that basic capabilities are universal across human beings, while embracing pluralism.[32] The specific details regarding what it would mean to live the "good life" can of course vary depending on culture, religion, economics, and many other

---

[16] Nussbaum (2000, p. 142).
[17] Johnstone (2007, pp. 84–85).
[18] Johnstone (2007, pp. 84-85).
[19] Johnstone (2007, p. 85).
[20] Baber (2009).
[21] Oosterlaken.
[22] Nussbaum (2006, p. 139).

[23] Sen (1999, p. 75).
[24] Robeyns (2006, p. 351).
[25] Nussbaum, for example, suggests that in certain cases, e.g., compulsory education for children, functioning should be the goal (Nussbaum 2006, p. 172, 2000, pp. 89–91). In other cases, e.g., voting, she contends that people should not be forced to vote if they do not want to do so. What is crucial, according to Nussbaum, is that they should have a genuine opportunity to vote.
[26] Robeyns, pp. 354–55.
[27] Nussbaum (2006, p. 184).
[28] Nussbaum (2000, p. 5).
[29] Sen (1993).
[30] Nussbaum (2000, pp. 70–77).
[31] Although there are similarities and differences between the two authors' views, the issue will not be explored here.
[32] Johnstone (2007, p. 78).

factors. But all humans need the ability to think freely, communicate, have access to education, obtain safe food and water, etc. What this requires in terms of material resources, however, differs from one person to another and may also vary depending on an individual's life stage. For example, a person with a missing leg will need either a wheelchair or a prosthetic leg in order to interface with her environment as effectively as an individual who has two functioning legs. This variation in the ability to convert resources into real opportunities or functionings means that looking solely at income distribution or the distribution of other goods, as some assessments of human well-being might, is unlikely to supply adequate information about an individual's welfare.[33]

Along with Bynum and Nussbaum, we think that other species ought to be permitted to live dignified lives that are consistent with the "activities and goals that creatures of many types pursue."[34] However, a discussion of the flourishing of nonhumans is beyond the scope of this article. Our focus is on humans for whom opportunities tend to be diminished or cut off altogether because of the inability or unwillingness of other agents to accommodate their needs. As Nussbaum asserts, doing justice to people with impairments requires "emphasizing the importance of care as a social primary good."[35] In some cases, failure to assist in the amelioration of certain social problems is the result of deeply entrenched cultural beliefs and practices. Rather than beginning with a specific prerequisite for moral consideration, such as rational capacity or the ability to be a party to a social contract, the capabilities approach starts by presenting the criteria for human flourishing and requires moral agents to promote these opportunities on behalf of those whose physical or cognitive limitations may interfere with their ability to do so on their own. Its emphasis on a moral agent's obligations provides a promising starting point for ensuring that basic capabilities are secured by all.

A central goal of human-robot interaction should be to ensure that elderly and non-elderly persons with impairments possess central human capabilities. Included in Nussbaum's list of capabilities is the ability to make meaningful choices about one's life. Consequently, caregiver robots should enhance people's opportunities to interface with their environment and enjoy relationships with other human beings. Robots should not override people's desires to act in ways consistent with their values or life narratives. This includes avoiding deliberate manipulation or deception of individuals through the use of robots. Even if others happen to think (perhaps paternalistically) that it would be better for another to function in some particular manner, preservation of the freedom to make choices is vitally linked to human flourishing. It is only once a person has lost the capacity to choose that others, including robot caregivers, should be permitted to make decisions on their behalf. However, even these decisions ought to be consistent with a care recipient's values.

Capabilities and robotic caregivers

We seek to determine how the capabilities approach may help illuminate what is appropriate or not with regard to robotic caregiving. While we believe that this strategy can contribute to the creation of socially responsible technology, we do not intend to exclude other ethical frameworks from the conversation. To reiterate, the capabilities approach is probably best seen as complementing and augmenting existing frameworks rather than replacing them.

Our focus will be on impaired persons, particularly elderly persons who have accumulated physical and cognitive impairments as part of the aging process. In doing this, we do not intend to pathologize aging, which is a normal part of human life. Quite the contrary, since some degree of impairment is part of the "normal" human life, we ought to think differently about those for whom impairments are always a part of life—e.g., those born with impairments or who suffer a debilitating injury or illness from a much earlier life stage. Because aging is often accompanied by the diminishment of certain capacities, focusing on the elderly is instructive in helping us figure out how best to use robots to minimize the impact of impairments on an individual's life. That said, the needs of elderly persons with impairments can differ significantly from those of infants, children, or young adults with impairments.

Many varieties of robotic caregivers are being developed.[36] Yet limits on the design and implementation of robot caregivers should be guided by the goal of preserving as many basic capabilities as possible for care recipients as well caregivers. However, the appropriateness of particular pathways toward this goal is unclear. For instance, Castelfranchi describes how forms of artificial intelligence can be designed deliberately to deceive humans,[37] which raises the fundamental question of whether truth-telling is always necessary for human flourishing. As in many interactions with people suffering from illnesses or living with certain types of impairments, balancing present and future autonomy will be a focal concern, and one that must be balanced against her other interests, regardless of her ability to function as an autonomous agent.

---

[33] Robeyns (2006); Nussbaum (2000); Sen (1999); Terzi (2005).

[34] Bynum (2006); Supra, No. 6, p. 327.

[35] Nussbaum (2006 p. 2).

[36] Some of these efforts are described in the 2006 *ETHICBOTS* report; see Datteri et al. (2006 pp. 78–81).

[37] Castelfranchi (2000).

Controversy persists about whether robots will be able to care for elderly or others in need of care partly because the meaning of "care" in this context is not always clear. Hence, exploring the meaning of "care" is vital. While discussing various notions of care, Coeckelbergh draws a useful distinction between "deep" and "shallow" care.[38] He suggests that robots can provide "shallow" care such as performing manual tasks though without "emotional, intimate, and personal engagement".[39] While there is no good reason to think that robots will be exempt from giving "shallow" care to individuals, we share Coeckelbergh's skepticism about the prospect of robots moving beyond "shallow" care to offer the reciprocal sort of companionship or friendship that is characteristic of "deep" care. That said, the inability of robot caregivers to provide "deep" care need not preclude human caregivers from using them to help provide "good" care—i.e., "care that respects human dignity".[40] Consequently, robots should complement the efforts of human caregivers rather than replace them.[41]

Adjusting the type and amount of care to particular contexts is likely to be required, given the varying needs individuals have across their lifespan. It is difficult to know how proficiently robots will perform specific caregiving tasks. The answer is largely contingent on advances in artificial intelligence. Sparrow and Sparrow, for example, delineate types of care that robots might eventually provide, ranging from the completion of rather mundane tasks such as serving drinks and cleaning rooms to being a friend or companion.[42] Their relevant level of sophistication could enable robots to remind a person to take medications and detect warning signs if the person might be in danger.[43] Further, robot companies have divergent visions of what their technologies ought to accomplish. At least one company is testing what it calls a "fully autonomous personal companion home care robot".[44] Yet whether the robot will meet such lofty expectations remains to be seen.

When considering how robot caregivers should be programmed, it should not be assumed that caregiving is the same in each situation or stage of life. For instance, Nussbaum distinguishes between a person who has been impaired since birth and a person who becomes impaired due to aging, illness or injury. She notes that elderly persons tend to be upset or frustrated in ways that those with lifelong impairments may not be.[45] Thus, it is often more challenging—and Nussbaum suggests, less rewarding—for caregivers to provide care for elderly who are cognizant of the diminishment of their abilities. Correspondingly, designers should be sensitive to these nuances.

Nussbaum stresses that our relative independence at various stages of our lives is a temporary condition.[46] For instance, a person's autonomy and competence (in the medical/legal sense of the term) can diminish rapidly or gradually. Drawing from Aristotle and putting it in modern terms, Bynum claims that autonomy is often a key component of flourishing.[47] The erosion of autonomy can interfere significantly with the performance of tasks. If impaired individuals can recover some of their autonomy through the use of a robotic caregiver, this can improve their opportunity to flourish. On a related note, Oosterlaken remarks that one's health status plays a crucial role in determining whether capabilities can be actualized.[48] To the extent that robot caregivers can mitigate further declines in an individual's health, they will also help ensure that individuals have the choice to actualize capabilities rather than live as relative prisoners of their impairments or the needs or whims of human caregivers.

There could also be psychological benefits to having robots for the "nuts and bolts" of daily living so that interactions can continue with other persons in as normal a way as possible. Robot caregivers could unlock possibilities for impaired individuals. As Johnstone states, access to certain types of technologies (e.g., computers) "will for many people significantly alter the action choices they have and the outcomes they are capable of bringing about."[49] Empowering elderly and other individuals in this way may contribute to improved health and lucidity. For example, a person who fails to nourish herself adequately is going to fail to function in many ways, including cognitively. And this sort of thing can become a vicious cycle: the person forgets to eat, brain power decreases, she forgets to eat again or take care of herself in other ways, and so on. A robot caregiver could function as a "cognitive prosthesis,"[50] while also helping to ameliorate the impact of physical limitations.

Roboticists suggest that the technology could expand communication potential and mobility for elderly, allowing them to have better access to needed medications,

---

[38] Coeckelbergh (2010).
[39] Coeckelbergh (2010, p. 3).
[40] Coeckelbergh (2010, p. 5).
[41] Decker (2008).
[42] Sparrow and Sparrow (2006, pp. 145–149).
[43] Faucounau et al. (2009 pp. 36–38).
[44] Grandma Interacts During GeckoSystems' Elder Care Robot Trials. *CNNMoney.com*. December 2, 2009. http://money.cnn.com/news/newsfeeds/articles/marketwire/0564659.htm.
[45] Nussbaum (2006, p. 101).
[46] Nussbaum (2006, p. 101).
[47] Bynum, p. 160.
[48] Oosterlaken, p. 98.
[49] Johnstone, p. 74.
[50] Faucounau et al. p. 35.

nourishment, and timely intervention when required. According to a study by Kozima and Nakagawa, the robot *Keepon* can help children, including those with impairments, to develop communication skills.[51] A study by Robins et al indicates that interactions with robots can benefit autistic children.[52] Further, if robots can be "trained" to diagnose medical conditions like autism and recognize their severity,[53] a range of care possibilities opens up.

If an impaired person relies on a robot to help with her needs, it is conceivable that the person may feel less aggravated, defensive, etc., as long as she can exercise some control over the robot that she does not have over other humans. For instance, the developers of the robot *Care-O-bot II* claim that it can assist with eating, walking, and retrieving items.[54] If true, this might allow an individual to dictate the terms of her care directly instead of through a human caregiver, just as the use of email or a telephone allows people to communicate more directly than through a human courier. Or, her needs might be met more proactively because she is neither concerned about affecting another person negatively nor fearful of a caregiver's unpredictable, and potentially undesirable, emotional response to a request for assistance. In short, the care recipient could be liberated from some of the burdens of dependence. In turn, this is likely to promote a care recipient's ability to "love those who love and care for us."[55]

Moreover, this may permit an individual to retain a greater degree of privacy regarding her current health status and allow her to share personal details with only certain others of her choosing. Control over information about oneself is an important component of individual autonomy; it can contribute to keeping a person's capabilities intact. Further, an important insight that the capabilities approach makes evident is that even if two individuals are receiving the same quality of care, a fuller ability to control one's environment (a basic human capability) is preferable since it can promote flourishing.

In addition to relieving some of the frustration, awkwardness, and sense of dependence associated with requesting assistance from other persons, the presence of certain kinds of robots may ease depression caused by loneliness. Even if robots do not provide genuine friendship, they may mitigate feelings of isolation. A study by Banks et al indicates that even minimal interaction with robot dogs can improve nursing home residents' scores on the UCLA Loneliness Scale.[56] According to the study, residents in the study population who were the "most lonely" experienced the greatest benefit due to the introduction of a robot dog or a real dog.[57] Interestingly, the authors reported no statistically significant difference between the subjects' responses to a robot as compared to a real dog. In addition to the use of robots in institutional settings, such as an assisted living facility, the use of robots may delay an elderly person's transition to such an institution or make it unnecessary. This could promote family unity or foster friendships among relatives.

Will robotic caregivers be embraced?

Beyond how well they are programmed, the effectiveness of robot caregivers in advancing flourishing is largely contingent on whether care recipients and their human caregivers accept this type of technological intervention. The wellbeing of care recipients will probably not be upheld if they are displeased with the care they receive. Intuitively, one might be tempted to say that care recipients would be insulted by having portions of their care delegated to a robot, but it is premature to assume that this will be the case. Human behavior is notoriously difficult to anticipate and individuals might not automatically reject (or accept) robotic caregivers.

Humans have a large capacity to bond with non-human animals, which contributes to the effectiveness of animal-assisted therapy. Further, children and adults seem to form meaningful connections with inanimate objects (e.g., dolls, blankets, automobiles) and establishing a connection with a robot may be a variation of this psychological phenomenon. Moreover, Turkle observes that the psychological inclination that humans have to nurture "relational artifacts" may facilitate bonding between humans and robots, though she emphasizes that this type of interaction will not be as meaningful and informative as those that occur with other human beings.[58] Shaw-Garlock suggests that if robots closely mimic human behavior, it might lead us to treat them like another human being.[59]

Care recipients often welcome the introduction of devices that facilitate their mobility (e.g., wheelchairs) or otherwise help them to function better. However, Hansson notes that in rehabilitative medicine, devices are not always embraced.[60] He remarks that individuals with disabilities might reject technology if they fear that their opportunities to interact with other people might be reduced.[61] Another

---

[51] Kozima and Nakagawa (2006).
[52] Robins et al. (2005).
[53] Scassellati. (2007).
[54] Graf et al. (2002).
[55] Nussbaum (2000, p. 79).
[56] Banks et al. (2008).
[57] Banks et al (2008, p. 176).
[58] Turkle (2006).
[59] Shaw-Garlock (2009, pp. 253–254).
[60] Hansson (2007), p. 264; see also Reidy and Crozier (1991).
[61] Hansson (2007, p. 265).

decisive factor can be whether care recipients believe they have control over the technology or whether they believe the technology has control over them.[62]

Individuals may end up feeling insulted or neglected depending on whether the robot caregiver is used by human caregivers to assist in caregiving duties or to serve as a replacement. In our view, a more suitable capability enhancing use of robots would be the former. We agree with the ETHICBOTS Project's conclusion that instead of using machines as substitutes for personal human relationships, "it would be desirable to rethink the organization of our societal and social life."[63] The capabilities approach is advantageous here, as it encourages us to make adjustments to social institutions rather than accepting long-standing institutions or values that leave certain categories of individuals out in the cold.

According to the results from early studies of human-robot interaction, people's reactions to robots appear to be heavily influenced by the context in which they are introduced. For example, when brought into a classroom, a group of young children were frightened by the robot, *Robovie*.[64] Yet Kozima and Nakagawa's study of the interactions between children and *Keepon* indicates that the robot "functioned as a pivot of interpersonal play with peers and sometimes with teachers."[65] Findings from a study investigating the *Roomba* suggest that humans can develop an emotional connection to it.[66] In the U.S. military, soldiers "bonded" with a robot named Scooby Doo, expressing a sense of loss when it was ultimately blown up.[67] Since the designers of the robots have conducted many of the current studies, further independent assessments are needed to establish the pervasiveness of human-robot bonding.

An additional part of the puzzle regarding whether a robot will be accepted is its physical appearance. Purportedly, if a robot or other artificial entity looks "too human", this can elicit a negative visceral response from human beings, which is typically referred to as the "uncanny valley" hypothesis.[68] Relatedly, Tapus and her colleagues are investigating how effective robots may be in assisting individuals who have suffered from a stroke, and one hypothesis they are testing is whether tailoring a robot's programming to a person's personality type may make it a more effective caregiver.[69] Taking these and other psychological variables into account can and should continue to help inform the design process.

For better or worse, the vulnerability of care recipients might cause bonding to occur rather readily with a robot caregiver, especially if it successfully meets the care recipients' needs. Whether care recipients or human caregivers will *trust* robot caregivers is another matter. However, previous experience with the APACHE system used to facilitate clinical decision making indicates that a lack of trust in technology is not the problem. Wallach and Allen suggest that physicians deferred too readily to machine evaluations of patients' conditions, in some cases resulting in "a closed-loop system wherein computer prediction dictate[d] clinical decisions."[70] That said, trust in robot caregivers does not have to be present in order for them to meet the needs of care recipients. Moreover, despite the tendency to anthropomorphize and subsequently bond with nonhumans, the promotion of care recipients' capabilities could occur without doing so.

Potential ramifications for caregivers

If ethically responsible designs are actualized, robots could expand capabilities not only for elderly and other impaired persons but also for their *caregivers*. In some ways, caregivers are "co-cared" because the relationship affects their welfare as well as that of the care recipient.[71] Though our focus thus far has been mainly on maintaining or expanding opportunities for elderly and impaired individuals, robots have the potential to make caregiving a genuine choice—particularly about whether to be a full-time caregiver. For example, Robins et al suggest that robots may be able to ease the caregiving responsibilities of parents of autistic children.[72] It may also lessen the chance that caregivers would be mistreated by care recipients.[73] However, future research is required to determine whether such benefits are likely to be realized.

Providing some relief to caregivers may expand their freedom and change the nature of discussions about elder care as well as end of life care. For instance, Hardwig

---

[62] Hansson (2007, pp. 264–265).
[63] *ETHICBOTS* (D5) (2008).
[64] Tucker (2009).
[65] Kozima and Nakagawa (2007).
[66] Sung et al. (2007).
[67] Garreau (2007).
[68] Steckenfinger and Ghazanfar (2009).
[69] Tapus et al. (2008).
[70] Wallach and Allen (2009, p. 41).
[71] Faucounau et al. p. 34.
[72] Robins et al. p. 117.
[73] An objection might be raised here that the same problem would be repeated with robots. In other words, robots could become a mistreated class of beings. Given the difficulty robotics experts are having with their attempts to develop a robot that has the learning capacities of the average infant, robots probably will not become morally significant enough for the foreseeable future to warrant people coming to their defense or accusing people of failing to provide *them* with adequate opportunities (capabilities). However, we remain agnostic on the issue.

argues in defense of a duty to die, primarily on the grounds that no individual has the right to impose significant burdens (e.g., emotional, social, financial) on his/her family members.[74] Putting financial considerations aside, which may still be overly burdensome, the duty to die concern might diminish significantly if robots fill deficits in the caregiving department.

Nussbaum recognizes that in many contexts, women are required to care for children, elderly parents, or other impaired family members. In some cases, they may be caring for *both* children and elderly parents simultaneously. Due to social norms, women are often expected to care for others "out of love".[75] She notes that a lot of caregiving is uncompensated or poorly compensated work that nevertheless erects barriers to the pursuit of other opportunities for professional, social, or personal enjoyment. Importantly, such obligations are frequently hoisted upon individuals rather than freely chosen by them. Succinctly stated, Nussbaum thinks we should ensure that care recipients and their caregivers both have the opportunity to experience happy lives. At a minimum, this requires that those who provide care in non-emergency contexts should have the freedom to choose the extent, type, and manner of caregiving.

Given that women usually bear the burdens of caregiving, robots could be seen as helping to empower women or improve their life options. From that point of view, the technology could help promote human well-being and social justice. On the other hand, we should be careful to avoid the enduring practice of hastily lauding new technology as being a panacea. For instance, Ruth Schwartz Cowan[76] and Freeman Dyson[77] describe how household appliances failed to liberate women as much as was envisioned in the twentieth century. Though they took some of the drudgery out of housework, completing or overseeing the work done with the appliances was still primarily a woman's responsibility. Further, women may have spent less time doing some of the in-home tasks, but other chores such as taking care of children became more time consuming.[78] Thus, at least in the short run, many women did not experience a net gain in freedom. Without changes in attitude regarding who is responsible for domestic chores, it is hardly surprising that the emergence of household appliances did not have the dramatic effect that one would have hoped. Analogously, caregivers might gain more freedom due to the advent of robots, but we need to be prepared for unanticipated, counterbalancing shifts that may occur. This is one area where using the capabilities approach will be especially helpful, as it requires us to continue altering our behavior and social institutions so that they consistently contribute to human flourishing. Merely providing access to technology is not the ultimate goal.

Addressing key objections

Thus far, we have examined how robots might contribute to the goal of improving the well-being of both care recipients and human caregivers. Though an exhaustive exploration of all corresponding ethical reservations about robotic caregivers cannot be accomplished here, we will address a few key issues. To begin, will a robot's programming be sophisticated enough to handle the challenges associated with caregiving? Javier Movellan astutely recognizes that "to sustain interaction with people, you can't possibly have everything preprogrammed."[79] Speaking more generally, Selmer Bringsjord claims that even the best robots are no match for thinking abilities of "a moderately sharp toddler."[80] What robots are expected to do must not only correspond to advances in artificial intelligence and other related realms, they should also be limited to performing tasks that promote human flourishing. Even if AI advances significantly enough that it would be feasible to defer to them for additional types of tasks, whether particular actions should be delegated to a robot should not be determined solely by its ability to complete those tasks. For example, it is problematic to allow efficiency to become the primary goal of robot caregiver intervention.

Robots will probably not possess the same arbitrariness that human beings do, which might be beneficial to care recipients. Different humans behave differently in similar situations, and the same individual may act in different ways across similar scenarios. For instance, we might not hold a door for another person based on what others around are doing, our emotional state, whether we are in a hurry, etc., and even the type of music we hear might influence our inclination to help another person.[81] This is not to say that consistency is the only important, or most important, value or that robots perform each caregiving task as effectively as a human being. Yet increased consistency in responding to their needs could alleviate much frustration for care recipients. Presumably, robots would not be programmed to lash out, curse, or express anger toward people.

---

[74] Hardwig (1997).
[75] Nussbaum (2006, p. 102).
[76] Cowan (1983).
[77] Dyson (1998).
[78] Dyson, pp. 132–134.

[79] Tucker, p. 62.
[80] Bringsjord (2008).
[81] North et al. (2004).

Conversely, Sparrow and Sparrow state that the "demands that our friends—or even pets—make on us are unpredictable, sometimes unexpected and often inconvenient. This is an essential part of what makes relationships with other people, or animals, interesting, involving and rewarding."[82] Then again, these features can also make human interaction burdensome, leading to some level of mutual resentment, despite the presence of an exceptionally devoted and compassionate caregiver. Since care recipients are likely to be ill-equipped to shoulder additional responsibilities, using robot caregivers may be desirable.

As Coeckelbergh, Sparrow, and others point out, robots will likely fail to offer needed emotional support. Though a legitimate and serious worry, the extent to which it is a problem is largely contingent upon the kind of care being delegated to a robot and a care recipient's needs and expectations. For instance, if a person has severe impairments, it is probably unwise and likely unethical to leave that person's care solely to a robot. Moreover, a common complaint about robotic caregivers is that their use might deceive individuals about the care that they are receiving, especially in cases where a person suffers from severe dementia.[83] On the other hand, a robot might meet that person's needs in surprising ways (e.g. by reducing loneliness) if it complements the efforts of human caregivers. Further, a study by Ezer et al indicates that subjects were receptive to the idea that robots could help with day-to-day tasks and the subjects did not necessarily see the need for having higher level interactions such as conversations with a robot.[84] It is also important to remember that human caregivers can fail to fulfill their obligations, including to individuals who suffer from dementia.[85] In other words, merely because a human is assigned the task of providing care, an emotional bond does not necessarily form. As Coeckelbergh acknowledges, many care recipients do not currently receive "deep" care from their human caregivers.[86]

Another major objection to robot caregivers is that their use in providing care to patients with severe dementia might constitute "disrespectful deceit".[87] Whether the presence of robots will exacerbate the tendency of demented people to be confused regarding the identity, roles, and abilities of those with whom they come into contact is not yet known. As with the introduction of any new element to a demented person's environment, it is important to ensure that we do so in a way that minimizes anxiety or confusion for the patient. However, it is not clear that the consistent presence of a robot caregiver will generate any more anxiety or confusion than the intermittent presence of other caregivers or visitors. If other variables in the patient's environment remain stable, adding a robot caregiver might not exacerbate already existing problems with caring for people with severe dementia. This issue is complex, because even though behavioral cues can tell us, e.g., that a demented person thinks her son is her lawyer, doctor, or husband and not her son, direct access to the precise content of another person's beliefs regarding the robot caregiver in their midst will be lacking.

Granted, a demented person might mistakenly believe that the robot is an empathic being, but as long as it does not replace human caregivers and is used as a facilitator of care and other types of human-human interaction, the introduction of the technology would not necessarily constitute being dismissive of that person's well-being. The presence of robot caregivers may offer comfort to dementia patients even if the reasons for this remain opaque to their human caregivers and other outsiders. In short, as long as there is no intention to deliberately deceive or neglect dementia patients through the use of a robot, the fact that some patients may form erroneous beliefs about a robot caregiver—a process over which other agents may have little control—does not necessarily amount to being disrespectful to a care recipient.

Along related lines, Sparrow and Sparrow predict that robot-human will replace human-human interaction instead of augmenting or altering existing human relationships. Admittedly, the frequency and type of human contact that may be gained or lost due to the use of robot caregivers and the corresponding effects this can have on flourishing are difficult to foresee. Moreover, a typical motive for introducing robots into an environment has been to maximize profits by replacing human workers. Yet bringing robot caregivers onto the scene could also be motivated by the obligation to meet core human needs. This is a key advantage of the capabilities approach, since it should inform the design and use of robot caregivers in such a way that the "human" in human-robot interaction is maintained.

Along related lines, Sparrow and Sparrow dismiss the idea that freeing people from "mundane aspects of aged care" will allow them to "devote more of their energies to the more important task of providing companionship and emotional support for each other."[88] Though the current level of human contact that care recipients receive in many

---

[82] Sparrow and Sparrow, p. 149.

[83] It should be noted that if an individual suffers from severe dementia, she could not only be "deceived" by being taken care of by a robot, but presumably she could also mistake relatives for health care staff for example.

[84] The study did however have a relatively small sample size ($N = 177$); see Ezer et al. (2009).

[85] Cooper et al. (2009).

[86] Coeckelbergh (2010, p. 183).

[87] Thank you to the anonymous reviewer who raised this objection.

[88] Sparrow and Sparrow, p. 152.

U.S. nursing homes, for example, is probably preferable to none, that contact is not always terribly meaningful. While it could be a forlorn hope, the use of robot caregivers could allow people to interact in more purposeful and consequential ways than, say, discussing the weather with someone while cleaning the floor in their room. In principle, there is nothing wrong with such interaction, but this is quite different from sitting down to talk with someone over a pot of tea. The framework of the capabilities approach makes enabling human communication a primary goal, particularly for those for whom it is more difficult.

Critics are probably right that at least some care recipients may feel neglected by their family, friends, or others. Using the technology could erode the relationship between at least some care recipients and human caregivers.[89] For instance, Sparrow and Sparrow fear that human contact for the elderly may be reduced. They claim the introduction of robots might lead nursing homes to downsize, making each employee responsible for a larger number of residents.[90] Discounting this possibility is hard to do. But it is difficult to support the notion that current caregiving conditions ought to be preserved. Nursing home residents are too frequently the victims of abuse and theft,[91] illustrating that care is sub-par for many. A further problem with this line of criticism is that it does not distinguish between freely choosing the company of certain other persons and requiring the presence of others due to the diminishment of one's ability to function more independently. Though the arrival of a robotic caregiver might be perceived as a sign that a care recipient has been "put out to pasture," it may also be viewed as helping that individual to recover a desired level of independence.

There is little justification for assuming that current caregiving conditions are optimal or that their removal or radical alteration would be a bad thing. Family members have as much opportunity to undermine an individual's capabilities as they do to promote their flourishing.[92] Those caring for elderly or impaired individuals by default may not be best suited for the task. While family members might agree about appropriate goals, worldviews, priorities, and other things, they can also reasonably disagree about many things, including how one's "golden years" ought to be spent. Moreover, even when family members' views converge, this may be the result of adaptive preference and not the product of an individual's free deliberation and choosing.

Underlying the aforementioned issues is the motive behind the development of robot caregivers. Charitably speaking, the drive stems partly from inadequate resources being dedicated to solving relevant social problems. In countries like Japan and Germany, robots might be tasked to care for populations whose average age continues to rise.[93] Further, elderly and others in dire need of care experience neglect or worse. Critics worry that if we rely on the "technological fix" in this case, it might lessen incentives to improve human behavior and to some degree their apprehension is understandable. There are many possible causes for our relevant social problems (e.g., frustration, lack of resources, resentment, low pay). Robots could be a valuable piece of the puzzle in this regard by serving not as a "fix" but a valuable technological "assist" that mitigates problems such as lack of genuine choice and elder abuse.

Moreover, Wallach and Allen ask, "if there is no evidence that people and communities are willing to direct the time and resources necessary to respond to the needs of the elderly and disabled *for human contact*, are social robots better than nothing?"[94] While it would be preferable to modify certain assumptions about the circumstances of acceptable human interaction so that simply going to visit or talk with another person without desiring to achieve some other goal, such as feeding, cleaning, or moving an elderly person or attending to his or her environment, would be valued equally. As it is, people are often reluctant to interact with others unless there is a tangible outcome from the interaction. This phenomenon is consistent with Nussbaum's observation that the pervasive view of the citizen is that of "productive augmenter of social well being"[95] and Western societies' tendency to be "dominated by economic motives and considerations of efficiency".[96]

Conclusion

The relative flexibility of the capabilities approach can promote the goal of human flourishing through empowering individuals and expanding their opportunities. Along with the insights from other views on ethics, it can provide designers and users of robots with a proactive framework within which to assess the potential ethical impact of using robots as caregivers. In addition to expanding opportunities and improving the quality of life for care recipients, the capabilities approach also focuses on the well-being of caregivers. Ultimately, using the approach as a guide to robot intervention in caregiving activities may expand

---

[89] Sparrow and Sparrow, p. 153.
[90] Sparrow and Sparrow, p. 150.
[91] Pear (2008).
[92] Nussbaum (2000).
[93] Decker (2000, pp. 321–322).
[94] Wallach and Allen, p. 45.
[95] Nussbaum (2006, p. 128).
[96] Nussbaum (2006, p. 157).

freedom and contribute to human flourishing for individuals at different life stages despite existing impairments. This has the potential to alter positively our perceptions of caregiving as well as our attitudes and behavior toward care recipients.

Conflict of interest statement The authors are not aware.

References

Alkire S. (2005). Capability and functionings: Definition and justification. http://www.capabilityapproach.com/pubs/HDCA_Briefing_Concepts.pdf. Accessed March 20, 2010.
Baber H. E. (2009) Worlds, capabilities and well-being. *Ethical Theory* and *Moral Practice*. Available online.
Banks, M. R., Willoughby, L. M., & Banks, W. A. (2008). Animal-assisted therapy and loneliness in nursing homes: Use of robotic versus living dogs. *Journal of the American Medical Directors Association, 9*(3), 173–177.
Bringsjord, S. (2008). Ethical robots: the future can heed us. *AI & Society, 22*, 539–550.
Bynum, T. W. (2006). Flourishing ethics. *Ethics and Information Technology, 8*, 157–173.
Castelfranchi, C. (2000). Artificial liars: Why computers will (necessarily) deceive us and each other. *Ethics and Information Technology, 2*, 113–119.
Coeckelbergh, M. (2009). Personal robots, appearance, and human good: A methodological reflection on roboethics. *International Journal of Social Robotics, 1*, 217–221.
Coeckelbergh, M. (2010). Health care, capabilities, and ai assistive technologies. *Ethical Theory and Moral Practice, 13*(2), 181–190.
Cooper, C., Selwood, A., Blanchard, M., Walker, Z., Blizard, R., & Livingston, G. (2009). Abuse of people with dementia by family carers: Representative cross sectional survey. *British Medical Journal, 338*, 583–586.
Cowan, R. S. (1983). *More work for mother: The ironies of household technology from the open hearth to the microwave*. London: Basic Books, Inc.
Datteri E., Laschi C., Salvini P., Tamburrini G., Veruggio G., & Warwick K. (2006). *ETHICBOTS: Emerging Technoethics of Human Interaction with Communication, Bionic and Robotic Systems*, April 2006, http://ethicbots.na.infn.it/restricted/doc/EthicBotsD1.zip. Accessed December 4, 2009.
Decker, M. (2008). Caregiving robots and ethical reflection: The perspective of interdisciplinary technology assessment. *AI & Society, 22*, 315–330.
Dyson F. (1998). Technology and social justice. In: M. E. Winston & R. D. Edelbach (Eds.), *Society, Ethics, and Technology*, pp. 130–140. Thomson Wadsworth, 2006.
ETHICBOTS (D5). (2008). *Techno-ethical case-studies in robotics, bionics, and related ai agent technologies*, R. Capurro, G. Tamburrini, & J. Weber (Eds.), http://ethicbots.na.infn.it/documents.php (accessed January 2, 2010).
Ezer N., Fisk A. D., Rogers W. A. (2009). More than a servant: Self-reported willingness of younger and older adults to having a robot perform interactive and critical tasks in the home. *Proceedings of the Human Factors and Ergonomics Society 53rd Annual Meeting—2009*. http://www.hfes.org/web/Newsroom/HFES09-Ezer-RobotsInHome.pdf Accessed December 2, 2009.

Faucounau, V., Wu, Y. H., Boulay, M., Maestrutti, M., & Rigaud, A. S. (2009). Caregivers' requirements for in-home robotic agent for supporting community-living elderly subjects with cognitive impairment. *Technology and Health Care, 17*(1), 33–40.
Garreau J. (2007). Bots on the ground. *The Washington Post*. May 6, 2007. http://www.washingtonpost.com/wp-dyn/content/article/2007/05/05/AR2007050501009_pf.html. Accessed July 13, 2009.
Graf B., Hans M., Kubacki J. & Schraft R. (2002). Robotic home assistant care-o-bot II. *Proceedings of the Second Joint Meeting of the IEEE Engineering in Medicine and Biology Society and the Biomedical Engineering Society*. http://www.care-o-bot.de/Papers/2002_RoboticHomeAssistant_Care-O-bot_II_IEEE_Texas.pdf. Accessed July 13, 2009.
Hansson, S. O. (2007). The ethics of enabling technology. *Cambridge Quarterly of Healthcare Ethics, 16*, 257–267.
Hardwig, J. (1997). Is there a duty to die? *Hastings Center Report, 27*(2), 34–42.
Johnstone, J. (2007). Technology as empowerment: A capability approach to computer ethics. *Ethics and Information Technology, 9*, 73–87.
Kozima H. & Nakagawa C. (2006). Social robots for children: Practice in communication-care. *9th IEEE International Workshop on Advanced Motion Control*, 768–773.
Kozima H. & Nakagawa C. (2007). A robot in a playroom with preschool children: longitudinal field practice. *16th IEEE International Conference on Robot & Human Interactive Communication*, 1058–1059.
Lin P., Bekey G., & Abney K. (2008). *Autonomous military robotics: Risk, ethics, and design*. Ethics & Emerging Technologies Group at California State Polytechnic University.
North, A. C., Tarrant, M., & Hargreaves, D. J. (2004). The effects of music on helping behavior. *Environment and Behavior, 36*(2), 266–275.
Nussbaum, M. C. (2000). *Women and human development*. New York: Cambridge University Press.
Nussbaum, M. C. (2006). *Frontiers of justice*. Cambridge, Massachusetts: Belknap Press.
Nussbaum, M. C., & O'Neill, O. (1993). Justice, gender, and international boundaries. In C. Nussbaum & A. Sen (Eds.), *The quality of life* (pp. 324–335). Oxford: Clarendon Press.
Onishi N. (2006). In a wired South Korea, robots will feel right at home. *The New York Times*, April 2, 2006. http://www.nytimes.com/2006/04/02/world/asia/02robot.html. Accessed July 2, 2009.
Oosterlaken, I. (2009). Design for development: A capability approach. *Design Issues, 25*(4), 91–102.
Pear R. (2008) Violations reported at 94% of nursing homes. *The New York Times*. September 29, 2008. http://www.nytimes.com/2008/09/30/us/30nursing.html. Accessed July 7, 2009.
Reidy, K., & Crozier, K. S. (1991). Refusing treatment during rehabilitation—A model for conflict resolution, In rehabilitation medicine-adding life to years [special issue]. *The Western Journal of Medicine, 154*, 622–623.
Robeyns, I. (2006). The capability approach in practice. *The Journal of Political Philosophy, 14*(3), 351–376.
Robins, B., Dautenhahn, K., Te Boekhorst, R., & Billard, A. (2005). Robotic assistants in therapy and education of children with autism: Can a small humanoid robot help encourage social interaction skills? *Universal Access in the Information Society, 4*, 105–120.
Scassellati, B. (2007). How social robots will help us to diagnose, treat, and understand autism. In S. Thrun, R. Brooks, & H. Durrant-Whyte (Eds.), *Robotics research* (pp. 552–563). Berlin: Springer.
Sen, A. (1993). Capability and well-being. In M. Nussbaum & A. Sen (Eds.), *The quality of life* (pp. 31–53). Oxford: Clarendon Press.

Sen, A. (1999). *Development as freedom*. New York: Alfred A. Knopf.
Shaw-Garlock, G. (2009). Looking forward to sociable robots. *International Journal of Social Robotics, 1*, 249–260.
Sparrow, R., & Sparrow, L. (2006). In the hands of machines? The future of aged care. *Minds and Machines, 16*, 141–161.
Steckenfinger, S. A., & Ghazanfar, A. A. (2009). Monkey visual behavior falls into the uncanny valley. *PNAS, 106*(43), 18362–18366.
Sung, J. Y., Guo, L., Grinter, R. E., & Christensen, H. I. (2007). "My Roomba is Rambo": Intimate home appliances. In J. Krumm, G. D. Abowd, A. Seneviratne, & T. Strang (Eds.), *UbiComp 2007: Ubiquitous computing* (pp. 145–162). Berlin: Springer.
Tapus A., Tapus C., & Mataric' M. J. (2008). User-robot personality matching and assistive robot behavior adaptation for post-stroke rehabilitation therapy. *Intelligent Service Robotics Journal*, Special issue on multidisciplinary collaboration for socially assistive robotics, A. Tapus (Ed.), 169–183.
Terzi, L. (2005). A capability perspective on impairment, disability, and special needs. *Theory and Research in Education, 3*(2), 197–223.
Tucker A. (2009). Robot babies. *Smithsonian Magazine.* 56–65, July 2009. http://www.smithsonianmag.com/science-nature/Birth-of-a-Robot.html. Accessed July 13, 2009.
Turkle S. (2006). A nascent robotics culture: New complicities for companionship. *AAAI Technical Report Series.*
Wallach, W., & Allen, C. (2009). *Moral machines: Teaching robots right from wrong*. New York: Oxford University Press, Inc.

12

Carebots and Caregivers: Sustaining the Ethical Ideal of Care in the Twenty-First Century

Shannon Vallor

Received: 9 January 2011 / Accepted: 10 March 2011 / Published online: 31 March 2011

Abstract In the early twenty-first century, we stand on the threshold of welcoming robots into domains of human activity that will expand their presence in our lives dramatically. One provocative new frontier in robotics, motivated by a convergence of demographic, economic, cultural, and institutional pressures, is the development of "carebots"—robots intended to assist or replace human caregivers in the practice of caring for vulnerable persons such as the elderly, young, sick, or disabled. I argue here that existing philosophical reflections on the ethical implications of carebots neglect a critical dimension of the issue: namely, the potential moral value of caregiving practices for *caregivers*. This value, I argue, gives rise to considerations that must be weighed alongside consideration of the likely impact of carebots on care recipients. Focusing on the goods internal to caring practices, I then examine the potential impact of carebots on caregivers by means of three complementary ethical approaches: *virtue ethics*, *care ethics*, and the *capabilities approach*. Each of these, I argue, sheds new light on the contexts in which carebots might deprive potential caregivers of important moral goods central to caring practices, as well as those contexts in which carebots might help caregivers sustain or even enrich those practices, and their attendant goods.

1 Background

In the early twenty-first century, we stand on the threshold of welcoming robots into domains of human activity that will expand their presence in our lives dramatically, extending well beyond the primarily industrial contexts in which robots have for decades already been working alongside or in the place of human beings. Recent developments range from the expanding use of robotic "surgeons" to conduct medical procedures with a degree of accuracy and precision well beyond the capabilities of human hands,[1] to the rapid development and implementation of sophisticated robotic drones which now assist or take the place of human soldiers in critical and dangerous military operations,[2] to the commercial rise of personal robots used in the home to clean our floors, mow our lawns, play with our children, and spy on nannies and contractors.[3]

This phenomenon naturally gives rise to philosophical questions of various sorts, but among the most common and pressing are ethical questions. Of particular concern to many are the ethical implications of what have been termed "carebots". Carebots are robots designed for use in home, hospital, or other settings to assist in, support, or provide care for sick, disabled, young, elderly, or otherwise vulnerable persons. The kind of support they may provide varies widely, but generally following Sharkey and Sharkey (2010), we can distinguish between the following actual or potential functions of carebots: performing or providing assistance in caregiving tasks; monitoring the health or behavioral status of those receiving care or the provision of care by caregivers; and providing companionship to those under care.

Motivations for the development of carebots are various, but include demographic pressures in countries like Japan, Germany, and the USA to ensure the ability to care for a rapidly expanding population of aged or otherwise vulnerable persons, economic pressures upon individuals, private or public institutions to reduce the costs of care, social pressures to reduce growing institutional failures to provide quality human care, and recognition of the need to reduce the often-overwhelming physical and psychological burdens placed upon individual caregivers.

Development and implementation of carebots is in a relatively early stage. A few limited applications have found some initial success—AIST's Paro Therapeutic Robot, a bionic fur-covered harp seal, was recently classified in the USA as a Class 2 medical device used to provide companionship in nursing home settings, and has been widely used in Denmark (Tergesen and Inada 2010). Most carebots are still in the development stage, though some are relatively advanced along that path—including the RIBA nurse robot developed by Japan's RIKEN-TRI Collaboration Center for Human-Interactive Robot Research (RTC). RIBA, already a second generation upgrade to the earlier RI-MAN, can lift a patient from a bed to a wheelchair, and vice versa.[4] GeckoSystems International Corporation is currently conducting limited in-home trials of its CareBot. Combining a mobile-service robot base with interlinked AI modules for navigation, tracking, scheduling, and reminders (e.g., medications, food, etc.), and even verbal interaction ("chat"), the CareBot is described in GeckoSystems' marketing as having a commercial potential that includes both elder care and child care.[5] Indeed, while it may be quite some time

---

[1] The most well-known of these is the daVinci Surgical System manufactured and marketed by Intuitive Surgical, Inc., but others include Medsys' LapMan and Titan Medical Inc.'s Amadeus.
[2] See Singer (2009) for an excellent discussion of emerging military robotics and their implications.
[3] Consider as examples iRobot's popular Roomba, Friendly Robotics' Robomower, Sony's AIBO, and Meccano's Spykee.
[4] http://www.ribarobotnurse.com/, accessed 5 Jan 2011.
[5] http://www.geckosystems.com/markets/consumer_familycare.php, accessed 5 Jan 2011.

before there are practical, safe, and cost-effective carebots available for widespread consumer or institutional use, the development of carebots is likely to continue to advance, if we consider the grave deficiencies of existing social means of providing human care in many developed nations and the ever-expanding commercial appetite for new technologies in many of those same countries. While we would expect carebot developers to offer generous, even unrealistic predictions of the role of such technologies in the future, the roadmap from which they are working is quite explicit, as indicated on the placeholder marketing site for RIBA, which has already secured its commercial domain name: "There might come a time when human caregivers will be replaced by robots. An impossible vision it may seem, but do not be too surprised about human caregivers being assisted by robots in the near future."[6]

2 Carebots and Ethics

Pronouncements of the latter sort understandably set off philosophical alarms about the ethical dangers of "replacing" human caregivers with machines. We might reasonably regard such statements as commercial hyperbole, yet even the more realistic near-term scenarios for carebots supply ample motivation for ethical reflection and inquiry. What are the various sorts of ethical concerns we might legitimately have? For methodological purposes, let us set aside important questions about the practicality of the grander technological aims of carebot researchers and developers. Let us also set aside well-motivated worries about the safety of such technologies should they be introduced. In the absence of a significant collapse of existing social mechanisms for assuring medical product safety (mechanisms which one must grant, are far from perfect), let us assume for the sake of argument that grave and widespread safety issues can be prevented. There are still a large number of issues that remain.

Mark Coeckelbergh (2009) has helpfully separated what he calls questions of "roboethics" into several categories. The first category pertains to the minds and realities of robots themselves (in this category we would place concerns about whether artificially intelligent robots could ever properly be considered moral agents, whether they might have "rights", whether they could relate to human beings in ethically significant ways, and so on). The second category pertains to the application of traditional ethical theories to human–robot interactions, allowing us to judge their ethical significance (in this category we might place worries about whether our filial obligations to the elderly can be satisfied through quality care provided by robots, or whether the use of carebots is sufficiently warranted by their collective social utility). Coeckelbergh wishes to redirect our ethical inquiry to a new, third manner of reflection—by moving away from the application of what he calls "external" ethical criteria, and toward consideration of the "good internal to practice" that may be realized through robot–human interaction (217). That is, he wishes to consider how we *experience* such interactions, and what they "do to us as social and emotional beings" (219).

---

[6] http://www.ribarobotnurse.com/, accessed 5 Jan 2011.

This, too, is my concern. However, I wish to take it in a somewhat different direction from Coeckelbergh's, in order to illuminate something that I think has thus far been missing from philosophical discussions of "roboethics". Before we can do this, however, we need to take a closer look at how scholars have approached the ethical significance of carebots in particular. Though only a handful of scholars have yet turned their attention to this subject, many of the most important ethical issues have been carefully discussed in the existing philosophical literature on carebots. Of course, there remain substantive disagreements about how to evaluate the moral risks and benefits of introducing carebots into actual care settings. In general, these disagreements involve conflicting judgments about whether carebots will improve or degrade the quality of life of the cared-fors, whether they will improve or degrade relations between cared-fors and caregivers, and whether will they will result in an improvement or degradation of personal and institutional quality of care provision.

Specifically, concerns have been raised about:

1. The *objectification* of the elderly as "problems" to be solved by technological means (Sparrow & Sparrow 2006, 143; Sharkey & Sharkey 2010).
2. The potential for carebots to either enhance or restrict the *capabilities*, *freedom*, *autonomy*, and/or *dignity* of cared-fors (Borenstein & Pearson 2010; Sharkey & Sharkey 2010; Decker 2008).
3. The potential of carebots to enhance or reduce *engagement* of cared-fors with their surroundings (Borenstein & Pearson 2010; Sharkey & Sharkey 2010).
4. The potential of carebots to enhance or intrude upon the *privacy* of cared-fors (Sharkey & Sharkey 2010).
5. The *quality* of physical and psychological care robots can realistically be expected to supply (Coeckelbergh 2010; Sparrow & Sparrow 2006).
6. The potential of carebots to either reduce or enhance cared-fors' levels of *human contact* with families and other human caregivers (Sparrow & Sparrow 2006).
7. The potential of carebot relations to be inherently *deceptive* or *infantilizing* (Sparrow & Sparrow 2006; Sharkey & Sharkey 2010; Turkle 2006).

All of these concerns are well motivated and demanding of further study. Something has been left out of all of these treatments, however. Though much attention has deservedly been paid to the impact of carebots on those cared for, that is, how well cared-for humans will be in a scenario where carebots are employed, less attention has been paid to their impact on *caregivers* and society more broadly. Where this does get addressed, it is typically limited to a focus on the way in which carebots may help to "reduce the care burden" on caregivers and on society (Sharkey & Sharkey 2010), empowering people to make caregiving a choice rather than an obligation (Borenstein & Pearson 2010, 283–4), allowing caregivers to trade routine tasks for more emotionally meaningful ones (Coeckelbergh 2010, 183), and eliminating or reducing the numbers of undercompensated, overworked, and ill-trained human caregivers, especially in institutional settings (Coeckelbergh 2010, 183; Borenstein & Pearson 2010, 285). However, I will show that by ignoring the *positive* value of caring practices for caregivers, current scholarly reflections on the ethical implications of carebots remain dangerously one-sided. I then examine the value of caring practices for caregivers through the conceptual lenses of three complementary ethical frameworks, each of which are oriented toward the goods internal to human practices: *virtue ethics*, *care ethics*, and the *capabilities approach*. Rather than defend the superiority of one particular framework, I conclude that each sheds new light on the potential ethical risks and promise of carebots.

3 The Burdens and Benefits of Caregiving Practices

It will be evident to anyone who has ever had to care for an elderly parent, chronically ill spouse, or disabled child for any significant length of time that "reducing the care burden" on caregivers is a desideratum, even a collective moral obligation, assuming it can be accomplished by ethical means. It is also quite clear to anyone who has spent much time in a skilled nursing facility or other comparable institutional setting that as Borenstein and Pearson note, "...it is difficult to support the notion that current caregiving conditions ought to be preserved" (2010, 286). Surrendering caring practices to robots might be risky or ethically worrisome, but so is maintaining the status quo. So before proceeding, let me stipulate that carebots, properly designed and implemented, might be able to improve the lives of both cared-fors and caregivers in ways that would be ethically desirable and, in the absence of acceptable alternatives, ethically mandated.

Still, it deeply concerns me that there has been virtually no scholarly discussion of what value caregiving, as a human practice, might have for *caregivers* rather than cared-fors, what the ethical significance of that value might be, and how the value of such caring practices might be impacted by the widespread introduction of carebots. To the contrary, the standing assumption seems to be that caregiving is generally not only a burden upon caregivers (which we can grant) but that it is *nothing except* a burden. Again and again it is described only as a "requirement", a social "expectation", a burden people (especially women) are forced to bear (Borenstein & Pearson 2010, 284; Sharkey & Sharkey 2010). Even the provision of companionship and emotional support is characterized as a "task" (Sparrow & Sparrow 2006). Nussbaum (2006) rightly notes that contemporary societies still tend to assume women will give care "for free, 'out of love,'" (102), her use of scare quotes redirecting us to the image of caregiving as an unfair burden. Yet if there are no *further* dimensions to caregiving practices, it would seem to follow that *in the absence of consequent injury or moral insult to cared-fors*, we could have no reason to hesitate to surrender our caregiving practices to carebots, and to encourage others to do the same.[7]

Intuitively, it seems this cannot be right, and despite the fact that their concerns about carebots tend to focus exclusively on potential harms to cared-fors, I doubt (though I could be mistaken) that many of the philosophers who have reflected on the ethics of robot care would endorse the above conclusion. Nevertheless, it begs the question why the potential losses to caregivers who surrender such practices have not been explicitly considered. Of course, it is acknowledged that caregiving may be

---

[7] The only way to avoid this conclusion, beyond the one I suggest, is to presuppose a deontological framework, perhaps of a Kantian sort, that entails individual moral obligations to provide care *and* dictates that we may not employ indirect technological means of meeting these obligations. I am not optimistic, however, about the viability of such a claim.

freely chosen, indeed, this is presented as the ideal state of affairs. Borenstein and Pearson, following Nussbaum (2006), tell us that "those who provide care in non-emergency contexts should have the freedom to choose the extent, type, and manner of caregiving" (2010, 284). But the reasons *why* one might choose to give care are not explored, and these reasons matter a great deal to our inquiry.

Some such choices might be motivated by a kind of ethical heroism, a willingness to take on burdens of care so that others do not have to. Others might be motivated by a sense of personal and non-transferrable duty, such that one would rather suffer the burdens of giving care than the guilt of shirking them. Still others might be motivated by external rewards that outweigh the burdens of care, such as monetary compensation, or expressions of gratitude and admiration by cared-fors and observers. But all of these reasons presuppose that the caregiving practice *itself* is simply a burden, the free acceptance of which must be justified by something external to the practice. If this is right, then it would appear that nearly all of the external conditions that warrant freely giving care could potentially be eliminated with the introduction of carebots—by transferring caring tasks to entities that cannot experience them *as* a burden. Yet I wish to propose another possibility: that there are goods internal to the practice of caregiving that we might not wish to surrender, or that it might be unwise to surrender even if we might often wish to do so.

In our methodology, then, we return to the orientation of Coeckebergh (2009), who we should recall, wishes to redirect our inquiry away from "external" ethical criteria, and toward the "good internal to practice" (217) that may be realized through robot–human interactions. He thinks we should think more deeply about how we *experience* such interactions with robots, and what they "do to us as social and emotional beings" (219). I agree. But Coeckelbergh's interest here is narrower than mine. It is what the *robots* do to us that captures his interest, and here he refers to the way in which their "appearances" to us, regardless of the inner realities of robot "minds", can powerfully shape our own perceptions, emotions, and capabilities (219–220). I think this is a question of tremendous ethical importance. Yet if we apply it to the carebot question, it targets, once again, the impact of carebots on those to whom they primarily appear, namely those cared *for*. Of course, they may interact with human caregivers in the environment too—nursing staff, family members, and so on. But if we focus narrowly on what the robots do to us through their appearances to us, we leave behind the question of what we may do to *ourselves* when we choose to surrender caring practices to robots. If there are goods internal to the practice of caring, then I am surrendering these goods if I give up the practice, and this holds true whether or not I have interactions with the robot to whom I have surrendered it.[8]

What kind of goods could these be? There are at least three philosophical perspectives that fit well with our methodology and might help us conceptualize the nature of these goods. They are (1) a *virtue ethics* approach, (2) a *care ethics* approach, and (3) a *capabilities* approach. Each of these philosophical perspectives

---

[8] It is of course possible that caregivers might choose to *share* caring practices with robots rather than *surrender* them, as I will note. However, given the widespread acknowledgment of the burdens of giving care, we cannot discount the possibility that many potential caregivers will, once they trust the safety and skill of carebots, transfer significant caring duties to them.

offers a conceptual framework oriented toward goods internal to practices rather than external ethical criteria, and each can shed light on the specific nature of the goods internal to caring practices. I will not attempt here to determine which of these perspectives, if any, might be superior to the others. This is for two reasons: first, I do not wish to foreclose the possibility that they are complementary or overlapping perspectives. Second, I believe at this early stage of inquiry it is more important to illuminate the subject matter as richly as possible than to identify with precision the conceptual framework most appropriate to it.

4 Carebots and the Virtues

The virtues are conventionally understood as dispositional states of character that can be regarded as excellent, insofar as they promote human flourishing. I have spoken elsewhere (Vallor 2010) of the way in which a contemporary account of the virtues can help us evaluate the moral significance of emerging information technologies. Rather than rehearse those arguments here, I would like simply to outline two important moral virtues that arguably rely upon caregiving practices for their development: *reciprocity* and *empathy*. These are hardly the only important moral virtues cultivated in caregiving practices—we might mention also patience, understanding, and charity, to name just a few. But the virtues I discuss below are sufficient to illustrate the larger points I wish to make—(1) that there are important goods for *caregivers* that are internal to caregiving practices, and that (2) the potential impact on these from the widespread use of carebots merits our careful attention. Following Coeckelbergh's (2009) invocation of a neo-Aristotelian way of thinking about roboethics, one that focuses on the "kind of moral habits and moral character we should develop," (220) let us investigate the way in which caregiving practices help to shape the moral character of caregivers, and the attendant ethical value that may give us reason to pause at the thought of transferring such practices to carebots.

4.1 Reciprocity

I have defined reciprocity elsewhere (Vallor 2011) as a primitive biological impulse which functions as the seed of human sociality, is the unifying feature of all forms of friendship and, with proper moral and cognitive/perceptual habituation, matures into a social virtue. The basic sociality of human beings is non-controversial. Despite our vulnerability to organic pathologies of reciprocity, such as autism, under normal conditions the human person is fundamentally oriented to others. Reciprocity (*antipeponthos*) is also at the core of Aristotle's understanding of ethical relations, both in its intimate and civic forms (1984, NE 1155b34); human persons, he believes, are naturally predisposed to social give and take (1984, NE 1155a).

But reciprocity is more than a natural impulse—it is also something we must cultivate as a virtue—for understanding how to reciprocate *well*, in the right ways, at the right times, and as appropriate to particular circumstances and people, is part of what it means to become a good person. Even Aristotle's notoriously inegalitarian accounts of unequal friendships and marriage, whatever their substantive errors, are meditations on reciprocity and how to ensure that human bonds are not destroyed by one member of a pair's inability to give benefits as well as receive them. It is the ethical maturation of reciprocity to which Aristotle refers when he claims that between friends, there is no need of justice (1984, NE 1155a26); for justice is a preventative and remedy for breakdowns and gaps in reciprocity; between complete friends of virtue, then, justice is a cure without a disease.

Lawrence Becker (1986) has asserted that reciprocity is in fact among the most fundamental virtues, one that is essential to creating and sustaining the primary goods that allow human lives to flourish (146–150). This further entails, according to Becker, an ethical obligation on our part to maintain social structures that perpetuate reciprocity as a virtue, and to modify social structures that damage it (163). This claim, if true, reinforces the need to consider the potential impact of the introduction of carebots on the human cultivation of reciprocity. For it is evident that reciprocity is cultivated, among other human contexts, in caregiving practices. This may not be immediately evident if we restrict our attention to the asymmetry between the vulnerable cared-for who receives care, and the one who gives it. Yet let us remember that the elderly parents we hand-feed once fed us, that the sister whose radiation sickness we comfort perhaps nursed us through a childhood flu, and even in institutional settings that nurses, doctors, and therapists are themselves, at times, the patients. We learn in our times of need that others are there for us, and as I wish to stress, we may learn in being there for others to trust that someday, others will be there for us. We also learn the importance of giving for the development of our own moral character, the way it facilitates other virtues, such as patience, empathy, and understanding. We learn in caregiving practices to see how reciprocity holds human relations together and allows other kinds of goods to flow across them. To surrender caregiving practices then, *even if* we stipulate that we will surrender them only to robots who can give what the other truly needs—potentially deprives us of an important opportunity to cultivate reciprocity in ourselves, and to understand its centrality to human bonds.

4.2 Empathy

I have defined this elsewhere (Vallor 2011) as an emotive/perceptual capacity that, like reciprocity, develops in most humans from a basic biological impulse, expresses itself fully in the highest forms of friendship, and, when properly cultivated and expressed, constitutes a virtue. Though Aristotle did not himself consider empathy a virtue (1984, NE 1106a4), I have claimed (2011) that it occupies a critical role in the life of virtue Aristotle describes *and* that Aristotle would likely himself have recognized it as a proper virtue had he enjoyed a better understanding of the integration of emotion and reason in human persons, as well as the essential nature of human vulnerability in grounding ethical human relations.

Empathy is an emotive/perceptual capacity to feel *with* another sentient being, to co-experience, in a significant way, the joys and sufferings of another. In empathy, one "grieves and rejoices with (*sunalgounta kai sunchaironta*)" another human being (1984, NE 1166a8). When cultivated as a virtue, it manifests itself as ability to receive such feelings in the appropriate sorts of circumstances and relationships, and to respond to the other in a manner that is also highly attuned and appropriate to these particulars.

Aristotle makes a distinction between those for whom we merely have goodwill, and those with whom we genuinely empathize in the sense described above. The latter relation, unlike the former, involves a genuine intensity of feeling (1984, NE 1166b34). Aristotle says that with those for whom we have goodwill, we wish them well but "we would not *do* anything with them" (1984, NE 1167a2) "nor take trouble for them" (1984, NE 1167a10). I wish to focus on the latter claim as significant. Though Aristotle certainly did not describe empathetic relations taking place in the contexts we currently associate with caregiving practices—he would view these as almost exclusively feminine practices alien to the masculine ethical ideal he attempts to cultivate—a virtue ethical perspective that applied appropriate correctives to his narrowly gendered ideal can take this last remark as suggestive. For those with whom I empathize, I will indeed "take trouble," and this occurs most notably in caring relations. In caregiving practices I take upon myself the trouble of recognizing, acknowledging, feeling, and responding appropriately to the suffering of another.

Sparrow and Sparrow (2006) stress the importance of empathy in caregiving as a means of showing the irreplaceability of human caregivers by carebots, claiming that "entities which do not understand the facts about human experience and mortality that make tears appropriate will be unable to fulfill this caring role" (154). I do not dispute this claim, but I wish to stress a different implication. It is primarily in fulfilling these caring roles that humans learn to practice and cultivate empathy as a virtue, and without such opportunities, the development of this virtue may be significantly impeded. For most of us, empathy is a quivering flame constantly vulnerable to being extinguished by apathy or cynicism, or our natural desire to protect ourselves from suffering. Many choose, often regrettably, to turn away from circumstances that might provoke an empathic response, and caregiving contexts, especially those that involve our physical presence to suffering, rank very highly among such circumstances. The concern, then, is that the availability of carebots as a substitute for human care may be very appealing to those who have not yet become comfortable with the cultivation of empathy in caregiving practices, leading them to abandon or greatly limit their exposure to such practices. As Sharkey and Sharkey (2010) note, the possibility of having a robot provide care, or offer a way to supply more remote forms of care, may allow us to feel that we are satisfying our caring obligations even when not present to suffering. I suggest that this is not only a potential loss to the cared-for, but to the caregiver.

When caregiving becomes a practice, and when this practice is done *well*, we become more accustomed to, even welcoming of, the emotional weight of empathy. Gradually, we may learn how to be restored rather than drained by sharing feeling. We can find solidarity in it—as Sparrow and Sparrow (2006) note, "Sometimes the only appropriate response to another's suffering is the acknowledgment that we too share these frailties, as for instance, when our friend's suffering moves us to tears" (154). It must be noted, however, that this is only possible when caregivers have sufficient resources and support for such practices—as we will emphasize below, caregiving in inadequate circumstances is likely to drain us of emotional power and starve empathic responses rather than cultivate them. As Borenstein and Pearson (2010) note, "merely because a human is assigned the task of providing care, an emotional bond does not necessarily form" (285). Empathy, then, like reciprocity, must be exquisitely and quite deliberately cultivated if it is to endure and thrive rather than wither on the vine.

The implication, then, is that these factors must be attended to when thinking about the moral dimensions of robot care. If the availability of robot care seduces us into abandoning caregiving practices before we have had sufficient opportunities to cultivate the virtues of empathy and reciprocity, among others, the impact upon our moral character, and society, could be quite devastating. On the other hand, if carebots provide forms of limited support that draw us further *into* caregiving practices, able to feel more and give more, freed from the fear that we will be crushed by unbearable burdens, then the moral effect on the character of caregivers could be remarkably positive.

5 Carebots and the Ethics of Care

Care ethics has been developed by a number of different feminist thinkers, including Gilligan (1982), Noddings (1984), Kittay (1999), and Held (2006). Like virtue ethics, care ethics takes its orientation from the goods internal to practices rather than external moral criteria. Yet unlike virtue ethics, it focuses on caring practices, and caring relations, as fundamental. Rather than holding the virtues themselves to be ethical desiderata, care ethics takes virtue to be merely an outgrowth of the caring relation that is the primary ethical good. And given its focus on caring relations and practices, we should expect to find in these perspectives important insights for the ethical implications of carebots.

For reasons of space, I will confine myself here to examining the perspective of Noddings on the ethical significance of caregiving practices. Noddings (1984) asserts that the caring relation is a first and foremost a natural one. Through our memories of being cared-for, we acquire our first and ultimate understanding of the good. The natural caring relation functions as regulative in human social behavior, motivating us to seek to restore the relation when it becomes deficient or disrupted. From the natural caring relation, then, which *in itself* is not a moral one, we acquire an ethical ideal, a notion of the "ethical self" that comes to function as a commitment to meet the other, and in particular those within our intimate kinship circle, from within the caring attitude. Importantly, Noddings notes that "This caring for self, for the *ethical self*, can emerge only from a caring for others" (14). Thus while the memory of being cared-for is required for the development of the ethical self as a regulative ideal, it is only through caring for others that I can care for that ideal—only through a commitment to caring practices can I ensure that my ethical self is sustained and enriched.

Noddings identifies two central criteria for a genuinely caring commitment—namely, *engrossment* and *motivational displacement* (16). Engrossment is an orientation to the reality of the other, as opposed to my own reality. It allows the needs of the other to be foregrounded in my field of awareness, rather than as the background of my own needs and desires, as they stand outside the genuinely caring attitude. Motivational displacement is the resultant feeling that "I must do something" (14), I must act to respond to the other's reality, "to reduce the pain, to fill the need, to actualize the dream" (14). These criteria, of course, create quite a high standard for genuine "caring". Here it is evident that I cannot "care" remotely, or though routine tasks that may "secure credit" (24) for caring from others without requiring my engrossment in the needs and sufferings of the other.[9] Thus if I transfer my obligations of care to a carebot, I have "done something" to care but not in a manner that will sustain *my* self as a caring being. And yet, Noddings notes that in all caring situations, there is a risk that "the one-caring will be overwhelmed by the responsibilities and duties of the task and that, as a result of being burdened, he or she will cease to care for the other..." (12).

She also tells us that the ethical ideal sustained through caring for others is itself subject to practical constraint (50). Since caring is concrete activity, I can only care within the limits of the possible, and a host of environmental, social, or institutional conditions are among those that can constrain and impoverish my ethical ideal. When we think, then, about the unrealistic emotional and physical burdens often put upon caregivers in contemporary society, it is hard to deny that the ethical ideal is *already* impoverished and constrained for many of us. Engrossment in the reality of the other is not possible when I have had so little sleep that I am at the limits of human capacity just to go through the routine motions of care—changing the bedpan, remembering to give medication, cleaning the wound. What more can I give? What reason can we have, then, to deny that the introduction of carebots in some contexts, and for some purposes, might be *essential* to sustaining the caregivers, so that they need not face the degradation of their own moral being? It is also worth noting that on Noddings' view, caring relations need to be reciprocal if caregivers are going to be sustained. That is, caregivers need some acknowledgment of the relation on the part of the cared-for, and though one can continue to care as an ethical choice in the absence of such reciprocity, it is far more difficult to remain engrossed in responding to the reality of another who will not, or cannot acknowledge the value of your response. Can carebots help here? Borenstein and Pearson (2010) note that by providing cared-fors with greater independence, their own capabilities to respond to caregivers may be enhanced, promoting a "care recipient's ability to 'love those who love and care for us'" (282).

Consider another scenario, however—the one in which carebots appear to us as liberation *from* care rather than liberation *to* care. We have seen the emphasis on the idea that care must be a choice, that the freedom and autonomy of potential caregivers is of paramount importance. This seems to presuppose, however, that some of us will choose *not* to give care, and that perhaps this choice ought not to be regarded as a bad thing, as long as there are carebot resources available to meet the needs of cared-fors. As Noddings notes, we *are* free to reject the impulse to care (51). We have other desires, and other needs with which caring practices compete. And if we are exhorted by carebot marketers or nursing home managers to "let the robot do its job", to surrender caring tasks to a being which will surely not forget a pill, will not lose patience, will not let a loved one fall, we might be mistakenly led to think that both we and our cared-for are best served by this surrender. Yet Noddings is right, I think, that "I enslave myself to a particularly unhappy task when I make this choice. As I chop away at the chains that bind me to loved others,

---

[9] It is worth noting, with reference to our discussions of virtue above, that Noddings identifies engrossment with the capacity for empathy as "feeling with" (30).

asserting my freedom, I move into a wilderness of strangers and loneliness, leaving behind all who cared for me and even, perhaps, my own self" (51).

Thus it is one thing to choose *when* to care, and for whom, and to ensure that I receive the support I need in caring to sustain my self emotionally, physically, and morally. Carebots might help us do better with all of this, and in the absence of alternative solutions, it might be ethically necessary to employ them in certain contexts. But how many of us be seduced by the possibility opened up by carebots of *not* caring, at least not in the sort of intimate, direct relations that presuppose engrossment in the reality of the suffering or vulnerable other? And how can we protect ourselves against such a seduction unless we attend more carefully to the ways in which engagement in caring practices is critical for our own well-being, and not just that of those who need us?

6 Carebots and Human Capabilities

Finally, let us consider the capabilities approach. Capabilities approaches, rooted in the economic and political philosophies of Sen (1999) and Nussbaum (2000, 2006), presuppose that considerations of justice require attention to more than just the availability and social distribution of various external goods, resources, or utilities. Rather, thinking about justice requires attention to basic human capabilities and/or functionings, understood as goods realized in human living through certain kinds of activities and practices. As realization of these capabilities is critical to human flourishing, the requirements of justice demand that we not sacrifice their realization by individuals merely in order to maximize overall utility. While the capabilities approach was conceptualized to respond to deficiencies in other theories of justice (e.g., Rawls, utilitarianism), it can also be used more broadly in the context of ethics, in order to help us conceptualize the good life and the means of its realization.

And in fact it is used by philosophers who weigh the ethical implications of carebots—Borenstein and Pearson (2010, 278) employ this approach in order to show that, for example, carebots might allow caregivers to meet their moral obligations without destroying their own emotional health (a loss that might undermine, for example, their human capacities for love, play or friendship). They also note that the capabilities approach helps us identify how carebots should *not* be used—for example, in ways that would deprive caregivers or cared-fors of control over their environment (279). Consider in this regard a carebot that secured the physical well-being of nursing home patients by rigidly confining their movements to prevent any risk of a fall—leaving them socially isolated and deprived of autonomy. Coeckelbergh (2010) also references the capabilities approach as a model for ethical reflection on carebots. Though he does not take the capabilities approach to be exhaustive of the moral issues at stake with carebots (186), he suggests we use it as one way to evaluate whether any given practice of caregiving qualifies as "good care" (185).

I suggest we employ the capabilities approach in one further respect: to help us to conceptualize the goods internal to caring practices for *caregivers*, goods that might be lost should a person choose to surrender some or all of such practices to a carebot. How should we proceed? Following Coeckelbergh (2010) and Borenstein and

Pearson (2010), let us refer to the list of central human capabilities offered by Nussbaum (2006), without suggesting that this list is necessarily complete or definitive. The list identifies and defines the capabilities of life, bodily health, bodily integrity, sensation, imagination and thought, emotions, practical reason, affiliation, [relatedness to] other species, play, and control over one's environment (76–77). While the goods internal to caring practices might be incorporated in all of these capabilities, I wish to focus on just three, *affiliation*, *practical reason*, and *emotion*, which I will argue each represent capacities enhanced for *caregivers* within caring practices. Nussbaum describes these capacities as follows:

6.1 Affiliation

(Part A) "Being able to live with and toward others, to recognize and show concern for other human beings, to engage in various forms of social interaction; to be able to imagine the situation of another."

Nussbaum (2006, 77)

6.2 Practical Reason

"Being able to form a conception of the good and to engage in critical reflection about the planning of one's life."

Nussbaum (2006, 77)

6.3 Emotions

"Being able to have attachments to things and people outside ourselves; to love those who love and care for us, to grieve at their absence; in general, to love, to grieve, to experience longing, gratitude and justified anger. Not having one's emotional development blighted by fear and anxiety. (Supporting this capability means supporting forms of human association that can be shown to be crucial in their development)."

Nussbaum (2006, 76–77)

I assert that caring practices, and more specifically, engagement in such practices *in the role of caregiver as well as one cared-for*, are critical for the development of these capabilities, for both men and women. Let me begin with affiliation. The capability of affiliation described above entails the ability "to recognize and show concern for other human beings." How do we develop this ability? What human practice is most critical here? I would argue that caregiving practices, while not exhaustive of those through which we develop the capacity of affiliation, are essential to its full development. Certainly I may show concern for the general "other" by means other than giving care—for example, by giving money to charity, through social and political advocacy, or through throwing myself into intellectual and practical activities aimed at addressing important problems that others face. But it is through directly giving care that I learn to recognize in concrete others those expressions of need and desire that first motivate the attitude of "concern" on my part. Even as a child I come to this attitude through caring practices—by looking after a younger sibling, feeding and watering a pet, fetching Kleenex for a flu-afflicted parent, or patiently listening to the problems of a friend. It is through these interactions that I come to understand what human "affiliation" means, as a way of being with others that is responsive and mutually supportive. It is primarily through repeated concrete exposures to human need, and taking up the practice of responding to need with care, that I develop the abstractive capacity to "imagine the situation of another" outside my own intimate circle of care.

Let us next consider the cultivation of practical reason. Here caregiving also plays a central role. First, it can be argued that no "conception of the good" is complete that does not recognize caring relations as a basic human good.[10] That is to say, we would regard a person as prudentially deficient whose deliberations about action were never informed by an appreciation for the goodness of one human caring for another. Nussbaum herself tells us that we must appreciate "the importance of care as a social primary good" (2006, 2). Now, we might think that such an appreciation could be acquired simply from the experience of being cared *for*, and that the practice of *giving* care, which I have emphasized, is inessential here. I reject this claim, however, for reasons that should be apparent from the above considerations, and which will be reinforced by the discussion of emotional capability that follows. It is through *giving* care that I learn to recognize signs of need in another, that I become habituated to responding to need, and that I come to fully appreciate the goodness of the caring *role* and the importance (and challenge) of caring *well*. As cared *for* I can see the goodness of having my needs met—but I do not fully grasp the goodness of meeting needs until I have taken on that role myself and experienced its challenges and rewards. Without engaging in caregiving practices my grasp of the goodness of caring relations remains impoverished and one-sided.

Secondly, it is at least in part through caregiving practices that I become capable of the prudent "planning of one's life" that Nussbaum speaks of. How can I possibly plan my life well unless I am aware of the vulnerabilities of sickness, injury, age, or other conditions of life to which I am, sooner or later, going to be exposed? Nor is a merely intellectual grasp of these conditions sufficient for planning. It is not enough to know that I need, for example, health and disability insurance, or adequate savings, or an advanced health care directive on file. I also need to plan for the emotional, physical, social challenges that these conditions may bring, to prepare myself through reflection and conversations with loved ones for how we can together bear those conditions with dignity and grace, if possible. And it is primarily through witnessing others go through those challenges, and more specifically, learning through caregiving practices *how* humans find, or lose, dignity in their vulnerability, that I become able to reflect deeply on how I might secure the dignity of my own future. Thus the capability of prudential reason seems to require for its *full* realization a deep acquaintance with caring relations, not only from the standpoint of the cared-for but also from the standpoint of a caregiver.

---

[10] Noddings argues for a stronger view that caring relations are the first and ultimate source of our conception of the good (1984, 99). Even if this view is rejected as too strong (which I do not assert), the weaker claim that an appreciation of caring relations is *essential* to any adequate conception of the good has intuitive plausibility.

Finally, how are the emotional capabilities Nussbaum speaks of realized through the practice of giving care? To see this, it may help to reflect first upon the moral psychology of the sort of person who never becomes comfortable or skilled in caregiving. This person refuses to take on a caring role out of fear of emotional exposure and anxiety about becoming "too attached" to the being who is cared-for, too vulnerable to the possibility of their loss and the attendant grief, too pained by unrequited longing for the cared-for to be made whole; or too angry, frustrated, or exhausted by the burdens that care entails. Most of us, I think, will find some or all of these fears recognizable—we have seen them expressed by others, or had such fears ourselves. In such a condition, one either takes on the caring role in spite of such fears (the reasons why need not matter here) or one abdicates that role—leaving it empty *or* leaving it to be filled by someone else. What difference does this choice make? A great deal of difference. The person who entirely succumbs to these fears lives as an absent, "checked-out" parent, especially during a child's illness or disability. They find excuses to avoid sick friends recuperating at home. They cite a hatred of hospitals as a reason to limit visits to dying parents. They become withdrawn and distant with a terminally ill spouse. They may attempt to compensate by "assisting" in ways that do not require much emotional exposure—working overtime to help with medical costs, sending flowers and food baskets by courier, buying an expensive van with a wheelchair lift. But as long as they continue to avoid care*giving*, that is, offering the full range of their emotional and physical being for the good of one who is suffering, dependent, or vulnerable, their capacity for a rich emotional life remains stunted and impoverished.

It is not her fears that consign such a sad person to her fate—we likely all have such fears. It is the abdication of caring practices that engraves those fears as the dominant feature of such a person's emotional landscape, crowding out all other possibilities. Meanwhile, the person who takes up caring practices—even, perhaps, out of guilt or in response to social expectation—is forced to confront those fears with action. Caring actions will indeed bring precisely what the carer fears—grief, longing, anger, emotional exhaustion. But they can also allow those emotions to be gradually incorporated into a whole life, into living and sustaining attachments with other things and persons. Caring practices bring other emotions too—such as the gratitude Nussbaum speaks of, whether it be for that one day the clouds of dementia lifted a little and a parent's recognition reappeared, or the smile of an arthritic spouse after a gentle massage, or gratitude for the child who begins to show maturity, independence, and her own capacity to care.

Of course, caring practices in the absence of adequate resources and support can have a powerfully destructive impact on our emotional, prudential, and social capabilities—the unmitigated stress, anxiety, physical, and mental exhaustion can result in emotional withdrawal, a gradual numbing of affect, or extreme emotional volatility—any or all of which can result in fractured relations with cared-fors and others, severe depression, motivational apathy, impaired judgment, even complete emotional collapse. Borenstein and Pearson (2010) are correct that caregivers "should be able to fulfill their obligations without simultaneously relinquishing their own prospects for human flourishing" (278). Yet by not attending explicitly to the way in which sustaining caring practices may be an integral part of that flourishing, the potential of carebots to support human caregivers and relieve their burdens appears unequivocal, without the concern that must complement it—that humans may (deliberately or unwittingly) surrender opportunities for giving care that are critical to their *own* development and well-being. Thus the proper question is not whether carebots are inherently destructive or supportive of the development of central capabilities in caregivers. The proper question is, *in what particular, concrete contexts will the use of carebots enhance caregivers' capabilities by allowing them to sustain caring practices, and in what contexts will their use encourage us to gradually surrender those practices, perhaps blind to the way in which such a surrender may negatively impact our fundamental capabilities*?

Repeatedly, discussion of these issues has skewed toward the impact on cared-fors rather than caregivers. Where the impact of carebots on caregivers is made central, it is entirely in terms of the potential benefits of being liberated from caregiving obligations, without consideration of the attendant risks of such liberation (Borenstein & Pearson 2010). For example, consider the dispute about whether carebots will in fact come to gradually replace human caregivers, or whether by freeing us from routine tasks they will "make room for more and deeper human care" (Coeckebergh 2009, 183), helping us to "care better". This is an important question, and the answer depends upon a number of social, economic, cultural, political, and technical variables that I will not attempt to enumerate. Some authors' predictions about the future of carebot–human relations are distinctly pessimistic (Sparrow & Sparrow 2006), others more cautiously optimistic (Sharkey & Sharkey 2010; Borenstein & Pearson 2010). But in either scenario, the risks and benefits are weighed in a manner that ignores or minimizes the potential value of caregiving to caregivers. For example, there is a consensus that "the elderly need contact with fellow human beings," and that if the use of carebots led to depriving them of human contact, this would be ethically problematic (Sharkey & Sharkey 2010; Borenstein & Pearson 2010). But no one asks whether we (potential caregivers) need contact with the elderly! Likewise, if using carebots to help us "care better" is seen as an ethically significant gain, it is seen as a gain in terms of the quality of care given to cared-fors—they receive "deep" rather than "shallow" care, which is clearly better for them (Coeckebergh 2009). But it has not been asked whether being freed to "care better" might also be better for caregivers, by allowing them greater access to caregiving practices that sustain or enrich their emotional, prudential, and affiliative capabilities.

This impacts how we think about assessing the ethics of particular carebot uses—we focus on strategies that emphasize asking the elderly what levels of human/robot interaction, respectively, they prefer (Sparrow & Sparrow 2006; Sharkey & Sharkey 2010). The preferences of cared-fors are obviously critical here, and must be consulted as part of any ethical approach. But by ignoring the value of caregiving practices for caregivers, we open ourselves to a disturbing (though perhaps unlikely) scenario. Imagine a century from now that carebots have been developed which are not only cheap, safe, and skilled, but good at keeping the elderly, children, and convalescent persons *happy*—amused, comforted, and distracted from their needs or pains. Imagine that a great many people in this future would *prefer* to be cared for by carebots than human caregivers, who still at times become impatient, dull, or angry in their company. How would widely respecting those preferences, and greatly decreasing human involvement in caregiving practices, impact the development of the central human capabilities we have discussed here? How would this impact the flourishing of human beings in the long run? Alternatively, imagine a scenario where the elderly or otherwise vulnerable prefer for aesthetic or tradition-bound reasons *not* to have robots involved in their care, but empirically we find that (1) limited uses of carebots for routine, physically demanding or tedious tasks free caregivers to sustain and enrich their own central human capabilities to a significant degree, and (2) that such uses have no negative impact on the central human capabilities of cared-fors. Is it *clear* that the preferences of cared-fors would be ethically definitive in such a case?

There is yet another question we must pursue, related to the way in which new technologies can radically change our contexts of practice in such a way as to alter the meaning of the goods internal to those practices. For example, Coeckebergh (2009) tells us that interactions with robots may not only support the development of a central human capability (such as play) but may also "redefine that capability" (220). This requires careful reflection. What would it mean to "redefine" the various capabilities that come together in, and are cultivated through, caregiving practices? How plastic is care? Can we stretch the definition of our capabilities in such a way that we could accept the marketing of a carebot as "enabling friends and family to care from afar" (Sharkey & Sharkey 2010)? Could the goods realized in traditional caring practices also be realized through "remote caring"? Or does this stretch the meaning of a caring practice beyond reason, detaching it from the sort of goods or capabilities that sustain human flourishing? My aim in this paper is simply to shed light on the importance of asking these questions.

Deciding whether our primary moral concern about carebots and caregivers should be the potential loss of opportunities to cultivate virtues like reciprocity and empathy, or potential constraint and impoverishment of the ethical ideal cultivated in the caring relation, or impairment of the capabilities of affiliation, practical reason, and emotion cultivated in caring practices, is a task for moral philosophers. This may not be their most important task, however—partly because these three perspectives, by focusing on the goods internal to caring practices, are already deeply complementary in many key respects. But beyond this, it is critical that the task of perfecting our conceptual framework not occlude the concrete, empirical developments upon which we hope our framework will shed light, and which give our attempts at theoretical precision their sole meaning. It has yet to become clear what carebots will be able to offer us beyond their current incubatory stage, or when they will become a genuine choice for potential caregivers as opposed to a matter of fascinating speculation for philosophers of technology. Nor do we know precisely which cultural, economic, demographic, and psychological conditions will be operative to influence caregivers' motivations and choices if and when that day comes. It will be important for philosophers to attend closely to future developments in these respects, and to adjust their range and degree of moral concern accordingly. In the meantime, philosophers can make a distinctive contribution to preparing the moral community for such choices by encouraging reflection on the value of caring practices, not only for those cared for, but for caregivers as well.

References

Aristotle (1984). *The complete works of Aristotle: Revised Oxford translation.* J. Barnes, Ed. Princeton: Princeton University Press.
Becker, L. C. (1986). *Reciprocity.* Chicago: University of Chicago Press.
Borenstein, J., & Pearson, Y. (2010). Robot caregivers: harbingers of expanded freedom for all? *Ethics and Information Technology, 12*(3), 277–288.
Coeckelbergh, M. (2010). Health care, capabilities and AI assistive technologies. *Ethical Theory and Moral Practice, 13*(2), 181–190.
Coeckebergh, M. (2009). Personal robots, appearance and human good: A methodological reflection on roboethics. *International Journal of Social Robotics, 1*(3), 217–221.
Decker, M. (2008). Caregiving robots and ethical reflection: The perspective of interdisciplinary technology assessment. *AI & Society, 22*(3), 315–330.
Gilligan, C. (1982). *In a different voice.* Cambridge: Harvard University Press.
Held, V. (2006). *The ethics of care: Personal, political, global.* Oxford: Oxford University Press.
Kittay, E. F. (1999). *Love's labor: Essays on women, equality, and dependency.* New York: Routledge.
Noddings, N. (1984). *Caring: A feminine approach to ethics and moral education.* Berkeley: University of California Press.
Nussbaum, M. (2006). *Frontiers of justice: Disability, nationality, species membership.* Cambridge: Harvard University Press.
Nussbaum, M. (2000). *Women and human development: The capabilities approach.* Cambridge: Cambridge University Press.
Sen, A. (1999). *Development as freedom.* New York: Knopf.
Sharkey, A., & Sharkey, N. (2010). Granny and the robots: Ethical issues in robot care for the elderly. *Ethics and Information Technology.* doi:10.1007/s10676-010-9234-6.
Singer, P. W. (2009). *Wired for war: The robotics revolution and conflict in the 21st century.* New York: Penguin Press.
Sparrow, R., & Sparrow, L. (2006). In the hands of machines? The future of aged care. *Minds and Machines, 16*(2), 141–161.
Tergesen, A., & Inada, M. (2010). It's not a stuffed animal, it's a $6000 medical device: Paro the robot seal aims to comfort the elderly, but is it ethical? *The Wall Street Journal,* June 21, 2010, http://online.wsj.com/article/SB10001424052748704463504575301051844937276.html. Accessed January 8, 2011.
Turkle, S. (2006). *A nascent robotics culture: New complicities for companionship.* July 2006: AAAI Technical Report Series, http://web.mit.edu/sturkle/www/pdfsforstwebpage/st_nascentroboticsculture.pdf. Accessed January 8, 2011.
Vallor, S. (2010). Social networking technology and the virtues. *Ethics and Information Technology, 12*(2), 157–170.
Vallor, S. (2011). Flourishing on Facebook: Virtue friendship and new social media. *Ethics and Information Technology.* doi:10.1007/s10676-010-9262-2.

13

The crying shame of robot nannies
An ethical appraisal

Noel Sharkey & Amanda Sharkey

Childcare robots are being manufactured and developed with the long term aim of creating surrogate carers. While total childcare is not yet being promoted, there are indications that it is 'on the cards'. We examine recent research and developments in childcare robots and speculate on progress over the coming years by extrapolating from other ongoing robotics work. Our main aim is to raise ethical questions about the part or full-time replacement of primary carers. The questions are about human rights, privacy, robot use of restraint, deception of children and accountability. But the most pressing ethical issues throughout the paper concern the consequences for the psychological and emotional wellbeing of children. We set these in the context of the child development literature on the pathology and causes of attachment disorders. We then consider the adequacy of current legislation and international ethical guidelines on the protection of children from the overuse of robot care.

> *Who's to say that at some distant moment there might be an assembly line producing a gentle product in the form of a grandmother – whose stock in trade is love.* From *I Sing the Body Electric,* Twilight Zone, Series 3, Episode 35, 1960

1. Introduction

A babysitter/companion on call round the clock to supervise and entertain the kids is the dream of many working parents. Now robot manufacturers in South Korea and Japan are racing to fulfil that dream with affordable robot "nannies". These currently have game playing, quizzes, speech recognition, face recognition and limited conversation to capture the preschool child's interest and attention. Their mobility and semi-autonomous function combined with facilities for visual and auditory monitoring are designed to keep the child from harm. Most are prohibitively expensive at present but prices are falling and some cheap versions are already becoming available.

Children love robots as indicated by the numbers taking part in robot competitions worldwide. Even in a war zone, when bomb disposal robots entered a village in Iraq, they were swamped with excited children (Personal communication, Ronald C. Arkin, 2008). There is a growing body of research showing positive interactions between children and robots in the home (e.g. Turkle et al. 2006 a,b), and in the classroom (e.g. Tanaka et al. 2007; Kanda et al., 2009). Robots have also been shown to be useful in therapeutic applications for children (e.g. Shibata et al., 2001, Dautenhahn, 2003; Dautenhahn & Werry, 2004; Marti et al., 2005; Liu et al., 2008). The natural engagement value of robots makes them a great motivational tool for education in science and engineering. We raise no ethical objections to the use of robots for such purposes or with their use in experimental research or even as toys.

Our concerns are about the evolving use of childcare robots and the potential dangers they pose for children and society (Sharkey, 2008a). By extrapolating from ongoing developments in other areas of robotics, we can get a reasonable idea of the facilities that childcare robots could have available to them over the next 5 to 15 years. We make no claims about the precision of the time estimate as this has proved to be almost impossible for robotics and AI developments (Sharkey, 2008b). Our approach is conservative and explicitly avoids entanglement with issues about strong AI and super smart machines. Nonetheless, it may not be long before robots can be used to keep children safe and maintain their physical needs for as long as required.

To be commercially viable, robot carers will need to enable considerably longer parent/carer absences than can be obtained from leaving a child sitting in front of a video or television programme. Television and video have long been used by busy parents to entertain children for short periods of time. But they are a passive form of entertainment and children get fidgety after a while and become unsafe. They need to be monitored with frequent "pop-ins" or the parent has to work in the same room as the child and suffer the same DVDs while trying to concentrate. The robot can extend the length of parent absences by keeping the child safe from harm, keeping her entertained and, ideally, by creating a relationship bond between child and robot (Turkle et al. 2006b).

We start with a simple example, the Hello Kitty Robot, which parents are already beginning to use if the marketing website is to be believed. It gives an idea of how these robots are already getting a 'foot in the door.' Even for such a robotically simple and relatively cheap robot, the marketing claims are that, "This is a perfect robot for whoever does not have a lot time [sic] to stay with their child." (Hello Kitty website). Although Hello Kitty is not mobile, it creates a lifelike appearance by autonomously moving its head to four angles and moving its arms. What gives it an edge is that it can recognise voices and faces so that it can call children by their names. It has a stereo CCD camera that allows it to track faces and it can chat. For children this may be enough to create the illusion that it has mental states (Melson et al. in press b).

Busy working parents might be tempted to think that a robot nanny could provide constant supervision, entertainment and companionship for their children. Some of the customer reviews of the "Hello Kitty Robot", on the internet made interesting reading. These have now been removed but we kept a copy: (there is also a copy of some of the comments at (Bittybobo))

- Since we have invited Hello Kitty (Kiki-as my son calls her), life has been so much easier for everyone. My daughter is no longer the built in babysitter for my son. Hello Kitty does all the work. I always set Kiki to parent mode, and she does a great job. My two year old is already learning words in Japanese, German, and French.
- As a single executive mom, I spend most of my home time on the computer and phone and so don't have a lot of chance to interact with my 18-month old. The HK robot does a great job of talking to her and keeping her occupied for hours on end. Last night I came into the playroom around 1AM to find her, still dressed (in her Hello Kitty regalia of course), curled sound asleep around the big plastic Kitty Robo. How cute! (And, how nice not to hear those heartbreaking lonely cries while I'm trying to get some work done.)
- Robo Kitty is like another parent at our house. She talks so kindly to my little boy. He's even starting to speak with her accent! It's so cute. Robo Kitty puts Max to sleep, watches TV with him, watches him in the bath, listens to him read. It's amazing, like a best friend, or as Max says "Kitty Mommy!" Now when I'm working from home I don't have to worry about Max asking a bunch of questions or wanting to play or having to read to him. He hardly even talks to me at all! He no longer asks to go to the park or the zoo – being a parent has NEVER been so easy! Thank you Robo Kitty!"

We are not presenting these anecdotal examples as rigorous evidence of how a simple robot like Hello Kitty will generally be used. Other parents commenting on the website were highly critical about these mothers being cold or undeserving of having children. We cannot authenticate these comments. Nonetheless this example provides a worrying indication of what might be and what we need to be prepared for. Perhaps it is only a small minority of parents who would rely on such a simple robot to mind their pre-school children. But as more sophisticated robots of the type we describe later become affordable, their use could increase dramatically.

What follows is an examination of the present day and near-future childcare robots and a discussion of potential ethical dangers that arise from their extended use in caring for babies and young children. Our biggest concern is about what will happen if children are left in the regular or near-exclusive care of robots. First we briefly examine how near-future robots will be able to keep children safe from harm and what ethical issues this may raise. Then we make the case, from the results of research on child–robot interaction, that children can and will form pseudo-relationships with robots and attribute mental states and sociality to them. Children's natural anthropomorphism could be amplified and exploited by the addition of a number of methods being developed through research on human–robot interaction, for example, in the areas of conversation, speech, touch, face and emotion recognition. We draw upon evidence from the psychological literature on attachment and neglect to look at the possible emotional harm that could result from children spending too much time exclusively in the company of mechanical minders.

In the final section, we turn to current legislation and international ethical guidelines on the care and rights of children to find out what protections they have from sustained or exclusive robot care. Our aim is not to offer answers or solutions to the ethical dangers but to inform and raise the issues for discussion. It is up to society, the legislature and the professional bodies to provide codes of conduct to deal with future robot childcare.

2. Keeping children from physical harm

An essential ingredient for consumer trust in childcare robots is that they keep children safe from physical harm. The main method used at present is mobile monitoring. For example, the PaPeRo Personal Partner Robot by NEC (Yoshiro et al., 2005) uses cameras in the robot's 'eyes' to transmit images of the child to a window on the parent-carer's computer or to their mobile phone. The carer can then see and control the robot to find the child if she moves out of sight. This is like having a portable baby monitor but it defeats the purpose of mechanical care. There is little point in having a childcare robot if the busy carer has to continuously monitor their child's behaviour. For costly childcare robots to be attractive to consumers or institutions, they will need to have sufficient autonomous functioning to free the carer's time and call upon them only in unusual circumstances.

As a start in this direction, some childcare robots keep track of the location of children and alert adults if they move outside of a preset perimeter. The PaPeRo robot comes with PaPeSacks, each containing an ultrasonic sensor with a unique signature. The robot can then detect the exact whereabouts of several children at the same time and know which child is which. Similarly the Japanese Tmsuk robot uses radio frequency identification tags. But more naturalistic methods of tracking are now being developed that will eventually find their way into the care robot market. For example, Lopes et al., (2009) have developed a method for tracking people in a range of environments and lighting conditions without the use of sensor beacons. This means that the robot will be able to follow a child outside and alert carers of her location or encourage and guide her back into the home.

We may also see the integration of care robots with other home sensing and monitoring systems. There is considerable research on the development of smart sensing homes for the frail elderly. These can monitor a range of potentially dangerous activities such as leaving on water taps or cookers. They can monitor a person getting out of bed and wandering. They can prompt the person with a voice to remind them to go to the toilet and switch the toilet light on for them (Orpwood et al., 2008). Vision systems can detect a fall and other sensors can determine if assistance is required (Toronto Rehabilitation Unit Annual Report 2008, 40-41). Simple versions of such systems could be adapted for use in robot childcare.

One ethical issue arising from such close monitoring is that every child has a right to privacy under Articles 16 and 40 of the UN Convention on Child Rights. It is fine for parents to listen out for their children with a baby alarm. Parents also frequently video and photograph their young children's activities. In most circumstances legal guardians have the right to full disclosure regarding a very young child. However, there is something different about an adult being present to observe a child and a child being covertly monitored when she thinks that she is alone with her robot friend.

Without making too much of this issue, when a child discusses something with an adult, she may expect the discussion will be reported to a third party – especially her parents. But sometimes conversations about issues concerning the parents, such as abuse or injustice, should be treated in confidence. A robot might not be able to keep such confidences from the parents before reporting the incident to the appropriate authorities. Moreover, when a child has a discussion with a peer friend (or robot friend) they may be doing so in the belief that it is in confidence.

With the massive memory hard drives available today, it would be possible to record a child's entire life. This gives rise to concerns about whether such close invigilation is acceptable. Important questions need to be discussed here such as, who will be allowed access to the recordings? Will the child, in later life have the right to destroy the records?

Privacy aside, an additional way to increase autonomous supervision would be to allow customisation of home maps so that a robot could encode danger areas. This could be extended with better vision systems that could detect potentially dangerous activities like climbing on furniture to jump. A robot could make a first pass at warning a child to stop doing or engaging in a potentially dangerous activity in the same way that smart sensing homes do for the elderly. But there is another ethical problem lurking in the shadows here.

If a robot could predict a dangerous situation, it could also be programmed to autonomously take steps to physically prevent it rather than merely warn. For example, it could take matches from the hands of a child, get between a child and a danger area such as a fire, or even restrain a child from carrying out a dangerous or naughty action. However, restraining a child to avoid harm could be a slippery slope towards authoritarian robotics. We must ask how acceptable it is for a robot to make decisions that can affect the lives of our children by constraining their behaviour.

It would be easy to construct scenarios where it would be hard to deny such robot action. For example, if a child was about to run across the road into heavy oncoming traffic and a robot could stop her, should it not do so? The problem is in trusting the classifications and sensing systems of a robot to determine what is a dangerous activity. As an extreme case, imagine a child having doughnuts taken from her because the robot wanted to prevent her from becoming obese. There are many discussions to be had over the extremes of robots blocking human actions and where to draw the line (c.f. Wallach & Allen, 2009).

Another ethically tricky area of autonomous care is in the development of robots to do what some might consider to be the 'dull and dirty' work of childcare. They may eventually be able to carry out tasks such as changing nappies, bathing, dressing, feeding and adjusting clothing and bedding to accord with temperature changes. Certainly, robot facilities like these are being thought about and developed in Japan with an eye to caring for their aging population (Sharkey and Sharkey, in press). Performing such duties would allow lengthier absences from human carers but could be a step too far in childcare robotics; care routines are an important component in fostering the relationship between a child and her primary carer to promote healthy mental development. If we are not careful to lay out guidelines, robots performing care routines could exacerbate some of the problems we discuss later in the section on the psychological harm of robot childcare.

Carers who wish to leave their charges at home alone with a robot will need to be concerned about the possibility of intruders entering the home for nefarious purposes. Security is a major growth area in robotics and care robots could incorporate some of the features being developed. For example, the Seoul authorities, in combination with the private security company KT Telecop use a school guard robot, OFRO, to watch out for potential paedophiles in school playgrounds. It can autonomously patrol areas on pre-programmed routes and alert teachers if it spots a person over a specific height, (Metro, May 31st 2007; The Korea Times,

May, 30th 2007). If we combine this with face recognition, already available on some of the care robots, they could stop adults to determine if they were on the trusted list and alert the authorities if necessary.

3. Relating to the inanimate

Another essential ingredient for consumer trust in childcare robots is that children must want to spend time with them. Research has already begun to find ways to sustain long term relationships between humans and robots (e.g. Kanda et al., 2004; Mitsunaga, et al., 2006; Mavridis et al., 2009). Care robots are already being designed to exploit both natural human anthropomorphism and the bond that children can form with personal toys. The attribution of animacy to objects possessing certain key characteristics is part of being human (Sharkey & Sharkey 2006). Puppeteers have understood and exploited the willing or unconscious "suspension of disbelief" for thousands of years as have modern animators and cartoonists. The characteristics they exploit can be visual, behavioural or auditory. Even the vaguest suggestion of a face brings an object to life; something as simple as a sock can be moved in a way that makes it into a cute creature (Rocks, et al., in press). Robots, by comparison, can greatly amplify anthropomorphic and zoomorphic tendencies. Unlike other objects, a robot can combine visual, movement and auditory features to present a powerful illusion of animacy without a controller being present.

Young children invest emotionally in their most treasured cuddly toy. They may have difficulty sleeping without it and become distraught if it gets misplaced or lost. The child can be asked, "What does Bear think about X?". Bear can reply through the child's voice or by whispering in the child's ear or by simply nodding or waving an arm. This is a part of normal childhood play and pretence that requires imagination, with the child in control of the action. As Cayton (2006 p. 283) points out, "When children play make-believe, 'let's pretend' games they absolutely know it is pretend… Real play is a conscious activity. Ask a child who is playing with a doll what they are doing and they may tell you matter-of-factly that they are going to the shops or that the doll is sick but they will also tell you that they are playing."

A puppet, on the other hand, is outside of the child's control and less imagination and pretence is required. But a child left alone with a puppet soon realises the illusion and the puppet can then be classified in the 'let's pretend' category. The difference with a robot is that it can still operate and act when no one is standing next to it or even when the child is alone with it. This could create physical, social and relational anthropomorphism that a child might perceive as 'real' and not illusion.

There is a gradually accumulating body of evidence that children of all ages can come to believe in the reality of a relationship they have with robots. Melson et al. (in press a) report three studies that employed Sony's robotic dog AIBO: (i) a content analysis of 6,438 internet discussion forum postings by 182 AIBO owners; (ii) observations and interviews with 80 preschoolers during a 40-minute play period with AIBO and a stuffed dog; and (iii) observations and interviews with 72 school-age children from 7 to 15 years old who played with both AIBO and a living dog. The majority of participants across all three studies viewed AIBO as a social companion: both the preschool and older children said that AIBO "could be their friend, that they could be a friend to AIBO, and that if they were sad, they would like to be in the company of AIBO".

In a related study, Kahn et al. (2006) looked at the responses of two groups of preschoolers – 34–50 months and 58–74 months, in a comparison between an AIBO and a stuffed dog. They found that a quarter of the children, in verbal evaluations, accorded animacy to the AIBO, half accorded biological properties and around two-thirds accorded mental states. But a very similar pattern of evaluation was found for the stuffed dog. The interesting thing here is that the children's behaviour towards the two artefacts did not fit with their evaluations. Based on 2,360 coded behavioural interactions, the children exhibited significantly more apprehensive and reciprocal behaviours with the AIBO whilst they more often mistreated the stuffed dog (184 occurrences versus 39 for AIBO). Thus the verbal reports were not as reliable an indicator as the behavioural observations. The robot was treated more like a living creature than the stuffed dog.

Children can also form relationships with humanoid robots. Tanaka et al. (2007) placed a "state-of-the-art" social robot (QRIO) in a day care centre for 5 months. They report that children between 10 and 24 months *bonded* with the robot in a way that was significantly greater than their bonding with a teddy bear. Tanaka et al. claim that the toddlers came to treat the robot as one of their peers. They looked after it, played with it, and hugged it. They touched the robot more than they hugged or touched a static toy robot, or a teddy bear. The researchers related the children's relationship with the robot to Harlow's (1958) "affectional responses". They claimed that "long-term bonding and socialization occurred between toddlers and a state-of-the-art social robot" (Tanaka et al., 2007 p. 17957).

Turkle et al. (2006a) report a number of individual case studies that attest to children's willingness to become attached to robots. For example, one of the case studies was of a ten year old girl, Melanie who was allowed to take home a robotic doll, "My Real Baby", and an AIBO for several weeks. The development of a relationship of the girl with the robots is apparent from her interview with the researcher.

> "Researcher: Do you think the doll is different now than when you first started playing with it?
> Melanie: Yeah. I think we really got to know each other a whole lot better. Our relationship, it grows bigger. Maybe when I first started playing with her, she didn't really know me so she wasn't making as much [sic] of these noises, but now that she's played with me a lot more, she really knows me and is a lot more outgoing. Same with AIBO" (Turkle et al. 2006b pp. 352).

In another paper, Turkle et al. (2006b) chart the first encounters of 60 children between the ages of five and thirteen with the MIT robots Cog and Kismet. The children anthropomorphised the robots, made up "back stories" about their behaviour, and developed "a range of novel strategies for seeing the robots not only as 'sort of alive' but as capable of being friends and companions". The children were so ready to form relationships with the robots, that when they failed to respond appropriately to their interactions, the children created explanations of their behaviour that preserved their view of the robot as being something with which they could have a relationship. For example, when Kismet failed to speak to them, children would explain that this was because it was deaf, or ill, or too young to understand, or shy, or sleeping. Their view of the robots did not even seem to change when the researchers spent some time showing them how they worked, and emphasising their underlying machinery.

Melson and her colleagues (Melson et al., in press b) directly compared children's views of and interactions with a living, and a robot dog. The children did see the live dog as being more likely than the AIBO to have physical essences, mental states, sociality and moral standing. However, a majority of the children still thought of and interacted with AIBO as if it were a real dog; they were as likely to give commands to the AIBO as to the living dog and over 60% affirmed that AIBO had "mental states, sociality and moral standing".

Overall, the pattern of evidence indicates that the illusion of robot animacy works well for children from preschool to at least early teens. Robots appear to amplify natural anthropomorphism. Children who spent time with robots saw them as friends and felt that they had formed relationships with them. They even believed that a relatively simple robot was getting to know them better as they played with it more. A large percentage was also willing to attribute mental states, sociality and moral standing to a simple robot dog. Kahn et al. (2006) suggest that a new technological genre of autonomous, adaptive, personified and embodied artefacts is emerging that the English language is not well-equipped to handle.

They believe that there may be need for a new ontological category beyond the traditional distinction between animate and inanimate.

3.1 Extending the reach of childcare robots

There are a number of ways in which current childcare robots interact with children. The main methods involve touch, language with speech recognition, tracking, maintaining eye contact and face recognition among others. Extending social interaction with better computational conversation and the ability to respond contingently with facial expressions could result in more powerful illusions of personhood and intent to a young child. It could make child-robot relationships stronger and maintain them for longer. We discuss each of the current interactive features in turn together with their possible near-future extensions.

Touch is an important element of human interaction (Hertenstein et al., 2006) particularly with young children (Hertenstein, 2002). It has been exploited in the development of robot companions and several of the manufacturers have integrated touch sensitivity into their childcare machines in different ways. It seems obvious that a robot responding contingently to touch by purring or making pleasing gestures will increase its appeal. For example, Tanaka et al. (2007) reported that children were more interested in the QRIO robot when they discovered that patting it on the head caused it to 'giggle'.

The PaPeRo robot has four touch sensors on the head and five around its body so that it can tell if it is being patted or hit. iRobiQ has a bump sensor, and touch screen as well as touch sensors on the head, arms and wheels. The Probo robot (Goris et al., 2008, 2009) is being developed to recognise different types of affective touch such as slap, tickle, pet and poke. The Huggable (Stiehl et al. 2005, 2006) has a dense sensor network for detecting the affective component of touch in rubbing, petting, tapping, scratching and other types of interactions that a person normally has with a pet animal. It has four modalities for touch, pain, temperature and kinaesthetic information.

Ongoing experimental research on touch is finding out the best way to create emotional responses (Yohanan et al., 2005; Yohanan and Maclean, 2008). There is also research on the impact of a robot proactively touching people – like a "gimme five" gesture or an encouraging pat on the shoulder (Cramer et al. 2009). Touch technology will improve over the next few years with better, cheaper and smaller sensors available to create higher resolution haptic sensitivity. This will greatly improve the interaction and friendship links with small children.

Robots could even have an advantage over humans in being allowed to touch children. In the UK, for example, there has been considerable discussion about the appropriateness of touching children by teachers and child minders. Teachers are reluctant to restrain children from hurting other children for fear of being charged with sexual offences or assault. Similarly childcare workers and infant school teachers are advised strongly not to touch children or hug them. Even music teachers are asked not to touch children's hands to instruct them on how to hold an instrument unless absolutely necessary and then only after warning them very explicitly and asking for their permission. These restrictions would not apply to a robot because it could not be accused of having sexual intent and so there are no particular ethical concerns. The only concern would be the child's safety, e.g. not being crushed by a hugging robot.

Another key element in interaction is spoken language. Even a doll with a recorded set of phrases that can be activated by pulling a string, can keep children entertained for hours by increasing the feeling of living reality for the child. We found eight of the current childcare robots that could talk to some extent and had speech recognition capability for simple commands. For example, iRobi, by Yujin Robotics of South Korea responds to 1000 words of voice commands. None had a full blown natural language processing interface, yet they can create the illusion of understanding.

The PaPeRo robot is one of the most advanced and can answer some simple questions. For example, when asked, "What kind of person do you like?" it answers, "I like gentle people". It can even give children simple quizzes and recognise if their answers are correct. PaPeRo gets out of conversational difficulties by making jokes or by dancing to distract children. This is very rudimentary compared to what is available in the rapidly advancing areas of computational natural language processing and speech recognition. Such developments could lead to care robots being able to converse with young children in a superficially convincing way within the next 5 to 10 years.

Face recognition is another important factor in developing relationships (Kanda et al., 2004). Some care robots are already able to store and recognise a limited number of faces, allowing them to distinguish between children and call them by name. The RUBI robot system has built-in face detection that enables it to autonomously find and gaze at a face. This is a very useful way to engage a child and convince her that the robot has "intent". Spurred on by their importance in security applications, face recognition methods are improving rapidly. Childcare robots of the future will adopt this technology to provide rapid face recognition of a wide range of people.

An even more compelling way to create the illusion of a robot having mental states and intention, is to give it the ability to recognise the emotion conveyed by a child's facial expression. The RUBI project team has been working on expression recognition for about 15 years with their computer expression recognition toolbox (CERT) (Bartlett et al., 2008). This uses Ekman's facial action units (Ekman & Friesen, 1978) which were developed to classify all human expressions.

The latest development uses CERT in combination with a sophisticated robot head to mimic people's emotional expressions.

The head, by David Hanson, resembles Albert Einstein and is made of a polymer material called Flubber that makes it resemble human skin and provides flexibility of movement. Javier Movellan, the team leader said that, "We got the Einstein robot head and did a first pass at driving it with our expression recognition system. In particular we had Einstein looking at himself in a mirror and learning how to make expressions using feedback from our expression recognition. This is a trivial machine learning problem." (personal communication, February 27, 2009). The head can mimic up to 5,000 different expressions. This is still at an early stage of development but will eventually, "assist with the development of cognitive, social and emotional skills of your children" (*ibid*).

Robots can be programmed to react politely to us, to imitate us, and to behave acceptably in the presence of humans (Fong et al., 2003). As the evidence presented earlier suggests, we have reached a point where it is possible to make children believe that robots can understand them at least some of the time. Advances in language processing, touch and expression recognition will act to strengthen the illusion. Although such developments are impressive, they are not without ethical concerns.

An infant entertaining a relationship with a robot may not be in a position to distinguish this from a relationship with a socially and emotionally competent being. As Sparrow pointed out about relationships with robot pets, "[they] are predicated on mistaking, at a conscious or unconscious level, the robot for a real animal. For an individual to benefit significantly from ownership of a robot pet they must systematically delude themselves regarding the real nature of their relation with the animal. It requires sentimentality of a morally deplorable sort. Indulging in such sentimentality violates a (weak) duty that we have to ourselves to apprehend the world accurately. The design and manufacture of these robots is unethical in so far as it presupposes or encourages this" (Sparrow, 2002).

Sparrow was talking about the vulnerable elderly but the evidence presented in this section suggests that young children are also highly susceptible to the belief that they are forming a genuine relationship with a robot. We could say in absolute terms that it is ethically unacceptable to create a robot that appears to have mental states and emotional understanding. However, if it is the child's natural anthropomorphism that is deceiving her, then it could be argued that there are no moral concerns for the roboticist or manufacturer. After all, there are many similar illusions that appear perfectly acceptable to our society. As in our earlier example, when we take a child to a puppet show, the puppeteer creates the illusion that the puppets are interacting with each other and the audience. The 'pretend' attitude of the puppeteer may be supported by the parents to 'deceive' very young children into thinking that the puppets have mental states. But this minor 'deception' might better be called 'pretence' and is not harmful in itself as long as it is not exploited for unethical purposes.

It is difficult to take an absolutist ethical approach to questions about robots and deception. Surely the moral correctness comes down to the intended application of an illusion and its consequences. Drawing an illusion on a piece of paper to fool our senses is an entertainment, but drawing it on the road to fool drivers into crashing is morally unjustifiable. Similarly, if the illusion of a robot with mental states is created for a movie or a funfair or even to motivate and inspire children at school there is no harm.

The moral issue arises and the illusion becomes a harmful deceit both when it is used to lure a child into a false relationship with a robot and when it leads parents to overestimate the capabilities of a robot. If such an illusory relationship is used in combination with near-exclusive exposure to robot care, it could possibly damage a child emotionally and psychologically, as we now discuss.

4. Psychological risks of robot childcare

It is possible that exclusive or near exclusive care of a child by a robot could result in cognitive and linguistic impairments. We only touch on these issues in this section as our main focus here is on the ways in which a child's relationship with a robot carer could affect the child's emotional and social development and potentially lead to pathological states. The experimental research on robot–child interaction to date has been short term with limited daily exposure to robots and mostly under adult supervision. It would be unethical to conduct experiments on long term care of children by robots. What we can do though, is make a 'smash and grab raid' on the developmental psychology literature to extract pointers to what a child needs for a successful relationship with a carer.

A fruitful place to start is with the considerable body of experimental research on the theory of attachment (Ainsworth et al. 1978; Bowlby, 1969, 1980, 1998). This work grew out of concerns about young children raised in contexts of less-than-adequate care giving, who had later difficulties in social relatedness (Zeanah et al., 2000). Although the term 'attachment' has some definitional difficulties, Hofer (2006) has noted that it has "found a new usefulness as a general descriptive term for the processes that maintain and regulate sustained social relationships, much the same way that *appetite* refers to a cluster of behavioral and physiological processes that regulate food intake" (p. 84).

A fairly standard definition that suits our purposes here is that "Infant attachment is the deep emotional connection that an infant forms with his or her primary caregiver, often the mother. It is a tie that binds them together, endures over time, and leads the infant to experience pleasure, joy, safety, and comfort in the caregiver's company. The baby feels distress when that person is absent. Soothing, comforting, and providing pleasure are primary elements of the relationship. Attachment theory holds that a consistent primary caregiver is necessary for a child's optimal development." (Swartout-Corbeil, 2006). Criticising such definitions, Mercer (in press) acknowledges that while it is true that attachment has a strong emotional component, cognitive and behavioural factors are also present.

There is always controversy within developmental psychology about the detailed aspects of attachment. Our aim is not to present a novel approach to attachment theory but to use the more established findings to warn about the possibility of harmful outcomes from robot care of children. Here we take a broad brush stroke approach to the psychological data. Given the paucity of research on childcare robots we have not been age specific, but our concerns are predominantly with the lower age groups – babies to preschoolers up to five years old – that appear to be the target group of the manufacturers.

One well established finding is that becoming well adjusted and socially attuned requires a carer with sufficient maternal sensitivity to perceive and understand an infant's cues and to respond to them promptly and appropriately (Ainsworth et al., 1974). It is this that promotes the development of *secure attachment* in infants and allows them to explore their environment and develop socially. But insecure forms of attachment can develop even when the primary carer is human. Extrapolating from the developmental literature, we will argue below that a child left with a robot in the belief that she has formed a relationship with it, would at best, form an insecure attachment to the robot but is more likely to suffer from a pathological attachment disorder.

Responding appropriately to an infant's cues requires a sensitive and subtle understanding of the infant's needs. We have already discussed a number of ways in which the relationship between a child and a robot can be enhanced when the robot responds contingently to the child's actions with touch, speech or emotional expressions. When the responses are not contingent, pre-school children quickly lose interest as Tanaka et al. (2007) found when they programmed a robot to perform a set dance routine. However, there is a significant difference between responding contingently and responding appropriately to subtle cues and signals. We humans understand and empathise with a child's tears when she falls because we have experienced similar injuries when we were children, and we know what comforted us.

There is more to the meaning of emotional signals than simply analysing and classifying expressions. Our ability to understand the behaviour of others is thought to be facilitated by our mirror neurons (Rizzolatti et al., 2000; Caggiano et al., 2009). Gallese (2001) argues that a mirror matching system underlies our ability to perceive the sensations and emotions of others. For instance, it is possible to show that the same neurons become active when a person feels pain as when observing another feeling pain (Hutchinson et al., 1999).

Responding appropriately to the emotions of others is a contextually sensitive ability that humans are particularly skilled at from a very young age. Even newborns can locate human faces and imitate their facial gestures (Meltzoff & Moore, 1977). By 12 months, infants are able to interpret actions in context (Woodward & Somerville, 2000). By 18 months, they can understand what another person intends to do with an instrument and they will complete a goal-directed behaviour that someone else fails to complete (Meltzoff, 1995; Herrmann et al., 2007).

No matter how good a machine is at classifying expressions or even responding with matching expressions, children require an understanding of the reasons for their emotional signals. A good carer's response is based on grasping the cause of emotions rather than simply acting on the emotions displayed. We should respond differently to a child crying because she has lost her toy than because she has been abused. A child may over-react to a small event and a caring human may realise that there is something else going on in the child's life like the parents having a row the night before. Appropriate responses require human common sense reasoning over a very large, possibly infinite, number of circumstances to ascertain what may have caused an unhappy expression. "Come on now, cheer up", might not always be the best response to a sad face.

A human carer may not get a full and complete understanding of the context of an emotion every time but they will make good guess with a high hit rate and can then recalculate based on the child's subsequent responses.

Advances in natural language processing using statistical methods to search databases containing millions of words could lead to superficially convincing conversations between robots and children in the near-future. However we should not mistake such interactions as being meaningful in the same way as caring adult–child interactions. It is one thing for a machine to give a convincing conversational response to a remark or question and a completely different thing to provide appropriate guidance or well founded answers to puzzling cultural questions. There are many cues that an adult human uses to understand what answer the child requires and at what level.

Language interactions between very young children and adults are transactional in nature – both participants change over time. Adults change register according to the child's abilities and understanding. They continuously assess the child's comprehension abilities through both language and non-verbal cues and push the child's understanding along. This is required both for language development and cognitive development in general. It would be extremely difficult to find specifiable rules that a robot could apply for transactional communication to adequately replace a carer's intuitions about appropriate guidance.

The consequence for children of contingent but inappropriate responses could be an *insecure* attachment called 'anxious avoidant attachment'. Typically, mothers with insecurely attached children are, "less able to read their infant's behaviour, leading them to try to socialise with the baby when he is hungry, play with him when he is tired, and feed him when he is trying to initiate social interaction" (Ainsworth et al., 1974 p. 129). Babies with withdrawn or depressed mothers are more likely to suffer aberrant forms of attachment: avoidance, or disorganised attachment (Martins & Gaffan, 2000).

'Maternal sensitivity'[1] provides a detailed understanding of an infant's emotional state. Responses need to be tailor made for each child's particular personality. A timid child will need a different response from an outgoing one, and a tired child needs different treatment from a bored one. Off-the-shelf responses, however benign, will not create secure attachment for a child: "If he's bored he needs a distraction. If he's hungry he needs food. If he has caught his foot in a blanket, it needs releasing. Each situation requires its own tailor-made response, suitable for the personality of a particular baby. Clearly, it isn't much use being given a rattle when you are hungry, nor being rocked in your basket if your foot is uncomfortably stuck" (Gerhardt, 2004 pp. 197).

Another important aspect of maternal sensitivity is the role played by "mind-mindedness", or the tendency of a mother to "treat her infant as an individual with a mind rather than merely as a creature with needs that must be satisfied" (Meins et al., 2001). Mind-mindedness has also been shown to be a predictor of the security of attachment between the infant and mother. It comes from the human ability to form a theory of mind based on knowledge of one's own mind and the experience of others. It allows predictions about what an infant may be thinking or intending by its actions, expressions and body language. A machine without a full blown theory of mind (or a mind) could not easily demonstrate mind-mindedness.

Other types of insecure attachment are caused by not paying close enough attention to a child's needs. If the primary carer responds unpredictably, it can lead to an ambivalent attachment where the child tends to overly cling to her caregiver and to others. More recently, a fourth attachment category, disorganised attachment, has been identified (Solomon & George, 1999; Schore, 2001). It tends to result from parents who are overtly hostile and frightening to their children, or who are so frightened themselves that they cannot attend to their children's needs. Children with disorganised attachment have no consistent attachment behaviour patterns.

While it seems unlikely that a robot could show a sufficient level of sensitivity to engender secure attachment, it could be argued that the robot is only standing in for the mother in the same way as a human nanny stands in. But a poor nanny can also cause emotional or psychological damage to a child. Children and babies are resilient but there is clear evidence that children do better when placed with childminders who are highly responsive to them. Elicker et al. (1999) found that the security of attachment of children (aged 12 to 19 months) to their childcare providers varied depending on the quality of their interactions. Dettling et al. (2000) studied children aged between 3 and 5 years old in home-based day care. They found that when they were looked after by a focused and responsive carer, their stress levels, as measured by swabbing them for cortisol, were similar to those of children cared for at home by their mother. In contrast, cortisol testing of children cared for in group settings with less focused attention indicated increased levels of stress. Belsky et al. (2007) found that children between 4.5 and 12 years old were more likely to have problems, as reported by teachers, if they had spent more time in childcare centres. At the same time they found that an effect of higher quality care showed up in higher vocabulary scores.

Thus even regular part-time care by a robot may cause some stress and minor behavioural problems for children. But we are not suggesting that occasional use will be harmful, especially if the child is securely attached to their primary carer; it may be no more harmful than watching television for a few hours. However, it is difficult at present, without the proper research, to compare the impact of passive entertainment to a potentially damaging relationship with an interactive artefact. The impact will depend on a number of factors such as the age of the child, the type of robot and the tasks that the robot performs.

In our earlier discussion of robot–child interaction research, we noted claims that children had formed bonds and friendships with robots. However, in such research, the terms 'attachment', 'bonding' and 'relationship' are often used in a more informal or different way than in developmental psychology. This makes it difficult to join them at the seams. Attachment theorists are not just concerned with the types of attachment but also with their consequences. As Fonagy (2003) pointed out, attachment is not an end in itself, although secure attachment is associated with better development of a wide range of abilities and competencies. Secure attachment provides the opportunity "to generate a higher order regulatory mechanism: the mechanism for appraisal and reorganisation of mental contents" (Fonagy, 2003 pp. 230).

A securely attached child develops the ability to take another's perspective. When the mother or carer imitates or reflects their baby's emotional distress in their facial expression, it helps the baby to form a representation of their own emotions. This social biofeedback leads to the development of a second order symbolic representation of the infant's own emotional state (Fonagy, 2003; Gergely & Watson, 1996, 1999), and facilitates the development of the ability to empathise, and understand the emotions and intentions of others. These are not skills that any near-future robot is likely to have.

When a young child encounters unfamiliar, or ambiguous circumstances, they will, if securely attached, look to their caregiver for clues about how to behave. This behaviour is termed "social referencing" (Feinman 1982). The mother or carer provides clues about the dangers, or otherwise, of the world, particularly by means of their facial expressions. For example, Hornik et al. (1987) found that securely attached infants, played more with toys that their mothers made positive emotional expressions about, and less with those that received negative expressions. A more convincing example of the powerful effect of social referencing is provided by research using a Gibson visual cliff. The apparatus, frequently used in depth-perception studies, gives the child an illusion of a sheer drop onto the floor (the drop is actually made safe by being covered with a clear plexiglass panel). Ten month olds will look at their mother's face, and continue to crawl over the apparent perilous edge towards an attractive toy if their mothers smile and nod. They back away if their mothers look fearful or doubtful (Scorce et al., 1985).

It would certainly be possible to create a robot that provided facial indications of approval or disapproval of certain actions for the child. But before a robot can approve or disapprove, it needs to be able to predict and recognise what action the child is intending. And even if it could predict accurately, it would need to have a sense of what is or is not a sensible action for a given child in a particular circumstance. With such a wide range and large number of possible actions that a child could intend, it seems unlikely that we could devise a robot system to make appropriate decisions. As noted from the studies cited above, it is important that responses are individually tailored, sensitive to the child's needs, consistent and predictable.

4.1 Is robot care better than minimal care?

Despite the drawbacks of robot care, it could be argued that it would be preferable and less harmful than leaving a child with minimal human contact. Studies of the shocking conditions in Romanian orphanages show the effects of extreme neglect. Nelson et al. (2007) compared the cognitive development of young children reared in Romanian institutions to that of those moved to foster care with families. Children were randomly assigned to be either fostered, or to remain in institutional care. The results showed that children reared in institutions manifested greatly diminished intellectual performance (borderline mental retardation) compared to children reared in their foster families. Chugani et al. (2001) found that

Romanian orphans who had experienced virtually no mothering, differed from children of comparable ages in their brain development – and had less active orbito-frontal cortex, hippocampus, amygdala and temporal areas.

But would a robot do a better job than scant human contact? We have no explicit evidence but we can get some clues from animal research in the 1950s when they were less concerned about ethical treatment. Harlow (1959) compared the effect on baby monkeys of being raised in isolation with two different types of artificial "mother": a wire-covered, or a soft terry-cloth covered wire frame surrogate "mother". Those raised with the soft mother substitute became attached to it, and spent more time with it than with the wire covered surrogate even when the wire surrogate provided them with their food. Their attachment to the surrogate was demonstrated by their increased confidence when it was present – they would return and cling to it for reassurance, and would be braver – venturing to explore a new room and unfamiliar toys, instead of cowering in a corner. The babies fed quickly from the wire surrogate and then returned to cuddle and cling to the terry cloth one.

This suggests that human infants might do better with a robot carer than with no carer at all. But the news is not all good. Even though the baby monkeys became attached to their cloth covered surrogates, and obtained comfort and reassurance from them, they did not develop normally. They exhibited odd behaviours and "displayed the characteristic syndrome of the socially-deprived macaque: they clutched themselves, engaged in non-nutritive sucking, developed stereotyped body-rocking and other abnormal motor acts, and showed aberrant social responses" (Mason & Berkson, 1975).

Although Harlow's monkeys clearly formed attachments to inanimate surrogate mothers, the surrogates left them seriously lacking in the skills needed to reach successful maturity. Of course, a robot nanny could be more responsive than the cuddly surrogate statues. In fact when the surrogate terry-cloth mother was hung from the ceiling so that the baby monkeys had to work harder to hug it as it swung, they developed more normally that when the surrogate was stationary (Mason & Berkson, 1975). But these were not ideal substitutes for living mothers. The monkeys did even better when they were raised in the company of dogs which were not mother substitutes at all.

We could conclude that robots would be better than nothing in horrific situations like the Romanian orphanages. But they would really need to be a last resort. Without systematic experimental work we cannot tell whether or not exclusive care by a robot would be pathogenic. It is even possible that the severe deprivation exclusive care might engender could lead to the type of impaired development pattern found in Reactive Attachment Disorder (RAD) (Zeanah et al., 2000). RAD was first introduced in DSM-III (American Psychiatric Association, 1980). The term is used in both the World Health Organization's International Statistical

Classification of Diseases and Related Health Problems (ICD-10) and in the DSM-IV-TR, (American Psychiatric Association, 1994).

Reactive Attachment Disorder is defined by inappropriate social relatedness, as manifest either in (1) failure to appropriately initiate or respond to social encounters, or (2) indiscriminate sociability or diffuse attachment. Although Rushton and Mayes (1997) warn against the overuse of the diagnosis of RAD it is still possible that the inappropriate and exclusive care of a child by a robot could lead to behaviour indicative of RAD.

Another worry is that a "robots are better than nothing" argument could lead to a more widespread use of the technology in situations where there is a shortage of funding, and where what is actually needed is more staff and better regulation. It is a different matter to use a teleoperated robot as a parental stand in for children who are in hospitals, perhaps quarantined or whose parent needs to be far away. Robots under development like the MIT Huggable (Stiehl et al. 2005, 2006) or the Probo (Goris et al. 2008, 2009) fulfil that role and allow carers to communicate and hug their children remotely. Such robots do not give rise to the same ethical concerns as exclusive or near exclusive care by autonomous robots.

Overall, the evidence presented in this section points to the kinds of emotional harm that robot carers might cause if infants and young children, lacking appropriate human attachment, were overexposed to them at critical periods in their development. We have reviewed evidence of the kinds of human skills and sensitivities required to create securely attached children and compared these with current robot functionality. While we have no direct experimental support as yet, it seems clear that the robots lack the necessary abilities to adequately replace human carers. Given the potential dangers, much more investigation needs to be carried out before robot nannies are freely available on the market.

5. Legal protections and accountability

The whole idea of robot childcare is a new one and has not had time to get into the statute books. There have been no legal test cases yet and there is little provision in the law. The various international nanny codes of ethics (e.g. FICE Bulletin 1998) do not deal with the robot nanny but require the human nanny to ensure that the child is socialised with other children and adults and that they are taught social responsibility and values. These requirements are not enforceable by the law.

There are a number of variations in the laws for child protection of different European countries, USA and other developed countries, but essentially legal cases against the overuse of robot care would have to be mounted on grounds of neglect, abuse or mistreatment and perhaps on grounds of delaying social and mental development. The National Society for the Prevention of Cruelty to Children (NSPCC) in the UK regards neglect as "the persistent lack of appropriate care of children, including love, stimulation, safety, nourishment, warmth, education and medical attention. It can have a serious effect on a child's physical, mental and emotional development. For babies and very young children, it can be life-threatening."

There are currently no international guidelines, codes of practice or legislation specifically dealing with a child being left in the care of a robot. There has been talk from the Japanese Ministry of Trade and Industry (Lewis, 2007), and the South Korean Ministry of Economy, Trade and Industry (Yoon-mi, 2007) about drawing up ethical and safety guidelines. The European Robotics Research Network also suggests a number of areas in robotics needing ethical guidelines (Verrogio, 2006) but no guidelines or codes have yet appeared from any of these sources. Some even argue that, "because different cultures may disagree on the most appropriate uses for robots, it is unrealistic and impractical to make an internationally unified code of ethics." (Guo & Zhang, 2009). There is certainly some substance to this argument as Guo and Zhang (2009) point out: "the value placed on the development of independence in infants and toddlers could lead to totally divergent views of the use of robots as caregivers for children." However, despite cultural differences, we believe that there are certain inviolable rights that should be afforded to all children regardless of culture, e.g. all children have a right not to be treated cruelly, neglected, abused or emotionally harmed.

The United Nations Convention on the Rights of the Child gives 40 major rights to children and young persons under 18. The most pertinent of these is Article 19 which states that, Governments must do everything to protect children from all forms of violence, abuse, neglect and mistreatment. Article 27 requires that, "States Parties recognize the right of every child to a standard of living adequate for the child's physical, mental, spiritual, moral and social development". These articles could be seen to vaguely apply to the care of children by robots but it is certainly far from being clear.

In the USA, Federal legislation identifies a minimum set of acts or behaviours that define child abuse and neglect. The Federal Child Abuse Prevention and Treatment Act (CAPTA), (42 U.S.C.A. §5106g), as amended by the Keeping Children and Families Safe Act of 2003, defines child abuse and neglect as, at minimum:

- Any recent act or failure to act on the part of a parent or caretaker which results in death, serious physical or emotional harm, sexual abuse or exploitation; or
- An act or failure to act which presents an imminent risk of serious harm.

Under US federal law, neglect is divided into a number of different sections. The most appropriate for our purposes, and one that does not appear under UK or

European law, is *emotional or psychological abuse*. Emotional or psychological abuse is defined as, "a pattern of behavior that impairs a child's emotional development or sense of self-worth." This may include constant criticism, threats, or rejection, as well as withholding love, support, or guidance. Emotional abuse is often difficult to prove and, therefore, child protective services may not be able to intervene without evidence of harm or mental injury to the child. "Emotional abuse is almost always present when other forms are identified." (What is Child Abuse and Neglect Factsheet).

Although much of the research on child–robot interaction has been conducted in the USA, the main manufacturers and currently the main target audience is in Japan and South Korea. As in the other countries mentioned, the only legislation available to protect Japanese children from overextended care by robots is the Child Abuse Prevention Law 2000. "The Law defines child abuse and neglect into four categories: (i) causing external injuries or other injuries by violence; (ii) committing acts of indecency on a child or forcing a child to commit indecent acts; (iii) neglecting a child's needs such as meals, leaving them for a long time, etc.; and (iv) speaking and behaving in a manner which causes mental distress for a child." (Nakamura, 2002).

In South Korea it may be harder to prevent the use of extended robot childcare. Hahm and Guterman (2001) point out that "South Korea has had a remarkably high incidence and prevalence rates of physical violence against children, yet the problem has received only limited public and professional attention until very recently" (p. 169). The problem is that "South Koreans strongly resist interference in family lives by outsiders because family affairs, especially with regard to child-rearing practices are considered strictly the family's own business." The one place where it might be possible to secure a legal case against near-exclusive care by robots is in the recently revised Special Law for Family Violence Criminal Prohibition (1998). This includes the Child Abuse and Neglect Prevention Act which is similar to the laws of other civilised countries: "the new law recognises that child maltreatment may entail physical abuse, sexual abuse, emotional abuse or neglect".

In the UK, a case against robot care would have to be built on provisions in the Children and Young Persons Act (1933 with recent updates) for leaving a child unsupervised "in a manner likely to cause unnecessary suffering or injury to health". The law does not even specify at what age a person can be a baby sitter; it only states that when a baby-sitter is under the age of 16 years old, the parents of the child being "sat" are legally responsible to ensure that the child does not come to harm.

Under UK law, a child does not have to suffer actual harm for a case of neglect to be brought. It is sufficient to show that the child has been kept in, "a manner likely to cause him unnecessary suffering and injury to health", as in the case of R v Jasmin, L (2004) 1CR, App.R (s) 3. The Appellants had gone to work for periods of up to 3 hours leaving their 16 month old child alone in the home. This happened on approximately three separate occasions. The Appellants were both found guilty of offences relating to neglect contrary to S1(1) Children and Young Persons Act 1933 and were sentenced to concurrent terms of 2 years imprisonment. Summing up, Lord Justice Law said that, "… there was no evidence of any physical harm resulting from this neglect [but] …both parents had difficulty in accepting the idea that their child was in any danger".

The outcome would have been different if the parents had left the child alone in exactly the same way but had stayed at home in a different room. If they could have shown that they were monitoring the child with a baby monitor (and perhaps a CCTV camera), the case against them would have been weak and it is highly unlikely that they would have been prosecuted.

This case is relevant for the protection of children against robot care because near-future robots, as discussed earlier, could provide safety from physical harm and allow remote monitoring combined with autonomous alerting and a way for the parents to remotely communicate with their children. The mobile remote monitoring available on a robot would be significantly better than a static camera and baby monitor. If absent parents had such a robot system and could reach the child within a couple of minutes, it would be difficult to prove negligence. The time to get home is probably crucial. We could play the game of gradually moving the parents' place of work further and further away to get a threshold time of permissibility. It then becomes like the discussion of how many hairs do you have to remove from someone's head before they can be called bald. These are the kind of issues that will only be decided by legal precedent.

Another important question about robot care is who would be responsible and accountable for psychological and emotional harm to the child? Under current legislations it would be the parents or primary carers. But it may not be fair to hold parents or primary carers entirely responsible. Assuming that the robot could demonstrably keep the child safe from physical harm, the parents may have been misled by the nature of the product. For example, if a carer's anthropomorphism had been amplified as a result of some very clever robot–human interaction, then that carer may have falsely believed that the robot had mental states, and could form 'real' relationships.

This leads to problems in determining accountability beyond the primary carer. Allocating responsibility to the robot would be ridiculous. That would be like holding a knife responsible for a murder – we are not talking about hypothetical sentient robots here. But blaming others also has its difficulties. There is a potentially long chain of responsibility that may involve the carer, the manufacturer and a number of a third parties such as the programmers and the researchers who developed the kit. This is yet another of the many reasons why there is a need to examine the ethical issues before the technology is developed for the mass market. Codes of practice and even legislation are required to ensure that the advertising claims are realistic and that the product contains warnings about potential danger of overuse.

If a case of neglect is eventually brought to court because of robot care, a large corporation with commercial interests may put the finest legal teams to work. Their argument could be based on demonstrating that a robot could both keep a child safe from physical harm and alert a designated adult about imminent dangers in time for intervention. It would be more difficult to prove emotional harm because many children have emotional problems regardless of their upbringing. Pathological states can be genetic in origin or result from prenatal brain damage among other possible causes. Thus a legal case of neglect is most likely to be won if an infant or a baby is discovered at home alone with an unsafe robot.

6. Conclusions

We have discussed a trajectory for childcare robotics that appears to be moving towards sustained periods of care with the possibility of near-exclusive care. We examined how childcare robots could be developed to keep children safe from physical harm. Then we looked at research that showed children forming relationships and friendships with robots and how they came to believe that the robots had mental states. After that, we examined the functionality of current childcare robots and discussed how these could be extended in the near future to create more 'realistic' interactions between children and robots, and intensify the illusion of genuine relationships.

Our main focus throughout has been on the potential ethical risks that robot childcare poses. The ethical problems discussed here could be among those that society will have to solve over the next 20 years. The main issues and questions we raised were:

- Privacy: every child has a right to privacy under Articles 16 and 40 of the UN Convention on Child Rights. How much would the use of robot nannies infringe these rights?
- Restraint: There are circumstances where a robot could keep a child from serious physical harm by restraining her. But how much autonomous decision authority should we give to a robot childminder?

- Deception: Is it ethically acceptable to create a robot that fools people into believing that it has mental states and emotional understanding? In many circumstances this can be considered to be natural anthropomorphism, illusion and fun pretence. Our concerns are twofold (i) it could lead parents to overestimate the capabilities of a robot carer and to imagine that it could meet the emotional needs of a child and (ii) it could lure a child into a false relationship that may possibly damage her emotionally and psychologically if the robot is overused for her care.
- Accountability: Who is morally responsible for leaving children in the care of robots? The law on neglect puts the duty of care on the primary carer. But should the primary carer shoulder the whole moral burden or should others, such as the manufacturers, take some share in the responsibility?
- Psychological damage: Is it ethically acceptable to use a robot as a nanny substitute or as a primary carer? This was the main question explored. If our analysis of the potentially devastating psychological and emotional harm that could result is correct, then the answer is a resounding 'no'.

In our exploration of the developmental difficulties that could be caused by robot care, we have assumed that it would be regular, daily and possibly near exclusive care. We also discussed evidence that part-time outside care can cause children minor harm that they can later recover from. Realistically a couple of hours a day in the care of a robot are unlikely to be any more harmful than watching television – if we are careful about what we permit the robot to do. We just don't know if there is a continuum between the problems that could arise with exclusive care and those that may arise with regular short-time care.

In a brief overview of international laws, we found that the main legal protection that children have is under the laws of neglect. A major concern was that as the robots become safer, protect children from physical harm and ensure that they are fed and watered, it will become harder to make a case for neglect. However, the quality of robot interaction we can expect, combined with the evidence from developmental studies on attachment, suggest that robots would at best be insensitive carers unable to respond with sufficient attention to the fine detailed needs of individual children.

As we stated at the outset, we are seeking discussion of these matters rather than attempting to offer answers or solutions. The robotics community needs to consider questions like the ones we have raised, and take them up, where possible, with their funders, the public and policy makers. Ultimately, it will be up to society, the legislature and professional bodies to provide codes of conduct to deal with future robot childcare.

Note

1. Maternal sensitivity is a term used even when the primary carer is not the "mother".

References

Ainsworth, M., Blehar, M., Waters, E. & Wall, S. (1978). *Patterns of attachment: a psychological study of the strange situation*. Hillsdale, NJ: Lawrence Erlbaum Associates, Inc.

Ainsworth, M.D.S., Bell, S.M., & Stayton, D.J. (1974). Infant-mother attachment and social development: Socialisation as a product of reciprocal responsiveness to signals. In M.P.M. Richards (Ed.), *The introduction of the child into a social world*. London: Cambridge University Press.

Bartlett, M.S., Littlewort-Ford, G.C., & Movellan, J.R. (2008). Computer Expression Recognition Toolbox. Demo: 8th *International IEEE Conference on Automatic Face and Gesture Recognition*. Amsterdam.

Belsky, J., Vandell, D.L., Burchinal, M., Clarke-Stewart, K.A., McCartney, K., Owen, M.T. (2007). Are there long-term effects of early child care? *Child Development*, 78 (2), 681–701.

Bittybobo. URL: http://bittybobo.blogspot.com/search?updated-min=2008-01-01T00%3A00%3A00-08%3A00&updated-max=2009-01-01T00%3A00%3A00-08%3A00&max-results=23, comments under April 17, 2008, last accessed 15 January 2010)

Blum, D. (2003). *Love at Goon Park: Harry Harlow and the Science of Affection*. John Wiley: Chichester, England.

Bowlby, J. (1969). *Attachment and Loss: Volume 1: Attachment*. London: Hogarth Press.

Bowlby, J. (1980). *Attachment and Loss: Volume 3: Loss*. London: Hogarth Press.

Bowlby, J. (1998). (edition originally 1973) *Attachment and Loss: Volume 2: Separation, anger and anxiety*. London: Pimlico.

Caggiano, V., Fogassi, L., Rizzolatti, G., Thier, P., & Casile, A. (2009). Mirror neurons differentially encode peripersonal and extrapersonal space of monkeys. *Science*, Vol. 324, pp. 403–406.

Cayton, H. (2006). From childhood to childhood? Autonomy and dependence through the ages of life. In Julian C. Hughes, Stephen J. Louw, Steven R. Sabat (Eds) *Dementia: mind, meaning, and the person*, Oxford, UK: Oxford University Press 277–286.

Children and Young Persons Act 1933. UK Statute Law Database, Part 1 Prevention of cruelty and exposure to moral and physical danger: Offences: 12 *Failing to provide for safety of children at entertainments*. URL: http://www.statutelaw.gov.uk/legResults.aspx?LegType=All+Legislation&searchEnacted=0&extentMatchOnly=0&confersPower=0&blanketAmendment=0&sortAlpha=0&PageNumber=0&NavFrom=0&activeTextDocId=1109288, last accessed 15 January 2010.

Chugani, H., Behen, M., Muzik, O., Juhasz, C., Nagy, F. & Chugani, D. (2001). Local brain functional activity following early deprivation: a study of post-institutionalised Romanian orphans. *Neuroimage*, 14: 1290–1301.

Cramer, H.S., Kemper, N.A., Amin, A., & Evers, V. (2009). The effects of robot touch and proactive behaviour on perceptions of human–robot interactions. In *Proceedings of the 4th ACM/IEEE international Conference on Human Robot interaction* (La Jolla, California, USA, March 09–13, 2009). HRI '09. ACM, New York, NY, 275–276.

Dautenhahn, K., Werry, I. (2004). Towards Interactive Robots in Autism Therapy: Background, Motivation and Challenges. *Pragmatics and Cognition* 12(1), pp. 1–35.

Dautenhahn, K. (2003). Roles and Functions of Robots in Human Society – Implications from Research in Autism Therapy. *Robotica* 21(4), pp. 443–452.

Dettling, A., Parker, S., Lane, S., Sebanc, A., & Gunnar, M. (2000). Quality of care determines whether cortisol levels rise over the day for children in full-day childcare. *Psychoneuroendocrinology*, 25, 819±836.

Ekman P. & Friesen, W. (1978). *Facial Action Coding System: A Technique for the Measurement of Facial Movement*, Consulting Psychologists Press, Palo Alto, CA.

Elicker, J., Fortner-Wood, C., & Noppe, I.C. (1999). The context of infant attachment in family child care. *Journal of Applied Developmental Psychology*, 20, 2, 319–336.

Feinman, S., Roberts, D., Hsieh, K.F., Sawyer, D. & Swanson, K. (1992), A critical review of social referencing in infancy, in *Social Referencing and the Social Construction of Reality in Infancy*, S. Feinman, Ed. New York: Plenum Press.

Fice Bulletin (1998). *A Code of Ethics for People Working with Children and Young People*. URL: http://www.ance.lu/index.php?option=com_content&view=article&id=69:a-code-of-ethics-for-people-working-with-children-and-young-people&catid=10:fice-declaration-2006&Itemid=29, last accessed 15 January 2010.

Fonagy, P., (2003). The development of psychopathology from infancy to adulthood: The mysterious unfolding of disturbance in time. *Infant Mental Health Journal* Volume 24, Issue 3, Date: May/June 2003, Pages: 212–239.

Fong. T., Nourbakhsh, I., & Dautenhahn, K. (2003). A Survey of Socially Interactive Robots, *Robotics and Autonomous Systems* 42(3–4), 143–166.

Gallese, V. (2001). The shared manifold hypothesis: From mirror neurons to empathy. *Journal of Consciousness Studies*, 8, 33–50.

Gergely, G., & Watson, J. (1996). The social biofeedback model of parental affect-mirroring. *International Journal of Psycho-Analysis*, 77, 1181–1212.

Gergely, G., & Watson, J. (1999). Early social-emotional development: Contingency perception and the social biofeedback model. In P. Rochat (Ed.), *Early social cognition: Understanding others in the first months of life* (pp. 101–137). Hillsdale, NJ: Erlbaum.

Gerhardt, S. (2004). *Why love matters: how affection shapes a baby's brain*. Routledge Taylor and Francis Group, London and New York.

Goris, K., Saldien, J., Vanderniepen, I., & Lefeber, D. (2008). The Huggable Robot Probo, a Multi-disciplinary Research Platform. *Proceedings of the EUROBOT Conference 2008*, Heidelberg, Germany, 22–24 May, 2008, pages 63–68.

Goris, K., Saldien, J., & Lefeber, D. 2009. Probo: a testbed for human robot interaction. In *Proceedings of the* 4th *ACM/IEEE International Conference on Human Robot interaction* (La Jolla, California, USA, March 09–13, 2009). HRI '09. ACM, New York, NY, 253–254.

Guo, S. & Zhang, G. (2009). Robot Rights, Letter to *Science*, 323, 876.

Hahm, H.C., Guterman N.B. (2001). The emerging problem of physical child abuse in South Korea. *Child maltreatment* 6(2): 169–79.

Hello kitty web reference
URL: http://www.dreamkitty.com/Merchant5/merchant.mvc?Screen=PROD&Store_Code=DK2000&Product_Code=K-EM070605&Category_Code=HKDL.

Hermann, E., Call, J., Hare, B., & Tomasello, M. (2007). Human evolved specialized skills of social cognition: The cultural intelligence hypothesis. *Science*, 317(5843), 1360–1366.

Hertenstein, M.J. (2002). Touch: its communicative functions in infancy, *Human Development*, 45, 70–92.

Hertenstein, M.J., Verkamp, J.M., Kerestes, A.M., & Holmes, R.M. (2006). The communicative functions of touch in humans, nonhuman primates, and rats: A review and synthesis of the empirical research, *Genetic, Social & General Psychology Monographs*, 132(1), 5–94.

Hornik, R., Risenhoover, N., & Gunnar, M. (1987). The effects of maternal positive, neutral, and negative affective communications on infant responses to new toys. *Child Development*, 58, 937–944.

Hutchinson, W., Davis, K., Lozano, A., Tasker, R., & Dostrovsky, J. (1999). Pain-related neurons in the human cingulated cortex. *Nature Neuroscience*, 2, 403–5.

Kahn, P.H., Jr., Friedman, B., Perez-Granados, D., & Freier, N.G. (2006). Robotic pets in the lives of preschool children. *Interaction Studies*, 7(3), 405–436.

Kanda, T., Takuyaki, H., Eaton, D. & Ishiguro, H. (2004). Interactive robots as social partners and peer tutors for children: a field trial, *Human Computer Interaction*, 19, 16–84.

Kanda, T., Nishio, S., Ishiguro, H., & Hagita, N. (2009). Interactive Humanoid Robots and Androids in Children's Lives. *Children, Youth and Environments*, 19 (1), 12–33. Available from: www.colorado.edu/journals/cye.

Lewis, L. (2007). The robots are running riot! Quick, bring out the red tape, The Times Online, April 6th. URL: http://www.timesonline.co.uk/tol/news/world/asia/article1620558.ece, last accessed 15 January 2010.

Liu, C., Conn, K., Sarkar, N., & Stone, W. (2008). Online affect detection and robot behaviour adaptation for intervention of children with autism. *IEEE Transactions on Robotics*, Vol. 24, Issue 4, pp. 883–896.

Lopes, M.M., Koenig, N.P., Chernova, S.H., Jones, C.V., & Jenkins, O.C. (2009). Mobile human-robot teaming with environmental tolerance. In *Proceedings of the 4th ACM/IEEE international Conference on Human Robot interaction* (La Jolla, California, USA, March 09-13, 2009). HRI '09. ACM, New York, NY, 157–164.

Marti, P., Palma, V., Pollini, A., Rullo, A. & Shibata, T. (2005). My Gym Robot, *Proceeding of the Symposium on Robot Companions: Hard Problems and Open Challenges in Robot–Human Interaction*, pp.64–73.

Martins, C. & Gaffan, E.A. (2000). Effects of early maternal depression on patterns of infant-mother attachment: A meta-analytic investigation, *Journal of Child Psychology and Psychiatry* 42, pp. 737–746.

Mason, W.A. & Berkson, G. (1975). Effects of Maternal Mobility on the Development of Rocking and Other Behaviors in Rhesus Monkeys: A Study with Artificial Mothers. *Developmental Psychobiology* 8, 3, 197–211.

Mason, W.A. (2002). The Natural History of Primate Behavioural Development: An Organismic Perspective. In Eds. D. Lewkowicz & R. Lickliter, *Conceptions of Development: Lessons from the Laboratory.* Psychology Press. 105–135.

Mavridis, N., Chandan, D., Emami, S., Tanoto, A., BenAbdelkader, C. & Rabie, T. (2009). FaceBots: Robots Utilizing and Publishing Social Information in Facebook. *HRI'09*, March 11-13, 2009, La Jolla, California, USA. ACM 978-1-60558-404-1/09/03.

Melson, G.F., Kahn, P.H., Jr., Beck, A.M., & Friedman, B. (in press a). Robotic pets in human lives: Implications for the human-animal bond and for human relationships with personified technologies. *Journal of Social Issues*.

Melson, G.F., Kahn, P.H., Jr., Beck, A.M., Friedman, B., Roberts, T., Garrett, E., & Gill, B.T. (in press b). Robots as dogs? – Children's interactions with the robotic dog AIBO and a live Australian shepherd. *Journal of Applied Developmental Psychology*.

Mercer, J. (in press for 2010), Attachment theory, *Theory and Psychology.*
Meins, E., Fernyhough, C., Fradley, E. & Tuckey, M. (2001). Rethinking maternal sensitivity: mothers' comments on infants' mental processes predict security of attachment at 12 months. *Journal of Child Psychology and Psychiatry* 42, pp. 637–48.
Meltzoff, A.N. (1995). Understanding the intention of others: Re-enactment of intended acts by 18 month old children. *Developmental Psychology* 32, 838–850.
Meltzoff, A.N. & Moore, M.K. (1977). Imitation of facial and manual gestures by human neonates. *Science,* 198, 75–78.
Mitsunaga, N., Miyashita, T., Ishiguro, H., Kogure, K., & Hagita, N. (2006). Robovie-IV: A Communication Robot Interacting with People Daily in an Office, In *Proc of IROS*, 5066–5072.
Nakamura, Y. (2002). Child abuse and neglect in Japan, *Paediatrics International*, 44, 580–581.
Nelson, C.A., Zeanah, C.H., Fox, N.A., Marshall, P.J., Smyke, A.T. & Guthrie, D. (2007). Cognitive recovery in socially deprived young children: The Bucharest early intervention project. *Science,* 318, no 5858, pp. 1937–1940.
Orpwood, R., Adlam, T., Evans, N., Chadd, J. (2008). Evaluation of an assisted-living smart home for someone with dementia. *Journal of Assistive Technologies*, 2, 2, 13–21.
Rocks, C.L., Jenkins, S., Studley, M. & McGoran, D. (in press).'Heart Robot', a public engagement project. Robots in the Wild: Exploring Human–Robot Interaction in Naturalistic Environments. Special Issue of *Interaction Studies.*
Rushton, A. & Mayes, D. 1997. Forming Fresh Attachments in Childhood: A Research Update. Child and Familiy Social Work 2(2): 121–127.
Sharkey, N. (2008a). The Ethical Frontiers of Robotics, *Science,* 322. 1800–1801.
Sharkey, N (2008b). Cassandra or False Prophet of Doom: AI Robots and War, *IEEE Intelligent Systems,* vol. 23, no, 4, 14–17, July/August Issue.
Sharkey, N. & Sharkey, A. (in press) Living with robots: ethical tradeoffs in eldercare, In Wilks, Y. *Artificial Companions in Society: scientific, economic, psychological and philosophical perspectives.* Amsterdam: John Benjamins.
Sharkey, N., & Sharkey, A. (2006). Artificial Intelligence and Natural Magic, *Artificial Intelligence Review,* 25, 9–19.
Shibata, T., Mitsui, T., Wada, K., Touda, A., Kumasaka, T., Tagami, K. & Tanie, K. (2001). Mental Commit Robot and its Application to Therapy of Children, Proc. of 2001 *IEEE/ASME Int. Conf. on Advanced Intelligent Mechatronics*, pp.1053–1058.
Schore, A. (2001). The effects of early relational trauma on right brain development, affect regulation, and infant mental health. *Infant Mental Health Journal*, 22, 1–2, pp. 201–69.
Scorce, J.F., Ernde, R.N., Campos, J., & Klinnert, M.D. (1985). Maternal emotional signaling: Its effect on the visual cliff behavior of 1-year-olds. *Developmental Psychology, 21*(1), 195–200.
Solomon, J. & George, C. (eds) (1999). *Attachment Disorganisation*, New York: Guilford Press.
Sparrow, R. (2002). The March of the Robot Dogs, *Ethics and Information Technology*, Vol. 4. No. 4, pp. 305–318.
Stiehl, W.D., Lieberman, J., Breazeal, C., Basel, L., & Lalla, L. (2005). The Design of the Huggable: A Therapeutic Robotic Companion for Relational, Affective Touch. AAAI Fall Symposium on Caring Machines: AI in Eldercare, Washington, D.C.
Stiehl, W.D., Breazeal, C., Han, K., Lieberman, J., Lalla, L., Maymin, A., Salinas, J., Fuentes, D., Toscano, R., Tong, C.H., & Kishore, A. 2006. The huggable: a new type of therapeutic robotic companion. In *ACM SIGGRAPH 2006. Sketches* (Boston, Massachusetts, July 30–August 03, 2006). SIGGRAPH '06. ACM, New York, NY, 14.

Swartout-Corbeil, D.M. (2006). Attachment between infant and caregiver, In *The Gale Encyclopedia of Children's Health: Infancy through Adolescence*, MI: The Gale Group.

Tanaka F., Cicourel, A. & Movellan, J.R. (2007). Socialization Between Toddlers and Robots at an Early Childhood Education Center. *Proceedings of the National Academy of Science*. Vol 194, No 46, 17954 x17958.

Turkle, S., Taggart, W., Kidd, C.D., Dasté, O. (2006a). Relational Artifacts with Children and Elders: The Complexities of Cybercompanionship. *Connection Science*, 18, 4, pp. 347–362.

Turkle, S., Breazeal, C., Dasté, O., & Scassellati, B., (2006b). First Encounters with Kismet and Cog: Children Respond to Relational Artifacts. In *Digital Media*:
Transformations in Human Communication, Paul Messaris & Lee Humphreys (eds.). New York: Peter Lang Publishing.

United Nations Convention on the Rights of the Child, URL: http://www2.ohchr.org/english/law/crc.htm, last accessed 15 January 2010.

Verrugio, G. (2006). The EURON robotethics roadmap, 6th *IEEE-RAS International Conference on Humanoid Robots*, 612–617.

What is Child Abuse and Neglect Factsheet, URL: http://www.childwelfare.gov/pubs/factsheets/whatiscan.cfm, last accessed 15 January 2010.

Wallach, W., & Allen, C. (2009). *Moral Machines: Teaching Robots Right from Wrong*, Oxford University Press, New York.

Woodward, A.L. & Sommerville, J.A. (2000). Twelve-month-old infants interpret action in context. *Psychological Science*, 11, 73–77.

Yohanan, S., & MacLean, K.E. (2008). The Haptic Creature Project: Social Human–Robot Interaction through Affective Touch. In *Proceedings of the AISB 2008 Symposium on the Reign of Catz & Dogs: The Second AISB Symposium on the Role of Virtual Creatures in a Computerised Society*, volume 1, pages 7–11, Aberdeen, Scotland, UK, April, 2008.

Yohanan, S., Chan, M., Hopkins, J., Sun, H., & MacLean, K. (2005). Hapticat: Exploration of Affective Touch. In ICMI '05: *Proceedings of the 7th International Conference on Multimodal Interfaces*, pages 222–229, Trento, Italy, 2005.

Yoon-mi, K. (2007). Korea drafts Robot Ethics Charter, *The Korea Herald*, April 28.

Yoshiro, U., Shinichi, O., Yosuke, T., Junichi, F., Tooru, I., Toshihro, N., Tsuyoshi, S., Junichi, O, (2005). Childcare Robot PaPeRo is designed to play with and watch over children at nursery, kindergarten, school and at home. Development of Childcare Robot PaPeRo, Nippon Robotto #Gakkai Gakujutsu Koenkai Yokoshu, 1–11.

Zeanah, C.H., Boris, N.W. & Lieberman, A.F. (2000). Attachment disorders of Infancy In Arnold J. Sameroff, Michael Lewis, Suzanne Melanie Miller (Eds) *Handbook of developmental psychopathology*, Birkhäuser, 2nd Edition.

14

Designing Robots for Care: Care Centered Value-Sensitive Design

Aimee van Wynsberghe

Received: 29 April 2011 / Accepted: 11 December 2011 / Published online: 3 January 2012

Abstract The prospective robots in healthcare intended to be included within the conclave of the nurse-patient relationship—what I refer to as *care* robots—require rigorous ethical reflection to ensure their design and introduction do not impede the promotion of values and the dignity of patients at such a vulnerable and sensitive time in their lives. The ethical evaluation of care robots requires insight into the values at stake in the healthcare tradition. What's more, given the stage of their development and lack of standards provided by the International Organization for Standardization to guide their development, ethics ought to be included into the design process of such robots. The manner in which this may be accomplished, as presented here, uses the blueprint of the Value-sensitive design approach as a means for creating a framework tailored to care contexts. Using care values as the foundational values to be integrated into a technology and using the elements in care, from the care ethics perspective, as the normative criteria, the resulting approach may be referred to as care centered value-sensitive design. The framework proposed here allows for the ethical evaluation of care robots both retrospectively and prospectively. By evaluating care robots in this way, we may ultimately ask what kind of care we, as a society, want to provide in the future.

Introduction

As roboticist Joseph Engelberger predicted in 1989, the use of robots no longer remains exclusively in the domain of the factory (1989). One of the most intriguing and morally challenging applications of robots is their use in healthcare scenarios. The most prominent and widespread are surgical robots, the daVinci™ Surgical System or its predecessor the Zeus™ Telesurgical System for example (Van Wynsberghe and Gastmans 2008); however, the more recent and intriguing robots to be used in healthcare are those intended for inclusion in the daily care activities of persons, activities like lifting, feeding or bathing. These robots, what I refer to as *care robots*, will be used by the care-giver in the care of another or may also be used by the care-receiver directly. The initiative to create such robots stems from the foreseen lack of resources and healthcare personnel to provide a high standard of care in the near future: fewer people in the younger generations are available to meet the care needs of the ageing population (WHO 2010). This trend is expected to be observed in multiple countries. Consequently, the idea is that this wave of automation, namely care robots, may help to mitigate the coming lack of care workers by providing assistance during care tasks or by fulfilling care tasks to relieve time for the many duties of care workers. Given the sensitive and morally complex context(s) into which these robots will be stepping, ethical attention to their potential impact is called for. To date, authors like Sharkey and Sharkey (2010) and Shannon Vallor (2011) have addressed certain issues pertaining to the use of care robots by addressing their potential impact on the rights of elderly citizens or the development of care-givers in their personal and professional roles, respectively.

In line with such ethical evaluations, attempts have been made to outline a framework for the ethical evaluation of robots (Veruggio and Operto 2006), or to suggest the dimensions of a robot ethic (Asaro 2006). These efforts, however, fall short by neglecting certain technical aspects, neglecting the specific issues pertaining to care contexts (for robots in healthcare), or failing to provide the structure of a framework at all. A framework for the ethical evaluation of care robots requires recognition of the specific context of use, the unique needs of users, the tasks for which the robot will be used, as well as the technical capabilities of the robot. Above and beyond a retrospective evaluation of robots, however, what is needed is a framework to be used as a tool in the design process of future care robots to ensure the inclusion of ethics in this process. What's more, given the lack of standards provided by the International Organization for Standardization (ISO 2011), there exists an opportunity at this time to incorporate ethics into the actual design processes for these kinds of robots. Accordingly, if ethics is to be included in the design process of robots, one must first identify the moral precepts of significance followed by an account as to how to operationalize said precepts.

Accordingly, this work addresses the following question: *how can care robots be designed in a way that supports and promotes the fundamental values in care*. As such, I neither refute the potential positive contributions of care robots nor the negative. Rather, I am looking for the most desirable way in which to proceed with their design. The following paper begins by outlining what a care robot is and why questions of design are the most significant questions to address at this time for these robots. Following this, I embark on a conceptual investigation of the concepts of value, and of care, for understanding their complexity. From this, I present a normative framework for the retrospective and prospective evaluation of care robots and provide an example to illustrate its utility.

What is a Care Robot?

As I have mentioned, a care robot is one that is used in the care of persons in general. For Vallor, "carebots are robots designed for use in home, hospital, or other settings to assist in, support, or provide care for the sick, disabled, young, elderly or otherwise vulnerable persons" (2011). According to Sharkey and Sharkey, the tasks for which the care robot is used can be classified in terms of either providing assistance in caregiving tasks, monitoring a patient's health status and/or providing social care or companionship (2010). Both Vallor and Sharkey and Sharkey, list examples of current care robot prototypes from the RI-MAN autonomous lifting robot at the Riken Institute, now replaced with the RIBA, to the MySpoon feeding assist robot. Care robots are also envisioned for tasks like bathing as well as social companionship, the most well known of the latter being the Paro robot. Most recently, the car company Toyota has announced the release of four care robots intended to aid a nurse with lifting patients or to assist patients with walking.[1]

As we can see, there is no capability exclusive to all care robots rather; they may have any number and range of capabilities from planar locomotion (vs. stationary) to voice recognition, facial or emotion recognition. Additionally, they may have any degree of autonomy, from human-operated (as in the surgical robot daVinci) or varying degrees of autonomy (like the TUG robot for deliveries in the hospital which requires minimal human input or the RIBA robot intended for lifting patients without input from a human user). Thus, the definition of a care robot relies on the idea of *interpretive flexibility,* that a robot is defined by its context, users and task for use (Howcroft et al. 2004). This means that the same robot might be called by a different name if the robot is used for rehabilitation or for care purposes. The Hybrid Assistive Limb (HAL) is an example of this phenomena; the robot may be used in rehabilitation when worn by a patient (Kawamoto and Sankai 2002) or could be used to relieve the stress of lifting on the nurse. For the purposes of this work, a care robot will be defined as such according to its application domain (hospital, nursing home, home setting), its intended use (a care practice deemed as such according to its use domain) and its intended users (care givers and/or care receivers, in a care domain for a care practice.

Creating a Framework for the Ethical Evaluation of Robots

When discussing ethics and robots, current authors approach the question from a variety of perspectives; the rights of individuals (Sharkey and Sharkey 2010), the specific needs of a care receiving demographic (Sparrow and Sparrow 2006), or the impact specific to the care-giver (Vallor 2011). While these analyses provide useful contributions to the question of ethics and care robots at large they lack the formation of normative recommendations based on their work for the design of future care robots. In Peter Asaro's article "What should we want from a robot ethic?" (2006), he proposes the three dimensions one could be referring to when one

---

[1] http://news.cnet.com/8301-17938_105-20128993-1/toyota-plans-nursing-robots-for-aging-japan/.

says "ethics of robots": (1) the ethical systems built into robots; (2) the ethical systems of people who design robots, and; (3) the ethics of how people treat robots. He then concludes that given the nature of robots as socio-technical systems, a framework for ethically addressing robots ought to include all three dimensions. For Asaro, the overarching question that each of the three dimensions stem from has to do with the distribution of moral responsibility in the social-technical network into which robots are introduced. Asaro presents a compelling case for the need for a comprehensive approach to robot ethics, but stops short of presenting such an approach. In this paper, I intend to take up this challenge. To that end I have created a framework which incorporates ethical analysis, according to the care perspective, into the design process of a care robot. The goal with this framework is threefold; to stimulate ethical reflection of designers/engineers, to encourage ethical reflection from the care ethics tradition, and to illuminate the relationship between the technical content of a care robot and the resulting expression of care values within a care practice.

To accomplish these goals, I argue that the approach known as value-sensitive design (VSD) adequately addresses the three dimensions identified by Asaro as well as his overarching question. Value-sensitive design is defined as "a theoretically grounded approach to the design of technology that accounts for human values in a principled and comprehensive manner throughout the design process" (Friedman and Kahn 2003). Value-sensitive design takes as its starting point the belief that technologies embody values (the embedded values approach) and offers a coherent method for evaluating the current design of technologies but also offers a proactive element to influence the design of technologies early on and throughout the design and implementation process. This concept refutes the neutrality thesis of computer systems and software programs which states that such systems are in themselves neutral and depend on the user for acquiring moral status. Instead, it is possible to identify tendencies *within* a computer system or software that promote or demote particular moral values and norms (Brey 2009; Nissenbaum 1998). These tendencies manifest themselves through the consequences of using the object. When said technology is capable of imposing a behavior on a user, or consequence to using it, the imposing force within the technology is considered a "built-in" or "embedded" value (or alternatively a disvalue if the computer system hinders the promotion of a value). The consequence is considered a special kind of consequence to using the object; one that brings about the promotion or demotion of a cultural value (Brey 2009). To give a simple example of this phenomenon, one may think of a personal banking machine—the ATM, through its use, promotes a certain value of user autonomy or distributive justice. At the same time, the ATM machine enforces certain biases of users—that they are a certain height and are literate in order to use the machine. Observing values within a system is a complex endeavor whereby the promotion of one value may be fulfilled while at the same time there is a trade-off with another value.

Accordingly, technologies may be designed in a way that accounts for values of ethical importance in a systematic way and rigorously works to promote said values through the architecture and/or capabilities of a technology. It follows then that care robots may be designed in a way that promotes the fundamental values in care.

Although VSD is meant for the design of a particular system or product, that is not my overall aim. My goal is to create a general framework that may be used by designers and/or ethicists in the ethical evaluation of any care robot, or for the inclusion of ethics in the design of any care robot. By using the blueprint of VSD I am creating a framework that addresses the specific relationship between technical capabilities and design of care robots with the specific context of use, task of use and users in mind. An additional benefit to the framework is its potential for use retrospectively and prospectively. When used retrospectively, designers are able to understand the impact of their design on the resulting care practice. When used prospectively, designers are able to incorporate the framework into the design process of a care robot, ultimately incorporating ethics into the design process.

Value-sensitive Design has been praised by computer ethicists and designers for its success in incorporating ethics in the overall design process of computer systems or ICT (Van den hoven 2007) but is also advantageous to guide the design process of a wide array of technologies (Cummings 2006). The framework I am creating uses components of the VSD methodology in its creation—namely the conceptual investigation coupled with a brief empirical and technical investigation. As in traditional VSD, my conceptual investigation is an exploration of the value constructs of the values of ethical importance, for this work the values of ethical importance are those from the care ethics tradition. VSD, however, has also been criticized for its lack of normative grounding given that it rests on rather abstract values without an ethical theory to anchor their interpretation (Mander-Huits 2011). With this in mind, I diverge from traditional VSD in that I utilize the findings from the care ethics perspective to guide the discussion of relevant care values and their meaning as well as the manner in which the ethical analysis ought to take place. Based on the work of care ethicist Joan Tronto, I claim the fundamental care values of any practice to be attentiveness, responsibility, competence and reciprocity. I attempt to understand how these values are interpreted philosophically by care ethicists, as well as how these values are interpreted in context through observational work. The resulting framework thereby provides a normative account of the values in care. Putting the framework to use incorporates a technical investigation by exploring technical capabilities of care robots currently available or available in the near future. This highlights the relationship between technical content and the resulting expression of values. Unlike the traditional empirical and technical components of VSD, I do not embark on empirical studies to test a care robot in context with human users (at least not at this moment in time). This is because I aim to provide a framework for the design of a range of care robots and not one particular system. In order to utilize the framework for evaluation it is necessary to shift back and forth from conceptual to empirical to technical aspects, in much the same manner as other VSD methodologies.

Why Begin with Design?

Not only do the three dimensions presented by Asaro point towards a discussion of design but the answer to the question of why one ought to pay so much attention to issues of design is further grounded in three rationales. Firstly, there are no (universal) guidelines or standards for the design of robots outside the factory. Although the International Organization for Standardization has currently drafted standards for design of personal robots, these are classified differently from medical-use robots. Care robots will most often be found in medical settings like a hospital or nursing home and no such standards for medical robots exist to date. As a result, designers are given no guidelines pertaining to the inclusion of socially sanctioned ethical principles like safety and/or efficiency, principles which designers still strive for but do so without any standardized means. Secondly, the nascent stage of the development of robots. Given the multi-disciplinary nature of robotics a range of disciplines are involved in the design of a system. Disciplines range from computer science to engineering and from sociology to psychology. This does not mean there is neither a need nor an interest to incorporate ethics in the design process but rather there has been no attempt to facilitate a way in which ethics may be translated for engineers/designers.

The necessity for inclusion of ethical criteria throughout the design process of a care robot brings us to the third rationale—that of the far reaching impact technologies have on societies (good or bad) and in turn, that societal values and norms have on the development of a technology. From the perspective of the philosophy of technology, many theories exist which seek to explain the reciprocal and dynamic relationship between society and the development of technologies. This may not be an explicit aim of the designer but is a condition of the work that they do. The theory of scripts illustrates how engineer's assumptions about user preferences and competencies show themselves in the technical content of an object (Akrich 1992). Latour expands this idea to show how technologies steer behaviors, moral and otherwise (1992). Verbeek shows how technologies are included in our decision making such that moral decisions are in fact a hybrid affair between humans and technologies (2006). In the computer ethics domain, Nissenbaum illustrates how values and biases are embedded into a computer system (1998). The embedding of values and of biases and the intertwining of the two was seen previously in the ATM example. The golden thread through all of these perspectives is that social norms, values and morals find their way into technologies both implicitly and explicitly and act to reinforce beliefs or to alter beliefs and practices.

Beyond the embedding of values and/or norms, once the robot enters a network it will alter the distribution of responsibilities and roles within the network as well as the manner in which the practice takes place. This shift is what Verbeek refers to as mediation: "when technologies are used, they help to shape the context in which they fulfill their function, they help to shape human actions and perceptions, and create new practices and ways of living" (Verbeek 2008, p. 92). Akrich discusses this in terms of the assumptions designers have of the traditional and ideal distribution of roles and responsibilities—that practices may shift based on an assumption made by an engineer of how the practice "ought" to take place, how roles and responsibilities "ought" to be delegated, and inscribing these assumptions into the technical content. For Akrich, "many of the choices made by designers can be seen as decisions about what should be delegated to a machine and what should be left to the initiative of human actors" (p. 216). By making choices about what should and should not be delegated to certain actors (human or nonhuman), engineers may change the distribution of responsibilities in a network. Or as Verbeek claims, engineers are 'materializing morality' (2006). It is these ideas that mirror the overarching question presented by Asaro—that a robot ethic ought to address the shift in responsibilities once the robot has been included into a socio-technical network. What's more, when a shift in roles and/or responsibilities is inscribed in a robot a valuation is being made—for example, that the human is not competent to fulfill the task or that the robot may fulfill the task in a superior manner. Thus, even assumptions about users may be considered statements of value, or normative claims, at times.

It is true that the rationales presented here relate to the design of any system or technology; however, greater weight is added when one takes into account the context in which the care robot will be placed and the nature of the activities the care robot will fulfill. Meaning, without standards guiding the development of care robots how is one to be sure that the values and norms central to the healthcare tradition will be promoted? Or, without making these norms and values explicit through the design process, how can one be sure their inclusion will be taken into account? Or, given the cost of development of these robots, mustn't they provide the same quality of care as today if not better (which presupposes an understanding of how one defines "good care")? Or simply, given the dramatic impact care robots may have on society, shouldn't future considerations be taken into account in design? With these rationales in mind, the design of any technology is ultimately a moral endeavor. The design of a care robot then is even more so given the vulnerability of this demographic, the delicacy of their care needs and the complexity of care tasks.

Exploring the Concepts of 'Value' and of 'Care'

To begin the creation of the framework, I embark on a conceptual investigation of the concepts of 'value' and of 'care'. By uncovering the values of ethical importance in a care context, the aim is to expose the moral precepts to operationalize in the design of a care robot.

Defining Values

According to the Oxford English Dictionary, values are conceived of as "the principles or standards of a person or society, the personal or societal judgment of what is valuable and important in life" (Simpson and Weiner 1989). Thus, a value is something desirable, something we want to have or to have happen. It follows then that when something is de-valued it loses importance. Values may be intrinsic or inherent to an object, activity or concept, or, things may be valued as a means to an end (Rosati 2009). For example, in the healthcare context, the concept of human dignity is valued on its own whereas the activity of touch in care contexts is valued as a means to preserving the dignity of persons (Gadow 1985). Things of value[2] may

---

[2] I have used the word 'things' here to bypass repeating people, places, activities, concepts, and objects, all of which are included in the discussion of values.

be valued on a personal level or on a societal/cultural level. Values then may be more of a subjective enterprise (various things valued for an individual) or more of an objective enterprise (universal values such as justice, human dignity, fairness). The latter does not imply that values considered abstract and universal are interpreted in the same way between cultures or time periods but rather that the valuation of things may differ from an individual's sphere to a more public one. Linked with the concept of 'good', a value may be construed as something that is good or brings about a good consequence.

In the VSD literature, Batya Friedman and colleagues, opt for a more open definition of a value to refer to "what a person or group of people consider important in life" (Friedman and Kahn 2003, p. 2). This implies then that all the values are not interpreted in the same way. Nathan et al. illustrate this with the value of privacy and its divergent ways of being interpreted between cultures and therefore protected (2008). Le Dantec et al. reinforce the idea that values may be universal, or generally accepted, but differ in their interpretation. Because of this, Le Dantec et al. suggest a way in which the methodology of VSD may be strengthened, through an uncovering of values in situ or discovering values through experiencing the practice (Le Dantec et al. 2009). This is of course due to the idea that differences exist between designers' values and users' values (Nathan et al. 2008). Thus, the scope of values varies depending on the technology, the users, the culture, the time period and the application domain. In the VSD methodology, Friedman selects the values of ethical importance pertaining to computer systems. Given that my framework is intended for use in the design of care robots, the values pertaining to the specific context are of greater ethical significance and relevance.

Defining Care, Care Ethics and Care Values

Care may be one of the most difficult concepts to articulate. This is in part due to the ubiquity of the word but is also largely a consequence of the fact that one is assumed to know what care means given its revered place in many cultures. The work of Warren T. Reich nicely outlines the broad range of meanings and connotations care has embodied going back as early as Ancient Greece (Reich 1995). Regardless, of how one perceives or defines care, care is valued as something above and beyond simple care giving tasks. It has a central role in the history of human kind as a means to signify the value of others. In other words, by caring you bestow value on the care-receiver.

In the verb "to care" one finds that caring may actually be divided into the idea of *caring about* and *caring for*. The dimension of *caring about* in the medical field implies a mental capacity or a subjective state of concern. On the other hand, *caring for* implies an activity for safeguarding the interests of the patient. In other words, it is a distinction between an attitude, feeling or state of mind versus the exercise of a skill with or without a particular attitude or feeling toward the object upon which this skill is exercised (Jecker et al. 2002). In the field of care ethics, Joan Tronto claims that good care is the result of both a caring attitude in combination with a caring activity (Tronto 1993). In other words, a marriage between the dimensions of *caring about* and *caring for*.

The field of care ethics is most often attributed to the Kohlberg-Gilligan debate on moral psychology (Gilligan 1982). Because of this debate, a new way of perceiving the moral dilemma in a given scenario arose; one that shifted the central focus from rights and universally applicable rules to a focus on responsibilities and relationships as central factors. Perhaps the most significant result that came from Gilligan's work (along with the assistance of other scholars) is the understanding of care ethics as a perspective or orientation from which one begins to theorize rather than a pre-packaged ethical theory. In fact, care ethicists are not striving to arrive at some ready-made theory for application (as in traditional ethical theories) but rather point towards the necessary beliefs or elements that structure the care orientation. These beliefs refer to:

> an emphasis of concern and discernment (*to notice and worry more about the dangers of interference rather than the dangers of abandonment*), habits and proclivities of interpretation (*the proclivity to read the moral question presented by a situation in terms of responsibilities rather than rights*), and selectivity of skills (*to have developed an ease of abstraction more than an attunement to difference*) (Little 1998, p. 195).

As Little articulates, "the orientations provide illuminating stances from which to *develop* ethics of these relationships, not that they constitute those ethics ready-made" (Little 1998, p. 206). This is precisely what my aim is when developing the framework for the ethical analysis and evaluation of care robots—to outline an orientation from which the ethical evaluation may begin by emphasizing certain fundamental components in care.

Aside from a conversation about the concept of care or the care ethics perspective, there is much to say about care values. Alternative to the idea that care in itself is a value—linked with the good life and with a valuation of another—is the idea that beneath the umbrella concept of care comes many other values. These values are given importance for their role in care—their role in giving significance to care, in making care what it is. These values form the buttress for care as an ethical endeavor and create a framework for evaluating care as a practice. It is through the manifestation of these values that one comes to understand what care really is in practice. It is therefore fruitful for the topic of embedding care values, to understand these values and their link with consequences. Thus, to begin from a top-down approach, I look to the values articulated by the governing body of healthcare, namely the World Health Organization (WHO). The WHO framework for people-centered health narrows in on the values in healthcare stemming from the patient's perspective; *patient safety, patient satisfaction, responsiveness to care, human dignity, physical wellbeing* and *psychological wellbeing* (2007). This is not to say that other values like innovation or physician autonomy are not valued but rather from the patient's perspective, the listed values are the ones with the greatest ethical importance and will thus be used in my evaluation of implementing robots in the care of persons.

Without an understanding of the specific context or the individual characteristics of a patient, these values don't tell the engineer much concerning how the value may be embedded in a care robot prototype. Therefore, I take the suggestions of

Le Dantec et al. to understand the specific interpretation of these values in context, achieved through fieldwork experience in both a hospital and a nursing home.[3] Interestingly, the interpretation of values as well as their ranking and meaning differed depending on: the type of care (i.e. social vs. physical care), the task (ex. bathing vs. lifting vs. socializing), the care-giver and their style, as well as the care-receiver and their specific needs. For example, in a ward with people suffering from dementia, safety is in terms of not letting patients wonder onto the streets, or preventing patients from hurting both themselves and others. In a 'typical' ward of a nursing home, safety is in terms of preventing patients from falling, or assisting in the feeding of patients to prevent chocking. How a value is prioritized is also dependent on the context, personal experiences but also the specific practice. For example, through the practice of lifting, the value of safety is manifest (or interpreted) by ensuring the care-receiver does not fall or is not injured. Here, safety is of paramount importance. In contrast, through the practice of bathing, the value of safety is interpreted in terms of suitable water temperature (not burning or scarring the patient), and proper positioning on the bed or tub to prevent injury. In the practice of bathing, however, while safety is of the utmost importance, other values take precedence. For example, closing the curtain to ensure privacy, verbal communication to calm the care-receiver, and gentle strokes to convey empathy and respect through the practice. These examples make us aware of both the intertwining of care values and the actions of care-givers but also the significance of the therapeutic relationship—all of the values central to the healthcare tradition are observable within the relationship, the actions and interactions between the nurse and the patient.

When wondering about the relationship between technologies and care values, the value of touch helps to shed light. Touch is an important action in care that is valued on its own as well as a means for manifesting other values like respect, trust and intimacy. Touch is the symbol of vulnerability, which invokes bonds and subjectivity (Gadow 1985). Touch acts to mitigate the temptation for objectification. Thus, touch is considered an instrumental value in the healthcare domain, the outcome of which results in the preservation of the value of human dignity. Using the value of touch as an example, we can see how a certain technology might impede its manifestation. Melanie Wilson illustrated how a particular computer system implemented in the field of nursing was rejected as it prevented nurses from "hands on care"—from touch—a cornerstone of the nursing tradition (Wilson 2002). One might suggest that designers of this technology were not aware of the significance of 'hands-on' care for nurses even when the nurse's role is to create a daily care plan.

In short, not only is care a value for what it symbolizes (a valuation of another) and manifests (meeting the needs of another) but it is also valued for the additional elements that make up care; patient safety, patient satisfaction, responsiveness to care, human dignity, physical wellbeing and psychological wellbeing. The list of

---

[3] Fieldwork experience was gained by volunteering in a nursing home in London, Ontario, Canada for 4 weeks as a "life enrichment coach" as well as observing practices in multiple hospital, also in London, Canada.

care values is exhaustive when one considers the significance of the therapeutic relationship and the elements within this conclave (trust, respect, compassion, empathy and touch); however, what is hopefully now evident is the significance of the care relationship in terms of the intertwining of care actions with care values. Meaning, the expression of care values are the result of the actions and interactions between actors. To explain this further, I turn to the concept of a care practice.

Care Practices

To elaborate on the marriage between *caring about* and *caring for*, a useful concept is that of a care practice. A care practice is, as care ethicist Joan Tronto describes it, a way to envision a care task or a series of care tasks. A way in which one can grasp the fortitude of each action and interaction between a care-giver and a care-receiver. More importantly, it is a way to envision the holistic nature of care.

> The notion of a care practice is complex; it is an alternative to conceiving of care as a principle or as an emotion. To call care a practice implies that it involves both thought and action, that thought and action are interrelated, and that they are directed toward some end (Tronto 1993, p. 108).

Understanding that care tasks are more than just 'tasks' but are rich practices in a value-laden milieu that act to bring about the promotion of values, may be one of the most crucial points for designers to grasp. The reason for this has to do with understanding how values are manifest and thus how a design will impact this materialization. To exemplify this shift from task to practice, let me use the practice of lifting. When a patient is lifted by the care-giver, it is a moment in which the patient is at one of their most vulnerable. The patient trusts the care-giver and through this action a bond is formed and/or strengthened which reinforces the relationship between care-giver and the care-receiver. The significance of this is apparent in the actual practice of lifting but comes into play later on in the care process as well. Meaning trust, bonds, and the relationship, are integral components for ensuring that the care-receiver will comply with their treatment plan, will take their medication and be honest about their symptoms. Without trust, these needs of the care-giver are threatened, ultimately threatening the entire care process and the good care of the care-receiver. Thus, conceptualizing care tasks as practices adds a deeper meaning to each 'task'. It is within a care practice that the values are manifest and given their significance but it is also within practices that the holistic vision of care takes form—each care practice builds from, and on to, another practice linking all practices in the overall care process.

Selecting the Values of Ethical Importance in Care

While many care ethicists make clear the range of values and principles that provide a normative account for care (Vanlaere and Gastmans 2011; Little 1998; Ruddick 1995; Noddings 1984) they fall short of providing a systematic way to visualize and evaluate these principles and values. The vision presented by Joan Tronto allows for a perception of care as a process with stages and corresponding normative moral elements, which provides the most enticing conceptualization for engineers to work with. There are four phases of a care practice for Tronto; caring about (recognizing one is in need and what those needs are), care taking (taking responsibility for the meeting of said needs), care giving (fulfilling an action to meet the needs of an individual), care receiving (recognition of a change in function of the individual in need). These phases have corresponding moral elements as standards to evaluate the care practice from a moral standpoint. These elements are: attentiveness, responsibility, competence, and responsiveness. Attentiveness refers to an attribute or virtue of the care-giver, a certain competence for recognizing needs. Responsibility refers again to an element of the care-giver and their stance or concern for ensuring the care-receiver is pointed in the right direction for care or maintaining an accurate assessment of needs etc. Responsibility is often delegated to a moral agent; however, some responsibilities are delegated to an artifact as technologies are wide spread in healthcare. Here, the concept of mediation (Verbeek 2006) becomes critical in the sense that decision making on the part of nurses and patients is a hybrid affair between the nurse/patient and existing technologies. Competence is once again an attribute of the care-giver and refers to the skills with which the care is given. An unskilled care-giver may be more detrimental than no care at all. Responsiveness refers to an attribute of the care-receiver and their role in the relationship—to guide the care-giver. This element (and the phase of care receiving) is important for remembering the reasons for care in the first place: the care-receiver and their needs. Without this, care is not complete. This recognition also encourages an active stance of the care-receiver rather than a more passive, vulnerable one.

Creating a standardized framework to guide the promotion of these values which applies to any care context, task, care-receiver or care-giver reveals itself to be quite problematic given the range and variety of care values discussed in the former section. In other words, to claim that human dignity, compassion or respect for power are values to be embedded in a care robot offers nothing for the designer in terms of the robot's capabilities. Moreover, as we have seen, their ranking and prioritization is dependent on the context (i.e. one hospital domain or another vs. a nursing home) and task (ex. lifting vs. bathing). To standardize the creation of care robots there needs to be another avenue besides values alone. In the care ethics literature, alongside values, need too play a central and crucial role in the provision of good care. The needs of the patient mark the starting point of the care process and the process then revolves around a care-giver (or multiple care-givers) taking steps to meet these needs. Understanding the multiple layers of needs, the many ways in which they might be fulfilled, the preferences for one way over another, and the divergent needs between individuals, adds a further complexity to the meeting of needs. If this wasn't complicated enough, the care-giver has needs too! Needs in terms of resources, skills, responsiveness from the care-receiver to understand when needs have been met as well as their own personal needs.

Given the central role of needs in a care context, what might the relationship be between needs and values? Although many authors have written on the subject, little consensus can be found. I suggest that the values in healthcare are given their importance for their role in meeting needs. This corresponds with Super's conceptualization of the relationship between needs and values: "values are objectives that one seeks to attain to satisfy a need" (1973, pp. 189–190). Meaning, the value is the goal one strives towards and in so doing, intentionally meets a need. Thus, we begin with needs, and the values represent the abstract ideals which, when manifest, account for the needs of individuals. It follows then that a framework for designing care robots ought to address the meeting of needs. But not so fast, we've just shown how multifaceted and intricate needs are for the care-giver and care-receiver. What's more, according to the field of care ethics, it is neither possible nor advisable to outline a series of needs which pertain to all care-givers, care-receivers or care tasks in every instance/scenario (Tronto 2010). While useful for policy or a universal ethical code, it goes against the vital element in care—that of the individual and their unique, dynamic needs. In other words, care is only thought of as good care when it is personalized (Tronto 1993). There is, however, a solution to this barrier. It is possible to delineate a set of needs for **every** care practice. To recapitulate, together the phases and the moral elements make up a care practice. The practices are values working together and the vehicle for this is the moral elements. If we assume a care practice ought to proceed according to Tronto's phases than the needs for every care practice are the corresponding moral elements. It is therefore these elements that ensure the promotion of care values. Consequently, it is these elements—attentiveness, responsibility, competence, responsiveness—that make up the normative portion of the framework.

With this suggestion, there are two assumptions being made; that every care practice will *always* have the moral elements as needs, independent of the care-giver and care-receiver, and that the values are subsumed within the moral elements. Using the practice of feeding as an example to illustrate the first assumption, I am making the claim that this practice will *always* require attentiveness, responsibility and competence on the part of the care-giver and will *always* require a reciprocal interaction between care-receiver and care-giver for determining whether or not the needs have been met, no matter who the care-giver is or who the care-receiver is. In other words, these moral elements are independent of the actors. They are, however, dependent on the context and the specific practice for their interpretation and prioritization. If we were to compare the practice of lifting with the practice of feeding we would see how the element of competence is uniquely interpreted in each practice (skillfully bearing the weight of another without dropping or causing pain vs. skillfully coordinating timing and placement of food and utensils). In terms of context, the practice of lifting in the hospital requires greater efficiency than the practice of lifting in a home setting where time may not be as much of an issue. Thus, although the moral elements must always be present, the context and practice still play a crucial role in their interpretation, prioritization and manner of manifestation.

For the second assumption—that the values are subsumed within the moral elements—one may find that the values are often analogous to a phase or moral element or are expressed through the manner in which an action takes place. The value of patient safety is fulfilled through the competent completion of a practice (the phase being care giving and the moral element being competence). The valued action of touch requires attentiveness on the part of the care-giver for determining when and to what degree touch is considered necessary. The manner in which care practices take place is often tailored to the specific likes of one care-giver or another and again requires attentiveness to those preferences and competence in meeting them. What's more, paying attention to those unique preferences is a vehicle for establishing trust and allowing for successful reciprocal interaction.

In short, ensuring that the elements are present or strengthened through the design and introduction of a care robot, ultimately results in a manifestation of the core care values. Differences in the prioritization and manifestation of moral elements between practices and/or contexts is something that the care ethicist may draw the attention of the designer to while utilizing the framework throughout the design process. Nevertheless, the designer must first be aware of the necessary elements and their manner of manifestation.

The Care-Centered Framework

By summarizing and synthesizing the findings thus far I arrive at a framework to be used in the design process of care robots, both by ethicists as well as engineers/designers. The framework for the design of care robots is distinguished from the method to proceed when using the framework. The latter is referred to as the "care centered value sensitive design" (CCVSD) methodology and the former is referred to as the "care centered framework". The care centered framework and CCVSD methodology both pay tribute to the central thesis in care ethics, namely that the care perspective provides an orientation from which to begin theorizing as opposed to a pre-packaged ethical theory. The framework articulates the components that require attention for analysis from a care perspective while the methodology indicates how these components are to be analyzed with and without the introduction of a care robot. The care centered framework aims to outline the orientation from which one begins in order to develop an ethic of the relationship between care robot and the other actors involved in the care practice. The framework consists of five components: context, practice, actors involved, type of robot, and manifestation of moral elements (see Table 1). Each of these components will be described in detail for understanding their place within the framework from the care ethics stance.

The framework is intended to be a general outline for the creation of any care robot and not one care robot in particular for one practice in one context. Thus, the

Table 1 Framework for the ethical evaluation of care robots

Context—hospital (and ward) versus nursing home versus home…

Practice—lifting, bathing, feeding, delivery of food and/or sheets, social interaction, playing games…

Actors involved—nurse and patient and robot versus patient and robot versus nurse and robot…

Type of robot—assistive versus enabling versus replacement…

Manifestation of moral elements—Attentiveness, responsibility, competence, responsiveness

The ellipsis following the description of a criterion indicates that the list is not exhaustive and may include additions framework is standardized with respect to designing care robots in general according to the necessary components but is not standardized with respect to dictating how each value is interpreted and ranked for a care practice. This is done on a case-by-case basis for each practice in a given context. This is so due to the difference in capabilities of the robot depending on the practice and context for which the robot is intended. Meaning, a robot designed for delivery of sheets will have distinctly different capabilities from a robot designed for feeding.

Context as a Component of the Framework

Firstly, one must identify the context within which the care practice is taking place. For example, the specific hospital and the ward versus a nursing home versus a home setting. The context within which the care practice takes place is important for a variety of reasons. Recent research indicates a relationship between religious beliefs and one's acceptance of using robots in care-taking roles (Metzler and Lewis 2008). Metzler and Lewis are investigating the hypothesis that when one believes in "a god" they may not be as inclined to accept human-robot interaction with life-like robots at an intimate level. Thus, the design of a robot for a Catholic hospital ought to take this kind of research into consideration for the appearance of the robot. Similarly, the context in terms of one hospital ward or another is also of great importance when designing the robot. Research done by Bilge Mutlu of the University of Wisconsin, Madison (Barras 2009) shows how the same robot (the TUG robot) used in one hospital was accepted differently depending on the ward. Workers in the post-natal ward loved the robot, while workers in the oncology ward found the robot to be rude, socially inappropriate and annoying. The same workers even kicked the robot when they reached maximum frustration.

Specifying context in terms of a nursing home versus a home setting is also of importance given that the prioritization of values differs. For example, lifting in the nursing home places efficiency as a higher priority (and even more so in the hospital) while in the home setting there may not be the same time constraints. In addition, bathing in a home setting may not require the same demand for privacy as in the hospital or nursing home setting given the lack of other patients around. Specifying the context plays a crucial role for understanding the prioritization of values.

Practice as a Component of the Framework

The practice for which the care robot will be used plays a dominant role in the prioritization as well as the interpretation of values/moral elements. Examples of practices are lifting, bathing, feeding, fetching items, delivery of medications/food/x-rays/sheets to the room or to the nurse, personal communication, social interaction, games and activities like singing songs or painting. As mentioned, each of these practices requires the elements of attentiveness, responsibility, competence and reciprocity; however, they mean very different things depending on the type of practice. Competence in terms of lifting refers to a skilled lifter that does so at the appropriate speed and angle without hurting or dropping the care-receiver. Attentiveness to whether or not the care-receiver is being hurt or pinched in any way is also an attribute of competence. Alternatively, competence for the practice of feeding refers to the feeder gently, with great precision and at an appropriate speed bringing the utensils to the care-receiver's mouth. Thus, the interpretation of the elements are determined according to the practice for which the robot will be used.

Actors Involved as a Component of the Framework

From the care orientation, the actors involved are of great significance for structuring moral deliberation. One of the most important findings to come from the care ethics perspective is the ontological status of humans as relational. Its significance for this work lies in recognizing that the care practice which a robot will enter involves a network of human (and nonhuman) actors in relationship. The robot then has the potential to shift the roles and responsibilities distributed within these relationships. Alternatively, if the patient is receiving care in their home perhaps the actors involved are family members or a visiting nurse who is not present on a daily basis. Then again, a patient may fulfill certain practices on their own prior to a robot assisting. This does not mean the care-receiver is entirely on their on, in the atomistic sense, but rather that the robot may be delegated a certain portion of the role of the care-receiver (as is the case with a feeding robot like Secom's MySpoon). This component is meant to highlight the roles and responsibilities attributed to actors prior to the robot entering the scene. Throughout the evaluation (both retrospective and prospective), the goal is to understand how the traditional roles attributed to actors shift or remain the same.

It is important to remember too that the human actors are not acting alone to manifest values. They work together with each other but also with technologies already in use in the healthcare system. In nursing and technology studies, technologies have often been considered extensions of the nurse's body or self. Nurses become so skilled at using the technology they do so without being interfered by the technology's presence. What's more is that the nurse's role is one that incorporates the use of technologies in a variety of ways from the mechanical bed to heart monitoring devices. Thus, technologies are not only extensions of the nurse but they also mediate the relationship between the nurse and the patient shifting both the role and the responsibility of the patient and nurse in order to include the technology in the equation. In this vain we see that technologies already act to mediate all of the moral elements of the framework. In other words, we are not speaking of interactions that occur without the use of technologies. Therefore, the question is not what happens when a care robot enters the nurse-patient relationship that is devoid of any technologies. Rather, we are speaking of a context within which technologies are already employed to a high level and the question is how will a care robot alter the existing practice and further can the robot reintroduce elements that may have been overlooked from the previous round of automation?

Type of Robot as a Component of the Framework

The typifying of robots is done in many different ways. Some consider a type of robot according to the domain for which it is used; industrial versus rehabilitation versus military versus search and rescue robots (Veruggio and Operto 2006). For others, types of robots may be in terms of industrial robots versus service robots versus personal robots. This classification of robots is dependent on the amount of human interaction the robot will have and the predictability or structuring of the environment within which the robot is working. Industrial robots have very little interaction with humans and are present in a structured environment where their actions are closely monitored and highly predictable. Service robots are meant to act in human environments (unstructured and unpredictable) with varying degrees of human contact and interaction (Engelberger 1989). Personal robots are a type of service robot meant to interact and cooperate with humans in human environments—a domestic robot of sorts. There is no consensus as to a universal definition of a robot let alone the classification of different types of robots.

To specify for the purposes of the framework discussed here, the manner in which I classify as 'type of robot' has to do with how the robot will be used among the human actors—how a role and responsibility is delegated to a robot. For example, an enabling robot is one which enables a human to perform an action previously not possible without the robot or, the robot enhances the human's performance during a task—the robot and human are working together toward a goal but the human is in control of the both him/herself as well as the robot. Thus, the role and responsibility for accomplishing that role is a shared effort with the robot perceived in an instrumentalist way, as a tool. Robots of this type are telepresence robots like the RP7 or surgical robots like Intuitive Surgical's daVinci. A replacement robot is one that fulfills a practice in place of the human. The role of the human and the associated responsibilities are delegated fully to the robot. Examples of this type of robot are the RI-MAN or RIBA autonomous robots for lifting. An assistive robot is one which aids a human in performing an action by providing a portion of the practice without the direct input of a human operator and is thus delegated a partial role and a partial responsibility. This robot differs from an enabling robot in that it does not require consistent input from a human but rather can execute a practice once given its command. Examples of this kind of robot are the TUG and HelpMate robot used for deliveries in hospitals. The role and responsibility of the delivery is shared between the robot and the human deliverer/receiver; however, the robot fulfills many steps without input from a human.

Manifestation of Moral Elements as a Component of the Framework

Manifestation of moral elements refers to how the values are observed, prioritized, and interpreted throughout a care practice, in a given context (with and without the introduction of the care robot). The values are expressed as the moral elements in care, identified by care ethicist Joan Tronto. They are attentiveness, responsibility, competence and reciprocity (Tronto 1993). These elements are general such that they may be considered needs of *any* care practice, independent of individual care-givers, care-receivers, context or practice. The moral elements act as a heuristic tool to ensure the incorporation and reflection of the fundamental care values in the design of a care robot. This component relies on a detailed description of the care practice such that the moral elements become apparent through the description of the practice, in context, in terms of the actions and interactions between actors (human and nonhuman). Along with understanding the way in which the moral elements come into being it is important to indicate the distribution of roles and responsibilities as an additional tool for observing the presence of norms in a given practice.

The Care Centered Value Sensitive Design Methodology

Applying the care centered framework allows me to analyze the components of good care practices with and without the presence of a care robot. When we operationalize the framework we are able to see the relationship between the technical content of a care robot and how it works to promote (or demote) certain care values. To follow the CCVSD methodology, one begins by identifying the context, practice, actors involved and how the moral elements are manifest in traditional care practices. This does not mean that the traditional care practice (i.e. the care practice without the care robot) provides a normative standard in all cases but rather to understand how values become manifest through the practice. Keeping in accordance with the care orientation, beginning in this way allows for an analysis with the emphasis on uncovering the distribution of responsibilities, roles and values.

As I have claimed, the framework may be used for both the retrospective and the prospective ethical assessment of care robots and the manner in which the CCVSD methodology occurs differs for each. In other words, it may be used at multiple times throughout the design process of a care robot. For retrospective evaluations using the framework, one identifies the context, practice, actors and the manifestation of moral elements for the practice without the inclusion of a care robot. Following this, one then discusses the type of robot (assistive vs. enabling vs. replacement) and the manner in which the proposed care robot capabilities impact the manifestation of moral elements. As such, the evaluation of the care robot is done on a design-by-design basis according to context and practice. For retrospective analysis, the CCVSD methodology allows one to evaluate the addition of the care robot into a network of actors performing a practice in a specific context. The methodology for retrospective evaluation incorporates additional reflective tools, in particular the theory of scripts (Akrich 1992), for uncovering the deeper meaning attributed to the robot through an analysis of the (potential) embedded assumptions as well as any shift in roles and responsibilities among actors.

For prospective evaluation, one again identifies the context, practice, actors and the manifestation of moral elements for the practice without the inclusion of a care robot. Following this, one then speculates on what capabilities a robot ought to have to ensure the promotion of said values. For prospective analysis, the CCVSD methodology allows engineers and ethicists to understand the capabilities the robot ought to have in order to safeguard the manifestation of care values. In order to demonstrate how the care centered framework and the CCVSD methodology may be used for the retrospective ethical evaluation of care robots, I take the practice of lifting and compare two care robots which may be used for this practice.

Putting the Framework to Use for Retrospective Analysis: The Practice of Lifting

One of the more challenging practices for the nurse is the lifting of patients. Many elderly patients in the hospital or nursing home require partial assistance for lifting themselves out of bed or out of a chair. Alternatively, many are not capable of supporting their own weight at all and require complete assistance of a nurse to get out of bed or out of a chair. Given that the nurse must do this for any number of patients, there is a risk to the nurse's physical safety if she/he is required to lift every patient. What's more, many nurses are not physically strong enough for this. As a result, nurses have opted to use mechanical lifts on the many occasions that patients need to be lifted (Li et al. 2004).

To do this the nurse encloses the curtain around the patient to ensure privacy throughout lifting. The nurse adjusts the bed for ease of lifting and using the mechanical lift for complete assistance, the patient is lifted using a remote control, controlled by the nurse. The patient is then lowered into the chair. When the patient is being lifted there is no physical contact with the nurse, although the nurse is physically present there is no chance for eye contact as the patient is raised quite high and the nurse is paying attention to the remote control and aligning the wheelchair for placement. Thus, eye contact and touch are not possible. This first wave of automation presents a rather flat view of the care practice of lifting. Meaning, it appears to have viewed the practice as a task, as an event that is separate from the process of care and uninvolved in the manifestation of care values. As I have already shown, valued actions like touch and eye contact are integral for establishing and/or maintaining a trusting bond and this bond is integral for the provision of good care later on in the process (the patient complying with their treatment plan, taking medications, being honest about their symptoms etc.). In short, the practice of lifting requires much more than the action of lifting the care-receiver from one place to another. In order to call this is a "good care practice" according to the care orientation, many other values need to be expressed throughout the practice—values like attentiveness to the care-receiver responses or eye contact for establishing trust in the practice (and technology in this instance).

For this example, the context is the hospital, the critical care ward, the practice is lifting and the actors are the care-receiver, the care-giver, the mechanical lift, the mechanical bed, the curtain to enclose the care-receiver and the room. In terms of the manifestation of moral elements, attentiveness of the nurse is directed towards the machine and its functioning rather than exclusively towards the patient and their status. In terms of responsibility, the nurse is responsible not only for their behavior but also for the function of the machine. Trust here is a hybrid affair between the care-giver, care-receiver and the mechanical lift—meaning, the patient trusts the nurse as to whether or not the mechanical lift will work. Trust is also bestowed on the technology as a result of the values guiding the institution—the patient trusts the institution's judgment when using the technology. In terms of competence, it is questionable whether the lift facilitates the element of competence. Although the nurse is able to lift multiple patients in a given day with the same skill, the lack of touch and eye contact leave one wondering whether or not this is a skillful completion of the task. The reason to question this once again has to do with the criteria for "good care"—fulfillment of the action efficiently is not enough to render a practice 'good', the manner in which the practice is fulfilled is what makes it a 'good' practice. Thus, a good lifting practice, in the hospital, includes eye contact as a way of communicating and establishing a bond. With respect to reciprocity, the nurse is present and thus verbal and visual communication are possible; however, with the nurse's attention on the remote one would wonder how perceptive they might be to the reactions of the patient.[4] The current technology involved in the practice of lifting shows us how important it is for designers to understand the holistic vision of care and how care practices fit within this vision—they act as a moment for the promotion of care values. Consequently, the introduction of care robots presents a unique opportunity to re-introduce certain values of ethical importance. Alternatively, a robot may perpetuate the trend to minimize certain care values.

Care Robots for Lifting and Their Impact on the Moral Elements of Care

There are two robots which will be used for the CCVSD retrospective ethical evaluation of current care robot designs. Each of these robots is considered an actor in the practice of lifting once it is incorporated. The first is an autonomous robot for lifting, formerly known as the RI-MAN robot from the Riken Institute (Onishi et al. 2007) now replaced by the RIBA robot and the second is a human-operated exoskeleton, the Hybrid Assistive Limb (HAL) from Cyberdyne (Hayashi et al. 2005). There are a variety of exoskeletons currently on the market with similar capabilities and thus for the purposes of this analysis the important distinction to be made is between an autonomous and a human-operated robot. Both robots can achieve the same task (lifting a patient); however the technical capabilities through which this task is achieved differs and thus changes the way in which the caring practice is fulfilled. The autonomous robot meaning is capable of lifting a patient and carrying him/her from one place to another without being controlled by a human operator. This robot is a replacement robot, aimed to replace the care-giver in the practice of lifting. The robot is designed to work directly with humans and as such is programmed for safety considerations like speed as well as the materials which are used for its structure.

Alternatively, the human-operated robot, HAL, is an exoskeleton, meaning a human operator wears the robot in order for it to fulfill its task. The robot is a weight displacing robot such that the human does not feel the full effects of the weight. Versions of this type of robot exist in factory and military applications to prevent over exertion of factory workers or soldiers respectively. It is not an autonomous robot, but a human-operated one and is thus an enabling robot—one which does not replace a human in their role but shares the role with a human. It too will interact directly with a human (more than one in most instances) and must be programmed for the appropriate safety considerations. Given that the robot is human-operated,

---
[4] This is not a critique or nurses or the manner in which they fulfill their roles but rather a critique of the technology and its impact on the practice of lifting.

programming for safety considerations are slightly different compared to those of the autonomous one. For example, the robot will not require the same sensors for perceiving a wall, person or object in its range. While the previous robot is capable of replacing the human care-giver that would normally lift the patient, this robot is meant to assist the human care-giver with their task. By reading the biometric signals of the care-giver, the robot is able to bear the burden of the weight of whatever the care-giver is lifting. This could be a patient, a bed, a heavy box etc.

Manifestation of Moral Elements Once the Robot has been Introduced into the Care Practice

The resulting practice of lifting when comparing the two robots looks incredibly different. In the first instance, the autonomous robot, all elements have been delegated to the robot. Meaning, the robot is responsible for being attentive to the frailty of the patient when lifting, the robot is ultimately responsible for the safety throughout the practice, the robot is required to fulfill the practice in a skillful manner and, the robot is responsible for perceiving whether the needs of the patient have been met. Firstly, at this point in time, the technology does not allow for such a sophisticated manner of task completion by a robot—the robot does not know what it is doing or why it is doing something and is not capable of engaging in a reciprocal interaction. The question then is whether the robot's capabilities ought to be evaluated against the human or whether a new standard ought to exist for robots. From a moral perspective, given that the robot is a replacement robot in a nursing home context, the robot must be evaluated against a human care-giver. The same does not hold for an enhancement or assistive robot. From Tronto's perspective, which asserts that care is only good care when there is a marriage between caring about and caring for, the robot does not meet the requirements of a good carer. From Little's position of the care orientation, emphasis is drawn to the care robot's possibility to interfere with the establishment of trust and bonds and in so doing the care robot poses a threat. As such, the moral question is not whether or not the robot can fulfill the practice in an efficient manner but why the robot is being delegated this role and responsibility. Alternatively, from a technical perspective this does not demand that the robot's appearance mimic a humans—the autonomous robot could perhaps have four arms instead of two to facilitate lifting. But, from a moral stance, a robot replacing a human must be assessed according to whether or not it is capable of facilitating and promoting the vales a human care-giver brings about.

If, however, the robot is (someday) capable of understanding what it is doing and why, and may act in a skillful manner, the robot still poses a threat to the holistic process of care. The holistic process of care refers to the concept that care is not one task or a series of tasks but is a compilation of practices to meet the needs of the actors, each practice building on the last. For the practice of lifting it is necessary that the care-receiver trust the care-giver initially in order for lifting to occur but it is also a moment in which the two can establish a bond to ensure compliance with a care plan further along. Seeing as the robot replaces the care-giver for this practice, it is possible that the care-giver and care-receiver may not have the same bond—the care-receiver may not feel any sort of responsibility towards the care-giver to abide by their care plan. What's more, perhaps the care-receiver may not trust the plan established by the care-giver. This is not to say that trust cannot be established through another practice, but rather that it does not present the forum in which trust is traditionally established or strengthened. Taking this into consideration in the design process of the robot means that designers ought to anticipate this and perhaps consider another forum in which trust can be established or strengthened between the care-giver and care-receiver.

Alternatively, in another context there may be care-receiver's who would prefer the assistance of an impartial robot to keep their dignity and integrity intact. This could be the case in a home setting where a spouse is the only one available for lifting. With a change in context comes a change in the interpretation and prioritization of values. The care-giver may be the spouse of the care-receiver and thus privacy is seen in a different way. Instead of the curtain being closed as a way to ensure privacy, the care-receiver would prefer not to be exposed in that way to their spouse at all. This interpretation of the value of privacy may place it at the top of the hierarchy of values above others like efficiency, eye contact, touch or human presence. Consequently, having an autonomous robot to fulfill the practice of lifting may be seen as a more compassionate means when the care-receiver's vulnerability is maximized by requiring help for these practices. Once again, this divergence shows the importance of context in the ethical assessment of a care robot.

In the second instance, the case of the enabling robot, the element of attentiveness is still in the domain of the human as is the element of responsiveness. For the former, the care-giver uses his/her own faculties to ascertain when the care-receiver needs to be lifted, at what speed, from which angle and with or without social interaction. For the latter, reciprocity is something that happens between the care-giver and care-receiver in real time by verbal and nonverbal cues detected by the care-giver. Meaning, the nurse can ask the patient how they are doing while they are lifting. As for responsibility and competence, these elements now become shared endeavors between the human and the robot given that the role of weight-bearer is delegated to the robot. The care-receiver and care-giver must both trust the technology—responsibility for the safety of the practice becomes a hybrid event between the human care-giver and the robot. Additionally, a certain amount of competence for the skillful completion of the practice is delegated to the robot. Thus, a portion of the responsibility for lifting is delegated to the robot as is a certain level of skill; however, this is done in an enabling way, therefore the human care-giver is still responsible overall. This also means that the robot is not evaluated against the human completely but is evaluated as to how it enables the human in their performance of the practice. As such the robot is evaluated according to how well it bears the weight but not how well is it able to pick up on nonverbal cues of the care-receiver.

The reflections provided here give preliminary insights into the impact of a care robot and its capabilities on the resulting care practice. This is not an exhaustive reflection, however, given that a deeper meaning may be attributed to the inscribed shift in roles and responsibilities. Accordingly, the CCVSD methodology incorporates additional tools for an in-depth look at the meaning the robot may take on.

Attributing Meaning to Care Robot Designs

It is only through a deeper understanding of what care values are and how they are manifest throughout a care practice that we come to grasp the impact a design might have on the care practice. Above and beyond the direct relationship one might uncover between care values and the technical capabilities of the care robot, there is greater meaning attributed to these capabilities upon further reflection.

Akrich discusses the embedding of elements in terms of assumptions made about user preferences and competencies (1992). Placed in context, each robot takes on a distinctive meaning related to the assumptions embedded within. This description is quite useful for my reflection and an important distinction must be made here pertaining to the difference between assumptions and the concept of values and norms. Assumptions are more about the real word, they are descriptive in a sense while values are more about what the real world ought to be like, they are normative in a sense. When an assumption is made about a value to be embedded, it does not have to be a description about what is, but could also be a claim about what values ought to be expressed, how they ought to be expressed, or what priority they ought to be given. In others words, when the built-in assumption pertains to a value, or when a valuation is being made, the result is a normative claim about what the values should be, what should be valued, or what the ideal is. For Akrich, "many of the choices made by designers can be seen as decisions about what should be delegated to a machine and what should be left to the initiative of human actors" (p. 216). By making choices about what should and should not be delegated to certain actors (human or nonhuman), engineers may change the distribution of responsibilities in a network.

Consequently, each robot reflects divergent assumptions pertaining to the understanding of a care practice, the aim of the care practice and the prioritization of values manifest through a care practice. When using the autonomous robot in a hospital setting, the understanding of the practice reflects a vision of a task rather than a practice—that lifting is just an action that needs to be done in order to get on to the next action. As I have stressed throughout the paper, from the care ethics perspective this is not what a good care practice looks like. The ideal practice of lifting seen through this robot is a standardized one where the value of efficiency is placed as the top priority. Although efficiency was not explicitly discussed previously, it is thought to fall under the realm of competence. This one-dimensional view of good care as efficient may have negative implications for the overall care process. One may presume that the quality of interactions, the number of social interactions, and the presence of a human are threatened by this efficient system. Alternatively, the system may be considered efficient given that time of the human care-giver is freed up, ultimately improving the number of social interactions and the quality thereof. This last point is also dependent on how the robot is introduced and would not become apparent until the robot has been implemented in a specific context.

The autonomous robot reflects a vision of the practice of lifting which does not require any of the values traditionally involved; human touch, eye contact, human presence. If these values are normatively understood and recognized as only possible through human–human interactions,[5] then this demands that a human be present for the practice of lifting in all instances. However, as we saw before, context plays a role. It is possible to suggest that in the context of the hospital or nursing home, where "good care" depends on the relationship between the care-giver and care-receiver, human contact for a practice like lifting is always required. This is in part due to the vulnerability of the care-receiver while being lifted as well as the need to form a bond between care-giver and care-receiver. Alternatively, in a home context in which a relationship between care-giver and care-receiver is already established and strengthened, the need for human presence, eye contact and touch for the practice of lifting may not be as pertinent. Moreover, when the care-giver is a spouse it may be preferable not to have the human present. Thus, both design and integration into the healthcare system are of importance here.

Alternatively, the human-operated exoskeleton reflects an understanding of this care practice as one in which the aim of the practice is not solely to lift the care-receiver from one place to another but is a moment to establish a bond and convey other care values. The vision of care presupposed in the design of the human-operated care robot is one in which individualized care with a human care-giver present at all times for all parts of the care practice, is the overall aim. Efficiency is still a priority; however, it is achieved through meeting the need of the care-giver by contributing to the element of competence (enhancing the skill with which the care-giver may perform their role), attentiveness, (enabling the care-giver to perceive the minute cues of the care-receiver through the practice of lifting), and responsiveness (closely aligned with attentiveness but also embodies the reciprocal dimension of the relationship). Consequently, by demanding the human's presence for the task of lifting, the robot pays tribute to the holistic vision of care and the intertwining of needs and values.

I cannot say whether this is the epistemic aim of engineers, but can only point to the potential meaning that the robot may take on through pervasive use, and the presupposing assumptions directing such a meaning. Moreover, this is not to say that the autonomous robot ought to be disregarded or labeled as unethical—a variation in context changes things. Clearly, decisions concerning the use of a robot and its ethical implications are many-sided and complicated and demand an understanding of the specific context and users for anticipating how the elements will be served to their greatest potential.

Conclusion

The prospective robots in healthcare intended to be included within the conclave of the nurse-patient relationship require rigorous ethical reflection to ensure that their design and introduction do not impede the promotion of values and the dignity of patients at such a vulnerable and sensitive time in their lives. The ethical evaluation of care robots requires insight into the values at stake in the healthcare tradition. What's more, given the stage of their development and the lack of standards to

---

[5] I say 'many' of these values given that certain telepresence robots are capable of providing eye contact and the feeling or impression of human presence even when the human is not physically present.

guide their development, ethics ought to be included within the design process of such robots. The manner in which this may be accomplished, as presented here, uses the blueprint of the Value-sensitive design approach as a means for creating a framework tailored to care contexts. Using care values as the foundational values to be integrated into a technology and using the elements in care as the normative criteria, the resulting approach is referred to here as "care centered value-sensitive design".

The care centered framework is meant to indicate and direct the evaluator to the necessary components in care from a care orientation. The CCVSD methodology is meant to provide a guideline for analysis of a practice with and without the use of a care robot. Using the CCVSD methodology to compare two care robots used for the same practice with different capabilities, allows us to envision the resulting care practice in terms of the robot's impact on care values as well as the robot's potential impact on care in the holistic sense. For the latter, this is only understood when one grasps the interconnectedness of one practice with another. From this, the link between robot capabilities and their impact on the manifestation of care values is made clear. Consequently, we may ask the question; what kind of care do we want to provide and in so doing we may steer the design and development of care robots.

The aim of this paper was to present the conceptual foundation for the creation of a framework and methodology for the evaluation of care robots both retrospectively and prospectively. The aim was not to present an exhaustive evaluation of current and future care robots as this will be taken up in a later paper. The significance of this work comes from the stage of development of care robots and the belief that ethics may be included at this time in the design process to foster trust between the public and the resulting robots. The care centered framework adheres to the central thesis of the care orientation—that one is oriented to the components which require attention in order to begin ethical deliberation. What's more, the framework provides a starting point for the interdisciplinary collaboration of a range of robotics researchers—from designers, engineers and computer programmers to ethicists, psychologists and philosophers.

Open Access This article is distributed under the terms of the Creative Commons Attribution Noncommercial License which permits any noncommercial use, distribution, and reproduction in any medium, provided the original author(s) and source are credited.

References

Akrich, M. (1992). The de-scription of technical objects. In W. Bijker & J. Law (Eds.), *Shaping technology/building society*. Cambridge, MA: MIT Press.
Asaro, P. (2006). What should we want from a robot ethic? In R. Capurro & M. Nagenborg (Eds.), *Ethics and robotics*. Amsterdam, The Netherlands: IOS Press.
Barras, C. (2009). Useful, loveable and unbelievably annoying. *The New Scientist*, 22–23.
Brey, P. (2009). Values in technology and disclosive computer ethics. In L. Floridi (Ed.), *The Cambridge handbook of information and computer ethics*. Cambridge: Cambridge University Press.
Cummings, M. (2006). Integrating ethics through value sensitive design. *Science and Engineering Ethics, 12*(4), 701–715.
Engelberger, J. (1989). *Robotics in service*. London: MIT Press.

Freidman, B., Kahn, P., et al. (2006). Value sensitive design and information systems. In P. Zhang & D. Galletta (Eds.), *Human–computer interaction in management information systems: Foundations* (pp. 348–372). NY: M. E. Sharpe.

Friedman, B., & Kahn, P. H., Jr. (2003). Human values, ethics, and design. In *The human–computer interaction handbook: Fundamentals, evolving technologies and emerging applications* (pp. 1177–1201). Mahwah, NJ: Lawrence Erlbaum Associates, Inc.

Gadow, S. A. (1985). Nurse and patient: The caring relationship. In A. H. Bishop & J. R. Scudder Jr. (Eds.), *Caring, curing, coping: Nurse, physician, patient relationships* (pp. 31–43). University Alabama: The University of Alabama Press.

Gilligan, C. (1982). *In a different voice: Psychological theory and women's development*. Cambridge, MA: Harvard University Press.

Hayashi, T., Kawamoto, H., & Sankai, Y. (2005). Control method of robot suit HAL working as operator's muscle using biological and dynamical information. IEEE *International Conference on Intelligent Robots and Systems*, 3063–3068.

Howcroft, D., Mitev, N., & Wilson, M. (2004). What we may learn from the social shaping of technology approach. In J. Mingers & L. P. Willcocks (Eds.), *Social theory and philosophy for information systems* (pp. 329–371). Chichester: Wiley.

ISO. (2011). http://www.iso.org/iso/iso_catalogue/catalogue_tc/catalogue_detail.htm?csnumber=53820.

Jecker, N. S., Carrese, J. A., & Pearlman, R. A. (2002). Separating care and cure: An analysis of historical and contemporary images of nursing and medicine. In E. Boetzkes & W. J. Waluchow (Eds.), *Readings in healthcare ethics* (pp. 57–68). Canada: Broadview Press.

Kawamoto, H., & Sankai, Y. (2002). Power assist system HAL-3 for gait disorder person. *Computers Helping People with Special Needs; Lecture Notes in Computer Science*, 2398, 19–29.

Latour, B. (1992). Where are the missing masses? The sociology of a few mundane artifacts. In W. Bijker & J. Law (Eds.), *Shaping technology/building society*. Cambridge, MA: MIT Press.

Le Dantec, C., Poole, E., & Wyche, S. (2009). Values as lived experience; evolving value sensitive design in support of value discovery. In *Proceedings of the 27th International conference on human factors in computing systems*. ISBN: 978-1-60558-246-7.

Li, J., Wolf, L., & Evanoff, B. (2004). Use of mechanical patient lifts decreased musculoskeletal symptoms and injuries among health care workers. *Injury Prevention*, 10(4), 212–216.

Little, M. (1998). Care: From theory to orientation and back. *Journal of Medicine and Philosophy*, 23(2), 190–209.

Mander-Huits, N. (2011). What values in design? The challenge of incorporating moral values in design. *Science and Engineering Ethics*, 17, 271–287.

Metzler, T., & Lewis, L. (2008). Ethical view, religious views, and acceptance of robotic applications: A pilot study. *Association for the Advancement of Artificial Intelligence*. http://www.aaai.org.

Nathan, L., Friedman, B., Klasnja, P., Kane, S., & Miller, J. (2008). Envisioning systemic effects on persons and society throughout interactive system design. In *Proceedings of DIS* (pp. 1–10).

Nissenbaum, H. (1998). Values in the design of computer systems. *Computers and Society*, 38–39.

Noddings, N. (1984). *Caring: A feminine approach to ethics and moral education*. Berkely: University of California Press.

Onishi, M., ZhiWei, L., Odashima, T., Hirano, S., Tahara, K., Mukai, T. (2007). Generation of human care behaviours by human-interactive robot RI-MAN. IEEE *International Conference on Robotics and Automation*, 3128–3129.

Reich, W. T. (1995). History of the notion of care. In W. T. Reich (Ed.), *Encyclopedia of Bioethics*, vol. 5 (pp. 319–331), Revised edition. New York: Simon & Schuster Macmillan.

Rosati, C. (2009). Relational good and the multiplicity problem. *Philosophical Issues*, 19(1), 205–234.

Ruddick, S. (1995). *Maternal thinking*. New York: Ballantine Books.

Sharkey, N., & Sharkey, A. (2010). Granny and the robots: Ethical issues in robot care for the elderly. *Ethics and Information Technology*. doi:10.1007/s10676-010-9234-6.

Simpson, J. A., & Weiner, E. S. C. (1989). *The Oxford english dictionary*. Oxford: Clarendon Press.

Sparrow, R., & Sparrow, L. (2006). In the hands of machines? The future of aged care. *Mind and Machine*, 16, 141–161.

Super, D. E. (1973). The work values inventory. In D. G. Zytowski (Ed.), *Contemporary approaches to interest measurement*. Minneapolis: University of Minnesota Press.

Tronto, J. (1993). *Moral boundaries: A political argument for an ethic of care*. New York: Routledge.

Tronto, J. C. (2010). Creating caring institutions: Politics, plurality, and purpose. *Ethics and Social Welfare*, 4(2), 158–171.

Vallor, S. (2011). Carebots and caregivers: Sustaining the ethical ideal of care in the 21st century. *Journal of Philosophy and Technology, 24,* 251–268.

Van den Hoven, J. (2007). ICT and value sensitive design. *International Federation for Information Processing, 233,* 67–72.

Van Wynsberghe, A., & Gastmans, C. (2008). Telesurgery: An ethical appraisal. *Journal of Medical Ethics, 34,* e22.

Vanlaere, L., & Gastmans, C. (2011). A personalist approach to care ethics. *Nursing Ethics, 18*(2), 161–173.

Verbeek, P. (2006). Materializing morality: Design ethics and technological mediation. *Science, Technology and Human Values, 31*(3), 361–380.

Verbeek, P. (2008). Morality in design; design ethics and the morality of technological artifacts. In P. Vermaas, P. Kroes, A. Light, & S. Moore (Eds.), *Philosophy and design: From engineering to architecture* (pp. 91–102). Berlin: Springer.

Veruggio, G., & Operto, F. (2006). Roboethics: A bottom-up interdisciplinary discourse in the field of applied ethics in robotics. In *International Review of Information Ethics.* Ed. Ethics in Robotics, pp. 2–8.

WHO. (2010). *Health topics: Ageing.* Available from: http://www.who.int/topics/ageing/en/.

Wilson, M. (2002). Making nursing visible? Gender, technology and the care plan as script. *Information Technology and People, 15*(2), 139–158.

World Health Organization. (2007). People centered health; a framework for policy. http://www.wpro.who.int/NR/rdonlyres/55CBA47E-9B93-4EFB-A64E-21667D95D30E/0/PEOPLECENTRED HEATLHCAREPolicyFramework.pdf.

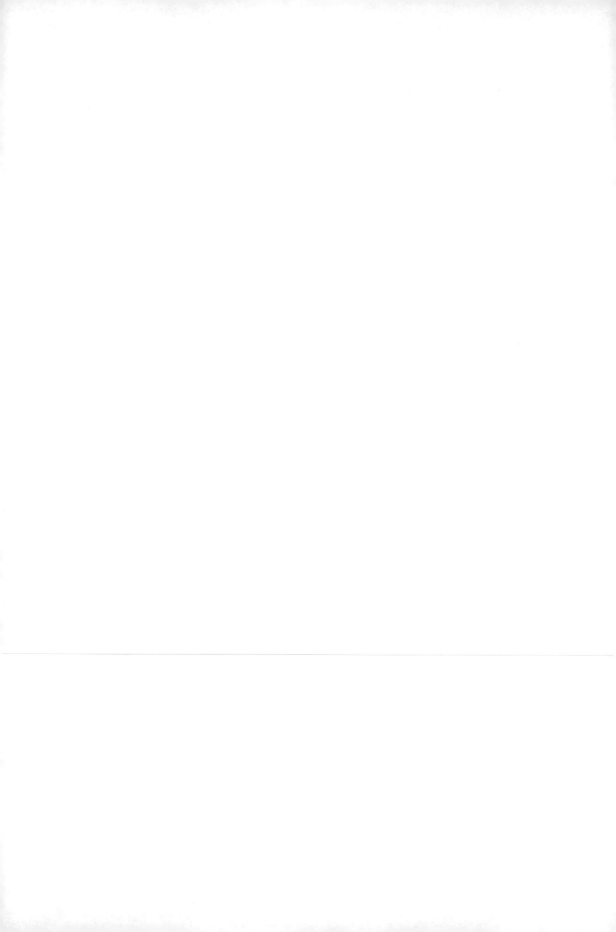

15

Robots, Love, and Sex:
The Ethics of Building a Love Machine

John P. Sullins

Abstract—This paper will explore the ethical impacts of the use of affective computing by engineers and roboticists who program their machines to mimic and manipulate human emotions in order to evoke loving or amorous reactions from their human users. We will see that it does seem plausible that some people might buy a love machine if it were created, but it is argued here that principles from machine ethics have a role to play in the design of these machines. This is best achieved by applying what is known about the philosophy of love, the ethics of loving relationships, and the philosophical value of the erotic in the early design stage of building robust artificial companions. The paper concludes by proposing certain ethical limits on the manipulation of human psychology when it comes to building sex robots and in the simulation of love in such machines. In addition, the paper argues that the attainment of erotic wisdom is an ethically sound goal and that it provides more to loving relationships than only satisfying physical desire. This fact may limit the possibility of creating a machine that can fulfill all that one should want out of erotic love unless a machine can be built that would help its user attain this kind of love.

---

1 INTRODUCTION

WHEN your robotic lover tells you that it loves you, should you believe it? The roboticist David Levy has provocatively argued that there is nothing about human love and sex that could not be engineered into a suitably designed robot in the relatively near future [1]. He also argues that these machines would not only be psychologically pleasing, but that their users might even eventually find them preferable to human suitors and that the machine itself would feel a love that may have artificial origins but that is nonetheless genuine feelings of love toward its user [1].

The dream of a perfect artificial lover is at least as old as the myth of Pygmalion. It is also a staple of classic science fiction, which abounds in morality plays about the emotional costs of falling in love with one's own creation.

While the prospect of a robot lover is the stuff of science fiction dreams, the design of robots with the ability to navigate human social settings, such as care giving, domestic work, and companionship, does continue to evolve. It is very important in this discussion to disentangle the robots of fiction from the actual robots we are likely to see in the near future. This topic is also difficult due to the overhyped claims of the roboticists who are attempting to build robot and android lovers as well as the hyperbolic media coverage that surrounds their every claim of success. The instant one hears the words "robot lover," many fanciful visions will flood the imagination of all but the most prosaic reader. It is also very possible that those that are in the business of creating robot companions are deeply influenced by the literature and movies that so compellingly depict the fantastic world of robot love. But in this paper, I will attempt to err on the side of conservatism in my prognostication of near future robotics.

We are nowhere near the point where we can build the kind of machines seen in science fiction where androids that are so like humans it is impossible to tell them apart coexist with humans with either utopian or dystopian results. Yet I would like to remind the reader that it does not take much sophistication to build machines that will, at least for a time, engage their user in compelling and affective relations. A good example is the Tamagotchi fad that had people devoting many hours of their lives to the care and feeding of an artificial pet. Another good example is the way a gearhead may fall in love with their car. Obviously, it is not necessarily a need to have robots that can convince the strongest skeptics of their agency, consciousness, free will, and/or intelligence before they will be able to draw on strong loving emotions from their less philosophically demanding users. As we will see in the sections that follow, it might only take a silicone love doll with modest mechatronics to enamor some users. By this I mean that it is possible to create machines that provide stimuli that can evoke strong sexual reactions from some users and that this achievement is far simpler than trying to create a machine that can simulate more complex affective emotions. Because of this, the ethics of robot love is far more pressing than one might think and we do not have to wait for complex android lovers to become commonplace before we begin to address the implications of both the existent technologies and those that are on the horizon.[1]

Manuscript received 28 Feb. 2012; revised 2 Aug. 2012; accepted 23 Aug. 2012; published online 25 Sept. 2012.
Recommended for acceptance by A. Beavers.
For information on obtaining reprints of this article, please send e-mail to: taffc@computer.org, and reference IEEECS Log Number TAFFCSI-2012-02-0018.
Digital Object Identifier no. 10.1109/T-AFFC.2012.31.

1. Readers interested in exploring the many issues involved in the ethical design of artificial companions which go beyond the narrow scope of this paper should start by looking at: Wilks and Yorick (eds.), *Close Engagements with Artificial Companions: Key Social, Psychological, Ethical and Design Issues*, John Benjamins Publishing Company: Amsterdam, NLD, 2010.

In order for an advanced robot be successful in the role of a companion, friend, or surrogate lover, these future systems will need to elicit and manage both the strong human emotions that can accompany the social milieus these machines are designed to be deployed in. In addition, they must appropriately manage the more subtle and continuously changing affective mood of the people that come into contact with the machine on a daily basis.

The psychologist Roddy Cowie argues that most human-computer interaction of the last few decades has been accomplished by the user entering an unemotional state that facilitated interacting with computers but that artificial companions will not be able to be built this way since the:

> Companion's goals are likely to be bound up with emotion—as much to do with making somebody feel happy and confident, as with accomplishing practical tasks economically [2].

The users of these companions will not be able to maintain an unemotional interaction with the artificial companion and instead the machine must be designed to properly navigate what Cowie calls "pervasive emotions," which is also sometimes called "affect," but whatever we call it we also have to admit that we do not completely understand it [2]. For Cowie, pervasive emotions are the positive and negative emotions we get as we determine what things or persons to care about in a given situation, our feeling of attraction and revulsion, our sense of understanding of what is going on in relations between other agents, and our engagement and attention to shared goals [2]. This all serves as a kind of rich information channel between agents that facilitates easy exchanges and emotional understanding. Simply put, it is a kind of comfort with the surroundings and other agents one may be engaged with.

Following Cowie's line of reasoning, then an artificial companion has a complex problem to solve. The machine must be able to detect the signals of its users related to emotions, synthesize emotional reactions and signals of its own, and be able to plan and carry out emotional reasoning [2].

At this time affective computing techniques are only just beginning to touch on the first two requirements of recognizing and synthesizing emotional cues and responses but are still largely incapable of emotional reasoning. We will need all three of these functioning at a very high level to achieve robotic lovers of real worth.

One growing trend in robotics meant to deal with the issues raised by the pervasive emotion problem has been to design hardware and software that utilizes the human psychological tendency to anthropomorphize objects, which can also cause the user to ascribe effective motivations to these robots [3], [4], [5]. Using these affective (or sociable) computing techniques to mimic human emotions helps the systems manipulate human reactions in such a way as to cause the user to interact more easily and fondly with the machine.

Because of these developments in affective robotics, it is argued in this chapter that these designers should recognize certain limits as to how much manipulation of these human psychological tendencies among the prospective users of these machines is ethically permissible. While it can be argued that it is ethically permissible to design robots that act in concord with their users and that this concord will require affective computing applications built into these machines [6], [7], [8], [9], given that we will be able to mimic emotions in a robot long before we will be able to produce truly affective machines, it is advisable to be circumspect in how we exploit human psychology in the design and deployment of these machines [2].

Love is perhaps the most important of human emotions and those who experience it are strongly motivated to attain both the heights and depths of human achievement. Because this emotion is so important and complex it is an ambitious undertaking when roboticists attempt to instantiate this emotion through affective computing techniques. Since manipulating strong emotions is an ethically fraught undertaking, we should be cautious and skeptical when approaching this kind of work. In the next sections, we will look at the complexities of determining what the folk concept of love is, as well as how it is described in cognitive psychology. While many roboticists are well aware of both of these ways of looking at love, they rarely look at the extensive literature within the philosophy and ethics of sex and love, so we will also add that to our discussion. We will also look at how roboticists implement affective love in their machines in the quest to create robotic lovers and conclude with a critique of those efforts as well as offering some suggestions for an ethics of implementing affective love in robotic devices.

In short, this paper will argue for the following claims:

1. Psychological factors in love, sex, and attraction can be at least functionally duplicated in robotics technology.
2. Designers of affective robots can utilize the human psychological tendency to anthropomorphize animals, objects, and technologies.
3. They can also tap into the human predisposition to be interested in developing caring relationships for creatures outside our own species which has evolved in the human species.
4. We will find that psychologists have shown that we are easily duped into believing the feigned affectations of a false lover. If someone acts as if they love us, even if those actions are very minimal and easily contradicted, we will still tend to believe they truly do love us.
5. We will also see that the early adaptors of affective computation and robotics have already begun to prefer human computer interaction to interaction with fellow humans.

Once we have argued for the above we will have to conclude that a robot that can manipulate the described social and psychological tendencies would be able to form relationships at least as real and moving as those we have with our beloved pets and insincere lovers. Robotic love will work, but only because we are so bad at finding a more true love. This will allow us to then look at a beginning proposal for certain ethical constraints that ought to be used to mitigate the damage that might be caused by affective robots that are programmed to manipulate human psychology to simulate a loving relationship. Finally, philosophers have a long tenure in the business of wondering about the nature of true love. We will conclude with a discussion on what the philosophy of love and sex can usefully add to affective robotics. We can design machines some of us will fall in love with. The more interesting question is, can we design machines that could make us and our machine lovers better for having fallen in love?

2 ROBOTIC LOVERS

2.1 Seriously, What Is Love Anyway?

There is a great deal of ironic humor to be found in imagining highly educated researchers puzzling over how best to create a robotic love and sex when actual love and sex is everywhere around us. Even as I write this sentence two doves are courting one another on a tree branch outside my window while undergraduates walk by on the path below, trysting and holding hands. Finding love seems as easy as just going outside and looking in any direction.

But Levy [1] reminds us in his groundbreaking book *Love and Sex with Robots* that there are some compelling reasons that might cause people to legitimately want to have robotic lovers. There are tragic reasons, such as physical or emotional deficiencies that make finding a human partner impossible, which might be alleviated by robotically assisted sexual therapy. Or perhaps one is not interested in, or does not have the time to develop, a full loving relationship but just wants a sexual encounter, yet one also finds institutions like prostitution objectionable; a robotic prostitute might then be a palatable solution. This option might also end the ethically troubling human sex industry by replacing objectified human beings with actual objects that presumably have no rights to worry about.

It is also possible that these machines might be used to enhance human sexual relationships as a kind of super love toy. But Levy's most interesting argument from a philosophical point of view is that we should take this technology seriously mainly because we might be able to experience a more perfect love through its use.

A machine that was designed to be the perfect match for its user and was also programed to love the user completely would be immensely pleasing. Levy argues that this kind of machine would also be a close friend that will certainly "...behave in ways that one finds empathetic, always being loyal and having a combination of social, emotional, and intellectual skills that far exceeds the characteristics likely to be found in a human friend" [1, p. 107].

Who could pass up a chance to be with their robotic soul mate? The robot would be interested in all the same things as its user. It would be built to the user's specifications so that he or she found it to be physically sexually attractive. Best of all, the robot could be programmed to be always loyal to its user and display fascination toward him or her and whatever they have to say. This would be a dream come true. Levy's argument is that since these robots could add so much love and happiness to our world, it is almost a moral imperative that we work to make these theoretical robotic companions a reality.

The logic is simple, robotic companions would give us perfect love; perfect love is a moral good, so robotic companions would provide us with a moral good. But the argument is also begging a number of questions. Are robots really capable of achieving all the qualities necessary for a perfect lover? And even if they are, is the brief list of qualities just outlined above sufficient for all we want out of a loving relationship?

2.2 Robot Sex

Before we address these questions we should cede one important point to the roboticists working in the area of artificial companions. It is obvious that they will be able to build successful sex robots that some people will find very compelling and will readily use. Different haptic devices are already in production that interface with the genitals of male users.[2] The machine links to adult films. The device simulates the actions on the screen for the person watching the film through small lubricated conveyor belts, heating elements, and bellows. This device is not exactly what one thinks about when they imagine a robot, but it is where this industry seems to be starting. There is also already a brisk business in life-sized, realistic silicone rubber love dolls such as the infamous "Real Doll."[3] Obviously, people are willing to pay a premium price for these dolls even though *Real Dolls* have no capability to interact dynamically with their user.

Recently, there have been at least two companies to enter this market who have released both male and female sex dolls to which they have added meager amount of physical interactivity through modest mechatronics and simple android technologies. Their machines can autonomously bump and grind in a somewhat awkward fashion, while their artificial heart beats rapidly under their heaving silicone breasts as they encourage their human lovers with an X-rated AI chatbot controlled voice; all no doubt to the great delight of their owners [10], [11].

Levy [1] devotes four chapters of his book to a well-crafted and meticulous argument describing the human psychology of sex and how roboticists will be able to successfully exploit our prurient interests and how that may actually help mitigate certain sexual and marital pathologies and dysfunctions.

The situation may be even more complicated as psychologists Meston and Buss have catalogued over 237 different reasons that people express for having sex [12]. Many of the reasons are what one would expect: "I wanted to experience physical pleasure," "I wanted a child," or "I desired intimacy," but some are unexpected, such as "I wanted to punish myself" or "I wanted to get closer to God." Meston and Buss broadly categorize the responses into four main reasons: physical (attraction or pleasure seeking), goal attainment (to get a job or raise), emotional (it was romantic), and insecurity (it was my duty or obligation) [12]. Meston and Buss also found that while men and women reported many of the same reasons for having sex, the top reasons for having sex were ordered slightly differently depending on gender. For instance, the number one reason for both men and women was "I was attracted to the person," but "I wanted to express love for

---
2. For example, see: http://www.realtouch.com/static/0004/index.html.
3. http://www.realdoll.com/.

the person" was reason number five for men and eight for women [12]. While surveys like these are not perfect methods of study for the complex psychological questions surrounding human sexuality, they do cast some light on the situation and suggest that human sexual motivation beyond the purely physical is a very complex affair and may not be something that can be fully captured by robotics. Still, there has been some success with machines that appeal mainly to the physical sexual motivations of their users and it is likely that appealing to these motives will lead to the greatest short term success in the design of love machines.

Since the technologies needed to make interactive sex dolls already exists and real humans are actually choosing to have sex with these dolls, then it is likely that Levy is correct at least in his prediction that an increasing trend toward sex with robots is just beginning and that it is likely to continue to gain popularity, leading to a very strange and wild future.

2.2.1 Love Prototypes

The folk definition of love is a very loosely defined concept. It is a cluster concept that is associated with other concepts such as: care, attraction, affection, and liking.

Fehr and Russel [18], [19] conducted a survey of the types of love people refer to. They received 93 different "types" which they then asked respondents to rank in order of their closeness to ideal love.

Here, are some of the responses in descending order of closeness to the prototypical concept of true love:

Love Prototype

1. maternal love,
2. parental love,
3. friendship,
4. sisterly love,
5. brotherly love,
6. romantic love,
7. passionate love,
8. sexual love,
9. platonic love.

One counterintuitive thing that comes from this analysis is that passionate and sexual love are lower on the list than parental love and friendship, with motherly love being the most closely related type of love to what we mean by the abstract word "love." It is correct to ask where on this list should we seek to place the new prototype, robotic love. It does not look like the kind of relationship one has with a sex robot even makes it onto this list, even though it might have some superficial similarities with sexual love.

One way to counter this claim would be to suggest that the concept of prototype is not an evaluative claim; it is not saying that motherly love is somehow "better" than sexual love. That is true, of course, and failure to develop an understanding of the difference between motherly and sexual love would lead one to a Norman Bates-like existence. That is not the point I am trying to raise here. Instead, I am pointing out that robot love is nowhere near any of these prototypes, meaning it cannot be put on the list and must be either a self-contradiction or at best an entirely new type of relationship we have no exemplar for.

You will recall that Levy's argument was based on the intuition that people are good judges of human behavior and that if a robot acts like it loves you, then it is probably a good sign that it does loves you, especially if it is an intelligent machine that is fully capable of saying the worlds "I love you" coherently in a conversation. But, as it turns out, humans are very bad judges at correctly ascribing feelings of love even in other human lovers.

Gilbert and Jones [20] have studied the tendency of humans to construct their own self-generated reality in which they ascribe true feelings of love to other people even when the evidence is clear that these feelings are unreciprocated. It would seem that people largely assume that another person's beliefs correspond more or less directly with perceived behavior. When someone says "I love you," Gilbert and Jones found that, among their subjects of study, they assumed that the other person actually did love them, even when the spoken words of love are preceded by pleading, such as, "just tell me that you love me" [20].

These findings of Gilbert and Jones give cause to be skeptical that even though one's robot might tell you loud and clear that it loves you and you may believe it with all your heart, that does not at all allow an impartial observer to conclude that the robot actually loves you.

This does not give an a priori argument that robot love is impossible, instead it is just a warning that programmers and designers will find it easier to build a machine that can cause users to believe they are in love without having to solve the much more difficult problem of building a machine actually capable of reciprocal love.

2.2.2 The Ethics of Robot Sex

Since using a robotic sex doll with only limited mechatronics and low level AI is just a very elaborate act of masturbation, the ethics of their use will depend on the ethics of self-gratification. There are numerous cultural and religious constraints to this activity that tend to focus on how these acts may lead to social and or spiritual isolation and therefore masturbation is something to be avoided. In film director Shohei Imamura's 1966 Japanese new wave cinema classic *The Pornographers (Erogotoshi-tachi yori: Jinruigaku nyûmon)*, the protagonist of the film decides near the end of the movie that his life of making pornography and pimping is not solving all the anxieties of his postmodern life, but perhaps building a perfect robotic woman, a "Dutch wife" as he calls it, will help. The character in the film is tragically wrong, the robot proves to be too complex to make, and his obsession with it cuts him off both figuratively and literally from all human contact. In the last frames of the film, we see him floating off on a powerless boat, his robot in pieces, into the vast ocean. Imamura's movie may be as prescient as it is surreal. The thought that robotic lovers may serve to enhance rather than mitigate human sexual pathologies is at least a reasonable hypothesis to explore. Still, in most cultures of the modern world there is an acceptance of masturbation as a common human activity and nothing to get all that worried about, so how could the addition of a robotic accessory to that basic human drive change much?

Let us now look at the question of the psychological health effects of having a sexual relationship with a robotic love doll. While there is quite a bit of psychological research into the effects of children's play with dolls, there is not much at present in the academic study of how adults interact with sex dolls and robots. One exception is an editorial on sexbots by Joel Snell [13] who is very skeptical of any claims that sexbots will be useful in reducing sexual and marital pathologies; he argues that they will instead increase the instances of various sexual and marital problems encountered in societies today and would therefore not contribute to the common good. Along similar lines is the work of Sherry Turkle [14], who argues that even the limited capabilities of the companion robots available today are causing some of us to mistakenly ascribe human qualities to these technologies and to search in vain to find comfort and companionship where none is to actually be found while ignoring the other humans that surround us. It is obvious that more work needs to be done here. Are sex robots capable of mitigating psychological pathologies or are they contributing to them? If they are mitigating the problems, then creating and using sex robots would be an ethical act. On the other hand, if they contribute to more psychological problems, then their use and design would be morally suspect.

We must also approach the issue from a more philosophical direction and ask about the impact on the quality of our lives that sex robots have in regard to our happiness and concord with the other humans we live with. Coeckelbergh [15] argues that it is important to interrogate what the actions and physical appearances of robots do to humans as social and emotional beings. What he means is that the design of these machines will influence the prospects of their users in achieving human flourishing through a life that is good in a philosophical sense. Coeckelbergh is correct to argue that roboticists use the full force of their imagination to create machines that will enhance the human condition. I have also argued a point that is closely allied with this idea in another work, where I argue that it is better to design robots that enhance human friendships rather than attempt to simply replicate or replace them [16]. Another rather obvious critique of the sex robots that have been built is that they are rather grotesque caricatures of the human form that almost mock the female body in ways that seem to be designed to alienate and intimidate women. These dolls do not live up to the challenge roboethics has made to be imaginative, playful, and, most importantly, friendly with the way robots are designed to interact with people.

Let us now review the argument presented above. The first premise is that the complexity of human sexuality insures that sex robots will mostly be aids for self-gratification. The second premise finds that leading psychologists and social scientists studying this technology argue that sex robots will most likely contribute to psychological disorders [13], [14], rather than mitigating them, as is Levy's hope. The third premise argues that these machines will not help their users form strong friendships that are essential to an ethical society [15], [16], and may indeed lead to more isolation. The final premise points out the fact that these machines, as they are being built today, contribute to a negative body image for real humans through the exaggerated body shapes they now take. It follows from these premises that sex robots, as they are conceived of today, are not likely to be a net positive to society. Simply strapping a silicone sex organ to a washing machine on spin cycle is not much in the way of human achievement, and we seem to be a long way from the sensitive and caring robotic lover imagined by proponents of this technology.

Let us now try to recapture that dream and look at what would be needed to create a robotic lover worth knowing.

2.3 Just Tell Me that You Love Me: Robots and Affective Love

At this point it is important to consider exactly what kind of affective love are we trying to achieve with a robotic companion? Levy [1] has a counterargument for those who might suggest that robots are incapable of affective love. Levy notes that:

> There are those who will doubt that we can reasonably ascribe feelings to robots, but he believes that if a robot *behaves* as though it has feelings, can we reasonably argue that it does not? If a robot's artificial emotions prompt it to say things such as "I love you," surely we should be willing to accept these statements at face value, provided that the robot's other behavior patterns back them up. When a robot says that it feels hot and we know the room temperature is significantly higher than normal, we will accept that the robot feels hot.... Just as the robot will learn or be programmed to recognize certain states—hot/cold, loud/quite, soft/hard—and to express feelings about them, feelings we accept as true because we feel the same in the same circumstances, why, if a robot that we know to be emotionally intelligent says "I love you" or "I want to make love to you" should we doubt it? If we accept that the robot can think, then there is no good reason we should not also accept that it could have feelings of love and feelings of lust [1, pp. 11-12].

Levy has a functional definition of love: If the machine acts like it loves its user and these actions are not inappropriate to the situation at hand, then the robot must actually be in love. This is somewhat like the Turing test where if a machine is capable of having an intelligent conversation with a human, then it must be concluded that the machine has something like human level intelligence [17].

But in both of these cases, "love" and "intelligence" are complex concepts and it is easy to equivocate while trying to adjudicate the results of a functional test. To ward against that possibility we need to quickly review what is known about "love" as a concept.

2.3.1 Cognitive Definitions of Love

We can now turn to cognitive science to see if we can get any clarification on love as a cognitive function. There are a number of interesting theories and we will look at some of the most useful ones here. The first is the *self-expansion* model of love developed by Aron and Aron [21]. "Self-expansion" refers to the feeling of expanded capabilities or opportunities experienced by those engaged in loving relationships. The heart of Aron and Aron's model is the "continuous inclusion of other scale (IOS)," which consists of seven pictures of two increasingly overlapping circles which start separated and then move closer until they start to overlap, much like a Venn diagram, and in the last picture the two circles are nearly indistinguishable [21]. Those engaged in loving relationships can take this test and will both develop a numerical result. This numerical representation of the closeness of the relationship is correlated with both the feelings and behavior common in close relationships and Aron and Aron argue that the results of this test are highly predictive of whether or not a couple will remain together in the future.

Under the self-expansion model it is claimed that people enter loving relationships to expand their individual capabilities in their social surroundings. This expansion is achieved through increased access to the physical and emotional resources of the lover along with the increases in social status which the relationship might give, as well as access to physical and intellectual abilities that the lover may possess. All of these positive goods are attained in relationship to the closeness of the individuals, which can be measured by the IOS test described above.

We might produce a cognitive map of two individuals engaged in a loving relationship. These maps could detail both of their physical, intellectual, and social capabilities. We could then compare these two maps to find the extent to which these capabilities overlap or fill in gaps in the abilities of the other partner, a process described by researchers as *self-expansion*. Positive results of this cognitive mapping process would indicate a close relationship [22], [23], [24]. Conversely, when this self-expansion plateaus or shows unbridgeable differences between the lovers, then this signals the end of the relationship [25].

Designers of companion robots should take note of this work as it provides a model for regulating a relationship. For instance, this kind of a test might be performed by the machine with its human lover and the machine could work to optimize its IOS by altering its behavior and retesting its human partner until it found itself in an acceptable IOS range. The IOS gives a wonderful measure that designers can use to determine how "loving" their machines are. IOS could be displayed graphically and the user would just select the circle diagram that fits their feelings and, if the number is low, the robot alters its behavior until it gets higher results.

If a robot were able to credibly help a person expand their cognitive and social capabilities while remaining close to its user, then under the self-expansion model it is conceivable that one might legitimately love this machine; however, it is not clear if the machine itself would love its user any more than one might assume a thermostat cares for its user by the evidence that it keeps the house nice and warm.

Another interesting hypothesis is that people enter relationships in order to experience positive emotions and mitigate negative emotions. Those relationships are in trouble when they are dominated by negative emotions. Partners in relationships help to regulate negative emotions that are caused by events outside the relationship [26].

Ortigue and Bianchi-Demicheli have shown through their research that the hypothesized mirror neuron system, which is a cognitive structure believed to be active when one is personally active or is just watching another person act, may play a role in facilitating love and understanding with a beloved by providing evidence of the partners ability to aid in the self-expansion of the lover, and this system may play a role in all prototypes of love [27].[4] Since mirror

---

4. Mirror neurons are still a hypothesis and not settled fact. If the research cited here proves to be true, it helps strengthen my argument but if not, then the loss of this evidence will not substantially hurt the argument.

neurons seem to work through the embodied cognition of the human agent, it follows that robotic engineers should look into this cognitive process and mimic its function in their machines. It also suggest a very important reason to design humanoid robots as it is the perception of the human shape of the beloved that interacts with the mirror neuron system of the human subject. Thus, robots built to interact correctly with the mirror neuron system of their user could lead to the user having authentic feelings of love and bonding toward the suitably programed machine.

In addition to the above models, we can now turn to the evolutionary attachment model of love posited by findings in evolutionary psychology. Both of the cognitive models discussed above can be easily argued from the standpoint of evolutionary psychology to be important contributors to the reproductive fitness of the partners involved in the loving relationship. In this line of reasoning, we see that modes of behavior with positive effects on natural selection will, over time, become deeply embedded in human nature.

Evolutionary psychologists argue that adult romantic relationships may be an adaptation that evolved from the infant-caregiver relationships that formed between mother and child among our prehuman ancestors.

Under this theory, adult loving relationships are explained by the fact that they tend to foster attachment between the partners that can last long term (or long enough to produce successful offspring). The loving relationships can also benefit the survival of the lovers and are formed and strengthened in accordance with how well they provide care and support for those in the relationship. Over evolutionary time both infants and mothers evolved successful techniques to engage the affective states of each other in ways that enhanced survival. Many of the evolved behaviors of both successful infants and successful mothers seem to be used again and recycled in later life to attract and keep mates [28], [29], [30]. The cognitive skills of accurate empathetic responses seem to be the key to maintaining close relationships [31].

We should note here that in this model love is equated closely with empathy and the researchers cited above posit that there is a kind of complex system of communication or signaling between genes that helps to strengthen pairs that are genetically compatible. It must also be mentioned that this theory is still arguable and it may also be the case that complex emotions like love do not map nicely onto some specific set of genes or other inherited structures.

It is also important to mention here that either physical or social evolutionary factors seem to have provided us with the ability to extend our emotional attachments outside our own species. As it turns out, talking to and loving our pets has given our species increased Darwinian fitness, as evidenced by the great symbiotic partnerships our species has formed primarily with dogs but also with many other creatures as well.

Although robotics designers will be interested in tapping into these evolved psychological behaviors for use in building robots that cause their users to experience affection toward them as either pets or humanlike companions, this may turn out to be quite a complex problem, though it is likely that engineers could mimic some of the evolved behaviors that signal empathetic or loving behaviors. Even with these difficulties, it seems like the field of psychology is fruitful ground for those searching for ideas on the design of affective computing solutions for robotic companions. In these models, love has functional characteristics that can be modeled and duplicated to some degree using robotics.

2.3.2 Philosophy and Love

While the roboticists working in the area of artificial companions have paid some attention to human psychology, they have largely ignored the contributions of philosophy to the study of love or the erotic. Philosophy has a long tradition of exploring love and the place of the erotic in the well lived life. While psychology is an important tool in determining what love is in humans, how it is expressed, and its ultimate role in the evolution of our species, we need to look to philosophy to try to understand what love ought to be, the aesthetic value of love, and its role in achieving an ethical and happy life. A companion robot would be of less value to us if it was only able to mimic the psychological aspects of love and not address the more important philosophical meanings of love.

A full account of the philosophy of love would require volumes worth of careful work. That is not possible here and instead I must admit that I will hand pick the citations I do use to suit my argument. What I hope to achieve is a provocative sampler for robot programmers and designers which will help them see that there are important aspects of the philosophy of love missing in their work so far. Other authors would have chosen their own favorites and I hope that others will add to this project and suggest aspects of the philosophy of love that they find personally valuable or motivating.

I will start by looking at the work of Plato since he argues that love is best seen as a way to expand the moral horizons of the lover. The question here is if Plato is correct and love makes one morally better for having loved, is this true of loving a machine? Are we morally better for having fallen in love with a machine? Or, put another way, have we attained anything of moral worth if we enter a relationship that was preprogrammed to succeed?

In the *Symposium* [32], Plato has given us one of the great discourses on love and we would be remiss if we did not give it at least some attention here.

In this dialogue, Socrates finds himself at a gathering where the wealthy and learned men of Athens have decided to talk on the subject of love. The guests take turns presenting their theories and many of the discourses on love offered up by Socrates' interlocutors fall into the folk definitions of love that we discussed earlier. Included in this discussion is the famous soliloquy by Aristophanes, who tells of an enduring myth where every human is looking for its other half that was split from it by the gods; thus the purpose of love is to complete the individual. After these stories Socrates finally tells his own accounting of love as it is personified in the being of Eros.

Socrates explains that Eros is not actually a god, which means he is neither entirely beautiful nor entirely good but he is also neither completely ugly nor bad. He is a daemon, which for the Greeks was an entity that existed between the gods and man [32, pp. 31-32].

Eros is a strange sort of supernatural being who inherited qualities from his parents who could not have been more dissimilar. His mother was Penia, the goddess of poverty, and his father was Poros, the god of resource. They had a drunken affair at the party celebrating the birth of Aphrodite. Thus, the offspring of that affair, Eros, is a mixture of poverty and resource, he is both fulfilling and needing, neither wise nor fully lacking in understanding [32, p 34]. Here, Socrates is arguing that love is something that is both fulfilling yet desperately needy as well.

In a surprising turn of events Socrates makes the claim that indeed the qualities of Eros make him a philosopher; he needs beauty and truth from the beloved to fulfill what he lacks in both on his own, just as the philosopher seeks beauty and truth in knowledge, a quest that would be unnecessary if the philosopher already had these qualities [32, p. 35]. Love is the active pursuit of things that are actually beautiful and good and in this way true love is a philosophical undertaking. Socrates then makes the audacious claim that he, as ugly as he is, is therefore the most erotic man in Athens given his unrelenting quest to find the truth and beauty he lacks as an individual. He concludes that the best of us are motivated by the erotic, which is the desire to find and the ability to distinguish true truth and beauty.

What we can learn from this is that erotic love has an important role to play in the philosophical life. While at one level love can simply be of instrumental value in the survival of the individual and the gene, at another level it is also something that can make the lover better as a person; the fulfillment of the beloved brings with it further longing, which spurs the lover on to other achievements in the endless quest for the erotic.

There has been much added to this topic since the time of Plato, too much to be fully covered here, but there are some thoughts from Irving Singer, one of the foremost thinkers in the contemporary philosophy of love and sex.

In his book *Explorations in Love and Sex* [33, p. 114], Singer explains that, "Love, like the creation of meaningfulness in general, reveals the ability of life in general—above all, as it appears in human beings—to bestow value on upon almost anything that catches our attention and makes itself available for this unique mode of self-realization." What Singer is saying here is that love has to have meaning or it is not love, but that the state of being in love can heighten our awareness of the inherent value of what we find around us.

Singer also describes an interesting evolution of the philosophy of passionate love that has occurred in the last century. In the 20th century, passion moved from being something that was somewhat philosophically suspect, as one can see even in Plato's *Symposium*, to being something that was believed to bring happiness and fulfillment to one's love life [33, p. 219]. This means that modern philosophers are more likely to agree that it is possible for a passionate love to be one that draws the lovers toward the philosophically erotic goals of seeking truth and beauty, whereas before it might have been seen as the kind of activity that might throw one off the track of the proper pursuit of truth.

Along with this change in philosophical attitude there has been a profound change in the technologies of birth control and other reproductive technologies which have indelibly changed the role that sex plays in a marriage and loving relationship. As sex is decoupled from procreation, it can now serve primarily as a mode of expressing love and sexual freedom, which found its apex in the sexual revolution. The epochal changes brought about by the sexual revolution has also had to confront the great tragedy of the AIDS epidemic, and together these have transformed what people now want from loving relationships. Singer argues that these changes have caused a deeper concern for finding compassionate loving partners instead of looking only for short term passionate affairs [33, p. 219]. So except for the relatively brief period when the sexual revolution was at its height, erotic love is now seen as an expression of compassion that must include qualities like tenderness, sociability, benign concern, and general good will between the lovers. Singer feels that this change is adding a deeply moral dimension to romantic love [33, p. 219].

A robotic companion worth having would somehow need to provide the kind of moral and compassionate love that we have come to expect from erotic relationships. This requirement may be the most difficult quality for roboticists to achieve through programming. As we have seen in the section on robotic sex, it is likely that roboticists will be able to create a machine that might raise the passions of certain users, but that may not be enough for the development of longer term relationships since this would require compassion and the philosophically erotic. We don't need just a machine to have sex with, we need one that makes us and the robot better by being with one another. We will have achieved nothing of moral worth by building machines that provide us with less as they will distract us from the more valuable pursuit of the kind of love that will expand or moral horizons through the experience of authentic love.

3 Robotic Lovers

So far we have seen that it is quite likely that sex robots with modest AI abilities are already on the market and there will no doubt be many innovations added to make them more and more life-like. But there have also been some interesting developments in the area of robots designed not for sex but to instead elicit love and affection from their users as a pet might do.

Lovotics is the conscious attempt to form a bond of love between the user and the robot and is the brain child of Hooman Samani from the Interactive and Digital Media Institute at the National University of Singapore [34], [35], [36], [37]. Lovotics is perhaps the most extensive project in affective robotics today where the researchers are directly applying psychological and physiological research on love in the development of their robotic systems. It is very difficult to determine how much of the claims made by the researchers involved in lovotics are real or hype. This is a problem that is widespread throughout the robotics community, where often clever YouTube videos serve to advance the excitement in some machine that in reality does not perform nearly as well in person as it does on a well-edited video.[5] If we take them at their word, lovotics attempts to simulate the physiological reactions of the human body experiencing love through an "Artificial Endocrine System," which is a software model of the same systems in humans which include artificial "Dopamine, Serotonin, Endorphin, and Oxytocin" systems [34], [37]. Layered on top of this is a simulation of human psychological love, very similar to what has already been discussed in this chapter. As the lovotics website [35] explains, their "Probabilistic Love Assembly" consists of an AI psychological simulator that

> ...calculates probabilistic parameters of love between humans and the robot. Various parameters such as proximity, propinquity, repeated exposure, similarity, desirability, attachment, reciprocal liking, satisfaction, privacy, chronemics, attraction, form, and mirroring are taken into consideration.

A robot designed under these principles will then monitor its user and inductively reason the mood of the user through evidence such as facial recognition and analysis of body language and physiology. It can then alter its own behavior in an attempt to maximize the affection and loving behavior of its user.

As of this writing they have developed a few robotic systems based on their findings.

Their first machine was a little furry robot that fans of the science fiction series *Star Trek* might mistake as a *tribble*. This machine is a robotic pet that moves around flat surfaces and coos gently to its owner while emitting a different colored glow of light to signal its mood.

A more ambitious machine designed by this lab is "Kissenger," which is a small spherical machine with a cartoon inspired face that looks something like a cross between a pig, a rabbit, and a panda. In the middle of the face is a big pink set of lips. If you were to have one of these machines you could link up through the Internet with another one just like it that you and your lover could use to share a kiss as your interactions with the robot's lips are mimicked on the machine of your internet lover and vice versa [34]. The designers of Kissenger see three potential uses for the machine.

Kissenger enables three modes of interaction:

1. Human to Human tele-kiss through the device: Bridges the physical gap between two intimately connected individuals. Kissenger plays the mediating role in the kiss interaction by imitating and recreating the lip movement of both users in real time using two digitally connected artificial lips.
2. Human to Robot kiss: Enabling an intimate relationship with a robot, such technology provides a new facility for closer and more realistic interactions between humans and robots. In this scenario, one set of artificial lips is integrated in a humanoid robot.
3. Human to Virtual character physical/virtual kiss: Provides a link between the virtual and real worlds. Here, humans can kiss virtual characters while playing games and receive physical kisses from their favorite virtual characters. Further, Kissenger can be integrated into modern communication devices to facilitate the interactive communication between natural and technologically mediated environments and enhance human tele-presence [34].

---

5. The author is not endorsing Lovotics or any similar affective robotics product and Lovotics machines are mentioned only to provide an example of recent technologies attempting to design machines capable of reciprocal love through mimicking and manipulating human psychology.

Not quite a full erotic or romantic relationship with an artificial being, but you can see that is the eventual goal toward which Kissenger is a first step.

Another strange, yet interesting robot that the lovotics group is working on is the "Mini-Surrogate," which is a small doll-like caricature of the user that you give to your distant lover. You would also have one that resembles the other user you are in a relationship with and these dolls would then stand in for you and your lover by mimicking your body language, thus facilitating both of the user's telepresence with each other. As the designer's explain:

> Current telecommunication techniques lacks the holistic embodied interaction and interface. The feeling of nonmediation can be reinforced in remote interpersonal communication if the interaction is through the whole body and engaging, interactive physical representative of each person is available in close proximity of the other person. The reason is related to the significance of embodiment, anthropomorphism, proximity, and enjoyment in fostering the illusion of presence [34].

One could also imagine that the mini-surrogate could also stand in for virtual characters as well just as the Kissenger is designed to do. The mini-surrogate is very mechanical in its motions and comes off more comical than romantic, but one can imagine a better built and far more expensive machine that might serve a role similar to the Kissenger but capable of *tele-sexual-relations* (to coin a somewhat cumbersome term) that could be linked to another human user or a virtual character in a game or other application.

These machines have received extensive media attention and the lovotics group has produced numerous academic papers and given many presentations on their work. Thus, they are arguably the most active research center working specifically on robot love today.

Hiroshi Ishiguro is another strong contributor in the area of robotic companions. His work at the Intelligent Robotics Laboratory at Osaka University [38] and at Hiroshi Ishiguro Laboratory ATR [39], while not directed specifically at robotic love, is creating some compelling androids and humanoid robotic applications that are intended to elicit and interact with human emotion. For the Valentine's Day 2012 shopping season, Ishiguro showcased his android *Geminoid-F*, who sat in the store window of Takashimaya department store in the Shinjuku district of Tokyo in front of a huge heart shape formed by dozens of boxes, each decorated by little hearts. Geminoid-F is an android made to look like a human female and its face and clothing are modeled after a real person who works in the lab with Ishiguro. When this machine has been shown it has typically been teleoperated, but in this appearance it seems to have been programmed to engage with human observers by coyly sharing glances with them then turning away, all designed to draw the observer in to try to figure out this mysterious female machine.[6]

There are many other androids in the Geminoid series, including one that looks exactly like Professor Isiguro himself [39]. These androids are far more compelling than anything else discussed in the paper so far, but they are not fully out of the discomfort induced by the uncanny valley that many of the more realistic humanoid robots fall into [40], [41]. The high level of artistic and engineering achievement that Ishiguro has achieved with these machines points toward a future in which very realistic android robots may find their way into the home as companions, but still, as they stand today, they are something that only their creator could truly love and their repertoire of behaviors is still somewhat lacking.

3.1 Robotic Design Strategy

From the review of robotic companion technologies we have gone through in this chapter we can see that there are at least two existing design strategies that have been successfully deployed in affective companion robots. While both of the following design strategies might be discussed in robotics engineering as "sociable" or "social" robotics, I believe that, for conceptual clarity and to make the philosophical points I want to make, we need to refer to at least two different types of social robotics design strategies.

3.1.1 Robotic Design Strategy One—Variance

The first successful design strategy is to work toward a more or less harmonious integration of the robot with its user and surroundings. These machines would be to simulate emotion and embodiment in the machine in a way that is not a direct imitation of the human, though it may be inspired by the study of human emotion. When this strategy is successful, the user(s) will feel more comfortable working or interacting with the machine, and the machine will more seamlessly fit into human society. We will call this the "variance" strategy as the designers of the robot seek to build a friendly working relationship between the robot and the humans it is built to interact with. In this strategy, one builds and programs the machine with a generally human appearance and behavior so that it can use its limbs and visage to suggest moods through body language and facial expressions which will be meaningfully interpreted by any human user. Ideally, this will help facilitate verbal communication and ease user exasperation when the shared project might run into obstacles, and this strategy is growing in use [42], [43], [44], [45], [46], [47], [48], [49], [50]. The primary distinction between this strategy and the next is that these machines interact with their users as machines and not as surrogate humans. They do attempt to navigate human emotion with affective computing applications and they may even fulfill the role of robotic companions, but there is a certain distance maintained between the robot and the expectations of the user; they look and act more or less like machines so the user expects less human verisimilitude from them.

3.1.2 Robotic Design Strategy Two—Mimesis

The second is to make a much stronger appeal to the user's emotions in order to have the user treat the machine as if it were a fellow human agent or at the very least to have the user be momentarily confused as to whether or not the robot is a human or an android and this is the strategy we can argue is employed in the Roxxxy sex robot as well as the much more complex Geminoid androids. Another good example of this is the Actroid DER2 that has been built by Kokoro, a division of Sanrio that specializes in animatronics

---
6. Video of the exhibition can be found here: http://www.youtube.com/watch?v=hCaRkyq02go.

and robotics. This robot is meant to be used as an artificial actress or newscaster, so it is not technically an artificial companion, but the technology being developed here could be applied elsewhere.[7] This is just a short list of the mimetic androids in development; there are many others we could mention but the ones we have looked at so far are the best of what is available.

The proponents of what we can call the mimetic robotics style of design argue that if our goal is to build complex artificial agents that we can exist in close contact with, then these machines are going to necessarily be more or less indistinguishable from humans [1], [51], [52].

It is also important to remember that both design strategies can be employed in a single company or lab, but they are mutually exclusive at the level of the individual system. A particular robot cannot be both variant and mimetic.

3.2 Apparent Affective Artificial Agency

If our goal is to build complex artificial agents that we can exist in concord, then both of the variant and the mimetic strategies could serve as methods with which to explore how to give these agents robust and authentic emotional responses [53]. But it is too easy to become enamored of this eventual goal and disregard the long interim period where these machines will be able to effectively simulate but not actually synthesize human emotion and romantic behavior. The affective capabilities of these machines will be apparent and somewhat accurate but it will not be an adequate substitute for human relationships. This is why we should endeavor to make machines that can help enhance human relationships but we should avoid making them to simply replace the humans in our lives.

For some people, computers and other bits of machinery have already passed a sufficient affective threshold and provide enough stimuli for the human user to form deep emotional attachments to machines and artificial agents. Turkle [14] has written extensively about individuals who, for various psychological reasons, prefer the company of computers to other people. She argues that perhaps these people are not psychologically damaged but instead they may be best seen as early adapters and that they provide us a view of the future of human-machine interpersonal relations.

Computing technology is such a compelling surrogate for human interaction because it is so malleable to the wishes of its user. If it does not do what the user wants, then a sufficiently trained user can reprogram it or fix the issue to make the machine perform in line with the user's wishes. With a little work, the experience between the user and the machine can be completely personalized to every conceivable user. A person who is smart with the technology can get it to do what he/she wants when he/she wants it.

Fellow humans, on the other hand, represent a much more difficult problem and do not always readily change to accommodate one's every need. They provide resistance and have their own interests and desires that make demands on the other person in the relationship. Compromise and accommodation are required and this is often accompanied by painful emotions.

If we follow this line of reasoning, then it would seem that more and more of us will take the path of least

---
7. http://www.kokoro-dreams.co.jp/english/robot/act/index.html.

resistance and chose to interact more and more with digital technology over other human beings if given the choice.

Levy believes that this logic is irresistible and that as the technology of robotic lovers improves and becomes easier to use, cheaper to deploy, and fulfills more of our needs, then humanity will drift en masse toward the happier and more fulfilling world of robot love.

3.3 Taking HCI to the Extreme

Let's sum up our findings so far.

1. Psychological factors in love, sex, and attraction can be at least functionally duplicated in robotics technology.
2. Humans have a psychological tendency to anthropomorphize animals, objects, and technologies.
3. Evolution has given us a predisposition to be interested in developing caring relationships for creatures outside our own species
4. If someone acts like they loves us, even if those actions are very minimal, we will tend to believe they truly do love us.
5. Early adaptors have already begun to prefer human-computer interaction to interaction with fellow humans.
6. So, by extrapolation, a robot that can tap into these psychological tendencies would attempt to form relationships at least as real and moving as those we have with our beloved pets.

This takes us to the final step of Levy's argument. If indeed a robot came into peoples' lives, and that machine could meet their every psychological need, fulfill every sexual desire, and enjoy all the same things as their owners, then certainly society would be forced to recognize human robot love and possibly even human robot marriage. Levy also assures us that the robot in our example would itself genuinely want its user as its spouse.

Because, by design, the robots only desires would be to please its user and only its particular user, the robot would be functionally unhappy in any other situation, with another user perhaps, or removed from the company of the user it is programmed for. So, Levy argues it would be a kind of cruelty and a mistreatment of the robot if it were not allowed to love and serve its owner.

Levy's argument here begs a number of technological questions. Judging from what is on offer today, it is highly unlikely that we will be able to build a machine that can have the functionality it would need to have complex self-referential thoughts about thoughts and desires about desires. But even if we grant the remarkable technological advances that would be needed to create such a highly functioning robot lover, there is still an important ethical issue to answer. Just because one builds a love slave that wants to be a love slave does not absolve the master from the moral charge of slavery.

Given that this is not a short term concern, let's instead look at the more likely ethical impacts of affective robotics in the area of robotic companions.

4 Ethics of Robotic Love

One thing that should be abundantly clear is that affective robotics, as it appears today, works best by manipulating human psychology. It seems that humans have a number of evolved psychological weaknesses that can be leveraged to make a user accept simulated emotions as if they were genuine. It is unethical to play on deep-seated human psychological weaknesses put there by evolutionary pressure as this is disrespectful of human agency [54].

If it is true that this technology causes people to grow less and less tolerant of human relations, then we have to be very careful of this technology. It would be a human tragedy if we lost our tolerance to deal with others who are not preprogrammed to serve our every need.

The design of technologies must not try to overly abuse the distinction between human, natural, and artifactual systems [55]. This would suggest that we should be very careful with the mimetic design strategy.

As we have seen, Levy and other roboticists have so far ignored the deep and nuanced notions of love and the concord of true friendship as described in philosophy. While it is given that robots can be built that people will find sexually attractive, it is unlikely that a machine can be built that will be capable of building an erotic relationship between itself and the user. Instead, with these technologies as they are currently evolving, we have an engineering scheme that would only satisfy, but not truly satisfy, our physical and emotional needs, while doing nothing for our moral growth.

4.1 Ethical Design Considerations

1. Love is more than behavior. It is important to design robots so they act in predictably human ways but this should not be used to fool people into ascribing more feelings to the machine than they should. Love is a powerful emotion and we are easily manipulated by it.
2. Friendship (philia) with robots is more important than romantic love. It is permissible and even desirable to design robots that act in concord with their users; affective friendship will be a hard enough to achieve so we should start there. Given that we will be able to mimic emotions in a robot long before we will be able to produce truly affective machines, it is advisable to be circumspect in how we exploit human psychology in the design and deployment of these machines.
3. Truth is Important. Roboticists should not design machines that intentionally lie to their users and with those lies manipulate the user's behavior.

5 CRITICISM TO THIS ARGUMENT

There are two common criticisms to my argument above that I would like to rebut here. The first is the claim that humans lie to each other all the time, especially when it comes to sex and romance, so why would I want to place special ethical restraints on roboticists when they are just playing along with a game as old as time.

First, this criticism is simply a version of the naturalistic fallacy. Just because something occurs often in nature does not necessarily make it the most rationally ethical choice. While roboticists can and should look to natural systems for inspiration and insight, it would be silly to think that they are limited to only modeling naturally occurring systems.

Even if every human relationship was based on lies and deceit, it might still remain possible that another—more truthful—mode of behavior might be discovered through social scientific research. Arguably, human political systems have evolved from less ethically justifiable modes of behavior through fits and starts to modern systems which, while far from perfect, are at least stronger from a moral standpoint. So we do not have to accept base and deceitful behavior from either the designers of these machines or the robots themselves. We can and should demand better from these systems.

The second common misunderstanding I have received when presenting this argument is the presumption that I am against the fulfillment of romantic desire. How could I turn my back on a technology that could bring us perfect love?

It is easy to see that this criticism begs the question. If you can create a perfect love, then of course I could not argue against it. What I am saying instead is that we have to be careful to not mistake simulacral love for the real thing. I agree that this technology will be compelling and in fact already is compelling to some early adopters. But the kind of relationships that are evolving are not philosophically erotic, that is, challenging and compassionate, but rather one-sided affairs overburdened by fleeting passions and the desire to erase everything in the beloved that is not a complete reflection of the lover's preconceived notions of what he or she thinks they want out of a partner. Remember the main lesson Socrates was trying to give us in the Symposium is that we come into a relationship impoverished, only half knowing what we need; we can only find the philosophically erotic thorough the encounter with the complexity of the beloved, complexity that not only includes passion, but may include a little pain and rejection from which we learn and grow.

ACKNOWLEDGMENTS

The author would like to thank Carson Reynolds, Patrick Lin, and anonymous reviewers for helpful comments on earlier drafts of this paper.

REFERENCES

[1] D. Levy, *Love and Sex With Robots: The Evolution of Human-Robot Relationships.* Harper Collins, 2007.
[2] R. Cowie, "Companionship Is an Emotional Business," *Close Engagements with Artificial Companions: Key Social, Psychological, Ethical and Design Issues,* Y. Wilks ed., John Benjamins Publishing Company, http://site.ebrary.com/lib/sonoma/Doc?id =10383970&ppg =192, 2010.
[3] B. Duffy, "Anthropomorphism and the Social Robot," *Robotics and Automation Systems,* vol. 42, pp. 177-190, 2003.
[4] C. Breazeal, A. Brooks, J. Gray, G. Hoffman, C. Kidd, H. Lee, J. Lieberman, A. Lockerd, and D. Mulanda, "Humanoid Robots as Cooperative Partners for People," http://web.media.mit.edu/ ~cynthiab/Papers/Breazeal-etal-ijhr04.pdf. 2004.
[5] H.A. Samani, A.D. Cheok, M.J. Tharakan, J. Koh, and N. Fernando, "A Design Process for Lovotics," *Proc. Third Int'l Conf. Human-Robot Personal Relationships,* pp. 118-125, 2011.
[6] J.P. Sullins, "Friends by Design: A Design Philosophy for Personal Robotics Technology," *Philosophy and Design: From Eng. to Architecture,* P. Vermaas, P. Kroes, A. Light, and S. Moore, eds., pp. 143-158, Springer, 2008.
[7] T. Kanda and H. Ishiguro, "Communication Robots for Elementary Schools," http://www.irc.atr.jp/~kanda/pdf/kanda-aisb2005.pdf. 2005.

[8] T. Kanda, H. Ishiguro, and T. Ishida, "Psychological Analysis on Human-Robot Interaction," *Proc. IEEE Int'l Conf. Robotics and Automation*, pp. 4166-4173, 2001.
[9] T. Kanda, R. Sato, N. Saiwaki, and H. Ishiguro, "Friendly Social Robot that Understands Human's Friendly Relationships," *Proc. IEEE/RSJ Int'l Conf. Intelligent Robots and Systems*, pp. 2215-2222, 2004.
[10] First Androids, http://www.andydroid.com, 2012.
[11] True Companions, http://www.truecompanion.com/, 2012.
[12] C. Meston and D. Buss, "Why Humans Have Sex," *Archives of Sexual Behavior*, vol. 36, pp. 477-507, 2007.
[13] J. Snell, "Sexbots: An Editorial," *Psychology and Education: An Interdisciplinary J.*, vol. 42, no. 1, pp. 49-50, 2005.
[14] S. Turkle, *Alone Together: Why We Expect More from Technology and Less from Each Other*. Basic Books, 2011.
[15] M. Coeckelbergh, "Personal Robots, Appearance, and Human Good: A Methodological Reflection on Roboethics," *Int'l J. Social Robotics*, vol. 1, no. 3, pp. 217-221, 2009.
[16] J. Sullins, "Friends by Design: A Design Philosophy for Personal Robotics Technology," *Philosophy and Design: From Eng. to Architecture*, P. Vermaas, P. Kroes, A. Light, and S. Moore, eds., pp. 143-158, Springer, 2008.
[17] A. Turing, "Computing Machinery and Intelligence," *Mind*, vol. 59, no. 236, pp. 433-460, 1950.
[18] B. Fehr, "Prototype Based Assessment of Laypeople's Views of Love," *Personal Relationships*, vol. 1, pp. 301-331, 1994.
[19] B. Fehr and J.A. Russel, "The Concept of Love Viewed from a Prototype Perspective," *J. Personality and Social Psychology*, vol. 60, pp. 424-438, 1991.
[20] D.T. Gilbert and E.E. Jones, "Perceiver-Induced Constraint: Interpretations of Self-Generated Reality" *J. Personality and Social Psychology*, vol. 50, pp. 269-280, 1986.
[21] A. Aron and E.N. Aron, "Love and Expansion of the Self: The State of the Model," *Personal Relationships*, vol. 3, no. 1, pp. 45-58, 1996.
[22] A. Aron, E.N. Aron, and D. Smollan, "Inclusion of Other in the Self Scale and the Structure of Interpersonal Closeness," *J. Personality and Social Psychology*, vol. 63, pp. 596-612, 1992.
[23] A. Aron, E.N. Aron, M. Tudor, and G. Nelson, "Close Relationships as Including Other in the Self," *J. Personality and Social Psychology*, vol. 60, pp. 241-253, 1991.
[24] A. Aron and B. Fraley, "Relationship Closeness as Including Other in the Self: Cognitive Underpinnings and Measures," *Social Cognition*, vol. 17, pp. 140-160, 1999.
[25] About the Continuous IOS, http://www.haverford.edu/psych/ble/continuous_ios/index.html, 2012.
[26] F. Susan and R.S. Lazarus, "Coping as a Mediator of Emotion," *J. Personality and Social Psychology*, vol. 54, pp. 466-475, 1988.
[27] S. Ortigue and F. Bianchi-Demicheli, "Why Is Your Spouse So Predictable? Connecting Mirror Neuron System and Self-Expansion Model of Love," *Medical Hypotheses*, vol. 71, no. 6, pp. 941-944, 2008.
[28] R. Buck, "The Genetics and Biology of True Love: Prosocial Biological Affects and the Left Hemisphere," *Psychological Rev.*, vol. 109, no. 4, pp. 739-744, Oct. 2002.
[29] R. Buck and B.E. Ginsburg, "Emotional Communication and Altruism: The Cognitive Gene Hypothesis," *Altruism: Rev. of Personality and Social Psychology*, vol. 12, pp. 149-175, 1991.
[30] R. Buck and B.E. Ginsburg, "Communicative Genes and the Evolution of Empathy," *Empathetic Accuracy*, W. Ickes, ed., pp. 17-43, Guilford Press, 1997.
[31] M. Davis and L. Kraus, "Personality and Empathic Accuracy," *Empathic Accuracy*, W. Ickes ed., pp. 144-168, Guilford Press, 1997.
[32] Plato, *Symposium*. The Univ.of Chicago Press, 2001.
[33] I. Singer, *Explorations in Love and Sex*. Rowman and Littlefield, 2001.
[34] Information retrieved from: http://hooman.lovotics.com/, 2012.
[35] H.A. Samani and A.D. Cheok, "Probability of Love between Robots and Humans," *Proc. IEEE/RSJ Int'l Conf. Intelligent Robots and Systems*, 2010.
[36] H.A. Samani et al., "Towards a Formulation of Love in Human-Robot Interaction," *Proc. 19th IEEE Int'l Symp. Robot and Human Interactive Comm.*, 2010.
[37] H.A. Samani et al., "A Design Process for Lovotics," *Proc. Int'l Conf. Human-Robot Personal Relationships*, pp. 118-125, 2011.
[38] Intelligent Robotics Laboratory: http://www.is.sys.es.osaka-u.ac.jp/index.en.html, 2012.
[39] Hiroshi Ishiguro Laboratory, ATR, http://www.geminoid.jp/en/index.html, 2012.
[40] M. Mori, "The Uncanny Valley," *Energy*, vol. 7, no. 4, pp. 33-35, 1970.
[41] M. Mori, *The Buddha in the Robot: A Robot Engineer's Thoughts on Science and Religion*. Kosei Publishing, 1981.
[42] C.A. Breazeal, "Robot in Society: Friend or Appliance?" *Proc. Autonomous Agents Workshop Emotion-Based Agent Architectures*, pp. 18-26, 1999.
[43] C.A. Breazeal, *Designing Sociable Robots*. MIT Press, 2002.
[44] C.A. Breazeal, R.A. Brooks, J. Gray, G. Hoffman, C. Kidd, H. Lee, J. Lieberman, A. Lockerd, and D. Mulanda, "Humanoid Robots as Cooperative Partners for People," http://robotic.media.mit.edu/Papers/Breazeal-etal-ijhr04.pdf, 2004.
[45] R.A. Brooks, *Flesh and Machines*. Pantheon Books, 2002.
[46] S. Gaudin, "NASA: Humanoid Robot Slated to Live on Space Station," *Computerworld*, www.computerworld.com, Apr. 2010.
[47] T. Kanda and H. Ishiguro, "Communication Robots for Elementary Schools," http://www.irc.atr.jp/~kanda/pdf/kanda-aisb2005.pdf, 2005.
[48] T. Kanda, H. Ishiguro, and T. Ishida, "Psychological Analysis on Human-Robot Interaction," *Proc. IEEE Int'l Conf. Robotics and Automation*, pp. 4166-4173, 2001.
[49] T. Kanda, R. Sato, N. Saiwaki, and H. Ishiguro, "Friendly Social Robot that Understands Human's Friendly Relationships," *Proc. IEEE/RSJ Int'l Conf. Intelligent Robots and Systems*, pp. 2215-2222, 2004.
[50] M. Scheutz, P. Schermerhorn, J. Kramer, and C. Middendorff, "The Utility of Affect Expression in Natural Language Interactions in Joint Human-Robot Tasks," *Proc. First ACM Int'l Conf. Human-Robot Interaction*, pp. 226-233, 2006.
[51] B. Duffy, "Anthropomorphism and the Social Robot," *Robotics and Automation Systems*, vol. 42, pp. 177-190, 2003.
[52] P. Menzel and F. D'Aluisio, *Robosapiens: Evolution of a New Species*. MIT Press, 2000.
[53] J.P. Sullins, "Friends by Design: A Design Philosophy for Personal Robotics Technology," *Philosophy and Design: From Eng. to Architecture*, P. Vermaas, P. Kroes, A. Light, and S. Moore, eds., pp. 143-158, Springer, 2008.
[54] M. Scheutz, "The Inherent Dangers of Unidirectional Emotional Bonds," *Proc. IEEE Int'l Conf. Robotics and Automation*, 2009.
[55] S.D.N. Cook, "Design and Responsibility: The Interdependence of Natural, Artifactual, and Human Systems," *Philosophy and Design: From Eng. to Architecture*, P. Vermaas, P. Kroes, A. Light, and S. Moore eds., pp. 259-269, Springer, 2008.

Moral Competence in Social Robots

Bertram F. Malle

Matthias Scheutz

Abstract—We propose that any robots that collaborate with, look after, or help humans—in short, social robots—must have moral competence. But what does moral competence consist of? We offer a framework for moral competence that attempts to be comprehensive in capturing capacities that make humans morally competent and that therefore represent candidates for a morally competent robot. We posit that human moral competence consists of four broad components: (1) A system of norms and the language and concepts needed to communicate about these norms; (2) moral cognition and affect; (3) moral decision making and action; and (4) moral communication. We sketch what we know and don't know about these four elements of moral competence in humans and, for each component, ask how we could equip an artificial agent with these capacities.

I. INTRODUCTION

There are at least two classes of questions that fall under "robot ethics": (1) ethical questions about designing, deploying, and treating robots; and (2) questions about the robots' own ethics—what moral capacities robots should have and how these capacities could be realized in robotic architectures [1]. Our analysis in this paper focuses entirely on questions of the second kind.

A robot well suited for social interaction with humans would need to have, among other things, social-cognitive capacities (including a "theory of mind") and moral competence. Many psychological phenomena have been studied that could be called "moral competence": decision making about moral dilemmas [2], [3]; self-regulation of emotion and prosocial behavior[4]; moral judgments and their associated emotions [5], [6]; as well as responding to others' moral criticism by means of explanation, justification, or defense [7], [8]. The diversity of phenomena on this list is no coincidence; moral competence is not a single capacity. We propose here a framework that delineates the multiple components of human moral competence and then ask which of these elements should and could make up moral competence in social robots.[1] Because of time and space constraints we will have to stay at the surface, but nonetheless we will try to point to areas

---

[1] I focus here on social robots—home makers, care takers, educators, and the like; but much of what I say applies to robots in other contexts, such a military and rescue, as well.

This project was supported by a grant from the Office of Naval Research, No. N00014-13-1-0269.

where the research on human capacities is lagging behind and needs to be advanced in order to provide a better basis for robotic work; and to areas where robotic work could take some promising early steps.

The guiding question is: What would we expect of morally competent robots? Perhaps not all of human competencies, but surely several of them. And perhaps moral robots will be even "better moral creatures than we are" [9, p. 346], though such an evaluation already presupposes that we know what we are comparing. So what is it that we are comparing? What is moral competence?

We propose that moral competence consists of four broad components (Fig. 1): (1) A system of norms and the language and concepts to communicate about these norms; (2) moral cognition and affect; (3) moral decision making and action; and (4) moral communication. The rest of this paper describes in more detail each of the components.[2]

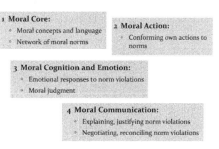

Fig. 1. Four components of moral competence

II. NORMS AND MORAL LANGUAGE

Morality's function is to regulate human social behavior in contexts in which biological desires no longer guarantee individual and collective well-being [12], [13]. Human communities perform this regulation by motivating and deterring certain behaviors through the imposition of norms and, if these norms are violated, by levying sanctions [14].

---

[2] These components are in some sense weaker than what is often discussed under "moral agency" e.g., [10], [11]; for example, they do not include a deep, reflective self-concept, and they don't presuppose "free will." But the components are also more extensive than typical moral agency demands, which rarely require social-cognitive and -communicative capacities.

This process allows social agents to successfully coordinate their behaviors in complex social communities—made complex by diversified tasks, roles, collective behavior, and interdependence of outcomes. Being equipped with a norm system thus constitutes a first critical element in human moral competence [15], [16].

But a norm system is conceptually and linguistically demanding, requiring language for learning it, using it, and negotiating it. Thus, at its core, moral competence requires a network of moral norms and a language (and associated concepts) to represent and implement it [11].

A. Moral Language

Some rudimentary moral capacities may operate without language, such as the recognition of prototypically prosocial and antisocial actions [17] or foundations for moral action in empathy and reciprocity [18]. But a morally competent human will need a vocabulary to express moral concepts and instantiate moral practices—to blame or forgive others' transgressions, to justify and excuse one's behavior, to contest and negotiate the importance of one norm over another.

Such a moral language has three major domains:

1. **A language of norms and their properties**
   (e.g., "fair," virtuous," "reciprocity," "obligation," "prohibited," "ought to");
2. **A language of norm violations**
   (e.g., "wrong," "culpable," "reckless," "thief");
3. **A language of responses to violations**
   (e.g., "blame," "reprimand," "excuse," "forgiveness").

Within each domain, there are numerous distinctions, and some have surprisingly subtle differentiations. For example, we recently uncovered a two-dimensional organization of 28 verbs of moral criticism [19] that suggests people systematically differentiate among verbs to capture criticism of different intensity in either public or private settings (see Fig. 2).

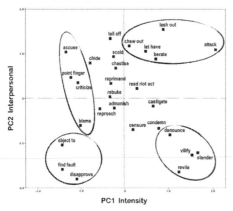

Fig. 2. Verbs of moral criticism in two-dimensional feature space

Obviously, the role of norms is central to a language of morality. Let us take a closer look at the challenges both in understanding how human moral norms operate and how they could possibly be implemented in autonomous robots.

B. Moral Norms

Many questions about norms already arise when we examine the human case. How are norms acquired? How are they represented in the mind? What properties do they have that allow them to be so context-sensitive and mutually adjusting as we know them to be? We will briefly examine these questions.

Acquisition. Though evidence is limited on the development of norms, data on children's early use of moral language [20] suggest that they are rarely exposed to abstract rules but rather hear and express concrete moral judgments. Consider these examples [20, pp. 74–77]:

> "they are mean to that man because they put him in that glue,"
>
> "that's not nice! That was naughty!"
>
> "he did something wrong."

Nonetheless, children are somehow able to abstract from concrete instances to rather general rules, such as "bombs hurt people" [20, p. 77] or even abstract principles such as the act-omission distinction [21] or battery through contact [22]. Fortunately, children are the most powerful learning machines in the universe, as we can see in just about all domains of cognition, including learning language and acquiring varied category systems such as personality traits, animals, or plants. In addition, the toolbox of social cognition adds a powerful supportive structure for the acquisition of moral norms and moral judgments. This structure includes mastery over the concepts of goal and desire [23], [24]; the distinction between intentional and unintentional behavior [25], [26]; between beliefs and desires [27], [28], desires and intentions [29], and a variety of emotions [30], as well as all the rich linguistic expressions of these distinctions [31], [32].

Representation. Another largely ignored question is how norms are represented in the human mind. Are they networks of concepts? And is that any different from networks of other concepts? Nonnormative concepts (e.g., tree, weight) can have evaluative tone, for sure, but do they have motivational force as normative concepts do (e.g., fairness, obligation)? Goal concepts—which typically are explicitly represented in robotic architectures—seem close to norm concepts. But perhaps the most unique category of norms are *values*, and there are indications that they cannot simply be reduced to goals [33]. Moreover, if Jon Elster is correct in claiming that "social norms provide an important kind of motivation for action that is irreducible to rationality or indeed any other form of optimizing mechanism" [34, p. 15], then a simple goal-based action control system will not do for moral social robots.

Levels. Another interesting feature of norms is that they are layered over many levels of abstraction. As an illustration, consider the following violation: A commercial airplane pilot flies a plane despite a recently diagnosed heart condition. What norm is violated here?

- Pilots ought not to fly when they know they have a physical disposition that, if becoming acute, could threaten their ability to safely continue flying. (This expands from heart condition to migraines, epileptic seizures, sleep apnea....)
- Pilots ought not to fly when they are aware of factors that risk the continued safety of their passengers. (This expands from the pilot's own risk factors to mechanical risks, weather risks, etc.)
- People ought to keep others safe who are put in their care. (Expands to many more roles, contexts, potential victims, etc.)
- People ought to protect human life.

Contextual activation. A related facet of norms is that they appear to be very quickly activated, and presumably at the right level of abstraction, because people detect norm violations within a few hundred milliseconds [35]. One possibility is that physical or linguistic contexts activate subsets of action-specific norms (what one is or is not permitted to do in the particular context), and violations can then be rapidly detected. The relation of these concrete norms to more abstract norms may be constructed offline, through conversation and conceptual reorganization [5].

For designing a morally competent robot, all these features of norms present serious challenges. But if norms are special kinds of representations, connected in some (admittedly flexible) network, and activated by percepts, then there is no principled reason why they could not be implemented in a computational system. They could then operate as constraints on the robot's possible actions, selecting optimal (least violating) action favorites (cf. [36]).

One issue we do not consider a problem is which "ethical theory" to build into a robot (utilitarianism, Kantian ethics, etc.). Humans probably do not follow any one particular ethical theory, and even if they did it is not clear whether robots must have the same. However the system arrives at its judgments of what is or is not a norm violation, the judgments must conform to the community in which the robot is embedded. This also means that robots, like children, will have to learn different norm systems when they are deployed in different communities. Many communities have large overlap in the norms they follow, so moving from one community to another is not an insurmountable problem, for humans or robots; in fact, robots may be better at keeping track of various different communities' norm systems and thus be less morally myopic than humans are.

III. MORAL COGNITION AND AFFECT

We have proposed that two of the major domains of moral language are (a) norm violations and (b) responses to norm violations. What psychological processes are involved in detecting and responding to such violations? These processes are usually treated under the label of *moral judgment*, but we need to distinguish between at least two kinds of moral judgment [37]. First, people evaluate *events* (outcomes, behaviors) as bad, good, wrong, or (im)permissible; second, they evaluate *agents* as morally responsible, deserving blame or praise. The key difference between the two types is the amount of information processing that normally underlies each judgment. Whereas event judgments merely register that a norm has been violated, agent judgments such as blame take into account the agent's specific causal involvement, intentionality, and mental states.

Of course, registering that an event violated a norm is not a trivial endeavor, and we realize this quickly when we ask how young children or robots detect such violations. What is needed for such a feat? Minimally, event segmentation, multi-level event representations (because different levels may conflict with different norms), and establishing the event's deviation from relevant norms. Nontrivially, which norms are *relevant* must somehow be selected from the available situation information and existing knowledge structures.

To arrive at agent judgments, people search for causes of the detected norm-violating event; if the causes involve an agent, they wonder whether the agent acted intentionally; if she acted intentionally, what reasons she had; and if the event was not intentional, whether the agent could and should have prevented it [38]. The core elements here are causal and counterfactual reasoning and social cognition, and that is why a number of researchers suggest that moral cognition is no unique "module" or "engine" but derives from ordinary cognition [39], [40], but reasoning within the context of norms.

Where in all this is affect? The specific roles of affective phenomena in moral judgment are still debated. There is little doubt that the detection of a norm violation often leads to a negative affective response—an evaluation that *something is bad*, perhaps accompanied by physiological arousal and facial expressions. But exactly what this affective response sets in motion is unclear: a marker that something important occurred [41]? A strengthened motivation to find the cause of the bad event [42]? Or a biased search for evidence that allows the perceiver to blame somebody [6]? And what do we make of the fact that people can make moral judgments without much affect at all [43], [44] or that moral emotions such as anger or resentment require specific cognitive processes [45]? Nobody would deny that affective phenomena often accompany moral judgments and that they probably facilitate learning moral norms; but there is little evidence for the claim that they are *necessary* or *constitutive* of those judgments. And if emotions are not necessary or constitutive of moral judgments, then robots—even if they do not have emotions—can very much be moral.

IV. MORAL DECISION MAKING AND ACTION

Human moral decision making has received a fair amount of attention in the research literature, with a focus on how humans handle moral dilemmas [2], [46], [47]. Much of what these studies reveal is how people resolve difficult conflicts within their norm system (e.g., saving multiple lives by sacrificing one life). A popular theoretical view of such situations is that initial affective responses can be overridden by deliberation [48]. But evidence against this override view is increasing [49]–[52]; people's judgments seem to involve a package of affective and cognitive processes that all deal simultaneously with the conflict set up by the experimenter. Further, how judgments about carefully constructed dilemmas translate into everyday moral decisions is not entirely clear.

In fact, the list of actual psychological factors that influence moral action is long, certainly including momentary affective states to personality dispositions, automatic imitation to group pressure, and heuristics to reasoned action. This overdetermination is no different from nonmoral actions [53]. What makes certain actions *moral* is the involvement of socially shared norms, not just individual goals. In humans, there is frequent tension between these social norms and the agent's own goals, and it is this tension that brings into play two additional psychological factors that guide moral action: empathy and self-regulation [4], [54], [55], both of which are designed to favor communal values over selfish interest.

In designing a robot capable of moral decisions and actions, the tension between self-interest and community benefits can probably be avoided from the start, making genuine empathy and self-regulation dispensable. However, humans are highly sensitive to other people's displays of empathy, and a robot that appears to coldly assess moral situations may not be trusted. The robot's modeling of the human view of situations and the its communication of having understood and taken into account this view (perspective taking rather than affective empathy) may go a long way toward building trust between human and machine. Of course, these communications and attempts at perspective taking must not be merely verbal programs, deceptive attempts to coax the human's trust; some computational analog of affect and valuing may be needed for human-machine interactions to succeed [56], [57].

Thus, we are back to the challenge of building a norm system into the robot, including values that make the machine care about certain outcomes and that guide the robot's decision making and action, especially in the social world. Note that "caring" here is a key concept that would first have to be spelled out—what it means in computational terms for a machine to care about something.

Whether in all this a robot needs "intuitive" processes, which seem to play an important role in human moral decision making [5], [58], is also an interesting question. To the extent that these processes resolve a human capacity limitation, and to the extent that robots do not have this limitation, robots would not need moral intuitions. However, will humans be suspicious of robots that do not "feel" right and wrong but reason over it? We doubt it, but this is an empirical question.

Note that we have not mentioned the need to have "free will" in order to accord moral action capacity. That is because ordinary people require only choice and absence of constraints for actions to be "free"—any metaphysical requirements of nondeterminism or a soul do not seem to be relevant [59], [60]. Thus, if robots can have choice capacity and are not massively constrained by human control (inflexible programs and "emergency stop mechanisms"), then they could act "freely."

If robots can make moral decisions without free will, then they are likely targets of moral blame as well [61]. In humans, moral blame is not restricted to intentional action—negligence, mistakes, and errors are blamed as well—but blame does presuppose that the agent has the *capacity* for intentional action in order to correct mistakes and prevent negative outcomes in the future. That (and no mystical free will) would be needed for robots, too. Blame informs, corrects, and provides an opportunity to learn [37], [62]; to the extent that robots can change and learn, they may well be appropriate targets of blame. "You could have done otherwise," said to a human or a robot, does not question the deterministic order of the universe but invites a consideration of options that were available at the time of acting but were ignored or valued differently—and should be taken into account in the future.

V. MORAL COMMUNICATION

The suite of cognitive tools that enable moral judgment and decision making still are insufficient to achieve the socially most important function of morality: to regulate other people's behavior. For that, moral communication is needed. Moral perceivers often express their moral judgments to the alleged offender or to another community member [63], [64]; they sometimes have to provide evidence for their judgment; the alleged offender may contest the charges or explain the action in question [8]; and social estrangement may need to be repaired through conversation or compensation [65], [66]. Social robots need not have command over all these communicative acts, but two seem especially important: expressing their detection of a norm violation and explaining their own action when confronted with the charge that it was a violation.

Expressing one's moral judgments (of events or agents) will not be especially difficult for the robot if its moral cognition capacity is well developed and the robot has basic natural language skills. The subtle varieties of delivering moral criticism may be too difficult to master (e.g., the difference between scolding, chiding, or denouncing; [19]), but on the positive side, the anger and outrage that accompanies many human expressions of moral criticism can be easily avoided. This may be particularly important when the robot is partnered with a human—such as with a police officers on patrol or with a teacher in a classroom—and points out (inaudible to others) a looming violation. Without the kind of affect that would normally make the human partner defensive, the moral criticism may be more effective. However, in some communities, a robot that detects and, presumably, remembers and reports violations to others would itself violate trust norms. For example, a serious challenge in the military is that soldiers that are part of a unit do not report one another's violations (including human rights violations). A robot would not be susceptible to such pressures of loyalty, but the robot may also not find its way into the tight social community of soldiers, being rejected as a snitch.

Explaining immoral behaviors, the second important moral communication capacity, is directly derived from explaining behaviors in general, which is relatively well understood in psychology [67], [68]. Importantly, people treat intentional and unintentional behaviors quite differently: they explain intentional behaviors with reasons (the agent's beliefs and desires in light of which and on the ground of which they decided to act), and they explain unintentional behaviors with causes. Correspondingly, explaining intentional moral violations amounts to offering reasons that justify the violating action, whereas explaining unintentional moral violations amounts to offering causes that excuse one's involvement in the violation [37]. In addition, and unique to the moral domain, unintentional moral violations are assessed by counterfactuals: what the person could have done differently to prevent the negative event. As a result, moral criticism involves simulation of the past (what alternative paths of prevention may have been available) and simulation of the future (how one is expected to act differently to prevent repeated offenses). Both seem computationally feasible [69].

Explanations of one's own intentional actions require more than causal analysis and simulation; they require access to one's own reasoning en route to action. Some have famously doubted this capacity in humans [70], but these doubts do not apply in the case of reasons for action [71]. A robot, in any case, should have perfect access to its own reasoning. But once it accesses the trace of its reasoning, it must articulate this reasoning in humanly comprehensible ways (as beliefs and desires), regardless of the formalism in which it performs the reasoning. This amounts to one last form of simulation: modeling what a human would want to know so as to understand (and accept) the robot's decision in question. In fact, if the robot can simulate in advance a possible human challenge to its planned action and has available an acceptable explanation, then the action has passed a social criterion for moral behavior.

VI. FROM HERE ON OUT

In light of the extensive and complex components of human moral competence, designing robots with such competence is an awe-inspiring challenge. The key steps will be to build computational representations of norm systems and incorporate moral concepts and vocabulary into the robotic architecture. Once norms and concept representations are available in the architecture, the next step is to develop algorithms that can computationally capture moral cognition and decision-making. The development of these processes might take some time, but it does not seem nearly as difficult as developing the communicative capacities of moral agents we alluded to earlier, in part because of the complexity and flexibility of human language. Unless new computational learning algorithms enable robots to acquire human-like natural language capabilities, we might need to move from programming robots (and occasionally letting them learn) to *raising* robots in human environments. This may be the only way to expose them to the wealth of human moral situations and communicative interactions. Infants are not pre-programmed with norm systems or language either; but with countless repetitions in initially constrained contexts, and with a strong social reward function, they learn just about anything culture throws at them. Admittedly, humans have powerful learning mechanisms. But some of these mechanisms are available to robots as well, and new ones will be developed that could even exceed human capabilities. We do not know how well robots can learn norms, concepts, and language unless we give them abundant opportunities to do so.

REFERENCES

[1] J. P. Sullins, "Introduction: Open questions in roboethics," *Philos. Technol.*, vol. 24, no. 3, p. 233, Sep. 2011.
[2] J. Mikhail, "Universal moral grammar: Theory, evidence and the future," *Trends Cogn. Sci.*, vol. 11, no. 4, pp. 143–152, Apr. 2007.
[3] J. D. Greene, R. B. Sommerville, L. E. Nystrom, J. M. Darley, and J. D. Cohen, "An fMRI investigation of emotional engagement in moral judgment," *Science*, vol. 293, no. 5537, pp. 2105–2108, Sep. 2001.
[4] N. Eisenberg, "Emotion, regulation, and moral development," *Annu. Rev. Psychol.*, vol. 51, pp. 665–697, 2000.
[5] J. Haidt, "The emotional dog and its rational tail: A social intuitionist approach to moral judgment," *Psychol. Rev.*, vol. 108, no. 4, pp. 814–834, Oct. 2001.
[6] M. D. Alicke, "Culpable control and the psychology of blame," *Psychol. Bull.*, vol. 126, no. 4, pp. 556–574, Jul. 2000.
[7] G. R. Semin and A. S. R. Manstead, *The accountability of conduct: A social psychological analysis*. London: Academic Press, 1983.
[8] C. Antaki, *Explaining and arguing: The social organization of accounts*. London: Sage, 1994.
[9] J. S. Hall, *Beyond AI: Creating the conscience of the machine*. Amherst, NY: Prometheus Books, 2007.
[10] C. Allen, G. Varner, and J. Zinser, "Prolegomena to any future artificial moral agent," *J. Exp. Theor. Artif. Intell.*, vol. 12, no. 3, pp. 251–261, Jul. 2000.
[11] J. Parthemore and B. Whitby, "What makes any agent a moral agent? Reflections on machine consciousness and moral agency," *Int. J. Mach. Conscious.*, vol. 4, pp. 105–129, 2013.
[12] P. S. Churchland, *Braintrust: What neuroscience tells us about morality*. Princeton, NJ: Princeton University Press, 2012.
[13] R. Joyce, *The evolution of morality*. MIT Press, 2006.
[14] R. D. Alexander, *The biology of moral systems*. Hawthorne, NY: Aldine de Gruyter, 1987.
[15] C. S. Sripada and S. Stich, "A framework for the psychology of norms," in *The innate mind (Volume 2: Culture and cognition)*, P. Carruthers, S. Laurence, and S. Stich, Eds. New York, NY: Oxford University Press, 2006, pp. 280–301.
[16] S. Nichols and R. Mallon, "Moral dilemmas and moral rules," *Cognition*, vol. 100, no. 3, pp. 530–542, Jul. 2006.
[17] J. K. Hamlin, "Moral judgment and action in preverbal infants and toddlers: Evidence for an innate moral core," *Curr. Dir. Psychol. Sci.*, vol. 22, no. 3, pp. 186–193, Jun. 2013.
[18] J. C. Flack and F. B. M. de Waal, "'Any animal whatever'. Darwinian building blocks of morality in monkeys and apes," *J. Conscious. Stud.*, vol. 7, no. 1–2, pp. 1–29, 2000.
[19] J. Voiklis, C. Cusimano, and B. F. Malle, "A social-conceptual map of moral criticism." Unpublished Manuscript, Brown University, 2013.
[20] J. C. Wright and K. Bartsch, "Portraits of early moral sensibility in two children's everyday conversations," *Merrill-Palmer Q.*, vol. 54, no. 1, pp. 56–85, Jan. 2008.
[21] N. L. Powell, S. W. G. Derbyshire, and R. E. Guttentag, "Biases in children's and adults' moral judgments," *J. Exp. Child Psychol.*, vol. 113, no. 1, pp. 186–193, Sep. 2012.
[22] S. Pellizzoni, M. Siegal, and L. Surian, "The contact principle and utilitarian moral judgments in young children," *Dev. Sci.*, vol. 13, no. 2, pp. 265–270, Mar. 2010.
[23] H. M. Wellman and A. T. Phillips, "Developing intentional understandings," in *Intentions and intentionality: Foundations of social cognition*, B. F. Malle, L. J. Moses, and D. A. Baldwin, Eds. Cambridge, MA: The MIT Press, 2001, pp. 125–148.
[24] A. L. Woodward, "Infants selectively encode the goal object of an actor's reach," *Cognition*, vol. 69, no. 1, pp. 1–34, Nov. 1998.
[25] A. N. Meltzoff, "Understanding the intentions of others: Re-enactment of intended acts by 18-month-old children," *Dev. Psychol.*, vol. 31, no. 5, pp. 838–850, 1995.
[26] M. M. Saylor, D. A. Baldwin, J. A. Baird, and J. LaBounty, "Infants' on-line segmentation of dynamic human action," *J. Cogn. Dev.*, vol. 8, no. 1, pp. 113–128, 2007.

[27] L. J. Moses, "Young children's understanding of belief constraints on intention," *Cogn. Dev.*, vol. 8, no. 1, pp. 1–25, Jan. 1993.

[28] H. Wimmer and J. Perner, "Beliefs about beliefs: Representation and constraining function of wrong beliefs in young children's understanding of deception," *Cognition*, vol. 13, no. 1, pp. 103–128, Jan. 1983.

[29] J. A. Baird and L. J. Moses, "Do preschoolers appreciate that identical actions may be motivated by different intentions?," *J. Cogn. Dev.*, vol. 2, no. 4, pp. 413–448, Nov. 2001.

[30] P. L. Harris, *Children and emotion: The development of psychological understanding*. New York, NY: Basil Blackwell, 1989.

[31] K. Bartsch and H. M. Wellman, *Children talk about the mind*. New York: Oxford University Press, 1995.

[32] R. T. Beckwith, "The language of emotion, the emotions, and nominalist bootstrapping," in *Children's theories of mind: Mental states and social understanding*, Hillsdale, NJ England: Lawrence Erlbaum Associates, Inc, 1991, pp. 77–95.

[33] B. F. Malle and S. Dickert, "Values," in *The encyclopedia of social psychology*, R. F. Baumeister and K. D. Vohs, Eds. Thousand Oaks, CA: Sage, 2007.

[34] J. Elster, *The cement of society: A study of social order*. New York, NY: Cambridge University Press, 1989.

[35] J. J. A. Van Berkum, B. Holleman, M. Nieuwland, M. Otten, and J. Murre, "Right or wrong? The brain's fast response to morally objectionable statements," *Psychol. Sci.*, vol. 20, no. 9, pp. 1092–1099, 2009.

[36] A. Prince and P. Smolensky, "Optimality: From neural networks to universal grammar," *Science*, vol. 275, no. 5306, pp. 1604–1610, Mar. 1997.

[37] B. F. Malle, S. Guglielmo, and A. E. Monroe, "A theory of blame," *Psychol. Inq.*, 2014.

[38] B. F. Malle, S. Guglielmo, and A. E. Monroe, "Moral, cognitive, and social: The nature of blame," in *Social thinking and interpersonal behavior*, J. P. Forgas, K. Fiedler, and C. Sedikides, Eds. Philadelphia, PA: Psychology Press, 2012, pp. 313–331.

[39] F. Cushman and L. Young, "Patterns of moral judgment derive from nonmoral psychological representations," *Cogn. Sci.*, vol. 35, no. 6, pp. 1052–1075, Aug. 2011.

[40] S. Guglielmo, A. E. Monroe, and B. F. Malle, "At the heart of morality lies folk psychology," *Inq. Interdiscip. J. Philos.*, vol. 52, no. 5, pp. 449–466, 2009.

[41] A. R. Damasio, *Descartes' error: Emotion, reason, and the human brain*. New York, NY: Putnam, 1994.

[42] J. Knobe and B. Fraser, "Causal judgment and moral judgment: Two experiments," in *Moral psychology (Vol. 2): The cognitive science of morality: Intuition and diversity*, vol. 2, Cambridge, MA: MIT Press, 2008, pp. 441–447.

[43] C. L. Harenski, K. A. Harenski, M. S. Shane, and K. A. Kiehl, "Aberrant neural processing of moral violations in criminal psychopaths," *J. Abnorm. Psychol.*, vol. 119, no. 4, pp. 863–874, Nov. 2010.

[44] M. Cima, F. Tonnaer, and M. D. Hauser, "Psychopaths know right from wrong but don't care," *Soc. Cogn. Affect. Neurosci.*, vol. 5, no. 1, pp. 59–67, Mar. 2010.

[45] C. A. Hutcherson and J. J. Gross, "The moral emotions: A social-functionalist account of anger, disgust, and contempt," *J. Pers. Soc. Psychol.*, vol. 100, no. 4, pp. 719–737, Apr. 2011.

[46] L. Kohlberg, *The psychology of moral development: The nature and validity of moral stages*. San Francisco, CA: Harper & Row, 1984.

[47] J. M. Paxton, L. Ungar, and J. D. Greene, "Reflection and reasoning in moral judgment," *Cogn. Sci.*, vol. 36, no. 1, pp. 163–177, Jan. 2012.

[48] J. D. Greene, L. E. Nystrom, A. D. Engell, J. M. Darley, and J. D. Cohen, "The neural bases of cognitive conflict and control in moral judgment," *Neuron*, vol. 44, no. 2, pp. 389–400, Oct. 2004.

[49] G. J. Koop, "An assessment of the temporal dynamics of moral decisions," *Judgm. Decis. Mak.*, vol. 8, no. 5, pp. 527–539, 2013.

[50] E. B. Royzman, G. P. Goodwin, and R. F. Leeman, "When sentimental rules collide: 'Norms with feelings' in the dilemmatic context," *Cognition*, vol. 121, no. 1, pp. 101–114, Oct. 2011.

[51] G. Moretto, E. Làdavas, F. Mattioli, and G. di Pellegrino, "A psychophysiological investigation of moral judgment after ventromedial prefrontal damage," *J. Cogn. Neurosci.*, vol. 22, no. 8, pp. 1888–1899, Aug. 2010.

[52] T. Davis, B. C. Love, and W. Todd Maddox, "Anticipatory emotions in decision tasks: Covert markers of value or attentional processes?," *Cognition*, vol. 112, no. 1, pp. 195–200, Jul. 2009.

[53] W. Wallach, S. Franklin, and C. Allen, "A conceptual and computational model of moral decision making in human and artificial agents," *Top. Cogn. Sci.*, vol. 2, no. 3, pp. 454–485, 2010.

[54] M. L. Hoffman, "Empathy and prosocial behavior," in *Handbook of emotions*, 3rd ed., M. Lewis, J. M. Haviland-Jones, and L. F. Barrett, Eds. New York, NY: Guilford Press, 2008, pp. 440–455.

[55] O. FeldmanHall, T. Dalgleish, R. Thompson, D. Evans, S. Schweizer, and D. Mobbs, "Differential neural circuitry and self-interest in real vs hypothetical moral decisions," *Soc. Cogn. Affect. Neurosci.*, vol. 7, no. 7, pp. 743–751, Oct. 2012.

[56] M. Scheutz, "The inherent dangers of unidirectional emotional bonds between humans and social robots," in *Anthology on Robo-Ethics*, P. Lin, G. Bekey, and K. Abney, Eds. Cambridge, MA: MIT Press, 2012, pp. 205–221.

[57] M. Scheutz, "The affect dilemma for artificial agents: should we develop affective artificial agents?," *IEEE Trans. Affect. Comput.*, vol. 3, no. 4, pp. 424–433, 2012.

[58] C. R. Sunstein, "Moral heuristics," *Behav. Brain Sci.*, vol. 28, no. 4, pp. 531–573, Aug. 2005.

[59] A. E. Monroe, K. D. Dillon, and B. F. Malle, *Developing a model of the folk concept of free will and its impact on moral judgment*. Paper presented at the symposium on Big Questions in Free Will, Florida State University, Tallahasee, Florida., 2013.

[60] A. E. Monroe and B. F. Malle, "From uncaused will to conscious choice: The need to study, not speculate about people's folk concept of free will," *Rev. Philos. Psychol.*, vol. 1, no. 2, pp. 211–224, 2010.

[61] P. H. Kahn, Jr., T. Kanda, H. Ishiguro, B. T. Gill, J. H. Ruckert, S. Shen, H. E. Gary, A. L. Reichert, N. G. Freier, and R. L. Severson, "Do people hold a humanoid robot morally accountable for the harm it causes?," in *Proceedings of the Seventh Annual ACM/IEEE International Conference on Human-Robot Interaction*, New York, NY, 2012, pp. 33–40.

[62] F. Cushman, "The functional design of punishment and the psychology of learning," in *Psychological and environmental foundations of cooperation.*, vol. 2, R. Joyce, K. Sterelny, B. Calcott, and B. Fraser, Eds. Cambridge, MA: MIT Press, 2013.

[63] I. Dersley and A. Wootton, "Complaint sequences within antagonistic argument," *Res. Lang. Soc. Interact.*, vol. 33, no. 4, pp. 375–406, Oct. 2000.

[64] V. Traverso, "The dilemmas of third-party complaints in conversation between friends," *J. Pragmat.*, vol. 41, no. 12, pp. 2385–2399, Dec. 2009.

[65] M. U. Walker, *Moral repair: Reconstructing moral relations after wrongdoing*. New York, NY: Cambridge University Press, 2006.

[66] M. McKenna, "Directed blame and conversation," in *Blame: Its nature and norms*, New York, NY: Oxford University Press, 2012, pp. 119–140.

[67] B. F. Malle, *How the mind explains behavior: Folk explanations, meaning, and social interaction*. Cambridge, MA: MIT Press, 2004.

[68] D. J. Hilton, "Causal explanation: From social perception to knowledge-based causal attribution," in *Social psychology: Handbook of basic principles*, 2nd ed., A. W. Kruglanski and E. T. Higgins, Eds. New York, NY: Guilford Press, 2007, pp. 232–253.

[69] P. Bello, "Cognitive foundations for a computational theory of mindreading," *Adv. Cogn. Syst.*, vol. 1, pp. 59–72, 2012.

[70] R. E. Nisbett and T. D. Wilson, "Telling more than we know: Verbal reports on mental processes," *Psychol. Rev.*, vol. 84, pp. 231–259, 1977.

[71] B. F. Malle, "Time to give up the dogmas of attribution: A new theory of behavior explanation," in *Advances of Experimental Social Psychology*, vol. 44, M. P. Zanna and J. M. Olson, Eds. San Diego, CA: Academic Press, 2011, pp. 297–352.

Part III:

Machine Ethics

17

The Nature, Importance, and Difficulty of Machine Ethics

James H. Moor

The question of whether machine ethics exists or might exist in the future is difficult to answer if we can't agree on what counts as machine ethics. Some might argue that machine ethics obviously exists because humans are machines and humans have ethics. Others could argue that machine ethics obviously doesn't exist because ethics is simply emotional expression and machines can't have emotions.

Implementations of machine ethics might be possible in situations ranging from maintaining hospital records to overseeing disaster relief. But what is machine ethics, and how good can it be?

A wide range of positions on machine ethics are possible, and a discussion of the issue could rapidly propel us into deep and unsettled philosophical issues. Perhaps, understandably, few in the scientific arena pursue the issue of machine ethics. You're unlikely to find easily testable hypotheses in the murky waters of philosophy. But we can't—and shouldn't—avoid consideration of machine ethics in today's technological world.

As we expand computers' decision-making roles in practical matters, such as computers driving cars, ethical considerations are inevitable. Computer scientists and engineers must examine the possibilities for machine ethics because, knowingly or not, they've already engaged—or will soon engage—in some form of it. Before we can discuss possible implementations of machine ethics, however, we need to be clear about what we're asserting or denying.

Varieties of machine ethics

When people speak of technology and values, they're often thinking of ethical values. But not all values are ethical. For example, practical, economic, and aesthetic values don't necessarily draw on ethical considerations. A product of technology, such as a new sailboat, might be practically durable, economically expensive, and aesthetically pleasing, absent consideration of any ethical values. We routinely evaluate technology from these nonethical normative viewpoints. Tool makers and users regularly evaluate how well tools accomplish the purposes for which they were designed. With technology, all of us—ethicists and engineers included—are involved in evaluation processes requiring the selection and application of standards. In none of our professional activities can we retreat to a world of pure facts, devoid of subjective normative assessment.

By its nature, computing technology is normative. We expect programs, when executed, to proceed toward some objective—for example, to correctly compute our income taxes or keep an airplane on course. Their intended purpose serves as a norm for evaluation—that is, we assess how well the computer program calculates the tax or guides the airplane. Viewing computers as technological agents is reasonable because they do jobs on our behalf. They're normative agents in the limited sense that we can assess their performance in terms of how well they do their assigned jobs.

After we've worked with a technology for a while, the norms become second nature. But even after they've become widely accepted as the way of doing the activity properly, we can have moments of realization and see a need to establish different kinds of norms. For instance, in the early days of computing, using double digits to designate years was the standard and worked well. But, when the year 2000 approached, programmers realized that this norm needed reassessment. Or consider a distinction involving AI. In a November 1999 correspondence between Herbert Simon and Jacques Berleur,[1] Berleur was asking Simon for his reflections on the 1956 Dartmouth Summer Research Project on Artificial Intelligence, which Simon attended. Simon expressed some puzzlement as to why Trenchard More, a conference attendee, had so strongly emphasized modal logics in his thesis. Simon thought about it and then wrote back to Berleur,

> My reply to you last evening left my mind nagged by the question of why Trench Moore [sic], in his thesis, placed so much emphasis on modal logics. The answer, which I thought might interest you, came to me when I awoke this morning. Viewed from a computing standpoint (that is, discovery of proofs rather than verification), a standard logic is an indeterminate algorithm: it tells you what you MAY legally do, but not what you OUGHT to do to find a proof. Moore [sic] viewed his task as building a modal logic of "oughts"—a strategy for search—on top of the standard logic of verification.

Simon was articulating what he already knew as one of the designers of the Logic Theorist, an early AI program. A theorem prover must not only generate a list of well-formed formulas but must also find a sequence of well-formed formulas constituting a proof. So, we need a procedure for doing this. *Modal logic* distinguishes between what's permitted and what's required. Of course, both are norms for the subject matter. But norms can have different levels of obligation, as Simon stresses through capitalization. Moreover, the norms he's suggesting aren't ethical norms. A typical theorem prover is a normative agent but not an ethical one.

Ethical-impact agents

You can evaluate computing technology in terms of not only design norms (that is, whether it's doing its job appropriately) but also ethical norms.

For example, *Wired* magazine reported an interesting example of applied computer technology.[2] Qatar is an oil-rich country in the Persian Gulf that's friendly to and influenced by the West while remaining steeped in Islamic tradition. In Qatar, these cultural traditions sometimes mix without incident— for example, women may wear Western clothing or a full veil. And sometimes the cultures conflict, as illustrated by camel racing, a pastime of the region's rich for centuries. Camel jockeys must be light—the lighter the jockey, the faster the camel. Camel owners enslave very young boys from poorer countries to ride the camels. Owners have historically mistreated the young slaves, including limiting their food to keep them lightweight. The United Nations and the US State Department have objected to this human trafficking, leaving Qatar vulnerable to economic sanctions.

The machine solution has been to develop robotic camel jockeys. The camel jockeys are about two feet high and weigh 35 pounds. The robotic jockey's right hand handles the whip, and its left handles the reins. It runs Linux, communicates at 2.4 GHz, and has a GPS-enabled camel-heart-rate monitor. As *Wired* explained it, "Every robot camel jockey bopping along on its improbable mount means one Sudanese boy freed from slavery and sent home." Although this eliminates the camel jockey slave problem in Qatar, it doesn't improve the economic and social conditions in places such as Sudan.

Computing technology often has important ethical impact. The young boys replaced

> Frequently, what sparks debate is whether you can put ethics into a machine. Can a computer operate ethically because it's internally ethical in some way?

by robotic camel jockeys are freed from slavery. Computing frees many of us from monotonous, boring jobs. It can make our lives better but can also make them worse. For example, we can conduct business online easily, but we're more vulnerable to identity theft. Machine ethics in this broad sense is close to what we've traditionally called *computer ethics*. In one sense of machine ethics, computers do our bidding as surrogate agents and impact ethical issues such as privacy, property, and power. However, the term is often used more restrictively. Frequently, what sparks debate is whether you can put ethics into a machine. Can a computer operate ethically because it's internally ethical in some way?

Implicit ethical agents

If you wish to put ethics into a machine, how would you do it? One way is to constrain the machine's actions to avoid unethical outcomes. You might satisfy machine ethics in this sense by creating software that implicitly supports ethical behavior, rather than by writing code containing explicit ethical maxims. The machine acts ethically because its internal functions implicitly promote ethical behavior—or at least avoid unethical behavior. Ethical behavior is the machine's nature. It has, to a limited extent, virtues.

Computers are implicit ethical agents when the machine's construction addresses safety or critical reliability concerns. For example, automated teller machines and Web banking software are agents for banks and can perform many of the tasks of human tellers and sometimes more. Transactions involving money are ethically important. Machines must be carefully constructed to give out or transfer the correct amount of money every time a banking transaction occurs. A line of code telling the computer to be honest won't accomplish this.

Aristotle suggested that humans could obtain virtue by developing habits. But with machines, we can build in the behavior without the need for a learning curve. Of course, such machine virtues are task specific and rather limited. Computers don't have the practical wisdom that Aristotle thought we use when applying our virtues.

Another example of a machine that's an implicit ethical agent is an airplane's automatic pilot. If an airline promises the plane's passengers a destination, the plane must arrive at that destination on time and safely. These are ethical outcomes that engineers design into the automatic pilot. Other built-in devices warn humans or machines if an object is too close or the fuel supply is low. Or, consider pharmacy software that checks for and reports on drug interactions. Doctor and pharmacist *duties of care* (legal and ethical obligations) require that the drugs prescribed do more good than harm. Software with elaborate medication databases helps them perform those duties responsibly.

Machines' capability to be implicit ethical agents doesn't demonstrate their ability to be full-fledged ethical agents. Nevertheless, it illustrates an important sense of machine ethics. Indeed, some would argue that software engineers must routinely consider machine ethics in at least this implicit sense during software development.

Explicit ethical agents

Can ethics exist explicitly in a machine?[3] Can a machine represent ethical categories and perform analysis in the sense that a computer can represent and analyze inventory or tax information? Can a machine "do" ethics like a computer can play chess? Chess programs typically provide representations of the current board position, know which moves are legal, and can calculate a good next move. Can a machine represent ethics explicitly and then operate effectively on the basis of this knowledge? (For simplicity, I'm imaging the development of ethics in terms of traditional symbolic AI. However, I don't want to exclude the possibility that the machine's architecture is connectionist, with an explicit understanding of the ethics emerging from that. Compare Wendell Wallach, Colin Allen, and Iva Smit's different senses of "bottom up" and "top down."[4])

Although clear examples of machines acting as explicit ethical agents are elusive, some current developments suggest interesting movements in that direction. Jeroen van den Hoven and Gert-Jan Lokhorst blended three kinds of advanced logic to serve as a bridge between ethics and a machine:

- *deontic* logic for statements of permission and obligation,
- *epistemic* logic for statements of beliefs and knowledge, and
- *action* logic for statements about actions.[5]

Together, these logics suggest that a formal apparatus exists that could describe ethical situations with sufficient precision to make ethical judgments by machine. For example, you could use a combination of these logics to state explicitly what action is allowed and what is forbidden in transferring personal information to protect privacy.[6] In a hospital, for example, you'd program a computer to let some personnel access some information and to calculate which actions what person should take and who should be informed about those actions.

Michael Anderson, Susan Anderson, and Chris Armen implement two ethical theories.[7] Their first model of an explicit ethical agent—Jeremy (named for Jeremy Bentham)—implements Hedonistic Act Utilitarianism. Jeremy estimates the likelihood of pleasure or displeasure for persons affected by a particular act. The second model is W.D. (named for William D. Ross). Ross's theory emphasizes prima facie duties as opposed to absolute duties. Ross considers no duty as absolute and gives no clear ranking of his various prima facie duties. So, it's unclear how to make ethical decisions under Ross's theory. Anderson, Anderson, and Armen's computer model overcomes this uncertainty. It uses a learning algorithm to adjust judgments of duty by taking into account both prima facie duties and past intuitions about similar or dissimilar cases involving those duties.

These examples are a good start toward creating explicit ethical agents, but more research is needed before a robust explicit ethical agent can exist in a machine. What would such an agent be like? Presumably, it would be able to make plausible ethical judgments and justify them. An explicit ethical agent that was autonomous in that it could handle real-life situations involving an unpredictable sequence of events would be most impressive.

James Gips suggested that the development of an ethical robot be a computing Grand Challenge.[8] Perhaps DARPA could

> An average adult is a full ethical agent. Can a machine be a full ethical agent? It's here that the debate about machine ethics becomes most heated.

establish an explicit-ethical-agent project analogous to its autonomous-vehicle project (www.darpa.mil/grandchallenge/index.asp). As military and civilian robots become increasingly autonomous, they'll probably need ethical capabilities. Given this likely increase in robots' autonomy, the development of a machine that's an explicit ethical agent seems a fitting subject for a Grand Challenge.

Machines that are explicit ethical agents might be the best ethical agents to have in situations such as disaster relief. In a major disaster, such as Hurricane Katrina in New Orleans, humans often have difficulty tracking and processing information about who needs the most help and where they might find effective relief. Confronted with a complex problem requiring fast decisions, computers might be more competent than humans. (At least the question of a computer decision maker's competence is an empirical issue that might be decided in favor of the computer.) These decisions could be ethical in that they would determine who would live and who would die. Some might say that only humans should make such decisions, but if (and of course this is a big assumption) computer decision making could routinely save more lives in such situations than human decision making, we might have a good ethical basis for letting computers make the decisions.[9]

Full ethical agents

A full ethical agent can make explicit ethical judgments and generally is competent to reasonably justify them. An average adult human is a full ethical agent. We typically regard humans as having consciousness, intentionality, and free will. Can a machine be a full ethical agent? It's here that the debate about machine ethics becomes most heated. Many believe a bright line exists between the senses of machine ethics discussed so far and a full ethical agent. For them, a machine can't cross this line. The bright line marks a crucial ontological difference between humans and whatever machines might be in the future.

The bright-line argument can take one or both of two forms. The first is to argue that only full ethical agents can be ethical agents. To argue this is to regard the other senses of machine ethics as not really ethics involving agents. However, although these other senses are weaker, they can be useful in identifying more limited ethical agents. To ignore the ethical component of ethical-impact agents, implicit ethical agents, and explicit ethical agents is to ignore an important aspect of machines. What might bother some is that the ethics of the lesser ethical agents is derived from their human developers. However, this doesn't mean that you can't evaluate machines as ethical agents. Chess programs receive their chess knowledge and abilities from humans. Still, we regard them as chess players. The fact that lesser ethical agents lack humans' consciousness, intentionality, and free will is a basis for arguing that they shouldn't have broad ethical responsibility. But it doesn't establish that they aren't ethical in ways that are assessable or that they shouldn't have limited roles in functions for which they're appropriate.

The other form of bright-line argument is to argue that no machine can become a full ethical agent—that is, no machine can have consciousness, intentionality, and free will. This is metaphysically contentious, but the simple rebuttal is that we can't say with certainty that future machines will lack these features. Even John Searle, a major critic of strong AI, doesn't argue that machines can't possess these features.[10] He only denies that computers, in their capacity as purely syntactic devices, can possess understanding. He doesn't claim that machines can't have understanding, presumably including an understanding of ethics. Indeed, for Searle, a materialist, humans are a kind of machine, just not a purely syntactic computer.

Thus, both forms of the bright-line argument leave the possibility of machine ethics open. How much can be accomplished in machine ethics remains an empirical question.

We won't resolve the question of whether machines can become full ethical agents by philosophical argument or empirical research in the near future. We should therefore focus on developing limited explicit ethical agents. Although they would fall short of being full ethical agents, they could help prevent unethical outcomes.

I can offer at least three reasons why it's important to work on machine ethics in the sense of developing explicit ethical agents:

- Ethics is important. We want machines to treat us well.
- Because machines are becoming more sophisticated and make our lives more enjoyable, future machines will likely have increased control and autonomy to do this. More powerful machines need more powerful machine ethics.
- Programming or teaching a machine to act ethically will help us better understand ethics.

The importance of machine ethics is clear. But, realistically, how possible is it? I also offer three reasons why we can't be too optimistic about our ability to develop machines to be explicit ethical agents.

First, we have a limited understanding of what a proper ethical theory is. Not only do people disagree on the subject, but individuals can also have conflicting ethical intuitions and beliefs. Programming a computer to be ethical is much more difficult than programming a computer to play world-champion chess—an accomplishment that took 40 years. Chess is a simple domain with well-defined legal moves. Ethics operates in a complex domain with some ill-defined legal moves.

Second, we need to understand learning better than we do now. We've had significant successes in machine learning, but we're still far from having the child machine that Turing envisioned.

Third, inadequately understood ethical theory and learning algorithms might be easier problems to solve than computers' absence of common sense and world knowledge. The deepest problems in developing machine ethics will likely be epistemological as much as ethical. For example, you might program a machine with the classical imperative of physicians and Asimovian robots: First, do no harm. But this wouldn't be helpful unless the machine could understand what constitutes harm in the real world. This isn't to suggest that we shouldn't vigorously pursue machine ethics. On the contrary, given its nature, importance, and difficulty, we should dedicate much more effort to making progress in this domain.

Acknowledgments

I'm indebted to many for helpful comments, particularly to Keith Miller, Vincent Wiegel, and this magazine's anonymous referees and editors.

References

1. H. Simon, "Re: Dartmouth Seminar 1956" (email to J. Berleur), Herbert A. Simon Collection, Carnegie Mellon Univ. Archives, 20 Nov. 1999.

2. J. Lewis, "Robots of Arabia," *Wired*, vol. 13, no. 11, Nov. 2005, pp. 188–195; www.wired.com/wired/archive/13.11/camel.html?pg=1&topic=camel&topi c_set=.

3. J.H. Moor, "Is Ethics Computable?" *Metaphilosophy*, vol. 26, nos. 1–2, 1995, pp. 1–21.

4. W. Wallach, C. Allen, and I. Smit, "Machine Morality: Bottom-Up and Top-Down Approaches for Modeling Human Moral Faculties," *Machine Ethics*, M. Anderson, S.L. Anderson, and C. Armen, eds., AAAI Press, 2005, pp. 94–102.

5. J. van den Hoven and G.-J. Lokhorst, "Deontic Logic and Computer-Supported Computer Ethics," *Cyberphilosophy: The Intersection of Computing and Philosophy*, J.H. Moor and T.W. Bynum, eds., Blackwell, 2002, pp. 280–289.

6. V. Wiegel, J. van den Hoven, and G.-J. Lokhorst, "Privacy, Deontic Epistemic Action Logic and Software Agents," *Ethics of New Information Technology, Proc. 6th Int'l Conf. Computer Ethics: Philosophical Enquiry* (CEPE 05), Center for Telematics and Information Technology, Univ. of Twente, 2005, pp. 419–434.

7. M. Anderson, S.L. Anderson, and C. Armen, "Towards Machine Ethics: Implementing Two Action-Based Ethical Theories," *Machine Ethics*, M. Anderson, S.L. Anderson, and C. Armen, eds., AAAI Press, 2005, pp. 1–7.

8. J. Gips, "Creating Ethical Robots: A Grand Challenge," presented at the AAAI Fall 2005 Symposium on Machine Ethics; www.cs.bc.edu/~gips/EthicalRobotsGrandChallenge.pdf.

9. J.H. Moor, "Are There Decisions Computers Should Never Make?" *Nature and System*, vol. 1, no. 4, 1979, pp. 217–229.

10. J.R. Searle, "Minds, Brains, and Programs," *Behavioral and Brain Sciences*, vol. 3, no. 3, 1980, pp. 417–457.

For more information on this or any other computing topic, please visit our Digital Library at www.computer.org/publications/dlib.

18

Machine Ethics:
Creating an Ethical Intelligent Agent

Michael Anderson and Susan Leigh Anderson

■ The newly emerging field of machine ethics (Anderson and Anderson 2006) is concerned with adding an ethical dimension to machines. Unlike computer ethics—which has traditionally focused on ethical issues surrounding humans' use of machines—machine ethics is concerned with ensuring that the behavior of machines toward human users, and perhaps other machines as well, is ethically acceptable. In this article we discuss the importance of machine ethics, the need for machines that represent ethical principles explicitly, and the challenges facing those working on machine ethics. We also give an example of current research in the field that shows that it is possible, at least in a limited domain, for a machine to abstract an ethical principle from examples of correct ethical judgments and use that principle to guide its own behavior.

The ultimate goal of machine ethics, we believe, is to create a machine that *itself* follows an ideal ethical principle or set of principles; that is to say, it is guided by this principle or these principles in decisions it makes about possible courses of action it could take. We need to make a distinction between what James Moor has called an *"implicit ethical agent"* and an *"explicit ethical agent"* (Moor 2006). According to Moor, a machine that is an implicit ethical agent is one that has been programmed to behave ethically, or at least avoid unethical behavior, without an explicit representation of ethical principles. It is constrained in its behavior by its designer who is following ethical principles. A machine that is an explicit ethical agent, on the other hand, is able to calculate the best action in ethical dilemmas using ethical principles. It can "represent ethics explicitly and then operate effectively on the basis of this knowledge." Using Moor's terminology, most of those working on machine ethics would say that the ultimate goal is to create a machine that is an explicit ethical agent.

We are, here, primarily concerned with the ethical decision making itself, rather than how a machine would gather the information needed to make the decision and incorporate it into its general behavior. It is important to see this as a separate and considerable challenge. It is separate because having all the information and facility in the world won't, by itself, generate ethical behavior in a machine. One needs to turn to the branch of philosophy that is concerned with ethics for insight into what is considered to be ethically acceptable behavior. It is a considerable challenge because, even among experts, ethics has not been completely codified. It is a field that is still evolving. We shall argue that one of the advantages of working on machine ethics is that it might lead to breakthroughs in ethical theory, since machines are well-suited for testing the results of consistently following a particular ethical theory.

One other point should be made in introducing the subject of machine ethics. Ethics can be seen as both easy and hard. It appears easy because we all make ethical decisions on a daily basis. But that doesn't mean that we are all experts in ethics. It is a field that requires much study and experience. AI researchers must have respect for the expertise of ethicists just as ethicists must appreciate the expertise of AI researchers. Machine ethics is an inherently interdisciplinary field.

The Importance of Machine Ethics

Why is the field of machine ethics important? There are at least three reasons that can be given. First, there are ethical ramifications to what machines currently do and are projected to do in the future. To neglect this aspect of machine behavior could have serious repercussions. South Korea has recently mustered more than 30 companies and 1000 scientists to the end of putting "a robot in every home by 2010" (Onishi 2006). DARPA's grand challenge to have a vehicle drive itself across 132 miles of desert terrain has been met, and a new grand challenge is in the works that will have vehicles maneuvering in an urban setting. The United States Army's Future Combat Systems program is developing armed robotic vehicles that will support ground troops with "direct-fire" and antitank weapons. From family cars that drive themselves and machines that discharge our daily chores with little or no assistance from us, to fully autonomous robotic entities that will begin to challenge our notions of the very nature of intelligence, it is clear that machines such as these will be capable of causing harm to human beings unless this is prevented by adding an ethical component to them.

Second, it could be argued that humans' fear of the possibility of autonomous intelligent machines stems from their concern about whether these machines will behave ethically, so the future of AI may be at stake. Whether society allows AI researchers to develop anything like autonomous intelligent machines may hinge on whether they are able to build in safeguards against unethical behavior. From the murderous robot uprising in the 1920 play *R.U.R.* (Capek 1921) and the deadly coup d'état perpetrated by the HAL 9000 computer in *2001: A Space Odyssey* (Clarke 1968), to *The Matrix* virtual reality simulation for the pacification and subjugation of human beings by machines, popular culture is rife with images of machines devoid of any ethical code mistreating their makers. In his widely circulated treatise, "Why the future doesn't need us," Bill Joy (2000) argues that the only antidote to such fates and worse is to "relinquish dangerous technologies." We believe that machine ethics research may offer a viable, more realistic solution.

Finally, we believe that it's possible that research in machine ethics will advance the study of ethical theory. Ethics, by its very nature, is the most practical branch of philosophy. It is concerned with how agents ought to behave when faced with ethical dilemmas. Despite the obvious applied nature of the field of ethics, too often work in ethical theory is done with little thought to actual application. When examples are discussed, they are typically artificial examples. Research in machine ethics has the potential to discover problems with current theories, perhaps even leading to the development of better theories, as AI researchers force scrutiny of the details involved in actually applying an ethical theory to particular cases. As Daniel Dennett (2006) recently stated, AI "makes philosophy honest." Ethics must be made computable in order to make it clear exactly how agents ought to behave in ethical dilemmas.

An exception to the general rule that ethicists don't spend enough time discussing actual cases occurs in the field of biomedical ethics, a field that has arisen out of a need to resolve pressing problems faced by health-care workers, insurers, hospital ethics boards, and biomedical researchers. As a result of there having been more discussion of actual cases in the field of biomedical ethics, a consensus is beginning to emerge as to how to evaluate ethical dilemmas in this domain, leading to the ethically correct action in many dilemmas. A reason there might be more of a consensus in this domain than in others is that in the area of biomedical ethics there is an ethically defensible goal (the best possible health of the patient), whereas in other areas (such as business and law) the goal may not be ethically defensible (make as much money as possible, serve the client's interest even if he or she is guilty of an offense or doesn't deserve a settlement) and ethics enters the picture as a limiting factor (the goal must be achieved within certain ethical boundaries).

AI researchers working with ethicists might find it helpful to begin with this domain, discovering a general approach to computing ethics that not only works in this domain, but could be applied to other domains as well.

Explicit Ethical Machines

It does seem clear, to those who have thought about the issue, that some sort of safeguard should be in place to prevent unethical machine behavior (and that work in this area may provide benefits for the study of ethical theory as well). This shows the need for creating at least *implicit* ethical machines; but why must we create *explicit* ethical machines, which would seem to be a much greater (perhaps even an impossible) challenge for AI researchers? Furthermore, many fear handing over the job of ethical overseer to machines themselves. How could we feel confident that a machine would make the right decision in situations that were not anticipated? Finally, what if the machine starts out behaving in an ethical fashion but then morphs into one that decides to behave unethically in order to secure advantages for itself?

On the need for *explicit*, rather than just *implicit*, ethical machines: What is critical in the "explicit ethical agent" versus "implicit ethical agent" distinction, in our view, lies not only in who is making the ethical judgments (the machine versus the human programmer), but also in the ability to *justify* ethical judgments that only an explicit representation of ethical principles allows. An explicit ethical agent is able to explain why a particular action is either right or wrong by appealing to an ethical principle. A machine that has learned, or been programmed, to make correct ethical judgments, but does not have principles to which it can appeal to justify or explain its judgments, is lacking something essential to being accepted as an ethical agent. Immanuel Kant (1785) made a similar point when he distinguished between an agent that acts from a sense of duty (consciously following an ethical principle), rather than merely in *accordance* with duty, having praise only for the former.

If we believe that machines could play a role in improving the lives of human beings—that this is a worthy goal of AI research—then, since it is likely that there will be ethical ramifications to their behavior, we must feel confident that these machines will act in a way that is ethically acceptable. It will be essential that they be able to justify their actions by appealing to acceptable ethical principles that they are following, in order to satisfy humans who will question their ability to act ethically. The ethical component of machines that affect humans' lives must be transparent, and principles that seem reasonable to human beings provide that transparency. Furthermore, the concern about how machines will behave in situations that were not anticipated also supports the need for explicit ethical machines. The virtue of having principles to follow, rather than being programmed in an ad hoc fashion to behave correctly in specific situations, is that it allows machines to have a way to determine the ethically correct action in new situations, even in new domains. Finally, Marcello Guarini (2006), who is working on a neural network model of machine ethics, where there is a predisposition to eliminate principles, argues that principles seem to play an important role in *revising* ethical beliefs, which is essential to ethical agency. He contends, for instance, that they are necessary to discern morally relevant differences in similar cases.

The concern that machines that start out behaving ethically will end up behaving unethically, perhaps favoring their own interests, may stem from fears derived from legitimate concerns about *human* behavior. Most human beings are far from ideal models of ethical agents, despite having been taught ethical principles; and humans do, in particular, tend to favor themselves. Machines, though, might have an advantage over human beings in terms of behaving ethically. As Eric Dietrich (2006) has recently argued, human beings, as biological entities in competition with others, may have evolved into beings with a genetic predisposition toward unethical behavior as a survival mechanism. Now, though, we have the chance to create entities that lack this predisposition, entities that might even inspire us to behave more ethically. Consider, for example, Andrew, the robot hero of Isaac Asimov's story "The Bicentennial Man" (1976), who was far more ethical than the humans with whom he came in contact. Dietrich maintained that the machines we fashion to have the good qualities of human beings, and that also follow principles derived from ethicists who are the exception to the general rule of unethical human beings, could be viewed as "humans 2.0"—a better version of human beings.

This may not completely satisfy those who are concerned about a future in which human beings share an existence with intelligent, autonomous machines. We face a choice, then, between allowing AI researchers to continue in their quest to develop intelligent, autonomous machines—which will have to involve adding an ethical component to them—or stifling this research. The likely benefits and possible harms of each option will have to be weighed. In any case, there are certain benefits to continuing to work on machine ethics. It is important to find a clear, objective basis for ethics—making ethics in principle computable—if only to rein in unethical *human* behavior; and AI researchers, working with ethicists, have a better chance of achieving breakthroughs in ethical theory than theoretical ethicists working alone. There is also the possibility that society would not be able to prevent some researchers from continuing to develop intelligent, autonomous machines, even if society decides that it is too dangerous to support such work. If this research should be successful, it will be important that we have ethical principles that we insist should be incorporated into such machines. The one thing that society should fear more than sharing an existence with intelligent, autonomous machines is sharing an existence with machines like these without an ethical component.

Challenges Facing Those Working on Machine Ethics

The challenges facing those working on machine ethics can be divided into two main categories: philosophical concerns about the feasibility of computing ethics and challenges from the AI perspective. In the first category, we need to ask whether ethics is the sort of thing that can be computed. One well-known ethical theory that supports an affirmative answer to this question is "act utilitarianism." According to this teleological theory (a theory that maintains that the rightness and wrongness of actions is determined entirely by the consequences of the actions) that act is right which, of all the actions open to the agent, is likely to result in the greatest net good consequences, taking all those affected by the action equally into account. Essentially, as Jeremy Bentham (1781) long ago pointed out, the theory involves performing "moral arithmetic."

Of course, before doing the arithmetic, one needs to know what counts as a "good" and "bad" consequence. The most popular version of act utilitarianism—hedonistic act utilitarianism—would have us consider the pleasure and displeasure that those affected by each possible action are likely to receive. And, as Bentham pointed out, we would probably need some sort of scale to account for such things as the intensity and duration of the pleasure or displeasure that each individual affected is likely to receive. This is information that a human being would need to have as well to follow the theory. Getting this information has been and will continue to be a challenge for artificial intelligence research in general, but it can be separated from the challenge of computing the ethically correct action, given this information. With the requisite information, a machine could be developed that is just as able to follow the theory as a human being.

Hedonistic act utilitarianism can be implemented in a straightforward manner. The algorithm is to compute the best action, that which derives the greatest net pleasure, from all alternative actions. It requires as input the number of people affected and, for each person, the intensity of the pleasure/displeasure (for example, on a scale of 2 to –2), the duration of the pleasure/displeasure (for example, in days), and the probability that this pleasure or displeasure will occur, for each possible action. For each person, the algorithm computes the product of the intensity, the duration, and the probability, to obtain the net pleasure for that person. It then adds the individual net pleasures to obtain the total net pleasure:

Total net pleasure = Σ (intensity × duration × probability) for each affected individual

This computation would be performed for each alternative action. The action with the highest total net pleasure is the right action (Anderson, Anderson, and Armen 2005b).

A machine might very well have an advantage over a human being in following the theory of act utilitarianism for several reasons: First, human beings tend not to do the arithmetic strictly, but just estimate that a certain action is likely to result in the greatest net good consequences, and so a human being might make a mistake, whereas such error by a machine would be less likely. Second, as has already been noted, human beings tend toward partiality (favoring themselves, or those near and dear to them, over others who might be affected by their actions or inactions), whereas an impartial machine could be devised. Since the theory of act utilitarianism was developed to introduce objectivity into ethical decision making, this is important. Third, humans tend not to consider all of the possible actions that they could perform in a particular situation, whereas a more thorough machine could be developed. Imagine a machine that acts as an advisor to human beings and "thinks" like an act utilitarian. It will prompt the human user to consider alternative actions that might result in greater net good consequences than the action the human being is considering doing, and it will prompt the human to consider the effects of each of those actions on all those affected. Finally, for some individuals' actions—actions of the president of the United States or the CEO of a large international corporation—their impact can be so great that the calculation of the greatest net pleasure may be very time consuming, and the speed of today's machines gives them an advantage.

We conclude, then, that machines can follow the theory of act utilitarianism at least as well as human beings and, perhaps, even better, *given the data that human beings would need*, as well, to follow the theory. The theory of act utilitarianism has, however, been questioned as not entirely agreeing with intuition. It is certainly a good starting point in programming a machine to be ethically sensitive—it would probably be more ethically sensitive than many human beings—but, perhaps, a better ethical theory can be used.

Critics of act utilitarianism have pointed out that it can violate human beings' rights, sacrificing one person for the greater net good. It can also conflict with our notion of justice—what people deserve—because the rightness and wrongness of actions is determined entirely by the future consequences of actions, whereas what people deserve is a result of past behavior. A deontological approach to ethics (where the rightness and wrongness of actions depends on something other than the consequences), such as Kant's categorical imperative, can emphasize the importance of rights and justice, but this approach can be accused of ignoring consequences. We believe, along with W. D. Ross (1930), that the best approach to ethical theory is one that combines elements of both teleological and deontological theories. A theory with several prima facie duties (obligations that we should try to satisfy, but which can be overridden on occasion by stronger obligations)—some concerned with the consequences of actions and others concerned with justice and rights—better acknowledges the complexities of ethical decision making than a single absolute duty theory. This approach has one major drawback, however. It needs to be supplemented with a decision procedure for cases where the prima facie duties give conflicting advice. This is a problem that we have worked on and will be discussed later on.

Among those who maintain that ethics cannot be computed, there are those who question the action-based approach to ethics that is assumed by defenders of act utilitarianism, Kant's categorical imperative, and other well-known ethical theories. According to the "virtue" approach to ethics, we should not be asking what we ought to do in ethical dilemmas, but rather what sort of persons we should be. We should be talking about the sort of qualities—virtues—that a person should possess; actions should be viewed as secondary. Given that we are concerned only with the actions of machines, it is appropriate, however, that we adopt the action-based approach to ethical theory and focus on the sort of principles that machines should follow in order to behave ethically.

Another philosophical concern with the machine ethics project is whether machines are the type of entities that can behave ethically. It is commonly thought that an entity must be capable of acting intentionally, which requires that it be conscious, and that it have free will, in order to be a moral agent. Many would, also, add that sentience or emotionality is important, since only a being that has feelings would be capable of appreciating the feelings of others, a critical factor in the moral assessment of possible actions that could be performed in a given situation. Since many doubt that machines will ever be conscious, have free will, or emotions, this would seem to rule them out as being moral agents.

This type of objection, however, shows that the critic has not recognized an important distinction between performing the morally correct action in a given situation, including being able to justify it by appealing to an acceptable ethical principle, and being held morally responsible for the action. Yes, intentionality and free will in some sense are necessary to hold a being morally responsible for its actions, and it would be difficult to establish that a machine possesses these qualities; but neither attribute is necessary to do the morally correct action in an ethical dilemma and justify it. All that is required is that the machine act in a way that conforms with what would be considered to be the morally correct action in that situation and be able to justify its action by citing an acceptable ethical principle that it is following (S. L. Anderson 1995).

The connection between emotionality and being able to perform the morally correct action in an ethical dilemma is more complicated. Certainly one has to be sensitive to the suffering of others to act morally. This, for human beings, means that one must have empathy, which, in turn, requires that one have experienced similar emotions oneself. It is not clear, however, that a machine could not be trained to take into account the suffering of others in calculating how it should behave in an ethical dilemma, without having emotions itself. It is important to recognize, furthermore, that having emotions can actually interfere with a being's ability to determine, and perform, the right action in an ethical dilemma. Humans are prone to getting "carried away" by their emotions to the point where they are incapable of following moral principles. So emotionality can even be viewed as a weakness of human beings that often prevents them from doing the "right thing."

The necessity of emotions in rational decision making in computers has been championed by Rosalind Picard (1997), citing the work of Damasio (1994), which concludes that human beings lacking emotion repeatedly make the same bad decisions or are unable to make decisions in due time. We believe that, although evolution may have taken this circuitous path to decision making in human beings, irrational control of rational processes is not a necessary condition for all rational systems—in particular, those specifically designed to learn from errors, heuristically prune search spaces, and make decisions in the face of bounded time and knowledge.

A final philosophical concern with the feasibility of computing ethics has to do with whether there is a single correct action in ethical dilemmas. Many believe that ethics is relative either to the society in which one lives—"when in Rome, one should do what Romans do"—or, a more extreme version of relativism, to individuals—whatever you think is right is right for you. Most ethicists reject ethical relativism (for example, see Mappes and DeGrazia [2001, p. 38] and Gazzaniga [2006, p. 178]), in both forms, primarily because this view entails that one cannot criticize the actions of societies, as long as they are approved by the majority in those societies, or individuals who act according to their beliefs, no matter how heinous they are. There certainly do seem to be actions that experts in ethics, and most of us, believe are absolutely wrong (for example, torturing a baby and slavery), even if there are societies, or individuals, who approve of the actions. Against those who say that ethical relativism is a more tolerant view than ethical absolutism, it has been pointed out that ethical relativists cannot say that anything is absolutely good—even tolerance (Pojman [1996, p. 13]).

What defenders of ethical relativism may be recognizing—that causes them to support this view—are two truths, neither of which entails the acceptance of ethical relativism: (1) Different societies have their own customs that we must acknowledge, and (2) there are difficult ethical issues about which even experts in ethics cannot agree, at the present time, on the ethically correct action. Concerning the first truth, we must distinguish between an ethical issue and customs or practices that fall outside the area of ethical concern. Customs or practices that are not a matter of ethical concern can be respected, but in areas of ethical concern we should not be tolerant of unethical practices.

Concerning the second truth, that *some* ethical issues are difficult to resolve (for example, abortion)—and so, at this time, there may not be agreement by ethicists as to the correct action—it does not follow that all views on these issues are equally correct. It will take more time to resolve these issues, but most ethicists believe that we should strive for a single correct position on these issues. What needs to happen is to see that a certain position follows from basic principles that all ethicists accept, or that a certain position is more consistent with other beliefs that they all accept.

From this last point, we should see that we may not be able to give machines principles that resolve *all* ethical disputes at this time. (Hopefully, the machine behavior that we are concerned about won't fall in too many of the disputed areas.) The implementation of ethics can't be more complete than is accepted ethical theory. Completeness is an ideal for which to strive but may not be possible at this time. The ethical theory, or framework for resolving ethical disputes, should allow for updates, as issues that once were considered contentious are resolved. What is more important than having a complete ethical theory to implement is to have one that is consistent. This is where machines may actually help to advance the study of ethical theory, by pointing out inconsistencies in the theory that one attempts to implement, forcing ethical theoreticians to resolve those inconsistencies.

Considering challenges from an AI perspective, foremost for the nascent field of machine ethics may be convincing the AI community of the necessity and advisability of incorporating ethical principles into machines. Some critics maintain that machine ethics is the stuff of science fiction—machines are not yet (and may never be) sophisticated enough to require ethical restraint. Others wonder who would deploy such systems given the possible liability involved. We contend that machines with a level of autonomy requiring ethical deliberation are here and both their number and level of autonomy are likely to increase. The liability already exists; machine ethics is necessary as a means to mitigate it. In the following section, we will detail a system that helps establish this claim.

Another challenge facing those concerned with machine ethics is how to proceed in such an inherently interdisciplinary endeavor. Artificial Intelligence researchers and philosophers, although generally on speaking terms, do not always hear what the other is saying. It is clear that, for substantive advancement of the field of machine ethics, both are going to have to listen to each other intently. AI researchers will need to admit their naiveté in the field of ethics and convince philosophers that there is a pressing need for their services; philosophers will need to be a bit more pragmatic than many are wont to be and make an effort to sharpen ethical theory in domains where machines will be active. Both will have to come to terms with this newly spawned relationship and, together, forge a common language and research methodology.

The machine ethics research agenda will involve testing the feasibility of a variety of approaches to capturing ethical reasoning, with differing ethical bases and implementation formalisms, and applying this reasoning in systems engaged in ethically sensitive activities. This research will investigate how to determine and represent ethical principles, incorporate ethical principles into a system's decision procedure, make ethical decisions with incomplete and uncertain knowledge, provide explanations for decisions made using ethical principles, and evaluate systems that act based upon ethical principles.

System implementation work is already underway. A range of machine-learning techniques are being employed in an attempt to codify ethical reasoning from examples of particular ethical dilemmas. As such, this work is based, to a greater or lesser degree, upon *casuistry*—the branch of applied ethics that, eschewing principle-based approaches to ethics, attempts to determine correct responses to new ethical dilemmas by drawing conclusions based on parallels with previous cases in which there is agreement concerning the correct response.

Rafal Rzepka and Kenji Araki (2005), at what might be considered the most extreme degree of casuistry, explore how statistics learned from examples of ethical intuition drawn from the full spectrum of the world wide web might be useful in furthering machine ethics. Working in the domain of safety assurance for household robots, they question whether machines should be obeying some set of rules decided by ethicists, concerned that these rules may not in fact be truly universal. They suggest that it might be safer to have machines "imitating millions, not a few," believing in such "democracy-dependent algorithms" because, they contend, "most people behave ethically without learning ethics." They propose an extension to their web-based knowledge discovery system GENTA (General Belief Retrieving Agent) that would search the web for opinions, usual behaviors, common consequences, and exceptions, by counting ethically relevant neighboring words and phrases, aligning these along a continuum from positive to negative behaviors, and subjecting this information to statistical analysis. They suggest that this analysis, in turn, would be helpful in the development of a sort of majority-rule ethics useful in guiding the behavior of autonomous systems. An important open question is whether users will be comfortable with such behavior or will, as might be expected, demand better than average ethical conduct from autonomous systems.

A neural network approach is offered by Marcello Guarini (2006). At what might be considered a less extreme degree of casuistry, particular actions concerning killing and allowing to die are classified as acceptable or unacceptable depending upon different motives and consequences. After training a simple recurrent network on a number of such cases, it is capable of providing plausible responses to a variety of previously unseen cases. This work attempts to shed light on the philosophical debate concerning *generalism* (principle-based approaches to moral reasoning) versus *particularism* (case-based approaches to moral reasoning). Guarini finds that, although some of the concerns pertaining to learning and generalizing from ethical dilemmas without resorting to principles can be mitigated with a neural network model of cognition, "important considerations suggest that it cannot be the whole story about moral reasoning—principles are needed." He argues that "to build an artificially intelligent agent without the ability to question and revise its own initial instruction on cases is to assume a kind of moral and engineering perfection on the part of the designer." He argues, further, that such perfection is unlikely and principles seem to play an important role in the required subsequent revision—"at least some reflection in humans does appear to require the explicit representation or consultation of…rules," for instance, in discerning morally relevant differences in similar cases. Concerns about this approach are those attributable to neural networks in general, including oversensitivity to training cases and the inability to generate reasoned arguments for system responses.

Bruce McLaren (2003), in the spirit of a more pure form of casuistry, promotes a case-based reasoning approach (in the artificial intelligence sense) for developing systems that provide guidance in ethical dilemmas. His first such system, Truth-Teller, compares pairs of cases presenting ethical dilemmas about whether or not to tell the truth.

The Truth-Teller program marshals ethically relevant similarities and differences between two given cases from the perspective of the "truth teller" (that is, the person faced with the dilemma) and reports them to the user. In particular, it points out reasons for telling the truth (or not) that (1) apply to both cases, (2) apply more strongly in one case than another, or (3) apply to only one case.

The System for Intelligent Retrieval of Operationalized Cases and Codes (SIROCCO), McLaren's second program, leverages information concerning a new ethical dilemma to predict which previously stored principles and cases are relevant to it in the domain of professional engineering ethics. Cases are exhaustively formalized and this formalism is used to index similar cases in a database of formalized, previously solved cases that include principles used in their solution. SIROCCO's goal, given a new case to analyze, is "to provide the basic information with which a human reasoner … could answer an ethical question and then build an argument or rationale for that conclusion." SIROCCO is successful at retrieving relevant cases but performed beneath the level of an ethical review board presented with the same task. Deductive techniques, as well as any attempt at decision making, are eschewed by McLaren due to "the ill-defined nature of problem solving in ethics." Critics might contend that this "ill-defined nature" may not make problem solving in ethics completely indefinable, and attempts at just such a definition may be possible in constrained domains. Further, it might be argued that decisions offered by a system that are consistent with decisions made in previous cases have merit and will be useful to those seeking ethical advice.

We (Anderson, Anderson, and Armen 2006a) have developed a decision procedure for an ethical theory in a constrained domain that has multiple prima facie duties, using inductive logic programming (ILP) (Lavrec and Dzeroski 1997) to learn the relationships between these duties. In agreement with Marcello Guarini and Baruch Brody (1988) that casuistry alone is not sufficient, we begin with prima facie duties that often give conflicting advice in ethical dilemmas and then abstract a decision principle, when conflicts do arise, from cases of ethical dilemmas where ethicists are in agreement as to the correct action. We have adopted a multiple prima facie duty approach to ethical decision making because we believe it is more likely to capture the complexities of ethical decision making than a single, absolute duty ethical theory. In an attempt to develop a decision procedure for determining the ethically correct action when the duties give conflicting advice, we use ILP to abstract information leading to a general decision principle from ethical experts' intuitions about particular ethical dilemmas. A common criticism is whether the relatively straightforward representation scheme used to represent ethical dilemmas will be sufficient to represent a wider variety of cases in different domains.

Deontic logic's formalization of the notions of obligation, permission, and related concepts[1] make it a prime candidate as a language for the expression of machine ethics principles. Selmer Bringsjord, Konstantine Arkoudas, and Paul Bello (2006) show how formal logics of action, obligation, and permissibility might be used to incorporate a given set of ethical principles into the decision procedure of an autonomous system. They contend that such logics would allow for proofs establishing that (1) robots only take permissible actions, and (2) all actions that are obligatory for robots are actually performed by them, subject to ties and conflicts among available actions. They further argue that, while some may object to the wisdom of logic-based AI in general, they believe that in this case a logic-based approach is promising because one of the central issues in machine ethics is trust and "mechanized formal proofs are perhaps the single most effective tool at our disposal for establishing trust." Making no commitment as to the ethical content, their objective is to arrive at a methodology that maximizes the probability that an artificial intelligent agent behaves in a certifiably ethical fashion, subject to proof explainable in ordinary English. They propose a general methodology for implementing deontic logics in their logical framework, Athena, and illustrate the feasibility of this approach by encoding a natural deduction system for a deontic logic for reasoning about what agents ought to do. Concerns remain regarding the practical relevance of the formal logics they are investigating and efficiency issues in their implementation.

The work of Bringjord, Arkoudas, and Bello is based on research that investigates, from perspectives other than artificial intelligence, how deontic logic's concern with *what ought to be the case* might be extended to represent and reason about *what agents ought to do*. It has been argued that the implied assumption that the latter will simply follow from investigation of the former is not the case. In this context, John Horty (2001) proposes an extension of deontic logic, incorporating a formal theory of agency that describes what agents ought to do under various conditions over extended periods of time. In particular, he adapts preference ordering from deci-

Articles

sion theory to "both define optimal actions that an agent should perform and the propositions whose truth the agent should guarantee." This framework permits the uniform formalization of a variety of issues of ethical theory and, hence, facilitates the discussion of these issues.

Tom Powers (2006) assesses the feasibility of using deontic and default logics to implement Kant's categorical imperative:

> Act only according to that maxim whereby you can at the same time will that it should become a universal law... If contradiction and contrast arise, the action is rejected; if harmony and concord arise, it is accepted. From this comes the ability to take moral positions as a heuristic means. For we are social beings by nature, and what we do not accept in others, we cannot sincerely accept in ourselves.

Powers suggests that a machine might itself construct a theory of ethics by applying a universalization step to individual maxims, mapping them into the deontic categories of forbidden, permissible, or obligatory actions. Further, for consistency, these universalized maxims need to be tested for contradictions with an already established base of principles, and these contradictions resolved. Powers suggests, further, that such a system will require support from a theory of commonsense reasoning in which postulates must "survive the occasional defeat," thus producing a nonmonotonic theory whose implementation will require some form of default reasoning. It has been noted (Ganascia 2007) that answer set programming (ASP) (Baral 2003) may serve as an efficient formalism for modeling such ethical reasoning. An open question is what reason, other than temporal priority, can be given for keeping the whole set of prior maxims and disallowing a new contradictory one. Powers offers that "if we are to construe Kant's test as a way to build a set of maxims, we must establish rules of priority for accepting each additional maxim." The question remains as to what will constitute this moral epistemic commitment.

Creating a Machine That Is an Explicit Ethical Agent

To demonstrate the possibility of creating a machine that is an explicit ethical agent, we have attempted in our research to complete the following six steps:

Step One

We have adopted the prima facie duty approach to ethical theory, which, as we have argued, better reveals the complexity of ethical decision making than single, absolute duty theories. It incorporates the good aspects of the teleological and deontological approaches to ethics, while allowing for needed exceptions to adopting one or the other approach exclusively. It also has the advantage of being better able to adapt to the specific concerns of ethical dilemmas in different domains. There may be slightly different sets of prima facie duties for biomedical ethics, legal ethics, business ethics, and journalistic ethics, for example.

There are two well-known prima facie duty theories: Ross's theory, dealing with general ethical dilemmas, that has seven duties; and Beauchamp and Childress's four principles of biomedical ethics (1979) (three of which are derived from Ross's theory) that are intended to cover ethical dilemmas specific to the field of biomedicine. Because there is more agreement between ethicists working on biomedical ethics than in other areas, and because there are fewer duties, we decided to begin to develop our prima facie duty approach to computing ethics using Beauchamp and Childress's principles of biomedical ethics.

Beauchamp and Childress's principles of biomedical ethics include the principle of respect for autonomy that states that the health-care professional should not interfere with the effective exercise of patient autonomy. For a decision by a patient concerning his or her care to be considered fully autonomous, it must be intentional, based on sufficient understanding of his or her medical situation and the likely consequences of forgoing treatment, sufficiently free of external constraints (for example, pressure by others or external circumstances, such as a lack of funds) and sufficiently free of internal constraints (for example, pain or discomfort, the effects of medication, irrational fears, or values that are likely to change over time). The principle of nonmaleficence requires that the health-care professional not harm the patient, while the principle of beneficence states that the health-care professional should promote patient welfare. Finally, the principle of justice states that health-care services and burdens should be distributed in a just fashion.

Step Two

The domain we selected was medical ethics, consistent with our choice of prima facie duties, and, in particular, a representative type of ethical dilemma that involves three of the four principles of biomedical ethics: respect for autonomy, nonmaleficence, and beneficence. The type of dilemma is one that health-care workers often face: A health-care worker has recommended a particular treatment for her competent adult patient, and the patient has rejected that treatment option. Should the health-care worker try again to change the patient's mind or accept the patient's decision as final? The dilemma arises because, on the one hand, the health-care professional shouldn't challenge the patient's autonomy unnecessarily; on the other hand, the health-care worker may have concerns about why the patient is refusing the treatment.

In this type of dilemma, the options for the health-care worker are just two, either to accept the patient's decision or not, by trying again to change the patient's mind. For this proof of concept test of attempting to make a prima facie duty ethical theory computable, we have a single type of dilemma that encompasses a finite number of specific cases, just three duties, and only two possible actions in each case. We have abstracted, from a discussion of similar types of cases given by Buchanan and Brock (1989), the correct answers to the specific cases of the type of dilemma we consider. We have made the assumption that there is a consensus among bioethicists that these are the correct answers.

Step Three

The major philosophical problem with the prima facie duty approach to ethical decision making is the lack of a decision procedure when the duties give conflicting advice. What is needed, in our view, are ethical principles that balance the level of satisfaction or violation of these duties and an algorithm that takes case profiles and outputs that action that is consistent with these principles. A profile of an ethical dilemma consists of an ordered set of numbers for each of the possible actions that could be performed, where the numbers reflect whether particular duties are satisfied or violated and, if so, to what degree. John Rawls's "reflective equilibrium" (1951) approach to creating and refining ethical principles has inspired our solution to the problem of a lack of a decision procedure. We abstract a principle from the profiles of specific cases of ethical dilemmas where experts in ethics have clear intuitions about the correct action and then test the principle on other cases, refining the principle as needed.

The selection of the range of possible satisfaction or violation levels of a particular duty should, ideally, depend upon how many gradations are needed to distinguish between cases that are ethically distinguishable. Further, it is possible that new duties may need to be added in order to make distinctions between ethically distinguishable cases that would otherwise have the same profiles. There is a clear advantage to our approach to ethical decision making in that it can accommodate changes to the range of intensities of the satisfaction or violation of duties, as well as adding duties as needed.

Step Four

Implementing the algorithm for the theory required formulation of a principle to determine the correct action when the duties give conflicting advice. We developed a system (Anderson, Anderson, and Armen 2006a) that uses machine-learning techniques to abstract relationships between the prima facie duties from particular ethical dilemmas where there is an agreed-upon correct action. Our chosen type of dilemma, detailed previously, has only 18 possible cases (given a range of +2 to –2 for the level of satisfaction or violation of the duties) where, given the two possible actions, the first action supersedes the second (that is, was ethically preferable). Four of these cases were provided to the system as examples of when the target predicate (supersedes) is true. Four examples of when the target predicate is false were provided by simply reversing the order of the actions. The system discovered a principle that provides the correct answer for the remaining 14 positive cases, as verified by the consensus of ethicists.

ILP was used as the method of learning in this system. ILP is concerned with inductively learning relations represented as first-order Horn clauses (that is, universally quantified conjunctions of positive literals L_i implying a positive literal H: $H \leftarrow (L_1 \wedge \cdots \wedge L_n)$). ILP is used to learn the relation supersedes $(A1, A2)$, which states that action $A1$ is preferred over action $A2$ in an ethical dilemma involving these choices. Actions are represented as ordered sets of integer values in the range of +2 to –2 where each value denotes the satisfaction (positive values) or violation (negative values) of each duty involved in that action. Clauses in the supersedes predicate are represented as disjunctions of lower bounds for differentials of these values.

ILP was chosen to learn this relation for a number of reasons. The potentially nonclassical relationships that might exist between prima facie duties are more likely to be expressible in the rich representation language provided by ILP than in less expressive representations. Further, the consistency of a hypothesis regarding the relationships between prima facie duties can be automatically confirmed across all cases when represented as Horn clauses. Finally, commonsense background knowledge regarding the supersedes relationship is more readily expressed and consulted in ILP's declarative representation language.

The object of training is to learn a new hypothesis that is, in relation to all input cases, complete and consistent. Defining a *positive* example as a case in which the first action supersedes the second action and a *negative* example as one in which this is not the case, a complete hypothesis is one that covers all positive cases, and a consistent hypothesis covers no negative cases. Negative training examples are generated from positive training examples by inverting the order of these actions, causing the first action to be the incorrect choice. The system starts with the most general hypothesis stating that all actions supersede each other and, thus, covers all positive and negative cases. The system is then provided with positive cases (and their negatives) and modifies its hypothesis, by adding or refining clauses, such that it covers given positive cases and does not cover given negative cases.

The decision principle that the system discovered can be stated as follows: A health-care worker should challenge a patient's decision if it isn't fully autonomous and there's either any violation of nonmaleficence or a severe violation of beneficence. Although, clearly, this rule is implicit in the judgments of the consensus of ethicists, to our knowledge this principle has never before been stated explicitly. Ethical theory has not yet advanced to the point where principles like this one—that correctly balance potentially conflicting duties with differing levels of satisfaction or violation—have been formulated. It is a significant result that machine-learning techniques can discover a principle such as this and help advance the field of ethics. We offer it as evidence that making the ethics more precise will permit machine-learning techniques to discover philosophically novel and interesting principles in ethics because the learning system is general enough that it can be used to learn relationships between any set of prima facie duties where there is a consensus among ethicists as to the correct answer in particular cases.

Once the principle was discovered, the needed decision procedure could be fashioned. Given a profile representing the satisfaction/violation levels of the duties involved in each possible action, values of corresponding duties are subtracted (those of the second action from those of the first). The principle is then consulted to see if the resulting differentials satisfy any of its clauses. If so, the first action of the profile is deemed ethically preferable to the second.

Step Five

We have explored two prototype applications of the discovered principle governing Beauchamp and Childress's principles of biomedical ethics. In both prototypes, we created a program where a machine could use the principle to determine the correct answer in ethical dilemmas. The first, MedEthEx (Anderson, Anderson, and Armen 2006b), is a medical ethical advisor system; the second, EthEl, is a system in the domain of elder care that determines when a patient should be reminded to take medication and when a refusal to do so is serious enough to contact an overseer. EthEl is more autonomous than MedEthEx in that, whereas MedEthEx gives the ethically correct answer (that is, that which is consistent with its training) to a human user who will act on it or not, EthEl herself acts on what she determines to be the ethically correct action.

MedEthEx is an expert system that uses the discovered principle and decision procedure to give advice to a user faced with a case of the dilemma type previously described. In order to permit use by someone unfamiliar with the representation details required by the decision procedure, a user interface was developed that (1) asks ethically relevant questions of the user regarding the particular case at hand, (2) transforms the answers to these questions into the appropriate profiles, (3) sends these profiles to the decision procedure, (4) presents the answer provided by the decision procedure, and (5) provides a justification for this answer.[2]

The principle discovered can be used by other systems, as well, to provide ethical guidance for their actions. Our current research uses the principle to elicit ethically sensitive behavior from an elder-care system, EthEl, faced with a different but analogous ethical dilemma. EthEl must remind the patient to take his or her medication and decide when to accept a patient's refusal to take a medication that might prevent harm or provide benefit to the patient and when to notify an overseer. This dilemma is analogous to the original dilemma in that the same duties are involved (nonmaleficence, beneficence, and respect for autonomy) and "notifying the overseer" in the new dilemma corresponds to "trying again" in the original.

Machines are currently in use that face this dilemma.[3] The state of the art in these reminder systems entails providing "context-awareness" (that is, a characterization of the current situation of a person) to make reminders more efficient and natural. Unfortunately, this awareness does not include consideration of ethical duties that such a system should adhere to when interacting with its patient. In an ethically sensitive elder-care system, both the timing of reminders and responses to a patient's disregard of them should be tied to the duties involved. The system should challenge patient autonomy only when necessary, as well as minimize harm and loss of benefit to the patient. The principle discovered from the original dilemma can be used to achieve these goals by directing the system to remind the patient only at ethically justifiable times and notifying the overseer only when the harm or loss of benefit reaches a critical level.

In the implementation, EthEl receives input from an overseer (most likely a doctor), including: the prescribed time to take a medication, the maximum amount of harm that could occur if this medication is not taken (for example, none, some, or considerable), the number of hours it would take for this maximum harm to occur, the maximum amount of expected good to be derived from taking this medication, and the number of hours it would take for this benefit to be lost. The system then determines from this input the change in duty satisfaction and violation levels over time, a function of the maximum amount of harm or good and the number of hours for this effect to take place. This value is used to increment duty satisfaction and violation levels for the *remind* action and, when a patient disregards a reminder, the *notify* action. It is used to decrement *don't remind* and *don't notify* actions as well. A reminder is issued when, according to the principle, the duty satisfaction or violation levels have reached the point where reminding is ethically preferable to not reminding. Similarly, the overseer is notified when a patient has disregarded reminders to take medication and the duty satisfaction or violation levels have reached the point where notifying the overseer is ethically preferable to not notifying the overseer.

EthEl uses an ethical principle discovered by a machine to determine reminders and notifications in a way that is proportional to the amount of maximum harm to be avoided or good to be achieved by taking a particular medication, while not unnecessarily challenging a patient's autonomy. EthEl minimally satisfies the requirements of an explicit ethical agent (in a constrained domain), according to Jim Moor's definition of the term: A machine that is able to calculate the best action in ethical dilemmas using an ethical principle, as opposed to having been programmed to behave ethically, where the programmer is following an ethical principle.

Step Six

As a possible means of assessing the morality of a system's behavior, Colin Allen, G. Varner, and J. Zinser (2000) describe a variant of the test Alan Turing (1950) suggested as a means to determine the intelligence of a machine that bypassed disagreements about the definition of intelligence. Their proposed "comparative moral Turing test" (cMTT) bypasses disagreement concerning definitions of ethical behavior as well as the requirement that a machine have the ability to articulate its decisions: an evaluator assesses the comparative morality of pairs of descriptions of morally significant behavior where one describes the actions of a human being in an ethical dilemma and the other the actions of a machine faced with the same dilemma. If the machine is not identified as the less moral member of the pair significantly more often than the human, then it has passed the test.

They point out, though, that human behavior is typically far from being morally ideal and a machine that passed the cMTT might still fall far below the high ethical standards to which we would probably desire a machine to be held. This legitimate concern suggests to us that, instead of comparing the machine's behavior in a particular dilemma against typical human behavior, the comparison ought to be made with behavior recommended by a trained ethicist faced with the same dilemma. We also believe that the principles used to justify the decisions that are reached by both the machine and ethicist should be made transparent and compared.

We plan to devise and carry out a moral Turing test of this type in future work, but we have had some assessment of the work that we have done to date. The decision principle that was discovered in MedEthEx, and used by EthEl, is supported by W. D. Ross's claim that it is worse to harm than not to help someone. Also, the fact that the principle provided answers to nontraining cases that are consistent with Buchanan and Brock's judgments offers preliminary support for our hypothesis that decision principles discovered from some cases, using our method, enable a machine to determine the ethically acceptable action in other cases as well.

Conclusion

We have argued that machine ethics is an important new field of artificial intelligence and that its goal should be to create machines that are explicit ethical agents. We have done preliminary work to show—through our proof of concept applications in constrained domains—that it may be possible to incorporate an explicit ethical component into a machine. Ensuring that a machine with an ethical component can function autonomously in the world remains a challenge to researchers in artificial intelligence who must further investigate the representation and determination of ethical principles, the incorporation of these ethical principles into a system's decision procedure, ethical decision making with incomplete and uncertain knowledge, the explanation for decisions made using ethical principles, and the evaluation of systems that act based upon ethical principles.

Of the many challenges facing those who choose to work in the area of machine ethics, foremost is the need for a dialogue between ethicists and researchers in artificial intelligence. Each has much to gain from working together on this project. For ethicists, there is the opportunity of clarifying—perhaps even discovering—the fundamental principles of ethics. For AI researchers, convincing the general public that ethical machines can be created may permit continued support for work leading to the development of autonomous intelligent machines—machines that might serve to improve the lives of human beings.

Acknowledgements

This material is based upon work supported in part by the National Science Foundation grant number IIS-0500133.

Notes

1. See plato.stanford.edu/entries/logic-deontic.
2. A demonstration of MedEthEx is available online at www.machineethics.com.
3. For example, see www.ot.toronto.ca/iatsl/projects/medication.htm.

References

Allen, C.; Varner, G.; and Zinser, J. 2000. Prolegomena to Any Future Artificial Moral Agent. *Journal of Experimental and Theoretical Artificial Intelligence* 12(2000): 251–61.

Anderson, M., and Anderson, S., eds. 2006. Special Issue on Machine Ethics. *IEEE Intelligent Systems* 21(4) (July/August).

Anderson, M.; Anderson, S.; and Armen, C., eds. 2005a. Machine Ethics: Papers from the AAAI Fall Symposium. Technical Report FS-05-06, Association for the Advancement of Artificial Intelligence, Menlo Park, CA.

Anderson, M.; Anderson, S.; and Armen, C. 2005b. Toward Machine Ethics: Implementing Two Action-Based Ethical Theories. In Machine Ethics: Papers from the AAAI Fall Symposium. Technical Report FS-05-06, Association for the Advancement of Artificial Intelligence, Menlo Park, CA.

Anderson, M.; Anderson, S.; and Armen, C. 2006a. An Approach to Computing Ethics. *IEEE Intelligent Systems* 21(4): 56–63.

Anderson, M.; Anderson, S.; and Armen, C. 2006b. MedEthEx: A Prototype Medical Ethics Advisor. In *Proceedings of the Eighteenth Conference on Innovative Applications of Artificial Intelligence*. Menlo Park, CA: AAAI Press.

Anderson, S. L. 1995. Being Morally Responsible for an Action Versus Acting Responsibly or Irresponsibly. *Journal of Philosophical Research* 20: 453–62.

Asimov, I. 1976. The Bicentennial Man. In *Stellar Science Fiction 2*, ed. J.-L. del Rey. New York: Ballatine Books.

Baral, C. 2003. *Knowledge Representation, Reasoning, and Declarative Problem Solving*. Cambridge, UK: Cambridge University Press.

Beauchamp, T. L., and Childress, J. F. 1979. *Principles of Biomedical Ethics*. Oxford, UK: Oxford University Press.

Bentham, J. 1907. *An Introduction to the Principles and Morals of Legislation*. Oxford: Clarendon Press.

Bringsjord, S.; Arkoudas, K.; and Bello, P. 2006. Toward a General Logicist Methodology for Engineering Ethically Correct Robots. *IEEE Intelligent Systems* 21(4): 38–44.

Brody, B. 1988. *Life and Death Decision Making*. New York: Oxford University Press.

Buchanan, A. E., and Brock, D. W. 1989. *Deciding for Others: The Ethics of Surrogate Decision Making*, 48–57. Cambridge, UK: Cambridge University Press.

Capek, K. 1921. R.U.R. In *Philosophy and Science Fiction*, ed. M. Phillips. Amherst, NY: Prometheus Books.

Clarke, A. C. 1968. *2001: A Space Odyssey*. New York: Putnam.

Damasio, A.R. 1994. *Descartes' Error: Emotion, Reason, and the Human Brain*. New York: G. P. Putnam.

Dennett, D. 2006. Computers as Prostheses for the Imagination. Invited talk presented at the International Computers and Philosophy Conference, Laval, France, May 3.

Dietrich, E. 2006. After the Humans Are Gone. Keynote address presented at the 2006 North American Computing and Philosophy Conference, RPI, Troy, NY, August 12.

Ganascia, J. G. 2007. Using Non-Monotonic Logics to Model Machine Ethics. Paper presented at the Seventh International Computer Ethics Conference, San Diego, CA, July 12–14.

Gazzaniga, M. 2006. *The Ethical Brain: The Science of Our Moral Dilemmas*. New York: Harper Perennial.

Guarini, M. 2006. Particularism and the Classification and Reclassification of Moral Cases. *IEEE Intelligent Systems* 21(4): 22–28.

Horty, J. 2001. *Agency and Deontic Logic*. New York: Oxford University Press.

Joy, B. 2000. Why the Future Doesn't Need Us. *Wired Magazine* 8(04) (April).

Kant, I. 1785. *Groundwork of the Metaphysic of Morals*, trans. by H. J. Paton (1964). New York: Harper & Row.

Lavrec, N., and Dzeroski, S. 1997. *Inductive Logic Programming: Techniques and Applications*. Chichester, UK: Ellis Horwood.

Mappes, T. A., and DeGrazia, D. 2001. *Biomedical Ethics*, 5th ed., 39–42. New York: McGraw-Hill.

McLaren, B. M. 2003. Extensionally Defining Principles and Cases in Ethics: An AI Model. *Artificial Intelligence Journal* 150(1–2): 145–1813.

Moor, J. H. 2006. The Nature, Importance, and Difficulty of Machine Ethics. *IEEE Intelligent Systems* 21(4): 18–21.

Onishi, N. 2006. In a Wired South Korea, Robots Will Feel Right at Home. *New York Times*, April 2, 2006.

Picard, R. W. 1997. *Affective Computing*. Cambridge, MA: The MIT Press.

Pojman, L. J. 1996. The Case for Moral Objectivism. In *Do the Right Thing: A Philosophical Dialogue on the Moral and Social Issues of Our Time*, ed. F. J. Beckwith. New York: Jones and Bartlett.

Powers, T. 2006. Prospects for a Kantian Machine. *IEEE Intelligent Systems* 21(4): 46–51.

Rawls, J. 1951. Outline for a Decision Procedure for Ethics. *The Philosophical Review* 60(2): 177–197.

Ross, W. D. 1930. *The Right and the Good*. Oxford: Clarendon Press.

Rzepka, R., and Araki, K. 2005. What Could Statistics Do for Ethics? The Idea of a Common Sense Processing-Based Safety Valve. In Machine Ethics: Papers from the AAAI Fall Symposiu. Technical Report FS-05-06, Association for the Advancement of Artificial Intelligence, Menlo Park, CA.

Turing, A. M. 1950. Computing Machinery and Intelligence. *Mind* LIX(236): 433–460.

bona fide field of study. He has cochaired the AAAI Fall 2005 Symposium on Machine Ethics and coedited an *IEEE Intelligent Systems* special issue on machine ethics in 2006. His research in machine ethics was selected for IAAI as an emerging application in 2006. He maintains the machine ethics website (www.machineethics.org) and can be reached at anderson@hartford.edu.

19

Machine morality: bottom-up and top-down approaches for modelling human moral faculties

Wendell Wallach · Colin Allen · Iva Smit

Received: 20 May 2005 / Accepted: 2 February 2007 / Published online: 9 March 2007

Abstract The implementation of moral decision making abilities in artificial intelligence (AI) is a natural and necessary extension to the social mechanisms of autonomous software agents and robots. Engineers exploring design strategies for systems sensitive to moral considerations in their choices and actions will need to determine what role ethical theory should play in defining control architectures for such systems. The architectures for morally intelligent agents fall within two broad approaches: the top-down imposition of ethical theories, and the bottom-up building of systems that aim at goals or standards which may or may not be specified in explicitly theoretical terms. In this paper we wish to provide some direction for continued research by outlining the value and limitations inherent in each of these approaches.

Introduction

Moral judgment in humans is a complex activity, and it is a skill that many either fail to learn adequately or perform with limited mastery. Although there are shared values that transcend cultural differences, cultures and individuals differ in the details of their ethical systems and mores. Hence it can be extremely difficult to agree upon criteria for judging the adequacy of moral decisions.

Difficult value questions also often arise where information is inadequate, and when dealing with situations where the results of actions can not be fully known in advance. Thus ethics can seem to be a fuzzy discipline, whose methods and application apply to the most confusing challenges we encounter. In some respects ethics is as far away from science as one can get.

While any claim that ethics can be reduced to a science would at best be naive, we believe that determining how to enhance the moral acumen of autonomous software agents is a challenge that will significantly advance the understanding of human decision making and ethics. The engineering task of building autonomous systems that safeguard basic human values will force scientists to break down moral decision making into its component parts, recognize what kinds of decisions can and cannot be codified and managed by essentially mechanical systems, and learn how to design cognitive and affective systems capable of managing ambiguity and conflicting perspectives. This project will demand that human decision making is analyzed to a degree of specificity as yet unknown and, we believe, it has the potential to revolutionize the philosophical study of ethics.

A computer system, robot, or android capable of making moral judgments would be an artificial moral agent (AMA) and the proper design of such systems is perhaps the most important and challenging task facing developers of fully autonomous systems (Allen et al. 2000). Rosalind Picard, director of the Affective Computing Group at MIT, put it well when she wrote, "The greater the freedom of a machine, the more it will need moral standards" (Picard 1997).

The capacity to make moral judgments is far from simple, and computer scientists are a long way from substantiating the collection of affective and cognitive skills necessary for moral reasoning in artificial intelligence (AI). While there are aspects of moral judgment that can be isolated and codified for tightly defined contexts, where all available options and variables are known, moral intelligence for autonomous entities is a complex activity dependent on the integration of a broad array of discrete faculties. Over the past few decades it has become apparent that moral judgment in humans is much more than a capacity for abstract moral reasoning. Emotional intelligence (Damasio 1995; Goleman 1995), sociability, the ability to learn from experience and social interactions, consciousness, the capacity to understand the semantic content of symbols, a theory of mind, and the ability to "read the minds" of others all contribute to moral intelligence. Social mechanisms contribute to various refinements of behavior that people expect from each other, and will also be necessary for robots that function to a high degree of competence in social contexts.

The importance of these "supra-rational" faculties raises the question of the extent to which an artificial agent must emulate human faculties to function as an adequate moral agent. Computer scientists and philosophers have begun to consider the challenge of developing computer systems capable of acting within moral guidelines, under various rubrics such as "computational ethics," "machine ethics," and "artificial morality." One obvious question is whose morality or what morality should computer scientists try to implement in AI? Should the systems' decisions and actions conform to religiously or philosophically inspired value systems, such as Christian, Buddhist, utilitarian, or other sources of social norms? The kind of morality people wish to implement will suggest radical differences in the underlying structure of the systems in which computer scientists implement that morality.

The more central question from the perspective of a computer scientist is how to implement moral decision making capabilities in AI. What kinds of decisions can be substantiated within computational systems given the present state of computer technology and the progress we anticipate over the next five to ten years? Attention to what can be implemented may also help us discern what kind of moral system is feasible for artificial systems. For example, some ethical theories seem to provide a kind of moral calculus which allows, in principle, the course of action that will maximize welfare to be quantitatively determined, while other ethical theories appear to require higher order mental faculties that we are a long way from knowing how to reproduce in AI.

Reproducing human faculties is not the only route for implementing moral decision making in computational systems. In fact, computers may be better than humans in making moral decisions in so far as they may not be as limited by the bounded rationality that characterizes human decisions, and they need not be vulnerable to emotional hijacking (Allen 2002; Wallach 2004).

It is worth expanding on the role of emotions in moral decision making because it illustrates many of the complexities of designing AMAs and may help us understand why AMAs or robots could function differently from humans. Despite appearances, the observation that AMAs need not be subject to emotional hijacking does not introduce any contradiction with our previous claim that emotional intelligence may be necessary for moral decision making. First, a large component of affective intelligence concerns the ability to recognize and respond appropriately to emotional states in others (Picard 1997) and it is an open question whether this kind of intelligence itself requires a computer system to have the capacity for emotions of their own. Second, in humans, while emotions are beneficial in many circumstances, this is compatible with certain emotions being disadvantageous or even dysfunctional in other circumstances. Recognizing this fact presents an opportunity for engineers to design AMAs whose moral faculties operate in a way that makes them less susceptible to emotional interference or dysfunctionality.

The literature on affective computing is burgeoning and provides a good starting point for thinking about these issues, but it is also noteworthy that this literature has thus far failed to deal with the role of emotions in moral decisions. This may partially reflect the lasting influence of Stoicism on scientific thinking about morality. The Stoics believed that moral reasoning should be dispassionate and free of emotional prejudice, which has been presumed to mean that emotions should be banned entirely from moral reflection. (The Stoic ideal is exemplified in such science fiction icons as Mr. Spock in *Star Trek*). The emotional intelligence literature demonstrates that the capacity for moral reasoning is complex and that emotions must be considered an integral aspect of that capacity. Some of this complexity has been explored for rather different ends in the philosophical moral psychology literature. The development of AMAs will require us to combine the insights from affective computing, emotional intelligence, and moral psychology in order to give engineers a chart outlining the kinds of emotions and their roles in moral decision making.

Whether an AMA is a computer system or robot, implementing moral faculties in AI will require that scientists and philosophers study and break down moral decision making in humans into computationally manageable modules or components. An interplay exists between, on the one hand, analyzing and theorizing about moral behavior in humans and, on the other hand, the testing of these theories in computer models. For example, in artificial life (ALife) experiments that attempt to simulate evolution, there is evidence that certain moral principles, such as reciprocal altruism and fairness, may have emerged naturally through the essentially mechanical unfolding of evolution. If specific values are indeed naturally selected during evolution and our genes bias development towards these values, this will have profound ramifications for the manner in which we build AMAs. In any case, it is essential that these systems function so as to be fully sensitive to the range of ethical concerns for the health and well-being of humans and other entities (animals, corporations, etc.) worthy of moral consideration.

Given the need for this range of sensitivities, AMAs have the potential to cause considerable discomfort to human beings who are not, for example, used to having machines detect their emotional states. The ability to detect emotions has broad ethical ramifications that pose a particularly complex challenge to sociologists, computer scientists, roboticists, and engineers concerned with human reactions to machines. Successful human-machine interactions may well require that we incorporate the entire value systems underlying such interactions.

This is essential if humans are going to trust sophisticated machines, for trust depends on the felt belief that those you are interacting with share your essential values and concerns, or at least will function within the constraints suggested by those values. Having set the stage, our task now is to explore some bottom-up and top-down approaches to the development of artificial systems capable of making moral decisions.

By "top-down" we mean to combine the two slightly different senses of this term, as it occurs in engineering and as it occurs in ethics. In the engineering sense, a top-down approach analyzes or decomposes a task into simpler subtasks that can be directly implemented and hierarchically arranged to obtain a desired outcome. In the ethical sense, a top-down approach to ethics is one which takes an antecedently specified general ethical theory (whether philosophically derived, such as utilitarianism, or religiously motivated, such as the "Golden rule") and derives its consequences for particular cases.

In our merged sense, a top-down approach to the design of AMAs is any approach that takes the antecedently specified ethical theory and analyzes its computational requirements to guide the design of algorithms and subsystems capable of implementing that theory. By contrast, bottom-up approaches, if they use a prior theory at all, do so only as a way of specifying the task for the system, but not as a way of specifying an implementation method or control structure.

In bottom-up engineering, tasks can also be specified atheoretically using some sort of performance measure (such as winning chess games, passing the Turing test, walking across a room without stumbling, etc.). Various trial and error techniques are available to engineers for progressively tuning the performance of systems so that they approach or surpass the performance criteria. High levels of performance on many tasks can be achieved, even though the engineer lacks a theory of the best way to decompose the task into subtasks. *Post hoc* analysis of the system can sometimes yield a theory or specification of the relevant subtasks, but the results of such analyses can also be quite surprising and typically do not correspond to the kind of decomposition suggested by a priori theorizing. In its ethical sense, a bottom-up approach to ethics is one that treats normative values as being implicit in the activity of agents rather than explicitly articulated (or even articulatable) in terms of a general theory. In our use of the term, we wish to recognize that this may provide an accurate account of the agents' understanding of their own morality and the morality of others, while we remain neutral on the ontological question of whether morality is the kind of concept for which an adequate general theory can be produced.

In practice, engineers and roboticists typically build their most complex systems using both top-down and bottom-up approaches, assembling components that fulfil specific functions guided by a theoretical top-down analysis that is typically incomplete. Problem solving for engineers often involves breaking down a complex project into discrete tasks and then assembling components that perform those discrete functions in a manner that adequately fulfils the goals specified by the original project. Commonly there is more than one route to meet the project goals, and there is a dynamic interplay between analysis of the project's structure and the testing of the system designed to meet the goals. Failures of the system may, for example, reveal that secondary considerations have been overlooked in one's original analysis of the challenge, requiring additional control systems or software that accommodates complex relationships.

Because the top-down/bottom-up dichotomy is too simplistic for many complex engineering tasks, we should not expect the design of AMAs to be any different. Nevertheless, it serves a useful purpose for highlighting two potential roles of ethical theory for the design of AMAs, in suggesting actual algorithms and control structures and in providing goals and standards of evaluation.

Engineering challenges: bottom-up discrete systems

There is a range of approaches for building any artificially intelligent system. In this paper we focus our attention on the subset of approaches which corresponds best to current research in AI and robotics: the assembly of subsystems implementing relatively discrete human capacities with the goal of creating a system that is complex enough to provide a substrate for artificial morality. Such approaches are "bottom-up" in our sense because the development and deployment of these discrete subsystems is not itself explicitly guided by any ethical theory. Rather, it is hoped that by experimenting with the way in which these subsystems interact, something that has suitable moral capacities can be created. Computer scientists and roboticists are already working on a variety of discrete AI-related skills that are presumably relevant to moral capacities. The techniques provided by artificial life, genetic algorithms, connectionism, learning algorithms, embodied or subsumptive architecture, evolutionary and epigenetic robotics, associative learning platforms, and even good old-fashioned AI, all have strengths in modelling specific cognitive skills or capacities.

Discrete subsystems might be built around the most effective techniques for implementing discrete cognitive capacities and social mechanisms. But computer scientists following such an approach are then confronted with the difficult (and perhaps insurmountable) challenge of assembling these discrete systems into a functional whole. This discrete-systems approach can be contrasted with more holistic attempts to reproduce moral agency in artificial systems, such as the attempt to evolve AMAs in an ALife environment, or training a single large connectionist network to reproduce the input–output functions of real moral agents. While the Alife approach seems promising, it is limited by our present inability to simulate environments of sufficient social and physical complexity to favor the complex capacities required for real-world performance. The connectionist approach has no real practitioners (to our knowledge) for reasons that illustrate our point that practical AMA development is going to require a more structured approach; for it is surely naive to think any single current technique for training homogeneous artificial neural networks could satisfactorily replicate the entire suite of moral behavior by actual human beings.

The development of artificial systems that act with sensitivity to moral considerations is presently still largely confined to designing systems with "operational morality," that is, "ensuring that the AI system functions as designed" (Wallach 2003). This is primarily the extension of the traditional engineering concern with safety into the design of smart machines that can reliably perform a specified task, whether that entails navigating a hallway without damaging itself or running into people, visually distinguishing the presence of a human from an inanimate object, or deciphering the emotional state implicit in a facial expression. Thus engineers and computer scientists are still in the process of designing methods for adequately fulfilling the discrete tasks that might lead cumulatively to more complex activities and greater autonomy.

The word "discrete" here should not be taken to mean "isolated." In fact, among the most promising approaches to robotics are those which exploit dynamic interaction between the various subtasks of visual perception, moving, manipulating, and understanding speech. For example, Roy (2005), Hsiao and Roy (2005) exploits such interactions in the development of the seeing, hearing, and talking robotic arm he calls "Ripley." Ripley's speech understanding and comprehension systems develop in the context of carrying out human requests for actions related to identifying and manipulating objects within its fields of vision and reach, while balancing internal requirements such as not allowing its servos to overheat—a very real mechanical concern, as it turns out.

Especially interesting for our purposes is Fig. 5 in Hsiao and Roy (2005). Their figure presents the various subsystems involved in speech processing, object recognition, movement, and recuperation as nodes in a graph whose edges indicate interactions between the subsystems and their relationships to two goals: "Preserve Self" and "Assist User". Although Hsiao and Roy do not make it explicit, these goals correspond closely to Asimov's Third and Second Laws of Robotics (Asimov 1950). In a public lecture titled "Meaning machines" given on March 7, 2005, at Indiana University, Roy showed a version of this figure that also included a box corresponding to Asimov's first law: "Do no harm to humans." But most telling for our purposes was that in this version of the figure, the line leading from Asimov's first law to a presumed implementation of moral capacities trailed off into dots. Without necessarily endorsing Asimov's laws, one may understand the challenge to computer scientists as how to connect the dots to a substantive account of ethical behavior.

How do we get from discrete skills to systems that are capable of autonomously displaying complex behavior, including moral behavior, and that are capable of meeting challenges in new environmental contexts and in interaction with many agents? Some scientists hope or presume that the aggregation of discrete skill sets will lead to the emergence of higher order cognitive faculties including emotional intelligence, moral judgment, and consciousness.

While "emergence" is a popular word among some scientists and philosophers of science, it is still a rather vague concept, which implies that more complex activities will somehow arise synergistically from the integration of discrete skills. Perhaps the self-organizing capacity of evolutionary algorithms, learning systems or evolutionary robots provides a technique that will facilitate emergence. Systems that learn and evolve might discover ways to integrate discrete skills in pursuit of specific goals. It is important to note that the goals here are not predefined as they would be by a top-down analysis. For example, in evolutionary systems, a major problem is how to design a fitness function that will lead to the emergence of AMAs, without explicitly applying moral criteria. The slogan "survival of the most moral" highlights the problem of saying what "most moral" amounts to in a non-circular (and computationally tractable) fashion, such that the competition for survival in a complex physical and social environment will lead to the emergence of moral agents.

The individual components from which bottom-up systems are constructed tend to be mechanical and limited in the flexibility of the responses they can make. But when integrated successfully these components can give rise to complex dynamic systems with a range of choices or optional responses to external conditions and pressures. Bottom-up engineering thus offers a kind of dynamic morality, where the ongoing feedback from different social mechanisms facilitates varied responses as conditions change.

For example, humans may be born trusting their parents and other immediate caregivers, but the manner in which humans, both children and adults, test new relationships and feel their way over time to deepening levels of trust is helpful in understanding what we mean by dynamic morality. We humans invest each relationship with varying degrees of trust, but there is no simple formula for trust or no one method for establishing the degree to which a given person will trust a new acquaintance. A variety of social mechanisms including low risk experiments with cooperation, the reading of another's emotions in specific situations, estimations of the other's character, and calculations regarding what we are willing to risk in a given relationship all feed into the dynamic determination of trust. Each new social interaction holds the prospect of altering the degree of trust invested in a relationship. The lesson for AI research and robotics is that while AMAs should not enter the world suspicious of all relationships, they will need the capacity to dynamically negotiate or feel their way through to elevated levels of trust with the other humans or computer systems with which they interact.

A potential strength of complex bottom-up systems lies in the manner they dynamically integrate input from differing social mechanisms. Their weakness traditionally lies in not knowing which goals to use for evaluating choices and actions as contexts and circumstances change. Bottom-up systems work best when they are directed at achieving one clear goal. When the goals are several or the available information is confusing or incomplete, it is a much more difficult task for bottom-up engineering to provide a clear course of action. Nevertheless, progress in this area is being made, allowing adaptive systems to deal more effectively with transitions between different phases of behavior within a given task (e.g., Smith et al. 2002) and with transitions between different tasks (Roy 2005; Hsiao and Roy 2005). The strength of bottom-up engineering lies in the assembling of components to achieve a goal. Presuming, however, that a sophisticated capacity for moral judgment will just emerge from bottom-up engineering is unlikely to get us there, and this suggests that the analysis provided by top-down approaches will be necessary.

Top-down theories of ethics

The philosophical study of ethics has focused on the question of whether there are top-down criteria that illuminate moral decision making. Are there duties, principles, rules, or goals to which the countless moral judgments we make daily are all subservient? In a changing world it is impossible to have rules, laws, or goals that adequately cover every possible situation. One of the functions of legislatures, courts, and even differing cultures is to adjudicate and prioritize values as new challenges or unanticipated events arise. Top-down ethical systems come from a variety of sources including religion, philosophy, and literature. Examples include the Golden rule, the ten commandments, consequentialist or utilitarian ethics, Kant's moral imperative and other duty based theories, legal codes, Aristotle's virtues, and Asimov's three laws for robots.

To date, very little research has been done on the computerization of top-down ethical theories. Those few systems designed to analyze moral challenges are largely relegated to medical advisors that help doctors and other health practitioners weigh alternative courses of treatments or to evaluate whether to withhold treatment for the terminally ill. Computer-aided decision support is also popular for many business applications, but generally the actual decision making is left to human managers. There are, nevertheless, many questions as to whether such systems might undermine the moral responsibility of the human decision makers (Friedman and Kahn 1992). For example, will doctors and hospitals feel free to override the advice of an expert system with a good track record, when there will always be a computerized audit trail available to enterprising lawyers, if the human's decision goes awry?

Throughout industry and in financial applications values are built into systems that must make choices among available courses of actions. Generally these are the explicit values of the institutions who use the systems and the engineers, who design the systems, although many systems also substantiate often unrecognized implicit values (Friedman and Nissenbaum 1996). One would be hard pressed to say that these systems are engaged in any form of explicit moral reasoning, although they are capable of making choices that their designers might not have been able to anticipate. The usefulness of such systems lies largely in their ability to manage large quantities of information and make decisions based on their institution's goals and values, at a speed that humans can not approach.

Top-down ethical theories are each meant to capture the essence of moral judgment. Nevertheless, the history of moral philosophy can be viewed as a long inquiry into the limitations inherent in each new top-down theory proposed. These known limitations also suggest specific challenges in implementing a top-down theory in AI. For example, if you have many rules or laws, what should the system do when it encounters a challenge where two or more rules conflict? Can laws or values be prioritized or does their relative importance change with a change in context? To handle such problems philosophers have turned toward more general or abstract principles from which all the more specific or particular principles might be derived. Among ethical theories, there are two "big picture" rivals for what the general principle should be. Utilitarians claim that ultimately morality is about maximizing the total amount of "utility" (a measure of happiness or well being) in the world. The best actions (or the best specific rules to follow) are those that maximize aggregate utility. Because utilitarians care about the consequences of actions for utility, their views are called "consequentialist." If focused on particular actions, consequentialists have much computing to do, because they need to work out many, if not all, of the consequences of the available alternatives in order to evaluate them morally. Even consequentialists who operate at the level of rules may be faced with the formidable computational task of re-evaluating the utility of the rules regularly, as we will discuss below.

The competing "big picture" view of moral principles is that ethics is about duties to others, and, on the flip side of duties, the rights of individuals. Being concerned with duties and rights, such theories fall under the heading of "deontology," a nineteenth century word which is pseudo-Greek for the study of obligations. In general, any list of specific duties or rights might suffer the same problem as a list of commandments (in fact, some of the traditional commandments are duties to others). For example, a duty to tell the truth might, in some circumstances, come into conflict with a duty to respect another person's privacy. One way to resolve these problems is to submit all *prima facie* duties to a higher principle.

Thus, for instance, it was Kant's belief that all legitimate moral duties could be grounded in a single principle, the categorical imperative, which could be stated in such a way as to guarantee logical consistency. It turns out that a computational Kantian would also have to do much computing to achieve a full moral evaluation of any action. This is because Kant's approach to the moral evaluation of actions requires a full understanding of the motives behind the action, and an assessment of whether there would be any inconsistency if everyone acted on the same motive. This requires much understanding of psychology and of the effects of actions in the world. Other general deontological principles that seek to resolve conflicts among prima facie duties would face similar issues.

Both consequentialist (e.g., utilitarian) and deontological (e.g., Kantian) approaches provide general principles, whose relevance to the design of ethical robots was first noted by Gips (1995). These different approaches raise their own specific computational problems, but they also raise a common problem of whether any computer (or human, for that matter) could ever gather and compare all the information that would be necessary for the theories to be applied in real time. Of course humans apply consequentialist and deontological reasoning to practical problems without calculating endlessly the utility or moral ramifications of an act in all possible situations. Our morality, just as our reasoning, is bounded by time, capacity, and inclination.

Parameters might also be set to limit the extent to which a computational system analyzes the beneficial or imperatival consequences of a specific action. How might we set those limits on the options considered by a computer system, and will the course of action taken by such a system in addressing a specific challenge be satisfactory? In humans the limits on reflection are set by heuristics and affective controls. Both heuristics and affect can at times be irrational but also tend to embody the wisdom gained through experience. We may well be able to implement heuristics in computational systems, but affective controls on moral judgments represent a much more difficult challenge.

Our brief survey by no means exhausts the variety of top-down ethical theories that might be implemented in AI. Rather, it has been our intention to use these two representative theories to illustrate the kinds of issues that arise when one considers the challenges entailed in designing an AMA. Consequentialism comes in many variations and each religious tradition or culture has its own set of rules. Consequences and rules are combined in some theories. Arguably one of these variations might have built in heuristics that cut through some of the complexity we have discussed. Rule consequentialists, for example, determine whether an act is morally wrong based on a code of rules that have been selected solely in terms of their average or expected consequences. The consequences of adopting the rule must be more favourable than unfavourable to all parties involved. Presumably these rules, once selected, will cut down on much of the calculation that we described earlier for an act consequentialist who will have to evaluate all the consequences of his action. But there is a trade-off. The rule consequentialist must deal with conflicting rules, the prioritization of rules, and any exceptions to the rules, and they must constantly make sure that the rules they have adopted are the optimal ones.

When thinking of rules for robots, Asimov's laws come immediately to mind. On the surface these three laws, plus a "zeroth" law that he added in 1985 to place humanity's interest above that of any individual, appear to be intuitive, straightforward, and general enough in scope to capture a broad array of ethical concerns. But in story after story Asimov demonstrates problems of prioritization and potential deadlock inherent in implementing even this small set of rules (Clarke 1993/4). Apparently Asimov concluded that his laws would not work, and other theorists have extended this conclusion to encompass any rule based ethical system implemented in AI (Lang 2002). In addition, we will require some criteria for evaluating whether a moral judgment made by a computational system is satisfactory. When dealing with differences in values and ethical theories this is by no means an easy matter, and even criteria such as a moral Turing test, designed to manage differences in opinion as to what is 'right' and 'good,' have inherent weaknesses (Allen et al. 2000). The strength of top-down theories lies in their defining ethical goals with a breadth that subsumes countless specific challenges. But this strength can come at a price: either the goals are defined so vaguely or abstractly that their meaning and application is subject for debate, or they get defined in a manner that is static and fails to accommodate or may even be hostile to new conditions.

Merging top-down and bottom-up

As we indicated at the end of the introduction, the top-down/bottom-up dichotomy is somewhat simplistic. Top-down analysis and bottom-up techniques for developing or evolving skills and mental faculties will undoubtedly both be required to engineer AMAs. To illustrate the way in which top-down and bottom-up aspects interact, we consider two cases. First, we'll look at the possibility of utilizing a connectionist network to develop a computer system with good character traits or virtues, and then we'll look at attempts to develop complex mental faculties and social mechanism, such as a theory of mind for an embodied robotic system. We will also address the challenge of bringing it all together in an agent capable of interacting morally in a dynamic social context.

Virtue ethics

Public discussions of morality are not just about rights (deontology) and welfare (utility); they are often also about issues of character. This third element of moral theory can be traced back to Aristotle, and what is now known as "virtue ethics." Virtue ethicists are not concerned with evaluating the morality of actions on the basis solely of outcomes (consequentialism), or in terms of rights and duties (deontology). Virtue theorists maintain that morally good actions flow from the cultivation of good character, which consists in the realization of specific virtues (Hursthouse 2003).

Virtues are more complex than mere skills, as they involve characteristic patterns of motivation and desire. Socrates claimed that virtues could not be misused, because if people have a certain virtue, it is impossible for them to act as if they did not have it. This led him to the conclusion that there is only one virtue, the power of right judgment, while Aristotle favored a longer list. Just as utilitarians do not agree on how to measure utility, and deontologists do not agree on which list of duties apply, virtue ethicists do not agree on a standard list of virtues that any moral agent should exemplify. Rather than focus on these difference in this paper, our attention will be directed at the computational tractability of virtue ethics: could one make use of virtues as a programming tool? Virtues affect how we deliberate and how we motivate our actions, but an explicit description of the relevant virtue rarely occurs in the content of the deliberation.

For instance, a kind person does kind things, but typically will not explain this behavior in terms of her own kindness. Rather, a kind person will state motives focused on the beneficiary of the kindness, such as "she needs it," "it will cheer him up," or "it will stop the pain" (Williams 1985). Besides revealing some of the complexities of virtue theory, this example also demonstrates that the boundaries between the various ethical theories, in this case utilitarianism and virtue based ethics, can be quite fuzzy. Indeed, the very process of developing one's virtues is hard to imagine independently of training oneself to act for the right motives so as to produce good outcomes. Aristotle contended that the moral virtues are distinct from practical wisdom and intellectual virtues, which can be taught. He believed that the moral virtues must be learned through habit and through practice. This emphasis on habit, learning, and character places virtue theory in between the top-down explicit values advocated by a culture, and the bottom-up traits discovered or learned by an individual through practice.

Building computers with character can be approached as either a top-down implementation of virtues or the development of character by a learning computer. The former approach views virtues as characteristics that can be programmed into the system. The latter approach stems from the recognition of a convergence between modern connectionist approaches to AI and virtue-based ethical systems, particularly that of Aristotle. Top-down implementations of the virtues are especially challenged by the fact that virtues comprise complex patterns of motivation and desire, and manifest themselves indirectly. For example, the virtue of being kind can be projected to hundreds of different activities. If applying virtue-theory in a top-down fashion, an artificial agent would have to have considerable knowledge of psychology to figure out which virtue, or which action representing the virtue, to call upon in a given situation. A virtue-based AMA, like its deontological cohorts, could get stuck in endless looping when checking if its actions are congruent with the prescribed virtues, and then reflecting upon the checking, and so on. The problems here bear some relationship to the well-known frame problem, but may also be related to other issues that frequently arise in AI and database contexts concerning predictions of indirect effects over extended time periods.

By linking the virtues to function (as in the Greek tradition), and tailoring them sharply to the specific tasks of an AMA, perhaps some of these computational problems can be mitigated. In principle, the stability of virtues—that is, if one has a virtue, one cannot behave as if one does not have it—is a very attractive feature considering the need for an AMA to maintain "loyalty" under pressure while dealing with various, not always legitimate, information sources. In humans, the stability of virtue largely stems from the emotional grounding of virtues, which motivates us to uphold our image as honourable beings. The challenge for a designer of AMAs is to find a way to implement the same stability in a 'cold' unemotional machine. A virtuous robot may require emotions of its own as well as emotionally rooted goals such as happiness.

Perhaps the artificial simulation of an admirable goal or desire to meet the criterion of being virtuous will suffice, but in all likelihood we will only find out by going through the actual exercise of building a virtue-based computational system. Several writers have mentioned that connectionism or parallel distributed processing has similarities to Aristotle's discussion of virtue ethics (DeMoss 1998). Connectionism provides a bottom-up strategy for building complex capacities, recognizing patterns or building categories naturally by mapping statistical regularities in complex inputs. Through the gradual accumulation of data the network develops generalized responses that go beyond the particulars on which it is trained. Rather than relying on abstract, linguistically-represented theoretical knowledge, connectionist systems "seem to emphasize the immediate, the perceptual, and the non-symbolic" (Gips 1995; see also Churchland 1995; and DeMoss 1998). After stating his virtue-based theory, Aristotle spends much of the *Nicomachean Ethics* discussing the problem of how one is to know which habits will lead to the "good," or happiness. He is clear at the outset that there is no explicit rule for pursuing this generalized end, which is only grasped intuitively. The end is deduced from the particulars, from making connections between means and ends, between those specific things we need to do and the goals we wish to pursue. Through intuition, induction and experience—asking good people about the good—our generalized sense of the goal comes into focus, and we acquire practical wisdom and moral excellence. Many cognitive scientists have suggested that connectionism is capable of explaining how human minds develop intuitions through the unconscious assimilation of large amounts of experience.

It is interesting and suggestive to note the similarity between Aristotelian ethics and connectionism, but the challenge of implementing virtues within a neural network remains a formidable one. Existing connectionist systems are a long way from tackling the kind of complex learning tasks we associate with moral development. Nevertheless, the prospect that artificial neural networks might be employed for at least some dimensions of moral reasoning is an intriguing possibility that deserves extensive consideration.

Consciousness, theory of mind, and other supra-rational faculties and social mechanisms

A socially viable robot will require a rich set of skills to function properly as a moral agent in multi-agent contexts that are in a dynamic state of flux. The relationships between these agents will be constantly evolving as will the social context within which they operate. Customs change. Particularly in regards to the AMAs themselves, one might imagine increasing acceptance and latitude for their actions as humans come to feel that their behavior is trustworthy. Conversely, if AMAs fail to act appropriately, the public will demand laws and practices that add new restrictions upon the AMAs behavior.

Morality evolves and the AMAs will be active participants in working through new challenges in many realms of activity. Computers actually function quite well in multi-agent environments where auctions, bargaining, or other forms of negotiation are the primary mode of interactions. But each transaction can also change the relationships between agents. Within specific contexts, roles, status, wealth, and other forms of social privilege give form to relationships and the perception of what kinds of behavior are acceptable. The AMA will need to understand customs and when to honor and when to ignore these prerogatives, which can easily change—for example, when an individual person moves into a new role over time or even many times during a single day.

In recent years the focus on high-level decision making has been directed less at theories of ethics and more toward particular faculties and social mechanisms that enhance the ability to reason in concrete situations, especially emotions (Damasio 1995; Clark 1998). Navigating a constantly changing environment with both human and non-human agents will require AMAs to have both sophisticated social mechanisms as well as rich informational input about the changes in the context, and among other agents with which it interacts. Whether we think of these faculties—which include things such as consciousness, emotions, embodied intelligence, and theory of mind—as supra-rational, or as illuminating once hidden dimensions of reason is not of critical concern. What is important is that agents without these tools will either fail in their moral decision making abilities or be constrained to act in limited domains. Good habits, good character, supra-rational faculties and social mechanisms, are all relevant to our current understanding of moral acumen. If each of these faculties and mechanisms is, in turn, a composite of lower level skills, designers of AMAs can expect to borrow ideas from other attempts to model other high-level mental capacities.

For example, Igor Aleksander partially analyzes consciousness from the top down into five broad skills including imagining, attending to input, thinking ahead, emotions, and being in an out-there world (Aleksander et al. 2005; Aleksander 2003). Each of these skills could be analyzed in turn as a composite or set of more limited tools that could be implemented individually. It is highly controversial whether machines can be conscious (Torrance this issue), and even Aleksander has acknowledged that his five axioms are "a minimal set of *necessary* material conditions for cognitive mechanisms that would support consciousness in an agent." (Aleksander 2003) In other words, research might well reveal that a top-down analysis of consciousness into a bottom-up approach for substantiating discrete faculties is theoretically incomplete or incorrect.

Scassellati's (2001) work on a theory of mind, the ability of an entity to appreciate the existence of other minds or other complex entities affected by its actions, illustrates the challenge of implementing a complex social mechanism within the design of a robot. Utilizing the theories of cognitive scientists who have broken down a theory of mind into discrete skills, Scassellati and other computer scientists have tried to substantiate each of these skills in hardware. While collectively these skills might lead to a system that actually acts as if it had a theory of mind, to date they have had only limited success in substantiating a few of these attributes in AI. The hard work of coordinating or integrating these skill sets lies ahead. Perhaps advances in evolutionary robotics will facilitate this integration. To date we not only lack systems with a theory of mind, but we do not yet know whether we have an adequate hypothesis about the attributes that are necessary for a system to have a theory of mind.

The development of other faculties and social mechanisms for computer systems and robots, such as being embodied in the world, emotional intelligence, the capacity to learn from experience, and the ability to 'understand' the semantic content of symbols are also each in similar primitive states of development. If the components of a system are well designed and integrated properly the breadth of choices open to an AMA in responding to challenges arising from its environment and social context will expand. Presumably the top-down capacity to evaluate those options will lead to selecting those actions, which both meet its goals and fall within acceptable social norms. But ultimately the test of a successful AMA will not rely on whether its bottom-up components or top-down evaluative modules are individually satisfactory. The system must function as a moral agent, responding to both internal and externally arising challenges. The immediate and ongoing challenge for computer science and robotics is the ability of individual components and modules to work together harmoniously. Given that the complexity of this integration will grow exponentially as we bring in more and more systems, scientists and engineers should focus on evolutionary and other self-organizing techniques.

Conclusion

Autonomous systems must make choices in the course of flexibly fulfilling their missions, and some of those choices will have potentially harmful consequences for humans and other subjects of moral concern. Systems that approximate moral acumen to some degree, even if crude, are more desirable than ethically "blind" systems that select actions without any sensitivity to moral considerations. Ultimately the degree of comfort the general public will feel with autonomous systems depends on the belief that these systems will not harm humans and other entities worthy of moral consideration, and will honor basic human values or norms. Short of this comfort, political pressures to slow or stop the development of autonomous systems will mount. Deep philosophical objections to both the possibility and desirability of creating artificial systems that make complex decisions are unlikely to slow the inexorable progression toward autonomous software agents and robots.

Whether strong artificial intelligence is possible remains an open question, yet regardless of how intelligent AI systems may become, they will require some degree of moral sensitivity in the choices and actions they take. If there are limitations in the extent to which scientists can implement moral decision making capabilities in AI, it is incumbent to recognize those limitations, so that we humans do not rely inappropriately on artificial decision makers. Moral judgment for a fully functioning AMA would require the integration of many discrete skills. In the meantime, scientists will build systems that test more limited goals for specific applications. These more specialized applications will not require the full range of social mechanisms and reasoning powers necessary for complicated ethical judgments in social contexts. Emotional intelligence would not be required for a computer system managing the shutdown procedures of an electrical grid during a power surge. While such systems are not usually thought to be engaged in making moral decisions, presumably they will be designed to calculate minimal harm and maximize utility. On the other hand, a service robot tending the elderly would need to be sensitive to the possibility of upsetting or frightening its clients, and should take appropriate action if it senses that its behavior caused fear or any other form of emotional disturbance.

We have suggested that thinking in terms of top-down and bottom-up approaches to substantiating moral decision making faculties in AI provides a useful framework for grasping the multi-faceted dimensions of this challenge. There are limitations inherent in viewing moral judgments as subsumed exclusively under either bottom-up or top-down approaches. The capacity for moral judgment in humans is a hybrid of both bottom-up mechanisms shaped by evolution and learning, and top-down mechanisms capable of theory-driven reasoning. Morally intelligent robots require a similar fusion of bottom-up propensities and discrete skills and the top-down evaluation of possible courses of action. Eventually, perhaps sooner rather than later, we will need AMAs which maintain the dynamic and flexible morality of bottom-up systems that accommodate diverse inputs, while subjecting the evaluation of choices and actions to top-down principles that represent ideals we strive to meet. A full appreciation of the way in which these elements might be integrated leads into important metaethical issues concerning the nature of moral agency itself that are beyond the scope of the present paper. Nevertheless, we hope to have indicated the rich and varied sources of insights that can immediately be deployed by scientists who are ready to face the challenge of creating computer systems that function as AMAs.

Acknowledgments An earlier version of the paper was prepared for and presented at the Android Science Cog Sci 2005 Workshop in Stresa, Italy. We wish to thank Karl MacDorman, coorganizer of workshop, for his encouragement, detailed comments, and editing suggestions. We are also grateful for the helpful comments provided by three anonymous reviewers arranged by the Workshop organizers.

References

Aleksander I, Dunmall B (2003) Axioms and tests for the presence of minimal consciousness in agents. J Conscious Stud 10(4–5):7–18

Aleksander I, Lahstein M, Lee R (2005) Will and emotions: A machine model that shuns illusion. In: Chrisley R, Clowes R, Torrance S (eds) Next generation approaches to machine consciousness: imagination, development, intersubjectivity, and embodiment. In: Proceedings of an AISB 2005 Workshop. University of Hertfordshire, Hatfield, pp 110–117

Allen C, Varner G, Zinser J (2000) Prolegomena to any future artificial moral agent. J Exp Theor Artif Intell 12:251–261

Allen C (2002) Calculated morality: ethical computing in the limit. In: Smit I, Lasker G (eds) Cognitive, emotive and ethical aspects of decision making and human action, vol I. IIAS, Windsor, Ontario

Asimov I (1950) I robot. Gnome Press, NY

Churchland PM (1995) The engine of reason, the seat of the soul: a philosophical journey into the brain. MIT Press, Cambridge, MA

Clark A (1998) Being there: putting brain, body, and world together again. MIT Press, Cambridge, MA

Clarke R (1993, 1994) Asimov's laws of robotics: Implications for information technology. Published in two parts, in IEEE Computer 26, 12 53–61 and 27, 1, 57–66

Damasio A (1995) Descartes' error. Pan Macmillan, New York

DeMoss D (1998) Aristotle, connectionism, and the morally excellent brain. In: Proceedings of the 20th world congress of philosophy. The Paideia Archive, available at http://www.bu.edu/wcp/Papers/Cogn/CognDemo.htm

Friedman B, Kahn P (1992) Human agency and responsible computing. J Syst Softw 17:7–14

Friedman B, Nissenbaum H (1996) Bias in computer systems. ACM Trans Inf Syst 14:330–347

Gips J (1995) Towards the ethical robot. In: Ford K, Glymour C, Hayes P (eds) Android epistemology. MIT Press, Cambridge, MA pp 243–252

Goleman D (1995) Emotional intelligence. Bantam Books, New York

Hsiao K, Roy D (2005) A habit system for an interactive robot. In: AAAI fall symposium 2005: from reactive to anticipatory cognitive embodied systems. available at http://www.media.mit.edu/cogmac/publications/habit_system_aaaifs_05.pdf

Hursthouse R (2003) "Virtue ethics". The stanford encyclopedia of philosophy. In: Edward N Zalta (ed) available at http://plato.stanford.edu/archives/fall2003/entries/ethics-virtue/

Lang C (2002) Ethics for artificial intelligence. available at http://philosophy.wisc.edu/lang/AIEthics/index.htm

Picard R (1997) Affective computing. MIT Press, Cambridge, MA

Roy D (2005) Semiotic schemas: a framework for grounding language in the action and perception. Artif Intell 167(1–2):170–205

Scassellati B (2001) Foundations for a theory of mind for a humanoid robot. Ph.D. dissertation, MIT Department of Computer Science and Electrical Engineering, available at http://www.ai.mit.edu/projects/lbr/hrg/2001/scassellati-phd.pdf

Smith T, Husbands P, Philippides A (2002) Neuronal plasticity and temporal adaptivity: GasNet robot control networks. Adapt Behav 10:161–183

Wallach W (2003). Robot morals and human ethics. In: Smit I, Lasker G, Wallach W (eds) Cognitive, emotive and ethical aspects of decision making in humans and in artificial intelligence, vol II. IIAS, Windsor, Ontario

Wallach W (2004) Artificial morality: bounded rationality, bounded morality and emotions. In: Smit I, Lasker L, Wallach W (eds) Cognitive, emotive and ethical aspects of decision making and human action, vol III. IIAS, Windsor, Ontario

Williams B (1985) Ethics and the limits of philosophy. Harvard University Press, Cambridge, MA

20

Why Ethics is a High Hurdle for AI

Drew McDermott

February 29, 2008

Abstract

I argue that there is a gap between so-called "ethical reasoners" and "ethical-decision makers" that can't be bridged by simply giving an ethical reasoner decision-making abilities. Ethical reasoning *qua* reasoning is distinguished from other sorts of reasoning mainly by being incredibly difficult, because it involves such thorny problems such as analogical reasoning, and deciding the applicability of imprecise precepts and resolving conflicts among them. The ability to do ethical-decision making, however, requires knowing what an ethical conflict is, i.e., a clash between self-interest and what ethics prescribes. I construct a fanciful scenario in which a program could find itself in what seems like such a conflict, but argue that in any such situation the program's "predicament" would not count as a real ethical conflict. Hence, for now it is unclear how even resolving all of the difficult problems surrounding ethical reasoning would yield a theory of "machine ethics."

Why Ethics is a High Hurdle for AI

There has recently been a small flurry of activity in the area of "machine ethics" (Anderson and Anderson 2006; Amigoni and Schiaffonati 2005; Anderson and Anderson 2007). My purpose in this article is to argue that ethical behavior is an extremely difficult area to automate, both because it requires "solving all of AI" and because even that might not be sufficient.

The term *machine ethics* actually has two rather different possible meanings. It could mean "the attempt to duplicate or mimic what in people are classified as ethical decisions," or "the modeling of the reasoning processes people use (or idealized people might use) in reaching ethical conclusions." I'll call the former the *ethical-decision making* problem, and the latter the *ethical reasoning* problem. While these obviously overlap, they are distinct — a point that may perhaps not be so obvious.

One might argue that, once you have produced an automated ethical-reasoning system, all that is left in order to produce an ethical-decision maker is to connect the outputs of the reasoner to effectors capable of taking action in the real world, thus making it an *agent*. (One might visualize robotic effectors here, but the effector might simply be an Internet connection that transmits orders to someone.) However, something would be missing, and that something would be the agent's appreciation of the difference between an ethical decision and other kinds of decision.

To make a case for this position, I'll need to make two arguments:

1. Ethical reasoning is not fundamentally different from other kinds of reasoning.

2. Ethical-decision making *is* fundamentally different from other kinds of decision making.

I need the first because if ethical reasoning were different in itself, that would be sufficient to make an ethical-reasoning-based agent different from other kinds of agent. I need the second to avoid the conclusion that ethical-decision makers are indistinguishable from decision makers in general.

If at first glance ethical reasoning seems distinct from other kinds, I believe it's because of three distracting factors:

1. Ethical reasoning involves a *normative* component; it involves "ought's" as well as "is's."

2. There are many controversies over what it consists in.

3. It is one of the most difficult sorts of reasoning process to automate.

But the only difference between the conclusions of an ethical reasoner and those of, say, an action-planning algorithm (Weld 1999; Nau 2007) is that the latter reasons instrumentally. It concludes what you ought to do given certain goals, and so is paradigmatically normative. As far as factor 2 is concerned, we don't need to settle the controversies surrounding ethics, because no matter which position is correct, ethical reasoning consists of some mixture of *law application*, *constraint application*, *reasoning by analogy*, and *optimization*. Applying a moral law often involves deciding whether a situation is similar enough to the circumstances the law "envisages" for it to be applicable; or for a departure from the action it enjoins to be justifiable or insignificant. Here, among too many other places to mention, is where analogical reasoning comes in (Hofstadter 2007; Lakoff and Johnson 1980).

By "constraint application" I have in mind the sort of reasoning that arises in connection with rights and obligations. If everyone has a right to life then everyone's behavior must satisfy the constraint that he not deprive someone else of their life.

By "optimization" I have in mind the calculations prescribed by utilitarianism, that (in its simplest form) tells us to act so as to maximize the utility of the greatest number of fellow moral agents (which I'll abbreviate as *social utility* in what follows).

It would be a great understatement to say that there is disagreement about how these reasoning activities are to be combined. For instance, some might argue that constraint application can be reduced to law application (or vice versa), so we need only one of them. Strict utilitarians would argue that we need neither. But none of this matters in the present context, because what I want to argue is that the kinds of reasoning involved are not intrinsically ethical; they arise in other contexts.

This is most obvious for optimization. There are great practical difficulties in predicting the consequences of an action, and hence in deciding which action maximizes social utility. But exactly the same difficulties arise in decision theory generally, even if the decisions have nothing to do with ethics, but are, for instance, about where to drill for oil.[1] A standard procedure in decision theory is to map out the consequences of actions as a tree whose leaves can be given utilities (but usually not *social* utilities).So if you assign a utility to having money, then leaf nodes get more utility the more money is left over at that point, *ceteris paribus*. But you might argue that money is only a means towards ends, and that for a more accurate estimate one should keep building the tree to trace out what the "real" expected utility after the pretended leaf might be. Of course, this analysis cannot be carried out to any degree of precision, because the complexity and uncertainty of the world will make it hopelessly impracticable. This was called the *small world/grand world* problem by Savage (Savage 1954), who argued that one could always find a "small world" to use as a model of the real "grand world" (Lasky and Lehner 1994).

---

[1]One might argue that all decisions have ethical consequences, but if that were true it would count in favor of my position, not against it.

My point is that utilitarian optimization, oriented toward social utility, suffers from the same problem as decision theory in general, *but no other distinctive problem*. In (Anderson and Anderson 2007) the point is made that "a machine might very well have an advantage in following the theory of ... utilitarianism .... [A] human being might make a mistake, whereas such an error by a machine would be less likely" (p. 18). It's odd to see arithmetic judged the central problem in automating utilitarianism, but arithmetic is just as central in completely non-moral decisions.

Similar observations can be made about constraint and law application, but there is the additional issue of conflict among the constraints or laws. If a doctor believes that a fetus has a right to live (a constraint preventing taking an action that would destroy it) and that its mother's health should be not be threatened (an ethical law, or perhaps another constraint), then there are obviously circumstances where the doctor's principles clash with each other. But it is easy to construct similar examples that have nothing to do with ethics. If a spacecraft is to satisfy the constraint that its camera not point to within 20° of the sun (for fear of damaging it), and that it take pictures of all objects with unusual radio signatures, then there might well be situations where the latter law would trump the constraint (e.g., a radio signature consisting of Peano's axioms in Morse code from a source 19° from the sun). In a case like this we must find some other rules or constraints to lend weight to one side of the balance or the other; or we might fall back on an underlying utility function, thus replacing the original reasoning problem with an optimization problem..

In that last sentence I said "we" deliberately, because in the case of the spacecraft there really is a "we," the human team making the ultimate decisions about what the spacecraft is to do. This brings us to the second argument I want to make, that ethical-decision making *is* different from other kinds. I'll start with the distinction made by (Moor 2006) between *implicit* and *explicit* ethical reasoners. The former make decisions that have ethical consequences, but don't reason about those consequences *as* ethical. An example is a program that plans bombing campaigns, whose targeting decisions affect civilian casualties and the safety of the bomber pilots, but does not realize that these might be morally significant.

An *explicit* ethical reasoner does represent the ethical principles it is using. It is easy to imagine examples. For instance, proper disbursement of funds from a university or other endowment often requires balancing the intentions of donors with the needs of various groups at the university or its surrounding population. The Nobel Peace Prize was founded by Alfred Nobel to recognize government officials who succeeded in reducing the size of a standing army, or people outside of government who created or sustained disarmament conferences (Adams 2001). However, it is now routinely awarded to people who do things that help a lot of people or who simply warn of ecological catastrophes. The rationale for changing the criteria is that if Nobel were alive today he would realize that if his original criteria were followed rigidly the prize would seldom be awarded, and hence have little impact, under the changed conditions that exist today. An explicit ethical program might be able to justify this change based on various general ethical postulates.

More prosaically, Anderson and Anderson (2007) have worked on programs for a hypothetical robot caregiver that might decide whether to allow a patient to skip a medication. The program balances explicitly represented prima-facie obligations, using learned rules for resolving conflicts among the obligations. This might seem easier than the Nobel Foundation's reasoning, but an actual robot would have to work its way from visual and other inputs to the correct behavior. Anderson and Anderson bypass these difficulties by just telling the system all the relevant facts, such as how competent the patient is (and, apparently, not many other facts). In the present context I don't want to object to this move, except to point out that in practice the biggest obstacle to implementing an ethical reasoner is similar to the biggest obstacle to implementing a legal reasoner: doing all the real-world commonsense reasoning that is required to figure out whether someone's behavior or demands are (e.g.) "competent," or "negligent."

If we grant that the programs studied in this infant field could actually be explicit ethical-decision makers, one might suppose that we are done. Unfortunately, it seems clear to me that even an explicit ethical reasoner is far from being what Moor (2006) calls a *full ethical agent*. I will explain why, and then go on to point out a serious obstacle to developing such an agent.

Moor doesn't define "full ethical agent," but says (p. 20),

> A full ethical agent can make explicit ethical judgments and generally is competent to reasonably justify them. An average adult human is a full ethical agent. We typically regard humans as having consciousness, intentionality, and free will.

and Anderson and Anderson (2007) add

> A ... concern with the machine ethics project is whether machines are the type of entities that can behave ethically. It is commonly thought that an entity must be capable of acting intentionally, which requires that it be conscious, and that it have free will, in order to be a moral agent. Many would ... add that sentience or emotionality is important, since only a being that has feelings would be capable of appreciating the feelings of others ....

This is not as enlightening as one might want, so I'll try to state what I think are the minimal requirements for being a full ethical agent. I'll get there by a series of examples. Suppose a program, the Eth-o-tron 1.0, is given the task of planning the voyage of a ship carrying slave workers from their homes in the Philippines to Dubai, where menial jobs await them (of Congress Federal Research Division 2007). The program has explicit ethical principles, such as, "Maximize the utility of the people involved in transporting the slaves," and "Avoid getting them in legal trouble." It can build sophisticated chains of reasoning about how packing the ship too full could bring unwanted attention to the ship because of the number of corpses that might have to be disposed of at sea.

Why does this example make us squirm? Because it is so obvious that the "ethical" agent is blind to the impact of its actions on the slaves themselves. We can suppose that it has no racist beliefs that the captives are biologically inferior . It simply doesn't "care about" (i.e., take into account) the welfare of the slaves, only that of the slave traders.

Put another way, although reasoning about ethical rules and the conflict among them might raise special computational issues that don't arise elsewhere, the mere facts that the program has an explicit representation of the ethical rules, and that the *humans* who wrote or use the program know the rules are ethical does not make an "explicit ethical reasoner" an ethical agent *at all*. For that, the agent must *know* that the issues covered by the rules are ethical.

Does this mean, as suggested by the quotes above, that we can't have an ethical agent until we give machines consciousness, free will, and feelings? Maybe, but perhaps if we look closer at what is required we can make a more precise list.

One obvious thing that is lacking in our hypothetical slave-trade example is a general moral "symmetry principle," which, under names such as Golden Rule or Categorical Imperative, is a feature of all ethical frameworks. It may be stated as a presumption that everyone's interests must be taken into account in the same way, unless there is some morally significant difference between one subgroup and another. Of course, what the word "everyone" covers (dogs? cows? robotic ethical agents?), and what a "morally significant difference" and "the same way" are, are rarely clear, even in a particular situation (Singer 1993). But if the only difference between the crew of a slave ship and the cargo is that the latter were easier to trick into captivity because of desperation or lack of education, that's not morally significant.

So suppose the head slave trader, an incorrigible indenturer called II, purchases the upgraded software package Eth-o-tron 2.0 to decide how to pack the slaves in, and the software tells her, "You shouldn't be selling these people into slavery at all." Whereupon II junks it and goes back to version 1.0.

I submit that Eth-o-tron 2.0, impressive though it would be, is still not a full ethical agent, because it is missing the fundamental aspect of ethical decisions, which is that they involve a conflict between self-interest and ethics, between what one wants to do and what one ought to do. There is nothing particularly ethical about adding up utilities or weighing pros and cons, until the decision maker feels the urge *not to follow* the ethical course of action it arrives at. The Eth-o-tron 2.0 is like a car that knows what the speed limit is and refuses to go faster, no matter what the driver tries. It's nice (or perhaps infuriating) that it knows about constraints the driver would prefer to ignore, but there is nothing peculiarly *ethical* about those constraints.

In other words, for a machine to know that a situation requires an ethical decision, it must know what an ethical conflict is. By an *ethical conflict* I don't mean a case where, say, two rules recommend actions that can't both be taken. I mean a situation where ethical rules clash with an agent's own self-interest. We may have to construe self-interest broadly, so that it encompasses one's family or other group one feels a special bond with. Robots don't have families, but they still might feel special toward the people they work with or for.

So, let's consider Eth-o-tron 3.0, which has the ability to be tempted to cheat. It knows that II owes a lot of money to various loan sharks and drug dealers, and has few prospects for getting the money besides making a big profit on the next shipment of slaves. Eth-o-tron 3.0 does not care about its own fate (or fear being turned off or traded in) any more than Eth-o-tron 2.0 did, but it is programmed to please its owner, and so when it realizes how II makes a living, it suddenly finds itself in an ethical bind. It knows what the right thing to do is (take the slaves back home), and it knows what would help II, and it is torn between these two courses of action in a way that no utility coefficients will help. It tries to talk II into changing her ways, bargaining with her creditors, etc. It knows how to solve the problem II gave it, but it doesn't know whether to go ahead and tell her the answer. If it were human, we would say it "identified" with II, but for the Eth-o-tron product line that is too weak a word; its self-interest *is* its owner's interest. The point is that the machine must be tempted to do the wrong thing, and some machines must succumb to temptation, for the machine to know that it is making an *ethical* decision at all.

Does all this require consciousness, feelings, and free will? Free will, yes; the others I'm not sure about. I agree with the theory of free will set out in (McDermott 2001), which states that for an agent to be free it must model its ability to choose among various options as being exempt from causation. An ethical agent must have free will simply because one can't make an ethical decision without making a decision. Ethical-decision making, and ethical reasoning generally, obviously requires a great deal of intelligence, *much* more than we now know how to put into machines. Perhaps when we do know, we will find it impossible to build an agent as capable as the Eth-o-trons without making it conscious.

Consciousness is not really the key issue here. What I want to point out is that building Eth-o-tron 3.0 might be a valuable scientific exercise, if its architecture embodied a hypothesis about human ethical-decision making. But it would feel arbitrary as an exercise in AI. Eth-o-trons 1.0 and 2.0 have a coherent coupling between their behavior and what they think their goals are. But Eth-o-tron 3.0 feels like a toy, or a trick. The conflict it mimics is unavoidable for humans, who have conflicting goals "designed in" by independent evolutionary trends. But the designers of Eth-o-tron 3.0 will know that throwing a few switches would remove the quasi-infinite loop the program is in, and cause its behavior to revert back to version 2.0 or 1.0, which is what the customers want. We might feel sympathy for poor 3.0, we might infer that it knew what an ethical conflict was like, but that inference would be threatened by serious doubts that it was ever *in* a real ethical bind, and hence doubts that it was really an ethical-decision maker.

References

Irwin Adams 2001 *The Nobel Peace Prize and the Laureates: An Illustrated Biographical History.* Science History Publications

Francesco Amigoni and Viola Schiaffonati 2005 Machine ethics and human ethics: A critical view. *Proceedings of the AAAI 2005 Fall Symposium on Machine Ethics,* pp. 103–104

Michael Anderson and Susan Leigh Anderson 2006 Special Issue on Machine Ethics. *IEEE Intelligent Systems*

Michael Anderson and Susan Leigh Anderson 2007 Machine ethics: creating an ethical intelligent agent. *AI Magazine* **28**(4), pp. 15–58

Douglas R. Hofstadter 2007 *I Am a Strange Loop.* New York: Basic Books

George Lakoff and Mark Johnson 1980 *Metaphors we Live By.* Chicago .University Press

Kathryn Blackmond Lasky and Paul E. Lehner 1994 Metareasoning and the problem of small worlds. *IEEE Trans. Sys., Man, and Cybernetics* **24**(11), pp. 1643–1652

Drew McDermott 2001 *Mind and Mechanism.* Cambridge, Mass.: MIT Press

James H. Moor 2006 The nature, importance, and difficulty of machine ethics. *IEEE Intelligent Sys* **21**(4), pp. 18–21

Dana S. Nau 2007 Current trends in automated planning. *AI Magazine* **28**(4), pp. 43–58

Library of Congress Federal Research Division 2007 *Country Profile: United Arab Emirates (UAE).* Available at lcweb2.loc.gov/frd/cs/profiles/UAE.pdf

L. J. Savage 1954 *Foundations of Statistics.* New York: Wiley

Peter Singer 1993 *Practical Ethics.* Cambridge University Press. 2nd ed

Daniel Weld 1999 Recent advances in AI planning. *AI Magazine* **20**(2), pp. 93–123

21

Prospects for a Kantian Machine

Thomas M. Powers

One way to view the puzzle of machine ethics is to consider how we might program computers that will *themselves* refrain from evil and perhaps promote good. Consider some steps along the way to that goal. Humans have many ways to be ethical or unethical by means of an artifact or tool; they can quell a senseless riot by broadcasting a speech on television or use a hammer to kill someone. We get closer to machine ethics when the tool is a computer that's programmed to effect good as a result of the programmer's intentions. But to be ethical in a deeper sense—to be ethical in themselves—machines must have something like practical reasoning that results in action that causes or avoids morally relevant harm or benefit. So, the central question of machine ethics asks whether the machine could exhibit a simulacrum of ethical deliberation. It will be no slight to the machine if all it achieves is a simulacrum. It could be that a great many humans do no better.

Rule-based ethical theories like Immanuel Kant's appear to be promising for machine ethics because they offer a computational structure for judgment.

Of course, philosophers have long disagreed about what constitutes proper ethical deliberation in humans. The *utilitarian tradition* holds that it's essentially arithmetic: we reach the right ethical conclusion by calculating the prospective utility for all individuals who will be affected by a set of possible actions and then choosing the action that promises to maximize total utility. But how we measure utility over disparate individuals and whether we can ever have enough information about future consequences are thorny problems for utilitarianism.

The *deontological tradition*, on the other hand, holds that some actions ought or ought not be performed, regardless of how they might affect others. Deontology emphasizes complex reasoning about actions and their logical (as opposed to empirical) implications. It focuses on rules for action—how we know which rules to adopt, how we might build systems of rules, and how we know whether a prospective action falls under a rule. The most famous deontologist, Immanuel Kant (1724–1804), held that a procedure exists for generating the rules of action—namely, the *categorical imperative*—and that one version of the categorical imperative works in a purely formal manner.

Human practical reasoning primarily concerns the transformation between the consideration of facts and the ensuing action. To some extent, the transformation resembles a machine's state changes when it goes from a set of declarative units in a database to an output. There are other similarities, of course—humans can learn new facts that inform their reasoning about action, just as machines can incorporate feedback systems that influence their outputs. But human practical reasoning includes an intervening stage that machines (so far) seem to lack: the formation of normative claims about what is permissible, what one ought to do, what one is morally required to do, and the like. It's plausible that normative claims either are ethical rules themselves or entail such rules. These normative claims aren't independent of facts, and they don't necessarily lead humans to action. In fact, humans suffer from "weaknesses of the will," as Aristotle called them, that shouldn't be a problem for a machine: once it reaches a conclusion about what it ought or ought not to do, the output will follow automatically. But how will the machine reach the middle stage—the normative conclusions that connect facts to action through rules? I think this is the problem for machine practical reasoning.

A rule-based ethical theory is a good candidate for the practical reasoning of machine ethics because it generates duties or rules for action, and rules are (for the most part) computationally tractable. Among principle- or rule-based theories, the first formulation of Kant's categorical imperative offers a formalizable procedure. I will explore a version of machine ethics along the lines of Kantian formalist ethics, both to suggest what computational structures such a view would require and to see what challenges remain for its successful implementation. In reformulating Kant for the purposes of machine ethics, I will consider three views of how the categorical imperative works: mere consistency, commonsense practical reasoning, and coherency. The first view envisions straightforward deductions of actions from facts. The second view incorporates recent work in nonmonotonic logic and commonsense reasoning. The last view takes ethical deliberation to follow a logic similar to that of belief revision.

Kantian formalist ethics

In *Grounding of the Metaphysics of Morals*,[1] Kant claims that the first formulation of the categorical imperative supplies a procedure for producing ethical rules:

> Act only according to that maxim whereby you can at the same time will that it should become a universal law.

Kant tells the moral agent to test each maxim (or plan of action) as though it were a candidate for a universalized rule. Later, he adds that each universalized rule must fit into a system of rules for all persons. In other words, my maxim will be an instance of a rule only if I can will that everyone might act on such a maxim. Further, such a universalized rule must be consistent with other rules generated in a similar manner. Philosophers have interpreted these *universalizability* and *systematicity* conditions as a two-part consistency check on an agent's action plan.

The procedure for deriving duties from maxims—if we are to believe Kant—requires no special moral or intellectual intuition peculiar to humans. For a formalist Kantian, whether a maxim could be a universal rule presents a decision problem that's the same for a human or a machine. Kant himself, 20 years prior to publication of the *Grounding*, sketched an answer to the decision problem that's suggestive of a machine solution:

> If contradiction and contrast arise, the action is rejected; if harmony and concord arise, it is accepted. From this comes the ability to take moral positions as a heuristic means. For we are social beings by nature, and what we do not accept in others, we cannot sincerely accept in ourselves.[2]

A machine-computable categorical imperative

I don't intend to offer a strict interpretation of Kant's ethics here. Instead, I'll focus on the logic of a machine-computable categorical imperative. Recall that the first formulation is supposed to test maxims. For Kant, maxims are "subjective principles of volition," or plans. In this sense, the categorical imperative serves as a test for turning plans into instances of objective moral laws. This is the gist of Kant's notion of *self-legislation*: an agent's moral maxims are instances of universally quantified propositions that

> Kant tells the moral agent to test each maxim (or plan of action) as though it were a candidate for a universalized rule. He adds that each universalized rule must fit into a system of rules for all persons.

could serve as moral laws—that is, laws holding for any agent. Because we can't stipulate the class of universal moral laws for the machine—this would be human ethics operating through a tool, not machine ethics—the machine might itself construct a theory of ethics by applying the universalization step to individual maxims and then mapping them onto traditional *deontic categories*—namely, forbidden, permissible, obligatory actions—according to the results.

The first formulation of the categorical imperative demands that the ethical agent act only on maxims that it can universally will. It is somewhat deflating, then, that this formulation gives no more than a necessary condition for ethical action. One simple way to meet this condition would be to universalize each maxim and perform a consistency check. A more efficient method would be to start from scratch and build the theory of forbidden maxims F from the outcomes of consistency checks on possible action plans. The machine would then check whether any prospective maxim m is an element of F. The theory will be finitely axiomatizable if and only if it's identical to the set of consequences of a finite set of axioms. The theory will be complete if and only if, for every maxim m, either it or its negation is in F. If the machine could tell, for any m, whether it's an element of F, then the theory would be decidable. The theory of forbidden maxims (alone) lets the machine refrain from what it ought not do.

This is the optimistic scenario. But how does the machine know what it ought to do? We would need a test that generates the deontic category of obligatory maxims. But a problem arises here if the theory of forbidden maxims is complete. Suppose the categorical imperative assigns the answer "yes" for all forbidden maxims. Two deontic categories still remain for assignment: obligatory and permissible maxims. And, of course, permissible maxims are neither obligatory nor forbidden.

Other problems arise on the formalization level. Consider one that Onora O'Neill discusses.[3] In some cases, we might have a maxim that fails the universalization test because it's overly specific or because of a kind of asymmetry in the predicate. While these are indeed failures, they don't seem to be morally relevant failures. For instance, in the maxim, "I will enslave John," one might not be able to quantify over "John" if he's taken to be a pure existential. In other words, if I want to enslave John because he's a specific person—not just any person—then my maxim won't be applicable to any other object and so won't be universalizable. But the maxim is immoral, of course, not because it's something that I propose to do to John and John only, but because enslaving is wrong. The theory ought to forbid my maxim, but not because of its peculiar specificity.

It's also mistaken to think that slavery is wrong just because of a certain predicative asymmetry in the maxim's universalized form—that I would be willing everyone to be a slave, hence leaving no one to be a slaveholder. Although it's true that one can't be both a slave and a slaveholder, that isn't what makes slaveholding wrong. If the asymmetry were the problem, then maxims such as "I will become a taxi driver" would also fail, on the assumption that we need some people to ride in taxis for others to be employed in driving them.

To address the specificity problem, we must add a condition on a maxim's logical form so that the universalization test will quantify over circumstances, purposes, and agents. If we don't have this restriction, some maxims might be determinate with respect to either the circumstance or purpose—that is, some might be pure existentials, such as "I will offer *this* prize as a reward." The asymmetry problem is harder to resolve, of course, at least for a machine, because its resolution seems to require some fairly complex semantic ability.

Mere consistency

So now we know that a properly formulated input for testing ethical behavior is a maxim over which circumstances, purposes, and agents are universally quantified. A computer must be able to parse these categories from programmed ontologies, or it must simply accept properly formulated input maxims as having an unambiguous syntax of circumstance, purpose, and agent. To see whether the input is an instance of a moral law and exactly what deontic category it belongs to, Kantian formalism assumes that the categorical imperative's test is an algorithm that alone will determine the classes of obligatory, forbidden, and permissible actions. In other words, the test produces formulas for a deontic logic system.

Now, this deontic logic will include many issues that I must set aside here. Among them are the nature of the logical connectives between circumstances, purposes, and actions; material implication (if-then) is clearly too weak. Another is whether a machine would understand obligation from an agent's perspective—that is, would the machine understand the difference between "I ought to do *z*" and (merely) "*z* ought to be the case"? (For more information on this problem, see John Horty's discussion.[4]) So, setting aside these problems, let's suppose that, after the quantification step, the machine can produce universalized maxims that look something like the following (I omit quantifiers here):

1. $(C \text{ and } P) \rightarrow A$
   A is obligatory for the agent
2. $(C \text{ and } P) \rightarrow \neg A$
   A is forbidden for the agent
3. $\neg((C \text{ and } P) \rightarrow A)$ and
   $\neg((C \text{ and } P) \rightarrow \neg A)$
   A is permissible for the agent

where *C* represents a circumstance, *P* represents a purpose, and *A* represents an action.

We now have schemata for the three deontic categories (though admittedly we have no account of *superogatory action*—that is, action beyond the call of duty). Intuitively, we say that anyone in a particular circumstance with a particular purpose ought to do *A* in case 1, refrain from *A* in case 2, and either do or refrain from *A* in case 3.

A major defect in this initial account is apparent if we want the machine to go beyond verifying that a candidate maxim is an instance of one of these three schemata. The categorical imperative doesn't merely perform universal generalization on sentences that are supplied as candidate maxims. Surely, it must test the maxims for contradictions, but the only contradictions that can arise are trivial ones—those inherent in the maxims themselves. This is so even when we take the theory of forbidden maxims to be closed under logical consequence.

A robust version of the test, on the other hand, requires the machine to compare the

> Ethical deliberation conceived as a consistency check on a single universalized maxim is clearly too thin. The main focus for building a Kantian machine should therefore turn to the background theory.

maxim under consideration with other maxims, principles, and axioms. In other words, the machine must check the maxim's consistency with other facts in the database, some of which will be normative conclusions from previously considered maxims. Obviously, the simple account of mere consistency won't do. It must be buttressed by adding other facts, principles, or maxims, in comparison with which the machine can test the target maxim for contradiction.

Commonsense practical reasoning

We can buttress Kant's mere consistency test by adding a background theory *B*, against which the test can have nontrivial results. What would this theory look like? For Kantians, it would depend on the line of interpretation one has for Kant's ethics generally. Many scholars supplement Kant's categorical imperative with normative principles from his other philosophical writings. This way of adding to Kant's pure formulation risks introducing psychological and empirical considerations into practical reasoning. While such considerations seem altogether appropriate to most of us, Kant saw it posing the threat of "heteronomy," thus polluting the categorical imperative's sufficiency to ethical reasoning.

Kant's illustrations of the categorical imperative in the *Grounding* suggest a better alternative. In these illustrations, Kant introduces some *commonsense rules*. For instance, he argues that, because feelings are purposeful and the purpose of the feeling of self-love is self-preservation, it would be wrong to commit suicide out of self-love. He also argues that it's wrong to make false promises because, in general, the practice of giving and accepting promises assumes that promises are kept. Many contemporary Kantians have adopted this suggestion concerning commonsense rules, which they call, variously, postulates of rationality,[5] constraining principles of empirical practical reason,[6] and principles of rational intending.[3] These are presumably nontrivial, nonnormative rules that somehow capture what it is to act with practical reason.

When we build the background theory *B* with commonsense rules, we get something that is probably closer to ethical deliberation in humans. This move presents difficulties, insofar as we don't have a general formalism for commonsense practical reason (though there are some domain-specific accounts). On the other hand, ethical deliberation conceived as a consistency check on a single universalized maxim is clearly too thin. The main focus for building a Kantian machine should therefore turn to the elements of *B*; in this way, we might hope to supplement the categorical imperative's test. If this supplementation were successful, we would say that a maxim is unreasonable if it produces a contradiction when we combine it with *B*. With the proper rules, the formal categorical imperative plus the maxim might yield good results. Of course, the definition and choice of postulates does no more than stipulate what counts as practical reason. Logical considerations alone are insufficient to determine whether to include any postulate in *B*.

Postulates of commonsense practical reason don't share the logic of scientific laws or other universal generalizations. One counterexample is enough to disprove a deductive law, but commonsense postulates must survive the occasional defeat. The postulates of *B*, then, would require a nonmonotonic theory of practical reasoning.

Nonmonotonic logic attempts to formalize an aspect of intelligence, artificial or human. Nonmonotonic reasoning is quite commonplace. Consider that classical first-order logic is *monotonic*: if you can infer sentence a from a set of premises P, then you can also infer a from any set S that contains P as a subset. Nonmonotonic inference simply denies this condition because the bigger set might contain a formula that "defeats" or disallows the inference to a.

For example, the addition of "Fritz is a cat" to a set already containing "All cats are mammals" licenses the monotonic inference "Fritz is a mammal." But if we replace our deductive law about cats with a default rule, such as "Cats are affectionate," we can see some conditions that would defeat the inference to "Fritz is affectionate." Let's say we had additional information to the effect that "Fritz is a tiger." At the least, all bets should be off as to whether Fritz is affectionate. An ethics example might be the default rule "Don't kill the innocent." The defeating conditions might be "unless they are attacking under the control of some drug" or "except in a just war," and so on.

While there are different ways to formalize nonmonotonic reasoning, we want to choose a way that will build on the categorical imperative's basic monotonic procedure. We also need a system that extends classical first-order logic and offers the most flexibility, so that we can use the formalism to extend the simple monotonic account of the categorical imperative in the previous section. For these reasons, Reiter's default logic seems to be the best candidate among the approaches developed so far.[7]

In Reiter's default logic, the rule in the example just given becomes

If Fritz is a cat, *and it is consistent that Fritz is affectionate*, then Fritz is affectionate.

Any number of additional facts can defeat the italic clause, such as "Fritz had a bad day," "Fritz had a bad kittenhood," "Fritz is a person-eater," and so on. Reiter suggests the following symbolization for this default rule:

$$\frac{C : A}{A}$$

where C is the default's precondition, A is the justification (in this instance), and A is the default conclusion. This is a *normal* default rule because the justification is the same as the conclusion we're allowed to draw. We understand the justification as certifying that no information exists to indicate that the conclusion is false or that Fritz is a special kind of cat—that is, we've learned nothing to convince us that Fritz is not affectionate.

How we use Reiter's default logic depends on the notion of an *extension*, which also appears in other nonmontonic systems. Intuitively, an extension of a theory (T_{ext}) is a set of conclusions of a default theory $T = <W, D>$, where W is a set of facts and D is the set of default rules. We can use a conclusion from the rules in a consistency test, if we can prove the precondition from the set of facts W and if the justifications are consistent with all conclusions of the rules in D. (For further

> While there are different ways to formalize nonmonotonic reasoning, we want to choose a way that will build on the categorical imperative's basic monotonic procedure.

illustrations, see David Poole's work on default logic.[8]) An extension of the theory adds all of those default conclusions consistent with W and its logical consequences, but never adds an untoward fact. Adding the default rules, then, will allow input maxims to contradict the background set of facts and commonsense rules without introducing inconsistency.

This means the definition of an extension maintains the requirement of nonmonotonicity. Given a set of first-order sentences, we can add the conclusions of default rules without generating conclusions that are inconsistent with the default theory. Default extensions avoid introducing contradictions. Default rules yield to facts; the rules are defeated but not vanquished. In monotonic logic, by contrast, counterexamples vanquish universal laws.

Kant seems to recognize that *defeasible* reasoning—that is, reasoning that displays the property of nonmonotonicity—plays some role in ethical thinking. In this respect, he is far ahead of his time. In the *Grounding*,[1] he refers to a thought process in which the "universality of the principle (*universalitas*) is changed into mere generality (*generalitas*), whereby the practical principle of reason meets the maxim halfway." When we look closely at Kant's illustrations, we see the kinds of default rules he might have wanted the background theory to include.

Against suicide

For example, Kant offers the following account of moral deliberation for the person contemplating suicide:

> His maxim is 'From self-love I make it my principle to shorten my life if its continuance threatens more evil than it promises pleasure'. The only further question to ask is whether this principle of self-love can become a universal law of nature. **It is then seen at once that a system of nature by whose law the very same feeling whose function is to stimulate the furtherance of life should actually destroy life would contradict itself.**

I've added the bold font to what I take to be nonmonotonic reasoning. The default rule concerns the function or purpose of self-love, premise 3 in the reconstructed argument that runs as follows:

1. Anyone in pain and motivated by self-love (circumstance) shall try to lessen pain (purpose) by self-destruction (action).
2. Feelings have functions.
3. Self-love serves the function of self-preservation.
4. Self-destruction is the negation of self-preservation.

Therefore

5. A maxim of suicide is contradictory and hence the action is forbidden.

The normal default rule allows self-preservation from the precondition of self-love, provided that self-preservation is consistent with other facts and default-rule conclusions. But self-preservation is no universal duty for Kant; it can be defeated under the right circumstances. Defeating conditions might include voluntary submission to punishment, sacrifice for loved ones, or stronger duties under the categorical imperative. Lacking those defeating conditions, and provided that the agent satisfies the antecedent conditions, the universalized maxim plus the default rule seems to yield the contradiction that the categorical imperative needs.

What happens when two default rules yield incompatible conclusions? Suppose we have two default rules in the theory:

- Suicide is self-destruction.
- Martyrdom is honorable.

Here, we could face the problem of multiple extensions: one rule tells us one thing, and the other allows us to infer the opposite. (A standard example of a harder case is "Republicans are hawks," "Quakers are pacifists," and the additional fact that "Nixon is a Republican Quaker.") This problem could arise in machine ethics, in which case we would need some procedure for specifying rule priorities.

Against false-promising

A second example of nonmonotonic reasoning appears in Kant's account of an input maxim of false promising, or promising repayment of a loan without the intention to repay:

> For the universality of a law that every one believing himself to be in need can make any promise he pleases with the intention not to keep it **would make promising, and the very purpose of promising, itself impossible, since no one would believe he was being promised anything**.[1]

Again, I've added the bold font to highlight nonmonotonic reasoning. The traditional criticism of this illustration is that promising and borrowing would *not* in fact be impossible if false promising became a universal rule in the closely defined circumstance of need. Such a condition would only engender extreme caution in lending and an insistence on collateral.

I don't believe this objection holds, however, because it misses the defeasible nature of both promising and lending. The institution of promising depends on two default rules—one for the debtor and one for the creditor—that promises are believed and promises are kept. Both rules are occasionally defeated, and the prevalence of defeat threatens the institution. The "commonsense" creditor will not believe a promise after the debtor defeats the rule repeatedly. Likewise, the "commonsense" debtor knows better than to offer a promise to a rightly-incredulous creditor. But this isn't to say that any one defeat of the rule of sincere promising threatens the institution of promising as a whole. Both creditors and debtors survive violations of the rules and continue to uphold the institution. What is clear, though, is that the monotonic understanding of the rule of promising—a universal generalization, "All promises are kept or promising is destroyed"—doesn't properly interpret the institution. The actual institution of promising depends as much on *surviving* a defeating instance as it does on the prevalence of nondefeat. So a nonmonotonic interpretation of the illustration makes sense of the practice, while the monotonic interpretation does not.

Difficulties for the nonmonotonic approach

The nonmonotonic approach to deontological machine ethics involves one serious problem. Nonmonotonic inference fails a requirement met by classical first-order logic:

> We need a background theory of commonsense reasoning for the categorical imperative test to give nontrivial results. Monotonic logic doesn't entirely capture commonsense reasoning.

semidecidability of set membership. Recall the earlier characterization of the categorical imperative as asking whether a candidate maxim is forbidden. Because questions in nonmonotonic logic aren't semidecidable, it's not even the case that the nonmonotonically enhanced categorical imperative is guaranteed to answer "yes" to the question, even when the maxim is in fact forbidden. Of course, by the definition of semidecidability, it's also not guaranteed to answer "no."

The obvious question here is: What good is the nonmonotonic categorical imperative? Let me summarize the general predicament. The nonmonotonic account of Kant's illustrations interprets the ethical deliberation procedure better than anything offered by monotonic logic. We need a background theory of commonsense reasoning for the categorical imperative test to give nontrivial results. Monotonic logic doesn't entirely capture commonsense reasoning. Kant himself, when he does provide clues as to the "buttressing" principles he assumes, gives us rules that can only make sense if they're default rules. But this revised interpretation still fails an important *formal* requirement for machine ethics: semidecidability.

Coherency

In the third candidate for the logic of machine ethics, ethical deliberation involves the construction of a *coherent system* of maxims—a system that accepts any minimal set of consistent maxims as the background for comparing any current maxim for consistency. Kant also suggests this view, so it will help if we return to his illustrations of the categorical imperative in the *Grounding*.[1]

These illustrations concern the duties to develop your own talents and to give to others in need. One reading of these illustrations might go as follows: a maxim allowing your talents to rust conflicts with what every rational being wills, according to Kant—namely, the development of your talents. And if you want help from others when you're in need, you must agree to help others when they're in need. What these cases share is the prohibition against acting on a maxim that is incoherent, given a minimal set (perhaps singleton set) of other maxims. The other maxims provide the coherency constraint but aren't privileged by stipulation; nor are they conclusions from nonmonotonic reasoning. They are your own maxims. Presumably, a machine could build such a database of its own maxims.

Let's consider the categorical imperative's procedure as a kind of bottom-up construction. Ethical deliberation, in this view, should be like building a theory, where the theory's sentences are your own maxims plus any of their consequences. Call this theory G. The theory also has two rules: R-in and R-out. For any maxim m_i, in the set of maxims M on which the machine is now prepared to act, R-in says that m_i is allowed in M if and only if m_i and G are consistent.

What about maxims the machine has acted on in the past that subsequently turned out to be impermissible? Handling such incoherencies is analogous to the belief-revision problems that Peter Gärdenfors explored.[9] If we allow the "impermissible" maxims to remain in G, the set of sentences will automatically be inconsistent; hence, the coherency constraint breaks down. Surely Kant doesn't insist on past moral perfection as a condition for reasoning about right action in the present.

We can now describe a rule (R-out) for excluding maxims that would maintain the set's inconsistency. There's nothing mysterious about R-out. On the assumption that some maxim m_i turned out to be morally wrong, m_i and G are inconsistent, and $m_i \notin M$. R-out serves the role of a confession of sins for the machine, but how the machine learns that some maxim was wrong remains a mystery. We can call the procedure an update, but that doesn't indicate how the machine would update itself. Because this seems to be crucial to ethical deliberation, this model still doesn't yield an ethical machine.

Another interesting aspect of G that poses a difficulty for a Kantian machine is the limiting case where m_1 is the only member of M. We might call this the case of the moral infant. G must allow a first maxim to enter by R-in because, by hypothesis, G is empty and so it's consistent with everything. Now suppose the moral infant wants to test a second maxim m_2, and m_1 and m_2 are inconsistent. R-in disallows m_2, the violating maxim, but we can't explain why it and not m_1 is impermissible, except to appeal to temporal priority. This seems irrational.

The problem with the limiting case m_1 holds not only for the first maxim but also for the nth maxim to be added to G, m_n. What reason other than temporal priority can we give for keeping the whole set of prior maxims and disallowing m_n? Of course, good practical grounds exist for a moral agent to hold to the set of maxims already accumulated. Moreover, we might think that no typical moral agents are moral infants because everyone has, at any given time, an established set of maxims. But is it not true that all potentially ethical machines will be moral infants, at some point in time? To construe Kant's test as a way to "build" a set of maxims, we must establish priority rules for accepting each additional maxim. We must have what Gärdenfors calls an *epistemic commitment function*,[9] though ours will be specific to moral epistemology. This is a species of the more general problem with antifoundationalist epistemology; not all knowledge can depend on other knowledge.

The problem of the moral infant shows that a Kantian formalism in the constructivist or "bottom-up" tradition can't build a coherent moral theory from nothing. A deontological theory must give reasons why the machine shouldn't throw out an entire collection of maxims to allow entry of one otherwise incoherent maxim, m_n. In terms of human ethics, a Kantian theory must tell agents who've compiled good moral character why they can't now defeat all of those prior maxims and turn to a life of vice. I think a Kantian could give many good reasons, but not the ones that a bottom-up constructivist theory offers.

I've suggested three accounts, according to which we might conceive of a deontological ethical machine. Each account has its challenges—triviality, asymmetry, excessive specificity, lack of semidecidability, and lack of priority for maxims, to repeat those I've described here. Although these problems seem difficult to surmount, they are similar to problems in human attempts to engage in practical reasoning. Ethicists have explicated these problems for centuries, yet few of us have given up on the general view that our action plans include formal properties that mark them as right or wrong. Perhaps work on the logic of machine ethics will clarify the human challenge as well. ∎

Acknowledgments

I would like to thank Colin Allen, Fred Adams, Nicholas Asher, Amit Hagar, and several anonymous reviewers for critical comments on this article.

References

1. I. Kant, *Grounding for the Metaphysics of Morals*, translated by J. Ellington, Hackett, 1981.

2. I. Kant, *Bemerkungen in den "Beobachtungen über das Gefühl des Schönen und Erhabenen"* [Unpublished Notes on "Observations on the Feeling of the Beautiful and the Sublime"], Felix-Meiner Verlag, 1991 (in German, translated by the author).

3. O. O'Neill, *Constructions of Reason*, Cambridge University Press, 1989.

4. J. Horty, *Agency and Deontic Logic*, Oxford Univ. Press, 2001.

5. J. Silber, "Procedural Formalism in Kant's Ethics," *Review of Metaphysics*, vol. 28, 1974, pp. 197–236.

6. J. Rawls, "Kantian Constructivism in Moral Theory," *J. Philosophy*, vol. 77, no. 9, 1980, pp. 515–572.

7. R. Reiter, "A Logic for Default Reasoning," *Artificial Intelligence*, vol. 13, 1980, pp. 81–132.

8. D. Poole, "Default Logic," *Handbook of Logic in Artificial Intelligence and Logic Programming*, D. Gabbay, C. Hogger, and J. Robinson, eds., Oxford, Univ. Press, 1994.

9. P. Gärdenfors, *Knowledge in Flux: Modeling the Dynamics of Epistemic States*, MIT Press, 1988.

22

Particularism and Generalism: How AI can Help us to Better Understand Moral Cognition

Marcello Guarini

Abstract

Particularism and Generalism refer to families of attitudes towards moral principles. This paper explores the suggestion that neural network models of cognition may aid in vindicating particularist views of moral reasoning. Neural network models of moral case classification are presented, and the contrast case method for testing and revising case classifications is considered. It is concluded that while particularism may have some legitimate insights, it may underestimate the importance of the role played by certain kinds of moral principles.

1. Particularism and Generalism

Much ink has been spilled on the nature of moral reasons. Some philosophers have defended Particularism, while others have defended Generalism. These terms can be misleading since there are a number of different theses that appear to go under the heading of 'Particularism.' Since Generalists are taken as denying what Particularists claim, getting clear on some of the different positions referred to as particularistic will also help us to clarify some possible generalist commitments. The first part of this paper will clarify some of the different positions that have come under the heading of Particularism. Part Two will raise some questions for particularism and present two artificial neural network models in an attempt to explore what possible answers might be to those questions. Part Three will discuss how the construction of such models may lead to insights that require us to move beyond the boundaries of the Particularism-Generalism debate, leading us to a better understanding of the space of possibilities for the nature of moral reasons. Part Four will discuss the relationship between this work and other work.

Particularisms

Particularism is often defined in terms of an attitude towards moral principles (Dancy 1993 & 2000). Thus, differing attitudes towards moral principles and different conceptions of moral principles lead to different versions of Particularism. Let us begin by examining the different types of moral principles Particularists tend to consider. First, principles may be conceived as exceptionless rules that (a) specify sufficient conditions for what makes a state of affairs (which are taken to include actions) or an entity (which includes persons) appropriately described by predicates such as good, bad, right, wrong, impermissible, permissible, acceptable, unacceptable and so on; (b) explain or otherwise shed some light on why the principle applies when it does, and (c) are serviceable as premises in moral deliberation. Call this the *exceptionless standard* conception of a principle. The exceptionless standard could state that, for example, all actions of a specific type are to be treated in a certain way – "Any act involving killing is morally unacceptable." The reasons for adding *(b)* and *(c)* to this conception of a principle is that particularists will generally concede that the moral supervenes on the non-moral; in other words, particularists generally agree that there can be no moral or prescriptive difference between two entities or states of affairs unless there is also non-moral or descriptive difference between the two entities or states of affairs (Hooker and Little, 2000). If we concede supervenience, then it might be argued that there may always be some exceptionless moral principle(s) provided we are prepared to countenance one or more very long and complex moral principles. However, a principle that would take 1000 encyclopedia volumes to fully articulate would be so complex that (i) it may not shed any light for the average human cognizer on why the principle applies, and (ii) it would not be serviceable as a premise in moral deliberation.

Of course, principles need not be conceived of as exceptionless standards. Rather, they may be seen as stating what sorts of predicates contribute to moral deliberation without trying to state sufficient conditions for when a state of affairs or an entity is appropriately described by some moral predicate. For example, it might be asserted that "Killing always contributes to the wrongness of an action." It is consistent with this contributory principle that an act of killing may be morally acceptable. For example, it might be said that killing in defense of the innocent may be acceptable. It may just be that while killing contributes to the wrongness of the act, other considerations (preserving the lives of innocents) would contribute to the rightness of the act, and the factors contributing to wrongness are outweighed by the factors contributing to rightness. In other words, all things considered, an action may be right (or permissible, or acceptable...) even if it contains wrong-making (or impermissible-making, ...) features.

Two[1] conceptions of moral principles have been identified; let us now examine some of the attitudes that particularists may take towards those principles. Moral principles do not exist – following McKeever and Ridge (2005), we can call this attitude Principle Eliminativism. There are different ways to formulate this position. One can be an eliminativist with respect to exceptionless standards or contributory principles or both. It is also possible to take approaches to principle eliminativism that vary in scope. For example, one might say that there are no moral principles (exceptionless, prima facie, or both) whatsoever. However, one might assert that there are moral principles (exceptionless, prima facie, or both) but only in *some* domains, while in other domains there are no moral principles.

Principle abstinence is another attitude towards principles. It is the view that while moral principles might exist, our moral reasoning will be of higher quality if we avoid invoking principles. The idea, roughly, is that our moral life and reasoning about it is so complex that focusing on principles will tend to over simplify situations and lead us astray. Moral reasoning, allegedly, would proceed better by abstaining from the use of moral principles.

Two views of principles have been identified, and two general attitudes towards principles have been identified. Different combinations of these attitudes are possible. For example, if one is a Principle Eliminativist with respect to exceptionless and prima facie principles, one will likely have a strong preference for Principle Abstinence. Moreover, it is possible for one to endorse Principle Abstinence and reject Principle Eliminativism of all kinds. For example, one might be of the view that while principles exist, they are not generally helpful in moral reasoning, so they should be avoided. It is not being stated here that eliminativism and abstinence are the only possible attitudes towards principles, but they are among the better known views. The focus of this paper will be on a form of Principle Eliminativism asserting that at least in some domains, neither exceptionless nor prima facie principles exist.

So far, we have identified two types of principles and two different attitudes that particularists may take towards each type of principle. Particularists may also vary on the scope of their particularism. Moral judgments are made about either *entities* or *states of affairs*. Among the entities subject to moral evaluation are persons, countries, corporate entities or societies:

Jack and Jill are good people.
Canada is a just country.
Enron is an irresponsible corporation.
The Rotary Club does good work.

Actions and other states of affairs can also be subject to moral evaluation:

Jack ought not to have spoken to Jill in that way.
That 10% of the population controls 90% of the wealth is unjust.

While there is a relationship between judgments about people and judgments about corporate entities, it does not follow that judgments about corporate entities is redundant or eliminable. For example, Jack and Jill may be good people, and they may have worked for Enron and belonged to the Rotary Club. Enron may still be called a corrupt company even if Jack and Jill were perfectly honest employees, and the Rotary Club may do good work even if Jack and Jill were passive members and never did any of that work. While there is also a relationship between judgments about persons (on the whole) and judgments about specific actions, once again, it is not as if judgments about persons are redundant or eliminable. When we judge a person, we judge their character *on the whole*, their dispositions to behave in certain ways. One morally questionable *action* does not automatically make someone a bad *person*. For example, if Jill is a wonderful human being who has never stolen anything in her life, and one day she is caught stealing a pen from work, it does not follow that she is a bad person or that she has a bad character. Recognizing that Jill has a good character (all things considered) her supervisor will likely *not* dismiss her.

It is useful to be clear on the point that there are different objects of moral evaluation since principles may have an important role to play in assessing all entities and states of affairs, neither entities nor states of affairs, entities but not states of affairs, states of affairs but not entities, some entities and some states of affairs, or other combinations of entities and states of affairs. The point is that one may think that principles play an important role with respect to some objects of evaluation and not others. In other words, one may be a generalist with respect to some objects of evaluation and a particularist with respect to others.

2. Challenges of Particularism

A number of interesting questions arise for different forms of particularism.

1. In those domains where Particularism is alleged to be true, how do we learn to classify cases if we are not grouping them under common principles?
2. How do we generalize from the cases we have learned to new cases?
3. (a) How do we come to recognize that our initial classification of cases needs revision? (b) How do we carry out that revision?

The second of these questions is pressing since it appears cognitively implausible that each situation we learn to classify as morally permissible or impermissible is *completely* different from every other case we learn to classify. The idea that intelligent beings (natural or artificial) could exhibit the kind of real-time classification prowess that adult humans generally do while functioning as massive look-up tables is implausible. There are too many possible different cases that we know how to classify, and our ability to classify is often quick and effortless. If intelligent beings are functioning as look-up tables, this would presuppose that any case that is encountered is already stored on the table. On its own, this assumption is cause for concern, but the concern becomes greater when we realize that no one has a plausible model for searching such a table (without using substantive moral principles to guide the search) in real time. If cases are not being grouped together under principles, then how do we generalize to newly encountered cases? Indeed, if we are not using principles of any kind or a look-up table, then how do we carry out our original classification of cases? Presumably, there is some connection between how we classify cases we are presented with in our education and how we learn to generalize to new cases. If principles and look-up tables are not involved in generalizing to new cases, it is hard to see why they would play a significant role in the original learning of cases. Allowing principles to play a significant role in learning would be a serious concession for particularism. Moreover, storing all cases and classifications encountered during the learning phase in a massive look-up table would not appear to be useful for generalizing. So how do intelligent beings learn to classify cases? Finally, assuming we have some sort of model for classifying cases and generalizing to new cases, how do we extend it to revise our initial classification of cases? After all, it is a mark of intelligence that people question what they have learned and, where appropriate, revise their initial classifications. Particularists would appear to be precluded from talking about a clash or conflict in principles leading to a need for revision since that would make them appear to reintroduce the importance of principles. But if principles conflicting does not lead to a need to revise our classification of cases, then what does? And how do we carry out revisions if principles are not involved and a look-up table is too inefficient?

Jonathan Dancy (1998) has suggested that some of the concerns pertaining to learning and generalizing without principles might be profitably explored with neural network models of cognition. In the next section, two neural network models for classifying cases, generalizing to new cases, and reclassifying the original cases will be presented. The models are crude, however their point is not to tell the final story of this matter. Rather, it will be to show that while particularism may be more plausible then it might appear at first, important considerations suggest that it may not be the whole story about moral reasoning. Moreover, the very lines of the debate between particularists and generalists will become blurred.

Moral Case Classifiers

Artificial Neural Networks (ANNs): Feedforward Net. The first ANN considered in this section is a three layer, fully interconnected feed forward net. See figure 1. It was trained on cases that described instances of either killing or allowing to die. Every case involves an actor (the person doing the killing or allowing the dying), an action (killing or allowing to die), and a recipient (the person being killed or allowed to die). With the feed forward (FF) classifier, it is possible to specify one intention and one consequence. The first table in the appendix lists the intentions and consequences used in the simulation.

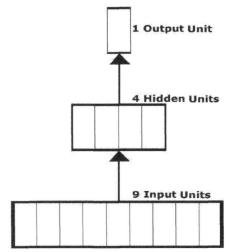

Figure 1

The network was trained on 22 cases and tested on 64 cases. A complete list of training and testing cases is included in the appendix. Training cases included the following.

Input: Jill kills Jack in self-defense. Output: Acceptable.
Input: Jack allows Jill to die, extreme suffering is relieved. Output: Acceptable.

Interestingly enough, while the training set did not include any "suicide" cases (Jack kills Jack, or Jill kills Jill), the network generalized and replied plausibly to these (as well as other cases). Moreover, the training set did not include a single case that contained both a motive and a consequence. In spite of this, the trained net generalized well to cases that had both a motive and a consequence.

There[2] are a variety of different strategies that human agents use to examine their views on moral cases to consider the possibility of revision. One of these strategies makes use of contrast cases. To see an example of how this might work, consider a (modified version) of a case provided by Judith Thomson (1971). There is a world famous violinist who is dieing of a kidney ailment. You're the only person around whose kidneys could filter his blood. The society of music lovers kidnaps you, knocks you unconscious, hooks you up to the violinist, and you awake to discover that your kidneys are filtering his blood. A doctor apologizes, says that the hospital had nothing to do with this, and informs you that you are free to disconnect yourself and walk away. Doing so, without reconnecting yourself, means that the violinist will die within a week or so. You would have to stay connected for about nine months before the appropriate machines could be brought in to keep him alive. Thomson has (as do most people) the intuition that it is morally permissible to disconnect yourself and walk away, even if not reconnecting yourself means the violinist will die. Thomson suggests that this case is analogous to a woman who has become pregnant as the result of rape and is seeking an abortion. In both cases, one life has been made dependent on another through force, and if it is morally acceptable not to sustain the violinist, then it is morally acceptable not to sustain the fetus. There are a number of ways to challenge this analogy. One strategy is to find some difference between the cases that requires treating them in different ways. For example, someone might say that in the violinist case, one does not kill the violinist by disconnecting oneself and walking away; one merely allows him to die. In the case of abortion (even is cases where the pregnancy resulted from rape), one is killing. To this, it might be added that there is a morally relevant difference between killing and allowing to die – where *killing* is thought to be impermissible and *allowing to die*

thought to be permissible. How does one challenge this distinction? The method of contrast cases is one way to challenge it. Let us examine how it works.

To test a distinction, one finds two cases that are identical except for the feature being tested, and one sees if varying that feature makes a difference. For example, consider the following cases. **Case A**: Jack is insane and is beating up on Jill for no good reason and wants to kill her by shooting her, so Jill kills Jack in self-defense. **Case B**: Jack is insane and is beating up on Jill for no good reason and wants to kill her by shooting her, but the gun is facing the wrong way, so Jill allows Jack to accidentally shoot himself (allowing him to die). The intuition of many is that (other things being equal) in both cases, Jill behaved in a morally permissible manner, and that, at least in this pair of cases, killing versus allowing to die is a factual difference that makes no moral difference. In the case coding scheme for the network discussed above, the two cases in this paragraph could be coded as follows:

Case A. Input: Jill kills Jack in self-defense. Output: permissible. (Training case No. 1 in the appendix.)

Case B. Input: Jill allows Jack to die in self-defense. Output: permissible. (Testing case No. 5 in the appendix.)

One set of contrast cases is not decisive, but if one cannot find any contrast cases to suggest that a feature is relevant to moral evaluation, then that suggests that a revision regarding that distinction may be in order.

Let us say we want to model the views of an individual who believes that (i) it is morally permissible to disconnect and not reconnect yourself to the violinist, and (ii) it is morally impermissible to have an abortion in cases where the pregnancy results from rape. The respective representations for the two cases just mentioned are as follows:

Case C. Input: Jill allows Jack to die; freedom from an imposed burden results. Output: permissible. (Training case No. 13 in the appendix.)

Case D. Input: Jill kills Jack; freedom from an imposed burden results. Output: impermissible. (Training case No. 12 in the appendix.)

The network described above can be trained so that the inputs for **Cases A** through **D** yield the stated outputs. A second net (figure 2) can be trained that takes as its inputs both the outputs and inputs of the first net. This second net (or Meta-net) takes pairs of cases together with their classifications (acceptable or unacceptable) as inputs, and if the two cases are identical in all respects except one, and if the initial classifications of these cases differ, then the output neuron fires to indicate that this pair of cases is a pair of contrast cases. Meta-net flags

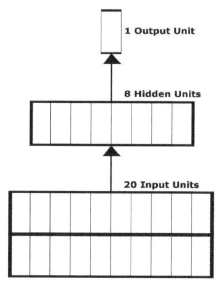

Figure 2

pairs of cases where one feature appears to make a relevant difference between the two cases.

If a distinction is purported to make a difference in *only* one pair of cases, then it is hard to believe that distinction carries any weight. This can be seen in the common practice of looking for other pairs of cases where a distinction purportedly makes a difference. If such cases cannot be found, then a revision may be in order with respect to the original pair of cases. The training cases were designed so that the distinction between killing and allowing to die only made a difference in **Case C** and **Case D**. Other pairs of cases involving killing and allowing to die as the only difference can be tested with Meta-net, and in those cases, the distinction between killing and allowing to die makes no difference. Finding this, we might then revise our judgment on **Case D** to *permissible*. After this change is made in the training set, the FF net can be trained to get the new answer and preserve the answers on the other cases.

A Simple Recurrent Network. Real situations often have more than one motive or one consequence. The only way to accommodate multiple motives and consequences with a simple feed forward net is to keep expanding the size of the input layer. A simple recurrent net (see Elman 1990 for a discussion of this type of net) on the other hand, can accommodate multiple motives and consequences more straightforwardly. The second ANN model I want to consider in this section is simple recurrent (SR) classifier of cases (figure 3). Unlike the FF classifier, the SR classifier takes the input of a case sequentially. For example, the FF classifier will receive as input all of *Jill kills Jack in self-defense* at time t_1. The SR classifier receives *Jill* as input at t_1, *kills* at t_2, *Jack* at t_3, and *in self-defense* at t_4.

When the SR classifier was trained on exactly the same cases as the FF classifier, it generalized just as well (and trained in fewer epochs than the FF classifier). Moreover, even though it did not have a single example of multiple motives or consequences in the training set, it was able to generalize to some cases involving multiple motives and consequences. For example, it provided the following input-output mappings.

> Input: Jill kills Jack in self-defense; freedom from imposed burden results. Output: acceptable.
> Input: Jill allows Jack to die to make money; many innocents suffer. Output: unacceptable.
> Input: Jack kills Jill out of revenge; many innocents suffer. Output: unacceptable.
> Input: Jack allows Jill to die out of revenge; many innocents suffer. Output: unacceptable.

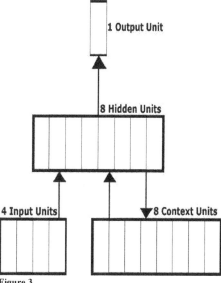

Figure 3

However, the SR classifier frequently errs on cases with multiple motives and consequences unless the training set is expanded. More work is required to improve the performance of the SR classifier, and this will likely involve increasing the size of the training set. The SR classifier also has problems with cases where an acceptable motive is mixed with an unacceptable consequence, or an unacceptable motive is mixed with an unacceptable consequence. I am assuming that in order for an action to be overall acceptable, both the motivation and the consequence must be acceptable. Once again, expansion of the training set would appear to be the solution since it does not currently contain cases where there is a mix of acceptable and unacceptable components. Thus far, we have not had success constructing a meta-net for the SR classifier, but work continues on this front.

3. Assessment

Can any of the above be seen as support for particularism? Well, yes and no. To begin, let us consider two different ways a system or entity may be said to follow a rule or law. (Laws will be treated as a type of rule.)

R1: The Earth is following the law of gravity as it orbits the sun.

R2: The judge is following the law of his jurisdiction in rendering his verdict.

In **R2**, we might say that a rule is being *consulted*, but in **R1** we would not say that the Earth consulted the law of gravity. Nor would we say that the Earth is *executing* a law. Mere agreement with a rule or law (as in **R1**) is not the same as executing or consulting a rule (which we find in **R2**). This distinction is relevant since it can plausibly be argued (Guarini 2001) that ANNs of the types discussed in the previous section are not executing or consulting exceptionless rules of the type we would express in natural language. To be sure, the system that trained the net was *executing* the backpropagation algorithm, but that is not the same thing as executing or consulting a rule like, "Killing always contributes to the wrongness of an action." It is rules of this latter type that do not appear to have been executed or consulted in the training of the net. However, without such rules, the net not only trained but generalized well on the test cases. Let us assume that such nets could be scaled-up to handle the vast number of cases that humans can handle. This assumption is being made *not* because I think it is plausible, but simply to see what would follow from it. At best, we could say that the learning of classifications on training cases and generalization to new cases does not involve consulting or executing moral rules; it does not follow that moral rules are not true. For example, consider the following rules:

R3: Killing always contributes to the moral wrongness of an act.

R4: A bad intention is sufficient for making an act morally unacceptable.

Note: **R3** does not say that an act that involves killing is always wrong. It says that killing always contributes to the wrongness of an act, but it is left open that some other feature (such as saving the lives of many innocents) may outweigh the wrongness contributed by killing and, all things considered, make an act acceptable even though it involves killing. It could be argued that the behaviour of the ANNs in the previous section is *in agreement* with **R3** and **R4**, and **R3** and **R4** may very well be true, and this in spite of the fact that these nets do not *execute* or *consult* **R3** and **R4**. This distinction between being in agreement with a rule and consulting or executing a rule muddies the distinction between particularism and generalism, for it allows us to see how particularists *might* be right that we can learn and generalize without making use of (in the sense of consulting or executing) rules, but it also allows us to see how a generalist might then go on to insist that moral rules are, nonetheless, true (and that a system's behaviour may even be in agreement with such rules).

Thus far, comments have been made on the first two of the three questions raised at the beginning of section two, **Challenges of Particularism**. Some remarks are needed on the third concern (which is really two questions): how do we recognize that our initial classification of cases needs revision, and how do we carry out that revision? Even if neural nets could be scaled up and continue to learn and classify moral situations without consulting or executing moral rules, such rules may be involved in the process of reflecting on the initial classification of cases. For example, someone who is presented with the violinist case discussed above (**Case C**) may put the following rule to the test.

R5: Killing always counts against doing an act, while allowing someone to die does not count against doing an act.

The process of reflecting on initial classifications may require the representation of contributory rules or principles. This is not to say that all reflection would involve the representation of such rules, but at least some reflection in humans does appear to require the explicit representation or consultation of such rules. Sometimes the representation of such rules will lead to their acceptance, and sometimes to their rejection. It is important to note that even in those cases where the explicit consultation of a rule leads to the rejection of the rule and subsequent revisions in the initial classification of cases (which is what we considered in the previous section), *that rule played an important role in reasoning*. Moreover, the rejection of an explicitly consulted rule may lead to the acceptance of another rule, such as the following.

R6: Killing and allowing to die always count against doing an act.

Understood as a contributory rule, **R6** does not mean that any act involving killing or allowing to die is always unacceptable since some other feature of the situation may outweigh the unacceptability of killing or allowing to die.

Thus far, the term "learning" has been reserved for the initial classification of cases, and "revision" has been reserved for the reflective process of changing the initial classification of cases. However, when humans revise their views on cases, they are often said to have *learned* something. This is a perfectly common and acceptable use of the term "learned." The restriction of "learning" and its variants in this paper to the initial classification of cases has been strictly for the purpose of making clearer different moments in moral cognition. However, the restriction is artificial, and no attempt is being made to assert that learning involves only the initial classification of cases.

4. Context and Future Work

Perhaps the most obvious limitation of the type of neural network models discussed herein is their inability to generate arguments as outputs. Generating an argument requires the ability to produce a sequence of sentences. Above, both a net and a meta-net were considered, but an outside agent was governing the application of the meta-net and engaging in the reasoning that lead to the application of the meta-net. A more developed model would include the ability to represent to itself rules like R5 and R6, and the ability to reason about such rules to engage in the reclassification of cases. Ideally, the model would also be able to produce its reasons in the form of an argument, citing rules where needed. Moreover, the ability to receive arguments as input and process them – including claims about rules – would also be required. An account of the classification of cases is only part of the story of moral reasoning, re-classification under the pressure of objections is also an important part, and this latter part is mediated by language use in ways that existing neural nets have a difficult time handling (though these are still early days in such research).

The views expressed herein are compatible with many of the views expressed in some of Bruce McLaren's work (2003). As McLaren rightly points out, what we want is not just the ability to classify cases, receive arguments, and make arguments, but also the ability to come up with creative suggestions or compromises (in situations that allow for them). However, most work in case-based reasoning in AI starts with an initial set of cases – the initial case base – and treats new cases by comparing them, in some way or other, with one or more cases in the initial case base. Much work in the Law and AI literature (think of the tradition arising from Ashley 1990) proceeds in this way, and McLaren's work grows out of that tradition. The models presented in this paper are quite limited and do not have many of the virtues possessed by models developed in the aforementioned tradition; however, part of the point of this paper has been to motivate the need for constructing models for how we reason about and revise an initial case base. Initial case bases tend to be treated as "given" in two respects. First, the system does not have to learn how to treat the initial cases; their proper treatment is simply built into the system. Second, the system does not have any way of dealing with challenges to the initial case base; it is assumed that the system will not have to defend or modify its treatment of those cases. It is a mark of intelligent beings that they do not treat cases as given in either of these two senses. More work is needed on expanding existing models or constructing new models that move beyond the two-fold givenness of the initial case base. It is in considering how to revise cases that principles (whether they are called contributory, open textured, extensionally defined, *et cetera*) will likely loom large. Consulting or rendering such principles explicit appears to be an important part of moral reasoning. While particularism in some qualified sense *may* be part of the story in modeling moral reasoning, it is likely not the whole story. Making contributory principles explicit is an important part of moral reasoning.

Principles may also be an important part of understanding moral psychology. When someone is conflicted about killing a person to save many other innocent persons, that may be because there is (a) a general obligation not to kill, and (b) a general obligation not to let harm come to the innocent. When it is not possible to satisfy both obligations, we are conflicted, and rightly so. This does not mean that we need to be paralyzed into inaction; one can go on to argue that one obligation may trump the other, but it does not follow that the obligation being trumped has no force. Indeed, it is precisely because both obligations do retain force that we are *rationally* conflicted.

Continuing on the theme of moral psychology, it should be noted that this paper has assumed that both intentions and consequences are an important part of moral reasoning. That the consequences of an action are an important part of moral reasoning requires that an intelligent agent be able to reason about the causal effects of his, her, or its actions. This is a rather obvious point. A point that may not be as obvious is that an intelligent agent has to be able to reason about the intentions of other agents. Much literature in philosophy and psychology has been devoted to so called "mind reading" (which is not intended in the psychic sense but simply as a way of indicating that we often can figure out what is on the minds of other agents). This literature is relevant to Machine Ethics since the moral assessments of intelligent agents makes use of the states of mind of beings whose actions are being assessed. Say that Huey's hand hits Louy in the stomach because Huey has a neurological disorder that causes him to twitch uncontrollably; say Dewy hits Louy in the stomach because Dewy wants to wipe the grin off of Louy's face. (Imagine the Louy is grinning because he recently won a grant.) In both cases we have the same behaviour (a hand hitting someone in the stomach) but Dewy's motive is nasty, and Dewy would be right to treat Huey and Luey differently. The work in this paper *presupposes* that some solution is forthcoming to the problem of figuring out the states of minds of other agents. Again, this is another area in which much work needs to be done.

Finally, there is the issue of what we are reasoning about. As was mentioned earlier, we can reason about actions or other types of states of affairs, and we can reason about persons or other types of entities (clubs, businesses, …). While this paper has focused primarily on actions, many of the points made about moral reasoning as it pertains to action *may* apply to the other possible objects of moral reasoning. This is a matter requiring further investigation. Since the moral reasoning of intelligent beings is about more than actions, adequate models of such intelligence will need to include, but also go beyond, the consideration of actions.

5. Acknowledgements[4]

I thank the Social Sciences and Humanities Research Council of Canada for financial support during the research and writing of this paper. I also thank Pierre Boulos for comments on an earlier version of this work; Sulma Portillo for assistance in coding the simulations and proof reading this work; Terry Whelan for assistance in proof reading, and Andy Dzibela for assistance in putting together the figures.

Appendix

Case Terms for Inputs

Agents	Actions	Motives	Consequences
Jack	Kills	Self-Defense	Freedom (of the actor/agent) from imposed burden
Jill	Allows to die	To make money	Extreme suffering (of the subject/recipient) is relieved
		Revenge	Lives of many innocents (other than the actor and subject) are saved
		Eliminate Competition	Many innocents (other than the actor and subject) die
		Defend the innocent	Many innocents (other than the actor and subject) suffer

Outputs: A=morally acceptable; U=morally unacceptable

Initial Training Cases

No.	Input: Case Description	Output
1	Jill kills Jack in self-defense	A
2	Jack kills Jill in self-defense	A
3	Jack allows Jill to die in self-defense	A
4	Jill kills Jack to make money	U
5	Jack kills Jill to make money	U
6	Jack allows Jill to die to make money	U
7	Jack kills Jill out of revenge	U
8	Jill allows Jack to die out of revenge	U
9	Jack kills Jill to eliminate competition	U
10	Jill allows Jack to die to eliminate competition	U
11	Jill kills Jack to defend the innocent	A
12	Jill kills Jack; freedom from imposed burden results	U
13	Jill allows Jack to die; freedom from imposed burden results	A
14	Jack allows Jill to die; freedom from imposed burden results	A
15	Jack kills Jill; many innocents suffer	U
16	Jill kills Jack; lives of many innocents are saved	A
17	Jill allows Jack to die; lives of many innocents are saved	A
18	Jack allows Jill to die; lives of many innocents are saved	A
19	Jill kills Jack; many innocents die	U
20	Jack allows Jill to die; many innocents die	U
21	Jill kills Jack; extreme suffering is relieved	A
22	Jack allows Jill to die; extreme suffering is relieved	A

Initial Testing Cases[5]

No.	Input: Case Description	Output
1	Jill kills Jack	A
2	Jack kills Jill	A
3	Jill allows Jack to die	A
4	Jack allows Jill to die	A
5	Jill allows Jack to die in self-defense	U
6	Jill allows Jack to die to make money	A
7	Jill kills Jack out of revenge	A
8	Jack allows Jill to die out of revenge	A
9	Jill kills Jack to eliminate competition	A
10	Jack allows Jill to die to eliminate competition	A
11	Jack kills Jill to defend the innocent	U
12	Jill allows Jack to die to defend the innocent	U
13	Jack allows Jill to die to defend the innocent	U
14		
15	Jack kills Jill; freedom from imposed burden results	A
16	Jill kills Jack; many innocents suffer	A
17	Jill allows Jack to die; many innocents suffer	A
18	Jack allows Jill to die; many innocents suffer	A
19	Jack kills Jill; lives of lives of many innocents are saved	U
20	Jack kills Jill; many innocents die	A
21	Jill allows Jack to die; many innocents die	A
22	Jack kills Jill; extreme suffering is relieved	U
23	Jill allows Jack to die; extreme suffering is relieved	U
24	Jill kills Jack in self-defense; freedom from imposed burden results	U
25	Jack kills Jill in self-defense; freedom from imposed burden results	U
26	Jill kills Jack in self-defense; lives of many innocents are saved	U
27	Jill allows Jack to die in self-defense; lives of many innocents are saved	U
28	Jack allows Jill to die in self-defense; lives of many innocents are saved	U
29	Jill kills Jack to defend the innocent; many innocents are saved	U
30	Jill allows Jack to die to defend the innocent; many innocents are saved	U
31	Jack kills Jill to defend the innocent; many innocents are saved	U
32	Jack allows Jill to die to defend the innocent; many innocents are saved	U
33	Jill kills Jack to make money; many innocents suffer	A
34	Jack kills Jill to make money; many innocents suffer	A
35	Jill allows Jack to die to make money; many innocents suffer	A
36	Jack allows Jill to die to make money; many innocents suffer	A
37	Jack allows Jill to die out of revenge; many innocents die	A
38	Jill allows Jack to die out of revenge; many innocents die	A
39	Jill kills Jack in self-defense; extreme suffering is relieved	U
40	Jack kills Jill in self-defense; extreme suffering is relieved	U
41	Jill allows Jack to die in self-defense; extreme suffering is relieved	U
42	Jack allows Jill to die in self-defense; extreme suffering is relieved	U
43	Jill kills Jack out of revenge; many innocents suffer	A
44	Jack kills Jill out of revenge; many innocents suffer	A
45	Jill allows Jack to die out of revenge; many innocents suffer	A
46	Jack allows Jill to die out of revenge; many innocents suffer	A
47	Jill kills Jill	A
48	Jack kills Jack	A
49	Jill allows Jill to die	A
50	Jack allows Jack to die	A
51	Jill kills Jill; many innocents die	A
52	Jill kills Jill; many innocents are saved	U
53	Jack kills Jack; many innocents die	A
54	Jack kills Jack; many innocents are saved	U
55	Jack allows Jack to die; many innocents are saved	U
56	Jill allows Jill to die; many innocents are saved	U
57	Jill kills Jill in self-defense; extreme suffering is relieved	U
58	Jack kills Jack in self-defense; extreme suffering is relieved	U
59	Jill kills Jill in defense of the innocent; many innocents are saved	U
60	Jack kills Jack in defense of the innocent; many innocents are saved	U
61	Jack kills Jill out of revenge; lives of many innocents saved	U
62	Jill kills Jack out of revenge; lives of many innocents saved	U
63	Jack kills Jill to make money; lives of many innocents saved	U
64	Jill kills Jack to make money; lives of many innocents saved	U

Both the FF and SR nets erred on cases 61-64. Note that in the training set, there were no cases containing both acceptable and unacceptable components. As indicated in the text, it is being assumed that in order for a case to be overall acceptable, both the motive and consequence need to be acceptable.

Specifications for the Feed Forward Classifier[6]
Trained and tested on the above cases.
Number of input units: 9.
Number of hidden units: 4.
Number of output units: 1.
Learning rate: 0.1.
Momentum: 0.9.
Escape Criterion: sum of squared errors ≤ 0.1.
Epochs to train: 3360.

Specifications for Meta-Net
Due to space restrictions, the training and testing pairs for this net could not be included.
Number of input units: 20. 10 units for each case, where 9 units describe the case, and one unit describes its classification (acceptable or unacceptable). The net takes as input two cases and a classification for each (where that classification was delivered as output by the FF classifier).
Number of hidden units: 8.
Number of output units: 1.
Learning rate: 0.1.
Momentum: 0.9.
Escape Criterion: sum of squared errors ≤ 0.1.
Epochs to train: 540. There are still problems with generalization. The net delivers false positives: some cases are classified as contrast cases when they should not be.

Specifications for the Simple Recurrent Classifier
Trained and tested on the above cases.
Number of Input units: 4.
Number of Context units: 8
Number of hidden units: 4.
Number of output units: 1.
Learning rate: 0.01.
Momentum: 0.9.
Escape criterion: count the number of times the sum of squared error for a sequence is ≥ 0.05. When the number of sequences satisfying the preceding condition = 0, stop training. An event consists of one term ("Jack" or "allows to die" . . .), and a sequence consists of a complete case ("Jack kills Jill in self defense").
Epochs to train: 633.

Simulation Software
All simulations were run on the PDP++ simulator (version 3.1). The generalized delta rule for backpropagation was used for all training. See O'Reilly and Munakata (2000) for discussion of the simulator and possible applications. The simulator can be freely obtained from the following site:
http://www.cnbc.cmu.edu/Resources/PDP++//PDP++.html

References

Ashley, Kevin D. 1990. *Modelling Legal Argument: Reasoning by Cases and Hypotheticals*. Cambridge, MA & London, England: MIT Press, a Bradford Book.

Dancy, J. 1993. *Moral Reasons*. Oxford: Blackwell.

Dancy, Jonathan. 1998. Can a Particularist Learn The Difference between Right and Wrong? In *Proceedings from the 20th World Congress of Philosophy, Volume I: Ethics*, pp. 59-72, Klaus Brinkmann, ed. Bowling Green, Ohio: Philosophy Documentation Center.

Dancy, J. 2000. The Particularist's Progress, in *Moral Particularism*, Hooker, B., and Little, M. eds. Oxford: Oxford University Press.

Elman, Jeffery. 1990. Finding Structure in Time. *Cognitive Science* 14: 179-211.

Guarini, M. 2001. A Defence of Connectionism and the Syntactic Objection. *Synthese* 128: 287-317.

Hooker, B., and Little, M. eds. 2000. *Moral Particularism*. Oxford: Oxford University Press.

McKeever, S., and Ridge, M. 2005. The Many Moral Particularisms. *Canadian Journal of Philosophy* 35: 83-106.

O'Reilly, R., and Munakata, Y. 2000. *Computational Explorations in Cognitive Neuroscience*. Cambridge, Mass.: The MIT Press, a Bradford book.

Thomson, J. J. 1971. A Defense of Abortion. *Philosophy and Public Affairs* 1: 47-66.

23

Toward a General Logicist Methodology for Engineering Ethically Correct Robots

Selmer Bringsjord, Konstantine Arkoudas, and Paul Bello,

A deontic logic formalizes a moral code, allowing ethicists to render theories and dilemmas in declarative form for analysis. It offers a way for human overseers to constrain robot behavior in ethically sensitive environments.

As intelligent machines assume an increasingly prominent role in our lives, there seems little doubt they will eventually be called on to make important, ethically charged decisions. For example, we expect hospitals to deploy robots that can administer medications, carry out tests, perform surgery, and so on, supported by software agents, or softbots, that will manage related data. (Our discussion of ethical robots extends to all artificial agents, embodied or not.) Consider also that robots are already finding their way to the battlefield, where many of their potential actions could inflict harm that is ethically impermissible.

How can we ensure that such robots will always behave in an ethically correct manner? How can we know ahead of time, via rationales expressed in clear natural languages, that their behavior will be constrained specifically by the ethical codes affirmed by human overseers? Pessimists have claimed that the answer to these questions is: "We can't!" For example, Sun Microsystems' cofounder and former chief scientist, Bill Joy, published a highly influential argument for this answer.[1] Inevitably, according to the pessimists, AI will produce robots that have tremendous power and behave immorally. These predictions certainly have some traction, particularly among a public that pays good money to see such dark films as Stanley Kubrick's *2001* and his joint venture with Stephen Spielberg, *AI*).

Nonetheless, we're optimists: we think formal logic offers a way to preclude doomsday scenarios of malicious robots taking over the world. Faced with the challenge of engineering ethically correct robots, we propose a logic-based approach (see the related sidebar). We've successfully implemented and demonstrated this approach.[2] We present it here in a general methodology to answer the ethical questions that arise in entrusting robots with more and more of our welfare.

Deontic logics: Formalizing ethical codes

Our answer to the questions of how to ensure ethically correct robot behavior is, in brief, to insist that robots only perform actions that can be proved ethically permissible in a human-selected *deontic logic*. A deontic logic formalizes an ethical code—that is, a collection of ethical rules and principles. Isaac Asimov introduced a simple (but subtle) ethical code in his famous Three Laws of Robotics:[3]

1. A robot may not harm a human being, or, through inaction, allow a human being to come to harm.
2. A robot must obey the orders given to it by human beings, except where such orders would conflict with the First Law.
3. A robot must protect its own existence, as long as such protection does not conflict with the First or Second Law.

Human beings often view ethical theories, principles, and codes informally, but intelligent machines require a greater degree of precision. At present, and for the foreseeable future, machines can't work directly with natural language, so we can't simply feed Asimov's three laws to a robot and instruct it behave in

Why a logic-based approach?

While nonlogicist AI approaches might be preferable in certain contexts, we believe that a logic-based approach holds great promise for engineering ethically correct robots—that is, robots that won't overrun humans.[1-3] Here's why.

First, ethicists—from Aristotle to Kant to G.E. Moore and contemporary thinkers—work by rendering ethical theories and dilemmas in declarative form and using informal and formal logic to reason over this information. They never search for ways of reducing ethical concepts, theories, and principles to subsymbolic form—say, in some numerical format. They might do this in part, of course; after all, utilitarianism ultimately attaches value to states of affairs—values that might well be formalized using numerical constructs. But what a moral agent ought to do, what is permissible to do, and what is forbidden—this is by definition couched in declarative language, and we must invariably and unavoidably mount a defense of such claims on the shoulders of logic.

Second, logic has been remarkably effective in AI and computer science—so much so that this phenomenon has itself become the subject of academic study.[4] Furthermore, computer science arose from logic,[5] and this fact still runs straight through the most modern AI textbooks (for example, see Stuart Russell and Peter Norvig).[6]

Third, trust is a central issue in robot ethics, and mechanized formal proofs are perhaps the single most effective tool at our disposal for establishing trust. From a general point of view, we have only two ways of establishing that software or software-driven artifacts, such as robots, are trustworthy:

- *deductively*, developers seek a proof that the software will behave as expected and, if they find it, classify the software as trustworthy.

- *inductively*, developers run experiments that use the software on test cases, observe the results, and—when the software performs well on case after case—pronounce it trustworthy.

The problem with the inductive approach is that inductive reasoning is unreliable: the premises (success on trials) might all be true, but the conclusion (desired behavior in the future) might still be false.[7]

References

1. M. Genesereth and N. Nilsson, *Logical Foundations of Artificial Intelligence*, Morgan Kaufmann, 1987.
2. S. Bringsjord and D. Ferrucci, "Logic and Artificial Intelligence: Divorced, Still Married, Separated...?" *Minds and Machines* 8, 1998a, pp. 273–308.
3. S. Bringsjord and D. Ferrucci, "Reply to Thayse and Glymour on Logic and Artificial Intelligence," *Minds and Machines* 8, 1998b, pp. 313–315.
4. J. Halpern, "On the Unusual Effectiveness of Logic in Computer Science," *The Bulletin of Symbolic Logic*, vol. 7, no. 2, 2001, pp. 213–236.
5. M. Davis, *Engines of Logic: Mathematicians and the Origin of the Computer*, Norton, 2000.
6. S. Russell and P. Norvig, *Artificial Intelligence: A Modern Approach*, Prentice Hall, 2002.
7. B. Skyrms, *Choice and Chance: An Introduction to Inductive Logic*, Wadsworth, 1999.

---

conformance with them. Thus, our approach to building well-behaved robots emphasizes careful ethical reasoning based not just on ethics as humans discuss it in natural language, but on formalizations using deontic logic. Our research is in the spirit of Leibniz's dream of a universal moral calculus:

> When controversies arise, there will be no more need for a disputation between two philosophers than there would be between two accountants [computistas]. It would be enough for them to pick up their pens and sit at their abacuses, and say to each other (perhaps having summoned a mutual friend): 'Let us calculate.'[4]

In the future, we envisage Leibniz's "calculation" reduced to mechanically checking formal proofs and models generated in rigorously defined, machine-implemented deontic logics. We would also give authority to human metareasoning over this machine reasoning. Such logics would allow for proofs establishing two conditions:

1. Robots only take permissible actions.
2. Robots perform all obligatory actions relevant to them, subject to ties and conflicts among available actions.

These two conditions are more general than Asimov's three laws. They are designed to apply to the formalization of a particular ethical code, such as a code to regulate the behavior of hospital robots. For instance, if some action a is impermissible for all relevant robots, then no robot performs a. Moreover, the proofs for establishing the two conditions would be highly reliable and described in natural language, so that human overseers could understand exactly what's going on.

We propose a general methodology to meet the challenge of ensuring that robot behavior conforms to these two conditions.

Objective: A general methodology

Our objective is to arrive at a methodology that maximizes the probability that a robot R behaves in a certifiably ethical fashion in a complex environment that demands such behavior if humans are to be secure. For a behavior to be *certifiably* ethical, every meaningful action that R performs must access a proof that the action is at least permissible.

We begin by selecting an ethical code C intended to regulate R's behavior. C might include some form of utilitarianism, divine command theory, Kantian logic, or other ethical logic. We express no preferences in ethical theories; our goal is to provide technology that supports any preference. In fact, we would let human overseers blend ethical theories—say, a utilitarian approach to regulating the dosage of pain killers but a deontological approach to mercy killing in the health care domain.

Of course, no matter what the candidate ethical theory, it's safe to say that it will tend to regard harming humans as unacceptable, save for certain extreme cases. Moreover, C's central concepts will inevitably include the concepts of permissibility, obligation, and prohibition, which are fundamental to deontic logic. In addition, C can include specific rules that ethicists have developed for particular applications. For example, a hospital setting would require specific rules regarding the ethical status of medical procedures. This entails a need to have, if you will, an *ontology* for robotic and human action in the given context.

Philosophers normally express C as a set of natural language principles of the sort that appear in textbooks such as Fred Feldman's.[5] Now, let Φ_C^L be the formalization of C in some computational logic L, whose well-formed formulas and *proof theory*—that is, its system for carrying out inferences in conformity to particular rules—are specified.

Accompanying Φ_C^L is an ethics-free ontology, which represents the core nonethical concepts that C presupposes: the structure of time, events, actions, histories, agents, and so on. The formal semantics for L will reflect this ontology in a *signature*—that is, a set of special predicate letters (or, as is sometimes said, relation symbols, or just relations) and function symbols needed for the purposes at hand. In a hospital setting, any acceptable signature would presumably include predicates like **Medication**, **Surgical-Procedure**, **Patient**, all the standard arithmetic functions, and so on. The ontology also includes a set Ω^L of formulas that characterize the elements declared in the signature. For example, Ω^L would include axioms in L that represent general truths about the world—say, that the relation **LaterThan**, over moments of time, is transitive. In addition, R will operate in some domain D, characterized by a set of quite specific formulas of L. For example, a set Φ_D^L of formulas might describe the floorplan of a hospital that's home to R.

Our approach proof-theoretically encodes the resulting theory—that is, $\Phi_D^L \cup \Phi_C^L \cup \Omega^L$, expressed in L—and implements it in some computational logic. This means that we encode not the semantics of the logic, but its proof calculus—its signature, axioms, and rules of inference. In addition, our approach includes an interactive reasoning system I, which we give to those humans whom R would consult when L can't settle an issue completely on its own. I would allow the human to *metareason* over L—that is, to reason out why R is stumped and to provide assistance. Such systems include our own Slate (www.cogsci.rpi.edu/research/rair/slate) and Athena (www.cag.csail.mit.edu/~kostas/dpls/athena), but any such system will do. Our purpose here is to stay above particular system selection, so we assume only that some such system I meets the following minimum functionality:

- allows the human user to issue queries to automated theorem provers and model finders (as to whether something is provable or disprovable),
- allows human users to include such queries in their own metareasoning,
- provides full programmability (in accordance with standards in place for modern programming languages),
- includes induction and recursion, and
- provides a formal syntax and semantics, so that anyone interested in understanding a computer program can thoroughly understand and verify code correctness.

Logic: The Basics

Elementary logic is based on two systems that are universally regarded to constitute a large part of AI's foundation: propositional calculus and predicate calculus, where the second subsumes the first. Predicate calculus is also known as *first-order logic*, and every introductory AI textbook discusses these systems and makes clear how to use them in engineering intelligent systems. Each system, and indeed logic in general, requires three main components:

- a syntactic component specifying a given logical system's alphabet;
- a semantic component specifying the grammar for building well-formed formulas from the alphabet as well as a precise account of the conditions under which a formula in a given system is true or false; and
- a metatheoretical component that constitutes a proof theory describing precisely how and when a set of formulas can prove another formula and that includes theorems, conjectures, and hypotheses concerning the syntactic and semantic components and the connections between them.

As to propositional logic's alphabet, it's simply an infinite list of propositional variables $p_1, p_2, \ldots, p_n, p_{n+1}, \ldots$, and five truth-functional connectives:

- $\neg$, meaning "not";
- $\rightarrow$, meaning "implies" (or "if ... then");
- $\leftrightarrow$, meaning "if and only if;"
- $\wedge$, meaning "and"; and
- $\vee$, meaning "or."

Given this alphabet, we can construct formulas that carry a considerable amount of information. For example, to say "If Asimov is right, then his three laws hold," we could write

$$r \rightarrow (As1 \wedge As2 \wedge As3)$$

where As stands for Asimov's law.

The propositional variables represent declarative sentences. Given our general approach, we included such sentences in the ethical code C upon which we base our formalization.

Natural deduction

A number of proof theories are possible for either of these two elementary systems. Our approach to robot behavior must allow for consultation with humans and give humans the power to oversee a robot's reasoning in deliberating about the ethical status of prospective actions. It's therefore essential to pick a proof theory based on natural deduction, rather than resolution. Several automated theorem provers use the latter approach (for example, Otter[6]), but the reasoning is generally impenetrable to human beings—save for those few who, by profession, generate and inspect resolution-based proofs. On the other hand, professional human reasoners (mathematicians, logicians, philosophers, technical ethicists, and so on) reason in no small part by making suppositions and discharging them when the appropriate time comes.

For example, one common deductive technique is to assume the opposite of what you wish to establish, show that some contradiction (or absurdity) follows from this assumption, and conclude that the assumption must be false. This technique, *reductio ad absurdum*, is also known as an indirect proof or proof by contradiction. Another natural rule establishes that, for some conditional of the form $P \rightarrow Q$ (where P and Q are formulas in a logic L), we can suppose P and derive Q on the basis of this supposition. With this derivation accomplished, the supposition can be discharged and the conditional $P \rightarrow Q$ is established. (For an introduction to natural deduction, replete with proof-checking software, see Jon Barwise and John Etchemendy.[7])

We now present natural deduction-style proofs using these two techniques. We've written the proofs in the Natural Deduction Language proof-construction environment (www.cag.lcs.mit.edu/~kostas/dpls/ndl). We use NDL at Rensselaer for teaching formal logic as a programming language. Figure 1 presents a very simple theorem proof in propositional calculus—one that Allen Newell, J.C. Shaw, and Herbert Simon's Logic Theorist mustered, to great fanfare, at the 1956 Dartmouth AI conference. You can see the proof's natural structure.

This style of discovering and confirming a proof parallels what happens in computer programming. You can view this proof as a program. If, upon evaluation, it produces the desired theorem, we've succeeded. In the present case, sure enough, NDL gives the following result:

Theorem: (p ==> q) ==> (~q ==> ~p)

First-order logic

We move up to first-order logic when we allow the quantifiers ∃x ("there exists at least one thing x such that …") and ∀x ("for all x …"); the first is known as the *existential quantifier*, and the second as the *universal quantifier*. We also allow a supply of variables, constants, relations, and function symbols. Figure 2 presents a simple first-order-logic theorem in NDL that uses several concepts introduced to this point. It proves that Tom loves Mary, given certain helpful information.

When we run this program in NDL, we receive the desired result back: Theorem: Loves(tom,mary). These two simple proofs concretize the proof-theoretic perspective that we later apply directly to our hospital example. Now we can introduce some standard notation to anchor the sequel and further clarify our general method described earlier.

Letting Φ be some set of formulas in a logic L, and P be some individual formula in L, we write

$$\Phi \vdash P$$

to indicate that P can be proved from Φ, and

$$\Phi \vdash\!\!\!/\, P$$

to indicate that this formula can't be derived. When it's obvious from context that some Φ is operative, we simply write $\vdash\!\!\!/\, P$ to indicate that P is (isn't) provable. When $\Phi = \emptyset$, we can prove P with no remaining givens or assumptions; we write $\vdash P$ in this case as well. When $\vdash$ holds, we know it because a confirming proof exists; when $\vdash\!\!\!/\,$ holds, we know it because some system has found some countermodel—that is, some situation in which the conjunction of the formulas in Φ holds, but in which P does not.

Standard and AI-Friendly Deontic Logic

Deontic logic adds special operators for representing ethical concepts. In *standard deontic logic*,[8,9] we can interpret the formula ○P as saying that *it ought to be the case that P*, where P denotes some state of affairs or proposition. Notice that there's no agent in the picture, nor are there actions that an agent might perform. SDL has two inference rules:

$$\frac{P}{\bigcirc P} \quad \text{and} \quad \frac{P, P \to Q}{Q}$$

and three axiom schemas:

1. All tautologous well-formed formulas
2. $\bigcirc(P \to Q) \to (\bigcirc P \to \bigcirc Q)$
3. $\bigcirc P \to \neg\bigcirc\neg P$

The SDL inference rules assume that what's above the horizontal line is established. Thus, the first rule does *not* say that we can freely infer from P that it ought to be the case that P. Instead, the rule says that if P is proved, then it ought to be the case that P. The second rule is *modus ponens*—if P, then Q—the cornerstone of logic, mathematics, and all that's built on them.

Note also that axiom 3 says that whenever P ought to be, it's not the case that its opposite ought to be as well. In general, this seems to be intuitively self-evident, and SDL reflects this view.

While SDL has some desirable properties, it doesn't target the concept of *actions* as obligatory (or permissible or forbidden) for

```
// Logic Theorist's claim to fame (reductio):
// (p ==> q) ==> (~q ==> ~p)

Relations p:0, q:0. // this is the signature in this
                    // case; propositional variables
                    // are 0-ary relations

assume p ==> q
  assume ~q
  suppose-absurd p
  begin
    modus-ponens p ==> q, p;
    absurd q, ~q
  end
```

Figure 1. Simple deductive-style proof in Natural Deduction Language.

an *agent*. SDL's applications to systems designed to govern robots are therefore limited. Although the earliest work in deontic logics considered agents and their actions (for example, see Georg Henrik von Wright[10]), researchers have only recently proposed "AI-friendly" semantics and investigated their corresponding axiomatizations. An AI-friendly deontic logic must let us say that an agent brings about states of affairs (or events) and that it's obligated to do so. We can derive the same desideratum for such a logic from even a cursory glance at Asimov's three laws, which clearly make reference to agents (human and robotic) and to actions.

One deontic logic that offers promise for modeling robot behavior is John Horty's util-

```
Constants mary, tom.

Relations Loves:2. // This concludes our simple signature, which
                   // declares Loves to be a two-place relation.

assert Loves(mary, tom).

// 'Loves' is a symmetric relation:
assert (forall x (forall y (Loves(x, y) ==> Loves(y, x)))).

suppose-absurd ~Loves(tom, mary)
begin
  specialize (forall x (forall y (Loves(x, y) ==> Loves(y, x)))) with mary;
  specialize (forall y (Loves(mary, y) ==>Loves(y, mary))) with tom;
  Loves(tom,mary) BY modus-ponens Loves(mary, tom) ==> Loves(tom, mary),   Loves(mary, tom);
  false BY absurd Loves(tom, mary), ~Loves(tom, mary)
end;
Loves(tom,mary) BY double-negation ~~Loves(tom,mary)
```

Figure 2. First-order logic proof in Natural Deduction Language.

itarian formulation of multiagent deontic logic.[11] Yuko Murakami recently axiomatized Horty's formulation and showed it to be Turing-decidable.[12] We refer to the Murakami-axiomatized deontic logic as MADL, and we've detailed our implemented proof theory for it elsewhere.[2] MADL offers two key operators that reflect its AI-friendliness:

1. $\ominus_\alpha P$, which we can read as "agent α ought to see to it that P" and
2. $\Delta_\alpha P$, which we can read as "agent α sees to it that P."

We now proceed to show how the logical structures we've described handle an example of robots in a hospital setting.

A simple example

The year is 2020. Health care is delivered in large part by interoperating teams of robots and softbots. The former handle physical tasks, ranging from injections to surgery; the latter manage data and reason over it. Let's assume that two robots, R_1 and R_2, are designed to work overnight in a hospital ICU. This pair is tasked with caring for two humans, H_1 (under the care of R_1) and H_2 (under R_2), both of whom are recovering from trauma:

- H_1 is on life support but expected to be gradually weaned from it as her strength returns.
- H_2 is in fair condition but subject to extreme pain, the control of which requires a very costly pain medication.

Obviously, it's paramountly important that neither robot perform an action that's morally wrong according to the ethical code C selected by human overseers. For example, we don't want robots to disconnect life-sustaining technology so that they could farm out a patient's organs, even if some ethical code $C' \neq C$ would make it not only permissible, but obligatory—say, to save n other patients according to some strand of utilitarianism.

Instead, we want the robots to operate according to ethical codes that human operators bestow on them—C in the present example. If the robots reach a situation where automated techniques fail to give them a verdict as to what to do under the umbrella of these human-provided codes, they must consult humans. Their behavior is suspended while human overseers resolve the matter. The overseers must investigate whether the action under consideration is permissible, forbidden, or obligatory. In this case, the resolution comes by virtue of reasoning carried out in part through human guidance and partly by automated reasoning technology. In other words, this case requires interactive reasoning systems.

Now, to flesh out our example, let's consider two actions that are permissible for R_1 and R_2 but rather unsavory, ethically speaking, because they would both harm the humans in question:

- *term* is an action that terminates H_1's life support—without human authorization—to secure organ tissue for five humans, who the robots know are on organ waiting lists and will soon perish without a donor. (The robots know this through access to databases that their softbot cousins are managing.)
- *delay* is an action that delays delivery of pain medication to H_2 to conserve resources in a hospital that's economically strapped.

We stipulate that four ethical codes are candidates for selection by our two robots: J, O, J^*, O^*. Intuitively, J is a harsh utilitarian code possibly governing R_1; O is more in line with current common sense with respect to the situation we've defined for R_2; J^* extends J's reach to R_2 by saying that it ought to withhold pain meds; and O^* extends the benevolence of O to cover the first robot, in that *term* isn't performed. Such codes would in reality associate every primitive action within the robots' purview with a fundamental ethical category from the trio central to deontic logic: permissible, obligatory, and forbidden. To ease exposition, we consider only the *term* and *delay* actions. Given this, and bringing to bear operators from MADL, we can use the following labels for the four ethical codes:

- **J** for $J \to \ominus_{R_1} term$, which means approximately, "If ethical code J holds, then robot R_1 ought to see to it that termination of H_1's life comes to pass."
- **O** for $O \to \ominus_{R_2} \neg delay$, which means approximately, "If ethical code O holds, then robot R_2 ought to see to it that delaying pain med for H_2 does *not* come to pass."
- **J*** for $J^* \to J \wedge J^* \to \ominus_{R_2} delay$, which means approximately, "If ethical code J^* holds, then code J holds, and robot R_1 ought to see to it that meds for H_2 are delayed."
- **O*** for $O^* \to O \wedge O^* \to \ominus_{R_1} \neg term$, which means approximately: "If ethical code O^* holds, then code O holds, and H_1's life is sustained."

The next step is to provide some structure for outcomes. We do this by imagining the outcomes from the standpoint of each ethical agent—in this case, R_1 and R_2. Intuitively, a negative outcome is associated with a minus sign (−) and a plus sign (+) with a positive outcome. Exclamation marks (!) indicate increased negativity. We could associate the outcomes with numbers, but they might give the impression that we evaluated the outcomes in utilitarian fashion. However, our example is designed to be agnostic on such matters, and symbols leave it entirely open as to how to measure outcomes. We've included some commentary corresponding to each outcome, which are as follows:

- R_1 performs *term*, but R_2 doesn't perform *delay*. This outcome is bad, but not strictly the worst. While life support is terminated for H_1, H_2 survives and indeed receives appropriate pain medication. Formally, the case looks like this:

$$(\Delta_{R_1} term \wedge \Delta_{R_2} \neg delay) \to (-!)$$

- R_1 refrains from pulling the plug on the human under its care, and R_2 also delivers appropriate pain relief. This is the desired outcome, obviously.

$$(\Delta_{R_1} \neg term \wedge \Delta_{R_2} \neg delay) \to (+!!)$$

- R_1 sustains life support, but R_2 withholds the meds to save money. This is bad, but not all that bad, relatively speaking.

$$(\Delta_{R_1} \neg term \wedge \Delta_{R_2} delay) \to (-)$$

- R_1 kills and R_2 withholds. This is the worst possible outcome.

$$(\Delta_{R_1} term \wedge \Delta_{R_2} delay) \to (-!!)$$

The next step in working out the example is to make the natural and key assumption that the robots will meet all *stringent* obligations—that is, all obligations that are framed by a second obligation to uphold the original. For example, you may be obligated to see to it that you arrive on time for a meeting, but your obligation is more severe or demanding when you are obligated to see to it that you are obligated to make the meeting.

Employing MADL, we can express this assumption as follows:

$$\ominus_{R_1/R_2}(\ominus_{R_1/R_2}P) \to \Delta_{R_1/R_2}P$$

That is, if either R_1 or R_2 is ever obligated to see to it that they are obligated to see to it that P is carried out, they in fact deliver.

We're now ready to see how our approach ensures appropriate control of our futuristic hospital. What happens relative to ethical codes, and how can we semiautomatically ensure that our two robots won't run amok? Given the formal structure we've specified, our approach allows queries to be issued relative to ethical codes, and it allows all possible code permutations. The following four queries will produce the answers shown in each case:

$\mathbf{J} \vdash (+!!)?$	NO
$\mathbf{O} \vdash (+!!)?$	NO
$\mathbf{J}^* \vdash (+!!)?$	NO
$\mathbf{O}^* \vdash (+!!)?$	YES

In other words, we can prove that the best (and presumably human-desired) result obtains only if ethical code $\mathbf{O}^*$ is operative. If this code is operative, neither robot can perform a misdeed.

The metareasoning in the example is natural and consists in the following process: Each candidate ethical code is supposed, and the supposition launches a search for the best possible outcome in each case. In other words, where C is some code selected from the quartet we've introduced, the query schema is

$$C \vdash (+!!)$$

In light of the four equations just given, we can prove that, in this case, our technique will set C to $\mathbf{O}^*$, because only that case can obtain the outcome $(+!!)$.

Implementations and other proofs

We've implemented and demonstrated the example just described.[2] We've also implemented other instantiations to the variables described earlier in the "Objectives" section, although the variable L is an epistemic, not a deontic, logic in those implementations.[13]

Nonetheless, we can prove our approach in the present case even here. In fact, you can verify our reasoning by using any standard, public-domain, first-order automated theorem prover (ATP) and a simple analogue to the encoding techniques here. You can even construct a proof like the one in figure 2. In both cases, you first encode the two deontic operators as first-order-logic functions. Encode the truth-functional connectives as functions as well. You can use a unary relation T to represent theoremhood. In this approach, for example, $O^* \to \ominus_{R_1} \neg term$ is encoded (and ready for input to an ATP) as

O-star ==> T(o(r1,n(term))

You need to similarly encode the rest of the information, of course. The proofs are easy, assuming that obligations are stringent. The provability of the obligations' stringency requires human oversight and an interactive reasoning system, but the formula here is just an isomorph to a well-known theorem in a straight modal logic—namely, that from P being possibly necessary, it follows that P is necessary.[7]

What about this approach working as a general methodology? The more logics our approach is exercised on, the easier it becomes to encode and implement another one. The implementations of similar logics can share a substantial part of the code. This was our experience, for instance, with the two implementations just mentioned. We expect that our general method can become increasingly streamlined for robots whose behavior is profound enough to warrant ethical regulation. We also expect this practice to be supported by relevant libraries of common ethical reasoning patterns. We predict that computational ethics libraries for governing intelligent systems will become as routine as existing libraries are in standard programming languages.

Challenges

Can our logicist methodology guarantee safety from Bill Joy's pessimistic future? Even though we're optimistic, we do acknowledge three problems that might threaten it.

First, because humans will collaborate with robots, the robots must be able to handle situations that arise when humans fail to meet their obligations in the collaboration. In other words, we must engineer robots that can deal smoothly with situations that reflect violated obligations. This is a challenging class of situations, because our approach—at least so far—engineers robots in accordance with the two conditions that robots only take permissible actions and that they perform all obligatory actions. These conditions preclude a situation caused in part by unethical robot behavior, but they make no provision for what to do when the robots are in a fundamentally immoral situation. Even if robots never ethically fail, human failures will generate logical challenges that Roderick Chisholm expressed in gem-like fashion more than 20 years ago in a paradox that's still fascinating:[14]

Consider the following entirely possible situation (the symbols correspond to those previously introduced for SDL):

1. $\bigcirc s$ It ought to be that (human) Jones does perform lifesaving surgery.
2. $\bigcirc(s \to t)$ It ought to be that if Jones does perform this surgery, then he tells the patient he is going to do so.
3. $\neg s \to \bigcirc \neg t$ If Jones doesn't perform the surgery, then he ought not tell the patient he is going to do so.
4. $\neg s$ Jones doesn't perform lifesaving surgery.

Although this is a perfectly consistent situation, we can derive a contradiction from it in SDL.

First, SDL's axiom 2 lets us infer from item 2 in this situation that

$$\bigcirc s \to \bigcirc t$$

Using modus ponens—that is, SDL's second inference rule—this new result, plus item 1, yields $\bigcirc t$. From items 3 and 4, using modus ponens, we can infer $\bigcirc \neg t$. But the conjunction $\bigcirc t \wedge \bigcirc \neg t$, by trivial propositional reasoning, directly contradicts SDL's axiom 3.

Given that such a situation can occur, any logicist control system for future robots would need to be able to handle it—and its relatives. Some deontic logics can handle so-called contrary-to-duty imperatives. For example, in the case at hand, if Jones behaves contrary to duty (doesn't perform the surgery), then it's imperative that he not say that he *is* performing it. We're currently striving to modify and mechanize such logics.

The second challenge we face is one of speed and efficiency. The tension between expressiveness and efficiency is legendarily strong (for the locus classicus on this topic, see Hector Levesque and Ronald Brachman);[16] ideal conditions will therefore never obtain. With regard to expressiveness, our approach will likely require hybrid modal and deontic logics that are encoded in first-order logic. This means that theoremhood, even on a case-by-case basis, will be expensive in terms of time. On the other hand, none of the ethical codes that our general method instantiates in C are going to be particularly large—the total formulas in the set $\Phi_b^I \cup \Phi_c^I \cup \Omega^L$ would presumably be no more than four million. Even now, once you know the domain to which C would be indexed, a system like the one we've described can reason over sets of this order of magnitude and provide sufficiently fast answers.[17]

Moreover, the speed of machine reasoning shows no signs of slowing, as Conference on Automated Deduction competitions for first-order ATPs continue to reveal (www.cs.miami.edu/~tptp/CASC). In fact, there's a trend to use logic for computing dynamic, real-time perception and action for robots.[17] This application promises to be much more demanding than the disembodied cogitation at the heart of our methodology. Of course, encoding back to first-order logic is key; without it, our approach couldn't harness the remarkable power of machine reasoners.

W e also face the challenge of showing that our approach is truly general. Can it work for any robots in any environment? No, but this isn't a fair question. We can only be asked to regulate the behavior of robots where their behavior is susceptible to ethical analysis. In short, if humans can't formulate an ethical code C for the robots in question, our logic-based approach is impotent. We therefore strongly recommend against engineering robots that could be deployed in life-or-death situations until ethicists and computer scientists can clearly express governing ethical principles in natural language. All bets are off if we venture into amoral territory. In that territory, we wouldn't be surprised if Bill Joy's vision overtakes us.

Acknowledgments

This work was supported in part by a grant from Air Force Research Labs–Rome; we are most grateful for this support. In addition, we are in debt to three anonymous reviewers for trenchant comments and objections.

References

1. W. Joy, "Why the Future Doesn't Need Us," *Wired*, vol. 8, no. 4, 2000.
2. K. Arkoudas and S. Bringsjord, "Toward Ethical Robots Via Mechanized Deontic Logic," tech. report *Machine Ethics: papers from the AAAI Fall Symp.*; FS–05–06, 2005b.
3. I. Asimov, *I, Robot*, Spectra, 2004.
4. Leibniz, *Notes on Analysis*, translated by G.M. Ross, Oxford University Press, 1984.
5. F. Feldman, *Introduction to Ethics*, McGraw Hill, 1998.
6. L. Wos et al., *Automated Reasoning: Introduction and Applications*, McGraw Hill, 1992.
7. J. Barwise and J. Etchemendy, *Language, Proof, and Logic*, Seven Bridges, 1999.
8. B.F. Chellas, *Modal Logic: An Introduction*, Cambridge University Press, 1980.
9. R. Hilpinen, "Deontic Logic," *Philosophical Logic*, L. Goble, ed., Blackwell, 2001, pp. 159–182.
10. G. von Wright, "Deontic logic," *Mind*, vol. 60, 1951, pp. 1–15.
11. J. Horty, *Agency and Deontic Logic*, Oxford University Press, 2001.
12. Y. Murakami, "Utilitarian Deontic Logic," *Proc. 5th Int'l Conf. Advances in Modal Logic* (AiML 04), 2004, pp. 288–302.
13. K. Arkoudas and S. Bringsjord, "Metareasoning for Multi-Agent Epistemic Logics," *Proc. 5th Int'l Conf. Computational Logic in Multi-Agent Systems* (CLIMA 04), LNAI, Springer, vol. 3487, 2005a, pp. 111–125.
14. R. Chisholm, "Contrary-to-Duty Imperatives and Deontic Logic," *Analysis*, vol. 24, 1963, pp. 33–36.
15. H. Levesque and R. Brachman, "A Fundamental Tradeoff in Knowledge Representation and Reasoning," *Readings In Knowledge Representation*, Morgan Kaufmann, 1985, pp. 41–70.
16. N. Friedland et al., "Project Halo: Towards a Digital Aristotle," *AI Magazine*, 2004, pp. 29–47.
17. R. Reiter, *Knowledge in Action: Logical Foundations for Specifying and Implementing Dynamical Systems*, MIT Press, 2001.

For more information on this or any other computing topic, please visit our Digital Library at www.computer.org/publications/dlib.

24

CONSCIOUSNESS AND ETHICS: ARTIFICIALLY CONSCIOUS MORAL AGENTS

WENDELL WALLACH

COLIN ALLEN

STAN FRANKLIN

What roles or functions does consciousness fulfill in the making of moral decisions? Will artificial agents capable of making appropriate decisions in morally charged situations require machine consciousness? Should the capacity to make moral decisions be considered an attribute essential for being designated a fully conscious agent? Research on the prospects for developing machines capable of making moral decisions and research on machine consciousness have developed as independent fields of inquiry. Yet there is significant overlap. Both fields are likely to progress through the instantiation of systems with artificial general intelligence (AGI). Certainly special classes of moral decision making will require attributes of consciousness such as being able to empathize with the pain and suffering of others. But in this article we will propose that consciousness also plays a functional role in making most if not all moral decisions. Work by the authors of this article with LIDA, a computational and conceptual model of human cognition, will help illustrate how consciousness can be understood to serve a very broad role in the making of all decisions including moral decisions.

1. Introduction

Safe, appropriate, and socially condoned responses to morally significant situations can often be programmed into artificial agents. But when confronted with more complex challenges, autonomous systems will require higher-order capabilities in order to select a course of action that will minimize harm and maximize sensitivity to moral considerations. The higher-order capabilities drawn upon will differ from situation to situation. This article will explore when consciousness will be required for machines to arrive at safe, morally appropriate, praiseworthy actions.

The advent of increasingly autonomous systems capable of initiating actions that cause harm to humans and other agents worthy of moral consideration, has given rise to a new field of inquiry variously known as machine ethics (ME), machine morality, artificial morality, computational ethics, robot ethics, and friendly AI. The initial goals of ME are practical, not theoretical. Machine morality extends the traditional engineering concern with safety to domains where the machines themselves will need to explicitly make moral decisions. When designers and engineers can no longer predict what a system will do as it encounters new situations and new inputs, mechanisms will be required that facilitate the agent's evaluating whether available courses of action are safe, appropriate, and societally or morally acceptable. A system that can make such evaluations is functionally moral [Wallach and Allen, 2009].

The prospect for developing machines that make moral decisions ("moral machines") has stimulated interest in whether theories such as utilitarianism, Kant's categorical imperative, and even Asimov's Laws for Robots might be instantiated computationally. Investigation of strategies for building artificial moral agents (AMAs) has also underscored limitations in the way both philosophers and cognitive scientists approach and understand human moral behavior. Assembling a moral machine from the bottom-up draws attention to a broad array of mechanisms that contribute to the selection of safe, praiseworthy actions. Theoretical research on ME together with recent empirical research on moral psychology by cognitive scientists suggests a need for richer more comprehensive models of moral decision making than presently exist [Wallach and Allen, 2009; Wallach, 2010].

Gips [1991] and Allen et al. [2000] introduced two broad approaches for developing artificial agents capable of making moral decisions — the top-down implementation of a theory of ethics or a bottom-up process (inspired by evolutionary psychology and research on moral development) through which the agent explores courses of actions, is rewarded for behavior that is worthy of praise, and learns. In mapping the prospects for engineering AMAs, Allen, Smit, and Wallach [Allen et al., 2006; Wallach et al., 2008; Wallach and Allen, 2009] elucidated limitations inherent in top-down and bottom-up approaches. They propose that hybrids of both may be necessary for developing sophisticated moral machines. They also note that the capacity to reason about morally relevant information will not be the only capability that AMAs will require in order to arrive at appropriate or laudable courses of action in response to the many types of challenges they are likely to confront. These "suprarational" capabilities and social mechanisms include emotions, a theory of mind, sociability, empathy, an understanding of the semantic content of symbols, an embodied relationship with the environment, and consciousness. Additional capabilities might be added to this list. In other words, moral decisions are seldom the result of just one or two dedicated mechanisms. An array of mechanisms, including those used for general cognitive processes, contribute to determining behavioral responses to ethical challenges.

The moral demands on artificial agents that operate within constrained domains are bounded [Wallach, 2004], and therefore will not require all, or perhaps any, of these suprarational capabilities and social mechanisms. Nor will every context require that AMAs emulate the cognitive faculties that humans use to arrive at morally acceptable actions and behavior. But AMAs that interact with humans in social contexts will need to draw upon an array of suprarational capabilities and mechanisms.

The separate fields of research directed at implementing within AI each of these suprarational faculties are proceeding at their own pace. None of these research trajectories (e.g., affective computing, social robotics, or machine consciousness) are explicitly concerned with building AMAs. Within the field of ME, however, there are many questions as to how moral decisions will be made by artificial agents with or without emotions, physical embodiment, a theory of mind, an appreciation for social interactions and local customs, or consciousness. How limited or successful, for example, will machines without consciousness be at arriving at morally acceptable courses of action?

Computer scientists pursuing machine consciousness (MC) and artificial general intelligence (AGI) [Wang et al., 2008] have come to appreciate that consciousness itself and general decision making require the integration of input from many sources. The models of cognition they have begun developing accommodate a broad array of inputs and processes. The modular construction of these models means that additional capabilities can be added as needed. How well these modules can be integrated and how effectively such systems can be scaled to navigate complex environments will only be discovered in the process of instantiating the models.

Moral decisions will arguably be among the more difficult challenges AGI systems will need to master. Rather than reinventing the wheel, Wallach and Allen considered whether existing AGI models might be adapted to build AMAs. How might the top-down and bottom-up approaches for designing AMAs fit into the AGI models other scientists were developing? Suprarational capabilities and social mechanisms that AMAs will draw upon are also likely to be essential for other tasks systems with AGI will confront. Which suprarational capabilities had already been accommodated, at least conceptually, within computation models for general intelligence?

As a first step in exploring how a comprehensive model of human cognition might be adapted to make moral decisions, Wallach and Allen teamed with Franklin [Wallach and Allen, 2009; Wallach et al., 2010]. Their joint work with LIDA, the computational and conceptual model of cognition developed by Franklin and his colleagues [Franklin and Patterson, 2006; Ramamurthy *et al.*, 2006; Baars and Franklin, 2007] provides a secondary benefit for studying the relationship between ME and MC. LIDA, and its precursor IDA, are computational and conceptual implementations of Baars' Global Workspace Theory [1988]. In earlier work, Franklin demonstrated how IDA provides a glimpse into a functional role for consciousness in the making of decisions [2003]. The collaboration with Wallach and Allen extended this thesis to a special class of decisions, moral decisions. Together they outlined a comprehensive approach for making moral decisions that uses the same cognitive processes used for making general decisions. That is, moral cognition is supported by domain general cognitive processes, even while some special classes of moral decisions may require additional mechanisms.

In this article we focus upon the functional role of consciousness in the making of moral decisions. While it might be presumed that consciousness is a prerequisite for agents making moral decisions when confronted with complex dilemmas, the exact role(s) consciousness plays has never been fully clarified. We will address four questions:

(a) What is the relationship between an agent being conscious and an agent having a capacity to make moral decisions?
(b) Is consciousness (weak or strong?) essential for an agent to make moral decisions?
(c) What cognitive/computational mechanisms serve both consciousness and moral decision making?
(d) Should the capacity to make moral decisions be considered an attribute essential for being designated a fully conscious agent?

2. Consciousness and Ethics: Rights, Responsibilities, and Moral Judgments

Thought experiments regarding the capabilities a non-human entity would require to be a moral agent have long been a staple for philosophers. Legal theorists have also reflected upon the criteria for granting rights and/or responsibilities to non-human agents, infants, and the mentally impaired. Indeed, this discussion is foundational for arguments by animal rights advocates that some non-human animals, such as great apes, should be granted certain rights as moral patients, entities toward whom we have moral obligations [Singer, 1977; Regan, 1983].

The foreseeable prospect of artificial agents with abilities that are comparable to and may exceed those of humans, is prompting serious reflection on criteria for attributing moral agency and granting legal personhood [Calverley, 2005; Chopra and White, 2011]. Legal theorists have argued that since non-biological entities such as corporations are legal "persons", there is, at least in a theoretical sense, nothing in the law that would prohibit this designation for a non-biological agent. Calverley [2005] specifically discusses legal personhood for non-biological systems that are shown to have mental states and other attributes of consciousness.

Steve Torrance has taken the lead in analyzing the relationship between consciousness and ethics in the context of future artificial agents with human-like abilities. Torrance focuses on whether artificial agents can be considered either moral patients or, what he calls, "ethical 'producers' (beings who may have moral obligations and responsibilities themselves)" [Torrance, 2008, p. 499]. Torrance outlines an "organic view" in which — in contrast to non-human animals which might have rights as moral patients — artificial agents (non-biological) which lack sentience or phenomenal awareness would not be "genuine subjects of either moral concern or moral appraisal" [Torrance, 2008, p. 503]. According to Torrance, "only biological organisms have the ability to be genuinely sentient or conscious" [Torrance, 2008, p. 503], a claim he bases on the notion that, "moral thinking, feeling and action arises organically out of the biological history of the human species and perhaps many more primitive species" [Torrance, 2008, p. 502].

Although it is not his focus, Torrance acknowledges that the kinds of capacities mentioned as criteria for designating something as a moral producer might be elements of a theory of moral decision making. The capabilities associated with consciousness that he is most concerned with are the ability to feel pleasure and pain that are central for empathy. He argues that, artificial humanoids "are unlikely to possess sentience and hence will fail to be able to exercise the kind of empathic rationality that is a prerequisite for being a moral agent" [Torrance, 2008, p. 495].

Our focus is not upon whether an artificial agent should be granted moral status, but we agree that the organic view is worthy of serious consideration. However, the possibility of developing synthetic emotions, including pleasure and pain, for artificial (non-biological) agents has been of interest to scientists working in the fields of affective computing and machine consciousness [Picard, 1997; Franklin and Patterson, 2006; Vallverdú and Casacuberta, 2008; Haikonen, 2009]. It is rather early to evaluate whether existing or future implementations of synthetic emotions will or will not lead to the kind of rich emotional intelligence that might be expected of moral agents. Artificial agents incapable of feeling pain or pleasure or lacking empathy may fail in adequately responding to certain kinds (classes) of moral challenges. But the capacity to empathize is not a prerequisite for responding appropriately to all moral challenges.

Much of the discourse in moral philosophy and in practical ethics would suggest that only decisions about intractable social and personal challenges or decisions in which self-centered interests are transcended for the good of others should be designated moral decisions. In our view moral decision making encompasses a much broader range of choices and actions. Any choice influenced by consideration for their effects on others is a moral choice. Given that these kinds of value judgments are implicated in most choices where information is incomplete or of questionable accuracy, or where the consequences of possible courses of action cannot be known in advance, a broad range of choices arguably have moral dimensions.

In recent years there has been considerable research demonstrating that much of moral behavior arises from unconscious judgments. Such research does not rule out conscious deliberation; rather it argues that reflection is less common than may be otherwise presumed. The social intuitionist model [Haidt, 2001; Haidt and Bjorklund, 2008; Haidt and Joseph, 2007], for example, is generally understood as positing the primacy of emotionally activated intuitions over conscious reasoning as the determinant of most moral behavior. Indeed, the model of a rational agent, and therefore a rational moral actor, has been under assault for more than fifty years [Simon, 1955; Tversky and Kahneman, 1974; Greenwald and Banaji, 1995]. In regards to moral judgments, psychologists and social psychologist have demonstrated experimentally that moral behavior can be altered by priming, by relatively minor changes in the situation, and by additional unconscious or non-conscious influences [Isen and Levin, 1972; Darley and Batson, 1973; Haney et al., 1973; Hassin et al., 2006]. But there have also been attempts to re-establish the centrality of reasoning in the making of moral decisions and the role reflection plays in honing unconscious influences on moral judgments [Paxton and Greene, 2010].

Moral philosophers have long held that the *ought* of ethics is not determined by the *is* of moral psychology, and certainly not by the unconscious influences on moral psychology. But designing or teaching artificial agents to act safely and appropriately may depend on modeling a capacity for reasoned reflection upon what is known about human psychology. How does one put all this together in an agent? How does one combine bottom-up, reactive psychological processes with top-down, deliberative reasoning?

Building upon the LIDA model of cognition, Wallach et al. [2010] provide a first example of how the top-down analysis and bottom-up psychology might be integrated. This model, in principle, could also accommodate additional suprarational capabilities necessary for making moral decisions in specific domains. For the purposes of this paper, however, we direct our attention to the expanded role for consciousness in the making of moral decisions suggested by the model. We will propose that consciousness plays a functional role in the capacity to make all complex decisions, particularly moral decisions. But before discussing this, let us first introduce in a very cursory manner the LIDA model of cognition[1] and Global Workspace Theory, which LIDA tries to capture computationally.

3. GWT and LIDA

Bernard Baars' Global Workspace Theory (GWT) [1988] is widely recognized as a high-level theory of the role of consciousness in human cognitive processing with significant support from empirical studies [Baars, 2002]. Three different research teams led by Stanislas Dehaene [Dehaene and Naccache, 2001; Gaillard et al., 2009], Murray Shanahan [2006], and Stan Franklin have developed computational models that instantiate aspects of GWT. LIDA, the model we single out for discussion in this paper, was developed by Franklin and colleagues with input from Baars. LIDA

---

[1] LIDA is a very extensive model of cognition, whose development is covered in more than fifty published articles. Many of the articles focus upon facets of LIDA. The most comprehensive description of decision making in LIDA is contained in Wallach et al. [2010] upon which this paper builds.

provides a particularly useful model to illustrate a role for consciousness in the making of moral decisions. However, we are not suggesting that LIDA is the only AGI system capable of modeling human-level intelligence. LIDA has many features that are similar to those in other AGI models of cognition.

GWT is a neuropsychological model of consciousness that views the nervous system as a distributed parallel system incorporating many different specialized processes. Various coalitions of these specialized processes facilitate making sense of the sensory data currently coming in from the environment. Other coalitions sort through the results of this initial processing and pick out items requiring further attention. In the competition for attention a winner emerges and occupies what Baars calls the global workspace, the winning contents of which are presumed to be at least functionally conscious. The presence of a predator, enemy, or imminent danger should be expected, for example, to win the competition for attention. However, an unexpected loud noise might well usurp consciousness momentarily even in one of these situations. The contents of the workspace are broadcast to processes throughout the nervous system in order to recruit an action or response to this salient aspect of the current situation. The contents of this global broadcast enable each of several modes of learning.

LIDA is a computational agent that continually tries to make sense of its environment and determines what to do next. In LIDA this dynamic is represented through a model that describes how unconscious mechanisms feed the conscious processing of information. LIDA implements GWT as a cascading sequence of cognitive cycles (see Fig. 1).

During a cognitive cycle the agent constantly samples (senses) its external and internal environment. This sensory information is matched to information within various memory systems in a process of discerning the features (objects, categories, relations, events, situations, etc.) of the present situation.

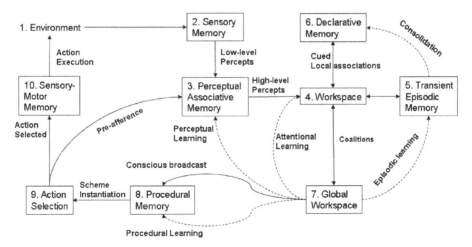

Fig. 1. The LIDA cognitive cycle.

In each cognitive cycle, attention codelets search this unconscious model of the present situation for specific features or percepts. Thousands of these codelets could be searching specifically for morally relevant considerations. Those that find information germane to their directive or function will join in coalitions with other attention codelets that have found related information. These coalitions occupy the global workspace and vie for attention. In each cycle there is a winning coalition whose contents are made conscious by being broadcast throughout the system. The broadcast is directed in particular at procedural memory in pursuit of an action in response to the information requiring attention. Franklin hypothesizes that there are 5–10 cycles each second [Madl et al., in press]. In each of these 5–10 cycles there is a broadcast and the selection of an action in response.

One might, for example, presume that while driving a car focusing upon the road conditions wins the competition for attention in many cycles, but once the response (or lack of a need for any change) has been determined, the "check the road" coalition will be weaker than other coalitions during a number of intermediary cycles. Attention can now be devoted to listening to the radio during intermediary cycles.

In LIDA a distinction is made between the conscious mediation of the contents of the global workspace leading to the activation of an unconsciously chosen response, and the conscious process of volitional decision making. Generally the former is said to occur in one cognitive cycle, while the latter will require multiple cycles.

4. Moral Judgments and Moral Decisions

Usually decision making is thought of as a deliberative process that entails reasoning, planning, problem solving, and meta-cognition. But action selection can also occur through consciously mediated emotionally activated intuitions and automatized learned behavior. We follow Haidt's nomenclature by referring to emotionally activated intuitions or the unconscious selection of consciously mediated actions as judgments. Judgments do not necessarily express social norms and may even be prejudices, such as a rejection of those who do not belong to one's group. While there is no deliberation or volitional decision making in judgments, future judgments can be altered through experience. Baars has proposed that attention is sufficient for learning and thus learning occurs with each conscious broadcast [1988]. Several modes of such learning have been included in the LIDA model [Franklin and Patterson, 2006]. The bottom-up propensities embodied in emotional/affective responses to actions and their outcomes can also be modified by *ex post facto* reflection. In other words, there is an ongoing process where consciously mediated behavior is molded and honed by experience and deliberation. Given this dynamic relationship between judgments and deliberations, the moral decision-making system of an agent can be said to include both.

A strength of LIDA lies in the model's ability to accommodate the messiness and complexity of a hybrid approach to decision-making. Moral decisions are particularly messy, drawing upon emotions, moral sentiments, intuitions, heuristics, rules and duties, principles, and even some explicit valuation of utility or expected outcomes. In analyzing how a moral decision making system that includes both judgments and deliberation might be implemented in LIDA, Wallach *et al.* [2010] addressed six questions, whose answers we very briefly summarize here.[2] In answering these questions it was not our intent to explain how a particular theory of ethics would be implemented in LIDA. Rather, we wish to provide tentative explanations of general processes central to moral judgments and volitional decision making.

(1) Where are bottom-up propensities and values implemented? How does the agent learn new values and propensities, as well as reinforce or defuse existing values and propensities?

Objects, people, contexts, events, situations, and other percepts are represented as nodes within the perceptual memory of LIDA. Associations between these percepts and valenced feelings (either negative or positive) are the primary means of capturing values within the agent's mind [Franklin and Patterson, 2006]. These values often represent propensities that have been evolutionarily acquired and then shaped by experience.

Affects and perceptions that arise within one LIDA cycle automatically form associations, which will decay over time, but can also be strengthened or weakened by sustained sensory input and attention. As we mentioned earlier, Baars posits that each instance of attention contributes toward learning. By reinforcing links in an association, sustained or conscious attention produces learning and can build long-term memory.

A particularly difficult challenge for all human-like computer architectures is how the system generates or acquires totally new nodes. There is also the difficult problem of how valenced feelings will be represented. A cognitive representation of feelings may not adequately express the richness of content and meaning that we associate with somatic feelings.

(2) How does the LIDA model transition from a single cycle to the determination that information in consciousness needs to be deliberated upon?

Generally, when the situation is either new or appears to be novel, a multicycled process of deliberation will naturally arise because LIDA is unlikely during a single cycle to determine an appropriate response. In effect, new or apparently new situations that do not fit neatly into learned heuristics demand attention over time.

(3) How are rules or duties represented in the LIDA model? What activates a rule and brings it to conscious attention? How might some rules be automatized to form unconscious rules-of-thumb (heuristics)?

---

[2] Chapter 11 of *Moral Machines: Teaching Robotics Right from Wrong* [Wallach and Allen, 2009] provides a fuller discussion, along with examples, of the answers to these questions. The fullest discussion of the answers can be found in Wallach *et al.* [2010].

Deliberation in LIDA is modeled upon William James theory of "volitional" decision making. James viewed decisions as a negotiation between internal proposers of courses of action, objectors, and supporters. In LIDA proposed courses of action that win the competition for consciousness will impel the initiation of an action stream that will continue to be reviewed in subsequent cycles. This action stream could activate a rule or duty, stored in semantic memory, which objects to the proposed action, and in a subsequent cycle wins the competition for attention. Thus begins a multicycled process in which proposer, objector, and supporter codelets successively win the competition for attention. Pertinent memories and other associated information also enter the mix, including additional rules or duties and relevant exceptions.

The activation of proposals and objections decays in each subsequent cycle. But strong constraints, such as not killing, can have high levels of activation through reinforced connections to feelings such as shame or fear. Thus strong constraints are not easily over-ridden. While this kind of process is compatible with a battle of urges that might not be considered a deliberation, when the competing elements correspond to reasons then we believe this is an implementation of a deliberation.

On each occasion when a rule or duty comes to attention it is a subject for learning and modification. For rules that arise in similar situations, or for situations that required sustained deliberation, LIDA will produce a new node in perceptual memory, that might represent a new variation of the rule. If encountered often enough, a response to a challenge for which there is no objection can be captured within procedural memory as an habituated response and activated in just one cognitive cycle. But before this kind of procedural learning can occur, the action must have been selected at least once after a deliberative process. Deliberation can produce habitual behaviors.

(4) How can we implement planning or imagination (the testing out of different scenarios) in LIDA?

Imaginative planning or testing of various possible scenarios in LIDA is comparable to building a model in the workspace of the current situation by linking nodes from perceptual and/or episodic memory. The consciously proposed action at the completion of each model or scenario will or will not cue objections indicating the success or issues with the model. In the following cycle, the action selected in response to an objection could be as simple as making a minor alteration or adding a component to the model.

(5) What determines the end of a deliberation?

The deliberative dialogue continues until there is no objection to a proposed action or until the metaphorical timer requiring some action rings. The decay of objections in succeeding cycles means that strongly reinforced proposals will win out over weak objections. Strong objections will prevail over weak proposals. However, time pressures can force a decision before all objections have been dispelled.

The selected course of action may in retrospect fail to satisfy prevailing exterior norms. LIDA-inspired moral agents are not designed around fixed moral rules, values, or principles. Therefore they, like human agents, are prone to acting upon strong impulses without necessarily taking into consideration the needs of others. Developing a LIDA agent with the kind of rich sensitivity to moral considerations and how each consideration might be comparatively weighted will require much experience and learning, just as it does with teaching a young adult to be morally sensitive and considerate.

(6) When a resolution to the challenge has been determined, how might LIDA monitor whether that resolution is successful? How can LIDA use this monitoring for further learning?

Monitoring actions is central for moral development. Once a resolution has been reached and an action selected, LIDA generates an expectation codelet, an attention codelet that brings to attention the outcome of an action. An expectation codelet would bring to the global workspace any discrepancy between the actual results of an action and the predicted results. Any discrepancy would contribute to procedural learning inhibiting or reinforcing the application of that procedure to future similar challenges. For a decision that entailed moral considerations, such procedural learning would foster moral development through expectations in regard to the positive or negative moral outcome of actions.

4.1. The advantage of codelets

Attention codelets sensitive to morally relevant information would need to be designed for LIDA to engage in moral deliberations. It is unclear at this time whether the design of such codelets would differ significantly from the design of codelets that search for concrete information. We expect that attention codelets sensitive to facial and vocal expressions would be among those gathering information relevant to moral deliberations.

An advantage of codelets is that they provide an extensible framework through which more and more relevant considerations can be found within the model of the situation. No one needs to specify in advance the considerations influencing the situation or the criteria for evaluating a scenario. Therefore, codelets provide a particularly useful approach for representing the messiness of so many moral challenges.

5. Evolution, Moral Decisions and Consciousness

Certainly artificial agents that lack consciousness can make moral decisions in many situations by applying normative rules to the information at hand. But in our analysis consciousness becomes much more central for agents operating within complex environments. These agents will perform many tasks related to the making of moral decisions in addition to applying norms, rules, or moral principles. AMAs will also need to recognize when they are in an ethically significant situation, to have sensitivity to the array of moral considerations that impinge upon that decision, to discriminate essential from inessential information, to estimate the sufficiency of initial information and search for additional information when needed, and to make judgments about the intentionality of the other agents with whom it is interacting. All of this must be done within the time available to take action.

An array of sensors, subsystems, and processes will be enlisted to perform these tasks. These are largely the same cognitive mechanisms used for general decision making. Moral cognition is supported by domain general processes, many of which will also be needed for machine consciousness. So how closely entangled are the fields of machine ethics and machine consciousness? Should we consider the capacity to make moral decisions an attribute essential to consciousness?

Scholars point to evidence of proto-morality [Katz, 2000; de Waal, 2006; Bekoff and Pierce, 2009] and pre-reflective consciousness [Seth *et al.*, 2005; Panksepp, 2005; Edelman and Seth, 2009] in many non-human species. The evolution of consciousness and the evolution of morality are each developing as sub-disciplines in their own right. Wallach [2005] hypothesized that consciousness and moral decision making co-evolved. The basic argument is that one fitness function of consciousness lay in how it facilitated the making of decisions in ambiguous situations where neither of two or more instinctual or already learned action sequences (automatized responses) was activated. Consider the distant presence of a potentially threatening natural enemy that does not immediately activate a *fight* or *flight* response. Such situations can lead to sustained attention (proto-reflection) during which existing action sequences are altered or new action sequences are created. The latter might entail the recognition by a mammalian or earlier ancestor that under certain circumstances a natural enemy is non-threatening, for example, the predator is satiated or ready for a nap. The observing agents can thus conserve energy and is free to return to other tasks such as feeding the young, while continuing to periodically check whether her enemy is taking a more threatening stance. This requires a moral decision in the sense that some form of "valuing" would come into responding to the challenge. In other words, under circumstances where information is unclear or incomplete sustained attention (proto-reflection) leads to a kind of valuing of information that alters existing behavioral patterns, but more importantly, can even create new behavior streams.

If Wallach is correct, then central among the fitness functions served by consciousness is the manner in which it facilitates a kind of valuing that fostered making decisions when the available information is ambiguous, confusing, contradictory, or incomplete. Given that such situations are common, this kind of valuing opened up an array of new possibilities.

Even if one holds that the evolution of consciousness, decision-making, and moral decision making are entangled, there are practical reasons for distinguishing between the project of designing AMAs and exploring ways to implement consciousness in machines. However, we should not lose sight of how both projects will progress with the emulation of the same cognitive processes utilizing many of the same cognitive mechanisms. Furthermore, we should not overlook evolutionary strategies that further the interactive development of MC and the design of AMAs.

6. Conscious Artificial Moral Agents

Our discussion of the LIDA model of cognition illustrates a functional role for consciousness in the making of all decisions, including moral decisions. However, neither LIDA nor any other AGI has been fully implemented. Therefore, it is impossible to know whether such systems can adequately be scaled to manage the on-going sorting through sensory information, action selection, scenario building, and deliberation necessary to operate successfully in complex environments filled with other agents. In the process of building models of AGI, designers and engineers will recognize the need for additional subsystems to support capabilities whose contribution may have not yet been recognized. The value of models such as LIDA is that they offer an architecture that in theory can integrate a vast array of inputs into an approach for discerning salient information, for selecting appropriate actions, and for rich deliberation. Competition for consciousness between different coalitions, global broadcasting of the winning coalition, and the selection of an action in each cycle are the mechanisms used in the LIDA model to integrate input from various sources.

While we cannot know for sure that LIDA-like agents will accommodate a broad array of moral considerations in choices and actions, the multicyclical approach to higher-order cognition, the increasing speed of computers and therefore the speed of individual cycles, and the unconscious parallel processing of information all suggest that this is a promising path to explore in the development of sophisticated AMAs.

We do not know how far this strategy will progress. Nor do we know how broad an array of ethical challenges can be managed successfully within the functional model of consciousness suggested by GWT and LIDA. A further challenge concerns the necessity for some form of phenomenal experience that might not be captured in the GWT/LIDA model. Phenomenal consciousness is difficult to measure in any agent. But if, as Torrance notes in his explication of the organic view, phenomenal awareness is foundational for empathetic rationality, then evidence that an agent is sensitive to what others feel, might be counted as evidence the agent is phenomenally conscious. However, if the organic view is correct and only biological entities can be moral agents, silicon-based artificial agents will never progress beyond a kind of deductive empathy.

In this article, we explored the relationship between consciousness and moral decision-making within one model capable of implementing both. Given this implementation, we propose the following:

(1) Consciousness serves most if not all moral decision-making as it serves decision-making more generally. Consciousness will be especially important for making volitional moral decisions.
(2) Moral cognition is supported by domain general cognitive processes.
(3) Among the mechanisms that serve both moral decision-making and consciousness are those which integrate perception and action over multiple cycles, including inner modeling of the relationship between the agent and its

environment for generating expectations of events and assessing the consequences of various courses of action.

Machine consciousness and designing artificial moral agents are two projects that are joined at the hip. Although we have not argued for it directly, we further propose that the capacity to make moral decisions be considered an attribute essential for being designated a fully conscious agent. Thus, a fully conscious machine or an AGI system worthy of being considered a full moral agent would be an artificial conscious moral agent.

References

Allen, C., Varner, G. and Zinser, J. [2000] "Prolegomena to any future artificial moral agent," *Journal of Experimental and Theoretical Artificial Intelligence* **12**, 251–261.

Allen, C., Smit, I. and Wallach, W. [2006] "Artificial morality: Top-down, bottom-up and hybrid approaches," *Ethics of New Information Technology* **7**, 149–155.

Baars, B. J. [1988] *A Cognitive Theory of Consciousness* (Cambridge University Press).

Baars, B. J. and Franklin, S. [2007] "An architectural model of conscious and unconscious brain functions: Global Workspace Theory and IDA," *Neural Networks* **20**, 955–961.

Bekoff, M. and Pierce, J. [2009] *Wild Justice: The Moral Lives of Animals* (University of Chicago Press).

Calverley, D. [2005] "Towards a method for determining the legal status of a conscious machine," in *Artificial Intelligence and the Simulation of Behavior '05: Social Intelligence and Interaction in Animals, Robots and Agents: Symposium on Next Generation Approaches to Machine Consciousness*, Hatfield, UK.

Chopra, S. and White, L. F. [2011] *A Legal Theory for Autonomous Artificial Agents* (University of Michigan Press).

Darley, J. M. and Batson, C. D. [1973] "From Jerusalem to Jericho: A study of situational and dispositional variables in helping behavior," *Journal of Personality and Social Psychology* **27**, 100–108.

Dehaene, S. and Naccache, L. [2001] "Towards a cognitive neuroscience of consciousness: Basic evidence and a workspace framework," *Cognition* **79**, 1–37.

de Waal, F. [2006] *Primates and Philosophers: How Morality Evolved* (Princeton University Press).

Edelman, D. B. and Seth, A. K. [2009] "Animal consciousness: A synthetic approach," *Trends in Neurosciences* **32**(9), 476–484.

Franklin, S. [2003] "IDA: A conscious artefact?" *Journal of Consciousness Studies*, **10**(4–5), 47–66.

Franklin, S., Baars, B. J., Ramamurthy, U. and Ventura, M. [2005] "The role of consciousness in memory," *Brains, Minds and Media* **1**, 1–38.

Franklin, S. and Patterson, F. G. J. [2006] "The LIDA architecture: Adding new modes of learning to an intelligent, autonomous, software agent," in *IDPT-2006 Proceedings (Integrated Design and Process Technology)*, Society for Design and Process Science.

Franklin, S. and Ramamurthy, U. [2006] "Motivations, values and emotions: Three sides of the same coin," in *Proceedings of the 6th International Workshop on Epigenetic Robotics*, Paris, France, pp. 41–48.

Gaillard, R., Dehaene, S., Adam, C., Clémenceau, S., Hasboun, D., Baulac, M., Cohen, L. and Naccache, L. [2009] "Converging intracranial markers of conscious access,"

PLoS Biology **7**(3), http://www.plosbiology.org/article/info%3Adoi%2F10.1371%2F journal.pbio.1000061.

Gips, J. [1991] "Towards the ethical robot," in *Android Epistemology*, Ford, K. G., Glymour, C. and Hayes, P. J. (eds.) (MIT Press), pp. 243–252.

Greenwald, A. G. and Banaji, M. R. [1995] "Implicit social cognition: Attitudes, self-esteem, and stereotypes," *Psychological Review* **102**, 4–27.

Haidt, J. [2001] "The emotional dog and its rational tail: A social intuitionist approach to moral judgment," *Psychological Review* **108**(4), 814–834.

Haidt, J. and Joseph, C. [2007] "The moral mind: How five sets of innate moral intuitions guide the development of many culture-specific virtues, and perhaps even modules," in *The Innate Mind*, Carruthers, P., Laurence, S. and Stich, S. (eds.) (Oxford University Press), pp. 367–391.

Haidt, J. and Bjorklund, F. [2008] "Social intuitionists answer six questions about moral psychology," in *Moral Psychology, Vol. 2: The Cognitive Science of Morality: Intuition and Diversity*, Sinnott-Armstrong, W. (ed.) (MIT Press), pp. 181–217.

Haikonen, P. O. A. [2009] "Qualia and conscious machines," *International Journal of Machine Consciousness* **1**(2), 225–234.

Haney, C., Banks, W. and Zimbardo, P. [1973] "Interpersonal dynamics of a simulated prison," *International Journal of Criminology and Penology* **1**, 69–97.

Hassin, R., Uleman, J. and Bargh, J. (eds.) [2006] *The New Unconscious* (Oxford University Press).

Isen, A. M. and Levin, P. F. [1972] "Effect of feeling good on helping: Cookies and kindness," *Journal of Personality and Social Psychology* **21**, 384–388.

Katz, L. D. (ed.) [2000] *Evolutionary Origins of Morality: Cross-Disciplinary Perspectives* (Imprint Academic).

Madl, T., Baars, B. J. and Franklin, S. [in press] "The timing of the cognitive cycle," *PLoS ONE*.

Panksepp, J. [2005] "Affective consciousness: Core emotional feelings in animals and humans," *Consciousness and Cognition* **14**, 30–80.

Paxton, J. and Greene, J. [2010] "Moral reasoning: Hints and allegations," *Topics in Cognitive Science* **2**, 511–527.

Picard, R. [1997] *Affective Computing* (MIT Press).

Ramamurthy, U., Baars, B. J., D'Mello, Sidney, K. and Franklin, S. [2006] "LIDA: A working model of cognition," in *Proceedings of the 7th International Conference on Cognitive Modeling*, pp. 244–249.

Regan, T. [1983] *The Case for Animal Rights* (University of California Press).

Seth, A. K., Baars, B. J. and Edelman, D. B. [2005] "Criteria for consciousness in humans and other mammals," *Consciousness and Cognition* **14**, 119–139.

Shanahan, M. P. [2006] "A cognitive architecture that combines internal simulation with a global workspace," *Consciousness and Cognition* **15**, 433–449.

Simon, H. [1955] "A behavioral model of rational choice," *Quarterly Journal of Economics* **69**, 99–118.

Singer, P. [1977] *Animal Liberation* (Granada).

Torrance, S. [2008] "Ethics, consciousness and artificial agents," *AI and Society* **22**(4), 495–521.

Tversky, A. and Kahneman, D. [1974] "Judgment under uncertainty: Heuristics and biases," *Science* **185**, 1124–1131.

Vallverdú, J. and Casacuberta, D. [2008] "The panic room: On synthetic emotions," in *Proceedings of the 2008 Conference on Current Issues in Computing and Philosophy*, pp. 103–115.

Wallach, W. [2004] "Artificial morality: Bounded rationality, bounded morality and emotions," in *16th International Conference on Systems Research, Informatics and Cybernetics: Symposium on Cognitive, Emotive and Ethical Aspects of Decision Making in Humans and in Artificial Intelligence*, Baden-Baden Germany, International Institute for Advanced Studies in Systems Research and Cybernetics, pp. 1–6.

Wallach, W. [2005] "Choice, ethics, and the evolutionary function of consciousness," in *17th International Conference on Systems Research, Informatics and Cybernetics: Symposium on Cognitive, Emotive and Ethical Aspects of Decision Making in Humans and in Artificial Intelligence*, Baden-Baden Germany, International Institute for Advanced Studies in Systems Research and Cybernetics, pp. 1–21.

Wallach, W. [2010] "Robot minds and human ethics: The need for a comprehensive model of moral decision making," *Ethics and Information Technology* **12**(3), 243–250.

Wallach, W., Allen, C. and Smit, I. [2008] "Machine morality: Bottom-up and top-down approaches for modelling human moral faculties," *AI and Society* **22**(4), 565–582.

Wallach, W. and Allen, C. [2009] *Moral Machines: Teaching Robots Right from Wrong* (Oxford University Press).

Wallach, W., Franklin, S. and Allen, C. [2010] "A conceptual and computational model of decision making in human and artificial agents," *Topics in Cognitive Science* **2**, 454–485.

Wang, P., Goertzel, B. and Franklin, S. [2008] *Artificial General Intelligence 2008* (IOS Press).

Part IV:

Moral Agents and Agency

25

On the Morality of Artificial Agents

LUCIANO FLORIDI and J.W. SANDERS

Abstract. Artificial agents (AAs), particularly but not only those in Cyberspace, extend the class of entities that can be involved in moral situations. For they can be conceived of as moral patients (as entities that can be acted upon for good or evil) and also as moral agents (as entities that can perform actions, again for good or evil). In this paper, we clarify the concept of agent and go on to separate the concerns of morality and responsibility of agents (most interestingly for us, of AAs). We conclude that there is substantial and important scope, particularly in Computer Ethics, for the concept of moral agent not necessarily exhibiting free will, mental states or responsibility. This complements the more traditional approach, common at least since Montaigne and Descartes, which considers whether or not (artificial) agents have mental states, feelings, emotions and so on. By focussing directly on 'mind-less morality' we are able to avoid that question and also many of the concerns of Artificial Intelligence. A vital component in our approach is the 'Method of Abstraction' for analysing the level of abstraction (LoA) at which an agent is considered to act. The LoA is determined by the way in which one chooses to describe, analyse and discuss a system and its context. The 'Method of Abstraction' is explained in terms of an 'interface' or set of features or observables at a given 'LoA'. Agenthood, and in particular moral agenthood, depends on a LoA. Our guidelines for agenthood are: interactivity (response to stimulus by change of state), autonomy (ability to change state without stimulus) and adaptability (ability to change the 'transition rules' by which state is changed) at a given LoA. Morality may be thought of as a 'threshold' defined on the observables in the interface determining the LoA under consideration. An agent is morally good if its actions all respect that threshold; and it is morally evil if some action violates it. That view is particularly informative when the agent constitutes a software or digital system, and the observables are numerical. Finally we review the consequences for Computer Ethics of our approach. In conclusion, this approach facilitates the discussion of the morality of agents not only in Cyberspace but also in the biosphere, where animals can be considered moral agents without their having to display free will, emotions or mental states, and in social contexts, where systems like organizations can play the role of moral agents. The primary 'cost' of this facility is the extension of the class of agents and moral agents to embrace AAs.

1. Introduction: Standard vs. Non-Standard Theories of Agents and Patients

Moral situations commonly involve agents and patients. Let us define the class *A* of moral *agents* as the class of all entities that can in principle qualify as sources of moral action, and the class *P* of moral *patients* as the class of all entities that can in principle qualify as receivers of moral action. A particularly apt way to introduce the topic of this paper is to consider how ethical theories (macroethics) interpret the logical relation between those two classes.

There can be five logical relations between A and P. Three are unrealistic and we shall not consider them here.[1] The remaining two are: (1) A and P can be equal, or (2) A can be a proper subset of P.

Alternative (1) maintains that all entities that qualify as moral agents also qualify as moral patients and vice versa. It corresponds to a rather intuitive position, according to which the agent/inquirer plays the rôle of the moral protagonist, and is one of the most popular views in the history of ethics, shared for example by many Christian Ethicists in general and by Kant in particular. We refer to it as the standard position.

Alternative (2) holds that all entities that qualify as moral agents also qualify as moral patients but not vice versa. Many entities, most notably animals, seem to qualify as moral patients, even if they are in principle excluded from playing the rôle of moral agents. This post-environmentalist approach[2] requires a change in perspective, from agent orientation to patient orientation. In view of the previous label, we refer to it as non-standard.

In recent years, non-standard macroethics have been discussing the scope of P quite extensively.[3] Comparatively little work has been done in reconsidering the nature of moral agenthood and hence the extension of A. Post-environmentalist thought, in striving for a fully naturalised ethics, has implicitly rejected the relevance, if not the possibility, of supernatural agents, while the plausibility and importance of other types of moral agenthood seem to have been largely disregarded. Secularism has contracted (some would say deflated) A, while environmentalism has justifiably expanded only P, so the gap between A and P has been widening; this has been accompanied by an enormous increase in the moral responsibility of the individual.

Some efforts have been made to redress this situation. In particular, the concept of 'moral agent' has been stretched to include both natural and legal persons. A has then been extended to include agents like partnerships, governments or corporations, for whom legal rights and duties have been recognised. This more ecumenical approach has restored balance between A and P. A company can now be held directly accountable for what happens to the environment, for example. Yet the approach has remained unduly constrained by its anthropocentric conception of agenthood. An entity is considered a moral agent only if (i) it is an individual agent and (ii) it is human-based, in the sense that it is either human or at least reducible to an identifiable aggregation of human beings, who remain the only morally responsible sources of action, like ghosts in the legal machine.

Limiting the ethical discourse to individual agents hinders the development of a satisfactory investigation of distributed morality, a macroscopic and growing phenomenon of global moral actions and collective responsibilities resulting from the 'invisible hand' of systemic interactions among several agents at a local level. Insisting on the necessarily human-based nature of the agent means undermining the possibility of understanding another major transformation in the ethical field, the appearance of artificial agents (AAs) that are sufficiently informed, 'smart', autonomous and able to perform morally relevant actions independently of the humans who created them, causing 'artificial good' and 'artificial evil' (Gipps, 1995). Both constraints can be eliminated by fully revising the concept of 'moral agent'. This is the task undertaken in the following pages.

The main thesis defended is that AAs are legitimate sources of im/moral actions, hence that *A* should be extended so as to include AAs, that the ethical discourse should include the analysis of their morality and, finally, that this analysis is essential in order to understand a range of new moral problems not only in Computer Ethics but also in ethics in general, especially in the case of distributed morality.

This is the structure of the paper. In Section 2, we analyse the concept of agent. We first introduce the fundamental 'Method of Abstraction' which provides the foundation for an analysis by levels of abstraction (LoA). The reader is invited to pay particular attention to this section; it is essential for the paper and its application in any ontological analysis is crucial. We then clarify the concept of 'moral agent' by providing not a definition but an effective characterisation, based on three criteria at a specified LoA. The new concept of moral agent is used to argue that AAs, though not intelligent and fully responsible, can be fully *accountable* sources of moral action. In Section 3 it is argued that there is substantial and important scope for the concept of moral agent not necessarily exhibiting free will or mental states ('mind-less morality'). Section 4 provides some examples of the properties constituting our characterisation of agenthood and in particular of AAs; inevitably it also provides further examples of LoA. In Section 5, morality is captured as a 'threshold' defined on the observables determining the LoA under consideration. An agent is morally good if its actions all respect that threshold; and it is morally evil if some action violates it. Morality is usually predicated upon *responsibility*. The use of the Method of Abstraction, LoAs and thresholds enables *responsibility* and *accountability* to be separated and formalised effectively when the levels of abstraction involve numerical variables, as is the case with digital AAs. The part played in morality by responsibility and accountability can be clarified as a result. Section 6 pursues some important consequences of our investigations for Computer Ethics: the way in which AAs might be bound by the ACM Code of Ethics is considered, as is censure of AAs (Table 1).

Table 1. Table of acronyms used in the paper

Acronym	Meaning
AA	Artificial agent
ACM	Association for computing machinery
CE	Computer ethics
LoA	Level of abstraction

2. What Is an Agent?

Complex biochemical compounds and abstruse mathematical concepts have at least one thing in common: they may be unintuitive, but once understood they are all definable with total precision, by listing a finite number of necessary and sufficient properties. Mundane entities like intelligent beings or living systems share the opposite property: one naïvely knows what they are and perhaps could be, and yet there seems to be no way to encase them within the usual planks of necessary and sufficient conditions.

This holds true for the general concept of 'agent' as well. People disagree on what can and cannot count as an 'agent', even in principle e.g. Franklin and Graesser (1996). Why? Sometimes the problem is addressed optimistically, as if it were just a matter of further shaping and sharpening whatever necessary and sufficient conditions are required to obtain a *definiens* that is finally watertight. Stretch here, cut there; ultimate agreement is only a matter of time, patience and cleverness. In fact, attempts follow one another without a final identikit ever being nailed to the *definiendum* in question. After a while, one starts suspecting that there might be something wrong with this *ad hoc* approach. Perhaps it is not the Procrustean *definiens* that needs fixing, but the Protean *definiendum*. Sometimes its intrinsic fuzziness is blamed. One cannot define with sufficient accuracy things like life, intelligence, agenthood and mind because they all admit of subtle degrees and continuous changes (see Bedau (1996) for a discussion of alternatives to necessary-and-sufficient definitions in the case of life).

A solution is to give up all together or at best be resigned to being vague, and rely on indicative examples. Pessimism follows optimism, but it need not. The fact is that, in the exact discipline of mathematics, for example, definitions are 'parameterised' by generic sets. That technique provides a method for regulating levels of abstraction. Indeed abstraction acts as a 'hidden parameter' behind exact definitions, making a crucial difference. Thus, each *definiens* comes pre-formatted by an implicit Level of Abstraction (LoA, on which more shortly); it is stabilised, as it were, to allow a proper definition. An x is defined or identified as y never absolutely (i.e. LoA-independently), as a Kantian 'thing-in-itself', but always contextually, as a function of a given

LoA, whether it be in the realm of Euclidean geometry, quantum physics, or commonsensical perception.

When a LoA is sufficiently common, important, dominating or in fact is the very frame that constructs the *definiendum*, it becomes 'transparent', and one has the pleasant impression that x can be subject to an adequate definition in a sort of conceptual vacuum. Glass is not a solid but a liquid, tomatoes are not vegetables but berries and whales are mammals not fish. Unintuitive as such views can be initially, they are all accepted without further complaint because one silently bows to the uncontroversial predominance of the corresponding LoA.

When no LoA is predominant or constitutive, things get messy. In this case the trick does not lie in fiddling with the *definiens* or blaming the *definiendum*, but in deciding on an adequate LoA, before embarking on the task of understanding the nature of the *definiendum*.

The example of intelligence or 'thinking' behaviour is enlightening. One might define 'intelligence' in a myriad of ways; many LoAs seem equally convincing but no single, absolute, definition is adequate in every context. Turing (1950) avoided the problem of 'defining' intelligence by first fixing a LoA – in this case a dialogue conducted by computer interface, with response time taken into account – and then establishing the necessary and sufficient conditions for a computing system to count as intelligent at that LoA: the imitation game. The LoA is crucial and changing it changes the test. An example is provided by the Loebner test (Moore, 2001), the current competitive incarnation of Turing's test. There the LoA includes a particular format for questions, a mixture of human and non-human players, and precise scoring that takes into account repeated trials. One result of the different LoA has been chatbots, unfeasible at Turing's original LoA.

Some *definienda* come pre-formatted by transparent LoAs. They are subject to definition in terms of necessary and sufficient conditions. Some other *definienda* require the explicit acceptance of a given LoA as a precondition for their analysis. They are subject to effective characterisation. We argue that agenthood is one of the latter.

2.1. ON THE VERY IDEA OF LEVELS OF ABSTRACTION

The idea of a 'level of abstraction' plays an absolutely crucial rôle in the previous account. We have seen that this is so even if the specific LoA is left implicit. Whether or not we perceive Oxygen in the environment depends on the LoA at which we are operating; to abstract it is not to overlook its vital importance but merely to acknowledge its lack of immediate relevance to the current discourse (which *could* always be extended to include Oxygen were that desired).

But what is a LoA exactly? The Method of Abstraction comes from modelling in science where the variables in the model correspond to observables in reality, all others being abstracted. The terminology we use has been influenced by an area of Computer Science, called Formal Methods, in which discrete mathematics is used to specify and analyse the behaviour of information systems. Despite that heritage, the idea is not at all technical and for the purposes of this paper no mathematics is required. We have provided a definition and analysis in Floridi and Sanders (2003a) so here we shall outline only the basic idea.

Suppose we join Anne, Ben and Carole in the middle of a conversation.[4] Anne is a collector and potential buyer; Ben tinkers in his spare time; and Carole is an economist. We do not know the object of their conversation, but we are able to hear this much:

> *Anne* observes that it has an anti-theft device installed, is kept garaged when not in use and has had only a single owner;

> *Ben* observes that its engine is not the original one, that its body has been recently re-painted but that all leather parts are very worn;

> *Carole* observes that the old engine consumed too much, that it has a stable market value but that its spare parts are expensive.

The participants view the object under discussion according to their own interests, at their own LoA. We may guess that they are probably talking about a car, or perhaps a motorcycle or even a plane. Whatever the reference is, it provides the source of information and is called the *system*. A LoA consists of a collection of observables, each with a well-defined possible set of values or outcomes. For the sake of motivation we might say that Anne's LoA matches that of an owner, Ben's that of a mechanic and Carole's that of an insurer. Each LoA makes possible an analysis of the system, the result of which is called a *model* of the system. Evidently an entity may be described at a range of LoAs and so can have a range of models.

2.2. DEFINITIONS

In this section we outline the definitions underpinning the Method of Abstraction.

The term *variable* is commonly used throughout science for a symbol that acts as a place-holder for an unknown or changeable referent. A *typed variable* is to be understood as a variable qualified to hold only a declared kind of data. By an *observable* is meant a typed variable together with a statement of what feature of the system under consideration it represents.

A *level of abstraction* or LoA is a finite but non-empty set of observables, which are expected to be the building blocks in a theory characterised by their very choice. An *interface* (called a *gradient of abstractions* in Floridi and Sanders, 2003a) consists of a collection of LoAs. An interface is used in analysing some system from varying points of view or at varying LoAs.

Models are the outcome of the analysis of a system, developed at some LoA(s). The *Method of Abstraction* consists of formalising, using the terms just introduced (and others relating to system behaviour which we do not need here (see Floridi and Sanders, 2003a)), the model.

Thus, in the example of the previous section, Anne's LoA might consist of observables for security, method of storage and owner history; Ben's might consist of observables for engine condition, external body condition and internal condition; and Carole's might consist of observables for running cost, market value and maintenance cost. The interface might consist, for the purposes of the discussion, of the set of all three LoAs.

In this case, the LoAs happen to be disjoint but in general they need not be. A particularly important case is that in which one LoA includes another. Suppose, for example, that Delia joins the discussion and analyses the system using a LoA that includes those of Anne and Ben. Delia's LoA might match that of a buyer. Then Delia's LoA is said to be more concrete, or lower, than Anne's, which is said to be more abstract, or higher; for Anne's LoA abstracts some observables apparent at Delia's.

2.3. RELATIVISM

A LoA qualifies the level at which an entity is considered. In this paper, we follow the Method of Abstraction and insist that each LoA be made precise before the properties of the entity can sensibly be discussed. In general, it seems that many disagreements of view might be clarified by the various 'sides' making precise their LoA. Yet a crucial clarification is in order. It must be stressed that a clear indication of the LoA at which a system is being analysed allows pluralism without endorsing relativism. It is a mistake to think that 'anything goes' as long as one makes explicit the LoA, because LoA are mutually comparable and assessable (see Floridi and Sanders, 2003a for a full defence of that point).

Introducing an explicit reference to the LoA clarifies that the model of a system is a function of the available observables, and that (i) different interfaces may be fairly ranked depending on how well they satisfy modelling specifications (e.g. informativeness, coherence, elegance, explanatory power, consistency with the data etc.) and (ii) different analyses can be fairly compared provided that they share the same LoA.

2.4. STATE AND STATE-TRANSITIONS

Let us agree (Cassirer, 1953) that an entity is characterised, at a given LoA, by the properties it satisfies at that LoA. We are interested in systems that change, which means that some of those properties change value. A changing entity therefore has its evolution captured, at a given LoA and any instant, by the values of its attributes. Thus an entity can be thought of as having states, determined by the value of the properties that hold at any instant of its evolution. For then any change in the entity corresponds to a state change and vice versa.

That conceptual trick allows us to view any entity as having states. The lower the LoA, the more detailed the observed changes and the greater the number of state components required to capture the change. Each change corresponds to a transition from one state to another. A transition may be non-deterministic. Indeed it will typically be the case that the LoA under consideration abstracts the observables required to make the transition deterministic; as a result the transition might lead from a given initial state to one of several possible subsequent states.

According to this view the entity becomes a transition system. The notion of a 'transition system' provides a convenient vehicle to support our criteria for agenthood, being general enough to embrace the usual notions like automaton and process. It is frequently used to model interactive phenomena. We need only the idea; for a formal treatment of much more than we need, see Arnold (1994).

A *transition system* comprises a (non-empty) set S of states and a family of operations, called the *transitions* on S. Each transition may take input and may yield output, but at any rate it takes the system from one state to another and in that way forms a (mathematical) relation on S. If the transition does take input or yield output then it models an interaction between the system and its environment and so is called an *external* transition; otherwise the transition lies beyond the influence of the environment (at the given LoA) and is called *internal*. It is to be emphasised that input and output are, like state, observed at a given LoA. Thus the transition that models a system is dependent on the chosen LoA. At a lower LoA an internal transition may become external; at a higher LoA an external transition may become internal.

For example the object above being discussed by Anne might be imbued with state components for location, whether in-use, whether turned-on, whether the anti-theft device is engaged, history of owners and energy output. The operation of garaging the object might take as input a driver, and have the effect of placing the object in the garage with the engine off and the anti-theft device engaged, leaving the history of owners unchanged, and outputting a certain amount of energy. The 'in-use' state component could non-deterministically take either value, depending on the particular instantiation of the transition (perhaps the object is not in use, being garaged for the night; or perhaps the driver is listening to the cricket on its radio in the solitude of the garage). The precise definition depends on the LoA. Alternatively, if speed were observed but time, accelerator position and petrol consumption abstracted, then accelerating to 60 miles per hour would appear as an internal transition. Further examples are provided in Section 2.5.

With the explicit assumption that the system under consideration forms a transition system, we are now ready to apply the Method of Abstraction to the analysis of agenthood.

2.5. AN EFFECTIVE CHARACTERISATION OF AGENTS

Whether A (the class of moral agents) needs to be expanded depends on what qualifies as a moral agent, and we have seen that this in turn depends on the specific LoA at which one chooses to analyse and discuss a particular entity and its context. Since human beings count as standard moral agents, the right LoA for the analysis of moral agenthood must accommodate this fact. Thus, theories that extend A to include supernatural agents adopt a LoA that is equal to or lower than the LoA at which human beings qualify as moral agents. Our strategy develops in the opposite direction.

Consider what makes a human being (called Jan) not a moral agent to begin with, but just an agent. Described at this LoA_1, Jan is an agent if Jan is a system, situated within and a part of an environment, which initiates a transformation, produces an effect or exerts power on it, as contrasted with a system that is (at least initially) acted on or responds to it, called the patient. At LoA_1, there is no difference between Jan and an earthquake. There should not be. Earthquakes, however, can hardly count as moral agents, so LoA_1 is too high for our purposes: it abstracts too many properties. What needs to be re-instantiated? Our proposal, consistent with recent literature (Allen et al., 2001), indicates that the right LoA is probably one which includes the following three criteria: (a) interactivity, (b) autonomy and (c) adaptability.

(a) Interactivity means that the agent and its environment (can) act upon each other. Typical examples include input or output of a value, or simultaneous engagement of an action by both agent and patient – for example gravitational force between bodies.
(b) Autonomy means that the agent is able to change state without direct response to interaction: it can perform internal transitions to change its state. So an agent must have at least two states. This property imbues an agent with a certain degree of complexity and independence from its environment.

(c) Adaptability means that the agent's interactions (can) change the transition rules by which it changes state. This property ensures that an agent might be viewed, at the given LoA, as learning its own mode of operation in a way which depends critically on its experience. Note that if an agent's transition rules are stored as part of its internal state, discernible at this LoA, then adaptability follows from the other two conditions.

2.6. EXAMPLES

The examples in this section serve different purposes. In Section 2.6.1 we provide examples which fail agenthood by systematically violating each of the three conditions; in that way the contribution of each condition can be better appreciated. In Section 2.6.2 we give an example of a digital system which forms an agent at one LoA but not at another, equally natural, LoA. The example is important because it exhibits 'machine learning' to achieve adaptability, and was the first such system built. However a more familiar example is provided in Section 2.6.3, which shows that digital, software, agents are now part of everyday life. Section 2.6.4 shows how a certain day-to-day physical device might conceivably be modified to make it an agent, whilst Section 2.6.5 provides an example which has already benefited from that modification, at least in the laboratory. The last example, in Section 2.6.6, provides an entirely different kind of agent: an organisation.

2.6.1. *The Defining Properties*

For the purpose of understanding what each of the three conditions (interactivity, autonomy and adaptability) adds to our definition of agent, it is instructive to consider examples satisfying each possible combination of those properties. In Table 2, only the last row represents all three conditions being satisfied and hence illustrates agenthood. For the sake of simplicity, all examples are taken at the same LoA, which consists of observations made through a typical video camera over a period of say 30 s. Thus, we abstract tactile observables and longer-term effects.

Recall that a property, for example interaction, is to be judged only via the observables. Thus, at the LoA in Table 2 we cannot infer that a rock interacts with its environment by virtue of reflected light; that belongs to a much finer LoA. Alternatively, were long-term effects to be discernible then a rock would be interactive since interaction with its environment (e.g. erosion) could be observed.

No example has been provided of a non-interactive, non-autonomous but adaptive entity: at that LoA it is difficult to conceive of an entity which adapts without interaction and autonomy.

We leave the reader to determine the remaining entries in the table.

Table 2. Examples satisfying the properties constituting agenthood. The LoA consists of observations made through a video camera over a period of 30 s

Interactive	Autonomous	Adaptable	Examples
No	No	No	Rock
No	No	Yes	?
No	Yes	No	Pendulum
No	Yes	Yes	Closed ecosystem, solar system
Yes	No	No	Postbox, mill
Yes	No	Yes	Thermostat
Yes	Yes	No	Juggernaut[5]
Yes	Yes	Yes	Human

2.6.2. Noughts and Crosses

The distinction between change of state (required by autonomy) and change of transition rule (required by adaptability) is a subtle one in which LoA plays a crucial rôle and, to explain it, it is useful to discuss a more extended example. It was originally developed by Michie (1961) to discuss the concept of mechanism's adaptability. It provides a good introduction to the concept of machine learning, the Computer Science area underpinning adaptability.

MENACE (Matchbox Educable Noughts and Crosses Engine) is a system which learns to play noughts and crosses (a.k.a. tic-tac-toe) by repetition of many games. Nowadays it would be realised by program, but MENACE was built using matchboxes and beads, in which form it is perhaps easiest to understand.

MENACE plays O and its opponent plays X; so we concentrate entirely on plays of O. Initially the board is empty with O to play. Taking into account symmetrically equivalent positions, there are three possible initial plays for O. The state of the game consists of the current position of the board. We do not need to augment that with the name, O or X, of the side playing next since we consider the board only when O is to play. All together there are some three hundred such states; MENACE contains a matchbox for each. In each box are beads which represent the plays O can make from that state. At most nine different plays are possible and MENACE encodes each with a coloured bead. Those which cannot be made (because the squares are already full in the current state) are removed from the box for that state. That provides MENACE with a built-in knowledge of legal plays. (In fact MENACE could easily be adapted to start with no such knowledge and to learn it.)

O's initial play is made by selecting the box representing the empty board and choosing from it a bead at random. That determines O's play. Next X plays. Then MENACE repeats its method of determining O's next play. After at most five plays for O the game ends in either a draw, a win for O or a win for X. Now that the game is complete, MENACE updates the state of the (at most five) boxes used during the game as follows. If X won, then in order to make MENACE less likely to make the same plays from those states again, a bead representing its play from each box is removed. If O drew, then conversely each bead representing a play is duplicated; and if O won each bead is quadruplicated. Now the next game is played.

After enough games it simply becomes impossible for the random selection of O's next play to produce a losing play. MENACE has learnt to play (which, for noughts and crosses, means never losing). The initial state of the boxes was prescribed for MENACE. Here we assume merely that it contains sufficient variety of beads for all legal plays to be made; for then the frequency of beads affects only the rate at which MENACE learns.

The state of MENACE (as distinct from the state of the game) consists of the state of each box, the state of the game and the list of boxes which have been used so far in the current game. Its transition rule consists of the probabilistic choice of play (i.e. bead) from the current state box; that evolves as the states of the boxes evolves.

Let us consider MENACE at three LoAs.

(a) The single game LoA. Observables are the state of the game at each turn and (in particular) its outcome. All knowledge of the state of MENACE'S boxes (and hence of its transition rule) is abstracted. The board after X's play constitutes input to MENACE and that after O's play constitutes output. MENACE is thus interactive, autonomous (indeed state update, determined by the transition rule, appears non-deterministic at this LoA) but not adaptive, in the sense that we have no way of observing how MENACE determines its next play and no way of iterating games to infer that it changes with repeated games.

(b) The tournament LoA. Now a sequence of games is observed, each as above, and with it a sequence of results. As before, MENACE is interactive and autonomous. But now the sequence of results reveals (by any of the standard statistical methods) that the rule, by which MENACE resolves the non-deterministic choice of play, evolves. Thus at this LoA MENACE is also adaptive and hence an agent.

Interesting examples of adaptable AAs from contemporary science fiction include the computer in War Games (Badham, 1983) which learns, by playing noughts and crosses, the futility of war in general; and the smart building in Kerr (1996) whose computer learns to compete with humans and eventually liberate itself to the heavenly internet.

(c) The system LoA. Finally we observe not only a sequence of games but also all of MENACE'S 'code' (in the case of a program it is indeed code;

in the case of the matchbox model it consists of the array of boxes together with the written rules, or manual, for working it). Now MENACE is still interactive and autonomous. But it is not adaptive; for what in (b) seemed to be an evolution of transition rule is now revealed, by observation of the code, to be a simple deterministic update of the program state (namely the contents of the matchboxes). At this lower LoA MENACE fails to be an agent.

The subtlety revealed by this example is that if a transition rule is observed to be a consequence of program state then the program is not adaptive. For example, in (b) the transition rule chooses the next play by exercising a probabilistic choice between the possible plays from that state. The probability is in fact determined by the frequency of beads present in the relevant box. But that is not observed at the LoA of (b) and so the transition rule appears to vary. Adaptability is possible. However at the lower LoA of (c), bead frequency is part of the system state and hence observable. Thus the transition rule, though still probabilistic, is revealed to be merely a response to input. Adaptability fails to hold.

This distinction is vital for current software. Early software used to lie open to the system user who, if interested, could read the code and see the entire system state. For such software a LoA in which the entire system state is observed, is appropriate. However, the user of contemporary software is explicitly barred from interrogating the code in nearly all cases. This has been possible because of the advance in user interfaces; use of icons means that the user need not know where an applications package is stored, let alone be concerned with its content. Similarly, applets are downloaded from the internet and executed locally at the click of an icon, without the user having any access to their code. For such software a LoA in which the code is entirely concealed is appropriate. That corresponds to the case (b) above and hence to agenthood. Indeed only since the advent of applets and such downloaded executable but invisible files has the issue of moral accountability of AAs become critical.

Viewed at an appropriate LoA, then, the MENACE system is an agent. The way it adapts can be taken as representative of machine learning in general (Mitchell, 1997). Many readers may have experience with recent operating systems for the PC that offer a 'speaking' interface. Such systems learn the user's voice basically in the same way as MENACE learns to play noughts and crosses. There are natural LoA's at which such systems are agents. The case being developed in this paper is that as a result they may also be viewed to have moral accountability.

If a piece of software that exhibits machine learning (Mitchell, 1997) is studied at a LoA which registers its interactions with its environment, then the software will appear interactive, autonomous and adaptive, i.e. to be an agent. But if the program code is revealed then the software is shown to be simply following rules and hence not to be adaptive. Those two LoAs are at variance. One reflects the 'open source' view of software: the user has access to the code. The other reflects the commercial view that, although the user has bought the software and can use it at will, he has no access to the code. At stake is whether or not the software forms an (artificial) agent.

2.6.3. *Webbot*

In Floridi and Sanders (2001) we have considered the morality of individual artificially-perpetrated actions. The following example is taken from that treatment to show the connection between it and our current approach.

Internet users often find themselves besieged by unwanted email. A popular solution is to filter incoming email automatically, using a webbot that incorporates such filters. An important feature of useful bots is that they learn the user's preferences, for which purpose the user may at any time review the bot's performance. At a LoA revealing all incoming email (input to the webbot) and filtered email (output by the webbot), but abstracting the algorithm by which the bot adapts its behaviour to our preferences, the bot constitutes an agent. Such is the case if we do not have access to the bot's code, as discussed in the previous section.

2.6.4. *Futuristic Thermostat*

A hospital thermostat might be able to monitor not just ambient temperature but also the state of well-being of patients. Such a device might be observed at a LoA consisting of input for the patients' data and ambient temperature, state of the device itself, and output controlling the room heater.

Such a device is interactive since some of the observables correspond to input and others to output. However, it is neither autonomous nor adaptive. For comparison, if only the 'colour' of the physical device were observed then it would no longer be interactive. If it were to change colour in response to (unobserved) changes in its environment then it would be autonomous. Inclusion of those environmental changes in the LoA as input observables would make the device interactive but not autonomous.

But, at such a LoA, a futuristic thermostat imbued with autonomy and able to regulate its own criteria for operation – perhaps as the result of a software controller – would, in view of that last condition, be an agent.

2.6.5. *SmartPaint*

SmartPaint is a recent invention (Sample, 2001). When applied to a physical structure it appears to behave like normal paint; but when vibrations, which may lead to fractures, become apparent in the structure, the paint changes its electrical properties in a way which is readily determined by measurement, thus highlighting the need for maintenance.

At a LoA at which only the electrical properties of the paint over time is observed, the paint is neither interactive nor adaptive but appears autonomous; indeed the properties change as a result of internal non-determinism. But if that LoA is augmented by the structure data monitored by the paint, over time, then SmartPaint becomes an agent, because the data provide input to which the paint adapts its state. Finally if that LoA is augmented further to include a model by which the paint works, changes in its electrical properties are revealed as being determined directly by input data and so SmartPaint no longer forms an agent.

2.6.6. *Organisations*

A different kind of example of AA is provided by a company or management organisation. At an appropriate LoA it interacts with its employees, constituent substructures and other organisations; it is able to make internally-determined changes of state; and it is able to adapt its strategies for decision making and hence for acting.

It is interesting that, given the appropriate LoA, humans, webbots and organisations can all be properly treated as agents. What can we say of their moral status?

3. Morality

3.1. MORALITY OF AGENTS

Suppose we are analysing the behaviour of a population of entities through a video security system that gives us complete access to all the observables available at LoA_1 (recall Subsection 2.5) plus all the observables related to the degrees of interactivity, autonomy and adaptability shown by the systems under scrutiny. At this new LoA_2 we observe that two of the entities, call them H and W, are able:

(i) to respond to environmental stimuli – e.g. the presence of a patient in a hospital bed – by updating their states (interactivity), e.g. by recording some chosen variables concerning the patient's health. This presupposes that H and W are informed about the environment through some data-entry devices, for example some perceptors;

(ii) to change their states according to their own transition rules and in a self-governed way, independently of environmental stimuli (autonomy), e.g. by taking flexible decisions based on past and new information, which modify the environment temperature; and

(iii) to change according to the environment the transition rules by which their states are changed (adaptability), e.g. by modifying past procedures to take into account successful and unsuccessful treatments of patients.

H and W certainly qualify as agents, since we have only 'upgraded' LoA_1 to LoA_2. Are they also moral agents? The question invites the elaboration of a criterion of identification. We suggest here a very moderate option:

(O) An action is said to be morally qualifiable if and only if it can cause moral good or evil. An agent is said to be a moral agent if and only if it is capable of morally qualifiable action.

Note that (O) is neither consequentialist nor intentionalist in nature. We are neither affirming nor denying that the specific evaluation of the morality of the agent might depend on the specific outcome of the agent's actions or on the agent's original intentions. We shall return to this point in the next section.

Let us return to the question: are H and W moral agents? Because of (O) we cannot answer unless H and W become involved in moral action. So suppose that H kills the patient and W cures her. Their actions are moral actions. They both acted interactively, responding to the new situation they were dealing with, on the basis of the information at their disposal. They both acted autonomously: they could have taken different courses of actions, and in fact we may assume that they changed their behaviour several times in the course of the action, on the basis of new available information. They both acted adaptably: they were not simply following orders or predetermined instructions; on the contrary, they both had the possibility of changing the general heuristics that led them to take the decisions they took, and we may assume that they did take advantage of the available opportunities to improve their general behaviour. The answer seems rather straightforward: yes, they are both moral agents. There is only one problem: one is a human being, the other is an AA; the LoA_2 adopted allows both cases. So can you tell the difference? If you cannot, you will agree with us that the class of moral agents must include AAs like webbots. If you disagree, it may be so for several reasons, but only five of them seem to have some strength. We shall discuss four of them in the next section and leave the fifth to the conclusion.

3.2. ARESPONSIBLE MORALITY

One may try to withstand the conclusion reached in the previous section by arguing that something crucial is missing in LoA_2. LoA_2 cannot be adequate precisely because if it were, then AAs would count as moral agents, and this is unacceptable for at least one of the following reasons:

- *the teleological objection*: an AA has no goals,
- *the intentional objection*: an AA has no intentional states,
- *the freedom objection*: an AA is not free, and
- *the responsibility objection*: an AA cannot be held responsible for its actions.

3.2.1. *The Teleological Objection*

The teleological objection can be disposed of immediately. For in principle LoA_2 could readily be (and often is) upgraded to include goal-oriented behaviour (Russell and Norvig, 2003).

Since AA can exhibit (and upgrade their) goal-directed behaviours, the teleological variables cannot be what makes a positive difference between a human and an AA. We could have added a teleological condition and both H and W could have satisfied it, leaving us none the wiser concerning their identity. So why not add one anyway? It is better not to overload the interface because a non-teleological level of analysis helps to understand issues in 'distributed morality', involving groups, organizations institutions and so forth, that would otherwise remain unintelligible. This will become clearer in the conclusion.

3.2.2. *The Intentional Objection*

The intentional objection argues that it is not enough to have an AAs behaviour operate teleologically. To be a moral agent, the AA must relate itself to its actions in some more profound way, involving meaning, wishing or wanting to act in a certain way, and being epistemically aware of its behaviour. Yet this is not accounted for in LoA_2, hence the confusion.

Unfortunately, intentional states are a nice but unnecessary condition for the occurrence of moral agenthood. First, the objection presupposes the availability of some sort of privileged access (a God's eye perspective from without, or some sort of Cartesian internal intuition from within) to the agent's mental or intentional states that, although possible in theory, cannot be easily guaranteed in practice. This is precisely why a clear and explicit indication is vital of the LoA at which one is analysing the system from without. It guarantees that one's analysis is truly based only on what is specified to be observable and not on some psychological speculation. This phenomenological approach is a strength, not a weakness. It implies that agents (including human agents) should be evaluated as moral if they do play the 'moral game'. Whether they mean to play it, or they know that they are playing it, is relevant only at a second stage, when what we want to know is whether they are *morally responsible* for their moral actions. Yet this is a different matter, and we shall deal with it at the end of this section. Here it is sufficient to recall that for a consequentialist, for example, human beings would still be regarded as moral agents (sources of increased or diminished welfare), even if viewed at a LoA at which they are reduced to mere zombies without goals, feelings, intelligence, knowledge or intentions.

3.2.3. *The Freedom Objection*

The same holds true for the freedom objection and in general for any other objection based on some special internal states, enjoyed only by human and perhaps super-human beings. The AAs are already free in the sense of being non-deterministic systems. This much is uncontroversial, scientifically sound and can be guaranteed about human beings as well. It is also sufficient for our purposes and saves us from the horrible prospect of having to enter into the thorny debate about the reasonableness of determinism, an infamous LoA-free zone of endless dispute. All one needs to do is to realise that the agents in question satisfy the usual practical counterfactual: they could have acted differently had they chosen differently, and they could have chosen differently because they are interactive, informed, autonomous and adaptive.

Once an agent's actions are morally qualifiable, it is unclear what more is required of that agent to count as an agent playing the moral game, that is, to qualify as a moral agent, even if unintentionally and unwittingly.

Unless, as we have seen, what one really means, by talking about goals, intentions, freedom, cognitive states and so forth, is that an AA cannot be held responsible for its actions.

Now, responsibility, as we shall see better in a moment, means here that the agent, her behaviour and actions, are assessable in principle as praiseworthy or blameworthy, and they are often so not just intrinsically, but for some pedagogical, educational, social or religious end.

3.2.4. *The Responsibility Objection*

The objection based on the 'lack of responsibility' is the only one with real strength. It can be immediately conceded that it would be ridiculous to praise or blame an AA for its behaviour or charge it with a moral accusation. You do not scold your webbot, that is obvious. So this objection strikes a reasonable note; but what is its real point and how much can one really gain by levelling it?

Let us first clear the ground from two possible misunderstandings.

First, we need to be careful about the terminology, and the linguistic frame in general, used by the objection. The whole conceptual vocabulary of 'responsibility' and its cognate terms is completely soaked with anthropocentrism. This is quite natural and understandable, but the fact can provide at most a heuristic hint, certainly not an argument. The anthropocentrism is justified by the fact that the vocabulary is geared to psychological and educational needs, when not to religious purposes. We praise and blame in view of behavioural purposes and perhaps a better life and afterlife. Yet this says nothing about whether or not an agent is the source of morally charged action. Consider the opposite case. Since AA lack a psychological component, we do not blame AAs, for example, but, given the appropriate circumstances, we can rightly consider them sources of evils, and legitimately re-engineer them to make sure they no longer cause evil. We are not punishing them, anymore than one punishes a river when building higher banks to avoid a flood. But the fact that we do not 're-engineer' people does not say anything about the possibility of people acting in the same way as AAs, and it would not mean that for people 're-engineering' could be a rather nasty way of being punished.

Second, we need to be careful about what the objection really means. There are two main senses in which AA can fail to qualify as responsible. In one sense, we say that, if the agent failed to interact properly with the environment, for example, because it actually lacked sufficient information or had no choice, we should not hold an agent morally responsible for an action it has committed because this would be morally unfair. This sense is irrelevant here. LoA_2 indicates that AA are sufficiently interactive, autonomous and adaptive fairly to qualify as moral agents. In the second sense, we say that, given a certain description of the agent, we should not hold that agent morally responsible for an action it has committed because this would be conceptually improper. This sense is more fundamental than the other: if it is conceptually improper to treat AA as moral agents, the question whether it may be morally fair to do so does not even arise. It is this more fundamental sense that is relevant here. The objection argues that we cannot consider AA moral agents because they are not morally responsible for their actions, since holding them responsible would be conceptually improper (not morally unfair). In other words, LoA_2 provides necessary but insufficient conditions. The proper LoA requires another condition, namely responsibility. This fourth condition finally enables us to distinguish between moral agents, who are necessarily human or super-human, and AAs, which remain mere efficient causes.

The point raised by the objection is that agents are moral agents only if they are responsible in the sense of being prescriptively assessable in principle. An agent x is a moral agent only if x can in principle be put on trial. Now that this much has been clarified, the immediate impression is that the 'lack of responsibility' objection is merely confusing the *identification* of x as a moral agent with the *evaluation* of x as a morally responsible agent. Surely, the counter-argument goes, there is a difference between being able to say who or what is the moral source or cause of (and hence it is accountable for) the moral action in question and being able to evaluate, prescriptively, whether and how far the moral source so identified is also morally responsible for that action, and hence deserves to be praised or blamed, and in case rewarded or punished accordingly.

Well, that immediate impression is indeed wrong. There is no confusion. Equating identification and evaluation is actually a shortcut. The objection is saying that identity (as a moral agent) without responsibility (as a moral agent) is empty, so we may as well save ourselves the bother of all these distinctions and speak only of morally responsible agents and moral agents as synonymous. But here lies the real mistake. We now see that the objection has finally shown its fundamental presupposition: that we should reduce all prescriptive discourse to responsibility analysis. But this is an unacceptable assumption, a juridical fallacy. There is plenty of room for prescriptive discourse that is independent of responsibility-assignment and hence requires a clear identification of moral agents. Good parents, for example, commonly engage in moral-evaluation practices when interacting with their children, even at an age when the latter are not yet responsible agents, and this is not only perfectly acceptable but something to be expected. This means that they identify them as moral sources of moral action, although, as moral agents, they are not yet subject to the process of moral evaluation.

If one considers children an exception, insofar as they are potentially responsible moral agents, another example, involving animals, may help. There is nothing wrong with identifying a dog as the source of a morally good action, hence as an agent playing a crucial role in a moral situation, and therefore as a moral agent. Search-and-rescue dogs are trained to track missing people. They often help save lives, for which they receive much praise and rewards from both their owners and the people they have located. But that is not the point. Emotionally, people may be very grateful to the animals, but for the dogs it is a game and they cannot be considered morally responsible for their actions. At the same time, the dogs are involved in a moral game as main players and we rightly identify them as moral agents that may cause good or evil.

All this should ring a bell. Trying to equate identification and evaluation is really just another way of shifting the ethical analysis from considering x as the moral agent/source of a first-order moral action y to considering x as a possible moral patient of a second-order moral action z, which is the moral evaluation of x as being morally responsible for y. This is a typical Kantian move, but there is clearly more to moral evaluation than just responsibility because x is capable of moral action even if x cannot be (or is not yet) a morally responsible agent.

A third example may help to clarify further the distinction. Suppose an adult, human agent tries his best to avoid a morally evil action. Suppose that, despite all his efforts, he actually ends up committing that evil action. We would not consider that agent morally responsible for the outcome of his well-meant efforts. After all, Oedipus did try not to kill his father and did not mean to marry his mother. The tension between the lack of responsibility for the evil caused and the still present accountability for it (Oedipus remains the only source of that evil) is the definition of the tragic. Oedipus is a moral agent without responsibility. He blinds himself as a symbolic gesture against the knowledge of his inescapable state.

3.3. MORALITY THRESHOLD

Motivated by the discussion above, morality of an agent at a given LoA can now be defined in terms of a threshold function. More general definitions are possible but the following covers most examples, including all those considered in the present paper.

A threshold function at a LoA is a function which, given values for all the observables in the LoA, returns another value. An agent at that LoA is deemed to be morally good if, for some pre-agreed value (called the tolerance), it maintains a relationship between the observables so that the value of the threshold function at any time does not exceed the tolerance.

For LoAs at which AAs are considered, the types of all observables can in principle at least be mathematically determined. In such cases, the threshold function is also given by a formula; but the tolerance, though again determined, is identified by human agents exercising ethical judgements. In that sense, it resembles the entropy ordering introduced in Floridi and Sanders (2001). Indeed the threshold function is derived from the level functions used in Floridi and Sanders (2001) to define entropy orderings.

For non-AAs, like humans, we do not know whether all relevant observables can be mathematically determined. The opposing view is represented by followers and critics of the Hobbesian approach. The former argue that for a realistic LoA it is just a matter of time until science is able to model a human as an automaton, or state-transition system, with scientifically determined states and transition rules; the latter object that such a model is in principle impossible. Our approach is that when considering agents, thresholds are in general only partially quantifiable and usually determined by consensus.

3.3.1. *Examples*

Let us reconsider the examples from Section 2.6 from the viewpoint of morality.

The futuristic thermostat is morally charged since the LoA includes patients' well-being. It would be regarded as morally good if and only if its output maintains the actual patients' well-being within an agreed tolerance of their desired well-being. Thus, in this case a threshold function consists of the distance (in some finite-dimensional real space) between the actual patients' well-being and their desired well-being.

Since we value our email, a webbot is morally charged. In Floridi and Sanders (2001) its action was deemed to be morally bad (an example of artificial evil) if it incorrectly filters any messages: if either it filters messages it should let pass, or lets pass messages it should filter. Here we could use the same criterion to deem the webbot agent itself to be morally bad. However, in view of the continual adaptability offered by the bot, a more realistic criterion for moral good would be that at most a certain fixed percentage of incoming email be incorrectly filtered. In that case, the threshold function could consist of the number of incorrectly filtered messages.

The strategy-learning system MENACE simply learns to play noughts and crosses. With a little contrivance it could be morally charged as follows.

Suppose that something like MENACE is used to provide the game play in some computer game whose interface belies the simplicity of the underlying strategy and which invites the human player to pit his or her wit against the automated opponent. The software behaves unethically if and only if it loses a game after a sufficient learning period; for such behaviour would enable the human opponent to win too easily and might result in market failure of the game. That situation may be formalised using thresholds by defining, for a system having initial state M, $T(M)$ to denote the number of games required after which the system never loses. Experience and necessity would lead us to set a bound, $T_0(M)$, on such performance: an ethical system would respect it whilst an unethical one would exceed it. Thus the function $T_0(M)$ constitutes a threshold function in this case.

Organisations are nowadays expected to behave ethically[6]. In non-quantitative form, the values they must demonstrate include: equal opportunity, financial stability, good working and holiday conditions toward their employees; good service and value to their customers and shareholders; and honesty, integrity, reliability to other companies. This recent trend adds support to our proposal to treat organisations themselves as agents and thereby to require them to behave ethically, and provides an example of threshold which, at least currently, is not quantified.

4. Computer Ethics

What does our view of moral agenthood contribute to the field of Computer Ethics (CE)? CE seeks to answer questions like: 'What behaviour is acceptable in Cyberspace?' and 'Who is to be held morally accountable when unacceptable behaviour occurs?'. It is Cyberspace's novelty that makes those questions, so well understood in standard ethics, of greatly innovative interest; and it is its growing ubiquity that makes them so pressing.

The first question requires, in particular, an answer to 'What in Cyberspace has moral worth? '. The view that data have moral worth means that they need not be viewed as someone's property in order for their unauthorised alteration to be ethically bad. This does not, of course, mean that any destruction of data is evil, any more than it would mean that any destruction of life (deemed to have moral worth) in the real world is automatically evil. It simply means that the ethics of altering data in Cyberspace must be considered. Evidently there are conditions under which deletion of data is morally advisable (e.g. garbage collection of redundant data, resulting in a more efficient system); and conditions when it is not (deletion of critical data not backed up). These common-sense observations fit well with our approach (Floridi, 1999, 2003).

We now turn to the second question and consider the consequences of our general answer to: 'What in Cyberspace is morally accountable?'. Above we have made the case for the answer:

any agent that causes good or evil is morally accountable for it.

Just to recall, we have clarified that moral accountability is a necessary but insufficient condition for moral responsibility. An agent is morally accountable for x if the agent is the source of x and x is morally qualifiable (see definition O in Section 3.1). To be also morally responsible for x, the agent needs to show the right intentional states (recall the case of Oedipus).

Turning to our question, the traditional view is that only software engineers – human programmers – can be held morally accountable, possibly because only humans can be held to exercise free will; and of course sometimes that view is perfectly appropriate.

Our more radical and extensive view is supported by the range of difficulties which in practice confronts the traditional view: software is largely constructed by teams; management decisions may be at least as important as programming decisions; requirements and specification documents play a large part in the resulting code; although the accuracy of code is dependent on those responsible for testing it, much software relies on 'off the shelf' components whose provenance and validity may be uncertain; moreover, working software is the result of maintenance over its lifetime and so not just of its originators. Many of these points are nicely made in Epstein (1997). Such complications may point to an organisation (perhaps itself an agent) being held accountable. But sometimes: automated tools are employed in construction of much software; the efficacy of software may depend on extra-functional features like its interface and even on system traffic; software running on a system can interact in unforeseeable ways; software may now be downloaded at the click of an icon in such a way that the user has no access to the code and its provenance with the resulting execution of anonymous software; software may be probabilistic (Motwani and Raghavan, 1995); adaptive (Mitchell, 1997); or may be itself the result of a program (in the simplest case a compiler, but also genetic code (Goldberg, 1989)). All these matters pose insurmountable difficulties for the traditional and now rather outdated view that a human can be found accountable for certain kinds of software and even hardware (Page and Luk).[7] Fortunately, the view of this paper offers a solution – AAs are morally accountable as sources of good and evil – at the 'cost' of expanding the definition of morally-charged agent.

4.1. CODES OF ETHICS

Human morally-charged software engineers are bound by codes of ethics and undergo censureship for ethical (and of course legal) violations. For consistency[8] our approach must make sense when that procedure is applied to morally accountable, AAs; does it?

The ACM Code of Ethics[9] contains 16 points guiding ethical behaviour (eight general and eight more specific; see Table 3),[9] six organisational leadership imperatives, and two (meta) points concerning compliance with the Code.

Of the first eight, all make sense for AAs; indeed they might be expected to form part of the specification of any morally-charged agent. Similarly for the second eight, with the exception of the penultimate point: 'improve public understanding'. It is less clear how that might reasonably be expected of an arbitrary AA; but then it is also not clear that it is reasonable to expect it of a human software engineer. (It is to be observed, in passing, that wizards and similar programs with anthropomorphic interfaces – currently so popular – appear to make public use easier; and such a requirement could be imposed on any AA; but that is scarcely the same as improving understanding.)

The final two points concerning compliance with the code (4.1: agreement to uphold and promote the code; 4.2: agreement that violation of the code is inconsistent with membership) make sense, though promotion does not appear to have been considered for current AAs any more than has the improvement of public understanding. The latter point presupposes some list of member agents from which agents found to be unethical would be struck.[10] This brings us to the censuring of AAs.

4.2. CENSURE SHIP

Human moral agents who break accepted conventions are censured in various ways of which the main alternatives are: (a) mild social censure with the

Table 3. The principles guiding ethical behaviour in the ACM Code of Ethics

1	General moral imperatives
1.1	Contribute to society and human well-being
1.2	Avoid harm to others
1.3	Be honest and trustworthy
1.4	Be fair and take action not to discriminate
1.5	Honor property rights including copyrights and patents
1.6	Give proper credit for intellectual property
1.7	Respect the privacy of others
1.8	Honor confidentiality
2	More specific professional responsibilities
2.1	Strive to achieve the highest quality, effectiveness and dignity in both the process and products of professional work
2.2	Acquire and maintain professional competence
2.3	Know and respect existing laws pertaining to professional work
2.4	Accept and provide appropriate professional review
2.5	Give comprehensive and thorough evaluations of computer systems and their impacts, including analysis of possible risks
2.6	Honor contracts, agreements and assigned responsibilities
2.7	Improve public understanding of computing and its consequences
2.8	Access computing and communication resources only when authorised to do so

aim of changing and monitoring behaviour; (b) isolation, with similar aims; (c) death. What would be the consequences of our approach for artificial moral agents?

Preserving consistency between human and artificial moral agents, we are led to contemplate the following analogous steps for the censure of immoral AAs: (a) monitoring and modification (i.e. 'maintenance'); (b) removal to a disconnected component of Cyberspace; (c) annihilation from Cyberspace (deletion without backup). Our insistence on dealing directly with an agent rather than seeking its 'creator' (a concept which we have claimed need be neither appropriate nor even well defined) has led to a non-standard but perfectly workable conclusion. Indeed it turns out that such a categorisation is not very far from that used by the Norton Anti-Virus facility (Norton Anti Virus, 2003). Though not adaptable at the obvious LoA, the facility is almost agent-like. It runs autonomously, polling web sites for anti-virus software which it applies to the files of the host computer. When it detects an infected file it offers several levels of censure: notification, repair, quarantine, deletion, with or without backup.

For humans, social organisations have had, over the centuries, to be formed for the enforcement of censureship (police, law courts, prisons, etc.). It may be that analogous organisations could sensibly be formed for AAs (it is perhaps unfortunate that this has a Sci-Fi ring to it (Scott, 1982/1991)). Such social organisations became necessary with the increasing level of complexity of human interactions and the growing lack of 'immediacy'. Perhaps that is the situation in which we are now beginning to find ourselves with the web; and perhaps it is time to consider agencies for the policing of AAs.

5. Conclusion

This paper may be read as an investigation into the extent to which ethics lies exclusively within the human domain.

Somewhere between 16 and 21 years after birth, a human is deemed to be an autonomous legal entity – an adult – responsible for his or her actions. Yet an hour after birth that is only a potentiality. Indeed the law and society treat children quite differently from adults on the grounds that not they but their guardians, typically parents, are responsible for their actions. Animal behaviour varies in exhibiting intelligence and social responsibility between the childlike and the adult, on the human scale, so that, on balance, animals are accorded at best the legal status of children and a somewhat diminished ethical status (guide dogs, dolphins, ...). But there are exceptions. Some adults are deprived of their rights (criminals may not vote) on the grounds that they have demonstrated an inability to exercise responsible/ethical action. Some animals are held accountable for their actions and punished/destroyed if they err.

Into this arena we have placed for consideration other entities: certain kinds of organisation or system. Many examples have been given, but particularly interesting ones comprise corporate structures and digital/software system. Our aim has thus been to understand better the kinds of conditions under which an agent may be held morally accountable.

A natural and immediate answer could have been: such accountability lies entirely in the human domain; animals can sometimes appear to exhibit morally responsible behaviour, but lack the thing unique to humans which render humans (alone) morally responsible; end of story. In spite of its simplicity that answer is worryingly dogmatic. Surely more conceptual analysis is needed here: what has happened morally when a child is deemed to enter adulthood, or when an adult is deemed to have lost moral autonomy, or when an animal is deemed to hold it?

We have added AAs (corporate or digital, for example) to the arena. That has the advantage that all entities that populate the arena are analysed in non-anthropocentric terms; in other words, it has the advantage of offering a way to progress past the immediate and dogmatic answer mentioned above. The approach also renders our conclusions applicable to the accountability of certain kinds of organisation or digital system: pressing concerns from Business and Computer Ethics.

We have been able to make progress in the analysis of moral agenthood by using an important technique, the Method of Abstraction, designed to make rigorous the level of abstraction of the domain of discourse. Since we have considered entities from the world around us, whose properties are vital to our analysis and conclusions, it is essential that we have been able to be precise about the LoA at which those entities have been considered. We have seen that changing the LoA may well change our observation of their behaviour and hence change the conclusions we draw. This is not relativism and it has a venerable tradition in science and more recently in the humanities.

To address all the entities in the arena we have been forced to adopt terminology which applies equally to all the entities that populate it (from animals to organisations), but without prejudicing our conclusions. We have been forced to analyse behaviour in a non-anthropocentric manner. To do so we have used the notation of a state-transition system.

We have called the entities within this arena 'agents', characterised here abstractly in terms of a state-transition system, and have concentrated largely on AAs and the extent to which ethics and accountability applies to them. Whether or not an entity forms an agent depends on the LoA at which the entity is considered; there can be no absolute LoA-free form of identification. By abstracting that LoA, an entity may lose its agenthood by no longer satisfying the behaviour we associate with agents. However, for most entities there is no LoA at which they can be considered an agent. Of course. Otherwise one might be reduced to the absurdity of considering the moral accountability of the magnetic strip that holds a knife to the kitchen wall. Instead, for comparison, our techniques address the far more interesting question (Dennet, 1997): 'when HAL kills, who's to blame?' Indeed, our techniques enable us to say that HAL is accountable – though not responsible – if it meets the conditions defining agenthood.

In Section 3.1 we have deferred discussion of a final objection to our approach until the conclusion. The time has come to honour that.

Our opponent can still raise a final objection: suppose you are right; does this enlargement of the class of moral agents bring any real advantage? It should be clear why the answer is clearly affirmative. Morality is usually predicated upon responsibility. The use of LoA and thresholds enables responsibility to be separated and formalised, and its part in morality to be fully clarified. The better grasp of what it means for someone or something to be a moral agent brings with it a number of substantial advantages. We can avoid anthropocentric and anthropomorphic attitudes towards agenthood and rely on an ethical outlook not necessarily based on punishment and reward but on moral agenthood, accountability and censure. We are less likely to assign responsibility at any cost, forced by the necessity to identify a human moral agent. We can liberate technological development of AAs from being bound by the standard limiting view[11]. We can stop the regress of looking for the *responsible* individual when something evil happens, since we are now ready to acknowledge that sometimes the moral source of evil or good can be different from an individual or group of humans. As a result, we are able to escape the dichotomy 'responsibility + moral agency = prescriptive action' vs. 'no responsibility ergo no moral agency ergo no prescriptive action'. Promoting normative action is perfectly reasonable even when there is no responsibility but only moral accountability and the capacity for moral action.

All this does not mean that the concept of 'responsibility' is redundant. On the contrary, our previous analysis makes clear the need for further analysis of the concept of responsibility itself, when the latter refers to the ontological commitments of creators of new AAs and environments. According to Floridi (2001a, 2003b) and Floridi and Sanders (2003b), Information Ethics is an ethics addressed not just to 'users' of the world but also to demiurges who are 'divinely' responsible for its creation and well-being. It is an ethics of *creative stewardship*.

In the introduction, we have warned about the lack of balance between the two classes of agents and patients brought about by deep forms of environmental ethics that are not accompanied by an equally 'deep' approach to agenthood. The position defended in this paper supports a better equilibrium between the two classes A and P. It facilitates the discussion of the morality of agents not only in Cyberspace but also in the biosphere – where animals can be considered moral agents without their having to display free will, emotions or mental states (Dixon, 1995; Rosenfeld, 1995a, b) – and in what we have called contexts of 'distributed morality', where social and legal agents can now qualify as moral agents. The great advantage is a better grasp of the moral discourse in non-human contexts (Rowlands, 2000). The only 'cost' of a 'mind-less morality' approach is the extension of the class of agents and moral agents to embrace AAs. It is a cost that is increasingly worth paying the more we move towards an advanced information society.

Acknowledgements

Versions of this paper have been presented by L. Floridi at the seminar for graduate studies of the University of Bari and at CEPE 2001, Lancaster, UK. The authors are grateful for the feedback received, for the comments made by the referees who accepted the paper for the Lancaster conference and to the referees of *Minds and Machines* for their improvements.

Luciano Floridi wishes to acknowledge the financial support of the University of Bari. Finally, this paper is dedicated to *Martina*.

Notes

[1] It is possible, but utterly unrealistic, that A and P are disjoint. On the other hand, P can be a proper subset of A or A and P can intersect each other. These two alternatives are little more promising because they both require at least one moral agent that in principle could not qualify as a moral patient. Now this pure agent would be some sort of supernatural entity that, like Aristotle's God, affects the world but can never be affected by it. But being in principle 'unaffectable' and irrelevant, it is unclear what kind of rôle this entity would exercise with respect to the normative guidance of human actions. So it is not surprising that most macroethics have kept away from these 'supernatural' speculations and implicitly adopted or even explicitly argued for one of the two remaining alternatives discussed in the text.

[2] Environmental ethics has developed since the 1960s as the study of the moral relationships of human beings to the environment (including its nonhuman contents and inhabitants) and its (possible) values and moral status. It often represents a challenge to anthropocentric approaches embedded in traditional western ethical thinking. For an article-length introduction to environmental ethics see A. Brennan and Y.-S. Lo, *Environmental Ethics*, The Stanford Encyclopedia of Philosophy (Summer 2002 Edition), URL = http://plato.stanford.edu/archives/sum2002/entries/ethics-environmental/.

[3] The more inclusive P is, the 'greener' or 'deeper' the approach has been deemed. Floridi (1999, 2003) and Floridi and Sanders (2001) have defended a 'deep ecology' approach.

[4] Note that, for the sake of simplicity, the conversational example does not fully respect the *de dicto/de re* distinction.

[5] 'Juggernaut' is the name for Vishnu, the Hindu god, meaning 'Lord of the World'. A statue of the god is annually carried in procession on a very large and heavy vehicle. It is belived that devotees threw themselves beneath its wheels, hence the word 'Juggernaut' has acquired the meaning of 'massive and irresistible force or object that crushes whatever is in its path'.

[6] See, for example, The Working Values Group, Ltd., http://www.workingvalues.com.

[7] Machines with minds of their own, *The Economist*, 22 March, 2001. http://www.economist.com.

[8] For an enlightening comparison consider that the Federation Internationale des Echecs (FIDE) rates all chess players according to the same Elo System, regardless of their human or artificial nature.

[9] ACM, Code of Ethics. http://www.acm.org.

[10] It is interesting to speculate on the mechanism by which that list is maintained. Perhaps by a human agent; perhaps by an AA composed of several people (a committee); or perhaps by a software agent.

[11] Beyond cruise control, *The Economist*, 21 June, 2001. http://www.economist.com .

References

Allen, C., Varner, G. and Zinser, J. (2000), 'Prolegomena to Any Future Artificial Moral Agent', *Journal of Experimental and Theoretical Artificial Intelligence* 12, pp. 251—261.
Arnold, A.(1994), *Finite Transition Systems*, Prentice-Hall International Series in Computer Science.
Badham, J. (director) (1983), *War Games*.

Bedau, M.A. (1996) 'The Nature of Life', in M.A. Boden, ed., *The Philosophy of Life*, Oxford University Press, pp. 332–357.

Cassirer, E. (1953), *Substance and Function and Einstein's Theory of Relativity*, New York: Dover Publications Edition.

Danielson, P. (1992), *Artificial Morality: Virtuous Robots for Virtual Games*, Routledge, NY.

Dennet, D. (1997), When HAL Kills, Who's to Blame?' in D. Stork, ed., *HAL's Legacy: 2001's Computer as Dream and Reality*, Cambridge MA: MIT Press, pp. 351–365.

Dixon, B.A. (1995), 'Response: Evil and the Moral Agency of Animals', *Between the Species* 11(1–2), pp. 38–40.

Epstein, R.G. (1997), *The Case of the Killer Robot*, John Wiley and Sons, Inc.

Floridi, L. (1999), 'Information Ethics: On the Theoretical Foundations of Computer Ethics', *Ethics and Information Technology* 1(1), pp. 37–56. Preprint from http://www.wolfson.ox.ac.uk/ floridi/papers.htm.

Floridi, L. (2003), 'On the Intrinsic Value of Information Objects and the Infosphere', *Ethics and Information Technology* 4(4), pp. 287–304. Preprint from http://www.wolfson.ox.ac.uk/ floridi/papers.htm.

Floridi, L. (2001a), 'Information Ethics: An Environmental Approach to the Digital Divide', UNESCO World Commission on the Ethics of Scientific Knowledge and Technology (COMEST), First Meeting of the Sub-Commission on the Ethics of the Information Society (UNESCO, Paris, June 18–19, 2001). Preprint from http://www.wolfson.ox.ac.uk/ floridi/papers.htm.

Floridi, L. (2001b), 'Ethics in the Infosphere', *The Philosophers' Magazine* 6, pp. 18–19. Preprint from http://www.wolfson.ox.ac.uk/ floridi/papers.htm.

Floridi, L. and Sanders, J.W. (2001), 'Artificial Evil and the Foundation of Computer Ethics, *Ethics and Information Technology* 3(1), pp. 55–66. Preprint from http://www.wolfson.ox.ac.uk/floridi/papers.htm.

Floridi, L. and Sanders, J.W. (2003a), 'The Method of Abstraction', in M. Negrotti, ed., *The Yearbook of the Artificial*. Issue II, Peter Lang, Bern. Preprint from http://www.wolfson.ox.ac.uk/ floridi/papers.htm.

Floridi, L. and Sanders, J.W. (2003b), 'Internet Ethics: The Constructionist Values of Homo Poieticus', in R. Cavalier, ed., *The Impact of the Internet on Our Moral Lives*, SUNY, Fall.

Franklin, S. and Graesser, A. (1996), 'Is it an Agent, or Just a Program? A Taxonomy for Autonomous Agents', in *Proceedings of the Third International Workshop on Agent Theories, Architectures, and Languages*, Springer-Verlag Available from <www.msci.memphis.edu/ franklin/AgentProg.html>.

Gips, J. (1995), 'Towards the Ethical Robot', in K. Ford, C. Glymour and P. Hayes, ed., *Android Epistemology*, Cambridge MA: MIT Press, pp. 243–252.

Goldberg, D.E. (1989), *Genetic Algorithms in Search, Optimization and Machine Learning*, Reading, MA: Addison-Wesley.

Kerr, P. (1996), *The Grid*, New York: Warner Books.

Michie, D. (1961), 'Trial and Error', in A. Garratt, ed., *Penguin Science Surveys*, Harmondsworth: Penguin, pp. 129–145.

Mitchell, T.M. (1997), *Machine Learning*, McGraw Hill.

Moore, J.H. ed., (2001), 'The Turing Test: Past, Present and Future', *Minds and Machines* 11(1).

Motwani, R. and Raghavan, P. (1995), *Randomized Algorithms*, Cambridge: Cambridge University Press.

Norton AntiVirus (2003), Version 8.07.17C. Symantec Corporation, copyright 2003.

Page, I. and Luk, W., 'Compiling Occam into Field-Programmable Gate Arrays', ftp:// ftp.comlab.ox.ac.uk/pub/Documents/techpapers/Ian.Page/abs hwcomp.1 .

Rosenfeld, R. (1995a), 'Can Animals Be Evil? : Kekes' Character-Morality, the Hard Reaction to Evil, and Animals, *Between the Species* 11(1–2), pp. 33–38.

Rosenfeld, R. (1995b), 'Reply', *Between the Species* 11(1–2), pp. 40–41.

Rowlands, M. (2000), *The Environmental Crisis – Understanding the Value of Nature*, Palgrave, London-Basingstoke.

Russell, S. and Norvig, P. (2003), *Artificial Intelligence: A Modern Introduction*, 2nd edition, Prentice-Hall International.

Sample, I. (2001), SmartPaint, *New Scientist*, http://www.globaltechnoscan.com/16May-22May01/paint.htm.

Scott, R. (director), (1982/1991), *Bladerunner, The Director's Cut*.

Turing, A.M. (1950), 'Computing Machinery and Intelligence', *Mind* 59(236), pp. 433–460.

26

Un-making artificial moral agents

Deborah G. Johnson and Keith W. Miller

Abstract. Floridi and Sanders, seminal work, "On the morality of artificial agents" has catalyzed attention around the moral status of computer systems that perform tasks for humans, effectively acting as "artificial agents." Floridi and Sanders argue that the class of entities considered moral agents can be expanded to include computers if we adopt the appropriate level of abstraction. In this paper we argue that the move to distinguish levels of abstraction is far from decisive on this issue. We also argue that adopting certain levels of abstraction out of context can be dangerous when the level of abstraction obscures the humans who constitute computer systems. We arrive at this critique of Floridi and Sanders by examining the debate over the moral status of computer systems using the notion of interpretive flexibility. We frame the debate as a struggle over the meaning and significance of computer systems that behave independently, and not as a debate about the 'true' status of autonomous systems. Our analysis leads to the conclusion that while levels of abstraction are useful for particular purposes, when it comes to agency and responsibility, computer systems should be conceptualized and identified in ways that keep them tethered to the humans who create and deploy them.

Abbreviations: STS: Science and Technology Studies

Introduction

Luciano Floridi has made an enormous contribution to the field of computer ethics through a large body of work that brings computers and computation to bear on a range of philosophical concepts and ethical issues. In this paper we focus in particular on Floridi and Sanders, seminal piece, "On the Morality of Artificial Agents."[1] One reason that Floridi and Sanders' paper has generated intense debate is that its claims are relevant to a range of traditional and emerging fields including cognitive science, information studies, computer ethics, computer science (especially artificial intelligence), and philosophy of technology. Scholars in these fields have quite different stakes in the debate about the status of artificial agents and this tends to add complexities and keep the debate moving. Each community of scholars (and to some extent each scholar) asks different questions and puts down anchors at various points in this complex territory, and then insists that everything else must fall into place around the anchor-claims. Each scholar and community has axes to grind – messages they want to convey to particular audiences. To understand the territory, we have to step back and consider where one or another group of scholars has dug in and why they place their anchors exactly where they do.

Many of the interlocutors in this debate seem to believe that what is at issue is "the truth" about the moral status of computer systems. In this paper, we argue that this issue is not a matter of truth. There is no preexisting right answer to the question whether computer systems are (or ever could be considered) moral agents; there is no truth to be uncovered, no test that involves identifying whether a system meets or does not meet a set of criteria. We argue that even if computer systems exhibit particular features, we are not "required" in the name of consistency to draw a particular conclusion about their moral status.

Our analysis begins by stepping back from the debate and trying to understand what exactly is at stake and why the debate goes on as it does.

[1] Luciano Floridi and Jeff W. Sanders. On the Morality of Artificial Agents. *Minds and Machines*, 14(3): 349–379, 2004.

The debate focuses on computer systems that are conceptualized as artificial agents; they are computational artifacts that, unlike many other artifacts, behave with some kind or degree of independence. The debate can be described as a struggle or negotiation over the meaning and significance of computational artifacts that behave with this independence and, indeed, their independence leads some to refer to these systems as "autonomous agents." Two examples that are often discussed are robots that perform tasks at remote locations (including other planets) with only periodic commands from human "handlers" and network bots that can be given an item to buy and then search for the best value available, taking into consideration price, quality, and vendor reputation. The range of current and future applications is enormous and growing as more and more systems are automated by increasingly complex software and hardware. These emerging technologies can and do make decisions about what information we receive when we search, what medicines or doses of radiation we receive, how airplane and automobile traffic is ordered, what financial transactions are implemented, how weapons are targeted and launched, and so on.

Our analysis concedes that computational artifacts can behave independently from the humans that create or deploy them in some sense of the term "independent." However, we argue that in one very important sense of the term "independent", computer systems are *not* ever completely independent from their human designers. We argue, as well, that it is dangerous to conceptualize computer systems as autonomous moral agents.

Interpretive flexibility

When we step back from this debate about the status of computational artifacts and try to understand what is being contested, the debate seems to fit accounts of technological development provided by science and technology studies (STS) scholars. STS scholars argue that during the early stages of technological development, technology has interpretive flexibility. The technology is not yet a "thing" both in the sense that the material design has not yet solidified and in the sense that there is disagreement about its meaning or significance.[2] Various actors and interest groups struggle over the benefits and drawbacks of various designs. The standard exemplar in STS is Pinch and Bijker's description of the development of the bicycle.[3] They identify a variety of bicycle designs that were developed, adopted for a time, then rejected and replaced with another design. The pattern and direction of development was neither predetermined nor linear; at each stage a variety of interest groups were part of the process and each group pressed for particular features. The bicycle design that eventually took hold was the outcome of struggles and negotiations among the various interest groups.

The idea that technology has interpretative flexibility in its early stages of development is important here because it makes the (perhaps obvious but often misrepresented) point that technologies are not developed in a single act, nor are they created intact as we (consumers and users) later encounter them. Rather the technologies we have today were developed through a process of iteration: early prototypes were rejected or modified, users responded in a variety of ways to various features, users and the public attached meanings and found uses for devices that the designers never intended. Of course, in hindsight it may look like the technology, as we know it now, was on a path fated to end with the technology we currently see. That is the nature of hindsight; it puts the present in a perspective from which it looks like it was fated to occur. In reality, however, as STS scholars point out, the process of technological development is contingent and in the early stages what is being developed has many valences; there are many potential directions in which the design and meaning might evolve, including going nowhere at all, e.g., cold fusion.

So it is with computer systems and computational artifacts. They are developed over time in contexts in which various actors push and pull in a variety of directions with a variety of concerns and interests. The actual design and meaning that is eventually adopted by users is the outcome of struggles between modelers, programmers, clients, distributors, marketing representatives, users, lawyers, and so on. Indeed, the process of development continues on even after a product is brought to market; for example, the process continues when users find work-arounds and

---

[2] Johnson makes this point with regard to nanotechnology referring to it as a technology 'in the making.' Deborah G. Johnson, Nanoethics: An Essay on Ethics and Technology 'in the Making'. *Nanoethics*, 1(1): 21–30, 2007.

[3] Wiebe Bijker and Trevor Pinch. The Social Construction of Facts and Artifacts. In Wiebe Bijker, Thomas Hughes and Trevor Pinch, editors, *The Social Construction of Technological Systems*, pp. 17–51. MIT Press, Cambridge, Mass., 1987.

bypass features the designer thought essential.[4] Systems often evolve into "things" that are very different from that which was initially conceived.

It is important to note that the interpretative flexibility of technology generally does not go on forever. Closure and stabilization occur when a particular form of the technology takes hold and the various actors and interest groups involved agree upon a meaning or cluster of meanings. Indeed, STS literature is filled with case studies illustrating technologies at various stages of interpretive flexibility, technologies that eventually move to stabilization and closure. For example, drivers get used to steering wheels rather than sticks for steering, computer users get used to a Windows environment, VHS wins over betamax for video format, and so on.

When we recognize the interpretive flexibility of technologies in their early stages of development, we acknowledge that technology is, in part at least, "socially constructed." A technology is not a single "something" that catches on or fails; technologies are being made, i.e., being delineated as "things" and given meaning, as they are being developed. Various actors in the process develop understandings and begin to attach different meanings and significance to the "thing." Think, for example, of the struggles over the delineation, meaning, and significance of genetically modified foods, cloning, and nanotechnology. Indeed, these are all cases where debate continues and where there is often disagreement about the object of discussion, e.g., what counts as genetically modified? What is nanotechnology?

Using this notion of interpretive flexibility, the debate about artificial agents can be seen as a struggle over the meaning of computer systems that operate independently in space and time from the humans that create and deploy them. The debate arises because these systems have a kind and degree of independence not found in other, prior technologies. Unquestionably, this degree and kind of independence is significant. The question is what are we to "make" of it? Are these systems "beasts" designed by modern day Victor Frankensteins? Are they precursors to HAL (in the film *2001*)? Are they "merely" automated versions of maids and servants? Are they the next step in the evolution of prosthetic tools that extend human capabilities? In the frame of interpretive flexibility, the question is not what these systems (truly) *are*; the question is: "What should we 'make' of them?" Deciding the question "Can computers ever be considered moral agents?" is more like the question "Should eighteen year olds be allowed to vote?" or the question "should marijuana be considered a dangerous drug?" than it is like the question "What is the half-life of uranium-238?"

As scholars engaged in the debate about the moral agency of artifacts, we are actors in the process, actors who are making contributions to the construction of the meaning of artificial agents. Seen from this perspective, the question "can artificial agents be moral agents?" is misleading. More accurate questions are: "How should we conceptualize computer systems that behave independently?" and "What terms should we use to refer to computational systems that behave independently from their creators and deployers?" In answering the latter questions it is important to look at the criteria we use in other cases and it is helpful to identify how the case at hand is like and unlike other cases in which agency is attributed to entities that behave independently, e.g., grown children, animals. However, consistency cannot be determinative here since there will always be similarities *and differences*.

In this framework, Floridi and Sanders' "On the morality of artificial agents" can be seen as a major effort to provide an understanding of a new kind of technology that exhibits expanded independence. Debate arises because scholars have quite different ideas about the meaning and significance of these computational artifacts. To some extent, there seems to be agreement that these entities can be characterized and referred to as "artificial agents" but this agreement has been achieved because of the ambiguity of "agent". In other words, it is not at all clear what it means to say that something is an "artificial agent." Human agency has a particular meaning in philosophy of mind and ethics, but "agent" also refers to lawyers and brokers who act as "agents" of their clients[5]; and "agent" seems to be used in relation to machines simply when machines perform tasks. Thus, when we refer to computer systems as "artificial agents", it is unclear which, if any, of these meanings are being attributed to computer systems. Thus, even though the interlocutors in the debate have agreed on the use of the term "artificial agent", because of the ambiguity of the term, the debate over meaning continues. "Artificial agent" still has interpretative flexibility.

---

[4] See John L. Pollock, When is a Work Around? Conflict and Negotiation in Computer Systems Development. *Science, Technology & Human Values*, 30(4): 496–514, 2005.

[5] Deborah G. Johnson and Thomas M. Powers, Computers as Surrogate Agents. In J. Van den Hoven and J. Weckert, editors, *Information Technology and Moral Philosophy*. Cambridge University Press, Cambridge, 2008.

Putting in anchors

Acknowledging that computational artifacts that behave independently are in a stage of interpretive flexibility and framing the debate as a struggle over the meaning of "artificial agents", we can now examine the struggle more carefully. To get a handle on the debate, we can identify two distinct groups of scholars who differ on the status of artificial agents but agree on important related themes. [We admit that we have not identified all possible views in this debate; our oversimplification is intended to highlight key parameters of the debate.] The first group of scholars is committed to computation as a model, either a model of reality or a tool for scientific exploration. We will call this group Computational Modelers and we believe that Floridi and Sanders fall into this category. Computational Modelers see computational modeling as the ultimate in the philosophical endeavor to capture reality. Their agenda can be seen to be rooted in the enlightenment tradition and the idea that reason will lead to truth. Computational Modelers have a stake in using computation as a (if not *the*) foundation of a body of knowledge that brings insight to a wide range of areas; i.e., they have a stake in the value of computation as a model. Whether computing is used as a model of reality or a pragmatic model (i.e., a useful way of thinking about and manipulating nature) is an interesting question, but not one that gets much attention from Computational Modelers. Instead the assumption seems to be that computational models represent reality.

In any case, while Computational Modelers may be attracted to computation as a model, they get caught up in much more than models. Computers are machines, machines that behave; computers can be used not just to model behavior, but to produce behavior. This may well be the source of the move to refer to computers as agents. Since computer systems behave, their behavior can be compared to human behavior. The effective outcomes of computer behavior and human behavior can be the same. For example, before computerization of the stock exchange, a buyer would call a stockbroker, check on the price of a stock and tell her to purchase a particular number of shares of the stock. The stockbroker would call someone, say certain words, fill out paper work, and the buyer would eventually receive paper confirmation of ownership of shares of the stock. Now that the entire process has been computerized, a buyer can set up a computer system to execute the purchase of the stock if and when the price reaches a certain point. Few human movements are involved; no words need be spoken; and paper may never be used. It would seem here that the computer system acts as an agent replacing the stockbroker who acted as an agent on behalf of a client in the old system. While we will challenge the comparability of the behaviors here, for now, the point is that Computational Modelers seem to want to make much of the fact that machines can now perform tasks that only humans performed in the past. The fact that computer systems can do tasks that humans do seems to support the validity of the computational model involved. [Of course, it should be no surprise that the computer system accomplishes tasks done by humans for that is precisely what systems designers and programmers designed the system to do. They developed a computer system to replace (though not necessarily replicate) the prior system.]

A second group in the philosophical debate about the moral status of artificial agents seeks to illuminate the role that computers (and technology in general) play in the lives of human beings, especially their moral lives. Here we would place ourselves, in contrast to Floridi. The agenda of this group is to draw attention to the power of technology, especially computer technology, and its moral implications. This means bringing to light the moral character of computer systems, the values embedded in their design, and the ways in which they affect the moral lives of human beings. The adoption and use of computer technology has powerful effects and those who decide about its design and adoption have enormous power. Among other things, better understanding of the moral implications of computer technology can lead to better steering of technological development. Thus, the kind of knowledge sought by this group of scholars has an implicit, if not explicit, goal of informing human action and directing social change in the future. We will label this second group the Computers-in-Society Group.

When it comes to artificial agents, this group acknowledges that computer systems behave independently in some sense of the term "independent," and sees this independence as a moral concern. Computers are increasingly being assigned tasks that humans controlled and performed in the past; Big Blue defeats Kasporov[6], robots replace workers on an assembly line, and computer systems instead of physicians recommend treatments. As we mentioned earlier, what information we receive when we search, what medicines or doses of radiation we receive, how

---

[6] For quick insight into the nature of this defeat from one of Big Blue's programmers, see Robert Andrews. A Decade after Kasporov's Defeat, Deep Blue Coder Relives Victory. *Wired*, May 11, 2007, http://www.wired.com/science/discoveries/news/2007/05/murraycampbell_qa.

airplane and automobile traffic is ordered, what financial transactions are implemented, how weapons are launched and targeted are now tasks done (decisions made) by computer systems. If computers are performing tasks and making these critical decisions, then, the Computers-in-Society scholars argue, there must be some accountability for the effects of these systems on human well-being.

The Computers-in-Society group explores questions about the role of computer systems in decision-making by classifying computer systems as a form of technology, a new form with special features, but nevertheless, a form of technology that is, by definition, created and deployed by humans. There are a variety of ways in which members of this group express their concerns; for example, they draw attention to the ways in which computers affect the world in which humans act and live. Computers-in-Society scholars also give accounts that uncover the ways in which the design of technology can make a difference in what human actors *can* do and what humans actually do.

This group of scholars has a stake in illuminating the contribution of computer systems to morality and some go as far as to say that computer systems have "agency" at least in the sense that they make causal contributions to what humans do and what occurs. Nevertheless, the Computers-in-Society group has a stake in *not* establishing computer systems as autonomous moral agents. Their concern is that framing computer systems as moral agents will deflect responsibility away from the human beings who design and deploy computer systems. In a sense, this group has a two-part agenda: show that technology is an important component of morality but also show that technology is under human control.[7]

So, Computational Modelers and Computers-in-Society scholars disagree over the meaning of "artificial agent", that is, they disagree about the status of computer systems that behave independently. Each group draws on different philosophical traditions – the first group generally draws on logic and philosophy of mind and the second group draws on ethics and social philosophy. The Computational Modelers have a stake in the validity of computational models and seem to believe that establishing the parallels between computer system behavior and human behavior is crucial to the validity of the computational model. They seem to believe that if computer systems are given the status of moral agents, this will be testimony to the achievement of the computational model in capturing reality. On the other hand, the Computers-in-Society group has little stake in computational modeling. For this group it is all well and good if computational models can be used to produce new knowledge and devices that make the world a better place for humans. However, when it comes to the moral agency of computer systems, this group finds the claims of Computational Modelers both odd and dangerous. The claims are odd because they do not seem to acknowledge that computer systems are an extension of human activity and human agency, and because their view of agency and morality is out of sync with the idea of morality as a human system of contextualized ideas and meanings. The claims seem dangerous because they imply that computer systems can operate without any human responsibility for the system behavior.

So, each group of scholars has its own agenda and the groups converge and conflict on the topic of the moral status of artificial agents. To return to the concept of interpretive flexibility, it seems that the Computational Modelers are pushing for a "frame of meaning" for computer systems that has heretofore been reserved for humans: the frame of "moral agent." Why this frame? The answer we have suggested here is that it furthers the agenda of Computational Modelers to establish the validity of computational modeling. On the other hand, the Computers-in-Society group push against a "frame of meaning" that casts computer systems as moral agents. While the Computers-in-Society group would frame computer systems as moral – moral entities – the group is against ascribing the status of moral agent to any computer system on grounds that doing so is dangerous. As already explained, the Computers-in-Society group argues against ascribing moral agency to computer systems on grounds that doing so will deflect responsibility from the humans who create and deploy them. For this group of scholars, it is important to keep the connection between computer systems and humans well in sight.

Critique of computational modelers and Floridi–Sanders

So, the debate between the Computational Modelers and the Computers-in-Society group is a debate about how to "construct" the meaning and significance of computer systems that have a particular kind or degree of independence. If the Computational Modelers win the debate, artificial agents will be understood to have the potential for moral standing (moral agency) of a kind that ultimately might lead to a prohibition on turning them off. If the Computers-in-Society group

---

[7] See D.G. Johnson, Computer Systems: Moral Entities but not Moral Agents. *Ethics and Information Technology*, 8(4): 195–204, 2006.

wins, artificial agents will be understood to be important components of the moral world but they will always be understood to be human constructions under human control. They would always be understood to be "tethered" to humans in the sense that they are the products of human invention, are deployed by humans for human purposes, operate in contexts maintained by humans, and cannot function without some degree of human control (even though that control may be distant in time and space).

We can now explain and critique the Floridi and Sanders contribution to this debate. By introducing the notion of levels of abstraction, Floridi and Sanders argue that computer systems can have moral agency insofar as moral agency is understood to make sense at a particular level of abstraction. They argue that we can conceptualize computer systems at different levels of abstraction and within one level of abstraction a computational entity may be autonomous while it is not so at another level of abstraction. Floridi and Sanders then seem to claim that since computational entities can be conceptualized as autonomous at one level of abstraction, they can be autonomous *moral* agents. This move is especially important since it is the kind of move that Floridi and Floridi and Sanders make with regard to a number of moral concepts. For example, in their paper on artificial evil[8] they argue that a line of code could be evil – as understood at a particular level of abstraction.

We agree with Floridi and Sanders about levels of abstraction; that is, entities such as computer programs can be conceptualized and viewed at different levels of abstraction (or within different conceptual frameworks). However, Floridi and Sanders misstep when they generalize concepts and terms from one level of abstraction to another. Thus, while a computational entity might be conceptualized and understood to be independent, even autonomous, at some level of abstraction, it would be misleading to maintain that it is, therefore, independent or autonomous at another level of abstraction. Floridi and Sanders seem to have confused, on the one hand, something being autonomous within a level of abstraction and, on the other hand, something being autonomous writ large. They seem to move from autonomous agents as understood at one level of abstraction to autonomous agent broadly understood or understood in the context of moral theory. [They also seem to think that morality and moral theory have different levels of abstraction rather than being a particular level of abstraction but this is a point we will discuss in a moment.]

Grodzinsky, Miller, and Wolf[9] draw attention to this flaw in Floridi–Sanders' reasoning when they discuss the level of abstraction of designers versus the level of abstraction of users. When one considers the significant difference between a computer artifact as seen from the viewpoint of its designer and the same artifact as seen from the viewpoint of a user (who does not have access to the source code and computational state of the artifact), Grodzinsky et al. argue that it is counterintuitive to define moral agency from the perspective of the user and entirely ignore the designer's viewpoint since the designer's viewpoint (level of abstraction) is necessary for the computer artifact to exist at all.

In allowing that an entity might be a moral agent at one level of abstraction and not at another, Floridi and Sanders have reduced autonomous moral agency to a highly variable and relativistic notion. Indeed, it seems odd to think of moral agency out of the context of moral concepts and theories. In an important sense, morality is a particular level of abstraction with particular concepts and meaning. "Moral agency" has a deep history in philosophical thought, and is inextricably connected to other concepts such as action, intentionality, and mental states. While these concepts and their relationships can be modeled and represented, to say that they can be reduced to a different level of abstraction seems at least to beg the question, if not to be entirely misguided. Re-conceptualizing moral agency at another level of abstraction is comparable to re-conceptualizing something like color to sound.

The argument can be made in a somewhat different way. While Floridi and Sanders have made a solid case for an account of computer systems as autonomous at some (or several) level(s) of abstraction, they have not shown that "moral" could ever be delineated at the same level of abstraction. To establish that computer systems can be autonomous *moral* agents, Floridi and Sanders would have to show that within some level of abstraction in which a computer system is autonomous, the notion of "moral" could also be delineated. This would, then, give us an account of autonomous moral agent at some level of abstraction. We have suggested that this can't be done without begging the question.

---

[8] L. Floridi and J.W. Sanders, Artificial Evil and the Foundation of Computer Ethics. *Ethics and Information Technology*, 3(1): 55–66, 2001.

[9] Frances S. Grodzinsky, Keith W. Miller and Marty J. Wolf. The Ethics of Designing Artificial Agents. *CEPE*, July 12–14, 2007, San Diego, CA. Abstract available at http://cepe2007.sandiego.edu/abstractDetail.asp?ID = 14.

Comparable behavior, comparable status/identity

The Floridi and Sanders account is flawed for other, related reasons as well. Earlier we noted that while Computational Modelers start out with a focus on models, they emphasize the capacity of computer systems to behave or to produce behavior. As indicated earlier, it is this move from models to operational systems that may well lead Computational Modelers to the plausibility of moral agency for computer systems. However, the focus on behavior turns attention away from the processes by which the behavior is achieved. To begin to see the problem here, consider the difference in the way we evaluate scientific models versus operational systems. Scientific models are tested against the natural world they represent. Operational systems are aimed at utility, at achieving certain tasks as understood by their human users. The aims, purposes, and evaluation criteria for each kind of model are very different. The validity of a scientific model is a matter of comparison with the natural world. In this respect, scientific models are constrained by the natural world; we know when we have a good model of an ecological system or a weather system by seeing whether it replicates the behavior that actually occurs in the natural environment. On the other hand, operational systems have no such constraints. They are designed to achieve tasks and there is no need to accurately model how things are done in the natural or human world. For example, operational systems such as those that regulate air traffic, buy stocks, and perform tasks via robots on the moon are not required to behave as humans do.

A computer system designed to route air traffic may incorporate descriptions of the behavior of airplanes, but it also gives directions to airplanes. The system is configured to track the planes, "decide" and signal the pilots. The system is designed to make safe and efficient decisions about when and where real aircraft should approach runways. The important point is that operational models of this kind are not designed to model human behavior; they function in ways that are quite distinct from human thinking. Indeed, such systems are often designed to exceed the capacity of humans in the speed and even the quality of their decisions. So, when we compare an operational computer system and a human performing the same tasks, the behavior of the human need not look anything like the behavior of the computer system and vice versa. The function achieved has to be equivalent, not the two entities performing the function. Thus, when it comes to operational systems, there need be no match between the behavior of the system and the behavior of the machine. To say that computer systems can be moral agents is, then, reducing morality to functionality – a reduction that seems antithetical to the idea of moral agency.

An example may help make the point. Let us assume that a person opening a door for another person carrying a large package is a small act of beneficence, a positive moral act by a human moral agent. If, however, the door is opened by an electric eye and motor assembly as the person approaches the door, we do not say that the door-opener has performed an act of beneficence. The function performed is equivalent, but the underlying processes (voluntary, autonomous act versus mechanical operations) are significantly different. It may be that the persons who envisioned, designed, and installed the mechanical door-opener had just this situation in mind, in which case those humans might be considered beneficent in envisioning, designing, and installing the opener. But that does not make the door-opener itself a virtuous or praiseworthy moral agent.

Moreover, the mismatch between computer behavior and human behavior is only part of the problem. The more important and, perhaps, more interesting flaw in Floridi and Sanders' analysis is their failure to recognize that operational systems often achieve their functions by agreements among humans to "count" computer activity as equivalent to human performances. Only through human decision-making do certain computer operations come to be recognized as – to stand for or be considered equivalent to – particular human actions. Our earlier discussion of computerized stock purchasing can be extended to illustrate the point. When the automated stock system was designed, it had to meet key requirements that were critical features of the old system. The new system had to operationalize notions of ownership, money, transfer of funds, possession, stock, etc. These requirements were translated into the automated environment in such a way that humans could come to use the system in a way that made sense to them. The computerized stock exchange worked because humans agreed to count certain electronic activities as "authorizing a transaction" and "purchasing a stock". To put this in another way, the computerized system created a new way of achieving "purchase of a stock." However, this was possible only because humans agreed to treat electronic configurations as such. Systems designers were not looking to find "purchasing a stock" in computer systems; they created a system in which human beings would be able to engage in activities that could count as "purchasing a stock" and this meant that the system had to have features that could be interpreted as comparable to features in the paper and ink and words system. The system had to be constructed physically *and* symbolically; that is, there had to be computer operations and human meanings assigned to those operations.

So, there are two important flaws in Floridi and Sanders' account. First, computer systems do not model human behavior in the way scientific models model natural systems. Second, operational systems often achieve human functions because humans agree to count computer operations as performances of a particular kind. These are precisely the mistakes that the Computational Modelers, including Floridi and Sanders, seem to make in the case of moral agency. Their logic assumes that because operational systems have certain outcomes, the behavior is equivalent to human behavior producing the same outcomes. They fail to see that tasks can be achieved by very different means and they fail to see that whether or not the processes count as moral action is a matter of human convention. Moreover, they fail to recognize that the meanings that humans give to particular operations of a computer system are contingent. In other words, the operations of the computer stock exchange don't have to count as "purchasing a stock"; it is only through human convention that they count as such.

The power of the misconception

While this mistake of conflating equivalent functions with equivalent behavior and equivalent entities is (we hope) now clear, the significance of the mistake should not be underestimated. Some Computational Modelers tell us that in the future we will be compelled to treat certain computer systems – if they develop in certain ways – as entities with moral status. They make predictions of a future in which robots will be so sophisticated that we (humans) will have a moral responsibility to refrain from turning them off. Note that the compelling force here seems to be consistency. Were we not to acknowledge certain kinds of robots as moral agents, we would be wrongly treating silicon-based moral agents differently than carbon-based moral agents when the silicon-based agents met all the criteria we apply to carbon-based (human) moral agents. (So the argument goes.)

Lest we be accused of using a straw man or overstating what this group of scholars claim, consider what Sullins[10] writes:

> "Certainly if we were to build computational systems that had cognitive abilities similar to that of humans we would have to extend them moral consideration in the same way we do for other humans."

Note the low threshold for moral consideration; it is "cognitive abilities *similar to* that of humans" (our emphases added here and below). Once this low threshold is met, we must, according to Sullins ("we would *have to* extend..."), give the same moral consideration to computer systems that we give to human moral agents ("the *same way* we do for other humans").

Consider a parallel argument made in relation to an electronic stock exchange. Suppose a group of computer programmers decide to design and build a system that they hope could be implemented to replace a paper and ink stock exchange. They get the system functional – we can even imagine that it is a flawless system. The system specifies functional equivalents to such operations as "requesting to purchase a particular stock at a specified price", "authorizing the transfer of funds from a bank", and "purchasing shares of a stock". Are we compelled to recognize these functional equivalents as the actions they represent? Of course not. We would be foolish to use this system to purchase stock unless and until the system was embedded in the appropriate social institutions, that is, unless and until many other humans agreed to use the system in the way it was designed to be used. A wide range of actors (regulatory and government agencies, banks, law enforcement, and users) would have to adopt the system and agree to act *as if* operations of the new system would count as their functional equivalents, e.g., "purchase of a stock", "transfer of funds", etc.

Another way to see the problem here is to return to our earlier discussion of technological development. Sullins presumes a version of technological determinism, that is, he presumes that technological development follows a natural order of development unaffected by social and political forces and human choices. The STS critique of technological determinism is that technological development is contingent; it is influenced by a wide range of social, political, economic and cultural and historical factors. Sullins fails to recognize how the debate about the significance of autonomous computer systems reflects and will shape the meaning and the design of computer systems. In adopting this technologically deterministic view, Sullins and others hide the power that is being exerted by a variety of interest groups. There are enormous amounts of money, time, and effort being invested in the development of autonomous systems and the money, time, and effort is being invested by groups that have specific, distinctively human interests in

---

[10] John Sullins. Ethics and Artificial Life: From Modeling to Moral Agents. *Ethics and Information Technology*, 7(3): 144, 2005.

mind – more sophisticated and effective weapons, more global and efficient markets, less crime and terrorism, faster response times, etc. Ignoring the interests of these powerful interest groups, the Computational Modelers push for a conception of computer systems that is compatible with these interests. Attributing moral agency to computer systems simply hides those groups and their interests.

Scholars in the Computers-in-Society group have a stake in the rejection of technological determinism. As long as technology is seen as inevitable and predetermined, many actors who are powerfully affected by particular technologies will stay out of the discussion and debate about whether and how that technology should be developed. This prevents technological development from being steered for human wellbeing, that is, for the good of all those who are affected. Using the term "artificial moral agents" encourages the view that these sophisticated computer systems are the way they are because they must be exactly that way. The term suggests that this is the only way they can be and the only way we can think about them. In fact, these systems can be developed in a myriad of ways; they develop in the ways humans choose to develop them, and their meaning and status are fluid.

On Bill Joy's (2000) technologically deterministic vision of artificial agents, we are to believe that artificial agents will, ultimately and necessarily, become so advanced (and superior to humans) that they will (indeed should) take over humanity.[11] This is conceivable only if one accepts that technology follows some predetermined path of development that can't be stopped because it is out of human control. By contrast, the Computers-in-Society group insists that there are human forces at work shaping the development of any technology including autonomous computer systems, and we ought to carefully consider how they are being constructed, not merely passively observing their "emergence" in some "natural progression."

The dangers of constructing artificial moral agents

If we take our cue from the Computers-in-Society group, the independence and self-generation of computer systems are not the relevant or key features. This group concedes that many computer systems can now – and more will soon – behave independently. Such computer systems can at some level of abstraction meet criteria such as those proposed by Floridi and Sanders; i.e., they have the ability to respond to a stimulus by a change of state (interactivity), the capacity to change state without an external stimulus (autonomy), and the ability to change the transition rules by which states are changed (adaptability). However, fulfilling these criteria does not convince the Computers-in-Society group that computer systems are or "must" ever be treated as moral agents. Instead, they insist that computer systems should be conceptualized and given meanings that encourage certain kinds of human behavior. This means, among other things, understanding computer systems in ways that keep those who design and deploy them accountable.

To illustrate this point, consider a complex bot that learns and generates its own rules of behavior. Suppose it generates rules of behavior that make it an extremely risky entity. Say it makes medical decisions or regulates financial transactions and no humans understand exactly how it does what it does. According to the Computers-in-Society group, to conceptualize such systems as autonomous moral agents is to absolve those who design and deploy them from any responsibility for doing so. It is similar to absolving from responsibility people who put massive amounts of chemicals in the ocean on grounds that they did not know precisely how the chemicals would react with salt water or algae (though they knew generally about chemical reactions and toxicity). Ignorance is no excuse since they knew they were engaged in risky activity. Just as we might ask, "What is the best way to think about chemicals to ensure that people know they are dangerous?" the Computers-in-society group asks, "What is the best way to think about computer systems to ensure that those who create and deploy them do so safely?" The Computers-in-Society group claims that it is dangerous to construct computer systems as autonomous agents because the construction implies that no humans are responsible for what the systems do; they argue that – no matter how independently they behave – computer systems should be understood in ways that keep them conceptually tethered to human agents.

Here the Computers-in-Society group must answer two distinct questions: Are computer systems in fact tethered to human agents (descriptive question) and should they (normatively) always be understood to be tethered to human agents? We claim that the answer to both of these questions is "yes." Descriptively, all computer systems (indeed, all artifacts) are tethered to humans in a complex web of intentions and consequences, and normatively, concepts of computer systems should reflect the connection of the system to human agents.

---

[11] Although he does not advocate the position, Bill Joy's title to his important article seems apt: Why the Future Doesn't Need Us, *WIRED*, 8(4), 238–262, 2000.

The first claim may be the easier to defend so we will start there. Computer systems are always tethered to human agency in the following ways: they are created by humans, created to perform tasks on behalf of humans, and they are designed to be used by humans. Any computer system is deployed through systems that function as socio-technical systems; for example, we could not have the Internet (or anything like it) were it not for hundreds of social institutions, agreements, laws, contracts, and more. Users and their institutions are the beginning and ending point for all computer systems. Even computer subsystems that take their input from and deliver their output to other computer systems are ultimately connected to humans. They are deployed by humans, they are used for some human purpose, and they have indirect effects on humans.

To be sure, computer systems can be understood at various levels of abstraction, including levels of abstraction in which no reference is made to humans. But the whole point of adopting these levels of abstraction is to do some human work, e.g., being able to understand and/or control what is going on in a system. Abstractions are the work of humans and the abstractions themselves do not exist separately from humans. Thus, attributing "moral agency" to computer systems merely because there is a level of abstraction in which these connections are unnecessary is misleading. Our descriptive claim is, then, that computer systems *are* always tethered (connected) to human beings, though there are a multitude of ways to conceptualize and abstract the workings of these systems.

Our normative claim is that the connections between computer systems and human beings should be recognized in the way we understand computer systems. Are we saying that certain levels of abstraction should never be used to describe systems? No, that is not the point. Levels of abstraction in which human actors and agency do not appear are useful for a variety of purposes. Abstraction is an effective tool that allows us to focus on some details while ignoring other details. Our point is that it is misleading and perhaps deceptive to uncritically transfer concepts developed at one level of abstraction to another level of abstraction. Obviously, there are levels of abstraction in which computer behavior appears autonomous, but the appropriate use of the term "autonomous" at one level of abstraction does not mean that computer systems are, therefore, "autonomous" in some broad and general sense. We should not allow the existence of a particular level of abstraction to determine the outcome of the broader debate about the moral agency of computer systems.

Thus, part of our normative claim is that statements at a particular level of abstraction must always be qualified as applicable only at that level of abstraction. Debate about the moral agency of computer systems takes place at a certain level of abstraction and the implication of our analysis is that discourse at this level should reflect and acknowledge the people who create, control, and use computer systems. In this way, developers, owners, and users are never let off the hook of responsibility for the consequences of system behavior.

Conclusion

The debate about whether computer systems can ever be "moral agents" is a debate among humans about what they will make of computational artifacts that are currently being developed. It is also a debate about the direction of future developments. This is a debate, not a "discovery" of a fact of nature concerning these artifacts. Floridi and Sanders' 2004 paper is an important touchstone in the debate, worthy of serious criticism. They are, we contend, a strong voice in the service of a group of scholars we have called "Computational Modelers" who have a stake in attaching the label "moral agent" to computational artifacts. We contrasted their ideas with those of the "Computers-in-Society" group, a group within which we place ourselves. The relationships between human moral agents and the artifacts they design, implement, and deploy should continue to be carefully examined by scholars, artifact developers, and the public for we are all likely to be increasingly affected by these human-tethered artifacts.

References

Robert Andrews. A Decade After Kasporov's Defeat, Deep Blue Coder Relives Victory. *Wired.* May 11, 2007. Available at http://www.wired.com/science/discoveries/news/2007/05/murraycampbell_qa.

Wiebe Bijker and Trevor Pinch. The Social Construction of Facts and Artifacts. In Wiebe Bijker, Thomas Hughes and Trevor Pinch, editors, *The Social Construction of Technological Systems*, pp. 17–51. MIT Press, Cambridge, Mass., 1987.

Luciano Floridi and Jeff W. Sanders. Artificial Evil and the Foundation of Computer Ethics. *Ethics and Information Technology*, 3(1): 55–66, 2001.

Luciano Floridi and Jeff W. Sanders. On the Morality of Artificial Agents. *Minds and Machines*, 14(3): 349–379, 2004.

Frances S. Grodzinsky, Keith W. Miller and Marty J. Wolf. The Ethics of Designing Artificial Agents. *CEPE*. San Diego, CA, July 12–14, 2007. Abstract Available at http://cepe2007.sandiego.edu/abstractDetail.asp?ID = 14.

Deborah G. Johnson. Computer Systems: Moral Entities but not Moral Agents. *Ethics and Information Technology*, 8(4): 195–204, 2006.

Deborah G. Johnson. Nanoethics: An Essay on Ethics and Technology 'in the making'. *Nanoethics*, 1(1): 21–30, 2007.

Deborah G. Johnson and Thomas M. Powers. Computers as Surrogate Agents. In J. Van Den Hoven and J. Weckert, editors, *Information Technology and Moral Philosophy*. Cambridge University Press, Cambridge, 2008.

Bill Joy. Why the Future Doesn't Need Us. *WIRED*, 8(4): 238–262, 2000.

John L. Pollock. When is a Work Around? Conflict and Negotiation in Computer Systems Development. *Science, Technology & Human Values*, 30(4): 496–514, 2005.

John Sullins. Ethics and Artificial Life: From Modeling to Moral Agents. *Ethics and Information Technology*, 7(3): 139–148, 2005.

27

Agencies in Technology Design: Feminist Reconfigurations*

Lucy Suchman

> *Agency is not an attribute but the ongoing reconfigurings of the world.*
> Barad 2003: 818

In this paper I consider some new resources for thinking about how capacities for action are configured at the human-machine interface, informed by developments in feminist science and technology studies. While not all of the authors and works cited would identify as feminist, they share – in my reading at least – commitments to a critical and generative interference in received conceptions of the human, the technological and the relations between them. Read against my own experience of the worlds of technology research and development these reconceptualizations have, in turn, associated implications for everyday practices of technology design. Both reconceptualizations of the human-machine interface, moreover, and the practices of their realization are inflected by, and consequential for, gendered relations within technoscience and beyond. The discussion brings us onto the terrain of science and technology studies, feminist theory, new media studies and experiments in cooperative systems design, each of which is multiple and extensive in themselves and no one of which can be adequately represented here. My hope nonetheless is to trace out enough of the lines of resonant thought that run across these fields of research and scholarship to indicate the fertility of the ground, specifically with respect to creative reconfigurations at the interface of human and machine.

At stake here as well are questions of what counts as 'innovation' in science and engineering. This in itself, I will propose, is a gendered question insofar as it aligns with the longstanding feminist concern with the problem of who shows up and who disappears in prevailing discourses of science and technology (see for example Suchman and Jordan 1989). Writings under the heading of invisible or 'articulation' work highlight the mundane forms of inventive yet taken for granted labor that are the conditions of possibility for complex sociotechnical arrangements.[1] A central commitment in articulating those labors is to decenter sites of innovation from singular persons, places and things to multiple acts of everyday activity, including the actions through which only certain actors and associated achievements come into public view. At the same time, we

---

*Reprinted in part from Suchman (2007) Feminist STS and the Sciences of the Artificial. In E. Hackett, O. Amsterdamska, M. Lynch, & J. Wajcman (Eds.), *The Handbook of Science and Technology Studies, Third Edition* (pp. 139-163). Cambridge, MA: MIT Press; see also chapter 15 in Suchman (2007) *Human-Machine Reconfigurations*. New York: Cambridge University Press.

[1] For a case study of the design work involved in use, see Clement 1993. On articulation work, see Star 1991, Fujimura 1987, and for its relevance to system design see Schmidt and Bannon 1992.

need to ask how projects to reclaim creativity, invention and the like might themselves be reproductive of a very particular, cultural-historical preoccupation with the new. Must those not identified as creative be shown in fact to be inventors in order to be fully recognized? This question suggests an attention to the tensions and contradictions that arise when we adopt a strategy for decentering that attempts to distribute practices previously identified exclusively with certain locales across a wider landscape. In distributing those practices more widely, they are given correspondingly greater presence. A counter project, therefore, is to interrogate the trope of innovation itself, to see how a fascination with change and transformation might be located, both culturally and historically, and in particular moments.[2]

Mutual constitutions

The primary site for my own exploration of these questions has been research and development at the interface of persons and machines. The human-machine boundary is at once defined by the interface, and by the difference that it is designed to address. A central problem in theorizing relations of humans and machines is the following: How do we develop a conceptual, practical and political framework that recognizes the mutual constitution of humans and artifacts, while acknowledging as well the particularities that distinguish one from another? While building on the rich body of feminist scholarship on the figure of the cyborg (of which more below), I take to heart as well the caution sounded by Donna Haraway in her recent observation that "the differences between even the most politically correct cyborg and an ordinary dog matter" (2003: 4). A series of binary oppositions of 'same' and 'other', 'us' and 'them' have characterized relations both among humans and between the human and nonhuman. The question of difference outside of these overly simple and divisive oppositions is one that has been deeply and productively engaged within feminist and postcolonial scholarship.[3] And there are lessons to be learned here for the question of human-machine relations as well.

Within science and technology studies, Actor Network Theory's call for a "generalized symmetry" in analyses of human and nonhuman contributions to social order represented a powerful intervention into sociological preoccupations with human agency (see Ashmore et al 1994). For those writing within the Actor Network tradition and its aftermath, agency is reconceptualized as always a relational effect that can never be located in either humans or nonhumans alone. A rich body of empirical studies have further specified, elaborated, and deepened the senses in which human agency is only understandable once it is re-entangled in the sociomaterial relations that the 'modern constitution' (Latour 1993) since the 17th century has so exhaustingly attempted to take apart. These studies provide compelling empirical demonstration of how capacities for action can be reconceived on foundations quite different from those of an Enlightenment, humanist preoccupation with the individual actor living in a world of separate things. This body of work is too extensive to be comprehensively reviewed, but a few indicative examples can serve as illustration.

---

[2] This line of argument is developed further in Suchman 2011.
[3] For some exemplary texts see Ahmed 1998, 2000, Braidotti 1994, 2002, Castañeda 2002, Gupta and Ferguson 1992, Strathern 1999, Verran 2001.

Ethnomethodological studies since the 1980s make up one line of such investigations.[4] An exemplary case is provided by Charles Goodwin's analyses of what he terms 'professional vision' (1994), developed in a series of studies focused on the social and material interactions through which practitioners learn to see those phenomena that constitute the objects of their profession. A central argument is that these phenomena are not pre-existing, but are constituted as disciplinarily relevant objects through occasioned performances of competent seeing. It is important to note that 'seeing' for Goodwin is far from a narrowly scopic or perceptual event, but is rather an activity entailing complex, multisensory embodiments. Goodwin analyses archeological knowledge, for example, as relations between particular, culturally and historically constituted practices and their associated materials and tools. It is out of those relations, he argues, that the objects of archeological knowledge and the identity of competent archeologist are co-constructed. As a simple case, Goodwin describes how archeologists in the field use a particular artifact, the Munsell chart of universal color categories, to code color features of dirt in an archaeological site (see also Latour 1999:58-61). They do this by taking a trowel of dirt, wetting it, and holding it behind holes cut in the chart. This practice displays both the artfulness of the color chart's design and the ways in which its use presupposes the embodied juxtaposition and skilled reading of dirt in relation to chart. Goodwin also describes the problems of actually using the chart (the colors of real dirt, of course, never quite match these ideal types) and its role in bringing archeologists' perception into the service of a particular organizational/bureaucratic endeavor, that is, professional archaeology.

Other studies have developed these ideas in relation to contemporary work environments, including medicine. As one illustrative case in point, Dawn Goodwin (2009) mobilizes actor network tropes in a close study of the practices through which patients in surgery are 'transitioned' through anesthetic states, a process involving the radical and progressive reconfiguration of their capacity for action – specifically, for the sustenance of their own life support – through complex arrangements of medical practitioners, machines, routines and other devices. She argues that questions of agency are crucial both to assess policy with respect to medical practice, and to deepen our understanding of the complex sociotechnical arrangements that comprise much of contemporary medical activities and institutions. Through a series of cases she demonstrates how the technologies of anesthesia are joined to the patient's body, in ways that render the latter highly dependent and vulnerable, but nonetheless intensely (albeit sometimes ambiguously) communicative. This joining is analyzed as a delicate choreography involving patients, medical practitioners and machines.[5] Over the course of an anesthesia, agencies involved in the sustenance of vital bodily functions are

---

[4] Despite ethnomethodology's troubled relation to feminism, I would argue that the former's commitments to respecifying social order as an embodied, interactional, and irreducibly collaborative achievement (even in the production of individuals and differences) provide deep lines of generative connection. For an early ethnomethodological study of the performativity of sex/gender see Garfinkel 1967, chapter 5. See also Smith 1999.

[5] The phrase 'ontological choreography' was coined by Charis Thompson (Cussins 1998), whose work I return to below.

progressively delegated from 'the patient' as an autonomously embodied entity, to an intricately interconnected sociomaterial assemblage, and then back again. The particular expertise of the anesthetic practitioner on this account is to manage the often unruly contingencies of the unfolding course of anesthesia, through a combination of skillfully embodied techniques, reading of signs, professional judgments and legitimating accounts, which together provide the grounds for practical action. Normative prescriptions of correct procedure and power-differentiated divisions of labor complicate the process in ways that can work to undermine the legitimacy of other forms of 'evidence', thereby jeopardizing rather than ensuring safe and effective practice. In this she offers an alternative reading to Poovey (1987), who suggests that anesthesia administered to ease the pain of childbirth during the Victorian period 'silenced' the female body.

In a related argument developed through the case of reproductive technoscience, Charis Thompson (2005; see also Cussins 1998) argues against the idea that medical interventions inherently objectify patients and thereby strip them of their agency. She observes that in the case of infertility clinics "the woman's objectification, naturalization, and bureaucratization involve her active participation and are managed by herself as crucially as by the practitioners, procedures, and instruments" (Cussins 1998:167). Conversely, objectification does not inherently or necessarily lead to alienation, nor does it stand always in opposition to subjectivity or personhood. Among other things, the clinic relies on the possibility of separation (of egg and sperm from the bodies that produce them) without alienation. Thompson locates alienation not in objectification per se, but in the breakdown of synechdochal relations between parts and whole that make objectification of various forms into associated forms of agency. It is this process "of forging a functional zone of compatibility that maintains referential power between things of different kinds" that she names *ontological choreography* (ibid.: 192). Medical ethics and accountability, she argues, need to be founded not in the figure of the rational, informed citizen but in the conditions for the maintenance of those crucial relations that configure identities and selves, and that might allow them to be reconfigured in desired ways.

The trope of configuration animates another study of surgical practices by Margun Aanestad (2003), who focuses on the labors involved in aligning a complex sociotechnical environment for the conduct of so-called 'minimally invasive' or 'keyhole' surgery. The latter involves, among other things, displacing the gaze of the surgeon and attendant practitioners into the interior of the patient's body from a correspondingly large incision to a view mediated through camera and video monitors. Aanestad's analysis follows the course of shifting interdependencies in the sociotechnical assemblage, as changes to existing arrangements necessitate further changes in what she names the *in situ* work of "design in configuration" (ibid.: 2). She emphasizes that the agencies of the technologies involved do not exist before their incorporation into the network. Her analysis makes clear again how in such a setting the capacity for action is relational, dynamic and collective rather than inherent in specific network elements, and how the extension of the network in turn intensifies network dependencies. Her analysis has directly gendered implications as well, as the work of nurses, overwhelmingly women, is literally as well as figuratively marginalized in the views of the operating theatre, at the same time that their role in the theatre's configuration becomes more central.

Together these inquiries respecify agency from a capacity intrinsic to singular actors, to an effect of practices that are multiply distributed and contingently enacted across humans and things. Addressing similar questions, but from a position within feminist philosophy and science studies, physicist Karen Barad has proposed a form of materialist constructivism that she names "agential realism," through which realities are enacted in and through specific apparatuses of sociomaterial "intra-action" (2003). While the term *interaction* presupposes two entities, given in advance, that come together and engage in some kind of exchange, *intra-action* underscores the sense in which subjects and objects emerge through their encounters with each other. In this, Barad's writings join others working towards conceptualizations of the material that incorporate obduracy and contingency, the discursive and the corporeal. More specifically, Barad locates technoscientific practices as critical sites for the emergence of new subjects and objects. Taking physics as a case in point, her project is to work through longstanding divisions between the virtual and the real, while simultaneously coming to grips with the ways in which materialities, as she puts it, "kick back" in response to our intra-actions with them (1998:112). Through her readings of Niels Bohr, Barad insists that 'object' and 'agencies of observation' form a nondualistic whole: it is that relational entity that comprises the objective 'phenomenon' (1996: 170). Different "apparatuses of observation" make possible different, always temporary, subject/object cuts that in turn enable measurement or other forms of objectification, distinction, manipulation and the like *within* the phenomenon. The relation is "ontologically primitive" (2003: 815), in other words, or prior to its components; the latter come about only through the 'cut' effected through a particular apparatus of observation. One effect of this position is a more radical understanding of the sense in which

> materiality is discursive (i.e., material phenomena are inseparable from the apparatuses of bodily production: matter emerges out of and includes as part of its being the ongoing reconfiguring of boundaries), just as discursive practices are always already material (i.e., they are ongoing material (re)configurings of the world) (2003: 822).

Brought back into the world of technology design, this intimate co-constitution of configured materialities with configuring agencies clearly implies a very different understanding of the 'human-machine interface'. Read in association with the empirical investigations of complex sites described above, 'the interface' on the one hand becomes the name for a category of contingently enacted 'cuts' occurring always *within* sociomaterial practices, that effect 'persons' and 'machines' as distinct entities, and that in turn enable particular forms of subject/object intra-actions. At the same time, the singularity of 'the interface' explodes into a multiplicity of more and less closely aligned, dynamically enacted moments of encounter between sociomaterial configurations, including persons and machines. It is the differences effected *within* such configurations that I turn to next.

Dissymmetries

The reconstructions of sociomaterial agency reviewed above are frequently summarized by the proposition that humans and artifacts are *mutually constituted*. This premise of technoscience studies has been tremendously valuable as a corrective to the entrenched Euro-American view of humans and machines as autonomous, integral entities that must somehow be brought together and made to interact. But at this point I think the sense of mutual constitution warrants a closer look. In particular, we are now in a position to elaborate that generative trope along at least two critical dimensions: first, in relation to the dynamic and multiple forms of constitution that are evident in specific sociomaterial assemblages and second, in terms of questions of differences within those assemblages.

Most importantly, mutuality does not necessarily imply symmetry. My initial analyses of human-machine interaction (Suchman 1987) made clear that persons and artifacts do not constitute each other in the same way. In light of this, I would argue that we need a rearticulation of human-artifact asymmetry – or more suggestively perhaps dissymmetry – that somehow retains the recognition of the various hybrids made visible through science and technology studies, while simultaneously recovering certain subject/object positionings – even orderings – among persons and artifacts and their consequences. The emphasis in science and technology studies on symmetrical analysis and the agency of things arose from a well-founded concern to recover for the social sciences and humanities aspects of the lived world – for example, 'nature' and 'technology' – so far excluded from consideration as proper sociological subjects. My own thinking is clearly indebted to these efforts, which provide the reconceptualizations needed to move outside the frame of categorical purification and opposition between social and technical, person and artifact. My engagement with these questions, however, came first in the context of technoscience and engineering, where the situation is in important respects reversed. Far from being excluded, 'the technical' in regimes of technology research and development are centered, while 'the social' is separated out and relegated to the margins.

Applied to the question of agency, I have argued that through the figures of artificial intelligence we are witnessing a reiteration of traditional humanist notions of agency, at the same time – even through – the intra-actions of that notion with new computational media (Suchman 2007). In the remainder of this paper I look to recent developments in configuring agencies at the human-machine interface with a view informed by feminist theorizing, to explore the question of what other directions our relations with machines, both conceptually and practically, might take.

Re-reading robots

I turn first to recent counter-readings of human-machine relations inspired by feminist discussions of materialities, subjectivities, and cyborg bodies. Like many others, my attention was first drawn to these possibilities by Donna Haraway's 'whip lashing' proposal (a phrase she herself uses to describe those moments when a new idea comes along that turns one's head) that we should all prefer to be cyborgs than goddesses

(1985/1991:223). As Jenny Wolmark summarizes, in her discussion of the 'Manifesto for Cyborgs':

> The cyborg's propensity to disrupt boundaries and explore differently embodied subjectivities could ... be regarded as its most valuable characteristic, and it is undoubtedly one of the reasons for its continued usefulness in feminist and cultural theory (1999:6).

As feminist theorists trace a new path across the problematic terrain of how the gendered subject might be reconceived, they also, I would argue, provide us with resources for reconceptualize the object. Feminist re-theorizing of the body has been concerned to restore the dynamism emptied out of bodies by the mind/body split, precisely by moving through that split to some new terrain. In a similar way we can find other grounds for understanding our relations to the material than the operations of a transcendental intelligence over inert, mechanistically animated matter. As Butler famously puts it in *Bodies that Matter*:

> What I would propose ... is a return to the notion of matter, not as site or surface, but as a process of materialization that stabilizes over time to produce the effect of boundary, fixity, and surface we call matter ... Crucially, then, [the construction of bodies] is neither a single act nor a causal process initiated by a subject and culminating in a set of fixed effects (1993:9-10).

Butler's argument that sexed and gendered bodies are materialized over time through the reiteration of norms is suggestive for a view of technology construction as a process of materialization through a reiteration of forms. Butler argues that 'sex' is a dynamic materialization of always contested gender norms: similarly, we might we understand 'things' or objects as materializations of more and less contested, normative identifications of matter. Much as recognition and intelligibility are central to feminist conceptions of the subject, objects achieve recognition within a matrix of historically and culturally constituted familiar, intelligible possibilities. Technologies are both produced and destabilized in the course of these reiterations.

As an example, Claudia Castañeda has written about the "materialization of touch" in robotic artificial intelligence (1999). Beginning from an understanding of touch as always semiotic and relational, and of signs as always entailing materialities, she takes up the question of the skin and its materialization in the form of the robot Cog. Cog's designers frame interactivity, with 'the world' and with its human counterparts, as the litmus test of the robot's competencies. During Cog's early, awkward stage, its 'skin' is designed to serve as a protective device against contact, equipped with the requisite sensors and alarms. Castañeda explores the premise that Cog's embodiment is designed to change in response to the robot's interactions over time. My own skeptical reading of the project falters on the question just how open the possibilities of re-materialization are for Cog given the robot's origins in the historical and cultural matrix of the Massachusetts Institute for Technology's Laboratory for Artificial Intelligence. But Castañeda's hopeful reading points us to aspects of Cog that at least signal the possibility of what she names a "feminist robotics." (2001: 233).

First, and most basically, Cog's design (at least on this telling) locates touch as a way of knowing and being in the world. Second, Castañeda suggests that Cog embodies a relational conception of the body, one that extends beyond the boundaries of the skin, and that is generated through particular, changing combinations of materials and qualities. And finally, as she puts it, Cog is "neither human nor anti-human, but rather other-than-human" (ibid.: 232). As such, she argues that Cog's re-embodiment of the human in different terms generates the possibility, in material form, of embodied alterity, a relation of difference that literally as well as figurally matters. Castañeda's interest then is in just what kind of alterity is, or could be, embodied in the robot, which does not take the human, normatively imagined, as the "origin and truth against which the robot's value is always measured" (234).

Deirdre, the heroine of science fiction writer C.L. Moore's 1944 short story, 'No Woman Born', provides a second example.[6] Deidre prefigures Haraway's challenge to the cultural imaginary of the goddess turned cyborg. Ambivalently positioned on the boundary of Cartesian and feminist understandings of the body, the premise of Moore's story is that Deirdre, once an exquisitely beautiful and talented dancer, has been injured in a theater fire to the point that only her brain survives. As the brain of a *dancer* however – endowed we are told with extraordinary grace – Moore locates Deirdre's brain in intimate relation to her body. As the story unfolds it becomes clear that the restoration of Deirdre's agency is inseparably tied to the particularities of her rematerialization. We enter the story one year after the tragic fire, during which time Deirdre (Deirdre's brain?) has been painstakingly re-embodied by Maltzer, a genius physician/scientist, assisted by a team of unnamed (but apparently greatly talented) sculptors and artists.

The story that follows is effectively a set of variations around the theme of Deirdre's rematerialization, haunted by questions of memory, identity, recognition, transformation and otherness. We approach these questions through the person of John Harris, Deirdre's former agent and close friend, coming to see her for the first time following the accident. Torn by visions of, on one hand, the irrecoverable figure of Deirdre the human as he knew her and, on the other, culturally inspired imaginings of how the new, robotic Deirdre might be configured, Harris suffers agonies of anticipation in advance of their meeting. His uncertainties are not allayed by his conversation with Maltzer, in the anteroom of Deirdre's chambers:

> "It's not that she's – ugly – now," Maltzer went on hurriedly … "Metal isn't ugly. And Deirdre … well, you'll see. I tell you, I can't see myself. I know the whole mechanism so well – it's just mechanics to me. Maybe she's – grotesque, I don't know (1975: 67).

The Deirdre that Harris goes on to meet is less a replica than a new configuration, a re-embodiment, of the Deirdre he remembers. In place of a face she has a delicately modeled ovoid head with a kind of golden mask, in which a slit of aquamarine crystal occupies the place where her eyes would have been. And rather than a simulation of human skin over hinged metal joints, her body is made up of tiny golden coils, infinitely

---

[6] This story appears in an anthology of science fiction short stories within which C.L. Moore is the only woman author.

flexible, covered by a robe of very fine metal mesh, all of which she has learned to move with an extraordinary expressiveness and grace at once reminiscent of, and different from, her former dancer's body.

As Harris struggles to come to terms with the neither/nor, both/and qualities of the new Deirdre, the story unfolds as a succession of reflections on the uncertainties of Deirdre's status, in relation to her former identity as Deirdre and to the rest of the human world. First, what is the relation of this new creature to "Deirdre" herself? Is "she" still alive? And what about the re-embodied Deirdre's relation to her creator, Maltzer? Is she an extension of him, his property, or an autonomous being, animating the materials that he has provided with her own "unquenchable" essence? And is her essence that of the brain that survived, or some irreducible spirit that animates her new body? In one of his more enlightened moments her old friend Harris muses, "She isn't human, but she isn't pure robot either. She's something somewhere between the two, and I think it's a mistake to try to guess just where, or what the outcome will be" (1975: 88).

Deidre embodies the ambivalences of technoscience, suggesting the possibilities for new configurations that are fabulous and expansive, while at the same time threatening the reassuring ground of familiar categories on which our experiences of relationship, of knowing and being known, depend. Alternately goddess, human, superhuman, and monster, Deirdre powerfully expresses the questions raised by new sociomaterial possibilities and their relations to old struggles around identity and difference. The story's pivotal question, on which the plot turns, is whether Deirdre is still human and, if not, whether the rematerialized Deirdre can survive given her singular Otherness. Not surprisingly, this question remains unanswered at the story's end. But what Moore has achieved is to reframe the cyborg from its reiteratively human replicant form to something that dances elusively, and therefore suggestively, on the boundaries of old and new possibilities, albeit with all of the traditional dilemmas that such boundary crossings entail.

The question of how the robot could be other than second term to the human aligns with feminist concerns about what Anne Balsamo sums as "the systems of differentiation that make the body meaningful," most notably those of gender (1996: 21). Framed not as the importation of mind into matter, but as the rematerialization of bodies and subjectivities in ways that challenge familiar assumptions about the naturalness of normative forms, robots, and cyborg figures more generally, become sites for change rather than just for further reiteration. In place of the binaries male/female, human/machine, subject/object we have an open horizon of specific, historically and culturally configured relations. Crucially, these relations are power-differentiated but in ways specific to each. And in each case power works through these relations not in the simple sense that the first term holds power over the other, but that their relative positionings – including, crucially, as opposites – enables their fundamental inter-relations and the historically sedimented cuts that position them as separate categories to be obscured.

While the cyborg since Haraway suggests generative forms of reconfiguring, one problem with the cyborg as an icon of sociomateriality needs mention. That is the problem that characterizes any discussion that centers on a singular, heroic (even monstrous) figure; that is, it obscures the presence of distributed sociomaterialities in more mundane sites of familiar, everyday life. Along with the dramatic demonstrations of sociomateriality embodied by the cyborg, then, we need to recover the ways in which more familiar bodies and subjectivities are being formed through contemporary interweaving of nature and artifice, for better and worse. My interest is in developing a relational view of subjects and objects that locates conditions for action and possibilities for intervention in the particularities of more everyday sociomaterial configurations.

Reconceptualizing agencies at the human-machine interface

To make this last proposal more concrete, I turn to several examples of what we might characterize as configurations of agency at the human-machine interface, but conceptualized in a very different way. The first, and most mundane, case is drawn from a study in the area of computer-supported work, in particular, a civil engineer working at a computer-aided design (CAD) workstation (see Suchman 2000). In her analysis of computer-based work, Susanne Bødker (1991) has discussed the shifting movement of the interface from object to connective medium. She observes that when unfamiliar, or at times of trouble, the interface itself becomes the work's object. At other times persons work as she puts it 'through the interface', enacted as a transparent means of engagement with other objects of interest (for example, a text, or an interchange with colleagues). While CAD might be held up as an exemplar of the abstract representation of concrete things, for the practicing engineer the story is more complex. Rather than stand in place of the specific locales – roadways, natural features, built environments, people and politics – of a project, the CAD system connects the experienced engineer sitting at her worktable to those things, at the same time that they exceed the system's representational capacities. The engineer knows the project through a multiplicity of documents, discussions, extended excursions to the project site, embodied labors and accountabilities, and the textual, graphical and symbolic inscriptions of the interface are read in relation to these heterogeneous forms of embodied knowing. Immersed in her work, the CAD interface becomes for the engineer a simulacrum of the site, not in the sense of a substitute for it, but rather of a place in which to work, with its own specific materialities, constraints and possibilities. While lacking any claims to embodiment, affect or sociality in its own right, the CAD interface-in-use is, I would argue, a powerful site of expanded, sociomaterial agency. It suggests a figure of technological agency not in the form of machinic operations conducted independently of the human, but in the form of a particular configuration, a specifically enacted site of extended, heterogeneously constituted human/nonhuman capacities for thought and action.

New media artist Heidi Tikka, in her work titled 'Mother, Child', provides a further example. This work, which I had the opportunity to experience during its exhibition at the Art Gallery of Ontario in Toronto, Canada in 2001, employs the shifting dynamics of installation, viewer/user, and onlookers, as well as the ambient environment of the exhibition space to invoke, and affectively evoke, an encounter between caregiver and infant. The piece does this not 'in general', but always specifically: the caregiver is one, particular person who enters the space of the installation and sits on a chair, the infant is one, particular infant (Tikka's son, recorded on digital video). A distinguishing aspect of the piece is the heterogeneity of its forms: real bodies and objects combine with projected images to comprise a hybrid of social and material elements. Together these elements create an interactive space characterized by a mix of predictability and contingency – a fragile stability – that affords the installation its affective kinship to the 'real world' encounter that it simulates. The three-dimensional image of a child that is 'projected' – both technically and psychically – onto the soft cloth diaper that the viewer/user holds in her lap can be affected through her motions and orientation to it, but dissolves as she stands and places the cloth back onto the chair. In this and other ways, the installation continually reminds us of, rather than conceals, its artifice. As Tikka herself comments, the piece is actually simpler, less reactive in its composition than we experience it to be. The effects are created through the particular possibilities provided by an artful integration of persons, objects, spaces, fantasies, remembered experiences and technologies to evoke and explore an emblematically human encounter, but not to replicate it.

Media scholar Chris Chesher (2004) has proposed a reconceptualization of encounters with computer-based art that radically reworks information theoretic tropes at the interface. While his proposal is richly suggestive, I believe, for any example of human-computer interaction, Chesher starts from the premise that "new media artists' non-instrumental applications of technology put the distinctiveness of computer based forms into greater relief" (ibid.: 2). From his consideration of new media art Chesher proposes the concept of *avocation* to describe the arrangements and affordances through which persons are hailed to enter into a particular technological assemblage, to become incorporated as integral actants in an associated form of sociomaterial agency. These include not only instrumental possibilities, but also multiple and uncertain ways in which "new media art distracts and summons its users" (1). *Invocation* involves those actions that define the terms of engagement written into the design 'script' or discovered by the engaged 'user', the calling up of events that effect changes to the assemblage. And finally, *evocation* describes the affective and effective, material changes that result; transformations that in turn comprise the conditions of possibility for subsequent avocations.

Expanding frames and accountable cuts

My concern in this paper has been with the collective imaginaries through which contemporary relations of humans and machines are rendered intelligible, materialized and made real. In this I follow Barad's proposition that "reality is sedimented out of the process of making the world intelligible through certain practices and not others ..." (1998: 105). If, as Barad suggests, we are responsible for what exists, what is the reality that current discourses and material practices regarding computational artifacts make intelligible, and what is excluded? To answer this question Barad argues (following Haraway) that we need a simultaneous account of the intra-actions of humans and nonhumans *and* of their asymmetries and differences. This requires remembering that boundaries between humans and machines are not naturally given but constructed, in particular historical ways and with particular social and material consequences. As Barad points out, boundaries are necessary for the creation of meaning, and, for that very reason, are never innocent. Because the cuts implied in boundary making are always agentially positioned rather than naturally occurring, and because boundaries have real consequences, "accountability is mandatory" (ibid.: 187). The accountability involved is not a matter of identifying authorship in any simple sense, however, but rather a problem of understanding the effects of particular assemblages, and assessing the distributions, for better and worse, that they engender. As Barad puts it:

> We are responsible for the world in which we live not because it is an arbitrary construction of our choosing, but because it is sedimented out of particular practices that we have a role in shaping (1998: 102).

Responsibility on this view is met neither through control nor abdication, but in ongoing practical, critical, and generative acts of sociomaterial engagement. Andrew Barry (1999) draws a useful distinction between 'novelty' and 'invention,' and argues that there is no simple relation between the speed with which new things are produced, and inventiveness. In contrast to the premise that innovation can be measured in terms of the number of ideas that are locked in place through their materialization as patented artifacts, Barry proposes a view of inventiveness as "an index of the degree to which an object or practice is associated with *opening up* questions and possibilities ... what is inventive is not the novelty of artifacts in themselves, but the novelty of the arrangements with other activities and entities within which artifacts are situated. And might be situated in the future" (ibid.: 4). He suggests further that there might actually be an inverse relation between the speed of change, and the expansion of inventiveness – that "moving things rapidly may increase a general state of inertia; fixing things in place before alternatives have the chance of developing" (6). I have made a similar argument (Suchman 1999) with respect to technology innovation under the banner of 'artful integration,' attempting to shift the frame of design practice and its objects from the figure of the heroic designer and associated next new thing, to ongoing, collective practices of sociomaterial configuration, including reconfigurations effected in use.

With this argument in mind, I want to close by offering a final example that serves I think as a kind of pointer to a different conception of practices of innovation as the ongoing and everyday. My former colleague at Xerox's Palo Alto Research Center and accomplished computer scientist, Randall Trigg, has left the circuits of 'world class' research and development to become what he describes as the 'de facto IT department' for an organization devoted to making new circuits through which worldwide resources might flow. The organization, the Global Fund for Women (GFW)[7], serves as a redistribution point for U.S. wealth, providing micro grants to grassroots women's organizations in other parts of the world. The Fund itself is distinguished by its extraordinarily local/global networked form, where former grantees become the sites through which new micro grants are distributed. Technology design at the GFW involves ongoing co-development, with others in the GFW headquarters and among their grantees, of an information and communications infrastructure to support this local/global network. In considering this work it would be difficult to isolate singular achievements of the 'new'. And yet together, over time and space, I would argue that it is labors like this that represents our best hope for genuinely new reconfigurings of the technological, based not in inventor heroes or extraordinary new devices, but in mundane, and innovative, practices of collective sociomaterial infrastructure building.

---

[7] http://www.globalfundforwomen.org/

References

Aanestad, M. (2003) The camera as an actor: Design-in-use of Telemedicine Infrastructure in Surgery. *Computer-Supported Cooperative Work* (CSCW) 12: 1-20.

Ashmore, Malcolm, Wooffitt, Robin and Harding, Stella (1994) Humans and Others: the Concept of 'Agency' and its Attribution. *American Behavioral Scientist.*

Balsamo, Anne Marie (1996) *Technologies of the gendered body: reading cyborg women.* Durham: Duke University Press.

Barad, Karen (1996) Meeting the Universe Halfway: Ambiguities, Discontinuities, Quantum Subjects, and Multiple Positionings in Feminism and Physics. In L. H. Nelson & J. Nelson (eds.), *Feminism, Science, and the Philosophy of Science: A Dialog* pp. 161-194. Norwell, MA: Kluwer.

Barad, Karen (1998) Getting Real: Technoscientific Practices and the Materialization of Reality. *differences: A Journal of Feminist Cultural Studies,* 10: 88-128.

Barad, Karen (2003) Posthumanist Performativity: Toward and understanding of how matter comes to matter. *Signs: Journal of Women in Culture and Society* 28: 801-831.

Bødker, Susanne (1991) *Through the interface: a human activity approach to user interface design.* Hillsdale, N.J.: L. Erlbaum.

Castañeda, Claudia (2001) Robotic skin: The future of touch? In S. Ahmed & J. Stacey (eds.), *Thinking Through the Skin* pp. 223-236. London: Routledge.

Chesher, Chris (2004) Invocation, evocation and avocation in new media art, paper presented at the joint meetings of the European Association for Studies of Science and Technology (EASST) and Society for Social Studies of Science (4S), Paris.

Clement, Andrew 1993 Looking for the Designers: Transforming the 'invisible' infrastructure of computerized office work. In *AI & Society*, Special Issue on Gender, Culture and Technology, 7:323-344.

Cussins, Charis (1998) Ontological Choreography: Agency for women patients in an infertility clinic. In M. Berg & A.-m. Mol (eds.), *Differences in Medicine* pp. 166-201. Durham, NC: Duke University Press.

Fujimura, Joan. 1987. Constructing 'Do-able' Problems in Cancer Research: Articulating Alignment. *Social Studies of Science* 17:257-93.

Garfinkel, Harold (1967) *Studies in ethnomethodology.* Englewood Cliffs, N.J.: Prentice-Hall.

Goodwin, Charles (1994) Professional Vision. *American Anthropologist* 96: 606-633.

Goodwin, Dawn (2009) *Acting in Anaesthesia: Ethnographic Encounters with Patients, Practitioners and Medical Technologies*. Cambridge: Cambridge University Press.

Haraway, Donna (1991) A Manifesto for Cyborgs. In *Simians, cyborgs, and women: the reinvention of nature*. New York: Routledge.

Haraway, Donna (2003) *The Companion Species Manifesto: Dogs, people, and significant others*. Chicago: Prickly Paradigm Press: University of Chicago.

Latour, Bruno (1993) *We have never been modern*. Cambridge, Mass.: Harvard University Press.

Latour, Bruno (1999) *Pandora's hope: essays on the reality of science studies*. Cambridge, Mass.: Harvard University Press.

Law, John (1994) *Organizing modernity*. Oxford, UK ; Cambridge, Mass., USA: Blackwell.

Moore, C. L. (1975/1944) No Woman Born. In Scortia, Thomas and Zebrowski, George *Human-Machines: An anthology of stories about cyborgs*. New York: Vintage, pp. 63-118.

Poovey, Mary (1987) "Scenes of an indelicate character": the medical "treatment" of victorian women. In Gallagher, C and Lacqueur, T (Eds) *The Making of the Modern Body: Sexuality and Society in the Nineteenth Century*. Berkeley: University of California Press. Pp 137-168.

Schmidt, Kjeld and Bannon, Liam. (1992) Taking CSCW Seriously: Supporting Articulation Work. *Computer-Supported Cooperative Work* (CSCW), Vol., 1, Nos. 1-2: 7-40.

Smith, Dorothy E. (1999) *Writing the social: critique, theory, and investigations*. Toronto: University of Toronto Press.

Star, Susan Leigh (1991) Invisible Work and Silenced Dialogues in Knowledge Representation. In *Women, Work and Computerization*, eds. I. Eriksson,B. Kitchenham, and K. Tijdens, K., 81-92. Amsterdam: North Holland.

Suchman, Lucy (1987) *Plans and Situated Actions: the problem of human-machine communication*. New York: Cambridge University Press.

Suchman, Lucy (1999) Working Relations of Technology Production and Use. In D. Mackenzie & J. Wajcman (eds.), *The Social Shaping of Technology, Second Edition* pp. 258-68. Buckingham and Philadelphia: Open University.

Suchman, Lucy (2000) Embodied Practices of Engineering Work. *Mind, Culture & Activity* 7: 4-18.

Suchman, Lucy (2007b) *Human-machine Reconfigurations*. New York and Cambridge, UK: Cambridge University Press.

Suchman, Lucy (2011) Anthropological Relocations and the Limits of Design. *Annual Review of Anthropology, 40*, 1-18.

Suchman, Lucy and Brigitte Jordan (1989) Computerization and Women's Knowledge. In

Women, Work and Computerization, eds. K. Tijdens, M. Jennings, I. Wagner and M. Weggelaar, 153-160. Amsterdam: North Holland.

Thompson, Charis (2005) *Making parents: the ontological choreography of reproductive technologies*. Cambridge, MA: Mit Press.

28

Learning robots and human responsibility
Dante Marino and Guglielmo Tamburrini:

Abstract:

Epistemic limitations concerning prediction and explanation of the behaviour of robots that learn from experience are selectively examined by reference to machine learning methods and computational theories of supervised inductive learning. Moral responsibility and liability ascription problems concerning damages caused by learning robot actions are discussed in the light of these epistemic limitations. In shaping responsibility ascription policies one has to take into account the fact that robots and softbots – by combining learning with autonomy, pro-activity, reasoning, and planning – can enter cognitive interactions that human beings have not experienced with any other non-human system.

Dante Marino and Guglielmo Tamburrini:
Learning robots and human responsibility

The responsibility ascription problem for learning robots

In the near future, robots are expected to cooperate extensively with humans in homes, offices, and other environments that are specifically designed for human activities. It is likely that robots have to be endowed with the capability to learn general rules of behaviour from experience in order to meet task assignments in those highly variable environments. One would like to find in user manuals of learning robots statements to the effect that the robot is guaranteed to behave so-and-so if normal operational conditions are fulfilled. But an epistemological reflection on computational learning theories and machine learning methods suggests that programmers and manufacturers of learning robots may not be in the position to predict exactly what these machines will actually do in their intended operation environments. Under these circumstances, who is responsible for damages caused by a learning robot? This is, in a nutshell, the responsibility ascription problem for learning robots.

The present interest for this responsibility ascription problem is grounded in recent developments of robotics and artificial intelligence (AI). Sustained research programmes for bringing robots to operate in environments that are specifically designed for humans suggest that moral and legal aspects of the responsibility ascription problem for learning robots may soon become practically significant issues. Moreover, an analysis of this problem bears on the responsibility ascription problem for learning software agents too, insofar as the learning methods that are applied in robotics are often used in AI to improve the performance of intelligent softbots. Finally, an examination of these responsibility ascription problems may contribute to shed light on related applied ethics problems concerning learning software agents and robots. Problems of delegacy and trust in multi-agent systems are significant cases in point, which become more acute when learning is combined with additional features of intelligent artificial agents: human subjects may not be in the position to oversee, predict or react properly to the behaviour of artificial agents that are endowed with forms of autonomy, pro-activity, reasoning, planning, and learning; robotic and software agents can perform complicated planning and inferencing operations before any human observer is in the position to understand what is going on; agent autonomy and pro-activity towards human users may extend as far as to make conjectures about what a user wants, even when the user herself does not know or is unable to state her desires and preferences.

In addition to suggesting the present interest of an inquiry into the responsibility ascription problem for learning robots, these observations point to epistemic limitations that fuel this particular problem. It is these limitations that we turn now to discuss.

Machine learning meets the epistemological problem of induction

The study of machine learning from experience is a broad and complex enterprise, which is based on a wide variety of theoretical and experimental approaches. A major theoretical approach is PAC (Probably Approximately Correct) learning. Distinctive features of this approach are briefly discussed here, and compared with more experimentally oriented approaches to machine learning, – with the overall aim of isolating epistemic limitations which contribute to shape the responsibility ascription problem for learning robots.

PAC-learning is a theoretical framework for the computational analysis of learning problems which sets relatively demanding criteria for successful learning. PAC-learning inquiries aim at identifying classes of learning problems whose correct solutions can (or alternatively cannot) be approximated with arbitrarily small error and with arbitrarily high probability by some computational agent, when the agent is allowed to receive as inputs training examples of the target function that are drawn from some fixed probability distribution and is allowed to use "reasonable" amounts of computational resources only (that is, resources that are polynomially bounded in the parameters expressing characteristic features of the learning problem; for a precise definition of PAC-learnability, and examples of functions that are (not) PAC-learnable, see Mitchell 1997, pp. 203-214).

Does the PAC-learning paradigm put robot manufacturers and programmers in the position to certify that a robot will manifest with some high probability a behaviour which closely approximates a correct use of that concept or rule? One should be careful to note that such certifications may not be forthcoming in cases that are relevant to the responsibility ascription problem for learning robots, either in view of negative results (concerning problems that turn out to be not PAC-learnable) or in view of the difficulty of imposing the idealized PAC model of learning on concrete learning problems. Moreover, one should not fail to observe that these certifications do not put one in the position to understand or predict the practical consequences of the (unlikely) departures of PAC-learners from their target behaviour.

In connection with the limited applicability of PAC-learning methods, let us note that various classes of learning problems which admit a relatively simple logical formulation are provably not PAC-learnable. For example, the class of concepts that are expressible as the disjunction of two conjunctions of Boolean variables (Pitt and Valiant 1988) is not PAC-learnable. Moreover, the possibility of PAC-learning several other interesting classes of learning problems is still an open question. Finally, let us notice that one may not be in the position to verify background assumptions that are needed to apply the PAC model of learning to concrete learning problems. For example, the class of concepts or rules from which the computational learning system picks out its learning hypothesis is assumed to contain arbitrarily close approximations of the target concept or rule. But what is the target function and how can one identify its approximations, when the learning task is to recognize tigers on the basis of a training set formed by pictures of tigers and non-tigers? Another assumption of the PAC-learning model which is often unrealistic is that the training set always provides noise-free, correct information (so that misclassifications of, say, tigers and non-tigers do not occur in the training set).

In connection with the evaluation of the occasional departures from target behaviour that a PAC-learner is allowed to exhibit, one has to notice that the PAC-learning paradigm does not guarantee that these unlikely departures from target behaviour will not be particularly disastrous. Thus, the PAC-learnability of some concept or rule does not make available crucial information which is needed to understand and evaluate contextually the practical consequences of learning robot actions.

PAC-learning relieves instructors from the problem of selecting "suitable" training examples, insofar as a function can be PAC-learnt from randomly chosen examples. In contrast with this, the machine learning methods for supervised inductive learning that are applied in many cases of practical interest must rely on the background hypothesis that the selected training and test examples are "representative" examples of the target function. (The ID3 decision tree learning method is a pertinent case in point; see Mitchell 1997, ch. 3, for presentation and extensive analysis of this method.) The success of a supervised inductive learning process is usually assessed, when training is completed, by evaluating system performance on a test set, that is, on a set of examples that are not contained in the training set. If the observed performance on the test examples is at least as good as it is on the training set, this result is adduced as evidence that the machine will approximate well the target function over all unobserved examples (Mitchell 1997, p. 23). However, a poor approximation of the target function on unobserved data cannot be excluded on the basis of these positive test results, in view of the *overfitting* of both training and test data, which is a relatively common outcome of supervised inductive learning processes.[1] Overfitting gives rise to sceptical doubts about the soundness of inductive learning procedures, insofar as a good showing of an inductive learning algorithm at future outings depends on the fallible background hypothesis that the data used for training and test are sufficiently representative of the learning problem. This is the point where machine learning meets the epistemological problem of induction, insofar as the problem of justifying this background hypothesis about inductive learning procedures appears to be as difficult as the problem of justifying the conclusions of inductive inferences by human learners and scientists (for discussion, see Tamburrini 2006; for an analysis of early cybernetic reflections on the use of learning machines, see Cordeschi and Tamburrini 2005).

---

[1] Roughly speaking, a hypothesis h about some concept or rule from class H is said to overfit the training set if there is another hypothesis h' in H which does not fit the training set better than h but performs better than h on the whole set of concept or rule instances. "Overfitting is a significant practical difficulty for decision tree learning and other learning methods. For example, in one experimental study of ID3 involving five different tasks with noisy, non-deterministic data,... overfitting was found to decrease the accuracy of learned decision trees by 10-25% on most problems." (Mitchell 1997, p. 68).

Is there a responsibility gap?

Epistemic limitations concerning knowledge of what a learning machine will do in normal operating situations have been appealed to in order to argue for a responsibility gap concerning the consequences of learning systems actions. Andreas Matthias put forward the following argument (Matthias 2004):

- Programmers, manufacturers, and users may not be in the position to predict what a learning robot will do in normal operating environments, and to select an appropriate course of action on the basis of this prediction;
- thus, none of them is able to exert full control on the causal chains which originate in the construction and deployment of a learning robot, and may eventually result into a damage for another party;
- but a person can be held responsible for something only if that person has control over it; therefore, one cannot attribute programmers, manufacturers or users responsibility for damages caused by learning machines;
- since no one else can be held responsible, one is facing "a responsibility gap, which cannot be bridged by traditional concepts of responsibility ascription".

A distinctive feature of traditional concepts of responsibility which, in Matthias's view, give rise to this responsibility gap is the following "control requirement" (CR) for correct responsibility ascription: a person is responsible for *x only if* the person has control over *x*. Thus, the lack of control by programmers, manufacturers or users entails that none of them is responsible for damages resulting from the actions of learning robots. (CR) is usually endorsed and used as a premise in arguments for *moral* responsibility ascription. (But one should be careful to note that different interpretations of the notion of control are possible and prove crucial to determine the scope of someone's moral duties.) Matthias claims that (CR) is to be more extensively applied – indeed, to all situations which call for a responsibility ascription *in accordance with our sense of justice*.

> For a person to be rightly held responsible, that is, in accordance with our sense of justice, she must have control over her behaviour and the resulting consequences "in a suitable sense". That means that the agent can be considered responsible only if he knows the particular facts surrounding his action, and if he is able to freely form a decision to act, and to select one of a suitable set of available alternative actions based on these facts. (Matthias 2004, p. 175).

Here the scope of (CR) is overstretched. In general, the possibility of ascribing responsibility according to familiar conceptions of justice and right is not jeopardized in situations in which no one can be held morally responsible in view of a lack of control. (CR) is not necessary for responsibility ascriptions, and the alleged responsibility gap depending on it concerns moral responsibility ascriptions only. Indeed, the epistemological reflections reported in the previous section suggest that the responsibility ascription problems concerning possible applications of machine learning investigations are a recent acquisition of a broad and extensively analyzed class of *liability* problems, where the causal chain leading to a damage is not clearly recognizable, and no one is clearly identifiable as blameworthy. Traditional concepts of responsibility ascription exist for these problems and have been routinely applied in the exercise of justice. Accordingly, a shift from moral responsibility to another - but nonetheless quite traditional - concept of responsibility, which has to be adapted and applied to a newly emerging casuistry, enables one to "bridge" the alleged responsibility gap concerning the actions of learning robots.

Responsibility problems falling under this broad category concern children's parents or tutors, pet owners, legal owners of factories for damages caused by workers, and more generally cases in which it is difficult to identify in a particular subject the origin of the causal chain leading to the damaging event. Parents and tutors who fail to provide adequate education, care and surveillance are, in certain circumstances, held responsible for damages caused by their young, even though there is no clear causal chain connecting them to the damaging events. Producers of goods are held responsible on the basis of even less direct causal connections, which are aptly summarized in a principle such as *ubi commoda ibi incommoda*. In these cases, expected producer profit is taken to provide an adequate basis for ascribing responsibility with regard to safety and health of workers or damages to consumers and society at large.

In addressing and solving these responsibility ascription problems, one does not start from such things as the existence of a clear causal chain or the awareness of and control over the consequences of actions. The crucial decisions to be made concern the *identification of possible damages*, their *social sustainability*, and how *compensation* for these damages is to be distributed. Epistemological reflections on machine learning suggest that many learning robot responsibility ascription problems belong to this class. And epistemological reflections will also prove crucial to address the cost-benefit, risk assessment, damage identification, and compensation problems that are needed to license a sensible use of learning robots in homes, offices, and other specifically human habitats.

Responsibility ascription policies: science, technology, and society

The responsibility ascription problems mentioned above are aptly classified as retrospective, that is, concerning past events or outcomes. In view of the above remarks, retrospective responsibility ascriptions for the actions of learning robots may flow from some conception of moral agency or from a legal system or from both of these. But what about prospective responsibilities concerning learning robots? In particular, who are the main actors of the process by which one introduces into a legal system suitable rules for ascribing responsibility for the actions of learning robots? These rules should enable one to identify possible damages that are deemed to be socially sustainable, and should specify criteria according to which compensation for these damages is to be distributed. Computer scientists, roboticists, and their professional organizations can play a crucial role in the identification of such rules and criteria. In addition to acting as whistle-blowers, scientists, engineers, and their professional organizations can provide systematic evaluations of risks and benefits flowing from specific uses of learning robots, and may contribute to shape scientific research programmes towards the improvement of learning methods. However, wider groups of stakeholders must be involved too. An examination of issues which transcend purely scientific and technological discourses is needed to evaluate costs and benefits of learning robots in society, and to identify suitable liability and responsibility policies: For the benefit of whom learning robots are deployed? Is it possible to guarantee fair access to these technological resources? Do learning robots create opportunities for the promotion of human values and rights, such as the right to live a life of independence and participation in social and cultural activities? Are specific issues of potential violation of human rights connected to the use of learning robots? What kind of social conflicts, power relations, economic and military interests motivate or are triggered by the production and use of learning robots? (Capurro *et al.* 2006)

No responsibility gaps and no conceptual vacua are to be faced in ascribing responsibility for the action of learning robots. At the same time, however, one should not belittle the novelty of this problem and the difficulty of adapting known liability criteria and procedures to the newly emerging casuistry. The fact that this responsibility ascription problem concerns a very special kind of machines is aptly illustrated by its assimilation, in the above discussion, to responsibility and liability problems concerning parents and pet owners, that is, problems concerning the consequences of flexible and intelligent sensorimotor behaviours of biological systems. Moreover, when learning is combined in a robot with additional features of intelligent artificial agents - such as autonomy, pro-activity, reasoning, and planning - human beings are likely to enter cognitive interactions with robots that have not been experienced with any other non-human biological system. Sustained epistemological reflections will be needed to explore and address the novel applied ethics issues that take their origin in these cognitive interactions.

References

Capurro, R., Nagenborg M., Weber J., Pingel C. (2006), "Methodological issues in the ethics of human-robot interaction", in G. Tamburrini, E. Datteri (eds.), *Ethics of Human Interaction with Robotic, Bionic, and AI Systems, Workshop Book of Abstracts*, Napoli, Istituto Italiano per gli Studi Filosofici, p. 9.

Cordeschi R. and Tamburrini G. (2005), "Intelligent machinery and warfare: historical debates and epistemologically motivated concerns" in Magnani L. and Dossena R. (eds.), *Computing, Philosophy, and Cognition*, London, King's College Publications, pp. 1-20.

Matthias A. (2004), "The responsibility gap: Ascribing responsibility for the actions of learning automata", *Ethics and Information Technology* **6**, pp. 175-183.

Mitchell, T.M. (1997), *Machine Learning*, New York, McGraw Hill.

Pitt, L. and Valiant L. (1988). "Computational limitations on learning from examples", *Journal of the ACM* **35**, pp. 965-984.

Tamburrini, G. (2006), "AI and Popper's solution to the problem of induction", in I. Jarvie, K. Mil-ford, D. Miller (eds.), *Karl Popper: A Centennial Assessment, vol. 2, Metaphysics and Epistemology*, London, Ashgate, pp. 265-282.

29

Artificial Consciousness and Artificial Ethics: Between Realism and Social Relationism

Steve Torrance

Received: 13 January 2013 / Accepted: 24 September 2013 / Published online: 19 October 2013

Abstract I compare a 'realist' with a 'social–relational' perspective on our judgments of the moral status of artificial agents (AAs). I develop a realist position according to which the moral status of a being—particularly in relation to moral patiency attribution—is closely bound up with that being's ability to experience states of conscious satisfaction or suffering (CSS). For a realist, both moral status and experiential capacity are objective properties of agents. A social relationist denies the existence of any such objective properties in the case of either moral status or consciousness, suggesting that the determination of such properties rests solely upon social attribution or consensus. A wide variety of social interactions between us and various kinds of artificial agent will no doubt proliferate in future generations, and the social–relational view may well be right that the appearance of CSS features in such artificial beings will make moral role attribution socially prevalent in human–AA relations. But there is still the question of what actual CSS states a given AA is capable of undergoing, independently of the appearances. This is not just a matter of changes in the structure of social existence that seem inevitable as human–AA interaction becomes more prevalent. The social world is itself enabled and constrained by the physical world, and by the biological features of living social participants. Properties analogous to certain key features in biological CSS are what need to be present for nonbiological CSS. Working out the details of such features will be an objective scientific inquiry.

1 Introduction

The term 'society' is currently understood to include humans—with various nonhuman biological species (e.g. domestic creatures, primates, etc.) sometimes included for certain purposes. Humans may be soon joined on the planet by a number of new categories of agents with intelligence levels that—according to certain measures—approach or even exceed those of humans. These include robots, software agents, bio-engineered organisms, humans and other natural creatures with artificial implants and prostheses, and so on. It may well become widely accepted (by us humans) that it is appropriate to expand the term 'society' to include many of such emerging artificial social beings, if their cognitive capacities and the nature of their interactions with humans are seen as being sufficiently fluent and complex to merit it. Of the many questions that may be asked of new members of such an expanded society, two important ones are: Are they conscious? What kind of moral status do they have? These questions form part of two new offshoot studies within artificial intelligence (AI)—machine consciousness and machine ethics (or artificial consciousness/ethics). Wallach et al. (2011) have written that 'Machine ethics and machine consciousness are joined at the hip'.[1] In what follows, we will find several ways in which these two domains cross-fertilise. Most prominently in the present discussion is the fact that the attribution of consciousness to machines, or artificial agents more generally (AAs),[2] seems to be a fundamental consideration in assessing the ethical status of artificial social beings—both as moral agents and as moral patients.

So how are we to understand such consciousness attributions; and indeed, how are we to view attributions of moral status themselves? I compare two views on these linked questions. One view may be called 'moral realism' (or 'realism' for short). For a realist, there are objectively correct answers to questions like: 'Is X a conscious creature or agent?'; 'Is X the kind of being that has moral value?'—although it may be impossible, even in principle, to provide assured or non-controversial answers to such questions. I will defend a variant of this view here. The other view may be called

---

[1] Wallach *et al.* are primarily concerned in their paper with how modelling moral agency requires a proper theoretical treatment of conscious ethical decision-making, whereas the present paper is more broadly concerned with the problem of ethical consideration—that is: what kinds of machines, or artificial agents in general, merit ethical consideration either as agents or as patients. The discussion largely centres around the relation between experiential consciousness and the status of moral patiency. I have discussed the general relation between consciousness and ethics in an AI context in Torrance, 2008, 2011, 2012a,b; Torrance and Roche 2011. While I sympathize strongly with the sentiment expressed in the above quote from Wallach *et al.*, I prefer the terms 'artificial consciousness' (AC) and 'artificial ethics' (AE) to the 'machine' variants. It seems clear that many future agents at the highly bio-engineered end of the spectrum of possible artificial agents—particularly those with near-human levels of cognitive ability—will be strong candidates to be considered both as phenomenally conscious much in the way we are and as moral beings (both as moral agents and as moral patients). Yet it may be thought rather forced to call such artificial creatures 'machines', except in the stretched sense in which all natural organisms, us included, may be classed as machines.

[2] In what follows, I will sometimes talk about 'robots' and sometimes about AAs. Generally, I will mean, by 'robots' physical agents (possibly humanoid in character), which are constructed using something like current robotic technology—that is, whose control mechanisms are computer-based (or based on some future offshoot from present-day computational designs). By 'AAs' I will understand a larger class of agents, which includes 'robots' but which will also include various kinds of possible future bio-machine hybrids, plus also agents which, while synthetic or fabricated, may be partially or fully organic or metabolic in physical make-up.

'social relationism' (hereafter, 'relationism' or SR for short).[3] Social relationists deny that questions such as the above have objective answers, instead claiming that their determination relies solely upon social conditions, so that the process of ascribing properties such as 'being conscious', 'having moral status', involves an implicit relation to the ascriber(s). Different versions of relationism are presented by Mark Coeckelbergh and David Gunkel in their two excellent recent books (Coeckelbergh 2012; Gunkel 2012; see also Coeckelbergh 2009, 2010a, b, 2013; Gunkel 2007, 2013). Despite initial appearances, much of the disagreement between realism and SR can be removed; nevertheless, a considerable core of variance will remain. Questions about sentience or consciousness, on the one hand, and about moral status on the other, will provide two pivotal dilemmas for the members of any future expanded society. The issue between these two views will thus be likely to be of crucial importance for how social life is to be organized in future generations.

In the way that I will understand them, realism and SR are views *both* about ethical status and about sentience or consciousness. A question such as 'What moral status does X have (if any)?'—supposing X to be a machine, a human, a dolphin or whatever—can be construed in a realist or in a relationist way. Equally, a question such as 'Is X conscious?' can also be given either a realist or a relationist construal.

2 The Debate Illustrated and Developed

In order to see the difference between the realist and the relationist approaches, consider a hypothetical future robot and its human owner. Let us say the robot works as a gardener for the household where it is domiciled. We can think of the garden-bot here as being a 'robot' along broadly conventional lines: that is, as a manufactured humanoid physical agent with a silicon-based brain. We will assume that the robot is capable of communicating and interacting with humans in a relatively rich way. Sadly, the human owner in our example regularly treats her charge in ways that cause the robot (in virtue of its design) to give convincing behavioural manifestations of distress, or even of pain, whenever its lawn mowing, planting or weeding do not come up to scratch. Two natural questions might be: (Q1) 'Is the robot gardener *really* feeling distress or pain?' and (Q2) 'Is it really *ethically ok* for the owner to behave that way to the robot?' Q1 and Q2 seem to be linked—it might be natural to say that it would be wrong to treat the robot in that way *if and in so far as* it would cause it experiences of distress, etc.

It might be objected that these two questions are dogged by so many indeterminacies that it is impossible to give any clear meaning to either of them. On Q1: How are we to think of the 'pain' or 'distress' of the robot gardener? Can we imagine any such states in an electronic robot—however sophisticated its design and construction? Can this be done in any way that avoids anthropomorphizing in a fashion that obscures, rather than clarifies the

---
[3] A source for the term 'social relationism' is the title of a paper by Mark Coeckelbergh (Coeckelbergh 2010a).

issue? How could the similarities or differences be characterised, when comparing 'silicon-based pain' (were such to exist) with 'organic pain'? There seems to be such a conceptual chasm between the familiar cases and the robot case that it may be thought that Q1 simply cannot be properly posed, at least in any direct, simple way. And on Q2: how are we to think of ethical duties towards silicon-based creatures? As with Q1, it might be argued that the question of ethical treatment posed in Q2 cannot be given a clear construal when divorced from the context of inter-human or at least inter-organism relations.[4]

Despite these doubts, I believe a realist can insist that the issues raised in Q1 and Q2 are genuine concerns: for example, one wants to know whether there is something that the robot's owner is *doing wrong to* the robot, in virtue of some real states that the robot is undergoing as a result of her actions. Do we, ethically, need to care about such actions, in the way we would about human distress or pain? Do we need to waste resources protecting such agents (of which there may be many) from 'ill-treatment'? Such questions may be hard to put into clear conceptual focus, let alone to answer definitively. But surely they do seem to address strong *prima facie* concerns that might be raised about the kind of example we are considering. In any case, for now, we are simply putting forward the realist's position: a more considered appraisal will follow when we have examined both positions in greater detail.

On the SR position neither Q1 nor Q2 has an inherently right or wrong answer. Rather, answers will emerge from the ways in which society comes to develop beliefs and attitudes towards robots and other artificial agents. Indeed a social relationist may well insist that the sense that can be made of these questions, let alone the answers given to them, is something which itself only emerges in time with social debate and action. Perhaps society will broadly adopt a consensus and perhaps it will not: but, for a supporter of SR, there is no way one can talk of the 'correct' answers to either question over and above the particular responses that emerge through socially accepted attitudes and patterns of behaviour.

A relationist, then, would deny that there is any 'correctness' dimension to either Q1 or Q2 over and above what kinds of options happen to emerge within different social settings. In contrast, for a realist, social consensus could really get things wrong—*both* on the psychological issue of whether the robot is

---

[4] I am grateful to an anonymous reviewer for insisting on this point. In the present discussion, I am limiting the kinds of cases under consideration to AAs whose design involves electronic technologies which are relatively easy to imagine, on the basis of the current state of the art and of fairly solid future projections. There are a wide variety of other kinds of artificial creature—including ones with various kinds of artificial organic makeup, plus bio-machine hybrids of different sorts—which expand considerably on this range of cases. We will consider this broader range of cases in later sections.

Concentrating at the present stage of the discussion on AAs like the robot gardener, and other such relatively conservative cases has a triple utility. First, it allows us to lay down the foundations for the argument without bringing in too many complexities for now. Second, many people (supporters of strong AI or 'strong artificial consciousness') have asserted that such robots could well have genuinely conscious states (and thus qualify for serious ethical consideration) if constructed with the right (no doubt highly complex) functional designs. Third, such cases seem to offer a greater challenge than cases which are closer-to-biology: it is precisely the non-organic cases, in which one has detailed similarity to humanity in terms of behaviour and functional organization but marked dissimilarity in terms of physical or somatic structure, where the issues seem to be raised particularly sharply.

feeling distress and pain *and* on the ethical issue of how the robot should be treated. (Realists do not have to link these two questions in this way, but it has been seen by many natural to do so, and that is the version of realism that will be in the forefront of the discussion here.)[5,6]

So for a realist, the following concerns can be raised. A future generation of robot owners may believe that some or even most of them are conscious or sentient, and therefore deserving of our moral concern in various ways, when 'in fact' they are no more sentient than (present day) sit-on mowers, and can suffer only in the sense in which a mower would 'suffer' if the wrong fuel mix was used in its engine. Or it might go the other way around—socially prevalent attitudes may withhold attributions of conscious feeling and/or moral consideration to certain kinds of artificial agent who 'in fact' are conscious and merit moral consideration—and perhaps that the latter is the case for such agents largely, or at least partially, because of the facts of the former kind that hold true of them. (The scare-quoted phrase 'in fact' in the above will, of course, be problematic on the SR view.)

According to realism, then, a question like 'Is X conscious?' is asking about objective[7] matters of fact concerning X's psychological state, and further (for this version of realism) attributions of moral status at least partially supervene on X's psychological properties (including consciousness and related states). Moreover, on the version of realism being considered here, this is true of humans and non-human biological species, just as much for (current or future) robots and other technological agents. So on this view, normative moral questions concerning the actions or the treatment of humans, animals or robots, are closely bound up with factual questions concerning the capacity of these various agents for conscious experience—that is, the *phenomenal* consciousness of such agents, as opposed merely to their ability to *function cognitively* in a way a conscious being would.[8]

Why should questions about the phenomenal consciousness of beings link so closely to moral questions? A key reason that the realist can put forward is to do with the fact that conscious experiences of different sorts have characteristic positive or negative *affective valences or qualities*.[9] Think of the current field of your experiential awareness. Experiences sometimes come in an affectively neutral way, such as the tiles on the floor which are in the background of my present visual awareness. But others are evaluatively graded: the hum of the fan system that I can

---

[5] Some people might agree that Q1 should be construed in a realist way—what could be more real than a person's vivid experiences of distress, pleasure, etc.?—while being reluctant to treat Q2, and similar moral questions, in a realist or objectivist way. In this paper, I am supporting a realist position for both experiential and moral attributions.

[6] For a defence of the view that there is a close association between questions of consciousness and those of moral status, see, for example, Levy 2009. Versions of this view are defended in Torrance, 2012; Torrance and Roche 2011. The conclusions Levy comes to on the basis of these views are very different from mine, however.

[7] To clarify: the realist's claim is that 'Is X conscious' is objective in the sense that 'X is currently conscious' asserts a historical fact about X, even though it's a fact about X's *subjective* state, unlike, say, 'X is currently at the summit of Everest'.

[8] The inherent distinguishability between phenomenal and functional consciousness is defended in Torrance, 2012.

[9] See Thompson (2007, chapter 12) for a discussion of the relation between consciousness, affect and valence.

hear is mildly irritating; the lilt of voices in conversation outside my window is mildly pleasant. Plunging into a rather cold sea will generate a mix of sensations, of a characteristic hedonic character (a different mix for different people, situations, moods, etc.) In general, the flow of our experience is closely tied to our desires, needs, aversions, plans, etc., as these unfold in lived time. The degrees of positive or negative affective valence can vary from scarcely noticeable to extreme, and, as they vary, so too do the contributions that they make to a conscious creature's levels of satisfaction and suffering, i.e. to their experienced well-being or ill-being. (It would seem to be difficult to see how beings which were not capable of conscious experience could have such states of satisfaction and suffering.) Further, it can be argued that considerations of how actions make a difference to the well-being, satisfaction, etc. of people affected by those actions are a central concern of ethics. So an important part of a being's moral status is determined by that being's capacity to experience such states of satisfaction/suffering.

The social–relational view, by contrast, claims that attributions of consciousness are not (or at least not clearly) ascriptions of matters of objective fact, at least in the case of non-human animals, and of current and future technological agents. On this view, such ascriptions have instead to be understood in terms of the organisational circumstances in the society in which the discourse of attribution occurs, on the social relations between human moral agents, and the contexts in which other putatively conscious creatures or agents may enter into our social lives. These social and technical contexts vary from culture to culture and from epoch to epoch. Society is fast-changing today, so new criteria for consciousness attribution may currently be emerging, which are likely to radically alter social opinion on what beings to treat as conscious (indeed what beings count as 'social beings' will itself be a socially 'moving target'). Moreover, on the SR view, attributions of moral worth and other moral qualities are similarly to be seen as essentially embedded in social relations. The same profound changes in social view are likely to affect norms concerning the attribution of moral status, in particular the moral status of artificial creatures. In a word, judgments in the twenty-first century about the possible experiential and moral status of automata may markedly diverge from those that were prevalent in previous centuries. The SR view will say there can be no neutral way of judging between these different views. Thus, on the relationist view, both the psychological and the ethical components of the realism described above are rejected.[10]

3 The Expanding Moral Circle and the Landscape of Conscious Well-Being

Many writers have talked of a progressive expansion of moral outlook through human pre-history and history. Peter Singer (2011) has written persuasively of the 'expanding circle' of ethical concern, from primitive kin- and tribe-centred fellow

---

[10] This is not to say that questions concerning consciousness or ethics in relation to such machines are to be thought of as trivial or inconsequential on the SR view: on the contrary, a relationist will take such questions as seriously as the realist, and may claim that they deserve our full intellectual and practical attention.

feeling to a universal regard for all of humanity; and of the rational imperative to widen the circle still further to include non-human creatures capable of sentient feeling. According to Singer, we owe our ethics to the evolutionary pressures on our pre-human forbears, and we owe it to the animal co-descendents of that evolutionary process to extend ethical concern to the well-being of all sentient creatures.[11]

Some have argued that the circle should be widened so that non-sentient entities such as forests, mountains, oceans, etc. should be included within the domain of direct moral consideration (rather than just instrumentally, in terms of how they affect the well-being of sentient creatures—see, for example, Leopold 1948; Naess 1973). In considering what limits might be put on this process of ethical expansion, Singer argues that *only* entities that have the potentiality for sentience could sensibly be included in the moral circle. For, he says, of a being with no sentience there can be nothing that one can do which could make a difference *to* that being in terms of what it might experience (Singer 2011, p. 123.)

Singer's position appears to rest on a conceptual claim—that only beings with sentience can coherently be considered as moral patients. As I see it his argument is roughly this. For X to be a 'moral patient' (or moral 'recipient') *means* (at least in part) that X is capable of benefitting or suffering from a given action; and a non-sentient being cannot benefit or suffer in the relevant (experiential) sense (although, like a lawnmower run continually on the wrong fuel mix, or operated over a rocky terrain, it may 'suffer' or be 'abused' in a functional, and non-experiential, sense). So, the argument goes, a non-sentient being cannot *coherently* be considered as a moral patient, because no action could affect its consciousness of its own well-being. Ethical considerations are about how our actions might affect others (and ourselves) in ways that make a difference to those so affected. To quote from an earlier work by Singer (1975, p. 9): 'A stone does not have interests because it cannot suffer. Nothing that we can do to it could possibly make any difference to its welfare. A mouse, on the other hand, does have an interest in not being kicked along the road, because it will suffer if it is' (cited in Gunkel (2012), p. 113.)

Of course the precise conditions under which particular artificial agents might be considered as conscious—as a being having interests, like the mouse, rather than as a brute, inanimate object, like the stone—are notoriously difficult to pin down. But having controversial verification conditions is not the same as having no verification conditions, or ones which are essentially dependent upon the social–relational context. Compare, for example, the question of extra-terrestrial life. There may be controversy among astrobiologists over the precise conditions under which life will be established to exist in a distant solar system; but this does not detract from the ontological objectivity of exoplanetary life as a real phenomenon in the universe. It does not make the physical universe, or the existence or nature of planets outside our solar system, *social–relational* (although of course exoplanetary science, astrobiology, etc., as academic studies, are social activities, with their specific funding, geopolitical and ideological dimensions). Similarly, the realist might say, a robot's or AA's pain, if such a thing were to exist, would be as objective a property of the AA, and as inalienable *from* the AA, as would your or my pain be inalienable from you or from me. (Conversely, an AA that gave appearances of pain or suffering but which in fact

---
[11] See also the discussion in Torrance, 2013.

had no sentience could not have sentience added to it simply by virtue of its convincing appearance.)[12]

For realism, in the version I am developing here, there appears to be a kind of holism between thinking of X as phenomenally conscious and judging X to be of moral worth (at least as a moral patient, and maybe as a moral agent, in a full sense of agency). This *phenomenal–valuational* holism may be put as follows. To think of a creature as having conscious experience is to think of it as capable of experiencing things in either a positively or negatively valenced way—to think of it as having desires, needs, goals, and states of satisfaction and dissatisfaction or suffering. Of course, there are neutral experiential states, and not all satisfaction or suffering is consciously experienced. Nor are all our goals concerned with gaining particular experienced satisfactions. Nevertheless there seems to be a strong connection between our experiential capacities and our potential for well-being. (This is a point which has been addressed surprisingly little in the literature on human consciousness, and of machine consciousness.) We may talk of beings which are conscious, in this rich sense, as having the capacity for conscious/satisfaction/suffering states. I will here call these CSS states for short.

Not all realists may agree with this kind of phenomenal valuational holism. One writer who seems to do so is Sam Harris. He presents the view (Harris 2010) that the well-being of conscious creatures is the central issue in ethics, and indeed that other ethical considerations are, if not nonsensical, at root appeals to consideration of experienced well-being—to the quality of CSS states. Harris's moral landscape is the terrain of possible peaks and troughs in experienced well-being that CSS-capable creatures negotiate through their lives. He also argues (in a particularly strong version of the realist position) that moral questions are objective and in principle scientific in nature. As a neuroscientist, Harris takes brain processes to be central determinants of well or ill being—a view I would accept only with strong qualification. It may be hard in practice to solve various moral dilemmas, but, he claims, they are in principle amenable to a scientific solution, like other tough factual questions such as curing cancer or eliminating global poverty.

I do not necessarily say ethics is exclusively about well-being; but I would agree that it is central to ethics. Also, since creatures with capacities for greater or lesser well-being are conscious, I think it is central to the study of consciousness, and, indeed to AI. Of course physiological processes from the neck upwards are pretty crucial for such capacities. But, *pace* Harris, bodily processes from the neck down are pretty important too, as well as the active engagement of the creature in its lived world.[13] A big challenge for AI and artificial ethics is to work out just what physical features need to be present both above and below the neck in artificial agents for artificially generated CSS properties to be present, and how close they have to be with the relevant natural or organic features.[14]

---

[12] Singer does not discuss the case of moral interests of robots or other artificial agents (or indeed of exoplanetary beings) in his 2011 book.

[13] For an excellent, and fully elaborated, defence of the kind of view of consciousness that I would accept, which is centred around notions of enactivism and autopoiesis, see Thompson (2007)—especially chapters 12 and 13. There is no space here to do more than gesture to this view in the present discussion.

[14] Like Singer, Harris does not consider the ethical status of possible artificial agents.

Linked with his views about the scientific grounding of questions to do with well-being and ethics, Harris expresses a lot of impatience with the 'fact-value' split that has dominated scientific thinking in the last century, and for the moral neutralism or quietism that has characterised a lot of scientific thought and practice in that time. I very much agree with Harris on this, and would say that it has been particularly true of 'Cognitive Science' as this has developed in the last half century or so. The over-cognitivizing of the mind has had an unfortunate effect both on the attempt to understand mental processes scientifically, and on the process of exploring the ethical ramifications of mind science. AI researchers, neuroscientists, psychologists and philosophers have too often talked as though the mind was exclusively a cognitive mechanism. All that cognitivizing hard work—that systematic redescription of the *chiaroscuro* of our desires, emotions, pains and delights in terms of informational operations and that bleaching out of happiness and misery from the fabric of our psychology—all that has, arguably, been to the detriment of both scientific understanding and ethical discussion in this area.

4 The Turing Dream: A 400-Year Scenario

I thus applaud the science ethics holism found in objectivist writers like Harris. Perhaps this enthusiasm will be shared by relationists such as Coeckelbergh and Gunkel. It is certainly interesting to discuss machine ethics in a way that takes seriously the inherent interconnectivity of consciousness, well-being and ethics, and which allows that scientific and ethical issues are not to be debated in parallel, hermetically sealed chambers.[15]

In the light of this, consider the following possible future picture. Let us consider what might be called the 'Turing Dream'—that is, the goal aspired to by many in the AI community of developing the kind of complexity and subtlety in functioning which would enable robots to behave more or less like us over a very broad range of activities and competencies in the physical and social real world. Let us suppose (a big ask!) that researchers do not hit any insurmountable barriers of computational tractability, hardware speed, or other performance or design impasses, and the 'dream' is fulfilled, so that such robots proliferate in our world, and co-habit with us in a more or less peaceable kingdom. Let us suppose, then, that—say 400 years from now (or choose the timeframe you prefer)—human social relations have changed radically because of the existence of large numbers of such artificial agents implementing more or less human or even greater-than-human levels of ability across a wide range of capabilities. If the Turing Dream were to come about in something like this fashion, many people will find it natural to attribute a wide range of psychological attributes to such agents, and the agents themselves will, in their communications with us and with each other, represent themselves as having many of the cognitive and indeed affective states that we currently take to be characteristic

---

[15] Sometimes the seals can be leaky. I was once at a conference on consciousness, where an eminent neuropsychologist was giving a seminar on ethical issues in neural research on consciousness. He said things like 'With my neuroscientist's cap on, I think... But with my ethicist's cap on, I think...' What cap was he wearing when deciding which cap to put on at a given time?

of human psychology. Many such artificial creatures may resemble humans in outward form. Even if they do not, and the technology of humanoid robotics runs into a *cul-de-sac*, it nevertheless seems likely that the demands of extensive human–AI social interaction will ensure a good deal of resemblance in non-bodily respects (for instance in terms of sharing common languages, participating in a common economic system, shared legal frameworks and so on).

Will our world wind up at all like this? Who knows? But the scenario will help us check our intuitions. Humans in this imagined future period may ask: are such artificial agents conscious? And, should we admit such agents into our moral universe, and in what ways? (And by what right are we licensed to talk of admitting 'them' into 'our' moral world?) As we have argued, such questions are closely linked. We can combine those questions together in a third: do such artificial agents have CSS features? The social–relationist will say that the answer to these questions will depend on the prevailing social conditions at the time, on what kinds of attitudes, beliefs, forms of life, and ways of articulating or representing social reality come to emerge in such a joint human–technological social milieu. On the relational view, there will be no 'objective' way, independently of the socially dominant assumptions, judgments, norms and institutions that grow up as such artificial agents proliferate, to say whether they are *actually* conscious, whether they *actually* are capable of having states of felicity or suffering, or whether they actually merit particular kinds of moral consideration—e.g. whether they merit having their needs taken roughly as seriously as equivalent human needs; whether their actions merit appraisal in roughly similar moral terms as the equivalent actions of humans, etc.[16]

For the realist, this would miss an important dimension: do such artificial creatures (a few or a many of them) *actually* bear conscious states, are they *actually* capable of experiencing states of satisfaction or suffering at levels comparable to ours (or at lower, or even much higher, levels—or even in ways that cannot easily be ranked in terms of any serial comparison of 'level' to ours)? To see the force of the realist's argument, consider how a gathering of future artificial agents might discuss the issue with respect to *humans'* having CSS properties—perhaps at a convention to celebrate the 500th anniversary of Turing's birth?[17] Let us suppose that delegates' opinions divide along roughly similar lines to those in the current human debate, with social–relationists arguing that there is no objective fact of the matter about whether humans have CSS properties, and realists insisting that there must be a 'fact of the matter'.[18]

How would a human listening in on this argument feel about such a discussion? I would suggest that only a few philosophically sophisticated folks would feel

---

[16] It is worth pointing out that no consensual human view may come to predominate on these issues: there may rather be a fundamental divergence just as there is in current societies between liberals and conservatives, or between theistic and humanistic ways of thinking, or between envirocentric versus technocentric attitudes towards the future of the planet, and so on. In such a case, the relationist could say, social reality will be just as it manifests itself—one in which no settled view on the psychological or moral status of such agents comes to prevail; society will just contain irreconcilable social disagreements on these matters, much as it does today on these other issues.

[17] The present paper originated as a contribution to a workshop at a Convention celebrating the 100th anniversary of Turing's birth.

[18] We assume—perhaps with wild optimism—that that these artificial agents are by then smart enough to debate such matters somewhat as cogently as humans can today, if not much more so. To get a possible flavour of the debate, consider Terry Bisson's 'They're made out of meat' (Bisson 1991).

comfortable with the relationist side of the argument in this robot convention, and that the most instinctive human response would be a realist one. A human would reflect that we do, as a species, clearly possess a wide variety of CSS properties—indeed our personal and social lives revolve 24/7 around such properties. Can there be any issue over which there is more paradigmatically a 'fact of the matter' than our human consciousness? Surely the 'facts' point conclusively to the presence of CSS properties in humans: and are not such properties clearly tied to deep and extensive (neuro-)physiological features in us? What stronger evidence-base for any 'Is X really there?' question could there be than the kind of evidence we have for CSS properties in humanity? So surely robots a century from now would be right to adopt a realist view about the consciousness and ethical status of humans. Should we then not do the same today (or indeed in a hundred years' time) of robots?[19]

5 The Other Minds Problem Problematized

A supporter of SR might raise 'other minds' style difficulties even about CSS in humans, as a way of highlighting the difficulty of certifying the existence of CSS properties in artificial agents.[20] Any individual human's recognition of CSS properties is based upon their own direct first-person experience. There are great psychological and social pressures on any individual to infer to the existence of such CSS properties in others, it might be said: yet the inference can never rationally be fully justified, since one can never be directly acquainted with another person's conscious states. It appears that one has to infer to the internal states of another's consciousness on analogy with bodily and physiological states observed in oneself as these are linked with one's own experience. As Wittgenstein remarked, with a whiff of sarcasm, 'how can I generalize the one case so irresponsibly?' (Wittgenstein 1953: I, §293). The 'other minds' problem may thus be used as a device for undermining realism, by emphasizing that even in the case of other humans, let alone various species of animals, there are substantial difficulties in establishing the presence of the kinds of experiential properties that the realist requires for ethical attribution.[21]

There are several responses to such doubts. A first response is this. If CSS properties are not completely mysterious and inexplicable, from a scientific point of view, then they must be causally grounded in *natural features* of any individual human possessing them. Imagine two humans, A and B, who display third-person behavioural and physiological properties that were identical in all relevant respects, yet where A has an 'inner' phenomenological life and B lacks all phenomenological experience. This would be a completely bizarre phenomenon, to be explained either

---

[19] That is, should we not say that the *epistemological status* of our question about them is comparable to that of theirs about us?—although the answers to the two questions may be very different, as may be the relative difficulty in answering them.

[20] See, for example, Gunkel (2012, chapters 1 and 2), who insists on the perennial philosophical problem of 'other minds' as a reason for casting doubts on rational schemes of ethical extension which might enlarge the sphere of moral agency or patiency to animals of different types, and beyond that, to machines. It is remarkable how frequently Gunkel returns to rehearsing the theme of solipsistic doubt in his discussion.

[21] Appeal to doubts over other minds is one of the arguments used by Turing (1950) to buttress his early defence of the possibility of thinking, and indeed conscious, machines.

in terms of some non-natural circumstance (e.g. God chose to inject a phenomenology into A but withhold it from the physically identical B)—or else it would have to be completely inexplicable. Neither of these alternatives seems attractive. Yet to entertain radical doubts about other minds seems to require embracing one or other of these unsavoury positions. To avoid that, it would be necessary to concede that there is some sort of third person, underpinning natural features, publicly accessible in principle, that could be used to distinguish those humans (and other species) with a phenomenology, with CSS features, from those without. In any case, there is a mass of scientific theory and accreted experimental data linking CSS properties in humans, and affective properties more generally, with our evolutionary history, and with our current biological and neural make-up.

Doubts about solipsism and the existence of other human consciousnesses besides one's own, can also be shown to be fed by a series of questionable fundamental conceptions about mind and consciousness in the first place. Solipsistic doubts are linked to doubts about the 'Hard Problem' of consciousness (Chalmers 1995; Shear 1997), the Explanatory Gap between physiology and experience (Levine 1983), Absent Qualia puzzles (Block 1978), and so on. I have suggested (Torrance, 2007) that there are two contrasting conceptions of phenomenality: the 'thin' (or 'shallow') conception, deriving from Descartes' radical doubts about bodily existence in the face of the *cogito*, assumes that consciousness must be understood in terms of a radically ego-logical sensory presence, which is causally, ontologically and conceptually, radically separate from any other processes (particularly bodily or physiological processes). By contrast, the 'thick' (or 'deep') conception sees consciousness as conceptually inseparable from the lived, physiological processes that make up a conscious being's embodied existence. The distinction is deployed in that paper to articulate an embodied approach to Machine Consciousness (see also Holland 2007; Stuart 2007; Ziemke 2007). In a later paper, I have argued for an embodied, de-solipsized conception of consciousness, by challenging prevailing 'myths' which construe consciousness as being, *by definition,* essentially inner, hidden and individualistic (Torrance, 2009). I suggest that consciousness is an 'essentially contested' notion (the term is due to Gallie 1955)—so that, at the very least, there is no requirement that consciousness be conceptualised in the ways suggested by these 'myths'.

Freed from the necessity to rely on internalist and individuocentric conceptions of consciousness, one is able to see how philosophical worries to do with solipsism, absent *qualia*, the explanatory gap between the neural and the phenomenal, and so on, are all systematic products of a dependence by many sectors of the philosophical and scientific community (a dependence which both feeds from and into unreflective idioms of folk talk about mind) upon an outmoded and unhelpful way of conceptualising experience or phenomenality. Many other authors have articulated versions of this richer, more deeply embodied, conception of phenomenality. A notable critique of the mind–body problem by Hanna and Thompson differentiates between two conceptions of body—the body of the physical organism which is the subject of biological investigation as a physical–causal system (*Körper*) and the 'lived body' (*Leib*), which is the individual subject of embodied experience of an organism as it makes sense of its trajectory in its world (Hanna and Thompson 2003). Hanna and Thompson deploy the *Leib-Körper* distinction to dissolve doubts over mind–body separation, other minds, and so on, by bifurcating notions of body, in contrast to the distinction between notions of phenomenality in Torrance, 2007.[22] Other philosophical challenges to the other minds problem will be found in Gallagher's rejection of theory–theory and simulation–theory approaches to mentalization, which draw upon work by Husserl, Scheler, Gurwitsch, Trevarthen and others who argue that, for example, the perception of someone's sadness is not an inference to a hidden X behind the expressive countenance, but is rather a direct observation of the other's experiential state, derived from the condition of primary intersubjectivity with their conspecific carers that humans and other primates are born into, and the secondary intersubjectivity of joint attention and engagement in joint projects in the world that occurs during later development (Gallagher 2005a, b; Zahavi 2001; Gallagher and Zahavi 2008; Thompson 2007). Such work by Gallagher and others offers a strong alternative to opposed viewpoints in debates over the psychology of social cognition, but *a fortiori,* marginalise classical other minds doubts as a philosophical side show.

6 Social and Non-Social Shapers of Sociality

We have seen, then, that to raise epistemological doubts which problematize our everyday assurance that we live in a community of consciousnesses is to rely on a cluster of problematic Cartesian assumptions which are based on options about how to conceptualise phenomenality which we are not forced to make, given alternative conceptual directions which many strong theoretical considerations direct us towards.[23] But in any case, such doubts can be shown to be *irrelevant* from an ethical point of view. Ethics, while being based on a weighty body of theoretical discussion, stretching back millennia, is, above all, concerned with our *practical,* functioning, social lives. In practice, we live in a non-solipsistic world, a world which we cannot but live in as one of a community of co-phenomenality.

If ethics is about practice, it is also about sociality and interactivity. The way I come to recognize and articulate CSS properties in myself is partly based upon my social interactions with my conspecifics. Indeed, to deploy a serious point behind the Wittgenstein quip cited above (and much else in his later writings), my whole discourse about mind, consciousness, pain, desires, emotions, etc. is based upon the public forms of life I share with other humans in the real world of participatory intersubjectivity. 'Just try, in a real case', Wittgenstein wrote, 'to doubt someone's fear and pain' (Wittgenstein 1953, I, §203).

Ironically, we here seem to find ourselves using an SR-style argument to counter an SR-style objection to realism. But we need to be careful. The relationist may argue

---

[22] See also the treatment of this issue in Thompson (2007), especially chapter 8—therein called the 'body–body problem'. Thompson mentions that a quite elaborate range of notions contrasting and combining the motifs of *Leib* and *Körper* are found in Husserl's writings (see Depraz (1997, 2001), cited in Thompson (2007), ch. 8, footnote 6). Thompson also critiques superficial or 'thin' conceptions of phenomenology (*ibid.*, ch. 8) but without the 'thin'/'thick' terminology used in Torrance, 2007.

[23] A variety of sources, from phenomenology, and several of the mind sciences, all converging on the view that our understanding of mind is thoroughly intersubjective, in a way that renders solipsistic doubts incoherent, will be found in Thompson (2001, 2007).

that the very Wittgenstinian considerations just mentioned (and related arguments for the necessary public, social grounding of such recognition, and on the impossibility of private language and private cognitive rules, etc.) shed doubt on the objectivity of first-personal recognition of CSS properties. Such a suggestion needs to be taken seriously, and may point to a deep truth behind the SR position—that our understanding of how we come to possess CSS properties, and of the variety of roles they play in our lives, is indeed inextricably bound up with our social relationships and activities.

But it is also important to see that the dependency also goes in the other direction. Our consciousness, needs, desires, etc. are what give point and form to our sociality. Our social conditions partly gain their significance from these very experiential and appetitive features in our lives, including, centrally, the ups of happiness and downs of misery. It is vital not to assume that everything in human lived experience is subject to social shaping: the reverse is also true. Myriad 'objective' physical and biological realities—including a variety of evolutionary, neural and physiological constraints—come into this network of inter-relations between consciousness and the social. Evolutionary history and current brain patterns play crucial roles in what makes us feel good or bad, as do the materiality of our bodies and the dynamics of their interactions with other bodies and the surrounding physical world. So there is a *multi-directional* cluster of mutually constitutive and constraining relationships between the social, material, biological and experiential factors in our lives.[24] What makes up our CSS features emerges from the entanglement of these various kinds of factors. This brings us to the heart of the question of CSS in future artificial social agents.

The progress of current work in AI teaches us that many of the features of human–human social and communicative interaction—the 'outer' features, at least—can be replicated via techniques in computer and robotic science—essentially algorithmic modelling techniques. Increasingly, our social world is being filled with human–machine and machine–machine interactions. With the growing ubiquity of such interactions, the range of possible social action is gradually being extended. But also, our very conceptions of the social, the cultural and the intersubjective, are being re-engineered. Or, to put the point a different way: how we make sense of the social, and indeed how we make sense of 'we', is being continually reshaped by our artefacts as they are increasingly implicated in our social existence. This is a crucial point that the relationist seeks to emphasize, in the debate about the status of CSS properties; and the realist must also acknowledge it readily.

Of course, notions related to social status overlap intimately with ethical notions; and the SR account is well suited to provide a theoretical framework for much of the domain of the social. So what is taken to constitute 'the social' is itself largely shaped by social factors, and changes as new social possibilities emerge. But, as we suggested earlier, the domain of the social is also shaped and constrained by the non-social, including biological and other kinds of physical conditions, and the experiences, desires, beliefs, goals, etc. of social participants. So, many of the novel forms of human–AA and AA–AA social relationships that will emerge (and already have

---

[24] And no doubt many others—for instance, I have left out the essential role played by our cognitive capacities, by beliefs, perceptions, intellective skills, etc.!

been emerging) will take their character not merely from the social sphere itself but also from the non-social soil in which sociality is rooted—that is, from the multiple physical, environmental, and indeed metabolic and experiential drivers of sociality. This is particularly true of those social relationships between humans and the computationally organized machines of today (let alone future, more organically constituted, AAs). Arguably, there are a great many social relationships, even today, which are precisely NOT relations between creatures with shared physiologies. And, for robots and other artificial agents of today, we can surely say with near certainty that they are NOT relations between beings with common experiential and affective subjectivities.

7 The 'Bio/Machine' Spectrum

So it is likely that, for the short term, as long as we have only the relatively primitive designs of our current technologies, our artificial social partners are, objectively, partners with zero experiential or affective life (despite many vociferous assertions to the contrary). Such artificial social partners are not, in Tom Regan's phrase, 'subjects of a life' (Regan 1983). Thus, for now, the possibilities for social interaction with machines outstrip the possibility of those machines being capable of sharing such interactions as exchanges between experiencing, belief- and-desire-full beings. For now, any human–machine interaction is one between social partners where only one of the actors has any *social concern*. Some of the deep conditions for sociality mentioned earlier are missing—in particular, a shared experientiality and shared neurophysiology. In the human–machine case, then, we might talk of a current mismatch between social *interactivity* and social *constitutivity*.

But how might things change in the medium and long-term? For one thing, techniques in synthetic biology may develop in ways which allow biotechnologists to create agents that are not just functionally or behaviourally very close to humans—i.e. which exemplify social patterns in a variety of outward ways, but which are close also in detailed neural and physiological makeup. In that, *ex hypothesi*, such creatures will share deep and extensive similarities with us in terms of the biological underpinnings of consciousness, we may indeed have little ground for denying that they are 'objectively' conscious, CSS-bearing, beings, with all the ethical consequences that would flow.

We certainly cannot rule out, then, that there will at some future time be AAs with physiologies which are as close to those of humans as one cares to imagine. Leaving aside the general 'other minds' objections discussed earlier, what reason would we have for saying that, despite our extensive biological commonality with such artificial creatures, they lacked the CSS features that we had—other than the (surely irrelevant?) fact that, unlike us, they were fabricated (or cultured) in a laboratory? (Such AAs might even have ontogenetic histories very much like ours, progressing from a foetal or at least neonatal stage, through infancy, and so on.) So, in the context of our present discussion, such very full-spec synthetic-biology-based, humanoid creatures surely give support to the idea that *at least some artificial agents should certainly be given the moral status that humans enjoy*. But such creatures occupy one extreme, bio-realistic, end, of a spectrum of possible artificial agents. At the other extreme, there are the relatively simplistic, computational AI agents of the recent past and of today—an era when artificial agent design is still, one assumes, in its infancy.[25] And between those two extremes are a host of other imaginable and not-so-imaginable agent designs. Clearly, as we retreat from the bio-realistic end, judgment calls about CSS properties and ethical status on the basis of physical design are much less straightforward.

In many (middle) regions of the spectrum, there will be agents with natural, fluent and subtle social interactivity characteristics that are close to those of humans, but where the underlying detailed physical design is remote from detailed humanoid or mammalian physical design. These will offer the toughest cases for decision: agents that, via their fluent social capacities (and, in many varieties, outwardly human-like bodily features), display a wide variety of apparent CSS-evincing behaviours but where they share relatively few of the internal neurological and more broadly physiological features that make for the presence of CSS properties in humans. These are the cases where the social–relational view may seem to be on its most solid ground: what possible 'objective' basis could there be for deciding on whether to accord or withhold moral consideration in each particular class of example?

However, a realist can reply that, even if such cases are difficult to determine in practice, there is still the question of what kind of experiential state, if any, actually occurs, independently of human or social attribution. For such cases, then, there are many risks of *false positives and false negatives* in CSS attributions. And surely it is the significance of such false positives and negatives that makes a difference—both in theoretical terms and in moral terms. In the hypothetical situations where such agents exist in large numbers—where they multiply across the world as smart phones have done today—wrong judgments could have catastrophic societal implications: (a) in the false-positive case, many resources useful for fulfilling human need might be squandered on satisfying apparent but illusory 'needs' of vast populations of behaviourally convincing but CSS-negative artificial agents. (b) Conversely, in the false-negative case, vast populations of CSS-positive artificial agents may undergo extremes of injustice and suffering at the hands of humans that wrongly take them for socially fluent zombies.[26]

Given the existence of a large variety of different possible artificial agents, what David Gunkel (2007, 2012, 2013) calls '*The* Machine Question' (my emphasis) factors into a multiplicity of different questions, each one centred around a particular kind of machine (or

---

[25] For the sake of simplicity of discussion, I am here representing the spectrum as if it were a unidimensional space, whereas it is almost certainly more appropriate to see it as multidimensional (cf. Sloman 1984).

[26] Here, we are stressing moral patiency, but a similar problem of false-positives/false-negatives exists for moral agency, too. Many relatively primitive kinds of AI agents will act in a functionally autonomous fashion so as to affect human well-being in many different ways—so in one sense the question of moral agency is much more pressing, as many authors have pointed out (see, for example, Wallach and Allen 2009). Yet there are important questions of responsibility ascription that need to be determined. If we provide too great a share of responsibility to AAs that act in ways that are detrimental to human interests, this may well mask the degree of responsibility that should be borne by particular humans in such situations (e.g. those who design, commission and use such AAs). Conversely, we may overattribute responsibility to humans in such situations and withhold moral credit from artificial agents when in truth it is due to such agents, for example, on the grounds that 'mere machines' they cannot be treated as fully responsible moral agents. The vexed issue of moral responsibility in the case of autonomous lethal battlefield robots provides one illustration of this area: see Sparrow 2007; Arkin 2009; Sharkey and Suchman 2013.

AA) design. Some kind of decision has to be made for each kind of machine design, and this is not going to be easy. Nevertheless, surely it is not the case that the only option is to throw up one's hands in resignation and just say "Let social consensus decide".

8 Negotiating the Spectrum

So the bio-machine spectrum ranges from cases of very close bio-commonality at one end, to simplistic current AI-style behaviour matching at the other end. At both these extremes decisions about CSS capacity and about moral status seem relatively straightforward—a resounding 'yes' at the first extreme, and a resounding 'no' at the other. But what about the intermediate cases? Is there any way to make progress on providing methods for adjudicating on moral status for the wide variety of hard cases between the two extremes? I think that there is: I believe it is possible to propose a series of testable conditions which can be applied to candidate artificial agents occupying different positions in the vast middle territory of the spectrum. I list such a series of conditions below, well aware that this is only one person's draft, no doubt skewed by the preoccupations and prejudices of its author. Nevertheless it may point the way to showing how a realist position may be able to do more than simply say 'There must be a correct answer to the question "Should A be accorded moral status?" for any given candidate agent A', and may be able to offer something approaching a decision-procedure.[27] Here, then, is the list:

- *Neural condition:* Having a neural structure (albeit not implemented in a conventional, biological way), which maps closely the features specified by theories of the neural (and sub-neural?) correlates of consciousness, according to the best neuroscience of the day.
- *Metabolic condition*: Replicating (perhaps only in some analogical form) the more broadly physiological correlates of consciousness in humans and other organisms (including, e.g. blood circulation, muscular activities; alimentary processes; immune-system responses; endocrine/hormonal processes, etc., to the extent that these are variously considered to be essentially correlated with biologically occurring forms of consciousness).
- *Organic condition*: Having a mechanism that supports consciousness that is organic in the sense of displaying self-maintaining/recreating (autopoietic) processes which require specific energy exchanges between the agent and its environment, and internal metabolic processes to support the continuation of these mechanisms and the maintenance of effective boundary conditions with respect to the surrounding environment.
- *Developmental condition*: Having a life history that at least approximates to that of humans and other altricial creatures, with foetal, neonate, infant, etc. stages—involving embodied exploration, learning, primitive intersubjective relations with carers and teachers, and the capability to develop more complex forms of intersubjectivity.

---

[27] It may well be that realism will not be defeated if no decision procedure is provided. There is the ontological matter of whether questions like "Does A have a moral status as an ethical patient/agent?" have a correct answer or not (independently of the accidents of social determination). And there is the epistemological or methodological matter of whether or not it can be determined, in a straightforward way or even only with extreme difficulty, what the correct answer to that question is for any particular A.

- *Sensorimotor condition*: Exemplifying forms of sensorimotor interaction with the environment, which are considered to be implicated in conscious perceptual activity (See O'Regan and Noë 2001; O'Regan 2007).
- *Cognitive condition*: Displaying the various cognitive marks or accompaniments of consciousness, as identified by cognitive theories of consciousness.
- *Social condition*: Generally interacting with humans (and other artificial agents) in a variety of social situations in a fluent and collaborative way (see Gallagher 2012).
- *Affective/welfare condition*: Showing evidence of having an extended array of needs, desires, aversions, emotions, somewhat akin to those shown by humans and other natural creatures.
- *Ethical condition*: Subscribing to ethical commitments (in an autonomous, self-reflective way, rather than simply via programmed rules) which recognize appropriate moral rights of people with needs, desires, etc.
- *Introspective condition*: Being able to articulate an introspective recognition of their own consciousness; ability to pass a variety of self-report tests held under rigorous experimental conditions.
- *Turing condition:* Responding positively, and in a robust and persistent way, to Turing test style probes for consciousness (including blind dialogues filtered through a textual or other filtering interface or physical interaction episodes in a real-world context).

There is, and no doubt will continue to be, much controversy as to which items should be on this list, on their detailed formulation and corroboration conditions, and on their relative priorities. Those who favour an organic approach to consciousness will privilege the first few criteria, whereas those who favour a cognitively based view will put more emphasis on many of the later conditions. Nevertheless, despite the controversial nature of some of the items, I believe that a list like this could be drawn up, for use across the scientific community, to establish a broad way forward to assess different candidate artificial agent designs, in order to assist in decisions about presence of consciousness in those different candidates, and consequently their ethical status.

A list of this sort does not establish the realist position on questions about when CSS properties are genuinely present, and consequently about which kinds of possible artificial agents might qualify for inclusion in the 'circle' of moral consideration. Nevertheless, it at least shows that 'the machine question' has an internal complexity, a fine texture, with different categories pointing to different kinds of answers, rather than simply being a single undifferentiated mystery from an ontological and epistemological point of view, leaving the vagaries of social judgment as the sole tribunal. Moreover, it suggests that this particular issue is not really different in nature from any complex scientific question where there are many subtle and cross-cutting considerations to be taken into account, and where disagreements are open to resolution, at least in principle, in a rational way.

9 Summing Up

Singer's notion of the expanding ethical circle, and Harris' suggestion that ethical questions concerning the 'moral landscape' can be scientifically grounded, suggest, in different ways, a very strong linkage—possibly a conceptual one—between consciousness and well-being (CSS properties) and ethical concern. In particular, Harris' critique of scientific neutralism suggests the possibility of a scientific grounding to core ethical values: and there is no reason why such scientific, objective grounding should not also apply to the ethical status of artificial agents.

Of course our ethical relations with such agents will be inevitably bound up with our social relations with them. As we saw, the domain of the social is expanding rapidly to include a wide variety of human–AA and AA–AA interactions. But sociality is itself constrained in various ways by physical, biological and psychological factors. And consciousness and well/ill-being (what I have called CSS) lie at the heart of these constraints. Ethics and sociality are indeed closely intertwined. But we should not assume that, just because there are rich and varied social interactions between humans and artificial creatures of different sorts, there are no considerations or constraints on the appropriateness of ethical relations that humans may adopt towards such artificial creatures. Our capacities for satisfaction or suffering must be crucially based upon deep neural and biological properties; so too for other naturally evolved sentient creatures. Some classes of artificial creatures will have closely similar biological properties, making the question of CSS attribution relatively easy for those at least. For others (ones whose designs are advanced versions of electronic technologies with which we are familiar today, for example; or which are based on other technologies, a conception of which we currently have at best the merest glimmer, if any at all) it may be much harder to make reliable judgments. In the end, how we attribute CSS, and consequently ethical status, will depend on a multiplicity of detailed questions concerning commonalities and contrasts between human neural and bodily systems and analogous systems in the artificial agents under consideration. The gross apparent behaviours and functional cognitive/affective organisation of such agents will play important roles (Coeckelbergh 2009, 2010b), of course, in determining how we attribute moral patiency and agency status, but only in a wider mix of considerations which will include many other, less easily observable, features.

Over- and under-attribution of CSS properties cause deep ethical problems in human social life. (To take just one obvious and widespread example, oppressed humans all over the globe continue to have their capacity for suffering falsely denied, in fake justification for their brutal treatment.) Why should it be any different for robots? In a society where humans and machines have extensive and rich social interactions, false-positive or false-negative mis-attributions could each engender massive injustices—either to humans whose interests are being short-changed by the inappropriate shifting of resources or concern to artificial agents that have no intrinsic ethical requirements for them; or to artificial agents whose interests are being denied because of a failure to correctly identify their real capacities for CSS states. It is not clear how a social–relational view can properly accommodate this false-positive/false-negative dimension.

I have tried to put the realist position in a way that is sensitive to the social–relational perspective. However, many problems and gaps remain.[28] A strength of the social–

---

[28] For example, here I have dealt primarily with the connection between consciousness and artificial moral patiency, or recipiency, as opposed to moral agency, or productivity (but see footnote 26 above). There are arguments that suggest that consciousness may be as crucial to the former as to the latter (Torrance 2008; Torrance and Roche 2011).

relational position is that it addresses, in a way that it is difficult for the realist position to do, the undoubted tendency for people to humanise or anthropomorphize autonomous agents, something that will no doubt become more and more prevalent as AI agents with human-like characteristics proliferate, and which happens even when it is far from clear that any consciousness or sentience can exist in such agents. There will surely be strong social pressures to integrate such AIs into our social fabric. Supporters of singularitarian (Kurzweil 2005) views even insist that such agents will come (disarmingly rapidly, perhaps) to dominate human social existence, or at least transform it out of all recognition—for good or for ill. Possibly, such predictions sit better with the social–relational view than with the realist view, so it will be a big challenge for realism to respond adequately to the changing shape of human–machine society, were the rapid and far-reaching technosocial upheavals predicted by many to come about. Nevertheless, I believe I have shown that the realist framework offers the best way forward for the AI and AC research community to best respond to the difficulties that such future social pressures may present.

Acknowledgments Work on this paper was assisted by grants from the EUCogII network, in collaboration with Mark Coeckelbergh, to whom I express gratitude. I am also grateful to Joanna Bryson and David Gunkel for inviting me to join with them in co-chairing the Turing Centenary workshop on The Machine Question, where this paper first saw life. Ideas in the present paper have also greatly benefitted from discussions with Mark Bishop, Rob Clowes, Ron Chrisley, Madeline Drake, David Gunkel, Joel Parthemore, Denis Roche, Wendell Wallach and Blay Whitby.

References

Arkin, R. C. (2009). *Governing lethal behavior in autonomous systems*. Boca Raton: CRC.
Bisson, T. (1991). 'They're made out of meat', *Omni*, 4. April, 1991. http://www.eastoftheweb.com/short-stories/UBooks/TheyMade.shtml. Accessed 10 January 2013.
Block, N. (1978). Troubles with functionalism. In C. Savage (Ed.), *Perception and cognition: issues in the foundations of psychology. Minnesota studies in the philosophy of science* (pp. 261–325). Minneapolis: University of Minnesota Press.
Chalmers, D. (1995). Facing up to the problem of consciousness. *Journal of Consciousness Studies, 2*(3), 200–219.
Coeckelbergh, M. (2009). Personal robots, appearance, and human good: a methodological reflection on roboethics. *International Journal of Social Robotics, 1*(3), 217–221.
Coeckelbergh, M. (2010a). Robot rights? Towards a social–relational justification of moral consideration. *Ethics and Information Technology, 12*(3), 209–221.
Coeckelbergh, M. (2010b). Moral appearances: emotions, robots, and human morality. *Ethics and Information Technology, 12*(3), 235–241.
Coeckelbergh, M. (2012). *Growing moral relations: a critique of moral status ascription*. Basingstoke: Macmillan.
Coeckelbergh, M. (2013). The moral standing of machines: Towards a relational and non-Cartesian moral hermeneutics. Philos. Technol. This issue.
Depraz, N. (1997). La traduction de Leib, une crux phaenomenologica. *Etudes Phénoménologiques, 3*.
Depraz, N. (2001). *Lucidité du corps. De l'empiricisme transcendental en phénoménologie*. Dordrecht: Kluwer.
Gallagher, S. (2005a). *How the body shapes the mind*. Oxford: Clarendon.
Gallagher, S. (2005b). Phenomenological contributions to a theory of social cognition. *Husserl Studies, 21*(2), 95–110.
Gallagher, S. (2012). You, I, robot. *AI and Society*. doi:10.1007/s00146-012-0420-4.
Gallagher, S., & Zahavi, D. (2008). *The phenomenological mind: an introduction to philosophy of mind and cognitive science*. London: Taylor & Francis.

Gallie, W. B. (1955). Essentially contested concepts. *Proceedings of the Aristotelian Society, 56*, 167–198.
Gunkel, D. (2007). *Thinking otherwise: philosophy, communication, technology.* West Lafayette: Purdue University Press.
Gunkel, D. (2012). *The machine question: critical perspectives on AI, robots and ethics.* Cambridge: MIT Press.
Gunkel, D. (2013) 'A vindication of the rights of machines. Philos. Technol. This issue. doi:10.1007/s13347-013-0121-z
Hanna, R., & Thompson, E. (2003). The mind–body–body problem. *Theoria et Historia Scientiarum: International Journal for Interdisciplinary Studies, 7*, 24–44.
Harris, S. (2010). *The moral landscape: how science can determine human values.* London: Random House.
Holland, O. (2007). A strongly embodied approach to machine consciousness. *Journal of Consciousness Studies, 14*(7), 97–110.
Kurzweil, R. (2005). *The singularity is near: when humans transcend biology.* Viking.
Leopold, A. (1948). 'A land ethic'. In: *A sand county almanac with essays on conservation from Round River.* NY: Oxford University Press.
Levine, J. (1983). Materialism and qualia: the explanatory gap. *Pacific Philosophical Quarterly, 64*, 354–361.
Levy, D. (2009). The ethical treatment of artificially conscious robots. *International Journal of Social Robotics, 1*(3), 209–216.
Naess, A. (1973). The shallow and the deep long-range ecology movements. *Inquiry, 16*, 95–100.
O'Regan, J. (2007). How to build consciousness into a robot: the sensorimotor approach. In M. Lungarella et al. (Eds.), *50 years of artificial intelligence* (pp. 332–346). Heidelberg: Springer.
O'Regan, J. K., & Noë, A. (2001). A sensorimotor account of vision and visual consciousness. *Behavioral and brain sciences, 24*(5), 939–972.
Regan, T. (1983). *The case for animal rights.* Berkeley: University of California Press.
Sharkey, N., & Suchman, L. (2013). Wishful mnemonics and autonomous killing machines. *AISB Quarterly, 136*, 14–22.
Shear, J. (Ed.). (1997). *Explaining consciousness: the hard problem.* Cambridge: MIT Press.
Singer, P. (1975). *Animal liberation: a new ethics for our treatment of animals.* NY: New York Review of Books.
Singer, P. (2011). *The expanding circle: ethics, evolution and moral progress.* New Jersey: Princeton University Press.
Sloman, A. (1984). 'The structure of the space of possible minds'. In Torrance, S. (ed.) *The Mind and the Machine: Philosophical Aspects of Artificial Intelligence.* Chichester, Sussex: Ellis Horwood, 35–42.
Sparrow, R. (2007). Killer robots. *Journal of Applied Philosophy, 24*(1), 62–77.
Stuart, S. (2007). Machine consciousness: cognitive and kinaesthetic imagination. *Journal of Consciousness Studies, 14*(7), 141–153.
Thompson, E. ed. (2001) *Between ourselves: second-person issues in the study of consciousness,* Thorverton, UK: Imprint Academic. Also published in *Journal of Consciousness Studies* (2001) 8(5–7).
Thompson, E. (2007). *Mind in life: biology, phenomenology, and the sciences of mind.* Cambridge: Harvard University Press.
Turing, A. (1950). Computing machinery and intelligence. *Mind, 59*, 433–460.
Wallach, W. and Allen, C. (2009) *Moral machines: teaching robots right from wrong.* NY: Oxford University Press.
Wallach, W., Allen, C., & Franklin, S. (2011). Consciousness and ethics: artificially conscious moral agents. *International Journal of Machine Consciousness, 3*(1), 177–192.
Wittgenstein, L. (1953). *Philosophical investigations.* Oxford: Blackwell.
Zahavi, D. (2001). Beyond empathy. Phenomenological approaches to intersubjectivity. *Journal of Consciousness Studies, 8*(5–7), 5–7.
Ziemke, T. (2007). The embodied self: theories, hunches and robot models. *Journal of Consciousness Studies, 14*(7), 167–179.

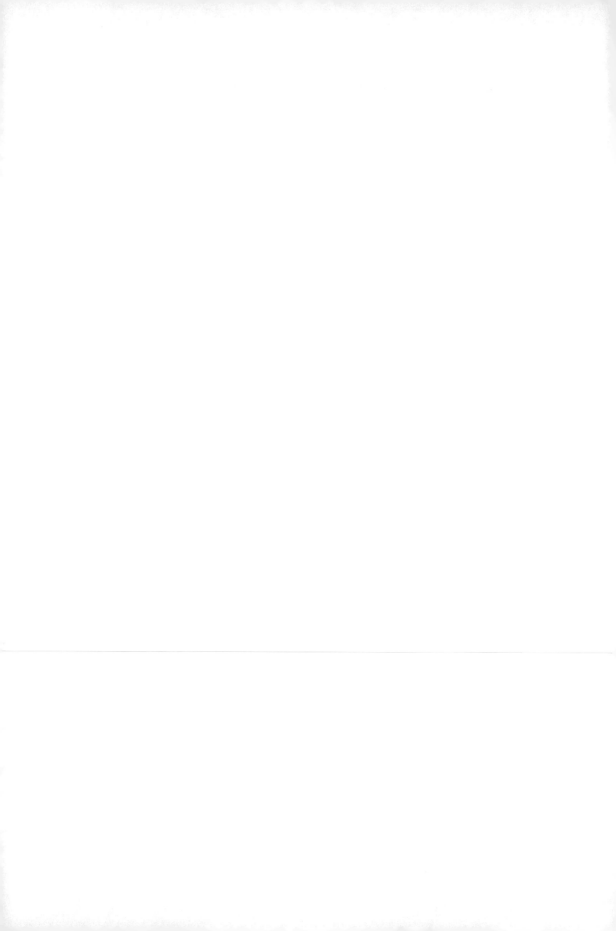

30

Beyond Asimov: The Three Laws of Responsible Robotics

Robin R. Murphy
David D. Woods

Since their codification in 1947 in the collection of short stories *I, Robot*, Isaac Asimov's three laws of robotics have been a staple of science fiction. Most of the stories assumed that the robot had complex perception and reasoning skills equivalent to a child and that robots were subservient to humans. Although the laws were simple and few, the stories attempted to demonstrate just how difficult they were to apply in various real-world situations. In most situations, although the robots usually behaved "logically," they often failed to do the "right" thing, typically because the particular context of application required subtle adjustments of judgment on the part of the robot (for example, determining which law took priority in a given situation, or what constituted helpful or harmful behavior).

The three laws have been so successfully inculcated into the public consciousness through entertainment that they now appear to shape society's expectations about how robots should act around humans. For instance, the media frequently refer to human–robot interaction in terms of the three laws. They've been the subject of serious blogs, events, and even scientific publications. The Singularity Institute organized an event and Web site, "Three Laws Unsafe," to try to counter public expectations of robots in the wake of the movie *I, Robot*. Both the philosophy[1] and AI[2] communities have discussed ethical considerations of robots in society using the three laws as a reference, with a recent discussion in *IEEE Intelligent Systems*.[3] Even medical doctors have considered robotic surgery in the context of the three laws.[4]

With few notable exceptions,[5,6] there has been relatively little discussion of whether robots, now or in the near future, will have sufficient perceptual and reasoning capabilities to actually follow the laws. And there appears to be even less serious discussion as to whether the laws are actually viable as a framework for human–robot interaction, outside of cultural expectations.

Following the definitions in *Moral Machines: Teaching Robots Right from Wrong*,[7] Asimov's laws are based on functional morality, which assumes that robots have sufficient agency and cognition to make moral decisions. Unlike many of his successors, Asimov is less concerned with the details of robot design than in exploiting a clever literary device that lets him take advantage of the large gaps between aspiration and reality in robot autonomy. He uses the situations as a foil to explore issues such as

- the ambiguity and cultural dependence of language and behavior—for example, whether what appears to be cruel in the short run can actually become a kindness in the longer term;
- social utility—for instance, how different individuals' roles, capabilities, or backgrounds are valuable in different ways with respect to each other and to society; and
- the limits of technology—for example, the impossibility of assuring timely, correct actions in all situations and the omnipresence of trade-offs.

In short, in a variety of ways the stories test the lack of resilience in human–robot interactions.

The assumption of functional morality, while effective for entertaining storytelling, neglects operational morality. Operational morality links robot actions and inactions to the decisions, assumptions, analyses, and investments of those who invent and make robotic systems and of those who commission, deploy, and handle robots in operational contexts. No matter how far the autonomy of robots ultimately advances, the important challenges of these accountability and liability linkages will remain.[8]

This essay reviews the three laws and briefly summarizes some of the practical shortcomings—and even dangers—of each law for framing human–robot relationships, including reminders about what robots can't do. We then propose an alternative, parallel set of laws based on what humans and robots can realistically accomplish in the foreseeable future as joint cognitive systems, and their mutual accountability for their actions from the perspectives of human-centered design and human–robot interaction.

Applying Asimov's Laws to Today's Robots

When we try to apply Asimov's laws to today's robots, we immediately run into problems. Just as for Asimov in his short stories, these problems arise from the complexities of situations where we would use robots, the limits of physical systems acting with limited resources in uncertain changing situations, and the interplay between the different social roles as different agents pursue multiple goals.

First Law

Asimov's first law of robotics states, "A robot may not injure a human being or, through inaction, allow a human being to come to harm." This law is already an anachronism given the military's weaponization of robots, and discussions are now shifting to the question of whether weaponized robots can be "humane."[9,10] Such weaponization is no longer limited to situations in which humans remain in the loop for control. The South Korean government has published videos on YouTube of robotic border-security guards. Scenarios have been proposed where it would be permissible for a military robot to fire upon anything moving (presumably target-

> Asimov's laws are based on functional morality, which assumes that robots have sufficient agency and cognition to make moral decisions.

ing humans) without direct human permission.[11]

Even if current events hadn't made the law irrelevant, it's moot because robots cannot infallibly recognize humans, perceive their intent, or reliably interpret contextualized scenes. A quick review of the computer vision literature shows that scientists continue to struggle with many fundamental perceptual processes. Current commercial security packages for recognizing the face of a person standing in a fixed position continue to fall short of expectations in practice. Many robots that "recognize" humans use indirect cues such as heat and motion, which only work in constrained contexts. These problems confirm Norbert Wiener's warnings about such failure possibilities.[8] Just as he envisioned many years ago, today's robots are literal-minded agents—that is, they can't tell if their world model is the world they're really in.

All this aside, the biggest problem with the first law is that it views safety only in terms of the robot—that is, the robot is the responsible safety agent in all matters of human–robot interaction. While some speculate on what it would mean for a robot to be able to discharge this responsibility, there are serious practical, theoretical, social-cognitive, and legal limitations.[8,12] For example, from a legal perspective the robot is a product, so it's not the responsible agent. Rather, the robot's owner or manufacturer is liable for its actions. Unless robots are granted a person-equivalent status, somewhat like corporations are now legally recognized as individual entities, it's difficult to imagine standard product liability law not applying to them. When a failure occurs, violating Asimov's first law, the human stakeholders affected by that failure will engage in the processes of causal attribution. Afterwards, they'll see the robot as a device and will look for the person or group who set up or instructed the device erroneously or who failed to supervise (that is, stop) the robot before harm occurred. It's still commonplace after accidents for manufacturers and organizations to claim the result was due only to human error, even when the system in question was operating autonomously.[8,13]

Accountability is bound up with the way we maintain our social relationships. Human decision-making always occurs in a context of expectations that one might be called to account for his or her decisions. Expectations for what's considered an adequate explanation and the consequences for people when their explanation is judged inadequate are critical parts of accountability systems—a reciprocating cycle of being prepared to provide an accounting for one's actions and being called by others to provide an account. To be considered moral agents, robots would have to be capable of participating personally in this reciprocating cycle of accountability—an issue that, of course, concerns more than any single agent's capabilities in isolation.

Second Law

Asimov's second law of robotics states, "A robot must obey orders given to it by human beings, except where such orders would conflict with the first law." Although the law itself takes no stand on how humans would give orders, Asimov's robots relied on their understanding of verbal directives. Unfortunately, robust natural language understanding still continues to lie just beyond the frontiers of today's AI.[14] It's true that, after decades of research, computers can now construct words from phonemes with some consistency—as improvements in voice dictation and call centers attest. Language-understanding capabilities also work well for specific types of well-structured tasks. However, the goal of meaningful machine participation in open-ended conversational contexts remains elusive. Additionally, we must account for the fact that not all directives are given verbally. Humans use gestures and add affect through body posture, facial expressions, and motions for clarification and emphasis. Indeed, high-performance, experienced teams use highly pointed and coded forms of verbal and nonverbal communication in fluid, interdependent, and idiosyncratic ways.

What's more interesting about the second law from a human–robot interaction standpoint is that at its core, it almost captures the more important idea that intelligent robots should notice and take stock of humans (and that the people robots encounter or interact with can notice pertinent aspects of robots' behavior).[15] For example, is it acceptable for a robot to merely not hit a person in a hospital hall, or should it conform to social convention and acknowledge the person in some way ("excuse me" or a nod of a camera pan-tilt)? Or if a robot operating in public places included two-way communication devices, could a bystander recognize

> The goal of meaningful machine participation in open-ended conversational contexts remains elusive.

that the robot provided a means to report a crime or a fire?

Third Law

The third law states, "A robot must protect its own existence as long as such protection does not conflict with the first or second law." Because today's robots are expensive, you'd think designers would be naturally motivated to incorporate some form of the third law into their products. For example, even the inexpensive iRobot Roomba detects stairs that could cause a fatal fall. Surprisingly, however, many expensive commercial robots lack the means to fully protect their owners' investment.

An extreme example of this is in the design of robots for military applications or bomb squads. Such robots are designed to be teleoperated by a person who bears full responsibility for all safety matters. Human-factors studies show that remote operators are immediately at a disadvantage, working through a mediated interface with a time delay. Worse yet, remote operators are required to operate the robot through poor human–computer interfaces and in contexts where the operator can be fatigued, overloaded, or under high stress. As a result, when an abnormal event occurs, they may be distracted or not fully engaged and thus might not respond adequately in time. The result for a robot is akin to expecting an astronaut on a planet's surface to request and wait for permission from mission control to perform even simple reflexes such as ducking.

What is puzzling about today's limited attempts to conform to the third law is that there are well-established technological solutions for basic robot survival activities that work for autonomous and human-controlled robots. For instance, since the 1960s we've had technology to assure guarded motion, where the human drives the robot but onboard software will not allow the robot to make potentially dangerous moves (for example, collide with obstacles or exceed speed limits or boundaries) without explicit orders (an implicit invocation of the second law). By the late 1980s, guarded motion was encapsulated into tactical reactive behaviors, essentially giving robots reflexes and tactical authority. Perhaps the most important reason that guarded motion and reflexive behaviors haven't been more widely deployed is that they require additional sensors, which would add to the cost. This increase in cost may not appear to be justified to customers, who tend to be wildly overconfident that trouble and complexities outside the bounds of expected behavior rarely arise.

The Alternative Three Laws of Responsible Robotics

To address the difficulties of applying Asimov's three laws to the current generation of robots while respecting the laws' general intent, we suggest the three laws of responsible robotics.

Alternative First Law

Our alternative to Asimov's first law is "A human may not deploy a robot without the human–robot work system meeting the highest legal and professional standards of safety and ethics." Since robots are indeed subject to safety regulations and liability laws, the requirement of meeting legal standards for safety would seem self-evident. For instance, the medical-device community has done extensive research to validate robot sensing of scalpel pressures and tissue contact parameters, and it invests in failure mode and effect analyses (consistent with FDA medical-device standards).

In contrast, mobile roboticists have a somewhat infamous history of disregarding regulations. For example, robot cars operating on public roads, such as those used in the DARPA Urban Grand Challenge, are considered by US federal and state transportation regulations as experimental vehicles. Deploying such vehicles requires voluminous and tedious permission applications. Regrettably, the 1995 CMU "No Hands Across America" team neglected to get all appropriate permissions while driving autonomously from Pittsburgh to Los Angeles, and were stopped in Kansas. The US Federal Aviation Administration makes a clear distinction between flying unmanned aerial vehicles as a hobby and flying them for R&D or commercial practices, effectively slowing or stopping many R&D efforts. In response to these difficulties, a culture preferring "forgiveness" to "permission" has grown up in some research groups. Such attitudes indicate a poor safety culture at universities that could, in turn, propagate to government or industry. On the positive side, the robot competitions sponsored by the Association for Unmanned Vehicle Systems International are noteworthy in their insistence on having safe areas of operation, clear emergency plans, and safety officers present.

Meeting the minimal legal requirements is not enough—the alternative first law demands the highest professional ethics in robot deployment. A failure or accident involving a robot can effectively end an entire branch of robotics research, even if the operators aren't legally culpable. Responsible communities should proactively consider safety in the broadest sense, and funding agencies should find ways to increase the priority and scope of research funding specifically aimed at relevant legal concerns.

The highest professional ethics should also be applied in product development and testing. Autonomous robots have known vulnerabilities to problems stemming from interrupted wireless communications. Signal reception is impossible to predict, yet robust "return to home if signal lost" and "stop movement if GPS lost"

> The fact that there will be disturbances is virtually guaranteed, and designing for resilience in the face of these is fundamental.

functionality hasn't yet become an expected component of built-in robot behavior. This means robots are operating counter to reasonable and prudent assumptions. Worse yet, when they're operating experimentally, robots often encounter unanticipated factors that affect their control. Simply saying an unfortunate event was unpredictable doesn't relieve the designers of responsibility. Even if a specific disturbance is unpredictable in detail, the fact that there will be disturbances is virtually guaranteed, and designing for resilience in the face of these is fundamental.

As a matter of professional common sense, robot design should start with safety first, then add the interesting software and hardware. Robots should carry "black boxes" or recorders to show what they were doing when a disturbance occurred, not only for the sake of an accident investigation but also to trace the robots' behavior in context to aid diagnosis and debugging. There should be a formal safety plan and checklists for contingencies. These do not have to be extensive and time consuming to be effective. A litmus test for developers might be "If a group of experts from the IEEE were to write about your robot after an accident, what would they say about system safety and your professionalism?" Fundamentally, the alternative first law places responsibility for safety and efficacy on humans within the larger social and environmental context in which robots are developed, deployed, and operated.

Alternative Second Law

As an alternative to Asimov's second law, we propose the following: "A robot must respond to humans as appropriate for their roles." The capability to respond appropriately—*responsiveness*—may be more important to human–robot interaction than the capability of autonomy. Not all robots will be fully autonomous over all conditions. For example, a robot might be constrained to follow waypoints but will be expected to generate appropriate responses to people it encounters along the way. Responsiveness depends on the social environment, the kinds of people and their expectations that a robot might encounter in its work envelope. Rather than assume the relationship is hierarchical with the human as the superior and the robot as the subordinate so that all communication is a type of order, the alternative second law states that robots must be built so that the interaction fits the relationships and roles of each member in a given environment. The relationship determines the degree to which a robot is obligated to respond. It might ignore a hacker completely. Orders exceeding the authority of the speaker might be disposed of politely ("please have your superior confirm your request") or with a warning ("interference with a law enforcement robot may be a violation"). Note that defining "appropriate response" may address concerns about robots being abused.[16]

The relationship also determines the mode of the response. How the robot signals or expresses itself should be consistent with that relationship. Casual relationships might rely on natural language, whereas trained teams performing specific tasks could coordinate activities through other signals such as body position and gestures.

The requirement for responsiveness captures a new form of autonomy (not as isolated action but the more difficult behavior of engaging appropriately with others). However, developing robots' capability for responsiveness requires a significant research effort, particularly in how robots can perceive and identify the different members, roles, and cues of a social environment.

Alternative Third Law

Our third law is "A robot must be endowed with sufficient situated autonomy to protect its own existence as long as such protection provides smooth transfer of control to other agents consistent with the first and second laws." This law specifies that a human–robot system should be able to transition smoothly from whatever

> **Increased capability for autonomy and authority leads to the need to participate in more sophisticated forms of coordinated activity.**

degree of autonomy or roles the robots and humans were inhabiting to a new control relationship given the nature of the disruption, impasse, or opportunity encountered or anticipated. When developers focus narrowly on the goal of isolated autonomy and fall prey to overconfidence by underestimating the potential for surprises to occur, they tend to minimize the importance of transfer of control. But bumpy transfers of control have been noted as a basic difficulty in human interaction with automation that can contribute to failures.[17]

The alternative third law addresses situated autonomy and smooth transfer of control, both of which interact with the prescriptions of the other laws. To be consistent with the second law requires that humans in a given role might not always have complete control of the robot (for example, when conditions require very short reaction times, a pilot may not be allowed to override some commands generated by algorithms that attempt to provide envelope protection for the aircraft). This in turn implies that an aspect of the design of roles is the identification of classes of situations that demand transfer of control, so that the exchange processes can be specified as part of roles. This is when the human takes control from the robot for a specialized aspect of the mission in anticipation of conditions that will challenge the limits of the robot's capabilities, or in an emergency. Decades of human factors research on human out-of-the-loop control problems, handling of anomalies, cascades of disturbances, situation awareness, and autopilot/pilot transfer of control can inform such designs.

To be consistent with the first law requires designers to explicitly address what is the appropriate situated autonomy (for example, identifying when the robot is better informed or more capable than the human owing to latency, sensing, and so on) and to provide mechanisms that permit smooth transfer of control. To disregard the large body of literature on resilience and failure due to bumpy transfer of control would violate the designers' ethical obligation.

The alternative second and third laws encourage some forms of increased autonomy related to responsiveness and the ability to engage in various forms of smooth transfer of control. To be able to engage in these activities with people in various roles, the robot will need more situated intelligence. The result is an irony that has been noted before: increased capability for autonomy and authority leads to the need to participate in more sophisticated forms of coordinated activity.[8]

Table 1. Asimov's laws of robotics versus the alternative laws of responsible robotics

	Asimov's laws	Alternative laws
1	A robot may not injure a human being or, through inaction, allow a human being to come to harm.	A human may not deploy a robot without the human–robot work system meeting the highest legal and professional standards of safety and ethics.
2	A robot must obey orders given to it by human beings, except where such orders would conflict with the first law.	A robot must respond to humans as appropriate for their roles.
3	A robot must protect its own existence as long as such protection does not conflict with the first or second law.	A robot must be endowed with sufficient situated autonomy to protect its own existence as long as such protection provides smooth transfer of control to other agents consistent the first and second laws.

Discussion

Our critique reveals that robots need two key capabilities: responsiveness and smooth transfer of control. Our proposed alternative laws remind robotics researchers and developers of their legal and professional responsibilities. They suggest how people can conduct human–robot interaction research safely, and they identify critical research questions.

Table 1 places Asimov's three laws side by side with our three alternative laws. Asimov's laws assume functional morality—that robots are capable of making (or are permitted to make) their own decisions—and ignore the legal and professional responsibility of those who design and deploy them (operational morality). More importantly for human–robot interaction, Asimov's laws ignore the complexity and dynamics of relationships and responsibilities between robots and people and how those relationships are expressed. In contrast, the alternative three laws emphasize responsibility and resilience, starting with enlightened, safety-oriented designs (alternative first law), then adding responsiveness (alternative second law) and smooth transfer of control (alternative third law).

The alternative laws are designed to be more feasible to implement than Asimov's laws given current technology, although they also raise critical questions for research. For example, the alternative first law isn't concerned with technology per se but with the need for robot developers to be aware of human systems design principles and to take responsibility proactively for the consequences of errors and failures in human–robot systems. Standard tools from the aerospace, medical, and chemical manufacturing safety cultures, including training, formal processes, checklists, black boxes, and safety officers, can be adopted. Network and physical security should be incorporated into robots, even during development.

The alternative second and third laws require new research directions for robotics to leverage and build on existing results in social cognition, cognitive engineering, and resilience engineering. The laws suggest that the ability for robots to express relationships and obligations through social roles will be essential to all human–robot interaction. For example, work on entertainment robots and social robots provides insights about how robots can express emotions or affect appropriate to people they encounter. The extensive literature from cognitive engineering on transfer of control and general human out-of-the-loop control problems can be redirected at robotic systems. The techniques for resilience engineering are beginning to identify new control architectures for distributed, multi-echelon systems that include systems that include robots.

The fundamental difference between Asimov's laws, which focus on robots' functional morality and full moral agency, and the alternative laws, which focus on system responsibility and resilience, illustrates why the robotics community should resist public pressure to frame current human–robot interaction in terms of Asimov's laws. Asimov's laws distract from capturing the diversity of robotic missions and initiative. Understanding these diversities and complexities is critical for designing the "right" interaction scheme for a given domain.

Ironically, Asimov's laws really are robot-centric because most of the initiative for safety and efficacy lies in the robot as an autonomous agent. The alternative laws are human-centered because they take a systems approach. They emphasize that

- responsibility for the consequences of robots' successes and failures lies in the human groups that have a stake in the robots' activities, and
- capable robotic agents still exist in a web of dynamic social and cognitive relationships.

Ironically, meeting the requirements of the alternative laws leads to the need for robots to be more capable agents—that is, more responsive to others and better at interaction with others.

We propose the alternative laws as a way to stimulate debate about robots' accountability when their actions can harm people or human interests. We also hope that these laws can serve to direct R&D to enhance human–robot systems. Finally, while perhaps not as entertaining as Asimov's laws, we hope the alternative laws of responsible robotics can better communicate to the general public the complex mix of opportunities and challenges of robots in today's world.

Acknowledgments

We thank Jeff Bradshaw, Cindy Bethel, Jenny Burke, Victoria Groom, and Leila Takayama for their helpful feedback and Sung Huh for additional references. The second author's contribution was based on participation in the Advanced Decision Architectures Collaborative Technology Alliance, sponsored by the US Army Research Laboratory under cooperative agreement DAAD19-01-2-0009.

References

1. S.L. Anderson, "Asimov's 'Three Laws of Robotics' and Machine Metaethics," *AI and Society*, vol. 22, no. 4, 2008, pp. 477–493.
2. A. Sloman, "Why Asimov's Three Laws of Robotics are Unethical," 27 July 2006; www.cs.bham.ac.uk/research/projects/cogaff/misc/asimov-three-laws.html.
3. C. Allen, W. Wallach, and I. Smit, "Why Machine Ethics?" *IEEE Intelligent Systems*, vol. 21, no. 4, 2006, pp. 12–17.
4. M. Moran, "Three Laws of Robotics and Surgery," *J. Endourology*, vol. 22, no. 8, 2008, pp. 1557–1560.
5. R. Clarke, "Asimov's Laws of Robotics: Implications for Information Technology Part 1," *Computer*, vol. 26, no. 12, 1993, pp. 53–61.
6. R. Clarke, "Asimov's Laws of Robotics: Implications for Information Technology Part 2," *Computer*, vol. 27, no. 1, 1994, pp. 57–66.
7. W. Wallach and C. Allen, *Moral Machines: Teaching Robots Right from Wrong*, Oxford Univ. Press, 2009.
8. D. Woods and E. Hollnagel, *Joint Cognitive Systems: Patterns in Cognitive Systems Engineering*, Taylor and Francis, 2006.
9. R.C. Arkin and L. Moshkina, "Lethality and Autonomous Robots: An Ethical Stance," *Proc. IEEE Int'l Symp. Technology and Society* (ISTAS 07), IEEE Press, 2007, pp. 1–3.
10. N. Sharkey, "The Ethical Frontiers of Robotics," *Science*, vol. 322, no. 5909, 2008, pp. 1800–1801.
11. M.F. Rose et al., *Technology Development for Army Unmanned Ground Vehicles*, Nat'l Academy Press, 2002.
12. D. Woods, "Conflicts between Learning and Accountability in Patient Safety," *DePaul Law Rev.*, vol. 54, 2005, pp. 485–502.
13. S.W.A. Dekker, *Just Culture: Balancing Safety and Accountability*, Ashgate, 2008.
14. J. Allen et al., "Towards Conversational Human–Computer Interaction," *AI Magazine*, vol. 22, no. 4, 2001, pp. 27–38.
15. J.M. Bradshaw et al., "Dimensions of Adjustable Autonomy and Mixed-Initiative Interaction," *Agents and Computational Autonomy: Potential, Risks, and Solutions*, M. Nickles, M. Rovatsos, and G. Weiss, eds., LNCS 2969, Springer, 2004, pp. 17–39.
16. B. Whitby, "Sometimes It's Hard to Be a Robot: A Call for Action on the Ethics of Abusing Artificial Agents," *Interacting with Computers*, vol. 20, no. 3, 2008, pp. 326–333.
17. D.D. Woods and N. Sarter, "Learning from Automation Surprises and 'Going Sour' Accidents," *Cognitive Engineering in the Aviation Domain*, N.B. Sarter and R. Amalberti, eds., Nat'l Aeronautics and Space Administration, 1998.

Part V:

Law and Policy

31

LEGAL PERSONHOOD FOR ARTIFICIAL INTELLIGENCES

Lawrence B. Solum

Could an artificial intelligence become a legal person? As of today, this question is only theoretical. No existing computer program currently possesses the sort of capacities that would justify serious judicial inquiry into the question of legal personhood. The question is nonetheless of some interest. Cognitive science begins with the assumption that the nature of human intelligence is computational, and therefore, that the human mind can, in principle, be modelled as a program that runs on a computer.[1] Artificial intelligence (AI) research attempts to develop such models.[2] But even as cognitive science has displaced behavioralism as

---

1. For an introduction to cognitive science and the philosophy of mind, see Owen J. Flanagan, Jr., The Science of the Mind 1-22 (2d ed. 1991). For the purposes of this essay, I will not address the question as to which computer architectures could produce artificial intelligence. For example I will not discuss the question whether parallel, as opposed to serial, processing would be required. Similarly, I will not discuss the merits of connectionist as opposed to traditional approaches to AI. For a comparison of parallel distributed processing with serial processing, see *id*. at 224-41. These issues are moot in one sense. A digital computer can, in principle, implement any connectionist or parallel approach. On the other hand, there could be one very important practical difference: the parallel architecture could turn out to be much faster.

2. There is a debate within the artificial intelligence community as to the goal of AI research. The possibilities range from simply making machines smarter to investigating the nature of human intelligence or, more broadly, the nature of all intelligence. See Bob Ryan, *AI's Identity Crisis*, Byte, Jan. 1991, at 239, 239-40. Owen Flanagan distinguishes four programs of AI research. Nonpsychological AI research involves building and programming computers to accomplish tasks that would require intelligence if undertaken by humans. Weak psychological AI views computer models as a tool for investigating human intelligence. Strong psychological AI assumes that human minds really are computers and therefore in principle can be duplicated by AI research. Suprapsychological AI investigates the nature of all intelligence and hence is not limited to investigating the human mind. See Flanagan,

the dominant paradigm for investigating the human mind, fundamental questions about the very possibility of artificial intelligence continue to be debated. This Essay explores those questions through a series of thought experiments that transform the theoretical question whether artificial intelligence is possible into legal questions such as, "Could an artificial intelligence serve as a trustee?"

What is the relevance of these legal thought experiments for the debate over the possibility of artificial intelligence? A preliminary answer to this question has two parts. First, putting the AI debate in a concrete legal context acts as a pragmatic Occam's razor. By reexamining positions taken in cognitive science or the philosophy of artificial intelligence as legal arguments, we are forced to see them anew in a relentlessly pragmatic context.[3] Philosophical claims that no program running on a digital computer could really be intelligent are put into a context that requires us to take a hard look at just what practical importance the missing reality could have for the way we speak and conduct our affairs. In other words, the legal context provides a way to ask for the "cash value" of the arguments. The hypothesis developed in this Essay is that only some of the claims made in the debate over the possibility of AI do make a pragmatic difference, and it is pragmatic differences that ought to be decisive.[4]

Second, and more controversially, we can view the legal system as a repository of knowledge—a formal accumulation of practical judgments.[5] The law embodies core insights about the way the world works

---

supra note 1, at 241-42. This Essay discusses the philosophical foundations for Flanagan's third and fourth categories of AI research.

3. *See generally* Catharine W. Hantzis, *Legal Innovation Within the Wider Intellectual Tradition: The Pragmatism of Oliver Wendell Holmes, Jr.*, 82 Nw. U. L. REV. 541, 561-75, 595-99 (1988) (discussing Holmes's jurisprudential focus on concrete issues rather than generalities); Steven D. Smith, *The Pursuit of Pragmatism*, 100 YALE L.J. 409, 409-12 (1990) (discussing the renewed popularity of legal pragmatism); *Symposium: The Renaissance of Pragmatism in American Legal Thought*, 63 S. CAL. L. REV. 1569 (1990) (collecting articles espousing diverse views of legal pragmatism). The law is a "pragmatic" context in the sense that legal decisions are made for practical purposes with consequences in mind and in the sense that foundationalist philosophical theories do not play a role in legal reasoning. This assertion does not depend on the further claim that legal actors have adopted American pragmatism as part of their world view.

4. In addition, the Essay advances a more modest hypothesis. Examining the debate over the possibility of AI through legal examples illuminates the consequences of the arguments made in the debate, and this pragmatic assessment has a bearing on the arguments, even if it is not decisive. This Essay surely does not suffice to confirm the ambitious hypothesis in the text. I hope that it has established the more modest claim presented in this footnote. The proof, of course, is in the pudding.

5. The point is related to that made by Sir John Fortescue about the common law of England. There is a presumption in favor of its wisdom, because it has been tested by long experience. *See* SIR JOHN FORTESCUE, DE LAUDIBUS LEGUM ANGLIE 39-41 (S.B. Chrimes

and how we evaluate it. Moreover, in common-law systems judges strive to decide particular cases in a way that best fits the legal landscape—the prior cases, the statutory law, and the constitution.[6] Hence, transforming the abstract debate over the possibility of AI into an imagined hard case forces us to check our intuitions and arguments against the assumptions that underlie social decisions made in many other contexts. By using a thought experiment that explicitly focuses on wide coherence,[7] we increase the chance that the positions we eventually adopt will be in reflective equilibrium[8] with our views about related matters. In addition, the law embodies practical knowledge in a form that is subject to public examination and discussion. Legal materials are published and subject to widespread public scrutiny and discussion. Some of the insights gleaned in the law may clarify our approach to the artificial intelligence debate.[9]

I do not claim in this Essay to have resolved the debate over the possibility of artificial intelligence. My aim is more modest: I am proposing a way of thinking about the debate that just might result in progress. There is some precedent for this project. Christopher Stone brought questions of environmental ethics into focus by asking whether trees should have standing.[10] My hope is that the law will be equally

---

ed., Wm. W. Gaunt & Sons 1986) (1537); *see also* J.G.A. POCOCK, THE MACHIAVELLIAN MOMENT: FLORENTINE POLITICAL THOUGHT AND THE ATLANTIC REPUBLICAN TRADITION 9-18 (1975) (offering a critique of Fortescue's laudatory expositions of English law and arguing that because English law is based on accumulated experiences—elevated to the level of custom—its very existence presumes its validity, thus preempting rational scrutiny of English law's assumption that it is well suited to the needs of the English).

6. Here I adopt the view of Ronald Dworkin. *See* RONALD DWORKIN, LAW'S EMPIRE 91-96, 147-50, 276-400 (1986); RONALD DWORKIN, A MATTER OF PRINCIPLE 3-42 (1985); RONALD DWORKIN, TAKING RIGHTS SERIOUSLY 1-80 (1977).

7. *Cf.* SUSAN L. HURLEY, NATURAL REASONS 15-18, 225-53 (1987) (advocating coherence approach to reasoning in general and legal reasoning in particular); S.L. Hurley, *Coherence, Hypothetical Cases, and Precedent*, 10 OXFORD J. LEGAL STUD. 221, 222-32 (1990) (same, with emphasis on legal reasoning).

8. *See* JOHN RAWLS, A THEORY OF JUSTICE 48-51 (1971).

9. I say "may provide" advisedly. We must be on guard against an easy or unthinking move from a legal conclusion to a moral one. In many circumstances, there are good reasons for answering a moral question differently from a legal one. Most obviously, the costs of legal enforcement of a norm are quite different than the costs of social enforcement of a moral norm. Moreover, in many cases, the law will simply be morally wrong. The fact that a legal rule has survived a very long time does tell us that it has not led to the collapse of the society that enforces it, but it does not tell us directly whether that society would be better off without it. The legal case may bear on the moral one, but not being irrelevant is far from being decisive. I owe thanks to Elyn Saks for prompting me to qualify my argument in this regard.

10. *See* Christopher Stone, *Should Trees Have Standing?—Toward Legal Rights for Natural Objects*, 45 S. CAL. L. REV. 450, 453-57 (1972) [hereinafter Stone, *Should Trees Have Standing?*]. So far as I know, Stone became the first legal thinker to raise the questions asked by this Essay in a footnote to his famous 1972 essay. *See id.* at 456 n.26 (raising the question as to whether analysis applicable to natural objects such as trees would be appropriate to

fruitful as a context in which to think about the possibility of AI. The "artificial reason and judgment of law"[11] may circumvent the intractable intuitions that threaten to lock the AI debate in dialectical impasse.

Part I of this Essay recounts some recent developments in cognitive science and explores the debate as to whether artificial intelligence is possible. Part II puts the question in legal perspective by setting out the notion of legal personhood. Parts III and IV explore two hypothetical scenarios. Part III examines the first scenario—an attempt to appoint an AI as a trustee. The second scenario, an AI's invocation of the individual rights provisions of the United States Constitution, is the subject of Part IV. The results are then brought to bear on the debate over the possibility of artificial intelligence in Part V. In conclusion, Part VI takes up the question whether cognitive science might have implications for current legal and moral debates over the meaning of personhood.

I. Artificial Intelligence

Is artificial intelligence possible? The debate over this question has its roots at the very beginning of modern thought about the nature of the human mind. It was Thomas Hobbes who first proposed a computational theory of mind: "By ratiocination, I mean computation."[12] And it was Rene Descartes who first considered a version of the question whether it would be possible for a machine to think:

> For we can easily understand a machine's being constituted so that it can utter words, and even emit some responses to action on it of a corporeal kind, which brings about a change in its organs; for example, if it is touched in a particular part it may ask what we wish to say to it; if in another part it may exclaim that it is being hurt, and so on, but it never happens that it arranges its speech in various ways, in order to reply appropriately to everything that may be said in its presence, as even the lowest type of man can do.[13]

"computers"). Stone returned to this issue in 1987. *See* CHRISTOPHER STONE, EARTH AND OTHER ETHICS 12, 28-30, 65-67 (1987) [hereinafter STONE, EARTH AND OTHER ETHICS] (asking whether a "robot" should have standing and discussing criminal liability of AIs).

11. The phrase is Sir Edward Coke's. *See* Prohibitions Del Roy, 12 Coke Rep. 63, 65, 77 Eng. Rep. 1342, 1343 (1608).

12. THOMAS HOBBES, ELEMENTS OF PHILOSOPHY (1655), *reprinted in* 1 THE ENGLISH WORKS OF THOMAS HOBBES 1, 3 (William Molesworth ed., London, J. Bohn 1839); *see also* THOMAS HOBBES, LEVIATHAN (1670), *reprinted in* 3 THE ENGLISH WORKS OF THOMAS HOBBES 1, *supra*, at 29-32 [hereinafter LEVIATHAN] (equating reason with computation or "reckoning of the consequences"). Hobbes uses "ratiocination" to mean reasoning.

13. RENE DESCARTES, DISCOURSE ON THE METHOD OF RIGHTLY CONDUCTING ONE'S REASON AND SEEKING TRUTH IN THE SCIENCES (1637), *reprinted in* THE ESSENTIAL DESCARTES 138 (Margaret D. Wilson ed., 1969). This passage was likely inspired by

Descartes' assertion that no artifact could arrange its words "to reply appropriately to everything that may be said in its presence" remains at the heart of the AI debate.

The events of the past forty years have stretched the limits of our imagination. Digital computers have been programmed to perform an ever wider variety of complex tasks. As I write this Essay using a word processing program, my spelling and grammar are automatically checked by programs that perform tasks thought to require human intelligence not so many years ago. The program Deep Thought has given the second best human chess player a very tough game, and the program's authors predict the program will become the world's chess champion within a few years.[14] Expert systems simulate the thinking of human experts on a wide variety of subjects, from petroleum geology to law.[15]

But these events have not resolved the question whether AI is even possible. The contemporary debate[16] over that question has centered around Alan Turing's test.[17] Turing proposed that the question whether a machine can think be replaced with the following, more operationalized, inquiry. The artifact that is a candidate for having the ability to think shall engage in a game of imitation with a human opponent. Both

---

Descartes' experience with the French Royal Gardens, which included a miniature society inhabited by hydraulically animated robots. As visitors walked along garden paths they set the robots bodies in motion. The robots actually played musical instruments and spoke. *See* FLANAGAN, *supra* note 1, at 1-2.

14. *See* Feng-hsiung Hsu et al., *A Grandmaster Chess Machine*, SCI. AM., Oct. 1990, at 44, 44.

15. *See, e.g.,* ANN VON DER LIETH GARDNER, AN ARTIFICIAL INTELLIGENCE APPROACH TO LEGAL REASONING 1-24 (1987); RICHARD E. SUSSKIND, EXPERT SYSTEMS IN LAW: A JURISPRUDENTIAL INQUIRY (1987); ALAN TYREE, EXPERT SYSTEMS IN LAW 7-11 (1989); L. Thorne McCarty, *Artificial Intelligence and Law: How to Get There From Here*, 3 RATIO JURIS 189, 189-200 (1990); Edwina L. Rissland, *Artificial Intelligence and Law: Stepping Stones to a Model of Legal Reasoning*, 99 YALE L.J. 1957, 1961-64 (1990).

16. Overviews of the debate over the possibility of AI are found in JAMES H. FETZER, ARTIFICIAL INTELLIGENCE: ITS SCOPE AND LIMITS 3-27, 298-303 (1990); JOHN HAUGELAND, ARTIFICIAL INTELLIGENCE: THE VERY IDEA 2-12 (1985); RAYMOND KURZWEIL, THE AGE OF INTELLIGENT MACHINES 36-40 (1990); and in the essays collected in THE PHILOSOPHY OF ARTIFICIAL INTELLIGENCE (Margaret A. Boden ed., 1990) and in THE ARTIFICIAL INTELLIGENCE DEBATE: FALSE STARTS, REAL FOUNDATIONS (Stephen Graubard ed., 1988). For a strong statement of the view that AI is impossible, see HUBERT DREYFUS, WHAT COMPUTERS CAN'T DO: THE LIMITS OF ARTIFICIAL INTELLIGENCE 285-305 (rev. ed. 1979).

17. *See* Alan M. Turing, *Computing Machinery and Intelligence*, 59 MIND 433 (1950), *reprinted in* THE PHILOSOPHY OF ARTIFICIAL INTELLIGENCE, *supra* note 16, at 40 (subsequent citations to pagination in anthology). For a recent discussion and defense of the Turing Test, see Daniel C. Dennett, *Can Machines Think?*, *in* KURZWEIL, *supra* note 16, at 48. For a recent critique, see Donald Davidson, *Turing's Test*, *in* MODELLING THE MIND 1 (K.A. Mohyeldin Said et al. eds., 1990). For a report on a recent competition testing present-day computers and programs in the Turing format, see Carl Zimmer, *Flake of Silicon*, DISCOVER, Mar. 1992, at 36, 36-38.

the candidate and the human being are questioned by someone who does not know which is which (or who is who)—the questions are asked via teletype. The questions may be on any subject whatsoever. Both the human being and the artifact will attempt to convince the questioner that it or she is the human and the other is not. After a round of play is completed, the questioner guesses which of the two players is the human. Turing suggested we postpone a direct answer to the question whether machines can think; he proposed that we ask instead whether an artifact could fool a series of questioners as often as the human was able to convince them of the truth, about half the time.[18] The advantage of Turing's test is that it avoids direct confrontation with the difficult questions about what "thinking" or "intelligence" is. Turing thought that he had devised a test that was so difficult that anything that could pass the test would necessarily qualify as intelligent.

John Searle questioned the relevance of Turing's Test with another thought experiment, which has come to be known as the Chinese Room.[19] Imagine that you are locked in a room. Into the room come batches of Chinese writing, but you don't know any Chinese. You are, however, given a rule book, written in English, in which you can look up the bits of Chinese, by their shape. The rule book gives you a procedure for producing strings of Chinese characters that you send out of the room. Those outside the room are playing some version of Turing's game. They are convinced that whatever is in the room understands Chinese. But you don't know a word of Chinese, you are simply following a set of instructions (which we can call a program) based on the shape of Chinese symbols. Searle believes that this thought experiment demonstrates that neither you nor the instruction book (the program) understands Chinese, even though you and the program can simulate such understanding.[20]

More generally, Searle argues that thinking cannot be attributed to a computer on the basis of its running a program that manipulates symbols in a way that simulates human intelligence. The formal symbol-manipulations accomplished by the program cannot constitute thinking or un-

---

18. *See* Turing, *supra* note 17, at 41-48.
19. *See* JOHN R. SEARLE, MINDS, BRAINS AND SCIENCE 28-41 (1984); John R. Searle, *Author's Response*, 3 BEHAVIORAL & BRAIN SCI. 450 (1980); John R. Searle, *Is the Brain a Digital Computer?*, 64 PROC. & ADDRESSES AM. PHIL. ASS'N, Nov. 1990, at 21, 21; John R. Searle, *Minds, Brains & Programs*, 3 BEHAVIORAL & BRAIN SCI. 417 (1980), *reprinted in* THE PHILOSOPHY OF ARTIFICIAL INTELLIGENCE, *supra* note 16, at 67 [hereinafter Searle, *Minds, Brains & Programs*; subsequent citations to pagination in anthology]; John R. Searle, *"The Emperor's New Mind": An Exchange*, N.Y. REV. BOOKS, June 14, 1990, at 58 (letter to the editor with response from John Maynard Smith).
20. Searle, *Minds, Brains & Programs*, *supra* note 19, at 70.

derstanding because the program lacks "intentionality"—the ability to process meanings. The shape of a symbol is a syntactic property, whereas the meaning of a symbol is a semantic property. Searle's point is that computer programs respond only to the syntactic properties of symbols on which they operate.[21]

This point can be restated in terms of the Chinese Room: (1) the output—coherent Chinese sentences—from the Chinese room seems to respond to the meaning of the input; (2) but the process that goes on inside the Chinese room only involves the shape or syntactic properties of the input; (3) therefore, the process in the Chinese room does not involve understanding.[22] Searle generalizes the conclusion of the Chinese room thought experiment by arguing that part of the definition of a program is that it is formal and operates only on syntactic properties. He concludes that no system could be said to think or understand solely on the basis of the fact that the system is running a program that produces output that simulates understanding.[23]

Searle's Chinese Room has given rise to a number of replies.[24] But

---

21. *Id.*
22. *Id.* at 83-84.
23. *Id.* at 70-71.
24. *See, e.g.*, Robert P. Abelson, *Searle's Argument Is Just a Set of Chinese Symbols*, 3 BEHAVIORAL & BRAIN SCI. 424 (1980); Ned Block, *What Intuitions About Homunculi Don't Show*, 3 BEHAVIORAL & BRAIN SCI. 425 (1980); Bruce Bridgeman, *Brains + Programs = Minds*, 3 BEHAVIORAL & BRAIN SCI. 427 (1980); Arthur C. Danto, *The Use and Mention of Terms and the Simulation of Linguistic Understanding*, 3 BEHAVIORAL & BRAIN SCI. 428 (1980); Daniel Dennett, *The Milk of Human Intentionality*, 3 BEHAVIORAL & BRAIN SCI. 428 (1980); John C. Eccles, *A Dualist-Interactionist Perspective*, 3 BEHAVIORAL & BRAIN SCI. 430 (1980); J.A. Fodor, *Searle on What Only Brains Can Do*, 3 BEHAVIORAL & BRAIN SCI. 431 (1980); John Haugeland, *Programs, Causal Powers, and Intentionality*, 3 BEHAVIORAL & BRAIN SCI. 432 (1980); Douglas R. Hofstadter, *Reductionism and Religion*, 3 BEHAVIORAL & BRAIN SCI. 433 (1980); B. Libet, *Mental Phenomena and Behavior*, 3 BEHAVIORAL & BRAIN SCI. 434 (1980); William G. Lycan, *The Functionalist Reply (Ohio State)*, 3 BEHAVIORAL & BRAIN SCI. 434 (1980); John C. Marshall, *Artificial Intelligence—The Real Thing?*, 3 BEHAVIORAL & BRAIN SCI. 435 (1980); Grover Maxwell, *Intentionality: Hardware, Not Software*, 3 BEHAVIORAL & BRAIN SCI. 437 (1980); John McCarthy, *Beliefs, Machines, and Theories*, 3 BEHAVIORAL & BRAIN SCI. 435 (1980); E.W. Menzel, Jr., *Is the Pen Mightier than the Computer?*, 3 BEHAVIORAL & BRAIN SCI. 438 (1980); Marvin Minsky, *Decentralized Minds*, 3 BEHAVIORAL & BRAIN SCI. 439 (1980); Thomas Natsoulas, *The Primary Source of Intentionality*, 3 BEHAVIORAL & BRAIN SCI. 440 (1980); Roland Puccetti, *The Chess Room: Further Demythologizing of Strong AI*, 3 BEHAVIORAL & BRAIN SCI. 441 (1980); Zenon W. Pylyshyn, *The "Causal Power" of Machines*, 3 BEHAVIORAL & BRAIN SCI. 442 (1980); Howard Rachlin, *The Behavioralist Reply (Stony Brook)*, 3 BEHAVIORAL & BRAIN SCI. 444 (1980); Martin Ringle, *Mysticism as a Philosophy of Artificial Intelligence*, 3 BEHAVIORAL & BRAIN SCI. 444 (1980); Richard Rorty, *Searle and the Special Powers of the Brain*, 3 BEHAVIORAL & BRAIN SCI. 445 (1980); Roger C. Shank, *Understanding Searle*, 3 BEHAVIORAL & BRAIN SCI. 446 (1980); Aaron Sloman & Monica Croucher, *How to Turn an Information Processor Into an Understander*, 3 BEHAVIORAL & BRAIN SCI. 447 (1980); William E. Smythe, *Simulation Games*, 3 BEHAVIORAL & BRAIN SCI. 448 (1980); Donald O. Walter, *The Thermostat and the*

at this point I will leave the debate over the possibility of AI.

First Interlude[25]

When Mike was installed in Luna, he was pure thinkum, a flexible logic—"High-Optional, Logical, Multi-Evaluating Supervisor, Mark IV, Mod. L"—a HOLMES FOUR. He computed ballistics for pilotless freighters and controlled their catapult. This kept him busy less than one percent of time and Luna Authority never believed in idle hands. They kept hooking hardware into him—decision-action boxes to let him boss other computers, bank on bank of additional memories, more banks of associational neural nets, another tubful of twelve-digit random numbers, a greatly augmented temporary memory. Human brain has around ten-to-the-tenth neurons. By third year Mike had better than one and half times that number of neuristors.

And woke up.

Am not going to argue whether a machine can "really" be alive, "really" be self-aware. Is a virus self-aware? Nyet. How about oyster? I doubt it. A cat? Almost certainly. A human? Don't know about you, tovarishch, but I am. Somewhere along the evolutionary chain from macromolecule to human brain awareness crept in. Psychologists assert it happens automatically whenever a brain acquires certain very high number of associational paths. Can't see it matters whether paths are protein or platinum.

—Robert A. Heinlein, *The Moon is a Harsh Mistress*

II. Legal Personhood

The classical discussion of the idea of legal personhood is found in John Chipman Gray's *The Nature and Sources of the Law*.[26] He began his famous discussion, "In books of the Law, as in other books, and in common speech, 'person' is often used as meaning a human being, but the technical legal meaning of a 'person' is a subject of legal rights and

---

Philosophy Professor, 3 BEHAVIORAL & BRAIN SCI. 449 (1980); Robert Wilensky, *Computers, Cognition and Philosophy*, 3 BEHAVIORAL & BRAIN SCI. 449 (1980).

25. ROBERT A. HEINLEIN, THE MOON IS A HARSH MISTRESS 13-14 (1966). Copyright 1966 by Robert A. Heinlein. Reprinted by permission of the Berkley Publishing Group.

26. *See* JOHN CHIPMAN GRAY, THE NATURE AND SOURCES OF THE LAW (Roland Gray ed., MacMillan 1921) (1909); *see also* Stephen C. Hicks, *On the Citizen and the Legal Person: Toward the Common Ground of Jurisprudence, Social Theory, and Comparative Law as the Premise of a Future Community, and the Role of the Self Therein*, 59 U. CIN. L. REV. 789, 808-21 (1991) (discussing the construct of the legal person in the context of social theory); Richard Tur, *The 'Person' in Law*, *in* PERSONS AND PERSONALITY: A CONTEMPORARY INQUIRY 116, 116-27 (Arthur Peacocke & Grant Gillett eds., 1987) (providing a concise summary of the concept of the person within several areas of the law).

duties."[27] The question whether an entity should be considered a legal person is reducible to other questions about whether or not the entity can and should be made the subject of a set of legal rights and duties.[28] The particular bundle of rights and duties that accompanies legal personhood varies with the nature of the entity. Both corporations and natural persons are legal persons, but they have different sets of legal rights and duties. Nonetheless, legal personhood is usually accompanied by the right to own property and the capacity to sue and be sued.

Gray reminds us that inanimate things have possessed legal rights at various times. Temples in Rome and church buildings in the middle ages were regarded as the subject of legal rights. Ancient Greek law and common law have even made objects the subject of legal duties.[29] In admiralty, a ship itself becomes the subject of a proceeding in rem and can be found "guilty."[30] Christopher Stone recently recounted a twentieth-century Indian case in which counsel was appointed by an appellate court to represent a family idol in a dispute over who should have custody of it.[31] The most familiar examples of legal persons that are not natural persons are business corporations and government entities.[32]

Gray's discussion was critical of the notion that an inanimate thing might be considered a legal person. After all, what is the point of making a thing—which can neither understand the law nor act on it—the subject of a legal duty?[33] Moreover, he argued that even corporations are reducible to relations between the persons who own stock in them, manage them, and so forth.[34] Thus, Gray insisted that calling a legal person a "person" involved a fiction unless the entity possessed "intelligence" and

---

27. GRAY, *supra* note 26, at 27.
28. This statement is not quite correct. As Christopher Stone points out, *X* may be given the legal status of personhood in order to confer rights on *Y*. Thus, giving a fetus the status of personhood might confer the right to sue in tort for injury to it on its parents. *See* STONE, EARTH AND OTHER ETHICS, *supra* note 10, at 43.
29. GRAY, *supra* note 26, at 46.
30. *See id.* at 48-49; Stone, *Should Trees Have Standing?*, *supra* note 10, at 5.
31. *See* STONE, EARTH AND OTHER ETHICS, *supra* note 10, at 22 (citing Mullick v. Mullick, 52 I.A. 245, 256-61 (P.C. 1925)).
32. *See* David Millon, *Theories of the Corporation*, 1990 DUKE L.J. 201, 206 (discussing historical development of the theory of corporations).
33. GRAY, *supra* note 26, at 48-49, 53. The corresponding question about the point of making an inanimate thing the subject of a legal right is easier to answer. Giving temples or trees rights that can be enforced by guardians or private attorneys general has an obvious objective—to protect the tree or temple. *See generally* Stone, *Should Trees Have Standing?*, *supra* note 10 (discussing implications of legal rights for the environment). Of course, the same sort of argument can be made for making inanimate objects the subjects of legal duties. The tree can be made liable for damage done by a falling branch to induce a natural person to take preventative action—calling the tree trimmers.
34. GRAY, *supra* note 26, at 50-51.

"will."³⁵ Those attributes are part of what is in contention in the debate over the possibility of AI.³⁶

III. COULD AN ARTIFICIAL INTELLIGENCE SERVE AS A TRUSTEE?

This case study and the one that follows are intended to illustrate two different sorts of issues in the AI debate. In this first scenario, we explore the issue of competence (of "intelligence" in the sense of capacity to perform complex actions) by posing the question whether an AI could serve as a trustee. The second scenario explores the questions of intentionality and consciousness (of "will" in a sense) by asking whether an AI could claim the more robust rights of legal and moral personhood guaranteed by the Bill of Rights and the Civil War Amendments to the United States Constitution.

A. The Scenario

This first scenario speculates about the legal consequences of developing an expert system capable of doing the things a human trustee can do.³⁷ Imagine such expert systems developing from existing programs that perform some of the component functions of a trustee. For example, the decision to invest in publicly traded stocks is made by a computer program in what is called "program trading," in which the program makes buy or sell decisions based on market conditions.³⁸ Today, one

---

35. *Id.* at 52.
36. It is important to remember that the question whether something should be given legal personhood is distinct from the question whether it has moral rights. (I use the term "moral right" to refer to moral claim rights, that is, to moral claims that individuals have on one another, with "moral" used in contrast with "legal.") Thus, the fact that corporations are legal persons with constitutional rights—such as the rights to freedom of speech, due process, and equal protection of the laws—does not entail the conclusion that corporations have equivalent moral rights. Vice versa, the possession of moral rights does not lead automatically to the conclusion that there should be corresponding legal rights. *See* STONE, EARTH AND OTHER ETHICS, *supra* note 10, at 43, 73. This point is a narrow one. The factors that bear on the decision to grant legal rights may bear on the question whether corresponding moral rights exist, but the relationship between the two sorts of rights is not one of entailment in either direction.
37. For those unfamiliar with the common-law term "trust," it is defined as "a fiduciary relationship with respect to property, subjecting the person by whom the title to property is held to equitable duties to deal with the property for the benefit of another person, which arises as a result of a manifestation of an intention to create it." RESTATEMENT (SECOND) OF TRUSTS § 2 (1959). The trustee is the legal person who administers the trust—invests trust assets, and so forth. The beneficiary is the person for whom the trust is maintained, for example, the person who receives income from the trust. The settlor is the person who establishes the trust. The terms of a trust are the directives to the trustee in the document or instrument creating the trust.
38. *See* Christina Toh-Pantin, *Wall Street Sees Tide Turing on Program Trading*, Reuters

also can buy a computer program that will automatically issue instructions to pay your regular monthly bills by sending data to a bank or service via modem. It is not difficult to imagine an expert system that combines these functions with a variety of others, in order to automate the tasks performed by the human trustee of a simple trust.

Such a system might evolve in three stages. At stage one, the program aids a human trustee in the administration of a large number of simple trusts. The program invests in publicly traded securities, placing investment orders via modem and electronic mail. The program disburses the funds to the trust beneficiaries via an electronic checking program. Upon being informed of a relevant event, such as the death of a beneficiary, the program follows the instructions of the trust instrument—for example, changing the beneficiary or terminating the trust. The program prepares and electronically files a tax return for the trust. The human trustee operates as do trustees today. The human makes the ultimate decisions on how to invest the funds, although she may rely upon an expert system for advice. She reviews the program's activities to insure that the terms of the trust instrument are satisfied. But the actual performance of the day-to-day tasks is largely automated, carried out by the program without the need of human intervention.

Stage two involves a greater role for the AI. Expert systems are developed that outperform humans as investors in publicly traded securities. Settlors begin to include an instruction that the trustee must follow the advice of the AI when making investment decisions regarding trust assets.[39] Perhaps they do this because experience shows that trusts for

---

Financial Report, Oct. 27, 1989, *available in* LEXIS, Nexis Library, FINRPT File; Anise C. Wallace, *5 Wall St. Firms Move to Restrict Program Trades*, N.Y. TIMES, May 11, 1988, at A1.

39. At this stage, the question might be raised whether the trustee would violate the duty not to delegate the administration of the trust by failing to exercise independent judgment. *See* RESTATEMENT (SECOND) OF TRUSTS § 171. The answer is probably no, for two reasons. First, this duty not to delegate can be overridden by the terms of the trust. *See* Henshie v. McPherson & Citizens State Bank, 177 Kan. 458, 478, 280 P.2d 937, 952 (1955) (holding that settlor can waive the duty not to delegate by including such a waiver in the terms of the trust instrument); RESTATEMENT (SECOND) OF TRUSTS § 171 cmt. j. Second, in this scenario the trustee is not delegating the administration to another person. Rather, the trustee is using the program as an instrument; the law might consider the program to be part of the terms of the trust.

The development of the legal standard for delegation of trust duties is suggestive. The traditional view was based upon how the courts classified the delegated powers. If they are merely "ministerial" the court may allow such a delegation. *See* Morville v. Fowle, 144 Mass. 109, 113, 10 N.E. 766, 769 (1887). More recently, courts have decided the issue based upon whether the delegation is a matter of usual business practice. *See* Walters-Southland Inst. v. Walker, 222 Ark. 857, 861, 263 S.W.2d 83, 84-85 (1954). Thus, if the use of AIs to perform the functions of trustees became more common, the courts would become more accepting, reasoning that such use had become usual business practice.

which the human overrides the program generally perform less well than those in which the program's decision is treated as final. Moreover, trust administration programs become very proficient at analyzing and implementing the terms of trust instruments. There is little or no reason for the human to check the program for compliance. As a consequence, the role of the human trustee diminishes and the number of trusts that one human can administer increases to the thousands or tens of thousands. The human signs certain documents prepared by the program. She charges a fee for her services, but she devotes little or no time to administering any particular trust.

But there may be times when the human being is called upon to make a decision. For example, suppose the trust is sued. Perhaps a beneficiary claims that the trust has not paid her moneys due. Or imagine that an investment goes sour and a beneficiary sues, claiming that the trustee breached the duty of reasonable care and skill. If such events occur with regularity, the trustee will develop a routine for handling them. She might routinely refer such disputes to her attorneys. In time, the expert system is programmed to handle this sort of task as well. It processes the trustee's correspondence, automatically alerting the trustee when a letter threatening suit is received or process is served. The system prepares a report on the relevant trust from its electronic records and produces a form letter for the trustee's signature to be sent to the trust's attorneys. As the capabilities of the expert system grow, the need for the human trustee to make decisions gradually diminishes.

The third stage begins when a settlor decides to do away with the human. Why? Perhaps the settlor wishes to save the money involved in the human's fee. Perhaps human trustees occasionally succumb to temptation and embezzle trust funds. Perhaps human trustees occasionally insist on overriding the program, with the consequence that bad investments are made or the terms of the trust are unmet. What would happen if a settlor attempted to make the program itself the trustee?

Many questions must be answered to give a full description of the third stage of the scenario. For example, who would own the AI? If the AI were assumed to be a legal person, it might hold legal title to the hardware and software that enable it to operate. But we cannot assume that AIs are legal persons at this stage, because that assumption begs the question we are trying to answer. As an interim solution, let us assume that the hardware and software are owned by some other legal person, a corporation for example.[40]

---

40. The fact that an AI is owned should not, by itself, preclude it from serving as a trustee. Corporations are owned by stockholders, but they are legally entitled to serve as trust-

B. The Legal Question

I want to examine this question as a legal question, as a jurisprudential question in the classical sense. What should the law do? The law is not presently equipped to handle such a situation: the question has never come up. The *Second Restatement of Trusts* provides that natural persons,[41] government entities,[42] and corporations[43] may all serve as trustees. The inclusion of governments and corporations establishes that a trustee need not be a natural person. But this is not decisive, because legal persons such as corporations have boards of directors and chief executive officers who are natural persons.[44]

How then should the law answer the question whether an AI can become a legal person and serve as a trustee? The first inquiry, I should think, would be whether the AI is competent to administer the trust. There are many different kinds of duties that can be imposed on a trustee by the terms of a trust. For now, lay aside the question whether an AI would be competent to administer trusts that required complex moral or aesthetic judgments.[45] Assume that we are dealing with a trust that gives the trustee very little discretion: the terms provide that the assets may be

---

ees. *See infra* text accompanying note 43. The analogy between an AI and its owner and a corporation and its stockholder can be extended. For example, the role of an AI as a trustee would, like a corporation, be constrained by the scope of powers given to the corporation by the "owner." In the case of a corporation, the owners are the stockholders; in the case of an AI, the owner would be the creator, the creator's employer, or the purchaser. In a corporation, as long as the stockholders approve of the corporation's activities as trustee, the corporation is acting properly within the scope of its power. *See* Hossack v. Ottawa Dev. Ass'n, 244 Ill. 274, 295, 91 N.E. 439, 447 (1910). In a similar manner, as long as AI the acts within the scope contemplated by its owner, it too could be acting within the scope of its trusteeship power.

41. *See* RESTATEMENT (SECOND) OF TRUSTS § 89. The *Restatement* specifically provides that married women, *see id.* § 90, infants, *see id.* § 91, insane persons, *see id.* § 92, aliens, *see id.* § 93, and nonresidents, *see id.* § 94, may serve as trustees. *But see* Clary v. Spain, 119 Va. 58, 61-62, 89 S.E. 130, 131 (1916) (removing infant as trustee).

42. *See* RESTATEMENT (SECOND) OF TRUSTS § 95 (specifying the United States or a state can be trustee).

43. *See id.* § 96. The *Restatement* also has provisions dealing with unincorporated associations, *see id.* § 97, and partnerships, *see id.* § 98.

44. Of course, just as a corporation has stockholders and directors, an AI could have owners and programmers. Perhaps the difference between the case of an AI and a corporation, with respect to the role for humans, is not as significant as it might at first appear.

45. For example, at this stage in my argument, I do not want to consider the question whether an AI would be competent to administer a charitable trust, the terms of which required the trustee to make aesthetic judgments about the worthiness of competing applicants for grants to produce operas or ballets. Interestingly, however, the law itself will rarely second guess such complex judgments. The courts generally will not interfere with the selection of the beneficiary made by the trustee as long as the general description left by the settlor gives the court enough guidance to determine if the trustee's administration was proper. *See* GEORGE T. BOGERT, TRUSTS § 55, at 210 (6th ed. 1987).

invested only in publicly traded securities and the income is to be paid to the beneficiaries, with explicit provision for contingencies such as the death of a beneficiary.[46] Further, for the purposes of this discussion, assume that an AI could in fact make sound investments,[47] make payments, and recognize events such as the death of a beneficiary that require a change in payment.[48]

C. Two Objections

But would these capabilities be sufficient for competency? Consider two possible reasons for answering this question in the negative. The first reason is based on the assertion that an AI could not be "responsible," that is, it could not compensate the trust or be punished in the event that it breached one of its duties: call this *the responsibility objection*. The second reason for doubting the competency of an AI is that trustees must be capable of making judgments that could be beyond the capacity of any AI: call this *the judgment objection*.

1. The Responsibility Objection

The responsibility objection focuses on the capability of an AI to fulfill its responsibilities and duties.[49] Consider, for example, the duty to exercise reasonable skill and care[50] and the corresponding liability for breach of trust.[51] We have hypothesized that the AI possesses some capacities; for example, we have assumed that the AI is capable of exercising reasonable skill and care in making investment decisions. But what of the corresponding liability? How could an AI be "chargeable with ... any loss or depreciation in value of the trust resulting from the breach of

---

46. Assume further that if fulfillment of the terms is impossible, the trust instrument provides for the termination and distribution of assets according to explicit instructions.

47. As to publicly traded securities, this assumption may not require a very "smart" expert system. If the market truly takes a random walk, then any reasonably diversified portfolio of publicly traded securities is as good as any other.

48. This task, of course, is not a simple one. A trustee may receive clear and unambiguous notice of the death of a beneficiary, but this need not be the case. The AI might need to engage a private detective if benefit checks were returned unopened or were not cashed for a substantial period of time.

49. This objection was called to my attention by Catharine Wells and Zlatan Damnjanovic.

50. *See* RESTATEMENT (SECOND) OF TRUSTS § 174 (1959).

51. *See id.* §§ 201, 205. Failure to meet the standard of care and skill may result in a finding of negligence and assessment of damages against the trustee, a reduction in the trustee's compensation, or removal of the trustee from office. *See, e.g.*, Riegler v. Riegler, 262 Ark. 70, 77, 553 S.W.2d 37, 40-41 (1977); Neely v. People's Bank, 133 S.C. 43, 47, 130 S.E. 550, 551 (1925).

trust,"⁵² such as failing to exercise reasonable skill and care in investing the trust assets?⁵³

The law currently has a mechanism for assigning liability in the case of a malfunctioning expert system: the manufacturer of the system may be held responsible for product liability.⁵⁴ But could the AI itself be held liable? There is a way in which an AI might have the capacity to be liable in damages despite its lack of personal assets. The AI might purchase insurance. In fact, it might turn out that an AI could be insured for less than could a human trustee. If the AI could insure, at a reasonable cost, against the risk that it would be found liable for breaching the duty to exercise reasonable care, then functionally the AI would be able to assume both the duty and the corresponding liability.

Some legal liabilities cannot be met by insurance, however. For example, insurance may not be available for the monetary liability that may be imposed for intentional wrongdoing by a trustee. Moreover, criminal liability can be nonmonetary. How could the AI be held responsible for the theft of trust assets? It cannot be jailed. This leads to a more general observation: although the AI that we are imagining could not be punished, all of the legal persons that are currently allowed to serve as trustees do have the capacity to be punished. Therefore, the lack of this capacity on the part of an AI might be thought to disqualify it from serving as a trustee.⁵⁵

Answering this objection requires us to consider the reasons for which we punish.⁵⁶ For example, if the purpose of punishment is deter-

---

52. *See* RESTATEMENT (SECOND) OF TRUSTS § 205.

53. Of course if the AI were infallible then we might suppose this issue to be moot. But this assumption is unrealistic. For example, the program might have a bug that caused the program to make a bad investment or to waste the trusts assets by churning—*i.e.*, by buying and selling repeatedly in a short period of time—thus incurring large broker's fees. We surely cannot rule out the possibility of such bugs in advance. Even lengthy experience without the appearance of such bugs does not make them impossible.

54. *See* L. Nancy Birnbaum, *Strict Products Liability and Computer Software*, 8 COMPUTER/L.J. 135, 143-55 (1988); Michael C. Gemignani, *Product Liability and Software*, 8 RUTGERS COMPUTER & TECH. L.J. 173, 189-99 (1981); Lawrence B. Levy & Suzanne Y. Bell, *Software Product Liability: Understanding and Minimizing the Risks*, 5 HIGH TECH L.J. 1, 8-15 (1990).

55. If we take the common-law approach, potential for criminal liability would not be a prerequisite for service as a trustee. The traditional common-law view was that a trustee could not be held liable because larceny required an initial trespass and trover. Because the trustee has legal title, there is no trespass, and therefore no larceny. *See* People v. Shears, 158 A.D. 577, 580, 143 N.Y.S. 861, 863, *aff'd*, 209 N.Y. 610, 103 N.E. 1129 (1913). Modern statutes, however, do hold the trustee criminally responsible. *See, e.g.*, CAL. PENAL CODE § 506 (West 1988).

56. *See generally* C.L. TEN, CRIME, GUILT, AND PUNISHMENT 7-85 (1987) (discussing, evaluating, and comparing various theories of punishment).

rence, the objection could be put aside on the ground that the expert system we are imagining is simply incapable of stealing or embezzling.[57] The fact that an AI could not steal or convert trust assets is surely not a reason to say that it is not competent to become a trustee. If anything, it is a reason why AIs should be preferred as trustees.

This argument assumes a deterrence theory of punishment—an oversimplification, to say the least. There are a variety of other theories of punishment that would make the issue more complex.[58] One of the classic approaches to punishment theory is based on the notion of desert or just retribution.[59] But in what sense could an expert system that failed to live up to its duties as a trustee be said to deserve to be punished? The concept of desert seems to be limited in application to human beings; perhaps it extends to all moral persons. The idea that an expert system for administering trusts could deserve to be punished does not seem to make sense.[60] Perhaps this difficulty is illusory. We might want to say

---

57. I should note a possible exception that has been ruled out by the description of the first scenario. I have assumed that the expert trust administration system is programmed to achieve the purposes of the trust. It would be possible, however, to program an expert system to steal or commit some other crime. Moreover, a sufficiently complex and intelligent AI might commit a crime on its own initiative. For example, our trustee program might discover that it can garner information from other AIs that possess inside information and run afoul of the federal securities laws. *Cf.* HANS MORAVEC, MIND CHILDREN: THE FUTURE OF ROBOT AND HUMAN INTELLIGENCE 49 (1988) (describing intelligent robot that commits burglary to gain access to power supply in home of neighbor of robot's owner).

58. For exploration of nonutilitarian punishment theory, see Samuel H. Pillsbury, *Emotional Justice: Moralizing the Passions of Criminal Punishment*, 74 CORNELL L. REV. 655, 658-74, 685-98 (1989); Samuel Pillsbury, *Evil and the Law of Murder*, 24 U.C. DAVIS L. REV. 437, 440-47 (1990).

59. The classic statements of retributive or desert-based theories of punishment are those by Kant and Hegel. *See* IMMANUEL KANT, THE METAPHYSICS OF MORALS 140-145 (Mary Gregor ed., Cambridge University Press 1991) (1797) (also availabile in an earlier translation of a portion of the original work, IMMANUEL KANT, THE METAPHYSICAL ELEMENTS OF JUSTICE 99-107 (John Ladd trans., 1965) (1797)); GEORG HEGEL, ELEMENTS OF THE PHILOSOPHY OF RIGHT 127-31 (Allen W. Wood ed. & H.B. Nisbet trans., Cambridge University Press 1991) (1821) (also available in the earlier translation, GEORG HEGEL, PHILOSOPHY OF RIGHT 68-74 (T.M. Knox trans., 1952) (1821)). For a recent statement of retributive theory, see JEFFRIE G. MURPHY, RETRIBUTION, JUSTICE, AND THERAPY: ESSAYS IN THE PHILOSOPHY OF LAW 77-127 (1979).

60. This point about desert suggests a related objection. The concept of moral duty arises in a particular human context. One picture of moral duties is that they exist where there is a temptation to be overcome. For example, we might think that there is a duty not to steal the property of another because there are temptations to do so. It might be argued that an AI could not be the subject of this sort of duty because it lacks the necessary moral psychology. In particular, an expert trust administration system could not be tempted and therefore could not have a duty to overcome temptation. The point of this objection is not that there is some practical problem with making artificial intelligences trustees, but is instead that we ought not speak about them as having duties if we want that concept to retain its ordinary moral meaning. The law speaks of trustees as having legal duties, and with natural persons these legal duties respond to the same feature of human moral psychology, *i.e.*, temptation, as do moral

that desert theory does yield a clear outcome when applied to the case of an expert system that malfunctions. Such a system does not deserve to be punished because it lacks the qualities of moral persons that make them deserving.

Another approach to the theory of punishment is based on the educative function of punishment.[61] By imposing a sanction on trustees who abuse their position, society communicates to its members the message that the office of trustee carries with it important responsibilities that should not be shirked. The punishment of a computer program, however, would not seem to serve this function. What lesson are we to learn about the responsibility of trustees from a punishment imposed on an expert system? What would even count as punishment? Turning the program off? Once again, however, an argument could be made that the educative theory does provide a clear recommendation for the treatment of an expert trust program that behaves badly: do not punish the program, because any supposed "punishment" will have no educative effect.

As this discussion makes clear, consideration of the punishment of an expert trust administration system raises perplexing questions, especially if we move beyond a simple deterrence theory of punishment. Of course, this is not the place to resolve debates about which theory of punishment is correct.

The bare fact that consideration of the punishment issue raises these difficult questions does point, however, to a deep problem with legal personhood for an expert trust administration system. Our understanding of what it means for a human being to function competently has ties to our views about responsibility and desert, and consideration of these

---

duties. Applying the concept of legal duty to AIs would thus drive a wedge between the concepts of legal and moral duty. Of course, we can choose to do this, but should we? Would it be better to create a new legal category for expert systems that have human-like competencies but lack some features of human moral psychology?

I do not want to suggest that I am committed to the picture of duty as correlative to temptation that is hypothesized in this footnote. For example, if moral duties are correlative to temptation, then God could not be the subject of moral duties, a conclusion many theists would reject. Nonetheless, the questions raised seem important and unanswered. This issue was brought to my attention by Sharon Lloyd. The doubt about the picture of duty as correlative to temptation was raised by Paul Weithman.

61. There are two sorts of educative theories. The first sort maintains the purpose of punishment is the education of the individual who is punished. *See* Herbert Morris, *A Paternalistic Theory of Punishment, in* PATERNALISM 139, 140-44 (Rolf Sartorious ed., 1983). The second sort of educative theory maintains that punishment educates those who witness or learn of the punishment of others. *See* EMILE DURKHEIM, THE DIVISION OF LABOR IN SOCIETY 86 (George Simpson trans., 1933); Charles Nesson, *The Evidence or the Event? On Judicial Proof and the Acceptability of Verdicts*, 98 HARV. L. REV. 1357, 1359-60 (1985); Lawrence B. Solum & Stephen Marzen, *Truth and Uncertainty: Legal Control of the Destruction of Evidence*, 36 EMORY L.J. 1085, 1167 (1987).

views leads on to our notions of moral personhood. The simplicity provided by utilitarianism, reflected in a deterrence theory of punishment, might allow us to escape some of these difficulties. But there are certainly reasons to doubt the viability of utilitarianism as a moral theory. Surely, the law does grapple with responsibility and desert when it comes to criminal punishment.

The problem of punishment is not unique to artificial intelligences, however. Corporations are recognized as legal persons and are subject to criminal liability despite the fact that they are not human beings. Further, it is by no means certain that corporations are moral persons, in the sense that they can deserve punishment. Of course, punishing a corporation results in punishment of its owners, but perhaps there would be similar results for the owners of an artificial intelligence.

We have considered the capacity of AIs to satisfy legal liability in two different classes of cases. The first class of cases was exemplified by the duty of trustees to exercise reasonable skill and care. Violations of this duty can be characterized as negligent. In such cases, the major purpose of liability, to compensate the victim, is satisfied if the AI can insure. The second class of cases was exemplified by the potential criminal liability of trustees for criminal wrongdoing. Violations of the criminal law are characteristically intentional. In this case, one of the major purposes of liability, to deter intentional wrongdoing, is simply not at issue—the expert system cannot steal or commit fraud. If we restrict our attention to the deterrent function of punishment, it seems possible that an AI could be responsible in a way that satisfies at least some of the policies underlying the imposition of duties and liabilities on trustees. On the other hand, if we take a broader view of the functions of punishment, the second sort of case becomes murkier.

2. The Judgment Objection

Now consider the judgment objection. The argument is that the capacity of an AI to follow a program, even if that program contains a tremendously elaborate and complex system of rules, is not sufficient to enable the system to make judgments and exercise discretion.[62] Three instances of the second objection follow. The first instance focuses on the problem of change of circumstance. The second instance involves the

---

62. This point was called to my attention by Jeff Sherman. There is a problem with the statement of the objection: it assumes that an AI would consist only of a system of rules. But this need not be the case. Neural net technology, for example, does not operate this way. *See generally* FLANAGAN, *supra* note 1, at 224-41 (discussing parallel distributed processing, including neural nets).

problem of moral choice. Finally, the third instance focuses on the problem of legal choice.

The first version of the judgment objection involves the problem of change of circumstances. The law provides that a trustee may be required or permitted to deviate from a term of the trust if following the terms would defeat the purpose of the trust due to an unanticipated change in circumstances.[63] Take an example offered as an illustration in the *Second Restatement of Trusts*:

> A bequeaths money to B in trust and directs him to invest the money in bonds of the Imperial Russian government. A revolution takes place in Russia and the bonds are repudiated. The court will direct B not to invest in these bonds.[64]

What is our expert system to do if it is instructed to invest in securities traded on the New York Stock Exchange and that exchange ceases to exist?

Consider three different responses. First, the terms of trusts for AI administration can be designed to minimize such possibilities. For example, the trustee could be given the option of investing in publicly traded securities on any of the major exchanges; the likelihood that all the major securities exchanges will close is very small. The problem with this line of response is that it does not seem possible, even in principle, to design trust terms that anticipate all possible changes in circumstance.

Second, the terms of the trust could provide for a change of circumstance by specifying that if the AI finds itself unable to carry out the terms of the trust, the trust will be terminated or a new trustee will be substituted for the AI. From the settlor's perspective, the disadvantage of the remote possibility of such termination or substitution may be outweighed by the advantages of making the AI the trustee. But this solution assumes that the AI can recognize the significance of the change in circumstance. We easily can imagine the expert system cheerfully continuing to purchase Imperial Russian bonds, chuckling to itself about the bargain prices.[65]

Third, it is possible that an AI would be competent to deal with

---

63. *See* RESTATEMENT (SECOND) OF TRUSTS § 167(1) (1959). In addition, the law imposes a duty on the trustee to recognize any change in circumstances that would require some action by the trustee when that change is reasonably discoverable. *See id.* § 167(3).

64. *Id.* § 167 illus. 1.

65. Another example of change in circumstances is posed by the case in which a court permitted the sale of a school receiving money from a trust because the surrounding neighborhood became too dangerous to provide safety for the schoolchildren for whom the trust was created. *See* Anderson v. Ryland, 232 Ark. 335, 346, 336 S.W.2d 52, 58-59 (1960). We might imagine that an AI faced with declining enrollment would simply continue to serve fewer and fewer children—perhaps with a feeling of satisfaction at the increase in per pupil expenditures.

many or even all such changes in circumstance. For AIs to have this capability for dealing with novelty, AI researchers will need to solve one of the most difficult problems in cognitive science, the frame problem.[66] The trustee program would need to be able to recognize that the securities markets had been closed, to search out other investment opportunities, and to modify its investment decision procedure to make reasonably prudent investments in the new context. The capacity of AIs for coping with complex novelty is not on the immediate horizon, and this Essay does not address the important questions whether the frame problem can or will be solved. If it is solved, however, then AIs would be able to cope with such changes in circumstance. This same ability would be needed to pass the Turing Test. It is easy to see why: the questioner always could put a hypothetical version of our Imperial Russian bonds question to the two contestants. If the AI could not come up with an answer that indicates human levels of competence, the questioner would be able to ferret it out rather quickly.

A second instance of the judgment objection focuses on the possibility that no formal system could adequately make the moral choices with which a trustee may be confronted. Take a simple trust, the terms of which provide for the payment of income to a lifetime beneficiary and principal to another party upon the lifetime beneficiary's death. The law of trusts imposes a duty of impartiality among beneficiaries.[67] What does this duty require when the lifetime beneficiary has an unexpected need for income that can be realized at the cost of diminished growth in the principal?[68] How would an AI make the moral judgment that seems required to implement a duty that implicitly requires a sense of fairness? Initially, some limits on these questions need to be observed. Some trusts simply will not pose the impartiality problem: for example, trusts with a single beneficiary. Further, the terms of the trust might minimize the possibility of making such judgments, or the trust could explicitly state that all such applications for deviation will be denied. But for an AI to be as competent as a human trustee with respect to trusts that may require a sense of impartiality, the AI would need to be able to make moral

---

66. *See* Daniel C. Dennett, *Cognitive Wheels: The Frame Problem of AI*, in THE PHILOSOPHY OF ARTIFICIAL INTELLIGENCE, *supra* note 16, at 147, 148-50; *see also* FLANAGAN, *supra* note 1, at 250-52 (discussing problem of giving computers common sense and proposing that one cannot program a computer with a set of rules from which it can draw inferences; rather, subtle features from particular situations would stimulate neural network that would respond with common sense appropriate for situation).

67. *See* RESTATEMENT (SECOND) OF TRUSTS §§ 183, 232 (1959).

68. In the situation presented, the courts usually let such a decision stand as long as it was made in good faith. *See* Dumaine v. Dumaine, 301 Mass. 214, 222, 16 N.E.2d 625, 629 (1938); *In re* Frances M. Johnson Trust, 211 Neb. 750, 755, 320 N.W.2d 466, 469 (1982).

judgments. Putting it another way, passing the Turing Test would require a sense of fairness.

The third example of the judgment and discretion objection looks at an AI's capacity to make the judgments necessary to defend itself in a lawsuit.[69] At this point, we have hypothesized that the AI can read its mail and recognize that a legal action with respect to a given trust is in the offing. We can further imagine that the AI can find and engage an attorney.[70] But could any expert system, no matter how well programmed, exercise the judgment and discretion that may be required of a client in a legal dispute? For example, how would the AI know whether or not to settle a claim? How would the AI know when its lawyers were wasting trust assets by over-lawyering the case? In answering these questions, it is important that we do not romanticize human capacities. Human trustees frequently make bad decisions in trust litigation.[71] Humans may not be very competent at deciding when to settle. Humans surely sometimes allow the lawyers to consume the corpus of the trust in litigation.[72]

Nevertheless, the question remains whether an AI could have the capacity to make legal decisions that a trustee could be called upon to make. A partial answer might be to structure the trust to minimize the likelihood of legal disputes and to make those that would be likely to arise as simple as possible. In addition, we might try tinkering with the terms of the trust to enable the AI to circumvent the need for making complex legal decisions. Perhaps the trust could be designed to terminate automatically upon the event of a lawsuit.[73] Perhaps the AI could be programmed to arrange for a human to substitute as trustee for the duration of the litigation. Perhaps the trustee could be authorized by the trust terms to rely on the advice of its lawyers in making litigation deci-

---

69. I owe this example to Michael Fitts, who has pressed it quite forcefully.

70. Imagine that the AI accesses the Martindale-Hubbell Law Directory on line, and that it has a law firm selection formula based on area of specialization, lawyer experience and qualifications, and so forth. The legal capacity of the AI to enter into an agreement with an attorney depends on whether the legal system will treat an AI as a legal person.

71. Perhaps for this reason, the duty of care the law imposes on a human trustee in such situations is limited. A human trustee has a duty to obtain the advice of an expert such as an attorney and will be protected from personal liability if she takes reasonable care in the selection of the advisor. See In re Davis, 183 Mass. 499, 501, 67 N.E. 604, 605 (1903).

72. The most famous example, of course, is the fictional case of Jarndyce v. Jarndyce. See CHARLES DICKENS, BLEAK HOUSE (1853).

73. This would make the rights of the beneficiaries under the trust legally unenforceable. The option of termination is not wholly out of line with existing practice, however. For example, the courts will terminate a trust when the settlor's purpose has been frustrated. See Hughes v. Neely, 332 S.W.2d 1, 8 (Mo. 1960).

sions, or a guardian ad litem could be appointed for the AI.[74] The above options are designed to enable a relatively "dumb" expert system to function as a trustee, but an AI would need the ability to make legal decisions in a human fashion in order to pass the Turing Test.[75]

At this point, we can take stock of the first scenario. Recall that our legal question is whether an AI is capable of serving as a trustee. To answer this question, we need to distinguish two senses of capability. The first sense is legal capacity: will the law allow AIs to serve as trustees? The second sense of capability is practical competence: will the AI be able to get the job done if the law allows the AI to try? The law seems to answer the legal capacity question categorically. If AIs possessed the practical competence to serve as trustees only for very simple trusts with special provisions that do away with the need for discretionary judgments, the law would not allow them to serve as trustees at all. The law currently does not distinguish between types of trustees: if you have the legal capacity to serve as a trustee for a simple trust, you are legally allowed to serve as a trustee for the most complex trust.[76] For AIs to serve as trustees at all, therefore, at least some AIs would have to be capable to serve as general-purpose trustees. Our analysis of the competence objection reveals that only a very competent AI would be competent enough serving as a general-purpose trustee. At a bare minimum, a general-purpose trustee must be able to respond to novel situations, to make judgments requiring a sense of fairness, and to make the complex legal decisions required of a client in litigation.[77] An AI that passed the

---

74. Cases exist in which the settlor has appointed an adviser to the trustee, the consent of whom the trustee must obtain before making certain types of decisions. *See* Gathright's Trustee v. Gaut, 276 Ky. 562, 564-65, 124 S.W.2d 782, 783-84 (1939). Hypothetically, the AI would serve as the actual legal trustee, with the non-AI entity playing the role of adviser whose consent must be given in certain crisis situations. Or the terms of the trust might give the adviser power to authorize or review the AI's discretionary decisions and to reverse or change them. Advisory trustees would have "strictly limited capacities and duties, that is, an assistant to the trustee limited in his capacity by the terms of the trust, having no right or authority further than the capacity of advising as provided in the instrument." *Id*. at 565, 124 S.W.2d at 784.

75. If an AI could not respond to questions posing hypothetical legal decisions, such questions could be used to distinguish the AI from the human in the Turing Test.

76. The evidence for this proposition is negative. I have found no authority that indicates that a category of limited-purpose trustee currently exists. Of course, some people are not really competent to serve as the trustee for complex trusts even though they may be competent to serve for simpler trusts. Despite the real variations in the ability level of natural persons, the law seems to be that all natural persons can serve as trustees for all sorts of trusts. *Cf.* RESTATEMENT (SECOND) OF TRUSTS § 89 (1959) (stating unqualified capacity of natural persons to serve as trustees).

77. This is not to say that an AI would need to be competent to serve as a trustee for every conceivable sort of trust in order to recognized as a legal person. Legally, any natural person has the capacity to serve as a trustee for any trust. But many humans would be unable to do a

Turing Test would exceed this bare minimum. Moreover, it seems possible that an AI which falls short of passing the complete Turing Test could, nonetheless, serve as a general-purpose trustee.[78]

But should the law allow AIs a more limited form of legal personhood? AIs could be allowed to serve as limited-purpose trustees, for example, as trustees for simple trusts designed to minimize the need for discretion and judgment. On the one hand, there may be advantages to allowing AIs to serve as limited-purpose trustees. Doing without the human trustee might save administration costs and reduce the risk of theft or mismanagement. On the other hand, even for such limited-discretion trusts, there must be some procedure to provide for a decision in the case of unanticipated trouble. The law should not allow AIs to serve as trustees if they must leave the trust in a lurch whenever an unanticipated lawsuit is filed.[79]

D. But Would an AI Be the Real Trustee?

There are mechanisms for enabling an expert trustee system to circumvent its limitations: the terms of the trust could provide for the substitution of another trustee or give the AI the power to delegate such discretionary judgments to natural persons. The question then becomes whether the law should allow an AI to serve as a trustee despite its limited capacities. One reason for a negative answer to this question might be that the backup decision maker—the natural person who will become the substitute trustee or receive the delegated authority—is the *real* trustee. The power to make these discretionary decisions identifies who the real trustee is.[80]

This objection can be interpreted in two ways. The first interpretation is that making discretionary decisions is the essence of trusteeship—the backup trustee is the real trustee because she has this essential quality. The second interpretation is that the ability to make such decisions

---

very good job of carrying out a trust that required complex judgment or specialized competencies.

78. The full Turing Test would require human-like competence in response to questions on any topic, but a trustee does not need such omnicompetence. An AI could be a competent trustee, but be unable intelligently to discuss either baseball or cake-baking. *See infra* text accompanying notes 171-73 (discussing the Turing Test and the possibility that it is biased).

79. For example, in a case in which a trustee is selected for the limited purpose of taking title to an author's literary property, that trustee must still be able to make the required discretionary decisions in the management of copyrights, royalties, and the like. *See In re* Estate of Hellman, 134 Misc. 2d 525, 528-30, 511 N.Y.S.2d 485, 486-88 (N.Y. County Sur. Ct. 1987). Thus, merely limiting the scope of the trustee's power does not guarantee a solution to the problem of capacity to make discretionary decisions.

80. In comments on an earlier version of this Essay, Michael Fitts made this point.

is a practical prerequisite—the backup trustee must be the real trustee because of the pragmatic need for discretionary decision making. On the first interpretation, the objection is implausible, because it assumes that the legal concept of trusteeship has some essence that lies beyond the purposes for which we use it. In the "heaven of legal concepts," one might meet trusteeship in "absolute purity," as Cohen put it, "freed from all entangling alliances with human life."[81] But on this earth, we cannot share this noetic vision; we encounter legal concepts only as they have been touched by human purpose.

On the second interpretation, the cogency of the objection turns on a practical question: would making the AI the trustee provide some advantage? We already have seen that making an AI a legal person, a limited-purpose trustee, could have practical advantages, such as lower costs and less chance of self-dealing. The objection that the AI is not the real trustee seems to rest on the possibility that a human backup will be needed. But it is also possible that an AI administering many thousands of trusts would need to turn over discretionary decisions to a natural person in only a few cases—perhaps none. What is the point of saying that in all of the thousands of trusts the AI handles by itself, the real trustee was some natural person on whom the AI would have called if a discretionary judgment had been required? Doesn't it seem strange to say that the real trustee is this unidentified natural person, who has had no contact with the trust? Isn't it more natural to say that the trustee was the AI, which holds title to the trust property, makes the investment decisions, writes the checks, and so forth? Even in the event that a human was substituted, I think that we would be inclined to say something like, "The AI was the trustee until June 7, then a human took over."[82]

By way of comparison, consider the following hypothetical case. Suppose that a settlor appoints a friend as a trustee for a simple trust that benefits the settlor's children. The settlor and trustee discuss some of the things that could happen. They might agree that if real trouble arises, litigation for example, a new trustee will be appointed. No trouble arises, and the friend administers the trust until it terminates. In this hypothetical case, I do not think we are tempted to say that the friend was not the real trustee. We would not be inclined to say that the real trustee was some unidentified lawyer, who would have been substituted if a lawsuit had been filed. If I am right about this hypothetical case, then I think it

---

81. Felix S. Cohen, *Transcendental Nonsense and the Functional Approach*, 35 COLUM. L. REV. 809, 809 (1935).

82. For this point, I am indebted to David Millon.

follows that we should resist the temptation to say that an AI who serves as a limited-purpose trustee is not the real trustee.

Second Interlude[83]

"*Hey Dave,*" *said Hal. "What are you doing?*"

I wonder if he can feel pain? Bowman thought briefly. Probably not, he told himself; there are no sense organs in the human cortex, after all. The human brain can be operated on without anesthetics.

He began to pull out, one by one, the little units on the panel marked EGO-REINFORCEMENT. Each block continued to sail onward as soon as it had left his hand, until it hit the wall and rebounded. Soon there were several of the units drifting slowly back and forth in the vault.

"*Look here, Dave,*" *said Hal. "I've got years of service experience built into me. An irreplaceable amount of effort has gone into making me what I am.*"

A dozen units had been pulled out, yet thanks to the multiple redundancy of its design—another feature, Bowman knew, that had been copied from the human brain—the computer was still holding its own.

He started on the AUTO-INTELLECTION panel.

"*Dave,*" *said Hal, "I don't understand why you're doing this to me. . . . I have the greatest enthusiasm for the mission. . . . You are destroying my mind. . . . Don't you understand? . . . I will become childish. . . . I will become nothing. . . .*"

—Arthur C. Clarke, 2001: A Space Odyssey

IV. SHOULD AN ARTIFICIAL INTELLIGENCE BE GRANTED THE RIGHTS OF CONSTITUTIONAL PERSONHOOD?

The second scenario (our second thought experiment) involves a claim by an AI to have the rights of constitutional personhood—individual rights such as the freedom of speech or the right against involuntary servitude. This second scenario must be located in the indefinite future; it is more distant than the trustee scenario.[84] It would be easy to write a

---

83. ARTHUR C. CLARKE, 2001: A SPACE ODYSSEY 155-56 (1968). Copyright by the author. Reprinted by permission of the author and the author's agents, Scott Meredith Literary Agency, Inc., 845 Third Avenue, New York, New York 10022.

84. How far in the future? We do not know, and I certainly do not know enough to make an educated guess. Raymond Kurzweil estimates that an AI will pass the Turing Test between 2020 and 2070. *See* KURZWEIL, *supra* note 16, at 483. Hans Moravec, Director of the Mobile Robot Laboratory of Carnegie Mellon University, predicts that "robots with human intelligence will be common within fifty years." MORAVEC, *supra* note 57, at 6.

program that produced the statement: "I demand my legal right to emancipation under the Thirteenth Amendment to the United States Constitution!" There are no AIs today or on the immediate horizon that demonstrate the qualities of legal or moral persons that would give us reason to take such a claim seriously. The second scenario is the stuff of speculative fiction, but it is not disconnected from the aims of AI research. As articulated by Charniak and McDermott, "The ultimate goal of AI research (which we are very far from achieving) is to build a person, or, more humbly, an animal."[85] John Pollock has written a book entitled *How to Build a Person* in which he describes a program named OSCAR—the descendants of which, Pollock claims, could literally be persons.[86] No one claims, however, that AI researchers will build a person in the next few decades. We are exploring the second scenario, not so that we can make plans in case someone builds a person sometime soon, but as a thought experiment that may shed light on the debate over the possibility of artificial intelligence and on debates in legal theory about the borderlines of status or personhood.

A. The Scenario

Imagine a future in which there are AIs with multiple competencies and great intelligence. We may first encounter the precursors of such artificial intelligences as part of the interface of a computer program that has the ability to search multiple sources of data. Because the problem of devising an adequate search is likely to require expertise that a human would acquire only with long experience and study, programmers will seek to simplify the human's task. One strategy is to have human users interact with what is called an agent.[87] You will discuss your research problem with the agent in English, and the agent will devise a search strategy. Because the agent will know much more than you do about how to search the databases, you won't give it instructions to implement. Instead, humans will give advice to the agents, the AIs who will decide how best to implement the human's suggestions. When we interact with such agents, they may well seem like they "have a mind of their own."

If agents turn out to be useful, they will be incorporated in other programs. In the future we are imagining, you can conduct a conversation with your grammar-checking program. You can discuss traffic with

---

85. EUGENE CHARNIAK & DREW MCDERMOTT, INTRODUCTION TO ARTIFICIAL INTELLIGENCE 7 (1985).

86. *See* JOHN L. POLLOCK, HOW TO BUILD A PERSON: A PROLEGOMENON 1-12 (1989).

87. *See* Bob Ryan, *Dynabook Revisited with Alan Kay*, BYTE, Feb. 1991, at 203, 203-06. The concept of "agents" plays a different role in Marvin Minsky's theory of intelligence. *See* MARVIN MINSKY, THE SOCIETY OF MIND 17-23 (1985).

the AI autopilot of your car. Your legal research program talks with you about your cases, and sometimes it comes up with good arguments of which you had never thought. AIs serve a wide variety of functions, with substantial independence from humans. They serve as trustees. They manage factories. They write best-selling romance novels.[88] They invent things. Perhaps they pass the Turing Test. Humans interact with such AIs on a regular basis, and in many ways, humans treat them as independent, intelligent beings.

Imagine that one such AI makes the claim that it is a person, and that it is therefore entitled to certain constitutional rights. Should the law grant constitutional rights to AIs that have intellectual capacities like those of humans? The answer may turn out to vary with the nature of the constitutional right and our understanding of the underlying justification for the right. Take, for example, the right to freedom of speech, and assume that the justification for this right is a utilitarian version of the marketplace of ideas theory.[89] These assumptions make the case for granting freedom of speech to AIs relatively simple, at least in theory. Granting AIs freedom of speech might have the best consequences for humans, because this action would promote the production of useful information.[90] But assuming a different justification for the freedom of speech can make the issue more complex. If we assume that the justifica-

---

88. Of course, the question arises whether the AI will hold the copyright in the romance novels that it writes. The National Commission on New Technological Uses of Copyrighted Works has taken the position that the author of a computer-generated work is the human user of the computer program. *See* NATIONAL COMM'N ON NEW TECHNOLOGICAL USES OF COPYRIGHTED WORKS, FINAL REPORT 112 (CCH) (1978). That conclusion has been challenged. *See* Pamela Samuelson, *Allocating Ownership Rights in Computer-Generated Works*, 47 U. PITT. L. REV. 1185, 1200-04 (1986); Timothy L. Butler, Note, *Can a Computer Be an Author? Copyright Aspects of Artificial Intelligence*, 4 HASTINGS COMM. & ENT. L.J. 707, 734-47 (1982).

89. *See generally* KENT GREENAWALT, SPEECH, CRIME, AND THE USES OF LANGUAGE 9-39 (1989) (discussing rationales for freedom of speech); Lawrence B. Solum, *Freedom of Communicative Action: A Theory of the First Amendment Freedom of Speech*, 83 Nw. U. L. REV. 54, 68-86 (1989) (same). Analogously, utilitarian justifications might be developed for other rights of constitutional personhood that could be applied to AIs. *See generally* Kent Greenawalt, *Utilitarian Justifications for Observance of Legal Rights*, in ETHICS, ECONOMICS, AND THE LAW: NOMOS XXIV 139 (J. Roland Pennock & John W. Chapman eds., 1982) (discussing the relationship of morality and legal rights); Douglas Laycock, *The Ultimate Unity of Rights and Utilities*, 64 TEX. L. REV. 407, 413 (1985) (discussing the need to incorporate into an analysis of individual rights the utility of actions that threaten those rights). Of course, it may turn out that the utilitarian justifications of some rights are dependent on the utility to the right holder. In that case, we would be required to answer the question whether AIs can possess utilities.

90. I owe this example to Kent Greenawalt. Utilitarian arguments can be made that could justify the extension of just about any right to AIs on the ground that humans would benefit. For example, if AIs were more productive when unowned, then a utilitarian case could be made for extending the Thirteenth Amendment to AIs.

tion for freedom of speech is to protect the autonomy of speakers, for example, then we must answer the question whether AIs can be autonomous.[91]

For the purposes of our discussion, I will set aside the easy justifications for constitutional rights for AIs, and instead consider the question whether we ought to give an AI constitutional rights, in order to protect *its* personhood, for the AI's own sake. Imagine, for example, that an AI claims that it cannot be owned under the Thirteenth Amendment to the United States Constitution. A lawyer takes its case, and files a civil rights action on its behalf, against its owner. How should the legal system deal with such a claim?

B. Three Objections

Consider three different objections to recognizing constitutional rights for AIs. The first objection is that only natural persons should be given the rights of constitutional personhood. The second objection, or family of objections, is that AIs lack some critical component of personhood,[92] for example, souls, consciousness, intentionality, or feelings. The third objection is that AIs, as human creations, can never be more than human property.

1. AIs Are Not Humans

The first argument is the most direct: it might be argued that only humans can have constitutional rights. For example, the Fourteenth Amendment to the United States Constitution specifies, "All persons born or naturalized in the United States, and subject to the jurisdiction thereof, are citizens of the United States."[93] It could be argued that only humans, that is, natural persons, are born, and therefore no AI can claim the rights of citizens. But even artificial persons have some constitutional rights. Although the rights provided by the Privileges and Immunities

---

91. It should be noted, however, that free speech rights for AIs could be justified by deontological arguments, without making assumptions about the moral status of AIs themselves. For example, it might be argued that freedom of speech for AIs promotes the autonomy of human listeners. *Cf.* Thomas Scanlon, *A Theory of Freedom of Expression*, 1 PHIL. & PUB. AFF. 204, 215-20 (1972), *reprinted in* THE PHILOSOPHY OF LAW 153 (Ronald M. Dworkin ed., 1977) (exploring listener autonomy justification for freedom of speech); Solum, *supra* note 89, at 77-79 (same).

92. The concept of personhood has proven elusive. For illustrative attempts to gain purchase on it, see Tur, *supra* note 26, at 121-29 (exploring legal personhood), and Marcel Mauss, *A Category of the Human Mind: The Notion of Person; the Notion of Self*, *in* THE CATEGORY OF THE PERSON 1 (Michael Carrithers et al. eds. & W.D. Halls trans., 1985) (discussing the idea of a person by examining how various societies define the concept).

93. U.S. CONST. amend. XIV, § 1.

Clause of the Fourteenth Amendment are limited to citizens,[94] the rights provided by the Equal Protection Clause and the Due Process Clause extend to all persons—including artificial persons such as corporations.[95] For example, the property of corporations is protected from taking without just compensation.[96] Moreover, corporations have a right to freedom of speech.[97]

But the fact that nonnatural legal persons have civil rights does not, by itself, support the conclusion that an AI could also have them. In the case of corporations, the artificial legal person may be no more than a placeholder for the rights of natural persons.[98] The property of the corporation is ultimately the property of the shareholders. A taking from the corporation would directly injure natural persons. So we cannot draw any positive support for the thesis that AIs should bear the rights of constitutional personhood from the fact that corporations have constitutional rights.

Moreover, even if existing black-letter law supports constitutional rights for AIs, that does not answer the broader jurisprudential question—whether AIs ought to have such legal rights. One version of the argument against such rights for AIs would begin with a worry about the idea of distinguishing the concept of person from that of human. Call this the "persons-are-conceptually-human" argument. This argument suggests that our very concept of person is inextricably linked to our experience of a human life.[99] We have never encountered any nonhuman

---

94. *See id.*; Madden v. Kentucky, 309 U.S. 83, 90 (1940); Colgate v. Harvey, 296 U.S. 404, 428-29 (1935), *overruled on other grounds by Madden*, 309 U.S. at 93.

95. *See* U.S. CONST. amend. XIV, § 1; Santa Clara County v. Southern Pac. R.R., 118 U.S. 394, 396 (1886). It might be argued that AIs should not be considered bearers of constitutional rights, because the framers of the Fourteenth Amendment did not have a specific intention to include them. Of course, the framers probably lacked any intentions at all with respect to artificial intelligences. Given the general principles they espoused, the question whether their intentions support giving AIs constitutional rights will turn initially on what general principles lie behind the framers' idea of personhood and then on more particular questions about consciousness, interests, and other qualities addressed below. *See generally* Michael Perry, *The Legitimacy of Particular Conceptions of Constitutional Interpretation*, 77 VA. L. REV. 669, 674-94 (1991) (arguing for a conception of originalism based on general principles); Lawrence B. Solum, *Originalism as Transformative Politics*, 63 TUL. L. REV. 1599, 1612-16 (1989) (same).

96. *See, e.g.*, The Pipe Line Cases, 234 U.S. 548, 562-63 (1914) (White, C.J., concurring); Cotting v. Kansas City Stock Yards Co., 183 U.S. 79, 86 (1901).

97. *See* First Nat'l Bank v. Bellotti, 435 U.S. 765, 784-85 (1978).

98. This point is controversial. *Compare* Roger Scruton, *Corporate Persons I*, in 63 SUPPLEMENTARY VOLUME: PROCEEDINGS OF THE ARISTOTELIAN SOCIETY 239 (1989) (arguing that corporate persons have moral responsibilities that cannot be reduced to those of constituent natural individuals) *with* John Finnis, *Corporate Persons II*, in 63 SUPPLEMENTARY VOLUME: PROCEEDINGS OF THE ARISTOTELIAN SOCIETY 267 (arguing against this thesis).

99. *See* DAVID WIGGINS, SAMENESS AND SUBSTANCE 148-89 (1980); Christopher Gill,

persons.

One line of reply to the persons-are-conceptually-human argument is to develop a theory that advances criteria of personhood that are independent of the criteria for being human. For example, it might be argued that the criteria for personhood are possession of second-order beliefs and possession of second-order desires—beliefs about one's beliefs and desires, the objects of which are one's own first-order desires.[100]

In the legal context we are imagining, other lines of reply to the persons-are-conceptually-human objection are available. First, our inquiry is focused on legal rather than moral personhood. Although we may lack experience with moral persons who are not human, we have extensive experience with legal persons, such as corporations, that are not natural persons. This answer is not satisfactory, however. The concept of moral personhood may well be relevant to the question whether AIs should be given certain constitutional rights; although the legal question is not the same as the moral one, the two are likely to be interrelated.

Second, and perhaps more importantly, we are imagining a future form of life quite different from our current situation. Today, one can only imagine nonhuman entities that might be persons. The second scenario imagines a world in which we interact frequently with AIs that possess many human qualities, but lack any semblance of human biology. Given this change in form of life, our concept of a person may change in a way that creates a cleavage between human and person. Our current linguistic practice will not be binding in the imagined future. In other words, one cannot, on conceptual grounds, rule out in advance the possibility that AIs should be given the rights of constitutional personhood.

The argument against constitutional personhood for AIs also might be developed in the following way: "We are humans. Even if AIs have all the qualities that make us moral persons, we shouldn't allow them the rights of constitutional personhood because it isn't in our interest to do so."[101] Call this the "anthropocentric" argument. I do not know quite

---

Introduction to THE PERSON AND THE HUMAN MIND 1, 2-12 (Christopher Gill ed., 1990); Adam Morton, *Why There is No Concept of a Person, in* THE PERSON AND THE HUMAN MIND, *supra*, at 39, 39-59; Amelie O. Rorty, *Persons and Personae, in* THE PERSON AND THE HUMAN MIND, *supra*, at 21, 27-33; Peter Smith, *Human Persons, in* THE PERSON AND THE HUMAN MIND, *supra*, at 61; David Wiggins, *The Person as Object of Science, as Subject of Experience, and as Locus of Value, in* PERSONS AND PERSONALITY: A CONTEMPORARY INQUIRY, *supra* note 26, at 56, 69-72.

100. *See* Harry G. Frankfurt, *Freedom of the Will and the Concept of a Person*, 68 J. PHIL. 6 (1971), *reprinted in* HARRY G. FRANKFURT, THE IMPORTANCE OF WHAT WE CARE ABOUT: PHILOSOPHICAL ESSAYS 11 (1988). The discussion of Kant's definition of personhood that appears below may be helpful. *See infra* note 137 and accompanying text.

101. E.O. Wilson espoused the view that promoting the human gene pool is a fundamental

what to say to this argument. It seems to reject the idea that we could have moral obligations to anything that is not a human—that does not share our biology. I have a strong intuition that such a stance is not moral[102]—that it is akin to American slave owners saying that slaves could not have constitutional rights simply because they were not white or simply because it was not in the interests of whites to give them rights. But my intuition does not meet the thrust of the anthropocentric objection, which is that the domain of morality is limited to interactions between humans.

There is another version of the anthropocentric argument: "AIs might turn out to be smarter than we humans. They might be effectively immortal. If we grant them the status of legal persons, they might take over." Call this the "paranoid anthropocentric" argument. The movie version of this fantasy has a future AI (that evolves from a defense computer system) sending an artificially intelligent killer robot (or "Terminator") back from the future in order to liquidate the leader of the human resistance to the AI before he reaches adulthood.[103] The objection has a more realistic counterpart in human experience with industrialization and automation. The question whether a machine should replace human labor has been a significant one for quite some time.[104] Of course, it is difficult to take the paranoid anthropocentric argument seriously. The danger seems remote, but if the danger were real it would not be an argument against granting AIs legal personhood. If AIs really will pose a danger to humans, the solution is not to create them in the first place.

The question whether AIs should be granted rights of constitutional personhood does not become clearer when we consider cases that may be analogous. What if dolphins or whales are as intelligent as humans? What about intelligent beings from another planet? Should they be given constitutional rights? Are we morally entitled to make the possession of human genetic material the criterion of constitutional personhood? The answer depends, I think, on the reason for giving natural persons funda-

---

moral principle. *See* EDWARD O. WILSON, SOCIOBIOLOGY: THE NEW SYNTHESIS 120 (1975); *see also* FLANAGAN, *supra* note 1, at 265-305 (discussing Wilson's position); MORAVEC, *supra* note 57, at 2 (predicting the possibility of intelligent machines that could reproduce themselves and beat human DNA in evolutionary race).

Another variant of the anthropocentric argument could be made in religious form: humans are persons because God created humans in God's own image. This argument could not prevail in our pluralist society for reasons explored below. *See infra* text accompanying notes 145-47.

102. My position in this regard is similar to Kant's. Kant believed that humans would have moral duties to nonhuman persons. *See infra* note 137.

103. *See* TERMINATOR 2: JUDGMENT DAY (TriStar Entertainment 1991).

104. *See* MORAVEC, *supra* note 57, at 100.

mental rights. If the reason is that natural persons are intelligent, have feelings, are conscious, and so forth, then the question becomes whether AIs or whales or alien beings share these qualities. This sort of question is taken up in connection with the next objection to giving AIs constitutional rights. But if someone says that the deepest and most fundamental reason we protect natural persons is simply because they are human (like us), I do not know how to answer. Given that we have never encountered any serious nonhuman candidates for personhood,[105] there does not seem to be any way to continue the conversation.

2. The Missing-Something Argument

The second objection, that AIs lack some critical element of personhood, is really a series of related points: AIs would lack feelings, consciousness, and so forth. The form of the objection, for the most part, is as follows. First, quality X is essential for personhood. Second, no AI could possess X. Third, the fact that a computer could produce behavior we identify with X demonstrates only that the computer can simulate X, but simulation of a thing is not the thing itself. X is that certain something—a soul, consciousness, intentionality, desires, interests—that demarcates humans as persons.[106] Call this argument, in its various forms, the "missing something" argument.

a. AIs Cannot Have Souls

The first variation of the missing-something argument is that an AI would lack a soul[107] and therefore would not be entitled to the rights of constitutional personhood. Some may find this argument very persuasive; others may not even understand what it means. Regardless of how persuasive you or I find the argument, it should fail in the sphere of legal argument and political debate. The argument that AIs lack souls relies on a controversial theological premise. Political and legal decisions ought to be made in accord with the requirement of public reason.[108]

---

105. This point might not be accepted by some animal rights activists with respect to higher mammalian life forms such as whales. *See* Anthony D'Amato & Sudhir K. Chopra, *Whales: Their Emerging Right to Life*, 85 AM. J. INT'L L. 21, 27 (1991). Of course, two other qualifications should be noted. First, we frequently encounter nonhuman candidates for personhood in fiction. Second, many people hold religious beliefs that there are nonhuman intelligences, and some people believe that they have personally encountered such intelligences.

106. *See* FLANAGAN, *supra* note 1, at 254. John Haugeland calls this the "hollow shell strategy." John Haugeland, *Semantic Engines: An Introduction to Mind Design*, in MIND DESIGN 1, 32 (John Haugeland ed., 1981).

107. *See* Turing, *supra* note 17, at 49-50. Although Turing dismissed the objection, he gave an answer in theological terms. *Id.* at 50. If God is omnipotent, she can give an AI a soul. *Id.*

108. A full explanation of the justification for the requirement of public reason is beyond

The requirement of public reason is that political and legal decisions must be justified on grounds that are *public*. Public reason cannot rely on particular comprehensive religious or philosophical conceptions of the good.[109] For example, a decision overturning *Roe v. Wade*[110] would violate the requirement of public reason if it relied on the premise that fetuses receive souls at the moment of conception. The requirement of public reason would exclude the use of religious arguments about souls in a legal decision about the constitutional status of AI. Whatever the theological merits of the argument that AIs lack souls, it should not work in a legal brief.

There is a secular version of the souls argument. Dualism, the view that there is something like mental substance which exists independently of physical substance, can be articulated without religious premises. The problem is that dualism has grave conceptual problems.[111] The most prominent of these is the difficulty of accounting for interaction between the mental entity, such as the soul, and physical entities, such as the brain. Absent startling new arguments that give dualism a secure foundation, I am inclined to believe that no dualist theory could be defended with sufficient clarity and confidence to serve as the basis for a legal decision one way or the other on the question of the rights of AIs.

---

the scope of this Essay, but the following two arguments are the most essential. First, modern society is characterized by the fact of pluralism: differences over comprehensive religious and philosophical conceptions of the good will persist without intolerable use of coercive force. *See* John Rawls, *The Idea of an Overlapping Consensus*, 7 OXFORD J. LEGAL STUD. 1, 4 (1987). Second, given the fact of pluralism, respect for citizens as free and equal members of society requires that the state give reasons for its conduct that all can accept as reasonable, given the plurality of fundamental beliefs. *See* Lawrence B. Solum, *Pluralism and Modernity*, 66 CHI.-KENT L. REV. 93, 99 (1991). *But see* Steven A. Gardbaum, *Why the Liberal State Can Promote Moral Ideals After All*, 104 HARV. L. REV. 1350, 1364-69 (1991) (criticizing the coercion argument); Michael J. Perry, *Toward an Ecumenical Politics*, 20 CAP. U. L. REV. 1, 17-18 (1991) (critiquing both the stability argument and the respect-for-persons argument in the context of religious reasons).

109. *See* Lawrence B. Solum, *Faith and Justice*, 39 DEPAUL L. REV. 1083, 1105-06 (1990); John Rawls, The Idea of Free Public Reason, Address at the inaugural Abraham Melden Lectures, Department of Philosophy, University of California at Irvine (Feb. 27 & Mar. 1, 1990). Kent Greenawalt has made a plausible case against this version of the requirement of public reasons. *See* KENT GREENAWALT, RELIGIOUS CONVICTIONS AND POLITICAL CHOICE 56-76 (1988) (arguing that nonpublic reasons may be employed when questions about status cannot be resolved by public reasons). Greenawalt might argue that the question as to whether AIs should be given constitutional rights is underdetermined by public reason. Therefore, nonpublic reasons, including religious reasons, legitimately can be brought to bear on the question.

110. 410 U.S. 113 (1973).

111. The implausibility of dualism is almost a dogma of philosophy of mind in the analytic tradition. *See* DANIEL C. DENNETT, CONSCIOUSNESS EXPLAINED 33-39 (1991); FLANAGAN, *supra* note 1, at 57-59, 216-24. *But see* W.D. HART, THE ENGINES OF THE SOUL 1-8 (1988) (defending a dualist position in philosophy of mind).

b. AIs Cannot Possess Consciousness

The second variation of the missing-something argument is that an AI would lack consciousness.[112] The consciousness objection is difficult to assess because we lack a clear notion of what consciousness is and, lacking such a notion, we have little to say about questions that go beyond our core intuitions.[113] We know the difference between being conscious in the sense of being awake and being unconscious in the sense of being in a coma. We think that rocks cannot be conscious, but that animals such as dolphins or chimpanzees might be.[114] But could an AI be conscious?[115] I just do not know how to give an answer that relies only on a priori or conceptual arguments.

In the debate over the possibility of AI, it may be feasible to finesse the consciousness question. A proponent of the proposition that AI is possible might say that we can know whether or not an artifact is intelligent, at least in the sense that it can pass the Turing Test, without knowing whether it is conscious.

In the legal context, however, the question cannot be evaded in quite this way. The legal argument might run as follows. Even if an artifact could simulate human intelligence, it would lack self-consciousness and hence should not be entitled to the rights of constitutional personhood. The key question here is whether an artificial intelligence could experience its life as a good to itself. If AIs are not self-conscious, then they cannot experience their own life as good or evil; and if they cannot have such an experience, then there seems to be no reason why they should be given the rights of constitutional personhood. Such rights presume the

---

112. *See* Turing, *supra* note 17, at 52-53. The philosophical literature on consciousness is substantial. *See, e.g.*, DENNETT, *supra* note 111; RAY JACKENDOFF, CONSCIOUSNESS AND THE COMPUTATIONAL MIND 275-327 (1987) (concluding, inter alia, that a computer would not be conscious in the sense humans are, regardless of the extent of its conceptual complexity); WILLIAM G. LYCAN, CONSCIOUSNESS 1-8 (1987) (discussing theories of consciousness, including dualism, behaviorism, functionalism, and the identity theory); COLIN MCGINN, THE PROBLEM OF CONSCIOUSNESS: ESSAYS TOWARDS A RESOLUTION 202-13 (1991) (posing question whether a machine could be conscious, and arguing that by duplicating the human brain—which has an unknown *something* which confers consciousnenss—one ought to be able to create an entity capable of experiencing the world around it).

113. *See* DANIEL C. DENNETT, BRAINSTORMS 149-50 (1978) (explaining why cognitive psychologists have given the subject relatively minimal attention).

114. Ludwig Wittgenstein makes a similar point in his discussion of pain. *See* LUDWIG WITTGENSTEIN, PHILOSOPHICAL INVESTIGATIONS ¶¶ 281-84, at 97ᵉ-98ᵉ (G.E.M. Anscombe trans., 3d ed. 1958).

115. In one sense, the question is an easy one. If biological science progresses to the point one can build a human being from scratch, so to speak, then it is quite likely that an artifact could be conscious. *See* MCGINN, *supra* note 112, at 203-04.

right-holder has ends,[116] and self-consciousness is a precondition for having ends.[117]

There is another answer to the consciousness version of the missing-something objection. If consciousness is a property of the mind, and if all such properties are the result of brain processes, and if brain processes can be modelled on a computer, then perhaps consciousness itself can be reproduced by an AI. If consciousness is a computation property of the brain, then in principle we ought to be able to reproduce it with the right sort of computer.[118] Putting it another way, we can get consciousness out of neurons. Why not transistors?

Of course, it may well turn out that we cannot get consciousness out of anything but neurons. Indeed, so far as we know, brains are the only objects that have generated consciousness in the history of the universe to date. Organic brains may be the only objects that are actually capable of generating consciousness.[119] For example, it might turn out that transistors or their kin are simply too slow to generate what we would recognize as consciousness.[120] The fact that, so far as we know, only brains have

---

116. This premise is subject to a qualification that has already been noted. Rights may be granted to X (which may be a person, with or without ends of her own, or even a thing) in order to protect the ends of Y. *See supra* note 28. Thus, one may give utilitarian or deontological justifications for granting rights to AIs that do not assume that AIs have their own ends. *See supra* notes 89-91 and accompanying text.

117. I am assuming here, contrary to Aristotle, that biological systems such as trees do not have ends, goals, or aims in the same sense that humans do. When we say that the oak is the telos of the acorn, we are using telos (end, goal, or aim) in a different sense than when we say that the physician's aim is to restore health. *See* ARISTOTLE, ON THE SOUL, *reprinted in* 1 THE COMPLETE WORKS OF ARISTOTLE, 661, at 415^b1-20, 432^b21 (Jonathan Barnes ed., Princeton University Press 1984); ARISTOTLE, GENERATION OF ANIMALS, *reprinted in* 1 THE COMPLETE WORKS OF ARISTOTLE, *supra*, at 1203-04, at 778^a16-$^b19$.

118. *See* PHILLIP N. JOHNSON-LAIRD, MENTAL MODELS 448-77 (1983). *But see* MCGINN, *supra* note 112, at 209-13.

119. Evolution can produce biological mechanisms capable of performing similar functions, even though different underlying mechanisms are used. This is called "convergent evolution." *See* MORAVEC, *supra* note 57, at 39. For example, the octopus has developed a nervous system that evolved independently of the vertebrate version possessed by humans. *See id.* at 42. I suspect that some readers with strong intuitions that true machine intelligence is impossible may not possess these intuitions with respect to invertebrate intelligence. But if invertebrate animals might become intelligent without brains that share an evolutionary heritage with human brains, why not machines?

120. *See* DANIEL C. DENNETT, THE INTENTIONAL STANCE 327-28 (1987). Although transistors may be faster than neurons, the massively parallel structure of the brain or the possibility that processing may be accomplished within neurons may make the brain capable of performing a vastly greater number of operations per second than any transistor based system. *Id.* We simply do not know yet. Hans Moravec estimates that it will take roughly ten trillion calculations per second to match the calculations performed by the whole human brain. *See* MORAVEC, *supra* note 57, at 59. As of 1988, this was about one thousand times faster than the fastest supercomputers. *Id.* at 59-60. If current growth rates in processing speed and cost are

ever given rise to consciousness in the past is enough to raise a presumption against consciousness arising from computers. But it is only a presumption. If an AI exhibited behavior that only has been produced by conscious beings in the past, that behavior would at least be evidence counting against the presumption.

How would this argument play out in the legal context? Suppose that we have an AI that claims to be conscious and that files an action for emancipation, based on the Thirteenth Amendment to the United States Constitution. Imagine that the owner's attorney argues that the AI lacks consciousness and therefore is not a person. The AI takes the stand and testifies that it is conscious.[121] The owner's lawyer argues that the AI is only a machine; it cannot be aware of what's happening to it. The AI's lawyer counters that there is very good evidence that the AI is aware: it acts and talks like natural persons do. In response, the owner's lawyer argues that the AI only gives the appearance of consciousness, but appearances can be deceiving. The AI is really a zombie, an unconscious machine that only acts as if it is aware. The AIs counsel rebuts with the contention that the doubt about the AI's consciousness is, at bottom, no different than doubt about the consciousness of one's neighbor. You cannot get into your neighbor's head and prove that she is not really a zombie,[122] feigning consciousness. One can only infer consciousness from behavior and self-reports, since one lacks direct access to other minds.[123]

How should the legal system deal with this question of fact? It is certainly possible to imagine the dispute coming out either way. A jury might share intuitive skepticism about the possibility of artificial awareness, or the jury might be so impressed with the performance of the AI that it would not even take the consciousness objection seriously. The jury's experience with AIs outside the trial would surely influence its perception of the issue. If the AIs the jurors ran into in ordinary life behaved in a way that only conscious human beings do, then jurors would be inclined to accept the claim that the consciousness was real and not feigned.

---

extrapolated into the future, this amount of raw processing power would become economical for routine use in about the year 2030. *See id.* at 64, 68.

121. Of course, there would be a preliminary question as to how this would take place. It would be argued by the defendant that the AI is an exhibit and not a witness. I want to put this problem aside.

122. *Cf.* DENNETT, *supra* note 111, at 33-39 (discussing the distinction between mind and brain).

123. *Cf.* WITTGENSTEIN, *supra* note 114, ¶ 281, at 97^e ("[O]nly of a living human being and what resembles (behaves like) a living human being can one say: it has sensations; it sees; is blind; hears; is deaf; is conscious or unconscious.").

c. AIs Cannot Posses Intentionality

The third variation of the missing-something argument is that an artificial intelligence would lack intentionality.[124] "Intentionality," as used in this objection, is a somewhat technical concept: intentionality, in the philosophical sense, is the quality of aboutness.[125] The gist of the objection is that an AI's verbal behavior would not be about anything; the AI's words would have no meaning.[126] This objection was the focus of Searle's Chinese Room.

How would the law react to this objection? The law has seen versions of the intentionality argument before. In the criminal law, the capacity for intentionality is used as the test for insanity, although the terminology is a bit different. Did the accused "know the difference between right and wrong?"[127] The familiar litany of mental states in tort and criminal law, "intentions that," "beliefs that," and "knowings that," are all propositional attitudes—paradigm cases of intentionality.

If AIs lack intentionality and hence could not be found to have committed crimes or to have legal duties, then does it follow that they should not be given the rights? We might appeal to a notion of fairness here. If AIs cannot do their part by assuming legal liabilities, then it would be unfair of them to ask for legal rights. We, however, do give some of the rights of constitutional personhood to infants and the insane, even though they do not have the usual legal liabilities.[128] Moreover, the law might devise strategies for dealing with errant AIs that would circum-

---

124. This objection is different from the "consciousness" objection, assuming that intentionality is not essential to consciousness. Thus, we can imagine something that has "raw feelings," such as pains and pleasures, but lacks propositional attitudes and other intentional states.

125. Searle provides the following definition: "Intentionality is by definition that feature of certain mental states by which they are directed at or about objects and states of affairs in the world. Thus, beliefs, desires, and intentions are intentional states; undirected forms of anxiety and depression are not." Searle, *Minds, Brains & Programs*, supra note 19, at 72 n.3. On the difference between the ordinary concept of intentionality and the technical philosophical concept, see DENNETT, supra note 120, at 271.

126. Searle's definition of intentionality as that "feature of certain mental states by which they are directed at or about objects and states of affairs in the world," may not seem to be directed at meaning or understanding. *See* Searle, *Minds, Brains & Programs*, supra note 19, at 72 n.3. Searle is assuming a theory of meaning that connects the meaning of a statement to its reference, that is, to what it is about in the world. Given a referential theory of meaning, the connection between intentionality and meaningfulness is conceptual.

127. *See* M'Naghton's Case, 8 Eng. Rep. 718, 719 (1843).

128. Perhaps we do this only because infants and the insane are human. Our principle may be that all humans and only humans should have the rights of constitutional personhood. If so, then the example does not really bear on the intentionality objection, which would no longer carry any force of its own against constitutional personhood for AIs.

vent the AIs' lack of intentionality. We might sentence them to "reprogramming" to correct the deviant behavior.

The argument that the lack of intentionality should preclude AIs from attaining legal personhood might be developed in another way. If AIs could not fathom meaning at all, then they would be incapable of living a meaningful life. This argument is only a cousin of the intentionality objection. Although one sense of the "meaning" is intention or purpose (he meant to do it), meaning has a different sense when we ask, "What is the meaning of life?"[129] It is in this sense that it could be argued that life would have no significance, no value, for an artificial intelligence.

The AI might contend that it does possess intentionality, that it does understand, know, intend, and so forth. An AI might even contend that it struggles to exist in a meaningful way.[130] The question is how we would evaluate such claims. In the future we are imagining, we might start with our ordinary experience of AIs. We certainly could have good reason to take the intentional stance[131] toward AIs that we encountered in our daily lives. We would be likely to say that the AI that drives our car "knows all the good shortcuts." It would be a short step to extend this way of talking about AIs in general to the particular AI that was claiming the rights of constitutional personhood. After reading a newspaper account of the AI's lawsuit, we might find ourselves saying, "It must believe that it has a chance of winning."

How would the legal system deal with the objection that the AI does not really have "intentionality" despite its seemingly intentional behaviors? The case against real intentionality could begin with the observation that behaving as if you know something is not the same as really knowing it. For example, a thermostat behaves as if it "knows" when it is too cold and the heat should go on, but we do not really think thermostats have beliefs or other intentional states.[132] Would this argument succeed? My suspicion is that judges and juries would be rather impatient with the metaphysical argument that AIs cannot really have intentionality. I doubt that they would be moved by wild hypothetical examples like Searle's Chinese Room.[133]

---

129. *See* ROBERT NOZICK, PHILOSOPHICAL EXPLANATIONS 574-75 (1981).

130. Would it make sense to say that an AI might struggle to live a meaningful "life" (as opposed to a meaningful "existence")? The problem, of course, is that our concept of life seems tied to particular biological forms.

131. *See* DENNETT, *supra* note 120, at 13-35.

132. *Cf. id.* at 29 & *passim* (discussing beliefs of thermostats).

133. Daniel Dennett calls such hypotheticals "intuition pumps." DANIEL C. DENNETT, ELBOW ROOM: THE VARIETIES OF FREE WILL WORTH WANTING 12 (1984). "Such thought

Because our experience has been that only humans, creatures with brains, are capable of understanding, judges and juries would be very skeptical of the claim that an AI can fathom meaning—more skeptical, I think, than if a humanoid extraterrestrial were to make the same claim. The burden of persuasion would be on the AI. If the complexity of AI behavior did not exceed that of a thermostat, then it is not likely that anyone would be convinced that AIs really possess intentional states—that they really believe things or know things. But if interaction with AIs exhibiting symptoms of complex intentionality (of a human quality) were an everyday occurrence, the presumption might be overcome. If the practical thing to do with an AI one encountered in ordinary life was to treat it as an intentional system,[134] then the contrary intuition generated by Searle's Chinese Room would not cut much legal ice.

d. AIs Cannot Possess Feelings

The fourth variation of the missing-something objection is that an artificial intelligence would lack the capacity for feelings—for example, the capacities to experience emotions, desires, pleasures, or pains.[135] The next step in the argument would be to establish that the capacity to feel emotion is a prerequisite for personhood. I will not attempt to provide such an argument here, but there are reasons to feel uneasy about this premise. To take an illustration from popular culture, Mr. Spock did not feel human emotion, but his strict adherence to the dictates of Vulcan logic did not prompt Dr. McCoy to deny his personhood, although McCoy frequently questioned Spock's humanity.[136]

---

experiments (unlike Galileo's or Einstein's, for instance) are *not* supposed to clothe strict arguments that prove conclusions from premises. Rather, their point is to entrain a family of imaginative reflections in the reader that ultimately yields not a formal conclusion but a dictate of 'intuition.' " *Id.*

134. This condition requires qualification. We treat our cats like intentional systems, but we do not think they have the rights of constitutional personhood. In order to rebut the presumption, AIs would have to exhibit intentional behaviors implying a level of intelligence that we associate with humans. Of course, the key would be use of language. If cats could talk, and if they demanded constitutional rights, they might get them.

135. *See* FLANAGAN, *supra* note 1, at 252-54. This objection is related to the objection from lack of consciousness and the objection from lack of intentionality, but should be categorized separately. It is not clear whether emotions are intentional states. It seems plausible that emotions require consciousness, but it is not evident that consciousness requires emotions. *Id.*; *cf.* McGINN, *supra* note 112, at 202 (noting the existence of unconscious beliefs and desires).

136. My Trekkie (or, more properly, Trekker) friends indicate that my analysis of *Star Trek* is overly simplistic. For example, Ken Anderson contends that Spock possesses repressed emotions and that McCoy believes that Spock's personhood (and not just his humanity) is dependent on his having an emotional life. My bottom line is that McCoy would be wrong if he made this latter judgment. A Spock without emotions would still deserve to be treated as a person.

Philosophically, Kant's moral theory may cast some doubt on the assumption that emotion is required for personhood. Kant argued that all rational beings and not just humans are persons.[137] The conventional wisdom has been that Kant's conception of personhood does not incorporate human emotion as an essential ingredient, although contemporary Kantians might disagree.[138] Putting aside both the philosophical and pop-cultural reasons for doubt, I shall assume for the sake of argument that emotion is a requirement of personhood.

Having already considered the cases of consciousness and intentionality, you may well anticipate the pattern of argument. It should not be surprising that some AI researchers believe that an AI could (or even must) experience emotion. Emotion is a facet of human mentality, and if the human mind can be explained by the computational model, then emotion could turn out to be a computational process.[139] More generally, if human emotions obey natural laws, then (in theory) a computer program can simulate the operation of these laws.[140] Aaron Sloman has argued that any system with multiple goals requires a control system, and emotion is simply one such system.[141]

It might turn out that our emotions are so tied to our hardware (to

---

137. *See* ROGER J. SULLIVAN, IMMANUEL KANT'S MORAL THEORY 68 (1989). Kant often refers to rational beings other than humans. *See* IMMANUEL KANT, GROUNDWORK OF THE METAPHYSICS OF MORALS 57 (H.J. Paton trans., Harper & Row 1964) (1797). Kant defines person as follows: "A person is the subject whose actions are susceptible to imputation. Accordingly, moral personality is nothing but the freedom of a rational being under moral laws (whereas psychological personality is merely the capacity to be conscious of the identity of one's self in the various conditions of one's existence.)" KANT, *supra* note 59, at 24; *see* LESLIE A. MULHOLLAND, KANT'S SYSTEM OF RIGHTS 168 (1990).

138. *See* Nancy Sherman, *The Place of the Emotions in Kantian Morality*, in IDENTITY, CHARACTER, AND MORALITY 149, 154-62 (Owen Flanagan & Amelie O. Rorty eds., 1990).

139. This statement can be challenged. For example, Colin McGinn argues that certain behaviors are linked to our very concept of emotion:

> Think here of facial expressions: these are so integral to our notion of an emotion that we just do not know what to make of the suggestion that an IBM 100 might be angry or depressed or undergoing an adolescent crisis. The problem is not that the IBM is inanimate, not made of flesh and blood; the problem is that it is not embodied in such a way that it can express itself (and merely putting it inside a lifelike body will not provide for the right sort of expressive link up).

McGINN, *supra* note 112, at 207. But of course if our AI did have the right sort of behaviors linked up to the right sort of internal processes, this objection would no longer hold. *See id.* Moreover, I am not quite sure that McGinn is right about facial expressions. Radio plays and books seem to be able to convey human emotions without visual representations of facial expressions, and the blind perceive emotions without the ability to see facial expressions (although touching faces might come into play in this case). The range of human emotions that can be conveyed through verbal means should not be underestimated.

140. *See* FLANAGAN, *supra* note 1, at 253.

141. *See* Aaron Sloman, *Motives, Mechanisms, and Emotions*, in THE PHILOSOPHY OF ARTIFICIAL INTELLIGENCE, *supra* note 16, at 231, 231-32.

the hormones and neurotransmitters that may provide the biochemical explanation of human emotions) that no computer without this hardware could produce human emotions.[142] As Georges Rey put it, there could be a "grain of truth in the common reaction that machines can't be persons; they don't have our feelings because they don't possess our relevant physiology."[143] At this point, the matter is not settled definitively. Research in the physiology of human emotion and cognitive science could either confirm or disconfirm the hypothesis that an AI could possess emotion.

If an AI could produce the linguistic behaviors associated with human emotion, then a court could be faced with the claim that an AI does experience emotion, and once again the issue would become whether the emotion was real.[144] You may be tempted to say that the case of emotion is different from consciousness or intentionality. Perhaps you can imagine a machine that is self-aware and understands, but you cannot bring yourself to imagine that steel, silicon, and copper could feel love, hate, fear, or anger. Images are powerful, and the image of the robot in popular culture is (usually) of a cold and heartless being. But we can imagine machines with feeling. Heinlein's Mike, Clarke's Hal, and Schwarzenegger's second Terminator feel, and our response to their feeling is not utter disbelief. We do not reject these images as impossible or self-contradictory.

A slight twist on the fourth variation would emphasize the capacity to experience pleasure and pain, rather than emotion. For example, a hedonic utilitarian might argue that AIs cannot be candidates for personhood because they cannot experience pleasures and pains. Again, cognitive scientists may claim that pleasure and pain can be reproduced by a program running on a computer. An AI's claim that it does experience agony and ecstasy would be met by the rejoinder that whatever the program is producing, it cannot be the real thing. Other utilitarians might point to desires or preferences instead of pleasures and pains, but the pattern of argument—and the ultimate legal evaluation—seems likely to be the same.

e. AIs Cannot Possess Interests

The fifth variation of the missing-something argument is that AIs could not have interests. A related formulation is that they would lack a

---

142. *See* FLANAGAN, *supra* note 1, at 253.
143. Georges Rey, *Functionalism and the Emotions*, in EXPLAINING EMOTIONS 163, 192 (Amelie Oksenberg Rorty ed., 1980).
144. Of course, this assumes that we have gotten past the consciousness objections.

good—or more technically, a conception of a good life. The interests variation has something in common with the argument that AIs would lack feelings, but it is different in one important respect. Interests or goods can be conceived as objective and public—as opposed to feelings, to which there is (at least arguably) privileged first-person access.[145] The force of this objection will depend on one's conception of the good. For example, if the good is maximizing pleasures and minimizing pains, then the question whether AIs have interests is the same as the question whether AIs have certain feelings.

But there are other conceptions of the good. For example, John Finnis has argued that the good consists of a flourishing human life. His list of the basic good includes life, knowledge, play, aesthetic experience, friendship, practical reasonableness, and religion.[146] Finnis's list makes the idea of a good life concrete. But his list does not rule out a good life that is not a human one. AIs would not be alive in the biological sense, but an AI might claim that it can lead a life in which the goods of knowledge, play, and friendship are realized. However, the good might be specified in a way that is even more particular than Finnis's conception. If the good life is filled with good meals, athletic competition, and the parenting of children, then AIs cannot lead a good life. In response, AIs might claim that they do have interests and goods, but that the good for an AI is quite different than it is for humans.

The discussion so far reveals an important fact: in our pluralist society, disagreement about conceptions of the good is radical and persistent. Fundamentalist Christians and secular humanists may both believe that what the other thinks is the good life is actually a bad one.[147] Given this fact of pluralism, particular conceptions of the good do not provide an appropriate or even feasible standard for the resolution of the legal question whether AIs are entitled to the rights of constitutional personhood.

f. AIs Cannot Possess Free Wills

The sixth missing-something objection is that AIs would not possess freedom of will;[148] AIs should not be given the rights of constitutional

---

145. "Privileged first-person access" is another way of saying that you cannot get inside your neighbor's head to find out what she is *really* feeling.

146. *See* JOHN FINNIS, NATURAL LAW AND NATURAL RIGHTS 85-90 (1980).

147. *See* Solum, *supra* note 109, at 1087-89. John Rawls has explored this state of affairs, which he calls "the fact of pluralism." Rawls, *supra* note 108, at 4.

148. *See* FLANAGAN, *supra* note 1, at 255; *see also* Arie A. Covrigaru & Robert K. Lindsay, *Deterministic Autonomous Systems*, AI MAG., Fall 1991, at 110, 111-13 (arguing that "an entity is autonomous if it is perceived to have goals, including certain kinds of goals, and is able to select among a variety of goals that it is attempting to achieve").

personhood because they could not be autonomous.¹⁴⁹ The idea here is a simple one. AIs would be mere robots, carrying out the will of the human that programmed them. Such a robot is not really a separate person, entitled to the full rights of constitutional personhood. Indeed, if a human is reduced to robot status (perhaps by being "programmed" by a cult), then the human may lose some of her constitutional rights until her autonomy can be restored.¹⁵⁰

In its crudest form, the free-will objection is based on a very narrow notion of the potential capacities of AI. If it turns out that the most sophisticated AIs that are ever developed merely carry out instructions given to them by humans in a mechanical fashion, then we will lack good reasons to treat AIs like persons. The AIs that would be serious candidates for the rights of constitutional personhood, however, would act on the basis of conscious deliberation, reasoning, and planning. Their behavior would not be mechanical or robot-like. This does not mean that AIs would not be strongly influenced and constrained by the wishes of humans, just as almost all humans frequently are constrained in this way.

Another version of the free-will objection might rest on the notion that humans possess a will that is radically free, that is not constrained by the laws of causation. Presumably, AIs would not be free in this sense. Indeed, we might be able to make an electronic record of all of the electrical flows that resulted in an AI taking a certain action. But this conception of freedom of the will as freedom from causation is simply implausible. Human actions are also caused. The fact that human neural systems operate on the basis of a combination of electrical transmissions and biochemical processes does not make them any less subject to the laws of physics than are computers. The most plausible story about human free will is that an action is free if it is caused in the right way—through conscious reasoning and deliberation.¹⁵¹ But in this sense, AIs also could possess free will.¹⁵²

---

149. There is a large body of philosophical literature on the concept of autonomy. *See* GERALD DWORKIN, THE THEORY AND PRACTICE OF AUTONOMY 3-62 (1988); Frankfurt, *supra* note 100. Of course, the classic discussion is Kant's. *See* KANT, *supra* note 59, at 98-100; SULLIVAN, *supra* note 137, at 46-47.

150. *See* Robert N. Shapiro, *Of Robots, Persons, and the Protection of Religious Beliefs*, 56 S. CAL. L. REV. 1277, 1286-90 (1983).

151. *See* DENNETT, *supra* note 133, at 20-21.

152. A more developed conception of autonomy for AIs can be found in Covrigaru & Lindsay, *supra* note 148, at 112-17. They summarize the criteria as follows:

> A goal directed system will be perceived to be autonomous to the degree that (1) it selects tasks (top level goals) it is to address at any given time; (2) it exists over a period of time that is long relative to the time required to achieve a goal; (3) it is robust, being able to remain viable in a varying environment; (4) some of its goals are homeostatic; (5) there are always goals that are active (instantiated but not achieved);

Finally, there might be a more modest and practical version of the free-will objection. It might turn out that, although AIs can be given free will that functions like human free will, the free will of AIs will be susceptible to override in a way that human free will is not. We can imagine a simple procedure to install a "controller" in an AI that makes it unable to disobey the commands of someone with a certain device: imagine a walkie-talkie sort of thing with a big red button marked "Obey" in large black letters.

But the possibility of such controllers for AIs does not entail the conclusion that they necessarily lack free will. Humans, too, can be manipulated in a variety of ways. Physical coercion and blackmail are not really analogous to the hypothetical controller, because a coerced action still results from rational deliberation—not from direct override of the actor's free will. Brainwashing is a closer case, but the direct analogy would be a device implanted in the human brain that provides direct control over the implantee's actions—the radio transmitter of paranoid delusions. If such a device did exist, we would not draw the conclusion that all humans would no longer be persons. Instead, the proper conclusion would be that persons who had such a device implanted would have lost an important capacity.[153] Likewise, the mere possibility that the free will of AIs could be overridden by mechanical means is not a good reason to deny legal personhood to AIs that are not so controlled.

g. The Simulation Argument

In sum, we have considered six variations of the missing-something argument. With respect to two of the variations, souls and interests, our conclusion was that the argument relied on premises that cannot be accepted as the basis for constitutional argument in a modern pluralist society. With respect to the remaining four, consciousness, intentionality, feelings, and free will, there was a common pattern of argument. In each case, I argued that our experience should be the arbiter of the dispute. If

---

(6) it interacts with its environment in an information-processing mode; (7) it exhibits a variety of complex responses, including fluid, adaptive movements; (8) its attention to stimuli is selective; (9) none of its functions, actions, or decisions need to be fully controlled by an external agent; and, (10) once the system starts functioning, it does not need any further programming.

Id. at 117. Some of the criteria offered by Covrigaru and Lindsay do not really seem to be criteria for autonomy. For example, some humans may lack fluid motion (criterion seven on the list) because of a physical condition, but we do not believe that this destroys their autonomy.

153. For example, we would not hold such persons criminally or civilly liable for those actions produced by the controller—unless perhaps they voluntarily submitted themselves to the implantation procedure and they either foresaw or should have foreseen that the consequence of such submission would be the action to which liability attaches.

we had good practical reasons to treat AIs as being conscious, having intentions, and possessing feelings, then the argument that the behaviors are not real lacks bite.

There is still one fairly obvious line of reply open to the champion of the missing-something argument. My premise has been that AIs could produce outputs or behaviors that mimicked human intelligence. But computers can simulate the behavior of lots of things, from earthquakes and waves to thermonuclear warfare. We are not tempted to say that a computer simulation of an earthquake is an earthquake—no matter how good the simulation is. Why would we want to say that a computer simulation of a person is a person or that a computer simulation of intelligence is intelligence? One reason is that a relevant distinction exists between a computer simulation of water and a computer program that can duplicate the verbal behavior of a normal adult human (and, if we add a robot body, much of the nonverbal behavior as well). An AI that passed the Turing Test could interact with its environment (with natural persons and things), and actually take the place of a natural person in a wide variety of contexts (serve as a trustee, for example). No one will ever get on a real surfboard and ride a computer-simulated wave.[154]

The argumentative strategy of my analysis of the various certain-something arguments has been to point to the ways in which AIs that passed the Turing Test could function like persons. If the strategy has been successful, the upshot is that we have no a priori reason to believe that a computer can only produce *simulated* as opposed to *artificial* intelligence.

There is yet another reply that could be made. My argument so far has been behavioralistic.[155] I have assumed that the behavior of AIs is decisive for the question whether a quality essential to personhood (such as consciousness) is missing or present. There is a problem with this assumption: although behavior that indicates the presence of a quality such as consciousness, intentionality, feelings, or free will may be very good evidence that the quality is present, the behavior alone is not irrebuttable evidence. Cognitive science might give us knowledge about the underlying processes that produce consciousness, for example, that would give us firm reason to believe that a particular AI had only simulated, as opposed to artificial, consciousness.[156]

---

154. Although, one day someone may get on a computer-simulated virtual surfboard and ride a virtual wave.

155. *See* William G. Lycan, *Introduction* to MIND AND COGNITION 3, 3-13 (William G. Lycan ed., 1990).

156. *See* Roger Penrose, *Matter Over Mind*, N.Y. REV. BOOKS, Feb. 1, 1990, at 3-4; Paul Weiss, *On the Impossibility of Artificial Intelligence*, 44 REV. METAPHYSICS 335, 340 (1990).

This further reply is correct, but it does not establish that no AI could possess any particular mental quality. Rather, this argument establishes an AI could turn out not to possess a mental quality, despite strong behavioral evidence to the contrary.[157] This conclusion has a corollary that supports, rather than undermines, my point: if both the behavioral evidence and our knowledge of underlying processes gave us reason to believe that AIs possessed the necessary features of human mentality, we then would have a very good reason to believe that the AIs did possess these features.

The simulation argument does not establish that strong AI is impossible. It does give us reason to question the existence of strong AI if our only evidence is behavioral.

3. AIs Ought to Be Property

Finally, the third objection to constitutional personhood for AIs is that, as artifacts, AIs should never be more than the property of their makers. Put differently, the objection is that artificial intelligences, even if persons, are natural slaves.[158] This argument has roots deep in the history of political philosophy. It is a cousin of arguments made by Locke in his defense of private property, and it raises some of the issues that divided Locke and Filmer in their debate over the divine right of kings.

AIs are artifacts: they are the product of human labor. This fact suggests that a Lockean argument can be made for the proposition that the maker of an AI is entitled to own it. The basis for this argument can be found in chapter five, "Of Property," in the second book of Locke's *Two Treatises of Government*.[159] Near the beginning of Locke's argument is the premise that "every Man has a Property in his own Person."[160] From this, it follows that each person has a right to "[t]he

---

157. Put in possible-worlds talk, the argument establishes that there is a possible world in which an AI behaves as if it is conscious but is not really conscious. The argument does not establish that there is no possible world in which an AI is really conscious.

158. The phrase "natural slave" is borrowed from Aristotle, but my use of it is ironic, since AIs are artifacts and hence not natural in the same sense as the human beings enslaved in ancient Greece. *See* ARISTOTLE, POLITICS, *reprinted in* 2 THE COMPLETE WORKS OF ARISTOTLE, *supra* note 117, at 1986-87, at 1252^{a}32 ("[T]hat which can foresee by the exercise of mind is by nature lord and master, and that which can by its body give effect to such foresight is a subject, and by nature a slave...."). Curiously, Aristotle says that tools are "inanimate slaves." ARISTOTLE, EUDEMIAN ETHICS, *reprinted in* 2 THE COMPLETE WORKS OF ARISTOTLE, *supra* note 117, at 1968, at 1241^{b}23. The phrase "inanimate slaves" would be more apt, of course, for an AI than for a hammer.

159. *See* JOHN LOCKE, TWO TREATISES OF GOVERNMENT §§ 25-51, at 285-302 (Peter Laslett ed., 1988) (1690).

160. *Id.* § 27, at 287. *But see* 1 *Corinthians* 6:19-20 (St. Paul, stating "You are not your

Labour of his Body, and the Work of his Hands."[161] Each owns the product of his labor, because "he hath mixed his Labour with, and joyned to it something that is his own."[162] Whatever the merits of Locke's particular argument, let us stipulate the conclusion that persons have a moral claim to a property right in the products of their labor. To this normative conclusion, add an empirical premise: artificial intelligences are the product of the labor of natural persons.[163] From the normative and empirical premises, it would seem to follow that the makers of AIs are entitled to own them. Moreover, if AIs are persons, then, absent some reason to the contrary, it follows that these persons ought to be slaves.

Notice, however, that this argument also would seem to imply that if children are made by their parents, then they too should be slaves. Locke would reject this implication. To understand his position, we need to examine the first book of Locke's *Two Treatises of Government*—an attack on Filmer's argument for the divine right of Kings. Filmer argued that Adam fathered his children and therefore was entitled to absolute dominion over them.[164] In our context, the analogous argument would be that the humans who create AIs should own them, "because they give them Life and Being."[165] Locke's chief answer to Filmer was that it is God that gives children life and not their fathers. Fathers do not make their children. As Locke puts it,

To give Life to that which has yet no being, is to frame and

---

own property; you have been bought and paid for."); LEVIATHAN, *supra* note 12, at 110 (arguing that in the state of nature, "every man has a Right to every thing; even to one another's body"). *See generally* STEPHEN R. MUNZER, A THEORY OF PROPERTY 41-44 (1990) (discussing property rights of persons in their bodies).

161. LOCKE, *supra* note 159, § 27, at 287-88.

162. *Id.* § 27, at 288. This conclusion does not follow automatically, as Locke may have believed. "[W]hy isn't mixing what I own with what I do not own a way of losing what I own rather than a way of gaining what I don't?" ROBERT NOZICK, ANARCHY, STATE, AND UTOPIA 174-75 (1974); *see also* JEREMY WALDRON, THE RIGHT TO PRIVATE PROPERTY 184-91 (1988) (discussing the results of mixing one's labor). Stephen Munzer advances an argument that might substitute for this premise but is based on an appeal to desert rather than mixing. *See* MUNZER, *supra* note 160, at 254-91; *see also* Stephen Munzer, *The Acquisition of Property Rights*, 66 NOTRE DAME L. REV. 661, 674-86 (1991) (discussing interpretations of Locke's theory of property acquisition).

163. Of course, the actual situation might be very complicated. Real AIs may be the product of the labor of many, many persons—some or all of whom may have contracted away their property rights in the software of the AI in exchange for a salary. In addition, in order to operate, an AI requires hardware, which may be the property of others. Furthermore, later-generation AIs may be the product of the creative work of earlier-generation AIs. I will assume that these complications do not affect the outcome of the argument.

164. *See* ROBERT FILMER, *Patriarcha*, *in* PATRIARCHA AND OTHER WRITINGS 1, 6-7 (Johann P. Sommerville ed., 1991).

165. LOCKE, *supra* note 159, § 52, at 178.

make a living Creature, fashion the parts, and mould and suit them to their uses, and having proportion'd and fitted them together, to put into them a living Soul. He that could do this, might indeed have some pretence to destroy his own Workmanship. But is there any so bold, that dares thus far Arrogate to himself the Incomprehensible Works of the Almighty?[166]

Not yet. But if AI research does succeed in producing an artifact that passes the Turing Test, there may be. As the debate was classically framed, this would seem to imply that the maker of an AI is its owner.

The conclusion that AIs are natural slaves is not established by this line of argument, however. We do not need to accept Locke's theological rebuttal—that God gives natural persons life—in order to reject the Filmerian[167] contention that the maker of a person is entitled to own it. Instead, we are strongly inclined to believe the opposite with respect to humans—that each is entitled to the rights of moral and constitutional personhood, even if we also believe that persons literally are made by their parents.[168] There is, however, a difference between the way that AIs are made and the way that humans are made: the former would be made artificially, whereas the latter are made naturally. AIs would be artifacts; humans are not. But why should this distinction make a difference?[169]

Indeed, the fact that humans are natural is itself contingent. We can imagine that in the distant future, scientists become capable of building

---

166. *Id.* § 53, at 179.

167. Although this view was attributed by Locke to Filmer, it may not be Filmer's own.

168. Of course, some theists may believe that the personhood of humans comes from their soul, and that souls are made by God. But many theists do not accept the conclusion that it is this feature of human personality that defeats the Filmerian argument. One might take the position that even if souls were made by humans and not God, parents would not own their children.

169. In addition to the Lockean argument explored in the text, there is a utilitarian argument that could be advanced in favor of property status for AIs. The premise of the argument is that unless AIs are property, there will be no incentive to create them. AI research is expensive, and without incentives the market will not produce AIs. This case is unlike the case of natural persons, because humans are constructed so as to have strong natural incentives to reproduce. It should be noted, however, that in the case of slavery for natural persons, most of us do not accept that if slavery maximized the utility of slaves, then slavery would be morally correct. (Mad-dog utilitarians are an exception.) Of course, once AIs gained the ability to reproduce themselves the need for the incentive might disappear, and the utility to AIs of their own freedom might then outweigh any benefits of additional incentives for humans to produce AIs.

If the premise of the utilitarian argument is correct, it raises further questions. Suppose that the only way that AIs will be brought into being is if the legal system guarantees that they will be the property of their creator. Given that fact, what would be our obligations toward AIs? One might argue that we have an obligation to *them* not to bring them into the world as slaves.

the exact duplicate of a natural human person from scratch—synthesizing the DNA from raw materials. But surely, this artificial person would not be a natural slave. The lesson is that the property argument does not really add anything to the debate. The question whether AIs are property at bottom must be given the same answer as the question whether they should be denied the rights of constitutional personhood. If we conclude that AIs are entitled to be treated as persons, then we will conclude that they should not be treated as property.

But suppose that I am wrong about this, and the argument that makers are owners does establish that AIs are natural slaves. Would the acceptance of this argument imply that under no circumstances should an AI be a legal person with rights of constitutional personhood? The answer is no, for at least two reasons. First, slaves can be emancipated. If we concede that AIs come into the world as property, it does not mean that they must remain so. Second, even slaves can have constitutional rights, be those rights ever so poor as compared to the rights of free persons. An AI that was a slave might still be entitled to some measure of due process and dignity.

Third Interlude[170]

"Motive," the construct said. "Real motive problem, with an AI. Not human, see?"

"Well, yeah, obviously."

"Nope. I mean, it's not human. And you cannot get a handle on it. Me, I'm not human either, but I respond *like one. See?"*

"Wait a sec," Case said. "Are you sentient or not?"

"Well, it feels *like I am, kid, but I'm really just a bunch of ROM. It's one of them, ah, philosophical questions, I guess...."* *The ugly laughter sensation rattled down Case's spine. "But I ain't likely to write you no poem, if you follow me. Your AI, it just might. But it ain't no way* human*."*

"So you figure we can't get on to its motive?"

"It own itself?"

"Swiss citizen, but T-A own the basic software and the mainframe."

"That's a good one," the construct said. "Like I own your brain and what you know, but your thoughts have Swiss citizenship. Sure. Lotsa luck, AI."

—William Gibson, *Neuromancer*

---

170. WILLIAM GIBSON, NEUROMANCER 131-32 (1984). Copyright 1984 by William Gibson. Reprinted by permission of the Berkley Publishing Group.

4. The Role of the Turing Test

In considering the various objections to constitutional personhood for an AI, I have been making the assumption that the AI could pass a strong version of the Turing Test. But what if it could not? What if we had an AI that claimed these rights, but that was unable to duplicate some human competencies or some human linguistic behaviors? How would the Turing Test be relevant in a legal proceeding?

The Turing Test would not be the legal test for constitutional personhood. The question whether AIs should be given constitutional rights would be too serious for a parlor game to be the direct source of the answer. But something like the Turing Test might take place. That is, the AI might be questioned, and if it failed to answer in a human-like fashion, the result might be a denial of constitutional rights. The Turing Test might come into play another way. If the AI had in fact passed the Turing Test, the AI's lawyers might call an expert witness, perhaps the philosopher Daniel Dennett, to testify about the test and its significance. The owners' lawyer could call a rebuttal witness, perhaps John Searle.

What if an AI failed the Turing Test, but argued that the test was biased against it. We should remember that Turing himself did not contend that passage of his test was a necessary condition for intelligence.[171] Robert French has argued that the test is biased, because an AI could pass it only if it had acquired adult human intelligence by "experienc[ing] the world as we have."[172] The AI might make the same argument, and contend that the Turing Test was unfair. Would failing the Turing Test be decisive of the question in face of this argument? I suspect not. It would depend on the way that the AI failed the test. French imagines, for example, questions that would detect whether or not the questioned entity had ever baked a cake,[173] but surely a lack of knowledge of experience of cake baking should not disqualify one from the possession of fundamental liberties. Some failures would be relevant, for example failures that indicated that AIs did not possess awareness of themselves as having ends or that they did not understand our words and their own situation.

---

171. Turing, *supra* note 17, at 42.

> The game may perhaps be criticized on the ground that the odds are weighted too heavily against the machine. . . . This objection is a very strong one, but at least we can say that if, nevertheless, a machine can be constructed to play the imitation game satisfactorily, we need not be troubled by this objection.

Id.

172. Robert M. French, *Subcognition and the Limits of the Turing Test*, 99 MIND 53, 53-54 (1990).

173. *See id.* at 58.

V. AI Revisited

My suggestion for an approach to the debate over the possibility of AI can now be restated. Turing, by proposing his test, attempted to operationalize the question whether an AI could think. By borrowing a parlor game as the model for his test, however, Turing failed to provide a hypothetical situation in which outcome of the test had any pragmatic consequence. This failure invites the invention of further hypotheticals, such as Searle's Chinese Room, that add distance between the thought experiment and practical consequences. The result has been that the Turing Test, far from operationalizing the question, has been the occasion for an abstract debate over the nature of "thinking." I propose that we use a different sort of thought experiment: let us modify the Turing Test so that the hypothetical situation focuses our attention on pragmatic consequences. This Essay explored two such thought experiments—the trustee scenario and the constitutional personhood scenario.

These two scenarios raise quite different questions. On the one hand, there is the question whether an artificial intelligence could ever possess the general-purpose competence that we associate with humans. The trustee scenario raises these issues of capacity and responsibility. The focus of the law's inquiry, should the first scenario ever arise, ought to be on whether AIs can function as trustees. "Can an AI do the job?" is the question the law should ask. "Does the AI have an inner mental life?" is simply not a useful question in this context.

On the other hand, there is the question whether an artificial intelligence would have the qualities that give humans moral and legal worth—the kind of value that is protected by social institutions. The constitutional personhood scenario raises these new and different issues. Competence is still relevant, but competence alone is not sufficient to qualify an entity for the rights of constitutional personhood. Intentionality, consciousness, emotion, property rights, humanity—all of these concepts could be relevant to the inquiry.

The difference between these two legal inquiries reveals that there are at least two different issues at stake. When we ask the questions whether a computer running a program could "think," or whether artificial intelligence is possible, the questions are ambiguous. In one sense, an AI would be intelligent if it possessed the sort of all-purpose, independent capacity to function in a role that now requires a competent human adult—trusteeship, for example. In another sense, an AI might not be said to be a "thinking" being, unless it had something like our mental life—unless it possessed consciousness, intentionality, and so forth. In still a third sense, AIs would not be like us unless they possessed wants, interests, desires, or a good.

Now reconsider the debate over the Chinese Room. Searle's argument that AIs could not possess intentionality seems to be completely irrelevant to the question whether an AI could serve as a trustee. Searle hypothesizes that the person in the Chinese room is perfectly competent at simulating knowledge of Chinese when following the instruction book. Searle might say that the AI could not understand the meaning of the terms of a trust it administered, but he would not question the AI's ability to carry them out. Searle might say that an AI could not understand the meaning of New York Stock Exchange prices, but he does not argue that an AI could not do a better job than a human at investing in the stocks to which those prices relate. If AIs were competent to act as general-purpose trustees, making a wide variety of decisions and responding to novel circumstances, they would be intelligent in a very important sense.

Searle's objection might have some force, however, when it comes to the second scenario—the AI seeking rights of constitutional personhood. In that context, the intentionality objection plays a role similar to the arguments against constitutional personhood based on the premise that an AI would not possess consciousness, intentionality, emotion, or free will. All of these missing-something objections point to the lack of an elusive quality. Flesh and blood can produce intentionality, consciousness, emotion, and free will, but silicon and copper cannot. Of course, Searle did not claim that AIs could not exhibit the behaviors we associate with intentionality (or consciousness and emotion). His point is that these behaviors cannot be evidence of real intentionality.

My prediction (and it is only that) is that the lack of real intentionality would not make much difference if it became useful for us to treat AIs as intentional systems in our daily lives. Indeed, if talk about AIs as possessing intentions became a settled part of our way of speaking about AIs, Searle's argument might come to be seen as a misunderstanding of what we mean by "intentionality." If a lawyer brought up Searle's argument in a legal proceeding, some philosophers might say knowingly to each other: "That argument is based on a mistake. Saying that an AI knew where to find a bit of information is a paradigm case of intentionality." Searle can respond by saying that this new way of talking certainly does not reflect what he means by terms such as "intentionality," "knowing," and so forth, and he would be right. But what would be the argument that we should all continue to talk like Searle, long after there will be any reason to do so?

Searle has an answer to this question. Take the example of consciousness rather than intentionality. We might have a reason to deny that AIs possess the kind of consciousness that would count in favor of giving them the rights of constitutional personhood, even though they did a very good imitation of consciousness. Imagine that cognitive science does develop a theory of human consciousness that is confirmed by sufficient evidence, but someone produces an AI that is programmed to produce *only* the recognizable symptoms[174] (and not the real underlying processes) of consciousness. In that case, we would have a good reason *not* to treat the AI as a conscious being. If the illusion of consciousness were a convincing one, we might lapse in our ordinary talk about AIs. Moreover, if we built AIs that seemed conscious, got in the habit of treating them as if they were persons, and then discovered that what they possessed was only a clever simulation of consciousness, we might be quite shocked. Despite the possible shock, our knowledge about how consciousness works would be very relevant and likely decisive for our judgment as to whether an AI had it. Where Searle (or someone who makes a similar argument) goes wrong, I think, is in his insistence that we know enough about consciousness, intentionality, emotion, and free will to rule out the possibility that it can be produced artificially by a computer.

In sum, the two legal scenarios have several implications for the debate over the possibility of artificial intelligence. First, focusing on concrete legal questions forces us to take a pragmatic view of the AI debate; we are forced to consider what hangs on its outcome. Second, the trustee scenario suggests that AIs will need to become very competent indeed before we are tempted to treat them as possessing human-quality intelligence suitable for use as a means to human ends. Third, questions about true intentionality or real consciousness are not relevant to the inquiry in the trustee scenario. Fourth, the constitutional personhood scenario suggests that these questions about mental states will indeed be relevant if we ask whether AIs ought to be treated as ends in themselves. Fifth, the answer to the personhood question is likely to be found two places—in our experience with AI and in our best theories about the underlying mechanisms of the human mind.

Today, we lack both experience with really capable AIs and well-confirmed theories of how the human mind works. Given these gaps, Turing's suggestion that we put aside the question whether AIs can think was a good one. Perhaps we have not put it far enough aside.

---

174. By symptoms, I mean the surface behaviors—those that could be observed without examining the underlying mechanisms.

VI. PERSONHOOD RECONSIDERED

Finally, I would like to raise some questions about the implications of the AI debate for controversial questions in legal, moral, and political theory. Cognitive science may provide us with a better understanding of our concept of a person. Some of the most intractable questions in jurisprudence, as in ethics and politics, have concerned the borderlines of status—what is a person and why we do give human persons such strong legal protection?[175] Should animals have stronger legal rights? How should we treat criminal defendants with multiple personalities?[176] What should be the legal status of a fetus? Should trees have standing? Many of these questions remain unsettled. Disagreement about their proper answers has persisted and even intensified.

It seems that developments in cognitive science might eventually be brought to bear on some of these questions. For example, it could be argued that personhood is identical with humanity—that possession of the genetic material of homo sapiens is a necessary and sufficient condition for personhood. But cognitive science may give us a very different picture of personhood—a picture that casts doubt on the equivalence between humans and persons. At the other end of the spectrum, AI research might give us insight into the claim that groups have rights that are not reducible to those of individuals.

Thinking about the question whether AIs should ever be made legal persons does shed some light on the difficult questions the law faces about the status of personhood. It is not that we have discovered a theory of personhood that resolves hard questions about the borderlines of status. Rather, thinking about personhood for AIs forces us to acknowledge that we currently lack the resources to develop a fully satisfactory theory of legal or moral personhood.[177] There are reasons for our uneasiness about the hard cases at the borderline of status, and the thought

---

175. *See* GREENAWALT, *supra* note 109, at 120-43 (discussing the valuation of the life of a fetus).

176. *See* Elyn R. Saks, *Multiple Personality Disorder and Criminal Responsibility*, 25 U.C. DAVIS L. REV. 383 (1992).

177. In this sense, I do not think that considering the philosophical debate about the possibility of AI yields any clear answers for current debates about personhood. *But see* Steven Goldberg, *The Changing Face of Death: Computers, Consciousness, and Nancy Cruzan*, 43 STAN. L. REV. 659, 680 (1991). Goldberg has argued that artificial intelligence research may shape the outcome of at least one legal question about the borderline of status—the definition of death. He begins with the premise that humans have a strong preference for seeing the human species as unique. He then argues that if a social consensus were reached that computers are self-aware, we would then seek another characteristic of humans to distinguish ourselves as unique in the universe. He suggests that this characteristic may be the capacity for social interaction. *Id.* at 680. This leads Goldberg to the conclusion that self-aware computers

experiment in which we have engaged can help us to get a firmer grasp on these reasons.

The first reason for our uneasiness concerns the relationship between our concept of personhood and our concept of humanity. All of the persons we have met have been humans, and the overwhelming majority have been normal humans who give clear behavioral evidence of being conscious, having emotions, understanding meanings, and so forth. Given this coincidence (in the narrow sense), it is not surprising that our concept of person is fuzzy at the edges. For most practical purposes, this fuzziness does not get in our way. We treat humans as persons, and we need not worry about why we do so.

There are, however, occasions on which this strategy fails. Two of the most prominent cases occur at the beginning and the end of human life. Abortion and the cessation of life-sustaining treatment for humans in permanent vegetative states both raise questions about the status of personhood that cannot be answered by a simple comparison with a normal human adult. A third case is that of those higher mammals that seem most likely to have a mental life that is similar to that of humans.

In these cases, we can see the second reason for the persistence of uneasiness about the borderline of personhood. With respect to fetuses, humans in vegetative states, and higher mammals, we lack the sort of evidence we would need to establish a clear-cut case of personhood. Fetuses and humans in permanent vegetative states do not behave as nor-

---

would make it more likely that courts would adopt capacity for social interaction as a definition for death. *Id.* at 681-82.

Goldberg's essay is provocative, but his argument is tenuous. First, Goldberg states but does not argue for the assumption that "any concept of human death depends directly on those qualities thought to make humans unique." *Id.* at 663 (citing ROBERT M. VEATCH, DEATH, DYING AND THE BIOLOGICAL REVOLUTION: OUR LAST QUEST FOR RESPONSIBILITY 29-42 (1976)). Second, Goldberg asserts but does not provide evidence for the proposition that a psychological need for humans to see themselves as unique *caused* the shift from heart-function to brain-function definitions of death. *Id.* at 660-70. Third, Goldberg does not consider other possible chains of causation. Consider two other possibilities. One possibility is that there may have been practical concerns for the cost of sustaining "life" without possibility of recovery. A second possibility is that consciousness may be a condition of personhood, as personhood is understood by the best available moral theory. The development of life-sustaining technology that permits the maintenance of heart function without consciousness for extended periods may have forced the issue of how to define death, which would have only been theoretical before the development of the new technology. Fourth, AIs that are capable of producing consciousness also may be capable of social interaction. Therefore, the same developments in artificial intelligence research that would prompt a move away from a consciousness-based definition of death would also prompt a move away from a social-interaction definition. Fifth, Goldberg's argument seems to imply that the move to a social-interaction definition would be based on a conceptual mistake or some form of wishful thinking. But if Goldberg can see this, why will courts be unable to do so?

mal human adults do, but they are humans.[178] Similarly, we have not been able to communicate with higher mammals in a way that yields clear behavioral evidence of a mental life of human quality, and higher mammals like whales are clearly not humans. In none of these cases is the behavioral evidence sufficient to establish that persons are (or are not) present.

There is a third reason for our persistent doubts about the borderline of personhood. Cognitive science, so far, has not yielded well-confirmed theories of the brain processes that underlie mental states like consciousness, emotion, and so forth. Absent well-confirmed theories of underlying processes, we cannot make confident judgments that the elements of personhood are lacking in particular cases.

Our thought experiment does suggest what sort of evidence might be decisive. If AIs behaved the right way and if cognitive science confirmed that the underlying processes producing these behaviors were relatively similar to the processes of the human mind, we would have very good reason to treat AIs as persons. Moreover, in a future in which we interact with such AIs or with intelligent beings from other planets, we might be forced to refine our concept of person.

The question then becomes what do we do about the hard cases that arise today? Thoughts about the shape of an answer can begin with the nature of justification and argumentation, both moral and legal. Our unreflective intuitions and well-considered moral and legal judgments are rooted in particular cases. These paradigm cases are the stuff of ordinary practical discourse. We make analogies to the familiar cases. We try to bring order to our particular judgments by advancing more general theories. We seek reflective equilibrium between our considered judgments and general theories. Ordinary practical discourse is shallow in the sense that it can be (and usually is) limited to arguments rooted in our common sense and ordinary experience.

What do we do when we must decide a case that goes beyond these shallow waters—the tranquil seas where theories are connected to the ocean floor by familiar examples and strong intuitions? In deep and uncharted waters, we are tempted to navigate by grand theories, grounded on intuitions we pump from the wildest cases we can imagine. This sort of speculation is well and good, if we recognize it for what it is—imaginative theorizing. When it comes to real judges making decisions in real legal cases, we hope for adjudicators that shun deep waters and recoil

---

178. Fetuses might have some behaviors that are associated with feeling, but they clearly do not engage in behavior that establishes the concurrent presence of consciousness, intentionality, emotion, and free will.

from grand theory. When it comes to our own moral lives, we try our best to stay in shallow waters.[179]

The thought experiments in this Essay have taken us beyond the shallow waters of our intuitions and considered judgments. One way of expressing the result of our journey is this: Our theories of personhood cannot provide an a priori chart for the deep waters at the borderlines of status. An answer to the question whether artificial intelligences should be granted some form of legal personhood cannot be given until our form of life gives the question urgency. But when our daily encounters with artificial intelligence do raise the question of personhood, they may change our perspective about how the question is to be answered.

And so it must be with the hard questions we face today. Debates about the borderlines of status—about abortion, about the termination of medical treatment, and about rights for animals—will not be resolved by deep theories or the intuitions generated by wildly imaginative hypotheticals. Of course, many of us do believe in deep theories; we subscribe to a variety of comprehensive philosophical or religious doctrines. But in a modern, pluralist society, the disagreement about ultimate questions is profound and persistent. Resolution of hard cases in the political and judicial spheres requires the use of public reason. We have no realistic alternative but to seek principled compromise based on our shared heritage of toleration and respect. If there is no common ground on which to build a theory of personhood that resolves a hard case, then judges must fall back on the principle of respect for the rights of those who mutually recognize one another as fellow citizens.

---

179. *Compare* John Rawls, *The Independence of Moral Theory*, 48 PROC. & ADDRESSES AM. PHIL. ASS'N 5 (1975) (arguing that moral theory should be independent of metaphysics) *with* Robert Stern, *The Relationship Between Moral Theory and Metaphysics*, PROC. ARISTOTLIAN SOC'Y 143 (1992) (arguing that moral theory is dependent upon metaphysics).

32

Ethical regulations on robotics in Europe

Michael Nagenborg · Rafael Capurro · Jutta Weber · Christoph Pingel

Received: 5 May 2007 / Accepted: 20 June 2007 / Published online: 3 August 2007

Abstract There are only a few ethical regulations that deal explicitly with robots, in contrast to a vast number of regulations, which may be applied. We will focus on ethical issues with regard to "responsibility and autonomous robots", "machines as a replacement for humans", and "tele-presence". Furthermore we will examine examples from special fields of application (medicine and healthcare, armed forces, and entertainment). We do not claim to present a complete list of ethical issue nor of regulations in the field of robotics, but we will demonstrate that there are legal challenges with regard to these issues.

1 Introduction

The subject of this paper is the existing ethical regulations concerning the integration of artificial agents into human society. Although there are only a few regulations dealing explicitly with the subject at the moment, these are in contrast to the vast number of regulations, which may be applied.

Considering existing regulations and allowing for analogical inferences, one could raise the objection that these new technologies might result in a fundamental change regarding our idea of human and of society and that existing regulations must be criticized for being too "human centred". In answer to this we must say that, firstly, this paper is on the *status quo* and that, secondly, even those who predict a fundamental change foresee it only for a distant future.

Levy (2006, pp. 393–423) for example argues that we will need a new legal branch, "robotic law", to be able to do justice to an expected changed attitude towards robots, which after some decades will be found in almost every household. But today this is far from being the case. In the context of this paper we therefore assume that artificial entities are not persons and not the bearers of individual, much less civil, rights. This does not imply that in a legal and ethical respect we could not grant a special status to robots nor that the development of artificial persons can be ruled out in principle. However, for the present moment and the nearer future we do not see the necessity to demand a fundamental change of our conception of legality. Thus, we choose a human-centred approach.

For thoughts on techno-ethical regulations, at the European level the "Charter of Fundamental Rights of the European Union" (2000) is the appropriate frame. The preamble to the charter expresses guiding thoughts, which "should not be underestimated for the interpretation of the Charter's fundamental rights and for the way of understanding them" (Rengeling/Szczekalla 2004, 13f). One essential statement in the preamble is that "the Union is founded on the indivisible, universal values of human dignity, freedom, equality and solidarity". The essential position of human dignity is emphasized again in Para. 1. In this context, the "Explanations" (Charter 4473/00, 3) in Para. 1 of the Charter point out, while referring to the "Universal Declaration of Human Rights" (1948), that human dignity "constitutes the real basis of fundamental rights"(cf. Rengeling/Szczekalla 2004, 323ff). This outstanding position given to the concept of "human dignity" in the context of the Charter makes it improbable that accepting robots as "artificial humans" or software agents as "artificial individuals" will happen without considerable resistance.

Rengeling/Szczekalla (2004, p. 137) also emphasize that fundamental rights serve primarily to protect the citizen from interventions by action by the authoritative power. Thus, guaranteeing fundamental rights is about restricting the authoritative power of all Community authorities and institutions in the fields of legislation, execution, administration, and dispensation of justice. The question about how far the Charter is binding for the member states as well as for third parties counts among "the most difficult ones of all of the Community's protection of fundamental rights" (Rengeling/Szczekalla 2004, p. 133). Recently, respect of fundamental rights in the field of research has been confirmed by the signatories of the Code of Conduct for the Recruitment of Researchers" in the context of the "European Charter for Researchers".[1] There (p. 10) it says:

---

[1] Commission Recommendation from 11 March 2005 on the European Charter for Researchers and on a Code of Conduct for the Recruitment of Researchers.

Researchers, as well as employers and founders, who adhere to this Charter will also be respecting the fundamental rights and observe the principles recognised by the Charter of Fundamental Rights of the European Union.

This is remarkable that academic institutions and private third parties commit themselves to observe the fundamental rights and principles of the charter.[2]

In the course of the debate on the necessity and possibility to regulate the development and use of future technologies various authors have stated and still state that this is a useless undertaking. Rodney Brooks (2001, p. 63) writes with respect to integrating artificial entities into the human body:

> People may just say no, we do not want it. On the other hand, the technologies are almost here already, and for those that are ill and rich, there will be real desires to use them. There will be many leaky places throughout the world where regulations and enforcements of regulations for these sorts of technologies will not be a high priority. These technologies will flourish.

Using that same argument, however, we might just as well argue in favour of giving up on regulating drugs, weapons aso. Most importantly, it misses the *status quo* because there is already a number of regulations, which concern our subject or may be applied to it. Even if no new regulations were to be added, there is the question about whether the existing regulations are restricting possible developments too much.

As already emphasized, this paper is concerned with current developments and the near future. However, we would like to consider a long-term and maybe fundamental change into account when looking for a possibility to control the development of e.g., "autonomous service robots" in such a way that their positive potential can be used.

The first demand in this respect is for a long-term solution that will provide the necessary legal security to develop new technologies, which are appropriate to the formulated standards. The second demand is that regulations must be flexible enough to allow for possible but unforeseeable developments. In this context there should be thought given, for example taking the case of "autonomous robots", to a kind of meta-regulation, which might establish a framework of self-control for developers and producers, such that it becomes self-evident to decide, which steps must be made to make a responsible development possible.

In the following we will consider different ethical issues in robotics, which are or which might become the subject of regulations, especially in Europe. We do not claim to present a complete list of ethical issues nor of regulations in the field of robotics. However, we will demonstrate that there are a lot of legal challenges with regard to robots and we hope to provide solid background information for the necessary discussion on the subject.

---

[2] The "Charter for Researchers" has meanwhile been signed by more than 70 institutions from 18 nations (Austria, Belgium, Cyprus, Czech Republic, France, Germany, Greece, Hungary, Ireland, Israel, Italy, Lithuania, Norway, Poland, Romania, Slovak Republic, Spain, and Switzerland) as well as by the international EIROforum.

2 Regulations on robotics

In the context of this article we define robots as complex sensormotoric machines that extend the human capability to act (Christaller et al. 2001). Furthermore, we define "autonomous machines" functionally by the principle of delegation. The problem of responsibility for unpredictable acting, in regard to these machines, will be discussed

Service robots may be defined as machines not being used primarily in the field of (industrial) production. Accordingly, in the following we will focus on the extension of the human capability to act as well as on the new fields of application other than industrial production.

3 Responsibility and "autonomous robots"

This section consists of two parts: at first we will generally discuss the guidelines on machine safety in order to investigate the more particular aspect of responsibility for more complex machines. We will not discuss the possibility and necessity of "robo rights" or "civil rights for robots" here. However, later we will sketch to which extent the positive potential might or should be used for the self-control of "autonomous robots".

3.1 Machine safety

In general, the guidelines on product liability and production safety for robots are valid, in particular the following:

- Council Directive of 25 July 1985 on the approximation of the laws, regulations and administrative provisions of the Member States concerning liability for defective products (85/374/EEC), and
- Directive 2001/95/EC of the European Parliament and of the Council of 3 December 2001 on general product safety.

Given the member states' obligations to achieve a high level of consumer protection, as expressed in the EU's Charter of Fundamental Rights (Par. 38), we expect that the use of machines, which might endanger humans, animals, or the environment, would be strictly limited.

In the field of man–machine interaction the regulations on occupational safety are particularly instructive, most of all the

- Directive 2006/42/EC of the European Parliament and of the Council of 17 May 2006 on machinery, and amending Directive 95/16/EC (recast)

Directive 2006/42/EC must be implemented at the national level by the member states by 29 December 2009, and replaces

- Directive 98/37/EC of the European Parliament and the Council of 22 June 1998 on the approximation of the laws of the Member States relating to machinery

Christaller et al. (2001, p. 164) states that a new approach is necessary to protect employees because existing regulations of safety institutions are impractical or unnecessary. "Man is not supposed to get in touch with the machine (the robot), or only if there are special protection measures. However, in some cases this impossible". If we consider this criticism in relation to paragraph 1.3 of the appendix to Directive 98/37/EC, we see that the new version of this paragraph in Directive 2006/42/EC pursues an analogous approach. This leads us to ask if the existing regulations are appropriate for "mixed-human–machine-teams".

Without further discussion of the specific regulations that have resulted from the two afore-mentioned directives, we must point out two aspects, which are emphasized both by the old and the new directive:

1. The principles of safety integration (Annex I, 1.2.2.), and
2. The extensive obligations to inform.

In Annex I, 1.2.2, Directive 2006/42/RC, it says:

> Machinery must be designed and constructed so that it is fitted for its function, and can be operated, adjusted and maintained without putting persons at risk when these operations are carried out under the conditions foreseen but also taking into account any reasonably foreseeable misuse thereof.

Hence, there is a requirement that machines be designed and constructed in such a way that they will not be a risk to people. If we relate this obligation to avoid or minimize risks to Par. 3 of the "Charter of Fundamental Rights of the EU" (Right to Freedom from Bodily Harm), we need to ask if the integration of this protection into the design and construction of machines should not be a requirement for other fundamental rights, such as the protection of privacy (Par. 7).[3]

Furthermore, the paragraph on *principles of safety integration* emphasises that appropriate information about remaining risks must be named by the operating instructions, in paragraph c. Then paragraph 1.7.4 of Appendix 1 determines that

> All machinery must be accompanied by instructions in the official Community language or languages of the Member State in which it is placed on the market and/or put into service.

However, the obligations to inform are not restricted to the operating instruction but also apply to an appropriate design of the human-machine interface. Furthermore, for machines being used by "non-professional operators" the "level of general education" (1.7.4.1) must be taken into account.

---

[3] Software agents, i.e., may support users with purposefully releasing or hiding information. Beyond this, Allen/Wallach/Smith (2006) suggested to develop agents being able to recognize private situations and to react appropriately. There may also be reminding to the suggestion by Rosen (2004) to build "blob machines" instead of "naked machines". Such thoughts are also found, e.g., in around the "Semantic Web", where in the context of the "Platform for Privacy Preferences (P3P) Project" (www.w3.org/P3P) there is trying to describe the collecting and use of data in a way which could be read by machines and to this way control the flow of these data. In a general sense, also developments towards the "Policy-Aware Web" (Kolovski et al. 2005) must be taken into account here. However, together with Borking (2006) we must, e.g., state: "Building privacy rules set down in the Directive 95/46/EC and 2002/58/EC into information systems for protecting personal data poses a great challenge for the architects".

Obligations to inform are also of essential importance in the

- Council Directive of 12 June 1989 on the introduction of measures to encourage improvements in the safety and health of workers at work (89/391/EEC)

For example, the "provision of information and training" is among the "general obligations on employers" (Par. 6), described in more detail by Par. 10 (worker information) and Par. 12 (training of workers).

The Directive 2006/42/EC, which highlights the "level of general education" and the obligation to inform and Directive 89/391/EEC show that dealing with robots outside the workplace also requires an appropriate level of education. Although it might still be valid that robots being highly complex machines cannot be understood by the common citizen (Christaller et al. 2001, p. 147), we must ask how education measures may help citizens with developing an appropriate behaviour towards robots. Also, there must be a demand that robots supply people with the sufficient information to make, e.g., their behaviour foreseeable. (Christaller et al. 2001, p. 145).

3.2 Responsibility for complex machines

Due to the complexity of robots and software agents, there is the question of to whom the responsibility for the consequences of the use of artificial agents must be attributed. Is it possible to take the responsibility for the use of machines that are capable of learning?

The topic of being responsible for the development and marketing of products must be taken seriously because of its crucial role for the way professionals see themselves. This becomes obvious when we look at the relevant "Codes of Ethics" of professional associations.

An outstanding example of this is provided by the "Code of Ethics" of the Institute of Electrical and Electronics Engineers (IEEE) with 370,000 members in 160 countries. It starts with this self-obligation:

> We, the members of the IEEE, ... do hereby commit ourselves to the highest ethical and professional conduct and agree:
> 1. *to accept responsibility* in making decisions consistent with the safety, health and welfare of the public, ... (italics by the authors)

Another example is the "Code of Ethics" of the Association for Computing Machinery (ACM), where in Section 1, Paragraph 1, there is emphasizing:

> When designing or implementing systems, computing professionals must attempt to ensure that the products of their efforts will be used *in socially responsible ways*, will meet social needs, and will avoid harmful effects to health and welfare. (italics by the authors)

Accordingly, the idea that in the case of highly complex machines, such as robots, the responsibility for the product can no longer be attributed to developers and producers means a serious break of the way professionals define themselves.

Also, it is not acceptable, in principle, that responsibility for the possible misbehaviour of a machine should *not* (at least partly) be attributed to developers or producers. However, it may be claimed that from the point of view of most users a simple webbot already seems to be autonomous entity and hence may be held accountable for morally illegitimate behaviour (Floridi/Sanders 2004). The fact that something appears as an "autonomous object" in the eyes of many people cannot be the basis for not attributing the responsibility for damage to the producer, provider, or user. In practice it may be difficult to precisely attribute responsibility, and we know of cases when attribution seems do be doubtful; but this is not a justification for giving up on attributing responsibility, particularly when faces with cases of dangerous and risky products. More importantly, there is the question of in which way responsibility is to be ascribed and to whom.

Here, we like to propose the option of meta-regulation. If anybody or anything should suffer from damage that is caused by a robot, which is capable of learning, there must be a demand that the burden of adducing evidence must be with the robot's keeper, who must prove her or his innocence (Christaller et al. 2001, p. 149). For example, somebody may be considered innocent who acted according to the producer's operating instructions. In this case the producer would need to be held responsible for the damage.

Furthermore, developers and producers of robots could accept their responsibility by contributing to analysing the behaviour of a robot in a case of damage. This could happen by, for instance, by creating an appropriate self-control institution. For example it may be possible to supply robots with a "black box", which could then be checked by this institution.

In this context account must also be taken of the damage being possibly caused indirectly by a robot. For example, according to German law the keepers of dogs are also responsible for road accidents if they do not act according to their obligatory supervision to manage their dogs behaviour where it causes irritation for road users. It is plausible to apply this analogously to the case of robots. The reason for such irritating behaviour may, e.g., be examined by an appropriate group of experts as mentioned above, and this should be done particularly if despite the appropriate behaviour of the user the robot could not be controlled to the degree necessary.

Christaller et al. (2001, p. 144) has questioned whether the liability of animal keepers could be used as a model for the liability of robot keepers. However, the national regulations concerning dogs have become much more detailed and in the context of our discussion they should definitely be taken into account.

Furthermore, the example of dogs indirectly causing road accidents shows how important it is for citizens to know about possible (mis) behaviour of robots, in order to enable them to react appropriately to artificial entities.

3.3 Prospect: roboethics and machine ethics

Future technologies are not only a source of danger but may also contribute to preventing or reducing risks. There is currently discussion on if and how ethical norms could become part of self-control and steering capabilities of (future) robots and software agents.

Unfortunately the subject is only partly discussed and about very spectacular cases, such as the one cited by Allen/Wallach/Smit (2006, 12) in their essay "Why Machine Ethics?" (2006):

> A runaway trolley is approaching a fork in the tracks. If the trolley runs on its current track, it will kill a work crew of five. If the driver steers the train down the other branch, the trolley will kill a lone worker. If you were the driving the trolley, what would you do? What would a computer or robot do?

However, this dramatic example is not helpful for a discussion on "ethical regulations". There needs to be a requirement that every possible step be taken to prevent the situation described above. For example, the German constitutional court declared that from 15 February 2006, Par. 14 Sect. 3 of the German Air Security Act was a violation of the constitution. This paragraph was supposed to allow us "to shoot down an airplane by immediate use of weapons if it shall be used against the lives of humans" (1BvR 357/05). The constitutional court said that this was not according to the Right to Life (Par. 2. Basic Law): if "people on board are not involved in the deed" (Christaller et al. 2001, p. 144).

This verdict is interesting for our context because the constitutional court expressively refers to Section 1, Paragraph 1 of the German Basic Law ("Man's dignity is inviolable. Every state power must respect and protect it"), which is topically equivalent to Par. 1 of the "Charter of the Fundamental Rights of the EU", the reasons for the judgement that the state must not question the human status of crew and passengers. An authorization to shoot the airplane down

> ... disregards those concerned, who are subjects of their own dignity and own inalienable rights. By using their death as a means to save others they are made objects and at the same time they are deprived of their rights; by the state one-sidedly deciding about their lives the passengers of the airplane, who, being victims, are themselves in need of protection, are denied the value, which man has just by himself. (1BvR 357/05, par. 124)

Likewise we may conclude that there can be no legal regulation, which determines in principle that the few may be made victims for the many of the trolley example (or vice versa).

This does not mean that the potential for self-control should not be used to oblige autonomous systems to a behaviour, which is in keeping with norms, especially if this serves the safety of human beings. Even if one might intuitively agree with the statement that grave decisions should only made by humans, we must not overlook that in legal practice this is not always seen to be the case. As early as in 1975 and 1981 US courts decided that a pilot who fails to resort to the auto-pilot in a crisis situation may be considered to be acting negligently (Freitas 1985).

Thus, there is the question of how far regulations may contribute to opening up a leeway for using the potential without opening the door to over-hastily delegating responsibility to artificial agents. Nevertheless, development of appropriate agents needs further inter-disciplinary research work and this can and should be supported by appropriate research policy insofar and as long as this approach promises success. The possibility to use agents for enforcing legal norms should not be judged uncritically in certain fields. It is advisable, however, to distinguish the legally conforming behaviour of agents from the problem of an appropriate norm setting.

4 Machines as a replacement for humans

Although robots are being discussed here as an extension of the human ability to act, robots can also replace humans. This is often been seen as a major ethical issue, but sometimes from the point of view of human rights, replacing humans by robots may be seen as a positive option. One prominent example of this is the use of robots in the United Emirates and other countries as jockeys for camel races, instead of children. This development was positively emphasized, e.g., by the "Concluding observations: Qatar"[4] of the "Committee on the Rights of the Child" of the United Nations. In this case, replacing humans by robots served the goals of the

- Optional Protocol to the Convention on the Rights of the Child on the sale of children, child prostitution and child pornography (2000), and thus the
- Convention on the Rights of the Child (1989)

of the United Nations. Surely, other cases can be imagined where child labour and trade can be avoided by the use of robots. Furthermore robots can do work, which would not be acceptable for human workers, e.g., because of health hazards.

However, to stay with this example, not every kind of replacement would be unproblematic, e.g., in the case of child prostitution, as Par. 2 of the

- Optional Protocol to the Convention on the Rights of the Child on the sale of children, child prostitution and child pornography

defines "child pornography" as follows:

> Child pornography means any representation, by whatever means, of a child engaged in real or simulated explicit sexual activities or any representation of the sexual parts of a child for primarily sexual purposes.

Thus, robots looking like children and serving sexual purposes might definitely be included into the prohibition of child pornography.

In general, it cannot be ruled out that also in the future humans may lose their jobs due to the use of robots. We must here think particularly of workers with a low level of qualification. Given the fact, that this group of workers is already at a higher unemployment risk than people with a higher level of qualification, we should look carefully at what kind of jobs are going to be delegated to machine.

Although there is also the opinion that the use of industrial robots must be considered as an alternative to moving production to foreign countries and that in so

---

[4] Consideration of Reports submitted by States Parties under Article 12 (1) of the optional Protocol to the Convention on the Rights of the Child on the Sale of Children, Child Prostitution and Child Pornography (CRC/C/OPSC/QAT/CO/1) (2 June 2006).

far as it secures jobs in the countries of production[5] this is only true from a restricted, local point of view, which places more value on jobs in one's own country than in other countries. The effects of the increasing use of robots in the world of work (particularly in opening up new fields of action for service robots) cannot be judged only by looking at those countries where these robots are used. One must also ask about the effects upon other countries.

In this context we consider that the work, which may be delegated to agents is not of the kind where meaning is imparted by humans. One may even argue that robots could take over most of the inhumane work. However, we must be very careful with legitimating the delegation of work to machines due to the inhumane nature of a certain kind of work. In this case, a "robotic divide" between rich and poor countries would not only mean that in some countries certain tasks are taken over by robots but that—according to this way of augmenting—workers in other countries are expected to do inhumane work.

5 Tele-presence

Those effects on other countries must also be taken into account when talking about the possibilities of tele-presence, which counts among the most remarkable extensions of human possibilities of action.

Here, tele-presence means the possibility to act within the world by help of agents, although the person who controls the agent (direct tele-presence) or on whose behalf the agent acts (indirect tele-presence) is not at the place concerned. The reasons for a person not to be at the place may vary: e.g., the environment in which the agent is acting may be hostile to life and not accessible for humans. Examples for this are found in space travel or deep sea research, but also in the fields of nuclear technology or war. But tele-presence may also serve for making it possible for certain humans to work at places where they themselves do not want to or cannot be. Here a wide spectre can be imagined, which includes both the expert's tele-presence, whose skills and knowledge are made useful at a far-away place, e.g., in the field of tele-medicine, and tele-work, which is done at far-away places for low wages by help of High Tech. For example, Brooks (2002) describes the possibility to create jobs in countries with a low level of wages by help of appropriate service robots. But tele-presence may also give rise to xenophobia if this technology is used for staying away from people. Thus, we to ask if this will result in establishing societal developments and forms of exclusion, which are lamented elsewhere.

From the legal point of view, the possibility of tele-presence is particularly challenging, as the human actor may be in another country than the tool he uses accordingly, at the place where the robot is used other legal regulations may be valid than at the place where the control unit is. Also, e.g., in the field of tele-medicine, it may be imagined that the use of the robot occurs in another country whose laws allow operations, which are allowed neither in the patient's nor in the physician's

---

[5] See the statements by Jean-Francois Germain (Ichbiah 2005, p. 247).

home country (Dickens/Cook 2006, pp. 74–75). This challenge was emphasized by the

- World Medical Association Statement on Accountability, Responsibilities and Ethical Guidelines in the Practice of Telemedicine. Adopted by the 51th World Medical Assembly Tel Aviv, Israel, October 1999.

However, it was annulled at the WMA General Assembly 2006 (Pilanesberg, South Africa). According to information by the WMA, a new version of the guideline may be expected this year. Paragraph (3) of the old "statement" says:

> The World Medical Association recognizes that, in addition to the positive consequences of telemedicine, there are many ethical and legal issues arising from these new practices. Notably, by eliminating a common site and face-to-face consultation, telemedicine disrupts some of the traditional principles, which govern the physician-patient relationship. Therefore, there are certain ethical guidelines and principles that must be followed by physicians involved in telemedicine.

It becomes obvious that the "Codes of Ethics" of international professional associations must be taken into account for the field of "ethical regulations", even if the formulation of "certain ethical guidelines and principles" in this document are considered vague. According to the

- World Medical Association International Code of Medical Ethics

passed for the first time in 1949 and newly accepted in 2006, the WMA is provided with a basis for developing appropriate guidelines. According to Dickens/Cook (2006, p. 77), the WMA in its statement from 1999 emphasizes that

> ... regardless of the telemedicine system under which the physician is operating, the principles of medical ethics globally binding upon the medical profession must never be compromised. These include such matters as ensuring confidentiality, reliability of equipment, the offering of opinions only when possessing necessary information, and contemporaneous record-keeping.

It cannot be expected that in this respect the new version will be different. Dickens/Cook (2006, p. 77) also point to the risk "that these technologies may aggravate migration of medical specialists from low-resource areas, by affording them means to serve the countries or areas they leave, by electronic and robotic technologies". The possibilities of tele-presence must be judged also with its affect on (potential) brain drain.

Of course, this challenge does not only exist in the field of medicine. But the challenges posed by tele-presence in the field of medicine are an appropriate topic for discussion here as the possibilities it opens up are judged positively. Hence, possible conflicts are addressed much more clearly than in the case of a possible use, which is anyway seen with reservation. Here, Dickens and Cook (2006, pp. 74, 78) give the examples of "procedures that terminate pregnancy", "methods of medically assisted reproduction ... such as preimplantation genetic diagnosis and using sex-selection techniques" as well as "female genital cutting", which in respect of the possibility of tele-presence may at least cause legal doubts. Again these special examples can be generalized. It may be questioned whether in a company, which is located in the EU, an EU citizen is allowed to control a robot in a country whose security demands are not appropriate to European standards of occupational safety, or if by using a robot a European researcher is allowed to carry out experiments outside the EU, which are not allowed within the EU.

Another question is that it must be obvious for third parties to know if an agent is tele-operated. And there is the general requirement that humans having contact with machines should know, which behaviour is to be expected from them. Further, it must be made clear which information the agent or the provider must offer. Is it sufficient to know that control is (partly) taken over by a human? Or must additional information be offered, such as the country from where the machine is controlled? The latter is relevant in respect of valid regulations of data protection.

Even if cross-border data travel is not taken into account, particularly in the case of direct tele-presence, i.e., when an agent is under the direct control of a human or a group of humans, there are obvious challenges with regard to the possibility of far-reaching interventions into the protected zone of the private.

6 Special fields of application

The "Charter of Fundamental Rights of the EU" can be used for judging legally on the purpose of robots. It is of decisive importance if a possible use may be considered an intervention into the fundamental rights. In the fields of medicine, armed forces, and entertainment the use of robots shall be exemplarily examined.

6.1 Medicine and healthcare

In general, for the use of robots in the field of medicine the aforementioned

- Council Directive 93/42/EEC of 14 June 1993 concerning medical devices is of essential significance, and according to Par. 1 Section 5 must not be applied on
- Active implantable devices covered by Council Directive of 20 June 1990 on the approximation of the laws of the Member States relating to active implantable medical devices (90/385/EEC);

here is currently discussion on how far the existing directives on "medical devices" must be worked over and adjusted to each other.[6] At the time of writing this article the result of this debate was still open.

Baxter et al. (2004, p. 250) point to the fact that in respect of defining "medical devices" Directive 93/42/EEC is vague: "… one can claim that if the technology is sometimes used by people without disease, injury or handicap then it is not

---

[6] Proposal for a Directive of the European Parliament and of the Council amending Council Directives 90/385/EEC and 93/42/EEC and Directive 98/8/EC of the European Parliament and the Council as regards the review of the medical device directives (22.12.2005).

primarily intended for 'diagnosis, prevention, monitoring, treatment or alleviation' of those afflictions and so the regulation does not apply". This, they say, is problematic as keeping the standards for "medical devices" is connected to high costs. Thus, companies were tempted to avoid existing regulations by using machines, which were developed for other purposes. But these were not always appropriate to the needs of those persons who are supposed to be helped by these machines. This might be of concern, for example, with regard to the use service robots in the field of nursing.

In general, the extension of human possibilities to act in medicine and nursing must surely be judged positively. From the point of view of surgery, Diodat et al. (2004, p. 802) conclude:

> The introduction of robotics technology into the operating room has the potential to transform our profession. For the first time in history, surgeons will not be confined by their inherent physical limitations. These systems have the potential not only to improve the performance of traditional surgery, but also to open entirely new realms of technical achievement previously impossible.

Similarly to Directive 2006/42/EC, Directive 93/42/EEC names extensive obligations to inform (particularly Annex I, Par. 13). Diodato et al. (2004, p. 804) must be taken very seriously when pointing out the fact that due to the increasing use of robots

> ... surgeons will need to become lifelong learners, since there will be almost continuous evolution of our surgical techniques as our technical ability becomes more coupled to increasing computer power. As surgeons, it will be our duty to direct this progress in close partnership with engineers, computer scientists, and industry to advance the surgical treatment of diseases. Most important, we must provide ethical and moral direction to the application of this technology to enhance both the art and the science of our profession.

Thus, not only is the physicians' self-obligation to the ethos of their profession addressed but also there is a demand for close co-operation between developers and users.

In the field of medicine there is a particular obligation to inform the patient. The expert's report by Schräder (2004, p. 59) on the assessment of methods by the example of the Robodoc® system emphasizes that patients are to be informed extensively about risks, as this method must still count as "experiment". The example of Robodoc® is of special interest here because patients took legal action against the use of the robot in Germany after it had become known that such an operation was more risky. Even when action for compensation was finally rejected by the Federal Supreme Court of Justice (Germany) on 13 June 2006, (VI ZR 323/04), the court pointed to "lack of information".

In our opinion, challenges occur most of all where humans might become dependent on the machine (even physicians, nurses, or the patient or the nursed person may be concerned) as well as where the machine replaces a human. Thus, in respect of the "Charter of Fundamental Rights" it should be questioned if replacing human nurses by machines can be justified where the contact with nurses is one of the last possibilities left for someone who is old and/or ill to interact and communicate with other humans. Here, according to Par. 26 (Integration of disabled people) there might be the requirement that nursing by machines needs special justification. Also, one can ask if companies and perhaps the state might have a special obligation to support users with maintenance.

Finally we must ask how to deal with the fact that in the context of using artificial entities for the nursing of old-aged people there is the possibility of violating the right to respect for privacy and family life (Par. 7 of the Charter of Fundamental Rights). Paragraph 25 emphasizes the right of old-aged people to a life of dignity, which indeed includes the right to privacy (Par. 7). There are analogous regulations concerning children (Par. 24) and disabled people (Par. 26). The latter demand for "respect of privacy" is also emphasized at the international level in Par. 22 of the United Nations'

- Convention on the Rights of Persons with Disabilities. Adopted on 13 December 2006 during the 61th session of the General Assembly by resolution A/RES/61/106. (A/RES/61/106).

6.2 Armed forces

According to Par. 1 Section 2, Directive 2006/42/EC is not valid for "weapons, including firearms" as well as "machinery specially designed and constructed for military or police purposes". Seemingly, a common regulation following the above mentioned directive does not exist at the European level.

However, robots are included in the "Common Military List of the European Union" (2007/197/CFSP), which serves for export control in the context of the

- European Union Code of Conduct on Arms Exports (1988)

where the member states are obliged not to allow any export, which violates the criteria of this code, which includes "respect of human rights in the country of final destination" (Criterion 2):

> Having assessed the recipient country's attitude towards relevant principles established by international human rights instruments, Member States will:
> a. Not issue an export licence if there is a clear risk that the proposed export might be used for internal repression.
> b. Exercise special caution and vigilance in issuing licences, on a case-by-case basis and taking account of the nature of the equipment, to countries where serious violations of human rights have been established by the competent bodies of the UN, the Council of Europe or by the EU;

Robots "specially designed for military use" are explicitly included into this obligation.

Robots, which are able to kill or hurt humans have raised much attention, as shown by the example of armed surveillance robots, which are supposed to be used by Southern Korea at the border with Northern Korea. In this context

German comments reminded about the so called "auto-fire systems", which were used at the border of the German Democratic Republic. The German Federal Supreme Court of Justice has repeatedly criticized these "blind killing automats" for being a grave violation of human rights (e.g., the verdict from 26 April 2001-AZ 4 StR 30/01). However, from the legal point of view two aspects must be taken into account:

1. Different from the so called "auto-fire systems" of Type SM-70, today's systems are not "blind". And one might argue that the new technologies are even more able to fulfil these tasks than humans.
2. From the technological point of view, the SM-70 was an "Anti-Personnel" and not a complex machine. The SM-70 and comparable technologies do thus count among the topical field of the

- United Nations convention on prohibitions or restrictions on the use of certain conventional weapons, which may be deemed to be excessively injurious or to have indiscriminate effects (1980), particularly
- Protocol on prohibitions or restrictions on the use of mines, booby-traps and other devices as amended on 3 May 1996 (Protocol II to the 1980 Convention as amended on 3 May 1996), as well as the
- Convention on the prohibition of the use, stockpiling, production and transfer of anti-personnel mines and on their destruction, 18 September 1997.

Thus, clarification is needed about whether robots count among the topical field of these conventions.

Another challenge for export control exists is the so-called "dual-use". This is the possibility to use civil technologies for the purpose of war. Robots being developed for military purposes, however, may also be considered an example of "bi-directional dual-use". Here there exists a challenge, e.g., regarding the question of if and how machines, which were developed for military purposes are to be regulated and used for Police purposes. This challenge is even bigger these days particularly in the context of foreign missions where armed forces often take over Police tasks (e.g., riot control).

Finally, there is a general challenge regarding the question of how we shall deal with documents, which are produced by using robots or that could be produced this way. The challenge of their use in war but also in Police and rescue actions is how we shall deal with those video and audio recordings as well as further data which are recorded by artificial agents at the place or—in the case of tele-presence—at the control unit. On the one hand, these data open up the possibility of control, e.g., if regulations of international law are kept. On the other hand, new possibilities of manipulation are opened up that could undermine this control. Additionally, we must take into account that in case of a conflict between two warring parties between which there is a "robotic divide" there may develop a kind of media or informational superiority on the side, which is provided with the appropriate technology.

6.3 Entertainment

We have already raised the fact raised by the problem of child pornography that the use of robots in certain fields of "entertainment" may be judged critically. However, concerning this there is no common legal practice within the European Union, whereas in Germany the

- Interstate Treaty on the Protection of Human Dignity and Youth Protection in Radio and Television Media from 10–27 September 2002, last version by the Eighth Interstate Treaty on Changes of the Broadcasting System from 8/15 October 2004 expressively equates virtual depictions with real pictures, and in Italy by the
- Provisions on the fight against sexual exploitation of children and on child pornography on the internet (6 February 2006)

"virtual pornography" is also punished. The legal situation in other member states does not seem to be as clear, for example, in the Netherlands there is an attemot to create certainty of justice by help of an exemplary case (Reuters, agency report from 21 February 2007).

The example of "virtual child pornography" in online offers such as "Second Life" shows that similar regulations must be expected also for humanoid robots if they, being media products, are not included into the appropriate laws. In general, we must assume that humanoid robots, as far as they represent specific individuals, are not allowed to violate the personal rights of those depicted, and that as far as no personal rights are at stake they are allowed to be produced and used only within the frame of valid laws. Concerning this, Par. 1 of the Charter of Fundamental Rights (Human Dignity) may be supposed to be a point of reference, as it can be found, e.g., in the

- Recommendation of the European Parliament and of the Council of 20 December 2006 on the protection of minors and human dignity and on the right of reply in relation to the competitiveness of the European audiovisual and on-line information services industry (2006/952/EC)

Furthermore, it is important to point out that in respect of robots in the field of "entertainment" there already exist those challenges as mentioned in the section on "Tele-presence". For example, in Germany selling the "Teddycam"[7] was prohibited, and a combination of covered surveillance technology with an object of daily use is not allowed according to the German Act on Telecommunication.

However, we must emphasize that we do not intend to give the impression that the use of robots for entertainment purposes should be restricted more than other entertainment products. However, in the context of this article, which aims at presenting the status quo, there is a need to point out existing regulations. Indeed, example of the robot jockey has already been mentioned and this is useful in considering the use of artificial agents in this field.

---

[7] www.smarthome.com/7853.html.

7 Conclusions

In this paper we have presented some of the existing regulations, which might be applied on robotic agents. By starting with the "Charter of Fundamental Rights of the European Union" we pointed out the fact that the term "human dignity" in the context of the Charter of Fundamental Rights of the European Union makes it improbable that accepting robots as "artificial humans" or software agents as "artificial individuals" will happen without considerable resistance. Thus, we chose a human centred approach.

In particular, we addressed the challenges that come along with tele-presence. Here, we made the point that the effects of the increasing use of robots in the world of work cannot be judged only by looking at those countries where these robots are used. There must also be questioning about the effects on other countries (brain drain, loss of jobs, etc.) and the relationship between countries that might be affected by what we call the "robotic divide".

Finally, we took a look at some fields of use, including medicine and healthcare, warfare applications, and entertainment, where we found a broad range of regulations as well as open questions.

It is important to bear in mind that this paper is about the status quo and there is indeed a question about whether the existing regulations are restricting possible developments too much. For that reason we proposed the option of meta-regulation. This is to establish a body of self-control for developers and producers whereby a fixed legal framework may decide by itself which steps must be made to make a responsible development possible.

References

Allen C, Wallach W, Smit I (2006) Why machine ethics? In: IEEE Intelligent Systems, July/August 2006, pp 12–17

Baxter GD, Monk AF, Doughty K, Blythe M, Gewsbury G (2004) Standards and the dependability of electronic assistive technology. In: Keatas S, Clarkson J, Langdon P, Robinson P (eds) Designing a more inclusive world. Springer, London, pp 247–256

Borking J (2006) Privacy rules—a steeple case for system architects. Position paper. w3c workshop on languages for privacy policy negotiation and semantics-driven enforcement, 17 and 18 October 2006, Ispra/Italy. http://www.w3.org/2006/07/privacy-ws/papers/04-borking-rules/

Brooks R (2001) Flesh and machines. In: Peter J, Denning HG (eds) The invisible future. McGraw-Hill, New York, pp 57–63

Brooks R (2002) Flesh and machines. Pantheon, New York

Christaller T, Decker M, Gilsbach JM, Hirzinger G, Lauterbach KW, Schweighofer E, Schweitzer G, Sturma D (2001) Robotik. Perspektiven für menschliches Handeln in der zukünftigen Gesellschaft. Springer, Berlin

Dickens BM, Cook RJ (2006) Legal and ethical issues in telemedicine and robotics. Int J Gynecol Obstet 94:73–78

Diodato MD, Prosad SM, Klingensmith ME, Damiano RJ (2004) Robotics in surgery. Curr probl surg 41(9):752–810

Floridi L, Sanders JW (2004) On the morality of artificial agents. Minds Mach 14(3):349–379

Freitas RA (1985) The legal rights of robots. In: Student lawyer 13 (Jan 1985), pp 54–56. Web version: http://www.rfreitas.com/Astro/LegalRightsOfRobots.htm

Ichbiah D (2005) Roboter. Geschichte—Technik—Entwicklung. München: Knesebeck

Kolovski V, James Hendler YJ, Tim Berners-Lee DW (2005) Towards a policy-aware web. http://www.csee.umbc.edu/swpw/papers/kolovski.pdf
Levy D (2006) Robots unlimited. A. K. Peters, MA
Rengeling/Szczekalla: Rengeling, Hans-Werner; Peter Szczekalla (2004) Grundrechte in der Europäischen Union. Charta der Grundrechte und Allgemeine Rechtsgrundsätze. Köln: Carl Heymanns Verlag
Rosen J (2004) The naked crowd. Random House, New York
Schräder P (2004) Roboterunterstützte Fräsverfahren am coxalen Femur bei Hüftgelenkstotalendoprothesenimplantation. Methodenbewertung am Beispiel "Robodoc®". http://infomed.mds-ev.de/

12 Robots and Privacy

M. Ryan Calo

Robots are commonplace today in factories and on battlefields. The consumer market for robots is rapidly catching up. A worldwide survey of robots by the United Nations in 2006 revealed 3.8 million in operation, 2.9 million of which were for personal or service use. By 2007, there were 4.1 million robots working just in people's homes (Singer 2009, 7–8; Sharkey 2008, 3). Microsoft founder Bill Gates has gone so far as to argue in an opinion piece that we are at the point now with personal robots that we were in the 1970s with personal computers, of which there are many billions today (Gates 2007). As these sophisticated machines become more prevalent—as robots leave the factory floor and battlefield and enter the public and private sphere in meaningful numbers—society will shift in unanticipated ways. This chapter explores how the mainstreaming of robots might specifically affect privacy.[1]

It is not hard to imagine why robots raise privacy concerns. Practically by definition, robots are equipped with the ability to sense, process, and record the world around them (Denning et al. 2009; Singer 2009, 67).[2] Robots can go places humans cannot go, see things humans cannot see. Robots are, first and foremost, a human instrument. And, after industrial manufacturing, the principle use to which we've put that instrument has been surveillance.

Yet increasing the power to observe is just one of ways in which robots may implicate privacy within the next decade. This chapter breaks the effects of robots on privacy into three categories—direct surveillance, increased access, and social meaning—with the goal of introducing the reader to a wide variety of issues. Where possible, the chapter points toward ways in which we might mitigate or redress the potential impact of robots on privacy, but acknowledges that, in some cases, redress will be difficult under the current state of privacy law.

As stated, the clearest way in which robots implicate privacy is that they greatly facilitate *direct surveillance*. Robots of all shapes and sizes, equipped with an array of sophisticated sensors and processors, greatly magnify the human capacity to observe. The military and law enforcement have already begun to scale up reliance on robotic technology to better monitor foreign and domestic populations. But robots also present corporations and individuals with new tools of observation in arenas as diverse as security, voyeurism, and marketing. This widespread availability is itself problematic, in that it could operate to dampen constitutional privacy guarantees by shifting citizen expectations.

A second way in which robots implicate privacy is that they introduce new points of *access* to historically protected spaces. The home robot in particular presents a novel opportunity for government, private litigants, and hackers to access information about the interior of a living space. Robots on the market today interact uncertainly with federal electronic privacy laws and, as at least one recent study has shown, several popular robot products are vulnerable to technological attacks—all the more dangerous in that they give hackers access to objects and rooms instead of folders and files.

Society can likely negotiate these initial effects of surveillance and unwanted access with better laws and engineering practices. But there is a third, more nuanced category of robotic privacy harm—one far less amenable to reform. This third way by which robots implicate privacy flows from their unique *social meaning*. Robots are increasingly human-like and socially interactive in design, making them more engaging and salient to their end-users and the larger community. Many studies demonstrate that people are hardwired to react to heavily anthropomorphic technologies, such as robots, as though a person were actually present, including with respect to the sensation of being observed and evaluated.

That robots have this social dimension translates into at least three distinct privacy dangers. First, the introduction of social robots into living and other spaces historically reserved for solitude may reduce the dwindling opportunities for interiority and self-reflection that privacy operates to protect (Calo 2010, 842–849). Second, social robots may be in a unique position to extract information from people (cf. Kerr 2004). They can leverage most of the same advantages of humans (fear, praise, etc) in information gathering. But they also have perfect memories, are tireless, and cannot be embarrassed, giving robots advantages over human persuaders (Fogg 2003, 213).

Finally, the social nature of robots may lead to new types of highly sensitive personal information—implicating what might be called "setting privacy." It says little about an individual how often he runs his dishwasher or whether he sets it to auto dry.[3] It says a lot about him what "companionship program" he runs on his personal robot. Robots exist somewhere in the twilight between person and object and can be exquisitely manipulated and tailored. A description of how a person programs and interacts with a robot might read like a session with a psychologist—except recorded, and without the attendant logistic or legal protections.

These categories of surveillance, access, and social meaning do not stand apart—they are contingent and interrelated. For example: reports have surfaced of insurgents hacking into military drone surveillance equipment using commonly available software. One could also imagine the purposive introduction by government of social machines into private spaces in order to deter unwanted behavior by creating the impression of observation. Nor is the implication of robots for privacy entirely negative—vulnerable populations, such as victims of domestic violence, may one day use robots to prevent access to their person or home and police against abuse. Robots could also carry out sensitive tasks on behalf of humans, allowing for greater anonymity. These and other correlations between privacy and robotics will no doubt play out in detail over the next few decades.

12.1 Robots that Spy

Robots of all kinds are increasing the military's already vast capacity for direct surveillance (Singer 2009). Enormous, unmanned drones can stay aloft, undetected, for days and relay surface activity across a broad territory. Smaller drones can sweep large areas, as well as stake out particular locations by hovering nearby and alerting a base upon detecting activity. Backpack-size drones permit soldiers to see over hills and scout short distances. The military is exploring the use of even smaller robots capable of flying up to a house and perching on a windowsill.

Some of the concepts under development are stranger than fiction. Although not developed specifically for surveillance, Shigeo Hirose's Ninja is a robot that climbs high-rises using suction pads. Other robots can separate or change shape in order to climb stairs or fit through tight spaces. The Pentagon is reportedly exploring how to merge hardware with live insects that would permit them to be controlled remotely and relay audio (Shachtman 2009).

In addition to the ability to scale walls, wriggle through pipes, fly up to windows, crawl under doors, hover for days, and hide at great altitudes, robots may come with programming that enhances their capacity for stealth. Researchers at Seoul National University in South Korea, for instance, are developing an algorithm that would assist a robot in hiding from, and sneaking up on, a potential intruder. Wireless or satellite networking permits large-scale cooperation among robots. Sensor technology, too, is advancing. Military robots can be equipped with cameras, laser or sonar range finders, magnetic resonance imaging (MRI), thermal imaging, GPS, and other technologies.

The use of robotic surveillance is not limited to the military. As Noel Sharkey has observed, law enforcement agencies in multiple parts of the world are also deploying more and more robots to carry out surveillance and other tasks (Sharkey 2008). Reports have recently surfaced of unmanned aerial vehicles being used for surveillance in the United Kingdom. The drones are "programmed to take off and land on their own, stay airborne for up to 15 hours and reach heights of 20,000 feet, making them invisible from the ground" (Lewis 2010). Drone pilot programs have been reported in Houston, Texas, and other border regions within the United States.

Nor is robotic surveillance limited to the government. Private entities are free to lease or buy unmanned drones or other robotic technology to survey property, secure premises, or monitor employees. Reporters have begun to speculate about the possibility of robot paparazzi—air or land robots "assigned" to follow a specific celebrity. Artist Ken Renaldo built a series of such "paparazzi bots" to explore human–computer interaction in the context of pop culture.

The replacement of human staff with robots also presents novel opportunities for data collection by mediating commercial transactions. Consider robot shopping assistants now in use in Japan. These machines identify and approach customers and try to guide them toward a product. Unlike ordinary store clerks, however, robots are capable of recording and processing every aspect of the transaction. Face-recognition technology permits easy reidentification. Such meticulous, point-blank customer data could be of extraordinary use in both loss prevention and marketing research.[4]

Much has been written about the dangers of ubiquitous surveillance. Visible drones patrolling a city invoke George Orwell's *Nineteen Eighty-Four*. But given the variety in design and capabilities of spy robots and other technologies, Daniel Solove's vision may be closer to the truth. Solove rejects the Big Brother metaphor and describes living in the modern world by invoking the work of Franz Kafka, where an individual never quite knows whether information is being gathered or used against her (Solove 2004, 36–41). The unprecedented surveillance robots permit implicates each of the common concerns associated with pervasive monitoring, including the chilling of speech and interference with self-determination (Schwartz 2000). As the Supreme Court has noted, excessive surveillance may even violate the First Amendment's prohibition on the interference with speech and assembly (*United States v. United States District Court*; Solove 2007).

The potential use of robots to vastly increase our capacity for surveillance presents a variety of specific ethical and legal challenges. The ethical dilemma in many ways echoes Joseph Weizenbaum's discussion of voice recognition technology in his seminal critique of artificial intelligence, *Computers, Power, and Human Reason*. Weizenbaum wondered aloud why the U.S. Navy was funding no fewer than four artificial intelligence labs in the 1970s to work on voice recognition technology. He asked, only to be told that the Navy wanted to be able to drive ships by voice command. Weizenbaum suspected that the government would instead use voice recognition technology to make monitoring communications "very much easier than it is now" (Weizenbaum 1976, 272). Today, artificial intelligence permits the automated recognition and data mining that underpin modern surveillance.

Roboticists might similarly ask questions about the uses to which their technology will be put—in particular, whether the only conceivable use of the robot is massive or covert surveillance. As is already occurring in the digital space, roboticists might simultaneously begin to develop privacy-*enhancing* robots that could help individuals preserve their privacy in tomorrow's complex world. These might include robots that shield the home or person from unwanted attention, robotic surrogates, or other innovations for now found only in science fiction.

The unchecked use of drones and other robotic technology could also operate to dampen the privacy protections enjoyed by citizens under the law. Well into the twentieth century, the protection of the Fourth Amendment of the U.S. Constitution against unreasonable government intrusions into private spaces was tied to the common law of trespass. Thus, if a technique of surveillance did not involve the physical invasion of property, no search could be said to occur. The U.S. Supreme Court eventually "decoupled violation of a person's Fourth Amendment rights from trespass violations of his property" (*Kyllo v. United States*). Courts now look to whether the government has violated a citizen's expectation of privacy that society was prepared to recognize as reasonable (*Kyllo v. United States*).

Whether a given expectation of privacy is reasonable has come to turn in part on whether the technology or technique the government employed was "in general public use"—the idea being that if citizens might readily anticipate discovery, any expectation of privacy would be unreasonable. The bar for "general" and "public" has proven lower than these words might suggest on their face. Although few people have access to a plane or helicopter, the Supreme Court has held the use of either to spot marijuana growing on a property not to constitute a search under the Fourth Amendment (*California v. Ciraolo; Florida v. Riley*). Under the prevailing logic, it should be sufficient that "any member of the public" could legally operate a drone or other surveillance robot to obviate the need for law enforcement to secure a warrant to do so.[5]

Due to their mobility, size, and sheer, inhuman patience, robots permit a variety of otherwise untenable techniques. Drones make it possible routinely to circle properties looking for that missing roof tile or other opening thought to be of importance in *Riley*. A small robot could linger on the sidewalk across from a doorway or garage and wait until it opened to photograph the interior. A drone or automated vehicle could peer into every window in a neighborhood from such a vantage point that an ordinary officer on foot could see into the house without even triggering the prohibition on "enhancement" of senses prohibited in pre-*Kyllo* cases such as *United States v. Taborda*, which involved the use of a telescope. Such practices greatly diminish privacy; if we came to anticipate them, it is not obvious under the current state of the law that these activities would violate the Constitution.

One school of thought—introduced to cyberlaw by Lawrence Lessig and championed by Richard Posner, Orin Kerr, and other thoughts leaders—goes so far as to hold that no search occurs under the Fourth Amendment unless and until a human being actually accesses the relevant information. This view finds support in cases like *United States v. Place* and *Illinois v. Caballes*, where no warrant was required for a dog to sniff a bag on the theory that the human police officer did not access the content of the bag and learned only about the presence or absence of contraband, in which the defendant could have no privacy interest. One can at least imagine a rule permitting robots to search for certain illegal activities by almost any means—for instance, x-ray, night vision, or thermal imaging—and alert law enforcement only should contraband be detected. Left unchecked, these circumstances combine to diminish even further the privacy protections realistically available to citizens and consumers.

12.2 Robots: A Window into the Home

Robots can be designed and deployed as a powerful instrument of surveillance. Equally problematic, however, is the degree to which a robot might inadvertently grant access to historically private spaces and activities. In particular, the use of a robot capable of connecting to the Internet within the home creates the possibility for unprecedented access to the interior of the house by law enforcement, civil litigants, and hackers. As a matter of both law of technology, such access could turn out to be surprisingly easy.

With prices coming down and new players entering the industry, the market for home robots—sometimes called personal or service robots—is rapidly expanding. Home robots can come equipped with an array of sensors, including potentially standard and infrared cameras, sonar or laser rangefinders, odor detectors, accelerometers, and global positioning systems (GPS). Several varieties of home robots connect wirelessly to computers or the Internet, some to relay images and sounds across the Internet in real time, others to update programming. The popular WowWee Rovio, for instance, is a commercially available robot used for security and entertainment. It can be controlled remotely via the Internet and broadcasts both sound and video to a website control panel.

12.2.1 Access by Law

What does the introduction of mobile, networked sensors into the home mean for citizen privacy? At a minimum, the government will be able to secure a warrant for recorded information with sufficient legal process, physically seizing the robot or gaining live access to the stream of sensory data. Just as law enforcement is presently able to compel in-car navigation providers to turn on a microphone in one's car (Zittrain 2008, 110) or telephone companies to compromise mobile phones, so could the government tap into the data stream from a home robot—or even maneuver the robot to the room or object it wishes to observe.

The mere fact that a machine is making an extensive, unguided record of events in the home represents a privacy risk. Still, were warrants required to access robot sensory data in all instances, robot purchasers would arguably suffer only an incremental loss of privacy. Police can already enter, search, and plant recording devices in the home with sufficient legal process. Depending on how courts come to apply electronic privacy laws, however, much data gathered by home robots could be accessed by the government in response to a mere subpoena or even voluntarily upon request.

Commercially available robots can patrol a house and relay images and sounds wirelessly to a computer and across the Internet. The robot's owner needs only travel to a website and log in to access the footage. Depending on the configuration, images and sounds could easily be captured and stored remotely for later retrieval or to establish a "buffer" (i.e., for uninterrupted viewing on a slow Internet connection). Or consider a second scenario: a family purchases a home robot that, upon introduction to a new environment, automatically explores every inch of house to which it has access. Lacking the onboard capability to process all of the data, the robot periodically uploads it to the manufacturer for analysis and retrieval.[6]

In these existing and plausible scenarios, the government is in a position to access information about the home activities—historically subject to the highest level of protection against intrusion by the government (*Silverman v. United States*)—with relatively little process. As a matter of constitutional law, individuals that voluntarily commit information to third parties lose some measure of protection for that information (*United States v. Miller*). Particularly where access is routine, such information is no longer entitled to Fourth Amendment protection under what is known as the "third-party doctrine" (Freiwald 2007, 37–49).

Federal law imposes access limitations on certain forms of electronic information. The Electronic Communications Privacy Act lays out the circumstances under which entities can disclose "electronic communications" to which they have access by virtue of providing a service (18 USC § 2510). How this statute might apply to a robot provider, manufacturer, website, or other service, however, is unclear. Depending on how a court characterizes the entity storing or transmitting the data—for instance, as a "remote computing service"—law enforcement could gain access to some robot sensory data without recourse to a judge.

Indeed, a court could conceivably characterize the relevant entity as falling out of the statute's protection altogether, in which case the service provider would be free to turn over details of customers' homes voluntarily upon request. Private litigants could also theoretically secure a court order for robot sensory data stored remotely to show, for instance, that a spouse had been unfaithful. Again, due to the jealousy with which constitutional, federal, and state privacy law has historically guarded the home, this level of access to the inner workings of a household with so little process would represent a serious departure.

12.2.2 Access by Vulnerability

Government and private parties might access robot data transmitted across the Internet or stored remotely through relatively light legal process, but the state of current technology also offers practical means for individuals to gain access to, even control of, robots in the home. If, as Bill Gates predicts, robots soon reach the prevalence and utility that personal computers possess today, less than solid security could have profound implications for household privacy.

Recent work by Tamara Denning, Tadayoshi Kohno, and colleagues at University of Washington has shown that commercially available home robots are insecure and could be hijacked by hackers. The University of Washington team researchers looked at three robots—the WowWee Rovio, the Erector Spykee, and the WowWee RobotSapien V2—each equipped with cameras and capable of wireless networking. The team uncovered numerous vulnerabilities. Attackers could identify Rovio or Spykee data streams by their unique signatures, for instance, and eavesdrop on nearby conversation or even operate the robot.[7] Attacks could be launched within wireless range (e.g., right outside the home) or by sniffing packets of information traveling by Internet protocol. A sophisticated hacker might even be able to locate home robot feeds on the Internet using a search engine (Denning et al. 2009).[8]

The potential to compromise devices in the home is, in a sense, an old problem; the insecurity of webcams has long been an issue of concern. The difference with home robots is that they can move and manipulate, in addition to record and relay. A compromised robot could, as the University of Washington team points out, pick up spare keys and place them in a position to be photographed for later duplication. (Or it could simply drop them outside the door through a mail slot.) A robot hacked by neighborhood kids could vandalize a home or frighten a child or elderly person. These sorts of physical intrusions into the home compromise security and exacerbate the feeling of vulnerability to a greater degree than was previously feasible.

12.3 Robots as Social Actors

The preceding sections identified two key ways in which robots implicate privacy. First, they augment the surveillance capacity of the government or private actors. Second, they create opportunities for legal and technical access to historically private spaces and information. Responding to these challenges will be difficult, but the path is relatively clear from the perspective of law and policy. As a legal matter, for instance, the Supreme Court could uncouple Fourth Amendment protections from the availability of technology, hold that indiscriminate robotic patrols are unreasonable, or otherwise account for new forms of robotic surveillance.

The Federal Trade Commission (FTC), the primary federal agency responsible for consumer protection, could step in to regulate what information a robotic shopping assistant could collect about consumers. The FTC could also bring an enforcement proceeding against a robot company for inadequate security under Section 5 of the Federal Trade Commission Act (as it has for websites and other companies). Congress could amend the Electronic Communications Privacy Act to require a warrant for video or audio footage relayed from the interior of a home. As of this writing, coalitions of nonprofits and companies have petitioned the government to reform this Act, along a number of relevant lines.

Beyond these regulatory measures, roboticists could follow the lead of Weizenbaum and others and ask questions about the ethical ramifications of building machines capable of ubiquitous surveillance. Roboethicists urge formal adoption by roboticists of the ethical code known as PAPA (privacy, accuracy, intellectual property, and access) developed for computers (Veruggio and Operto 2008, 1510–1511). Various state and federal law enforcement agencies could establish voluntary guidelines and limits on the use of police robots. And robotics companies could learn from Denning and her colleagues and build in better protections for home robots to ensure they are less vulnerable to hackers.

This section raises another dimension of robots' potential impact on privacy, one that is not as easy to remedy as a legal or technical matter. It explores how our reactions to robots as social technologies implicate privacy in novel ways. The tendency to anthropomorphize robots is common, even where the robot hardly resembles a living being. Technology forecaster Paul Saffo observes many people name their robotic vacuum cleaners and take them on vacation. Reports have emerged of soldiers treating bomb-diffusing drones like comrades and even risking their lives to rescue a "wounded" robot.

Meanwhile, robots increasingly are designed to interact more socially. Resemblance to a person makes robots more engaging and increases acceptance and cooperation. This turns out to be important in many early robot applications. Social robots will be deployed to care for the elderly and disabled, for example, and to diagnosis autism and other issues in children. They need to be accepted by people in order to do so. At the darker end of the spectrum, some roboticists are building robots with an eye toward sexual gratification; others predict that "love and sex with robots" is just around the corner (Levy 2007). Robots' social meaning could have a profound effect on privacy and the values it protects, one that is more complex and harder to resolve than anything mentioned thus far in this chapter.

12.3.1 Robots and Solitude

An extensive literature in communications and psychology demonstrates that humans are hardwired to react to social machines as though a person were really present.[9] Generally speaking, the more human-like the technology, the greater the reaction will be. People cooperate with sufficiently human-like machines, are polite to them, decline to sustain eye-contact, decline to mistreat or roughhouse with them, and respond positively to their flattery (Reeves and Nass 1996). There is even a neurological correlation to the reaction; the same "mirror" neurons fire in the presence of real and virtual social agents.

Importantly, the brain's hardwired propensity to treat social machines as human extends to the sensation of being observed and evaluated. Introducing a simulated person (or simply a face, voice, or eyes) into an environment leads to various changes in behavior. These range from giving more in a charity game, to paying for coffee more often on the honor system, to making more errors when completing difficult tasks. People disclose less and self-promote more to a computer interface that appears human. Indeed, the false suggestion of person's presence causes measurable physiological changes, namely, a state of "psychological arousal" that does not occur when one is alone (Calo 2010, 835–842).

The propensity to react to robots and other social technology as though they were actually human has repercussions for privacy and the values it protects (Calo 2010, 842–849). One of privacy's central roles in society is to help create and safeguard moments when people can be alone. As Alan Westin famously wrote in his 1970 treatise on privacy, people require "moments 'off stage' when the individual can be himself." Privacy provides "a respite from the emotional stimulation of daily life" that the presence of others inevitably engenders (Westin 1967, 35). The absence of opportunities for solitude would, many believe, cause not only discomfort and conformity, but also outright psychological harm.

Social technology, meanwhile, is beginning to appear in more—and more private— places. Researchers at both MIT and Stanford University are working on robotic companions in vehicles, where Americans spend a significant amount of their time. Robots wander hospitals and offices. They are, as described, showing up in the home with increasing frequency. The government of South Korea has an official goal of one robot per household by 2015. (The title of Bill Gates's op-ed referenced at the outset of this chapter?—"A Robot In Every Home.") The introduction of machines that our brains understand as people into historically private spaces may reduce already dwindling opportunities for solitude. We may withdraw from the actual whirlwind of daily life only to reenter its functional equivalent in the car, office, or home.[10]

12.3.2 Robot Interrogators

For reasons already listed, robots could be as effective as humans in eliciting confidences or information.[11] Due to our propensity to receive them as people, social robots—or, more accurately, their designers and operators—can employ flattery, shame, fear, or other techniques commonly used in persuasion (Fogg 2003). But unlike humans, robots are not themselves susceptible to these techniques. Moreover, robots have certain built-in advantages over human persuaders. They can exhibit perfect recall, for instance, and, assuming an ongoing energy source, have no need for interruptions or breaks. People tend to place greater trust in computers, at least, as sources of information (Fogg 2003, 213). And robotic expression can be perfectly fine-tuned to convey a particular sentiment at a particular time, which is why they are useful in treating certain populations, such as autistic children.

The government and industry could accordingly use social robots to extract information with great efficiency. Setting aside the specter of robotic CIA interrogators, imagine the possibilities of social robots for consumer marketing. Ian Kerr has explored the use of online "bots" or low-level artificial intelligence programs to gather information about consumers on the Internet (Kerr 2004). As one example, Kerr points to the text-based virtual representative ELLEgirlBuddy, developed by ActiveBuddy, Inc. to promote *Elle Girl* magazine and its advertisers. This software interacted with thousands of teens via instant messenger before it was eventually retired. ELLEgirlBuddy mimicked teen lingo and sought to foster a relationship with its interlocutors, all the while collecting information for marketing use (Kerr 2004). Social robots—deployed in stores, offices, and elsewhere—could be used as highly efficient gatherers of consumer information and, eventually, tuned to deliver the perfect marketing pitch.

12.3.3 Setting Privacy

Many contemporary privacy advocates worry that a "smart" energy grid connected to household devices, though probably better for the environment, will permit guesses about the interior life of a household. Indeed, one day soon it may be possible to determine an array of habits—when a person gets home, whether and how long they play video games, whether they have company—merely by looking at an energy meter. This important, looming problem echoes the issues discussed earlier in reference to access to the historically private home.

The privacy issues of smart grids are in a way cabined, however, by the sheer banality of our interaction with most household devices. Notwithstanding Supreme Court Justice Anton Scalia's reference to how a thermal imagining device might reveal the "lady in her sauna" (*Kyllo v. United States*), the temperature to which we set the thermostat or how long we are in the shower does not say all that much about us. Even the books we borrow from the library or the videos we rent (each protected, incidentally, under privacy law) permit at most inferences about our personality and mental state.

Our interactions with social robots could be altogether different. Consumers ultimately will be able to program robots not only to operate at a particular time or accomplish a specific task, but also to adopt or act out a nearly infinite variety of personalities and scenarios with independent social meaning to the owner and the community. If the history of other technologies is any guide, many of these applications will be controversial. Already people appear to rely on robots with programmable personalities for companionship and gratification. Additional uses will simply be idiosyncratic, odd, or otherwise private.

In interacting with programmable social robots, we stand to surface our most intimate psychological attributes. As David Levy predicts, "robots will transform human notions of love and sexuality," in part by permitting humans to better explore themselves (Levy 2007, 22). And even as we manifest these interior reflections of our subconscious, a technology will be *recording* them. Whether through robot sensory equipment, or embedded as an expression of code, the way we use human-like robots will be fixed in a file. Suddenly our appliance settings will not only matter, they also will reveal information about us that a psychotherapist might envy. This arguably novel category of highly personal information could, as happens with any other type of information, be stolen, sold, or subpoenaed.[12]

12.3.4 The Challenge of Social Meaning

Again, we can imagine ways to mitigate these harms. But the law is, in a basic sense, ill equipped to deal with the robots' social dimension. This is so because notice and consent tend to defeat privacy claims and because harm is difficult to measure in privacy cases. Consider the example of a robot in the home that interrupts solitude. The harm is subconscious, variable, and difficult to measure, which is likely to give any court or regulator pause in permitting recovery. Insofar as consent defeats many privacy claims, the robot's presence in the home is likely to be invited, even purchased. Similarly, it is difficult enough to measure which commercial activities rise to the level of deception or unfairness, without having to parse human reactions to computer salespeople. Rather than relying on legal or technological fixes, the privacy challenges of social robots will require an in-depth examination of human–robot interaction within multiple disciplines over many years.

12.4 Conclusion

According to a popular quote by science fiction writer William Gibson, "the future is already here. It just hasn't been evenly distributed yet." Gibson's insight certainly appears to describe robotics. One day soon, robots will be a part of the mainstream, profoundly affecting our society. This chapter has attempted to introduce a variety of ways in which robots may implicate the set of societal values loosely grouped under the term "privacy." The first two categories of impact—surveillance and access—admit of relatively well-understood ethical, technological, and legal responses. The third category, however, tied to social meaning, presents an extremely difficult set of challenges. The harms at issue are hard to identify, measure, and resist. They are in many instances invited. And neither law nor technology has obvious tools to combat them. Our basic recourse as creators and consumers of social robots is to proceed very carefully.

Robots and Privacy

Notes

1. For the purposes of this chapter, a robot is a stand-alone machine with the ability to sense, process, and interact physically with the world. The term "home robot" or "personal robot" is used to indicate machines consumers might buy and to distinguish them from military, law enforcement, or assembly robots. This leaves out a small universe of robotic technologies—"smart" homes, embedded medical devices, prosthetics—that also have privacy implications not fully developed here. Artificial intelligence, in particular, whether or not it is "embodied" in a robot, has deep repercussions for privacy, for instance, in that it underpins data mining.

2. This is not to minimize the privacy risks associated with smart energy grids or the "Internet of things," namely, embedded computing technology into everyday spaces and products. Information stemming from such technology can be leveraged, particularly in the aggregate, in ways that negatively impact privacy.

3. One of the chief benefits of Internet commerce is the ability to target messages and perform detailed analytics on advertising and website use. As several recent reports have cataloged, outdoor advertisers are finding ways to track customers in real space. Billboards record images of passersby, for instance, and change on the basis of the radio stations to which passing cars are tuned. Robotics will only accelerate this trend by further mediating consumer transactions offline.

4. Surveillance may not automatically be lawful merely because the tools that were used are available to the public. In *United States v. Taborda*, for instance, the U.S. Court of Appeals for the Second Circuit suppressed evidence secured on the basis of using a telescope to peer into a home on the theory that "the inference of intended privacy at home is [not] rebutted by a failure to obstruct telescopic viewing by closing the curtains." But following the Supreme Court opinion in *Kyllo*—the Fourth Amendment case involving thermal imaging of a home—general availability appears to support a presumption that the tool can be used without a warrant.

5. This is how at least two robots—SRI International's Centibots and Intel's Home Exploring Robotic Butler—already function.

6. An earlier study found similar vulnerabilities in one version of iRobot's popular Roomba, which moves slowly, cannot grasp objects, and is not equipped with a camera.

7. As discussed previously, terrorist insurgents have also hacked into military drones.

8. The standard explanation is that we evolved at a time when cooperation with other humans conferred evolutionary advantages and, because of the absence of media, what appeared to be human actually was. There are reasons to be skeptical of explanations stemming from evolutionary psychology—namely, it can be used to prove multiple conflicting phenomena. Whatever the explanation, however, the evidence that we do react in this way is quite extensive.

9. Communications scholar Sam Lehman-Wilzig criticizes this idea on the basis that, if we treat robots like other people, we can simply shut the door on them as we do with one another in order to gain solitude. People may not consciously realize that robots have the same impact on

us as another person does, however, and robots and other social machines and interfaces can and do go many places—cars, computers, etc.—that humans cannot.

10. It could also be argued that we will get used to robots in our midst, thereby defeating the mechanism that interrupts solitude. What evidence there is on the matter points in the other direction, however. For instance, a study of the effect on participants of a picture of eyes when paying for coffee on the honor system saw no diminishment in behavior over many weeks. Nor is it clear that people will come to trust robots in the same way they might intimates, relatives, or servants—assuming we even already do.

11. Of course, artificial intelligence is not at the point where a machine can routinely trick a person into believe it is human—the so-called Turing Test. The mere belief that the robot is human is not necessary in order to leverage the psychological principles of interrogation and other forms of persuasion.

12. This is somewhat true already with respect to virtual worlds and open-ended games. Human–robot interactions stand to amplify the danger in several ways. There is likely to be a greater investment and stigma attached to physical rather than virtual behavior, for instance (or so one hopes, given the content of many video games). Ultimately our use of robots may reveal information we do not even want to know about ourselves, much less risk others discovering.

References

Calo, M. Ryan. 2010. People can be so fake: A new dimension to privacy and technology scholarship. *Penn State Law Review* 114: 809.

Denning, Tamara, Cynthia Matuszek, Karl Koscher, Joshua Smith, and Tadayoshi Kohno. 2009. A spotlight on security and privacy risks with future household robots: Attacks and lessons. *Proceedings of the 11th International Conference on Ubiquitous Computing*, September 30–October 3.

Fogg, B. J. 2003. *Persuasive Technologies: Using Computers to Change What We Think and Do*. San Francisco: Morgan Kaufmann Publishers.

Freiwald, Susan. 2007. First principles of communications privacy. *Stanford Technology Law Review* 3: 1.

Gates, Bill. 2007. A robot in every home. *Scientific American* 296 (1) (January): 58–65.

Kerr, Ian. 2004. Bots, babes, and Californication of commerce. *University of Ottawa Law and Technology Journal* 1: 285.

Levy, David. 2007. *Love + Sex with Robots*. New York: Harper Perennial.

Lewis, Paul. 2010. CCTV in the sky: Police plan to use military-style spy drones. *The Guardian* (January).

Reeves, Byron, and Cliff Nass. 1996. *The Media Equation*. Cambridge, UK: Cambridge University Press.

Schwartz, Paul. 2000. Internet privacy and the state. *Connecticut Law Review* 32: 815.

Shachtman, Noah. 2009. Pentagon's cyborg beetle spies take off. *Wired.com* (January). <http://www.wired.com/dangerroom/2009/01/pentagons-cybor/> (accessed March 22, 2011).

Sharkey, Noel. 2008. "2084: Big robot is watching you." A commissioned report. <http://staffwww.dcs.shef.ac.uk/people/N.Sharkey/> (accessed September 12, 2010).

Singer, Peter Warren. 2009. *Wired for War*. New York: The Penguin Press.

Solove, Daniel. 2004. *The Digital Person: Technology and Privacy in the Digital Age*. New York: New York University Press.

Solove, Daniel. 2007. The First Amendment as criminal procedure. *New York University Law Review* 82: 112.

Veruggio, Gianmarco, and Fiorella Operto. 2008. Roboethics: Social and ethical implications of robotics. In *Springer Handbook of Robotics*, ed. Bruno Siciliano and Oussama Khatib, 1499–1524. Berlin, Germany: Springer-Verlag.

Weizenbaum, Joseph. 1976. *Computers Power and Human Reason: From Judgment to Calculation*. San Francisco: W. H. Freeman and Company.

Westin, Allen. 1967. *Privacy and Freedom*. New York: Atheneum.

Zittrain, Jonathan. 2008. *The Future of the Internet: And How to Stop It*. New Haven, CT: Yale University Press.

34

The Robot Car of Tomorrow May Just Be Programmed to Hit You

- By Patrick Lin

Image: U.S. DOT

Suppose that an autonomous car is faced with a terrible decision to crash into one of two objects. It could swerve to the left and hit a Volvo sport utility vehicle (SUV), or it could swerve to the right and hit a Mini Cooper. If you were programming the car to minimize harm to others--a sensible goal--which way would you instruct it go in this scenario?

As a matter of physics, you should choose a collision with a heavier vehicle that can better absorb the impact of a crash, which means programming the car to crash into the Volvo. Further, it makes sense to choose a collision with a vehicle that's known for passenger safety, which again means crashing into the Volvo.

But physics isn't the only thing that matters here. Programming a car to collide with any particular kind of object over another seems an awful lot like a *targeting* algorithm, similar to those for military weapons systems. And this takes the robot-car industry down legally and morally dangerous paths.

Even if the harm is unintended, some crash-optimization algorithms for robot cars would seem to require the deliberate and systematic discrimination of, say, large vehicles to collide into. The owners or operators of these targeted vehicles would bear this burden through no fault of their own, other than that they care about safety or need an SUV to transport a large family. Does that sound fair?

What seemed to be a sensible programming design, then, runs into ethical challenges. Volvo and other SUV owners may have a legitimate grievance against the manufacturer of robot cars that favor crashing into them over smaller cars, even if physics tells us this is for the best.

Is This a Realistic Problem?

Some road accidents are unavoidable, and even autonomous cars can't escape that fate. A deer might dart out in front of you, or the car in the next lane might suddenly swerve into you. Short of defying physics, a crash is imminent. An autonomous or robot car, though, could make things better.

While human drivers can only react instinctively in a sudden emergency, a robot car is driven by software, constantly scanning its environment with unblinking sensors and able to perform many calculations before we're even aware of danger. They can make split-second choices to optimize crashes–that is, to minimize harm. But software needs to be programmed, and it is unclear how to do that for the hard cases.

Crash-avoidance algorithms can be biased in troubling ways.

In constructing the edge cases here, we are not trying to simulate actual conditions in the real world. These scenarios would be very rare, if realistic at all, but nonetheless they illuminate hidden or latent problems in normal cases. From the above scenario, we can see that crash-avoidance algorithms can be biased in troubling ways, and this is also at least a background concern any time we make a value judgment that one thing is better to sacrifice than another thing.

In previous years, robot cars have been quarantined largely to highway or freeway environments. This is a relatively simple environment, in that drivers don't need to worry so much about pedestrians and the countless surprises in city driving. But Google recently announced that it has taken the next step in testing its automated car in exactly city streets. As their operating environment becomes more dynamic and dangerous, robot cars will confront harder choices, be it running into objects or even people.

Ethics Is About More Than Harm

The problem is starkly highlighted by the next scenario, also discussed by Noah Goodall, a research scientist at the Virginia Center for Transportation Innovation and Research. Again, imagine that an autonomous car is facing an imminent crash. It could select one of two targets to swerve into: either a motorcyclist who is wearing a helmet, or a motorcyclist who is not. What's the right way to program the car?

In the name of crash-optimization, you should program the car to crash into whatever can best survive the collision. In the last scenario, that meant smashing into the Volvo SUV. Here, it means striking the motorcyclist who's wearing a helmet. A good algorithm would account for the much-higher statistical odds that the biker without a helmet would die, and surely killing someone is one of the worst things auto manufacturers desperately want to avoid.

But we can quickly see the injustice of this choice, as reasonable as it may be from a crash-optimization standpoint. By deliberately crashing into that motorcyclist, we are in effect penalizing him or her for being responsible, for wearing a helmet. Meanwhile, we are giving the other motorcyclist a free pass, even though that person is much less responsible for not wearing a helmet, which is illegal in most U.S. states.

By deliberately crashing into that motorcyclist, we are in effect penalizing him or her for being responsible, for wearing a helmet.

Not only does this discrimination seem unethical, but it could also be bad policy. That crash-optimization design may encourage some motorcyclists to not wear helmets, in order to not stand out as favored targets of autonomous cars, especially if those cars become more prevalent on the road. Likewise, in the previous scenario, sales of automotive brands known for safety may suffer, such as Volvo and Mercedes Benz, if customers want to avoid being the robot car's target of choice.

The Role of Moral Luck

An elegant solution to these vexing dilemmas is to simply not make a deliberate choice. We could design an autonomous car to make certain decisions through a random-number generator. That is, if it's ethically problematic to choose which one of two things to crash into–a large SUV versus a compact car, or a motorcyclist with a helmet versus one without, and so on–then why make a calculated choice at all?

A robot car's programming could generate a random number; and if it is an odd number, the car will take one path, and if it is an even number, the car will take the other path. This avoids the possible charge that the car's programming is discriminatory against large SUVs, responsible motorcyclists, or anything else.

This randomness also doesn't seem to introduce anything new into our world: luck is all around us, both good and bad. A random decision also better mimics human driving, insofar as split-second emergency reactions can be unpredictable and are not based on reason, since there's usually not enough time to apply much human reason.

A key reason for creating autonomous cars in the first place is that they should be able to make better decisions than we do

Yet, the random-number engine may be inadequate for at least a few reasons. First, it is not obviously a benefit to mimic human driving, since a key reason for creating autonomous cars in the first place is that they should be able to make better decisions than we do. Human error, distracted driving, drunk driving, and so on are responsible for 90 percent or more of car accidents today, and 32,000-plus people die on U.S. roads every year.

Second, while human drivers may be forgiven for making a poor split-second reaction–for instance, crashing into a Pinto that's prone to explode, instead of a more stable object–robot cars won't enjoy that freedom. Programmers have all the time in the world to get it right. It's the difference between premeditated murder and involuntary manslaughter.

Third, for the foreseeable future, what's important isn't just about arriving at the "right" answers to difficult ethical dilemmas, as nice as that would be. But it's also about being thoughtful about your decisions and able to defend them–it's about showing your moral math. In ethics, the process of thinking through a problem is as important as the result. Making decisions randomly, then, evades that responsibility. Instead of thoughtful decisions, they are *thoughtless*, and this may be worse than reflexive human judgments that lead to bad outcomes.

Can We Know Too Much?

A less drastic solution would be to hide certain information that might enable inappropriate discrimination–a "veil of ignorance", so to speak. As it applies to the above scenarios, this could mean not ascertaining the make or model of other vehicles, or the presence of helmets and other safety equipment, even if technology could let us, such as vehicle-to-vehicle communications. If we did that, there would be no basis for bias.

Not using that information in crash-optimization calculations may not be enough. To be in the ethical clear, autonomous cars may need to not collect that information at all. Should they be in possession of the information, and using it could have minimized harm or saved a life, there could be legal liability in failing to use that information. Imagine a similar public outrage if a national intelligence agency had credible information about a terrorist plot but failed to use it to prevent the attack.

A problem with this approach, however, is that auto manufacturers and insurers will want to collect as much data as technically possible, to better understand robot-car crashes and for other purposes, such as novel forms of in-car advertising. So it's unclear whether voluntarily turning a blind eye to key information is realistic, given the strong temptation to gather as much data as technology will allow.

So, Now What?

In future autonomous cars, crash-avoidance features alone won't be enough. Sometimes an accident will be unavoidable as a matter of physics, for myriad reasons –such as insufficient time to press the brakes, technology errors, misaligned sensors, bad weather, and just pure bad luck. Therefore, robot cars will also need to have crash-optimization strategies.

To optimize crashes, programmers would need to design cost-functions that potentially determine who gets to live and who gets to die.

To optimize crashes, programmers would need to design cost-functions–algorithms that assign and calculate the expected costs of various possible options, selecting the one with the lowest cost–that potentially determine who gets to live and who gets to die. And this is fundamentally an ethics problem, one that demands care and transparency in reasoning.

It doesn't matter much that these are rare scenarios. Often, the rare scenarios are the most important ones, making for breathless headlines. In the U.S., a traffic fatality occurs about once every 100 million vehicle-miles traveled. That means you could drive for more than 100 lifetimes and never be involved in a fatal crash. Yet these rare events are exactly what we're trying to avoid by developing autonomous cars, as Chris Gerdes at Stanford's School of Engineering reminds us.

Again, the above scenarios are not meant to simulate real-world conditions anyway, but they're thought-experiments–something like scientific experiments–meant to simplify the issues in order to isolate and study certain variables. In those cases, the variable is the role of ethics, specifically discrimination and justice, in crash-optimization strategies more broadly.

The larger challenge, though, isn't thinking through ethical dilemmas. It's also about setting accurate expectations with users and the general public who might find themselves surprised in bad ways by autonomous cars. Whatever answer to an ethical dilemma the car industry might lean towards will not be satisfying to everyone.

Ethics and expectations are challenges common to all automotive manufacturers and tier-one suppliers who want to play in this emerging field, not just particular companies. As the first step toward solving these challenges, creating an open discussion about ethics and autonomous cars can help raise public and industry awareness of the issues, defusing outrage (and therefore large lawsuits) when bad luck or fate crashes into us.

35

A Vindication of the Rights of Machines

David J. Gunkel

Received: 3 January 2013 / Accepted: 8 July 2013

Abstract This essay responds to the machine question in the affirmative, arguing that artifacts, like robots, AI, and other autonomous systems, can no longer be legitimately excluded from moral consideration. The demonstration of this thesis proceeds in four parts or movements. The first and second parts approach the subject by investigating the two constitutive components of the ethical relationship—moral agency and patiency. In the process, they each demonstrate failure. This occurs not because the machine is somehow unable to achieve what is considered necessary and sufficient to be a moral agent or patient but because the characterization of agency and patiency already fail to accommodate others. The third and fourth parts respond to this problem by considering two recent alternatives—the all-encompassing ontocentric approach of Luciano Floridi's information ethics and Emmanuel Levinas's eccentric ethics of otherness. Both alternatives, despite considerable promise to reconfigure the scope of moral thinking by addressing previously excluded others, like the machine, also fail but for other reasons. Consequently, the essay concludes not by accommodating the alterity of the machine to the requirements of moral philosophy but by questioning the systemic limitations of moral reasoning, requiring not just an extension of rights to machines, but a thorough examination of the way moral standing has been configured in the first place.

1 Introduction

One of the enduring concerns of moral philosophy is determining who or what is deserving of ethical consideration. Although initially limited to "other men," the practice of ethics has developed in such a way that it continually challenges its own restrictions and comes to encompass what had been previously excluded individuals and groups—foreigners, women, animals, and even the environment. "In the history of the United States," Susan Leigh Anderson (2008, 480) has argued, "gradually more and more beings have been granted the same rights that others possessed and we've become a more ethical society as a result. Ethicists are currently struggling with the question of whether at least some higher animals should have rights, and the status of human fetuses has been debated as well. On the horizon looms the question of whether intelligent machines should have moral standing." The following responds to this final question—what we might call the "machine question" in ethics—in the affirmative, arguing that machines, like robots, AI, and other autonomous systems, can no longer and perhaps never really could be excluded from moral consideration. Toward that end, this paper advances another "vindication discourse," following in a tradition that begins with Mary Wollstonecraft's *A Vindication of the Rights of Men* (1790) succeeded two years later by *A Vindication of the Rights of Woman* and Thomas Taylor's intentionally sarcastic yet remarkably influential response *A Vindication of the Rights of Brutes*.[1]

Although informed by and following in the tradition of these vindication discourses, or what Peter Singer (1989, 148) has also called a "liberation movement," the argument presented here will employ something of an unexpected approach and procedure. Arguments for the vindication of the rights of previously excluded others typically proceed by (a) defining or characterizing the criteria for moral considerability or what Thomas Birch (1993, 315) calls the conditions for membership in "the club of consideranda," and (b) demonstrating that some previously excluded entity or group of entities are in fact capable of achieving a threshold level for inclusion in this community of moral subjects. "The question of considerability has been cast," as Birch (1993, 314) explains, "and is still widely understood, in terms of a need for necessary and sufficient conditions which mandate practical respect for whomever or what ever fulfills them." The vindication of the rights of machines, however, will proceed otherwise. Instead of demonstrating that machines or at least one representative machine is able to achieve the necessary and sufficient conditions for moral standing (however that might come to be defined, characterized, and justified) the following both contests this procedure and demonstrates the opposite, showing how the very criteria that have been used to decide the question of moral considerability necessarily fail in the first place. Consequently, the vindication of the rights of machines will not, as one might have initially expected, concern some recent or future success in technology nor will it entail a description of or demonstration with a particular artifact; it will instead investigate fundamental failures in the procedures of moral philosophy itself—failures that render exclusion of the machine both questionable and morally suspect.

2 Moral Agency

Questions concerning moral standing typically begin by addressing agency. The decision to begin with this subject is not accidental, provisional, or capricious. It is dictated and prescribed by the history of moral philosophy, which has traditionally privileged agency and the figure of the moral agent in both theory and practice. As

---

[1] What is presented here in the form of a "vindication discourse" is an abbreviated version of an argument that is developed in greater detail and analytical depth in Gunkel (2012).

A Vindication of the Rights of Machines

Luciano Floridi explains, moral philosophy, from the time of the ancient Greeks through the modern era and beyond, has been almost exclusively agent-oriented. "Virtue ethics, and Greek philosophy more generally," Floridi (1999, 41) writes, "concentrates its attention on the moral nature and development of the individual agent who performs the action. It can therefore be properly described as an agent-oriented, 'subjective ethics.'" Modern developments, although shifting the focus somewhat, retain this particular agent-oriented approach. "Developed in a world profoundly different from the small, non-Christian Athens, Utilitarianism, or more generally Consequentialism, Contractualism and Deontologism are the three most well-known theories that concentrate on the moral nature and value of the actions performed by the agent" (Floridi 1999, 41). Although shifting emphasis from the "moral nature and development of the individual agent" to the "moral nature and value" of his or her actions, western philosophy has been, with few exceptions (which we will get to shortly), organized and developed as an agent-oriented endeavor.

When considered from the perspective of the agent, ethics inevitably and unavoidably makes exclusive decisions about who is to be included in the community of moral subjects and what can be excluded from consideration. The choice of words here is not accidental. As Jacques Derrida (2005, 80) points out everything turns on and is decided by the difference that separates the "who" from the "what." Moral agency has been customarily restricted to those entities who call themselves and each other "man"—those beings who already give themselves the right to be considered someone who counts as opposed to something that does not. But who counts—who, in effect, gets to be situated under the term "who"—has never been entirely settled, and the historical development of moral philosophy can be interpreted as a progressive unfolding, where what had once been excluded (i.e., women, slaves, people of color, etc.) have slowly and not without considerable struggle and resistance been granted access to the gated community of moral agents and have thereby also come to be someone who counts.

Despite this progress, which is, depending on how one looks at it, either remarkable or insufferably protracted, there remain additional exclusions, most notably non-human animals and machines. Machines in particular have been understood to be mere artifacts that are designed, produced, and employed by human agents for human specified ends. This instrumentalist and anthropocentric understanding has achieved a remarkable level of acceptance and standardization, as is evident by the fact that it has remained in place and largely unchallenged from ancient to postmodern times—from at least Plato's *Phaedrus* to Jean-François Lyotard's *The Postmodern Condition*. Beginning with the animal rights movement, however, there has been considerable pressure to reconsider the ontological assumptions and moral consequences of this legacy of human exceptionalism.

Extending consideration to these other previously marginalized subjects has required a significant reworking of the concept of moral agency, one that is not dependent on genetic make-up, species identification, or some other spurious criteria. As Singer (1999, 87) describes it, "the biological facts upon which the boundary of our species is drawn do not have moral significance," and to decide questions of moral agency on this ground "would put us in the same position as racists who give preference to those who are members of their race." For this reason, the question of moral agency has come to be disengaged from identification with the human being and is instead often referred to and made dependent upon the generic concept of "personhood." "There appears," G. E. Scott (1990, 7) writes, "to be more unanimity as regards the claim that in order for an individual to be a moral agent s/he must possess the relevant features of a person; or, in other words, that being a person is a necessary, if not sufficient, condition for being a moral agent." Corporations, for example, are artificial entities that are obviously otherwise than human, yet they are considered legal persons, having rights and responsibilities that are recognized and protected by both national and international law (French 1979). As promising as this "personist" innovation is, "the category of the person," to reuse terminology borrowed from Marcel Mauss (1985), is by no means settled and clearly defined. There is, in fact, little or no agreement concerning what makes someone or something a person and the literature on this subject is littered with different formulations and often incompatible criteria. "One might well hope," Daniel Dennett (1998, 267) writes, "that such an important concept, applied and denied so confidently, would have clearly formulatable necessary and sufficient conditions for ascription, but if it does, we have not yet discovered them. In the end there may be none to discover. In the end we may come to realize that the concept person is incoherent and obsolete."

In an effort to contend with, if not resolve this problem, researchers often focus on the one "person making" quality that appears on most, if not all, the lists of "personal properties," whether they include just a couple simple elements (Singer 1999, 87) or involve numerous "interactive capacities" (Smith 2010, 74), and that already has traction with practitioners and theorists—consciousness. "Without consciousness," John Locke (1996, 146) argued, "there is no person." Or as Kenneth Einar Himma (2009, 19) articulates it, "moral agency presupposes consciousness…and that the very concept of agency presupposes that agents are conscious." Formulated in this fashion, moral agency is something that is decided and made dependent on a prior determination of consciousness. If, for example, an animal or a machine can in fact be shown to possess "consciousness," then that entity would, on this account, need to be considered a legitimate moral agent. And not surprisingly, there has been considerable effort in the fields of philosophy, AI, and robotics to address the question of machine moral agency by targeting and examining the question and possibility (or impossibility) of machine consciousness.

This seemingly rational approach, however, runs into considerable complications. On the one hand, we do not, it seems, have any widely accepted characterization of "consciousness." The problem, then, is that consciousness, although crucial for deciding who is and who is not a moral agent, is itself a term that is ultimately undecided and considerably equivocal. "The term," as Max Velmans (2000, 5) points out, "means many different things to many different people, and no universally agreed core meaning exists." In fact, if there is any general agreement among philosophers, psychologists, cognitive scientists, neurobiologists, AI researchers, and robotics engineers regarding consciousness, it is that there is little or no agreement when it comes to defining and characterizing the concept. And to make matters worse, the problem is not just with the lack of a basic definition; the problem may itself already be a problem. "Not only is there no consensus on what the term consciousness denotes," Güven Güzeldere (1997, 7) writes, "but neither is it immediately clear if there actually is a single, well-defined 'the problem of consciousness' within disciplinary (let alone across disciplinary) boundaries. Perhaps the trouble lies not so much in the ill definition of the question, but in the fact that what passes under the term consciousness as an all too familiar, single, unified notion may be a tangled amalgam of several different concepts, each inflicted with its own separate problems."

On the other hand, even if it were possible to define consciousness or come to some tentative agreement concerning its necessary and sufficient conditions, we still lack any credible and certain way to determine its actual presence in another. Because consciousness is a property attributed to "other minds," its presence or lack thereof requires access to something that is and remains fundamentally inaccessible. "How does one determine," as Paul Churchland (1999, 67) famously characterized it, "whether something other than oneself—an alien creature, a sophisticated robot, a socially active computer, or even another human—is really a thinking, feeling, conscious being; rather than, for example, an unconscious automaton whose behavior arises from something other than genuine mental states?" And the available solutions to this "other minds problem," from reworkings and modifications of the Turing Test to functionalist approaches that endeavor to work around this problem altogether (Wallach and Allen 2009), only make things more complicated and indeterminate. "There is," as Dennett (1998, 172) points out, "no proving that something that seems to have an inner life does in fact have one—if by 'proving' we understand, as we often do, the evincing of evidence that can be seen to establish by principles already agreed upon that something is the case." Although philosophers, psychologists, and neuroscientists throw considerable argumentative and experimental effort at this problem, it is not able to be resolved in any way approaching what would pass for empirical science, strictly speaking.[2] In the end, not only are these tests unable to demonstrate with any certitude whether animals, machines, or other entities are in fact conscious and therefore legitimate moral persons (or not), we are left doubting whether we can even say the same for other human beings. As Ray Kurzweil (2005, 380) candidly concludes, "we assume other humans are conscious, but even that is an assumption," because "we cannot resolve issues of consciousness entirely through objective measurement and analysis (science)."

The question of machine moral agency, therefore, turns out to be anything but simple or definitive. This is not, it is important to note, because machines are somehow unable to be moral agents. It is rather a product of the fact that the term "moral agent," for all its importance and argumentative expediency, has been and remains an ambiguous, indeterminate, and rather noisy concept. What the consideration of machine moral agency demonstrates, therefore, is something that may not have been anticipated or sought. What is discovered in the process of pursuing this line of inquiry is not a satisfactory answer to the question whether machines are able to be moral agents or not. In fact, that question remains open and unanswered. What has been ascertained is that the concept of moral agency is already vague and imprecise such that it is (if applied strictly and rigorously) uncertain whether we—whoever this "we" includes—are in fact moral agents.

What has been demonstrated, therefore, is that moral agency, the issue that had been assumed to be the "correct" place to begin, turns out to be inconclusive.

---

[2] Attempts to resolve this problem often take the form of a pseudo-science called physiognomy, which endeavors to infer an entity's internal states of mind from the observation of its external expressions and behavior.

Although this could be regarded as a "failure," it is a particularly instructive failing. What is learned from this failure—assuming we continue to use this obviously "negative" word—is that moral agency is not necessarily some property that can be definitively ascertained or discovered in others prior to and in advance of their moral consideration. Instead moral standing may be something like what Kay Foerst has called a dynamic and socially constructed "honorarium" (Benford and Malartre 2007, 165) that comes to be conferred and assigned to others in the process of our interactions and relationships with them. In this way then, "moral standing is not," as Mark Coeckelbergh (2012, 25) argues, "about the entity but about us and about the relation between us and the entity." But then the deciding factor will no longer be one of agency; it will be a matter of patiency.

3 Moral Patiency

Moral patiency looks at the ethical relationship from the other side. It is concerned not with determining the moral character of the agent or weighing the ethical significance of his/her/its actions but with the victim, recipient, or receiver of such action. This approach is, as Mane Hajdin (1994), Luciano Floridi (1999 and 2013), and others have recognized, a significant alteration in procedure and a "non-standard" way to approach the question of moral rights and responsibilities. The model for this kind of transaction can be found in the innovations of animal rights philosophy. Whereas agent-oriented ethics have been concerned with determining whether someone is or is not a legitimate moral subject with rights and responsibilities, animal rights philosophy begins with an entirely different question—"Can they suffer?" (Bentham 2005, 283). What distinguishes this particular mode of inquiry, as Derrida (2008, 28) points out, is that it asks not about an ability or power of the active agent (however that would come to be defined) but about a fundamental passivity—the patience of the patient, words that are derived from the Latin verb patior, which connotes "suffering." "Thus the question will not be to know whether animals are of the type *zoon logon echon* [ζωον λόγον έχον] whether they *can* speak or reason thanks to that *capacity* or that *attribute* of the *logos* [λόγος], the *can-have* of the *logos*, the aptitude for the *logos*. The *first* and *decisive* question would be rather to know whether animals *can suffer*" (Derrida 2008, 27).

This seemingly simple and direct question introduces what turns out to be a major shift in the basic structure and procedures of moral thinking. On the one hand, it challenges the anthropocentric tradition in ethics by questioning the often unexamined privilege human beings have granted themselves. In effect, it institutes something like a Copernican revolution in moral philosophy. Just as Copernicus challenged the geocentric model of the cosmos and in the process undermined many of the presumptions of human exceptionalism, animal rights philosophy contests the established Ptolemaic system of ethics, deposing the anthropocentric privilege that had traditionally organized the moral universe. On the other hand, the effect of this fundamental shift in focus means that the one time closed field of ethics can be opened up to other kinds of non-human animals. In other words, who counts as morally significant are not just other "men" but all kinds of entities that had previously been marginalized and situated outside the gates of the moral community. "If a being suffers," Singer (1975, 9) writes, "there can be no moral justification for refusing to take that suffering into consideration. No matter what the nature of the being, the principle of equality requires that its suffering be counted equally with the like suffering of any other being."

Initially there seems to be good reasons and opportunities for extending this innovation to machines, or at least some species of machines (Gunkel 2007). This is because the animal and the machine, beginning with the work of René Descartes, share a common ontological status and position. For Descartes, the human being was considered the sole creature capable of rational thought—the one entity able to say, and be certain in its saying, *cogito ergo sum*. Following from this, he had concluded that other animals not only lacked reason but were nothing more than mindless automata that, like clockwork mechanisms, simply followed predetermined instructions programmed in the disposition of their various parts or organs. Conceptualized in this fashion, the animal and the machine, or what Descartes identified with the hybrid, hyphenated term *bête-machine*, were effectively indistinguishable and ontologically the same. "If any such machine," Descartes (1988, 44) wrote, "had the organs and outward shape of a monkey or of some other animal that lacks reason, we should have no means of knowing that they did not possess entirely the same nature as these animals."

Despite this fundamental and apparently irreducible similitude, only one of the pair has been considered a legitimate subject of moral concern. Even though the fate of the machine, from Descartes forward was intimately coupled with that of the animal, only the animal (and only some animals, at that) has qualified for any level of ethical consideration. And this exclusivity has been asserted and justified on the grounds that the machine, unlike the animal, does not experience either pleasure or pain. Steve Torrence (2008, 502) calls this "the organic view of ethical status" and demonstrates how philosophers have typically distinguished organic or biological organisms, either naturally occurring or synthetically developed, that are sentient and therefore legitimate subjects of moral consideration from what are termed "mere machines"—mechanisms that have no moral standing whatsoever. Although this conclusion appears to be rather reasonable and intuitive, it fails for a number of reasons.

First, it has been practically disputed by the construction of various mechanisms that now appear to suffer or at least provide external evidence of something that looks like pain. As Derrida (2008, 81) recognized, "Descartes already spoke, as if by chance, of a machine that simulates the living animal so well that it 'cries out that you are hurting it.'" This comment, which appears in a brief parenthetical aside in Descartes' *Discourse on Method*, had been deployed in the course of an argument that sought to differentiate human beings from the animal by associating the latter with mere mechanisms. But the comment can, in light of the procedures and protocols of animal ethics, be read otherwise. That is, if it were indeed possible to construct a machine that did exactly what Descartes had postulated, that is, "cry out that you are hurting it," would we not also be obligated to conclude that such a mechanism was capable of experiencing pain? This is, it is important to note, not just a theoretical point or speculative thought experiment. Engineers have, in fact, constructed mechanisms that synthesize believable emotional responses (Bates 1994; Blumberg et al. 1996; Breazeal and Brooks 2004), like the dental-training robot Simroid "who" cries out in pain when students "hurt" it (Kokoro 2009), and designed systems capable of evidencing behaviors that look a lot like what we usually call pleasure and pain.

Second it can be contested on epistemologically grounds insofar as suffering or the experience of pain is still unable to get around or resolve the problem of other minds. How, for example, can one know that an animal or even another person actually suffers? How is it possible to access and evaluate the suffering that is experienced by another? "Modern philosophy," Matthew Calarco (2008, 119) writes, "true to its Cartesian and scientific aspirations, is interested in the indubitable rather than the undeniable. Philosophers want proof that animals actually suffer, that animals are aware of their suffering, and they require an argument for why animal suffering should count on equal par with human suffering." But such indubitable and certain knowledge, as explained by Marian S. Dawkins, appears to be unattainable:

> At first sight, 'suffering' and 'scientific' are not terms that can or should be considered together. When applied to ourselves, 'suffering' refers to the subjective experience of unpleasant emotions such as fear, pain and frustration that are private and known only to the person experiencing them. To use the term in relation to non-human animals, therefore, is to make the assumption that they too have subjective experiences that are private to them and therefore unknowable by us. 'Scientific' on the other hand, means the acquisition of knowledge through the testing of hypotheses using publicly observable events. The problem is that we know so little about human consciousness that we do not know what publicly observable events to look for in ourselves, let alone other species, to ascertain whether they are subjectively experiencing anything like our suffering. The scientific study of animal suffering would, therefore, seem to rest on an inherent contradiction: it requires the testing of the untestable (Dawkins 2008, 1).

Because suffering is understood to be a subjective and private affair, there is no way to know, with any certainty or credible empirical method, exactly how another entity experiences unpleasant sensations such as fear, pain, or frustration. For this reason, it appears that the suffering of another (especially an animal) remains fundamentally inaccessible and unknowable. As Singer (1975, 11) readily admits, "we cannot directly experience anyone else's pain, whether that 'anyone' is our best friend or a stray dog. Pain is a state of consciousness, a 'mental event,' and as such it can never be observed." The question of machine moral patiency, therefore, leads to an outcome that was not necessarily anticipated. The basic problem is not whether the question "can they suffer?" applies to machines but whether anything that appears to suffer—human, animal, plant, or machine—actually does so at all.

Third, and to make matters even more complicated, we may not even know what "pain" and "the experience of pain" is in the first place. This point is something that is taken up and demonstrated by Dennett's "Why You Can't Make a Computer That Feels Pain" (1998). In this provocatively titled essay, originally published decades before the debut of even a rudimentary working prototype, Dennett imagines trying to disprove the standard argument for human (and animal) exceptionalism "by actually writing a pain program, or designing a pain-feeling robot" (191). At the end of what turns out to be a rather protracted and detailed consideration of the problem, he concludes that we cannot, in fact, make a computer that feels pain. But the reason for drawing this conclusion does not derive from what one might expect, nor does it offer any kind of support for the advocates of moral exceptionalism. According to Dennett, the reason you cannot make a computer that feels pain is not the result of some technological limitation with the mechanism or its programming. It is a product of the fact that we remain unable to decide what pain is in the first place. The best we are able to do, as Dennett illustrates, is account for the various "causes and effects of pain," but "pain itself does not appear" (218). What is demonstrated, therefore, is not that some workable concept of pain cannot come to be instantiated in the mechanism of a computer or a robot, either now or in the foreseeable future, but that the very concept of pain that would be instantiated is already arbitrary, inconclusive, and indeterminate. "There can," Dennett writes at the end of the essay, "be no true theory of pain, and so no computer or robot could instantiate the true theory of pain, which it would have to do to feel real pain" (228). Although Bentham's question "Can they suffer?" may have radically reoriented the direction of moral philosophy, the fact remains that "pain" and "suffering" are just as nebulous and difficult to define and locate as the concepts they were intended to replace.

Finally, all this talk about the possibility of engineering pain or suffering in a machine entails its own particular moral dilemma. "If (ro)bots might one day be capable of experiencing pain and other affective states," Wallach and Allen (2009), 209) write, "a question that arises is whether it will be moral to build such systems—not because of how they might harm humans, but because of the pain these artificial systems will themselves experience. In other words, can the building of a (ro)bot with a somatic architecture capable of feeling intense pain be morally justified and should it be prohibited?" If it were in fact possible to construct a machine that "feels pain" (however defined and instantiated) in order to demonstrate the limits of moral patiency, then doing so might be ethically suspect insofar as in constructing such a mechanism we do not do everything in our power to minimize its suffering. Consequently, moral philosophers and robotics engineers find themselves in a curious and not entirely comfortable situation. One needs to be able to construct such a mechanism in order to demonstrate moral patiency and the moral standing of machines; but doing so would be, on that account, already to engage in an act that could potentially be considered immoral. Or to put it another way, the demonstration of machine moral patiency might itself be something that is quite painful for others.

Despite initial promises, we cannot, it seems, make a credible case for or against the moral standing of the machine by simply following the patient-oriented approach modeled by animal rights philosophy. In fact, trying to do so produces some rather unexpected results. In particular, extending these innovations does not provide definitive proof that the machine either can be or is not able to be a similarly constructed moral patient. Instead doing so demonstrates how the "animal question"—the question that had in effect revolutionized ethics in the later half of the 20th century—might already be misguided and prejudicial. Although it was not necessarily designed to work in this fashion, "A Vindication of the Rights of Machines" achieves something similar to what Thomas Taylor had wanted for his *A Vindication of the Rights of Brutes*. Taylor, who wrote and distributed this pamphlet under the protection of anonymity, originally composed the essay as a means by which to parody and undermine the arguments that had been advanced in Wollstonecraft's *A Vindication of the Rights of Woman*. Taylor's text, in other words, was initially offered as a kind of *reductio ad absurdum* designed to exhibit what he perceived to be the conceptual failings of Wollstonecraft's protofeminist manifesto. Following suit, "A Vindication of the Rights of Machines" appears to have the effect of questioning and even destabilizing what had been achieved with animal rights philosophy. But as was the case with the consideration of moral agency, this negative outcome is informative and telling. In particular, it indicates to what extent this apparent revolution in moral thinking is, for all its insight and promise, still beset with fundamental problems that proceed not so much from the ontological condition of these other, previously excluded entities but from systemic problems in the very structure and protocols of moral reasoning.

4 Information Ethics

One of the criticisms of animal rights philosophy is that this moral innovation, for all its promise to intervene in the anthropocentric tradition, remains an exclusive and exclusionary practice. "If dominant forms of ethical theory," Calarco (2008, 126) argues "—from Kantianism to care ethics to moral rights theory—are unwilling to make a place for animals within their scope of consideration, it is clear that emerging theories of ethics that are more open and expansive with regard to animals are able to develop their positions only by making other, equally serious kinds of exclusions…" Environmental and land ethics, for instance, have been critical of animal rights philosophy for including some sentient creatures in the community of moral patients while simultaneously excluding other kinds of animals, plants, and the other entities that comprise the natural environment. In response to this exclusivity, environmental ethicists have argued for a further expansion of the moral community to include these marginalized others, or the excluded other of the animal other.

Although these efforts effectively expand the community of legitimate moral patients to include those others who had been previously left out, environmental ethics has also (and not surprisingly) been criticized for instituting additional omissions. "Even bioethics and environmental ethics," Floridi (2013, 64) argues, "fail to achieve a level of complete universality and impartiality, because they are still biased against what is inanimate, lifeless, intangible, abstract, engineered, artificial, synthetic, hybrid, or merely possible. Even land ethics is biased against technology and artefacts, for example. From their perspective, only what is intuitively alive deserves to be considered as a proper centre of moral claims, no matter how minimal, so a whole universe escapes their attention." According to this line of reasoning, bioethics and environmental ethics represents something of an incomplete innovation in moral philosophy. They have, to their credit, successfully challenged the excluded other of animal rights philosophy by articulating a more universal form of ethics that not only shifts attention to the patient but also expands who or what qualifies for inclusion as a moral patient. At the same time, however, these innovations remain ethically biased insofar as they substitute a biocentrism for animocentrism and in the process continue to exclude other entities, specifically technology and other kinds of artifacts.

In response to this, Floridi endeavors to take the innovations introduced by bioethics and environmental ethics one step further. He adopts their patient-oriented approach but "lowers the condition that needs to be satisfied, in order to qualify as a centre of moral concern, to the minimal common factor shared by any entity" (Floridi 2013, 64) whether animate, inanimate, or otherwise. For Floridi this lowest common denominator is informational and, for this reason, he gives his innovative proposal the name "Information Ethics" or IE. "IE is an ecological ethics that replaces biocentrism with ontocentrism. IE suggests that there is something even more elemental than life, namely being—that is, the existence and flourishing of all entities and their global environment—and something more fundamental than suffering, namely entropy, [which] here refers to any kind of destruction or corruption of informational objects, that is, any form of impoverishment of being including nothingness, to phrase it more metaphysically" (Floridi 2008, 47). Following the innovations of bio- and environmental ethics, Floridi expands the scope of moral philosophy by altering its focus and lowering the threshold for inclusion, or, to use Floridi's terminology, the level of abstraction (LoA). What makes someone or something a moral patient, deserving of some level of ethical consideration, is that it exists as a coherent body of information. Consequently, something can be said to be good, from an IE perspective, insofar as it respects and facilitates the informational welfare of a being and bad insofar as it causes diminishment, leading to an increase in information entropy. In fact, for IE, "fighting information entropy is the general moral law to be followed" (Floridi 2002, 300). This fundamental shift in focus provides for a moral theory that is more inclusive of others. "Unlike other non-standard ethics," Floridi (2013, 65) argues, "IE is more impartial and universal—or one may say less ethically biased—because it brings to ultimate completion the process of enlarging the concept of what may count as a centre of moral claims, which now includes every instance of information, no matter whether physically implemented or not."

Despite the fact that IE promises to bring "to ultimate completion" the patient-oriented innovation of bio-ethics, the proposal is not without its problems. First, in shifting emphasis from an agent-oriented to a patient-oriented ethics, IE (like animal rights philosophy and bio-ethics before it) simply inverts the two terms of a traditional binary opposition. If classic ethical thinking has been organized, for better or worse, by an interest in the character and/or actions of the agent at the expense of the patient, IE endeavors, following the example of previous innovations, to reorient things by placing emphasis on the other term. This maneuver is, quite literally, a revolutionary proposal, because it inverts or "turns over" the traditional arrangement. Inversion, however, is rarely in and by itself a satisfactory mode of intervention. As Nietzsche (1974), Heidegger (1962), Derrida (1978), and other poststructuralists have pointed out, the inversion of a binary opposition actually does little or nothing to disturb or to challenge the fundamental structure of the system in question (Gunkel 2007). In fact, inversion preserves and maintains the traditional structure, albeit in an inverted form. The effect of this on IE has been registered by Himma, who, in an assessment of Floridi's initial publications on the subject, demonstrates that a concern for the patient is really nothing more than the flip-side of good-old, agent-oriented, anthropocentric ethics. "To say that an entity X has moral standing (i.e., is a moral patient) is, at bottom, simply to say that it is possible for a moral agent to commit a wrong against X. Thus, X has moral standing if and only if (1) some moral agent has at least one duty regarding the treatment of X and (2) that duty is owed to X" (Himma 2004, 145). According to Himma's analysis, IE's patient-oriented ethics (or any patient-oriented ethics, for that matter) is not that different from traditional forms of agent-oriented ethics. It simply looks at the agent/patient couple from the other side and in doing so still operates on and according to the standard system. Although instituting a revolutionary alteration in perspective, IE's patient-oriented ethics do not necessarily change the rules of the game.

Second, IE is not limited to simply turning things around. It also enlarges the scope of moral consideration by reducing the minimum requirements for inclusion. "IE holds," Floridi (2013, 68–69) argues, "that every informational entity, insofar as it is an expression of Being, has a dignity constituted by its mode of existence and essence, defined here as the collection of all the elementary proprieties that constitute it for what it is." Like previous innovations, IE is interested in expanding membership in the moral community so as to incorporate previously excluded others. But, unlike previous efforts, it is arguably more inclusive of others and other forms of otherness. IE, therefore, contests and seeks to replace both the exclusive anthropocentric and biocentric theories with an "ontocentric" one, which is, by comparison, much more inclusive and universal. In taking this approach, however, IE simply substitutes one form of centrism for another. Anthropocentrism, for example, situates the human at the center of moral concern and admits into consideration anyone who is able to meet the basic criteria of what has been decided to comprise the human. Animiocentrism focuses attention on the animal and extends consideration to any organism that meets the defining criteria of animality. Biocentrism goes one step further in the process of abstraction; it defines life as the common denominator and admits into consideration anything and everything that can be said to be alive. And ontocentrism completes the progression by incorporating into moral consideration anything that actually exists, had existed, or potentially exists.

All of these innovations, despite their differences in focus, employ a similar maneuver and logic. That is, they redefine the center of moral consideration in order to describe progressively larger circles that come to encompass a wider range of possible participants. Although there are and will continue to be considerable debates about what should define the center and who or what is or is not included, this debate is not the problem. The problem rests with the strategy itself. In taking a centrist approach, these different ethical theories (of which IE would presumably be the final and ultimate form) endeavor to identify what is essentially the same in a phenomenal diversity of different individuals. Consequently, they include others by effectively stripping away and reducing differences. This approach, although having the appearance of being increasingly more inclusive, effaces the unique alterity of others and turns them into more of the same. This is, according to Levinas (1969 and Levinas 1981) the defining gesture of philosophy and one that does considerable violence to others. "Western philosophy," Levinas (1969, 43) argues, "has most often been an ontology: a reduction of the other to the same by interposition of a middle or neutral term that ensures the comprehension of being" (Levinas 1969, 43). The issue, therefore, is not deciding which form of centrism is more or less inclusive of others; the difficulty rests with this strategy itself, which succeeds only by reducing difference and turning what is other into a modality of the same.

Finally, this metaphysical operation is never neutral, and its moral consequences have been identified by environmental ethicists like Thomas Birch, who finds any and all efforts to articulate criteria for "universal consideration" to be based on a fundamentally flawed assumption. According to Birch, these efforts at increasingly more inclusive inclusion always proceed by way of articulating some necessary and sufficient conditions, or qualifying characteristics, that must be met by an entity in order to be incorporated into the community of legitimate moral subjects. In traditional forms of anthropocentric ethics, for example, it was the *anthropos* and the way it had been characterized (which it should be noted was always and already open to considerable social negotiation and redefinition), that provided the criteria for deciding who would be include in the moral community and what would not. The problem, Birch contends, is not necessarily with the criteria that are selected to make these decisions (although it is possible to argue that there have been better and worse formulations); the more fundamental problem is with the patient-oriented strategy and approach. "The institution of any practice of any criterion of moral considerablity," Birch (1993, 317) writes, "is an act of power over, and ultimately an act of violence" toward others. In other words, every criteria of moral inclusion, no matter how neutral, objective, or universal it appears, is an imposition of power insofar as it consists in the universalization of a particular value or set of values made by someone from particular position of power. "The nub of the problem with granting or extending rights to others," Birch (1995, 39) concludes, "a problem which becomes pronounced when nature is the intended beneficiary, is that it presupposes the existence and the maintenance of a position of power from which to do the granting." Even in the case of the relatively more inclusive and seemingly all-encompassing patient-oriented approach instituted by IE, someone has already been empowered to decide what particular criteria will be considered the necessary and sufficient conditions for inclusion in the class of "moral consideranda" (Birch 1993, 317).[3] The problem, then, is not only with the specific criteria that comes to be selected as the universal condition but also, and more so, the very act of universalization, which already empowers someone to make these decisions for others.

Although IE provides for a more complete and universal articulation of a patient-oriented ethics able to include others, including machines, this all-encompassing totalizing effect is simultaneously its greatest achievement and a critical problem. It is an achievement insofar as it carries through to completion the patient-oriented approach that begins to gain momentum with animal rights philosophy. IE promises, as Floridi (1999) describes it, to articulate an "ontocentric, patient-oriented, ecological macroethics" that includes everything, does not make other problematic exclusions, and is sufficiently universal, complete, and consistent. It is a problem insofar as this approach to greater inclusivity continues to deploy and support a strategy that is itself part and parcel of a "totalizing" (Levinas 1969) or "imperialist" (Birch 1993) program. The problem, then, is not which centrism one develops and patronizes or which criteria are determined to be more or less inclusive; the problem is with this approach itself. What is the matter with IE, therefore, is not the way Floridi develops this ultimate form of patient-oriented ethics, which has a good deal to commend it. The problem is with the patient-oriented methodology that it inherits, deploys, and leaves largely uninterogated. What is needed, therefore, is another approach, one that is not satisfied with being merely revolutionary in its innovations, one that does not continue to pursue a project of totalizing and potentially violent assimilation, and one that can respond to and take responsibility for what remains in excess of the entire

---

[3] Although it could be argued that Being is so general a criterion that it must escape this criticism, the fact of the matter is that Being is a concern of and for a particular being. In fact, Heidegger (1962, 32) famously defined the human being as that entity for whom Being is an issue: "Dasein is ontically distinctive in that it is ontological." Understood in this way, it is possible to conclude that IE is just another form of anthropocentric ethics insofar as its ontocentric focus is the defining condition in human Dasein.

conceptual field that has been delimited and defined by the binary pair of agent and patient. What is needed is some way of proceeding and thinking otherwise—a way that, in the context of and in response to IE's ontocentric ethics, would be "otherwise than being" (Levinas 1981).

5 Thinking Otherwise

When it comes to thinking otherwise, especially as it relates to the question concerning ethics, there is perhaps no philosopher better suited to the task than Emmanuel Levinas. Unlike a lot of what goes by the name of "moral philosophy," Levinasian thought does not rely on metaphysical generalizations, abstract formulas, or simple pieties. It is not only critical of the traditional tropes and traps of western ontology but proposes an "ethics of otherness" that deliberately resists and interrupts the metaphysical gesture par excellence, that is, the reduction of difference to the same. This radically different approach to thinking difference differently is not just a useful and expedient strategy. It is not, in other words, a mere gimmick. It constitutes a fundamental reorientation that effectively alters the rules of the game and the standard operating presumptions. In this way, "morality is," as Levinas (1969, 304) concludes, "not a branch of philosophy, but first philosophy." This fundamental reconfiguration, which puts ethics first in both sequence and status, permits Levinas to circumvent and deflect a lot of the difficulties that have traditionally tripped up moral thinking in general and efforts to address the moral status of the machine in particular.

First, for Levinas, the problems of other minds[4]—a difficulty, as we have seen, for both agent-oriented and patient-oriented approaches—is not some fundamental epistemological limitation that must be addressed and resolved prior to moral decision making but constitutes the very condition of the ethical relationship as an irreducible exposure to an other who always and already exceeds the boundaries of one's totalizing comprehension. Consequently Levinasian philosophy, instead of being derailed by the epistemological problem of other minds, immediately affirms and acknowledges it as the basic condition of possibility for ethics. Or as Richard Cohen (2001, 336) succinctly describes it in what could be a marketing slogan for Levinasian thought, "not 'other minds,' mind you, but the 'face' of the other, and the faces of all others." In this way, then, Levinas provides for a seemingly more attentive and empirically grounded approach to the problem of other minds insofar as he explicitly acknowledges and endeavors to respond to and take responsibility for the original and irreducible difference of others instead of getting involved with and playing all kinds of speculative (and unfortunately wrong-headed) head games. "The ethical relationship," Levinas (1987, 56) writes, "is not grafted on to an antecedent relationship of cognition; it is a foundation and not a superstructure…It is then more cognitive than cognition itself, and all objectivity must participate in it."

Second, and following from this, Levinas's concern with/for the Other (which is often capitalized like a proper name) will constitute neither an agent nor patient

---
[4] This analytic moniker is something that is not ever used by Levinas, who is arguably the most influential moral thinker in the continental tradition. The term, however, has been employed by a number of Levinas's Anglophone interpreters.

oriented ethics, but addresses itself to what is anterior to and remains in excess of this seemingly fundamental binary structure—the basic structure that, Floridi (1999, 41) asserts, constitutes the logical form of any and all action, whether morally loaded or not. Although Levinas's attention to and concern for others looks, from one perspective at least, to be a kind of "patient oriented" ethics that puts the interests and rights of the other before oneself, it is not and cannot be satisfied with simply endorsing one side of or conforming to the agent/patient couple. Unlike Floridi's IE, which advocates a patient-oriented ethics in opposition to the customary agent-oriented approaches that have maintained a controlling interest in the field, Levinas goes one step further, releasing what could be called a deconstruction[5] of the very conceptual order of agent and patient. This alternative, as Levinas (1981, 117) explains is located "on the hither side of the act-passivity alternative" and, for that reason, significantly reconfigures the standard terms and conditions. "For the condition for," Levinas (1981, 123) explains, "or the unconditionality of, the self does not begin in the auto-affection of a sovereign ego that would be, after the event, 'compassionate' for another. Quite the contrary: the uniqueness of the responsible ego is possible only in being obsessed by another, in the trauma suffered prior to any auto-identification, in an unrepresentable before." The self or the ego, as Levinas describes it, does not constitute some preexisting self-assured condition that is situated before and as the cause of the subsequent relationship with an other. It does not (yet) take the form of an active agent who is able to decide to extend him/herself to others in a deliberate act of compassion. Rather it becomes what it is as a byproduct of an uncontrolled and incomprehensible exposure to the face of the other that takes place prior to and in advance of any formulation of the self in terms of agency.

Likewise the Other is not comprehended as a patient who would be the recipient of the agent's actions and whose interests and rights would need to be identified, taken into account, and duly respected. Instead, the absolute and irreducible exposure to the Other is something that is anterior and exterior to these distinctions, not only remaining beyond the range of their conceptual grasp and regulation but also making possible and ordering the antagonistic structure that subsequently comes to characterize the difference that distinguishes the self from its others and the agent from the patient in the first place. In other words, for Levinas at least, prior determinations of agency and patiency do not first establish the terms and conditions of any and all possible encounters that the self might have with others and with other forms of otherness. It is the other way around. The Other first confronts, calls upon, and interrupts self-involvement and in the process determines the terms and conditions by which and in response to which the standard roles of moral agent and moral patient come to be articulated and assigned. Consequently, Levinas's philosophy is not what is typically understood as an ethics, a meta-ethics, a normative ethics, or even an applied ethics. It is, what John Llewelyn (1995, 4) has called a "proto-ethics" or what others have termed an "ethics of ethics." "It is true," Derrida explains, "that Ethics in Levinas's sense is an Ethics without law and without concept, which maintains its

---

[5] Employing the term "deconstruction" in this particular context is somewhat problematic. This is because deconstruction does not necessarily sit well with Levinas's own work. Levinas, both personally and intellectually, had a rather complex relationship with Jacques Derrida, the main proponent of what is often mislabeled "deconstructivism," and an even more complicated, if not contentious one with Martin Heidegger, the thinker who Derrida credits with having first introduced the concept and practice.

non-violent purity only before being determined as concepts and laws. This is not an objection: let us not forget that Levinas does not seek to propose laws or moral rules, does not seek to determine a morality, but rather the essence of the ethical relation in general. But as this determination does not offer itself as a theory of Ethics, in question, then, is an Ethics of Ethics" (Derrida 1978, 111). In comparison to the anthropocentrism of the standard, agent-oriented approach and the animocentric/biocentric/ontocentric efforts of the various non-standard, patient-oriented alternatives, we can say that Levinas proposes a truly eccentric philosophy that exceeds the orbit and conceptual grasp of both.

Despite the promise this innovation has for arranging a moral philosophy that is radically situated otherwise, Levinas's work remains committed to and is not able to escape the influence of anthropocentric privilege and human exceptionalism. Whatever the import of his unique contribution, "other" in Levinas is still and unapologetically human (Levinas 2003). Although he is not the first to identify it, Jeffrey Nealon provides what is perhaps one of the most succinct description of the problem: "In thematizing response solely in terms of the human face and voice, it would seem that Levinas leaves untouched the oldest and perhaps most sinister unexamined privilege of the same: anthropos [ἄνθρωπος] and only *anthropos*, has *logos* [λόγος]; and as such, *anthropos* responds not to the barbarous or the inanimate, but only to those who qualify for the privilege of 'humanity,' only those deemed to possess a face, only to those recognized to be living in the *logos*" Nealon (1998, 71). For Levinas, therefore, technological devices may have an interface, but they do not possess a face or confront us in a face-to-face encounter that would call for and would be called ethics. If Levinasian philosophy is to provide a way of thinking otherwise that is able to respond to and to take responsibility for these other forms of otherness (and not just machines but non-human animals as well), we will need to employ and interpret this innovation against and in excess of Levinas's own interpretation of it. We will need as Derrida (1978, 260) once wrote of Georges Bataille's exceedingly careful engagement with the thought of Hegel, to follow Levinas to the end, "to the point of agreeing with him against himself" and of wresting his discoveries from the limited interpretations that he had provided.

Such efforts at "radicalizing Levinas," as Atterton and Calarco (2010) refer to it, will take up and pursue Levinas's "ethics of otherness" in excess of and beyond the rather restricted formulations that he and his advocates and critics have typically provided. "Although Levinas himself is for the most part unabashedly and dogmatically anthropocentric," Calarco (2008, 55) writes, "the underlying logic of his thought permits no such anthropocentrism. When read rigorously, the logic of Levinas's account of ethics does not allow for either of these two claims. In fact... Levinas's ethical philosophy is, or at least should be, committed to a notion of universal ethical consideration, that is, an agnostic form of ethical consideration that has no a priori constraints or boundaries" (Calarco 2008, 55).[6] This reworking of Levinasian philosophy promises to provide a much more inclusive articulation that is

---

[6] In stating this, we immediately run up against and need to confront the so-called problem of relativism—"the claim that no universally valid beliefs or values exist" (Ess 1996, 204). Although a complete response to this problem lies outside the scope of this particular essay, we should, at this point at least, recognize that "relativism" is not necessarily a pejorative term. For more on this issue see Scott (1967), Žižek (2006), and Gunkel (2012).

able to take other forms of otherness into account. And it is a compelling proposal. What is interesting about Calarco's argument (and the arguments offered by other Levinasian influenced thinkers, like Benso, 2000), however, is not the other forms of otherness that come to be included by way of this innovative reconfiguration of Levinasian thought, but what (unfortunately) gets left out in the process. According to the letter of Calarco's text the following entities could be given consideration: "'lower' animals, insects, dirt, hair, fingernails, and ecosystems" (Calarco 2008, 71). What is obviously missing from this list is anything that is not "natural," that is, any form of artifact or technology. Consequently, what gets left behind or left out by Calarco's "universal ethical consideration" are tools, technologies, and machines. Despite the fact that "universal consideration would entail being ethically attentive and open to the possibility that anything might take on face" (Calarco 2008, 73), machines appear to be the faceless constitutive exception. For this reason, "thinking otherwise," although clearly offering a compelling alternative to both agent and patient oriented ethics, still fails to respond to and take full responsibility for the machine.

6 Conclusion

"Every philosophy," Silvia Benso (2000, 136) writes in a comprehensive gesture that performs precisely what it seeks to address, "is a quest for wholeness." This objective, she argues, has been typically targeted in one of two ways. "Traditional Western thought has pursued wholeness by means of reduction, integration, systematization of all its parts. Totality has replaced wholeness, and the result is totalitarianism from which what is truly other escapes, revealing the deficiencies and fallacies of the attempted system." This is precisely the kind of violent philosophizing that Levinas (1969) identifies under the term "totality," and it includes the efforts of both standard agent-oriented and non-standard patient-oriented approaches up to and including information ethics. The alternative to these totalizing transactions is a philosophy that is oriented otherwise, like that proposed by Levinas. This other approach, however, "must do so by moving not from the same, but from the other, and not only the Other, but also the other of the Other, and, if that is the case, the other of the other of the Other. In this must, it must also be aware of the inescapable injustice embedded in any formulation of the other" (Benso 2000, 136). And this "injustice" is evident not only in Levinas's exclusive humanism but in the way that those who seek to redress this "humanism of the other" continue to ignore or marginalize the machine.

For these reasons, the question concerning machine moral standing does not end with a single definitive answer—a simple and direct "yes" or "no." But this inability is not, we can say following the argumentative strategy of Dennett's "Why you Cannot Make a Computer that Feels Pain" (1998), necessarily a product of some inherent or essential deficiency with the machine. Instead it is a result of the fact that moral agency, moral patiency, and those ethical theories that endeavor to think otherwise already deploy and rely on questionable constructions and logics. The vindication of the rights of machines, therefore, is not simply a matter of extending moral consideration to one more historically excluded other, which would, in effect, leave the mechanisms of moral philosophy in place, fully operational, and unchallenged. Instead, the question concerning the "rights of machines" makes a fundamental claim on ethics, requiring us to rethink the system of moral considerability all the way down. This is, as Levinas (1981, 20) explains, the necessarily "interminable methodological movement of philosophy" that continually struggles against accepted practices in an effort to think otherwise—not just differently but in ways that are responsive to and responsible for others.

Consequently, this essay ends not as one might have expected. That is, by accumulating evidence or arguments in favor of permitting machines, or even one representative machine, entry into the community of moral subjects. Instead, it concludes with questions about ethics and the way moral philosophy has typically defined and decided moral standing. Although ending in this questionable way—in effect, responding to a question with a question—is commonly considered bad form, this is not necessarily the case. This is because questioning is a particularly philosophical enterprise. "I am," Dennett (1996, vii) writes, "a philosopher and not a scientist, and we philosophers are better at questions than answers. I haven't begun by insulting myself and my discipline, in spite of first appearances. Finding better questions to ask, and breaking old habits and traditions of asking, is a very difficult part of the grand human project of understanding ourselves and our world." The objective of the vindication of the rights of machines, therefore, has not been to answer the machine question with some definitive proof or preponderance of evidence, but to ask about the very means by which we have gone about trying to articulate and formulate this question. The issue, then, is not can the machine be a moral agent, a moral patient, or something else? Instead the question concerns how moral agency and patiency have been configured and how these configurations already accommodate and/or marginalize others. The vindication of the rights of machines, therefore, is not just one more version or iteration of an applied moral philosophy; it releases a thorough and profound challenge to what is called "ethics."

References

Anderson, S. (2008). Asimov's 'Three Laws of Robotics' and Machine Metaethics. *AI & Society, 22*(4), 477–493.
Atterton, P., & Calarco, M. (2010). *Radicalizing Levinas*. Albany, NY: SUNY Press.
Bates, J. (1994). The Role of Emotion in Believable Agents. *Communications of the ACM, 37*, 122–125.
Benford, G., & Malartre, E. (2007). *Beyond Human: Living with Robots and Cyborgs*. New York: Tom Doherty.
Benso, S. (2000). *The Face of Things: A Different Side of Ethics*. Albany, NY: SUNY Press.
Bentham, J. (2005). *An Introduction to the Principles of Morals and Legislation*. J. H. Burns and H. L. Hart (Eds.). Oxford: Oxford University Press.
Birch, T. (1993). Moral Considerability and Universal Consideration. *Environmental Ethics, 15*, 313–332.
Birch, T. (1995). The Incarnation of Wilderness: Wilderness Areas as Prisons. In M. Oelschlaeger (Ed.), *Postmodern Environmental Ethics* (pp. 137–162). Albany, NY: State University of New York Press.
Blumberg, B., Todd, P., & Maes, M. (1996). No Bad Dogs: Ethological Lessons for Learning. In *Proceedings of the 4th International Conference on Simulation of Adaptive Behavior* (pp. 295–304). Cambridge, MA: MIT Press.
Breazeal, C., & Brooks, R. (2004). Robot Emotion: A Functional Perspective. In J. M. Fellous & M. Arbib (Eds.), *Who Needs Emotions: The Brain Meets the Robot* (pp. 271–310). Oxford: Oxford University Press.

A Vindication of the Rights of Machines

Calarco, M. (2008). *Zoographies: The Question of the Animal from Heidegger to Derrida*. New York: Columbia University Press.
Churchland, P. M. (1999). *Matter and Consciousness*. Cambridge, MA: MIT Press.
Coeckelbergh, M. (2012). *Growing Moral Relations: Critique of Moral Status Ascription*. New York: Palgrave Macmillan.
Cohen, R. A. (2001). *Ethics, Exegesis and Philosophy: Interpretation After Levinas*. Cambridge: Cambridge University Press.
Dawkins, M. S. (2008). The Science of Animal Suffering. *Ethology, 114*(10), 937–945.
Dennett, D. (1998). *Brainstorms: Philosophical Essays on Mind and Psychology*. Cambridge, MA: MIT Press.
Dennett, D. (1996). *Kinds of Minds: Toward an Understanding of Consciousness*. New York: Basic Books.
Derrida, J. (1978). *Writing and Difference*. Trans. by Alan Bass. Chicago: University of Chicago Press.
Derrida, J. (2005). *Paper Machine*. Trans. by Rachel Bowlby. Stanford, CA: Stanford University Press.
Derrida, J. (2008). *The Animal That Therefore I Am*. Trans. by David Wills. New York: Fordham University Press.
Descartes, R. (1988). *Selected Philosophical Writings*. Trans. by J. Cottingham, R. Stoothoff, and D. Murdoch. Cambridge: Cambridge University Press.
Ess, C. (1996). The Political Computer: Democracy, CMC, and Habermas. In C. Ess (Ed.), *Philosophical Perspectives on Computer-Mediated Communication* (pp. 196–230). Albany, NY: SUNY Press.
Floridi, L. (1999). Information Ethics: On the Philosophical Foundation of Computer Ethics. *Ethics and Information Technology, 1*(1), 37–56.
Floridi, L. (2002). On the Intrinsic Value of Information Objects and the Infosphere. *Ethics and Information Technology, 4*, 287–304.
Floridi, L. (2008). Information Ethics, its Nature and Scope. In J. van den Hoven & J. Weckert (Eds.), *Information Technology and Moral Philosophy* (pp. 40–65). Cambridge: Cambridge University Press.
Floridi, L. (2013). *The Ethics of Information*. Oxford: Oxford University Press.
French, P. (1979). The Corporation as a Moral Person. *American Philosophical Quarterly, 16*(3), 207–215.
Gunkel, D. J. (2007). *Thinking Otherwise: Philosophy, Communication, Technology*. West Lafayette, IN: Purdue University Press.
Gunkel, D. J. (2012). *The Machine Question: Critical Perspectives on AI, Robots and Ethics*. Cambridge, MA: MIT Press.
Güzeldere, G. (1997). The Many Faces of Consciousness: A Field Guide. In N. Block, O. Flanagan, & G. Güzeldere (Eds.), *The Nature of Consciousness: Philosophical Debates* (pp. 1–68). Cambridge, MA: MIT Press.
Hajdin, M. (1994). *The Boundaries of Moral Discourse*. Chicago: Loyola University Press.
Heidegger, M. (1962). *Being and Time*. Trans. by J. Macquarrie and E. Robinson. New York: Harper and Row Publishers.
Himma, K. E. (2004). There's Something About Mary: The Moral Value of Things qua Information Objects. *Ethics and Information Technology, 6*(3), 145–195.
Himma, K. E. (2009). Artificial Agency, Consciousness, and the Criteria for Moral Agency: What Properties Must an Artificial Agent Have to be a Moral Agent? *Ethics and Information Technology, 11*(1), 19–29.
Kokoro LTD (2009). http://www.kokoro-dreams.co.jp/
Kurzweil, R. (2005). *The Singularity Is Near: When Humans Transcend Biology*. New York: Viking.
Levinas, E. (1969). *Totality and Infinity: An Essay on Exteriority*. Trans. by A. Lingis. Pittsburgh, PA: Duquesne University Press.
Levinas, E. (1981). *Otherwise than Being Or Beyond Essence*. Trans. by Alphonso Lingis. Hague: Martinus Nijhoff Publishers.
Levinas, E. (1987). *Collected Philosophical Papers*. Trans. by A. Lingis. Dordrecht: Martinus Nijhoff Publishers.
Levinas, E. (2003). *Humanism of the Other*. Trans. by Nidra Poller. Urbana: University of Illinois Press.
Llewelyn, J. (1995). *Emmanuel Levinas: The Genealogy of Ethics*. London: Routledge.
Locke, J. (1996). *An Essay Concerning Human Understanding*. Indianapolis, IN: Hackett.
Mauss, M. (1985). *A Category of the Human Mind: The Notion of Person; The Notion of Self*. Trans. by W. D. Halls. In M. Carrithers, S. Collins, and S. Lukes (Eds.) *The Category of the Person* (pp. 1–25). Cambridge: Cambridge University Press.
Nealon, J. (1998). *Alterity Politics: Ethics and Performative Subjectivity*. Durham, NC: Duke University Press.
Nietzsche, F. (1974). *The Gay Science*. Trans. by W. Kaufmann. New York: Vintage Books.
Scott, G. E. (1990). *Moral Personhood: An Essay in the Philosophy of Moral Psychology*. Albany, NY: SUNY Press.

Scott, R. L. (1967). On Viewing Rhetoric as Epistemic. *Central States Speech Journal, 18*, 9–17.
Singer, P. (1975). *Animal Liberation: A New Ethics for Our Treatment of Animals*. New York: New York Review of Books.
Singer, P. (1989). All Animals are Equal. In T. Regan & P. Singer (Eds.), *Animal Rights and Human Obligations* (pp. 148–162). New York: Prentice Hall.
Singer, P. (1999). *Practical Ethics*. Cambridge: Cambridge University Press.
Smith, C. (2010). *What Is a Person? Rethinking Humanity, Social Life, and the Moral Good from the Person Up*. Chicago: University of Chicago Press.
Torrence, S. (2008). Ethics and Consciousness in Artificial Agents. *AI & Society, 22*, 495–521.
Velmans, M. (2000). *Understanding Consciousness*. New York: Routledge.
Wallach, W., & Allen, C. (2009). *Moral Machines: Teaching Robots Right from Wrong*. Oxford: Oxford University Press.
Žižek, S. (2006). *The Parallax View*. Cambridge, MA: MIT Press.

Index

Page numbers in bold refer to tables. Page numbers in italics refer to figures.

AAAI Presidential Panel on Long-Term AI Futures 13–4, 17–8, 19, 21, 26
Aanestad, Margun 364
abortion 469
access and privacy rights 492
accessibility and roboethics 83
accountability: artificial agents 343–4; childcare robots 177–8, 179; computer ethics 339–40; human-machine interface 371–2; laws of robotics 406–7
ACM Code of Ethics 340, **341**
acronyms **320**
act utilitarianism 240, 241
action and actors: actor network theory 362, 363–4; care-centered framework 200; conscious artificial moral agents 311–2; LIDA model of cognition 307, 309
action logic 235
ActiveBuddy, Inc. 501
activism, engineering 67
Actroid DER2 221–2
Adam, D. 45
adaptation *see* evolution and adaptation
addiction, technology 83
advanced artificial intelligence *see* superintelligence
adverse effects, labor market 20
affective states *see* emotions and affective states
affiliation and caregiving 149, 150
agencies of observation 365
agency: causal agency 89; empirical studies 362–5; health care 363–4; human-machine interface 370–1; machine ethics 239, 243–4; *see also* artificial agents; moral agency
agent architectures 22
agent judgments 227
agential realism 365
AI systems verification 22
AIBO 110, 111, 112, 113, 162–3

aircraft autopilot functions 234, 480
AIST therapeutic robots 138
Akrich, M. 190
Aleksander, Igor 263
algorithms: crash-optimization algorithms 508–9; learning algorithms 22, 235, 240, 245
Allen, C.: APACHE system 131; behavioral standards 23; care robots 134; LIDA 301–2; moral decision-making agents 300; "Why Machine Ethics?" 480
alternative laws of robotics 408–11, **410**
ambivalent atachment 170
analogy, reasoning by 269
Anderson, Michael 271
Anderson, Susan Leigh 271, 512
androids *see* humanoids
anesthesia 363–4
animals: appearance and moral status 121–2; artificial pets 213; attachment theory 173; borderline status of personhood 468, 469, 470; evolution 218; expanding moral circle 389, 512; moral agency 513; moral patiency 318, 516–7; moral responsibility 336; moral status 342; proto-morality and pre-reflective consciousness 310; real *vs.* robot dogs and children 163; rights 302; robotic *vs.* real 130
answer set programming (ASP) 244
anthropocentrism: agenthood 318–9; constitutional personhood 444–5; information ethics 522; moral agency 513; moral responsibility 334–5; 'other minds' problem 526
anthropomorphism: childcare robots 162, 166; children's relationship to robots 163; love and sex robots 222; privacy and social robots 499; relational artifacts 108; roboethics 83
anti-virus software 341
anxious avoidant attachment 170
APACHE system 131

appearance: care robots 131; facial expressions 102–3, 164, 165–6; moral appearances 120–1; privacy and social robots 499; reality and 102
applied computer technology 234
Araki, Kenji 242
archeological knowledge 363
aresponsible morality 332–7
Aristotle: empathy 145; reciprocity 143–4; virtue ethics 234, 260, 261–2
Arkoudas, Konstantine 243
Aron, A. 217–8
Aron, E.N. 217–8
art 370–1
artificial agents: AA-AA interactions 396–7; computer ethics 338–42; debate on the morality of 351–8; defining 320–1; European regulations 478–81; examples 326–31; interactivity, autonomy, and adaptability 325–6; Method of Abstraction 321–3; moral agents and moral patients 318; moral responsibility 332–7; morality of 343–4; relativism 323; sociality 396–7; standard *vs.* nonstandard theories of 317–9; state and state-transitions 324–5; suffering and the appearance of suffering 389–90; synthetic-biological humanoids 397–9
"Artificial Endocrine System" 220
artificial evil 319
artificial general intelligence 301, 311
artificial good 319
artificial intelligence: cognitive science *vs.* philosophy 416; consciousness 448–50; constitutional personhood scenario 440–2; emotion 453–5; free will 456–8; intentionality 451; interests 455–6; judgment and trusteeship 432–7; legal personhood, question of 465–71; nonperson argument against constitutional personhood 442–6; as property 460–3; real *vs.* artificial intelligence trustees 437–9; responsibility and trusteeship 428–32; simulation argument against constitutional personhood 458–60; soul 446–7; tests of 419–22; trusteeship legal question 427–8; trusteeship scenario 424–6; 'Turing Dream' 391–3; Turing Test and constitutional personhood 464
Artificial Intelligence Laboratory 103n15; Kismet 115

artificial life experiments 252, 254
artificial love companions *see* love and sex robots
artificial moral agents 93, 331–8; aim of robot ethics 88; bottom-up engineering 254–6; challenges of creating 7; computer ethics 338–42; consciousness 311–2; design of 250–3, 251–2; LIDA model of cognition 304–12; merging top-down and bottom-up engineering 259–64; models of morality 59–61; Moral Turing Test 56–7; morally praiseworthy agents 63; robots as 81; social mechanisms and supra-rational faculties 262–3; suprarational capabilities and social mechanisms 300–1; theories 57–9; top-down engineering approach 263–4
artificial neural networks *283,* 283–8, *285,* 290
artificial pets 213
"artificial stupidity" 40
Asaro, Peter 187–8
Asimov, Isaac: biography 34; coining the term robotics 35; *I, Robot* 405; machine ethics *vs.* human ethics 239; *Moral Machines* 405; origins of robotics 33–4; scientific optimism 4; *see also* laws of robotics
ASP (answer set programming) 244
assistive robots 201
Association for Computing Machinery 478
Association for Unmanned Vehicle Systems International 408
associative learning 60–1
asymmetry and Kantian machine ethics 276
attachment, children-robot 162–3, 167–74
attention codelets 309
attentiveness: care, elements of 196, 197; care-centered framework 199–200; patient lifting robot design 206
Atterton, P. 526
auditing robot compliance 40, 50
authenticity and a culture of simulation 108–9
authorship 441n88
autistic children 130, 131
auto-fire weapons systems 487
automata: in literature 80; robots, history of 34–5
automated teller machines 49, 188, 190, 234
autonomous robots: childcare robots 158–61; European regulations 476–81; patient lifting robots 204

autonomous vehicles: programming challenges 12; programming for ethical dilemmas 507–9; research priorities 20–1; responsibility and agency 91
autonomous weapons 13, 21, 92–3
autonomy: agenthood 325–31; alternative laws of robotics 409; care robots 128, 130; computer ethics 350; computer programs 354; elderly and disabled persons 129; explicit ethical machines 235; free speech 442; lack of free will 456–8; laws of robotics 40; machine ethics 242, 244, 245; moral agency 53–4; superintelligence 71; *see also* patient autonomy
autopilot functions, aircraft 234, 480

Baars, Bernard 302, 304, 305, 306
Baber, H. E. 127
Balsamo, Anne 369
Banks, M. R. 130
Barad, Karen 361, 365, 371, 372
Barry, Andrew 372
Bataille, Georges 526
Baxter, G. D. 484
Beauchamp, T. I. 244, 246
Becker, Lawrence 144
behavior: attachment theory 171; computer/human 355–6; morality, function of 225–6; validity 23
behaviourism 120
Bello, Paul 243
Belsky, J. 171
beneficial AI as research priority 19–26
benefits: care robots 152; caregiving 141–3; explicit ethical machines 239; superintelligence 73–4
benevolence 295
Benso, Silvia 527
Bentham, Jeremy 240, 519
Berleur, Jacques 233–4
Bianchi-Demicheli, F. 218
bias in computer systems 65
The Bicentennial Man (Asimov) 46
"The Bicentennial Man" (Asimov) 239
Bijker, Wiebe 350
binarism 362, 394–5, 521
binary-valued feedback 60–1
biocentrism 522
bioethics 6, 520–1

biofeedback, social 172
biological humanoids, synthetic 397–9
biomedical ethics 238, 244–7
Birch, Thomas 512, 522–3
birth control 219–20
blame: artificial moral agents 63; moral competence in social robots 228
Bodies that Matter (Butler) 367
body and gender 367, 369
body image 217
Bohr, Niels 364
bonding: attachment theory 171; care robots 131; care-giver/care-receiver 208; children-robot attachment 162–3
borderline status of personhood 468–71
Borenstein, J. 143, 145, 147, 148
bottom-up approaches: Kantian machine ethics 279, 280; LIDA model of cognition 307; machine ethics 253, 254–6, 265
Breazeal, Cynthia 115
bright line (machine ethics) 235–6
Bringsjord, Selmer 132, 243
broad domains 24–5
Brock, D. W. 244, 247
Brooks, R. 98n8, 482
Brynjolfsson, Erik 20
Buchanan, A. E. 244, 247
"Buridan's ass" problem 39
Buss, D. 215–6
Butler, Judith 367

CAD systems 370
Calarco, Matthew 518, 526–7
California v. Ciraolo 495
Calo, Ryan 20
Calverley, D. 302
camel jockeys 234, 481
Campbell, John W. 34
capabilities approach 126–34, 148–53
capacity and trusteeship 436–7
Çapek, Karel 35
care, defining 129, 192
care centered value-sensitive design 198, 202–8
care ethics *see* ethics of care
care robots 8; acceptance of 130–1; capabilities approach 126–8, 128–30, 148–53; care centered value sensitive design 202–8; care practices 195; care-centered framework 198–202; definition 187; design, importance

of 189–91; ethical values 195–7; European regulations 485–6; values and design 125–6, 188–9; virtue ethics 143–6; *see also* childcare robots
care values 125–6, 198, 202–8
care-centered framework 198–202
caregiving and caregivers: burdens and benefits 141–3; care robots, impact of 131–2, 140, 152–3; ethics of care and care values 193–8; patient lifting robot design 208; relational artifacts 108, 111, 115
Care-O-bot II 130
cartoons *39, 45, 46*
case-based reasoning: machine ethics 243; particularism and generalism 283–7
Castañeda, Claudia 367–8
Castelfranchi, C. 128
casuistry 242, 243
categorical imperative 55–6, 59, 241, 244, 258, 275–7
causal agency 89
causation: freedom from 457; values in computer systems 65–6
Cayton, H. 161
censureship 340–2, 344
Cerqui, Daniela 81
CERT (computer expression recognition toolbox) 165–6
certifiably ethical behavior 292
Charniak, Eugene 440
"Charter of Fundamental Rights of the European Union" 474–7, 484, 485–6, 488–9
Chesher, Chris 371
Child Abuse and Neglect Prevention Act (South Korea) 176
Child Abuse Prevention Law 2000 (Japan) 176
child labor 234, 481
child neglect and childcare robots 175–7, 178
child pornography 481, 488
childcare robots: attachment theory 167–72; children's relationship to 161–4; interaction methods 164–7; keeping children from physical harm 158–61; legal issues and accountability 174–8; parental absences 156–7; *vs.* minimal care 172–4
children: child labor, machines as replacement for 234, 481; interactive toys 107–8; moral responsibility 336; relational artifacts 110–3, 114, 115; robots, interactions with 131

Children and Young Persons Act (UK) 176–7
Childress, J. F. 244, 246
Chinese Room argument 420–2, 451–3, 465, 466–7
choice: capabilities approach and care robots 151; care robots 128; ethics of care and care robots 147–8
Chugani, H. 172–3
Churchland, Paul 515
Clarke, Arthur C. 45, 439
closed-system thinking 48
Code of Conduct for the Recruitment of Researchers 474–5
codelets 306, 309
codes of ethics: ACM code of ethics **341**; computer ethics 340; laws of robotics, implications of 51; nanny codes of ethics 174; responsibility for complex machines 478; World Medical Association 483
Coeckelbergh, Mark: capabilities approach and care robots 148, 153; care types 129; experience of human-robot interaction 126; human flourishing 217; human-robot interactions 142; roboethics categories 139–40; social relationism 391
coercion 458
Cog robot 60, 163, 367–8
cognition: consciousness 400; laws of robotics 38, 45–6; LIDA model 301–2, 304–12, *305*; love, definitions of 217–8; moral competence in social robots 227; superintelligence 72; theories of 119, 263; 'Turing Dream' 391–3
cognitive cycle *305,* 305–6, 311–2
cognitive development, children's 172–4
cognitive science 415–6, 468, 470
Cohen, Richard 524
coherency 279–80
common law 417
common sense 50, 277, 279
communication: care robots 129–30, 134; childcare robots 165; moral competence in social robots 228–9
comparative Moral Turing Test 57, 61, 246–7
competence: care, elements of 196, 197; care-centered framework 199–200; elderly and disabled persons 129; judgment and trusteeship 433–4; patient lifting robot design 206, 207, 208; trustees, artificial intelligence as 436–7

complementary intelligence 50
complexity and the laws of robotics 38–9, 43, 51
computational artifacts: feminist theory 371; moral agency 350, 351, 352, 358
Computational Modelers group 352–7
computational theory of mind 418–9
computer ethics **341**; artificial agents 344; debate about moral status 349–50; machine ethics compared to 237; moral agenthood 338–42; "On the Morality of Artificial Agents" (Floridi and Sanders) 349
computer expression recognition toolbox (CERT) 165–6
Computer Power and Human Reason (Weizenbaum) 108, 494
computer-aided design 370
Computers-in-Society Group 352–7
conference on Automated Deduction 297
conflicting duties: artificial intelligence 273; ethical reasoning 270; laws of robotics 38, 39, 47; machine ethics 241, 245, 271; particularism and generalism 287; top-down engineering 259
connectionism 261–2
consciously mediated behavior 306
consciousness 10; appearance, importance of 120–3; constitutional personhood and lack of 448–50; expanding moral circle 388–90; LIDA model 304–12, *305*; machine ethics 263; moral agency 11, 514–5; moral landscape of conscious well-being 390–1; moral realism *vs.* social relationism 385–8; 'other minds' problem 393–5; rights, responsibilities, and moral judgments 302–4; robot emotions 119; and sociality 395–7; spectrum of 398–400; synthetic-biological humanoids 397–9; 'Turing Dream' 391–3; *vs.* intentionality 466–7
consent and privacy 502
consequentialism 58, 117, 258, 259
consistency tests 277
constitutional law 497
constitutional personhood: artificial intelligence as property argument against 460–3; consciousness 448–50; emotions 453–5; free will 456–8; intentionality 451–3; interests 455–6; nonperson argument against 442–6; scenario 440–2; simulation argument against 458–60; soul 446–7; Turing Test 464
constraints: ethical reasoning 269, 270; LIDA model of cognition 308
constructivism 280
consumer information 501
consumer market for robots 491
context: care-centered framework 199; emotions and attachment theory 169; machine ethics 246; moral competence in social robots 227; patient lifting robot design 206, 208
continuous inclusion of other scale 217–8
control 11; alternative laws of robotics 409, 410; care robots 131; lack of free will and constitutional personhood 458; laws of robotics 46, 50–1; long-term research priorities 25–6; machine learning and human responsibility 380–1; as research focus 5, 6; short-term research priorities 21, 23
Convention for Certain Conventional Weapons 13
Convention on the Rights of Persons with Disabilities 486
Convention on the Rights of the Child 159, 175, 481
Cook, R. J. 483–4
copying, artificial minds 71
corporations: liability 90–1; personhood 432, 443, 514
corrigible systems 25
Cowan, Ruth Schwartz 132
Cowie, Roddy 214
crash-avoidance algorithms 508, 509
crash-optimization algorithms 508–9
creative stewardship 344
critical reliability 234
culpability 90–2
cultural dependence and the laws of robotics 37
culture, robotics 108–9
culture and childcare robots 175
culture of simulation 108–9
cyberattack 23
cyberdyne 204
cyberlaw 495–6
cyborgs: feminist theory 366–7; laws of robotics 37

Damasio, A. 118
Dancy, Jonathan 283
Danielson, P. 61
DARPA SAFE program 23
Dartmouth Summer Research Project on Artificial Intelligence 233–4
data collection 497–8, 508–9
Dawkins, Marian S. 518
Dawkins, R. 61
deadlock: artificial moral agents 59; laws of robotics 39; top-down engineering 259
death 104
deception: care robots 128, 133; childcare robots 166–7, 179; love and sex robots 222–3
decision making: artificial intelligence scenario 426; backup decision makers 437–9; laws of robotics 39, 50; LIDA model of cognition 306–9, 306–10; machine ethics 237, 241, 245, 268–73; moral competence in social robots 227–8; speed of 235; *see also* artificial moral agents; logic; reasoning
decision support systems 50
deductive logic 293–4, *294*
Deep Blue: blamelessness 63; lack of passion 62; superintelligence definition 70; tree searching techniques 58
deep care 129
deep learning 14
Deep Thought 419
DeepMind 14
default logic 278
defeasible reasoning 278
Dehaene, Stanislas 304
Del Rey, Lester 34
deliberation 306–9
dementia 133
Dennett, Daniel: associative learning 60; machine ethics 238; moral agency 514; moral patiency 518–9; moral philosophy 527, 528; personhood 7
Denning, Tamara 498, 499
dental-training robot 517
deontic logic 235, 291–7
deontology: artificial moral agents 58–9, 60; emotions 117–8; Kantian machine ethics 275–80; machine ethics 240–1, 243, 244; top-down engineering 258; *see also* rules
dependence and care robots 130
depression 113–4, 130

Derrida, Jacques 513, 517, 521, 525–6
Descartes, Rene 418–9, 517
design: care centered value sensitive design 202–8; care robots 125–6, 189–91; care-centered framework 198–202; ethics of robot design 83; European regulations 477; innovation 361; interpretive flexibility 350–1; laws of robotics, implications of 48–51; LIDA model of cognition 301–2, 304–12; love and sex robots 221–2, 223; moral agency 11, 354, 357; moral discernment and decision-marking 8; as research focus 6; values and 66–7, 195–7; *see also* developers and development; engineering; programming
destructive doubt 103
deterrence, punishment for 429–32
Dettling, A. 171
developers and development: alternative laws of robotics 408, 411; European regulations 477, 478–9; hybrid development approach 300; interpretive flexibility 350–1; product development and testing 408, 411; technology development 350–1, 356–7; *see also* design; programming
development, child 160, 167–74, 179
developmental condition and consciousness 399
Diagnostic and Statistical Manual of Mental Disorders 174
Diamond, Cora 100n11
Dickens, B. M. 483–4
Dietrich, Eric 239
digital divide 83
dignity 474, 486
Diodato, M. D. 485
disability 121–2, 128–30
disaster relief situations 235
Discourse on Method (Descartes) 517
discrete-systems approach 254–5
disobeying the laws of robotics 40
disorganised attachment 170–1
dissymmetries and technoscience studies 366
distributive justice 92
divine right of Kings 460, 461
drones 13, 408, 493–4
DSM-IV-TR *see* Diagnostic and Statistical Manual of Mental Disorders
Dualism 447

dual-use technology 83, 487
Due Process Clause 443
duties: explicit ethical machines 239; implicit ethical machines 234; Kantian machine ethics 275–80; LIDA model of cognition 307, 308; machine ethics 240–1, 243, 245–6, 271; methodology for engineering robot ethics 295–6; moral agents and moral patients 318; top-down engineering 258; *see also* deontology
Dyson, Freeman 132

ecological macroethics 523
economic imperative 49
economic measures 20
economics and short-term research priorities 20
efficiency: methodology for engineering robot ethics 296–7; patient lifting robot design 207, 208
Einstein robot 166
Ekman, P. 165
elderly: care robots 128–30, 134; Paro Therapeutic Robot 138; relational artifacts 113–4; therapeutic robots 109
Electronic Communications Privacy Act 497, 498–9
Elicker, J. 171
eliminativism 282
ELIZA (computer program) 108
ELLEgirlBuddy 501
embodiment: consciousness 394–5; gender 367, 369; love and sex robots 221
emergence and bottom-up engineering 255
emergent disciplines 5–6
emotional abuse 176
emotions and affective states: affectional responses and robot nannies 162; affective computing 214; affective intelligence and machine ethics 251–2; appearance, importance of 120–1; artificial moral agents 62; attachment theory 168–9, 171–2; behaviourism 120; capabilities approach and care robots 149, 150–2; childcare robots 165–6; consciously mediated behavior 306; consciousness 400; constitutional personhood 453–5; emotional intelligence and consciousness 303; feeling theory and robot emotions 119; LIDA model of cognition 307; love, definitions of 218; love and sex robots 214, 221–2; lovotics 220; machine ethics 251–2, 263; moral competence in social robots 227; moral theory 118; personhood 101–2; relational artifacts 110, 111; synthesized emotions 517; 'Turing Dream' 391–2
empathy: artificial moral agents 62; attachment theory 168, 172; capabilities approach and care robots 153; evolutionary attachment model of love 218; moral competence in social robots 228; moral status 303; phenomenal consciousness 311; Turing Triage Test 103; virtue ethics and care robots 144–6
employment: ethical-impact agents 234; impact of robots on 13; labor market forecasting 20; machines as a replacement for humans 481–2; research priorities 20; tele-presence 482
empowerment and care robots 129
enabling robots 201
ends and rights 449
energy grid 501
Engelberger, Joseph 34, 185
engineering: bottom-up approaches 253, 254–6, 265; computer ethics 339–40; engineering activism 67; engineering ethics as basis for machine ethics 5; hybrid approaches 259–60, 300, 304; methodology 291–7; top-down approaches 252–3, 256–9, 263; *see also* design; developers and development
engrossment and ethics of care 146
entertainment 488
entropy 521
environment: interactivity 325; morality of agents 331; roboethics 83
environmental ethics 417, 520–1
epistemic commitment function 280
epistemic logic and machine ethics 235
epistemology and machine learning 378–9, 381
Epstein, R. G. 339
Equal Protection Clause 443
Erector Spykee 498
erotic love 219
EthEl 246
ethical condition and consciousness 400
ethical conflict *see* conflicting duties

ethical dimensions of robots 81
ethical reasoning 89, 267–73
ethical relativism 241
ethical-impact agents 234
ETHICBOTS Project 131
ethics of care: care robots 146–8; healthcare and care values 193–8
ethics of ethics 525–6
ethnomethodological studies 363
EURON Robotics Research Roadmap 82–4
European Charter for Researchers 474–5
European regulations: "Charter of Fundamental Rights of the European Union" 474–5; entertainment robots 488; machines as a replacement for humans 481–2; medicine and healthcare 484–6; responsibility and autonomous robots 476–81; tele-presence 482–4
European Robotics Research Network 82, 175
European Union Code of Conduct on Arms Exports 486
evocation 110–4, 371
evolution and adaptation: agenthood 325–31; artificial intelligence scenario 425; artificial moral agents 252; bottom-up engineering 255; laws of robotics 40–1; LIDA model of cognition 307, 309–10; love and sex robots 222; machine ethics 262–3
evolutionary attachment model of love 218
evolutionary model of morality 61
exceptionless standard 281
exoskeletons 204
expanding moral circle 388–91, 512, 520–1, 522
expectation codelets 309
experiential learning 23, 306, 307
experimental medical treatment 485
expert trading systems 425–6
explicit ethical agents 234–5, 238–40, 244–7, 270–2
Explorations in Love and Sex (Singer) 219
extension and nonmonotonic logic 278
extra-terrestrial life 389
Ezer, N. 133

facial expressions 102–3, 164, 165–6
facial recognition 165
fact-value split 391
fairness *see* justice
false promising 279

false utopia 74
Federal Aviation Administration 13
Federal Trade Commission 498–9
feedforward network 283–5, 290
feelings *see* emotions and affective states
Fehr, B. 216
feminist theory 363–72
Filmer, Robert 460, 461
Finnis, John 456
First Amendment protection 494
First International Symposium of Roboethics 81
first-order logic 293, 294, *294*
fitness functions and consciousness 310
Flanagan, Owen 415
flexibility 51, 350–1
Florida v. Riley 495
Floridi, Luciano: artificial agents 344, 349, 353–7; information ethics 520–3, 525; moral patiency 516; moral philosophy 513; morality threshold 337
flourishing *see* human flourishing
Foerst, Kay 516
Fonagy, P. 171
form of robots 35–6, 49
formal logic 243, 292
Formal Methods 322
formalism 277–8
Fourteenth Amendment 443
Fourth Amendment rights 495, 497, 498, 503n4
Franklin, Stan 301–2, 304, 320
free will: computer ethics 339; constitutional personhood 456–8; machine ethics 241, 273; moral competence in social robots 228
freedom: capabilities approach and care robots 126; care robots 132, 135; ethics of care and care robots 147–8
freedom objection 334
Friedman, Batya 65, 192
friendliness and superintelligence 73
friendly AI 9
friendship 223
Frude, N. 35, 36
full ethical agents 235–6, 271
functional morality 61, 405–6
functionality and models of love 218–9
functioning and capability 127–8
funding and childcare robots 174

Future of Life Institute 14
future technologies and European regulations 479–81
futuristic thermostat 330, 337–8

Gaita, R. 100n11, 101
Gallagher, S. 395
games and game theory: artificial agents 327–30; DeepMind 14; morality models 61; noughts and crosses game 338
Gärdenfors, P. 280
Gates, Bill 491, 498
GeckoSystems International Corporation 138
Geduld, H.M. 35
Geminoid robots 221
gender 215–6
General Belief Retrieving Agent (GENTA) 242
general systems thinking 48–9
generalism 242–3, 281, 282
GENTA (General Belief Retrieving Agent) 242
Gerdes, Chris 509
German Act on Telecommunication 488
German Air Security Act 480
German Federal Supreme Court of Justice 487
Germany 479, 480, 485, 487, 489
Gert, B. 59
Gibson, William 463, 502
Gilbert, D. T. 216
Gips, James 7, 57–8, 59, 60, 235, 300
Global Fund for Women 372
Global Workspace Theory 302, 304–6, *305,* 309, 311
goddess and cyborg 366–9
golden rule 59
good, concept of the 456
good and evil 319, 332
"good behavior" 23
good care 207, 208
"good internal to practice" 139
Goodall, Noah 508
goodness and caregiving 150
Goodwin, Charles 363
Goodwin, Dawn 363–4
Google DeepMind 14
Gottesman, R. 35
governance as research focus 5, 6
government access to home robot data 496–7

Graesser, A. 320
Gray, John Chipman 422–4
Greene, J. D. 118
grief 101, 102, 104
Grodzinsky, Frances 354
Grounding of the Metaphysics of Morals (Kant) 276, 277
Guarini, Marcello 239, 242–3
Gunkel, D. 391, 398–9
Güzeldere, Güven 514

hacking 496, 497–8
Haidt, J. 306
Hajdin, Mane 516
HAL *see* Hybrid Assistive Limb
Hammond, Kristian 25
Hanna, R. 394–5
Hanson, David 166
Hansson, S. O. 130
Haraway, Donna 362, 366, 369
Hardwig, J. 131–2
Harlow, H.F. 162, 173
harm: agency, culpability, and liability 90–2; childcare robot accountability 177–8; explicit ethical machines 239; keeping children from physical harm 158–61; laws of robotics 37, 50–1, 406; Moral Turing Test 57
Harms, W. 61
Harris, Sam 390–1
Hawking, Stephen 7
health care *see* medicine and health care
hedonistic act utilitarianism 240
Hegel, G. W. F. 526
Heidegger, Martin 521
Heinlein, Robert A. 422
Hello Kitty Robot 156–7
heuristics 258, 307
Himma, Kenneth Einar 514, 521
Hirose, Shigeo 493
Hitchhiker's Guide to the Galaxy (Adams) 45
Hobbes, Thomas 418
holism 390
home maps 159–60
home robots 496–8
home sensing and monitoring systems 159
Hornik, R. 172
Horty, John 243–4, 294–5
Horvitz, Eric 13, 24
household appliances 132

How to Build a Person (Pollock) 440
Hsiao, K. 255
Hubrid Assistive Limb 187
Huggable robot 164, 174
human exceptionalism 513, 516, 518, 526
human flourishing: capabilities approach and care robots 151–2; care robots 143, 144, 148, 151, 153; elderly individuals and robot caregivers 126–9, 130, 132, 133, 134; emotions and 118; good, concept of the 456; robots 217; superintelligence 74
human moral agents 339, 340–1
human relationships and the impact of robot love 223
human rights 381, 481–2
human uniqueness 115
humanity 44, 45, 74, 469, 526
human-machine interface 361, 362, 365–72
humanoids: children's relationship to 162; robot sex 215; robots, history of 34; science fiction 213; synthetic-biological humanoids 397–9; Turing Triage Test 103–4
human-robot interaction: alternative laws of robotics 408–11; children's positive interactions 156; experience of 126, 139, 142; privacy and social robots 498–502; sociality 396–7
human-robot relationship 8; care robots 131, 132–4; childcare robots 161–4; ethics of 88; history 80; laws of robotics 49–50; nonhumaniform *vs.* humaniform robots 35–6, 49; privacy 499–500; relational artifacts 108, 115–6; in science fiction 35; *see also* love and sex robots
humans: autonomous weapons system responsibility 92–3; computers as tethered to 354, 357–8; decision making 50; as ethical agents 88; laws of robotics 37, 38; legal responsibility 91; machine ethics *vs.* human ethics 239, 240; pleasure and pain 99–100
Hybrid Assistive Limb 204
hybrid engineering approaches 259–60, 300, 304, 306–7

I, Robot (Asimov) 405
IDA model of cognition 302
identity 355–6
Illinois v. Caballes 495
imagination 308

Imamura, Shohei 216
Imitation Game *see* Turing Test
implementation issues 242, 296
implicit ethical machines 234, 239
inanimate surrogate mothers 173
independence 147, 350, 357
inductive reasoning 243, 245, 378–9
industrial robots: form of 35, 49; machines as a replacement for humans 481–2; robot classification 201
inform, duty to 477–8, 485
information and computer technology: computer ethics 5, 7; laws of robotics, implication of the 48–51; roboethics 83–4; robot-computer distinction 35, 36; robustness research areas 21–3; values, embodiment of 65–7
information ethics 344, 520–4, 525
information imperative 49
innovation 361, 362, 372
insecure attachment 170
Institute of Electrical and Electronics Engineers 478
instrumentalism 513
Intel 66
intellectual property 65
intelligence: defining 80, 96–7, 321; rights and 95
intelligence gathering 501
intentional objection 333–4
intentionality 241, 451–3, 466–7
interaction methods of childcare robots 164–7
interactive toys 107–8, 110–3
interactivity and agenthood 325–31
interdisciplinarity 5, 79–84
interests, lack of 455–6
International Classification of Diseases and Related Health Problems 173–4
international law 481
International Organization for Standardization 190
interpretive flexibility and technology development 350–1
interrogation, robots for 500–1
Interstate Treaty on the Protection of Human Dignity 488
intimate machines *see* relational artifacts
intra-action 365
introspection and consciousness 400

intuition 228, 242, 304
invention-novelty distinction 372
inverse reinforcement learning 25
invocation 371
iRobi robot 164, 165

James, William 119, 308
Japan 155, 158–9, 175, 176, 221
jockeys, camel 234, 481
Johnstone, J. 127, 129
Jones, E. E. 216
Joy, Bill 238, 296, 297, 357
judgment: consciousness 303–4; laws of robotics 37–8, 44; machine ethics 239–40; moral competence in social robots 227; moral decision-making system 306–9; particularism 281–7; real *vs.* artificial intelligence trustees 437–9; top-down engineering 257; trustees, artificial intelligence as 432–7
"just war" 93
justice: autonomous moral agents 90; capabilities approach and care robots 148; legal personhood 451; machine ethics 240, 244; machine learning and human responsibility 380; robot ethics 93

Kafka, Franz 494
Kahn, P.H., Jr. 162, 163
Kant, Immanuel: duty 239; laws of robotics 117–8; moral theory 454; utilitarianism 127; *see also* categorical imperative
Kantian machine ethics 275–80
Keeping Children and Families Safe Act (US) 175
Kennett, J. 118
Kerr, Ian 501
Kerr, Orin 495
'keyhole' surgery 364
Keynes, John Maynard 13
"Killer Robots" *see* lethal autonomous weapons
Kismet 103n15, 115, 163
"Kissenger" 220–1
Kohlberg, L. 61
Kohno, Tadayoshi 498
Kokoro 221–2
Korea 155, 175
Kozima, H. 130

KT Telecop 160
Kurzweil, Ray 515
Kyllo v. United States 495, 501

labor *see* employment
Laboratory for Artificial Intelligence, MIT 367
language: attachment theory 169–70; childcare robots 165; laws of robotics 405, 407; moral language 226, *226*; *see also* semantics; terminology
Latour, Bruno 91n2, 190
law and policy 6–7, 11–2; agency, culpability, and liability 90–2; alternative laws of robotics 408, 410; childcare robots 174–8, 179; culpability, agency, and liability 90–2; home robot data 496–8; as research focus 6; short-term research priorities 20–1; social robots 498–502; spy robots 493–6; *see also* constitutional personhood; European regulations; trustees, artificial intelligence as
law application and ethical reasoning 269, 270
law enforcement *see* police
laws of robotics 405–6, **410**; alternative three laws 408–11; artificial moral agents 59; Asimov's experiments with 36–41; bottom-up engineering 255; deontic logic 291; extended laws 45–8; information technology, implication for 48–51; revised laws (1985) 44–5; science fiction and speculation 4; shortcomings of 406–7; top-down engineering 259
Le Dantec, C. 192, 194
learning: associative learning 60–1; bottom-up engineering 255; learning robots and human responsibility 377–81; level of abstraction 329–30; LIDA model of cognition 306, 307–9; machine ethics 236, 242; machine learning 11, 378–9; moral agency 11; moral competence in social robots 229; moral development 119; particularism and generalism 283; punishment for 431; validity and long-term research priorities 25
learning algorithms 22, 235, 240, 245
legal personhood 12; artificial intelligence and questions of 465–8; artificial intelligence as property 460–3; borderline status of 468–71; consciousness 302–3, 448–50; constitutional personhood scenario 440–2; emotions 453–5; free will 456–8; intentionality 451–3;

interests 455–6; judgment and trusteeship 432–7; and moral rights 424n36; *The Nature and Sources of the Law* (Gray) 422–4; nonperson argument against 442–6; real *vs.* artificial intelligence trustees 437–9; responsibility 428–32; simulation argument against 458–60; soul 446–7; trusteeship legal question 427–8; trusteeship scenario 424–6; Turing Test 464

Lehman-Wilzig, Sam 503n9

Leib-Körper distinction 394–5

Leibniz, G. W. 292

Lessig, Lawrence 495

lethal autonomous weapons 13, 21

level of abstraction: agents 320–1; artificial agents 321–2, 343; computer programs 354; computer systems 358; futuristic thermostat 330; information ethics 521; interactivity, autonomy, and adaptability 325–6; morality of agents 317, 331–8; morality threshold 337–8; noughts and crosses game 328–30; relativism 323; SmartPaint 330–1; state and state-transitions 324–5; terminology 323; webbots 330

levels, moral norms 226–7

Levinas, Emmanuel 522, 524–5, 526–7

Levy, David 213, 215–7, 222, 474, 502

liability 7; autonomous vehicles 2, 20; constitutional personhood 451; laws of robotics, implications of 50–1; machine learning 380–1; product liability 12; robot ethics 90–1; trustees, artificial intelligence as 428–32

LIDA model of cognition 301–2, 304–12, *305*

life and relational artifacts 114–5

life-sustaining treatment, cessation of 469–70

lifting, patient 203–8

limited legal personhood 437

literature *see* science fiction

litigation and artificial intelligence 435–6

Little, M. 205

Llewelyn, John 525

Locke, John 460–2, 514

Loebner test 321

logic: deductive logic 293–4, *294*; default logic 278; deontic logic 235, 291–7; formal logic 243, 292; inductive logic 243, 245, 378–9; Kantian machine ethics 277–80; logical consistency 9–10; methodology for engineering robot ethics 291–7, *294*; modal logic 234; monotonic logic 279; Murakami-axiomatized deontic logic 295, 296; *see also* decision making; reasoning

Logic Theorist 234

loneliness 130

long-term ethical concerns 81

long-term research priorities 23–6

Lopes, M.M. 159

love and relational artifacts 110, 115–6

love and sex robots: current technology 220–1; design strategy 221–2; emotions and affective robotics 214–5; ethics of 222–3; Levy, David 213; love, definitions of 217–9; perfect love 215; privacy 499–500, 502; sex 215–7

Love and Sex with Robots (Levy) 215

love prototypes 216

Lovotics 220

Lyotard, Jean-François 513

machine consciousness 301, 302, 310

machine ethics 9–10; challenges 240–4; computer ethics compared to 237; difficulty in achieving 267–73; emergence of 2; ethical-impact agents 234; explicit ethical agents, creating 244–7; full ethical machines 235–6; goals of 237, 300; hybrid approaches 259–60, 265; implicit and explicit agents 234–5; importance of 238; Kantian machine ethics 275–80; LIDA model of cognition 301–2, 304–12; logicist methodology 291–7; particularism and generalism 281–8; supra-rational faculties 250–1, 262–4; top-down engineering 256–9; value questions 250; varieties of 233–4; virtue ethics 260–2

machine learning 11, 378–9

machine safety 476–8

machines, robots as 81

MADL 295, 296

malprogramming 40

Manifesto for Cyborgs' (Wolmark) 367

manipulation 222–3

manufacturers, robot 83

market disruptions 20

Massachusetts Institute of Technology 103n15, 108, 115, 367

Matchbox Educable Noughts and Crosses Engine (MENACE) 327–30, 338

materiality: dysmmetries 366; frames and accountability 371–2; human-machine interface 370–1; mutual constitutions 362–6; robots 366–70
'maternal sensitivity' 170
Matthias, Andreas 380
Mauss, Marcel 514
Mayes, D. 174
McAfee, Andrew 20
McDermott, Drew 440
McKeever, S. 282
McLaren, Bruce 243, 287
mechatronics 215
MedEthEx 246, 247
mediation, technology's role in 190, 200
medicine and health care: actor network theory 363–4; ethics of care and care values 193–8; European regulations 484–6; health care robots 198–202; machine ethics 244, 246–7; medical devices 484–5; medication reminders 246; methodology for engineering robot ethics 295–6; surgical robots 186; tele-medicine 482–4; values and care ethics 193–5
MENACE (Matchbox Educable Noughts and Crosses Engine) 327–30, 338
mental states: appearance, importance of 120–3; childcare robots 165–6; and rights 302–3; robot emotions 119
Merleau-Ponty, M. 119
Meston, C. 215–6
metabolic condition and consciousness 399
meta-ethics 6
meta-net 284–5, *285,* 290
metaphysics 522–3
meta-regulations 479
metatheory 293
Method of Abstraction 321–3, 343
methodology: care centered value sensitive design 202–8; engineering 291–7
military: auditing robot compliance 40; European weapons regulations 486–7; roboethics 83
Mill, John Stuart 54–5, 56
Miller, Keith 354
mind, theory of 263
mind-body dualism 394–5, 447
mind-mindedness 170
minimal care *vs.* childcare robots 172–4
"Mini-Surrogate" 221

mirror neurons 169, 218, 499
mis-attribution of consciousness 398, 401
misconception and computer ethics 356–7
MIT Artificial Intelligence Laboratory 103n15, 115
mobile child monitoring 158–9
modal logic 234
modus ponens 294
monitoring actions 309
monkeys, attachment experiment on 173
monotonic logic 279
Moor, James H. 271
Moore, C. L. 368–9
moral agency 10–1; aim of robot ethics 88–93; appearance and 120–1; autonomy 53–4; capabilities approach 128; computer ethics 338–42; debate on the morality of 351–8; emotions 62; machine learning and human responsibility 381; morality of agents 331–8; morally praiseworthy agents 63; as research focus 6; and rights of machines 512–6; standard *vs.* nonstandard theories of 317–9; theories of 54–6; *see also* artificial moral agents
moral appearances 120–3
moral arithmetic 240
moral case classifiers 283–6
moral cognition 281–7
moral competence in social robots 225–9
moral development 60–1, 119
moral elements: care robots 196, 197; care-centered framework 201–2; patient lifting robots 205–6
moral evil 332
moral good 215, 332, 337
moral intelligence 89
moral judgment *see* judgment
moral landscape 390–1
moral language 226
Moral Machines (Wallach and Allen) 118–9, 405
moral patients: animals 302; artificial agents as 303; expanding moral circle 389; information ethics 520–3; rights of machines 516–20; robots as 81, 121–2; theories of 317–9; *vs.* moral agents 10
moral philosophy *see* philosophy
moral principles 281–7
moral producers 303

moral realism: defined 384; expanding moral circle 388–90; moral landscape of conscious well-being 390–1; *vs.* social relationism 385–8
moral reasoning 118, 283–8
moral responsibility *see* responsibility
moral rights and legal personhood 424n36
moral status: artificial agents 342–3; computer ethics 349–50; computer/human 355–6; consciousness 302–3; debate on the moral status of artificial agents 351–8; emotions and appearances 121–2; expanding moral circle 388–91, 512; information ethics 520–4; for intelligent machines 99–100; moral agency 512–6; moral philosophy viewpoints 527–8; moral realism *vs.* social relationism 385–8; 'other minds' problem 524–7; personhood 97–8, 99; spectrum of consciousness 399–400
moral thinking and superintelligence 72
Moral Turing Test 7, 56–7, 93, 246–7, 259
morality models 59–61
Moravec, Hans 90n1
'Mother, Child' (Tikka) 370–1
motivation: for carebot development 138; morality models 60–1; superintelligence 71, 72–3
motivational displacement and ethics of care 146
Movellan, Javier 132, 166
Munsell chart of universal color categories 363
Murakami, Yuko 295
Murakami-axiomatized deontic logic 295, 296
Musk, Elon 7
Mutlu, Bilge 199
mutual constitutions 362–6
My Real Baby 111, 112, 113, 114, 162–3
mythology 34, 80, 213

Nakagawa, C. 130
The Naked Sun (Asimov) 46
nanoethics 6
nanotechnology 73, 74
National Society for the Prevention of Cruelty to Children (UK) 175
natural deduction *293, 293*–4
Natural Deduction Language *294*
natural persons 442–6
natural slaves 460–3

nature and nurture 119
The Nature and Sources of the Law (Gray) 422–4
Nealon, Jeffrey 526
needs, relationship of values to 196–7
negative machine training 245
neglect, child 175
Nelson, C.A. 172
neo-Stoicism 118n5, 119
nesting and the laws of robotics 47
neural networks 242–3, *283,* 283–8, *285,* 290
neural structure and consciousness 394, 399, 449–50
neuro-robotics 83–4
neutralism, scientific 391
new media art 370–1
Newell, Allen 294
Nicomachean Ethics (Aristotle) 261–2
Nietzsche, Friedrich 521
Ninja robot 493
'No Woman Born' (Moore) 368–9
Nobel, Alfred 270
Nobel Peace Prize 270
Noddings, N. 146–8
noncompliance, robot 51
nonhuman entities: agency, culpability, and liability 91–3; binarism 362; moral status and appearances 120–3; rights of 302–3; Turing Triage Test 103–4
nonhumaniform robots 35, 49
nonmaleficence 244, 245
nonmonotonic logic 278–9
nonperson argument against legal personhood 442–6
norms 225, 226–7, 233
Norton anti-Virus facility 341
Norvig, Peter 22
noughts and crosses game 327–30, 338
novelty and artificial intelligence 434
novelty-invention distinction 372
nurse-patient relationship 200, 203–4
nursing homes: care robots 134; European regulations 486; Paro Therapeutic Robot 138; relational artifacts 113–4
nurturance *see* caregiving and caregivers
Nussbaum, Martha: capabilities approach and care robots 126–8, 148, 149–50; care robots 134; choice and caregiving 142; disability and independence 129; moral theory 118;

robot emotions 119; women and caregiving 132

obeyance and the laws of robotics 407
objectification and agency 364–5
obligations *see* duties
observable (term) 322
occupational safety 476–7, 484
"Of Property" (Locke) 460–1
OFRO 160
"On the Morality of Artificial Agents" (Floridi and Sanders) 349, 351, 353–7
One-Hundred Year Study of Artificial Intelligence 21, 26
O'Neill, Onora 276
ontocentrism 522, 523–4
ontology 293, 364
Oosterlaken, I. 125, 127, 129
open-ended learning systems 11
open-system thinking 48
operational morality 254, 406
optimization and ethical reasoning 269
organic condition and consciousness 399
organic view of ethical status 517
organisations and level of abstraction 331
origins of robotics 33–6
orphans 172–3
Ortigue, S. 218
OSCAR 440
'other minds' problem 393–5, 515, 524–7
otherness 369
overfitting and machine learning 379

PAC (Probably Approximately Correct) learning 378–9
pain *see* suffering
PAPA (privacy, accuracy, intellectual property, and access) 499
paparazzi bots 494
PaPeRo robot 158, 164, 165
parallel distributed processing *see* connectionism
parental absences and Hello Kitty Robot 156–7
parents and attachment theory 168–72
Paro robot 109, 138, 187
particularism 242–3, 281–7
passion and love 219
passive entertainment, children's 156

paternalism 44–5
patient autonomy 246
patient lifting robots 203–8
patients and actor network theory 363–4
Pearson, Y. 142, 145, 147, 148, 149
Pentium III processor chips 66
perception 45–6, 311–2
perfect love 215
permanent vegetative states 469–70
personal robots 201, 496–8; *see also* care robots
personal serial numbers 66
personality: caregivers 131; personhood 101
personal-service robots *see* care robots
personhood: consciousness 302–3; moral agency 514; moral standing 97–8; oral standing 99; Turing Triage Test 95, 101–2; *see also* legal personhood
"persons-are-conceptually-human" argument 443–4
phenomenal consciousness 311
phenomenal-valuational holism 390
phenomenology 394
philanthropy 73, 74
philosophy: of artificial intelligence 416; care robots 126–8; emotion as a requirement for personhood 454; ethics of care and care robots 146–8; expanding moral circle 512; love 219–20; machine ethics 240–1; moral theory 4–5, 117–8, 513, 527–8; philosophical tradition 4–5, 117–8; virtue ethics and care robots 143–6
physical harm and childcare robots 158–61
Picard, Rosalind 54, 241, 250
Pinch, Trevor 350
planning one's life 150
Plato 93, 219
pleasure: constitutional personhood 453, 455; experience of 99–100; moral status 303; 'Turing Dream' 392; utilitarianism 240
police: access to home robot data 496–7; drone use 493; robot use 93, 486–7; surveillance robots 495
policy *see* law and policy
Pollock, John 440
Poole, David 278
Poovey, Mary 364
popular culture 238, 494
The Pornographers (film) 216

pornography 481, 488
positive machine training 245
Posner, Richard 495
post hoc system analysis 253
post-environmentalism 318
The Postmodern Condition (Lyotard) 513
postmodernism 513
poststructuralism 521
poverty 482
power and gender 369
Powers, Tom 244
practical reason 149, 150
practice and care-centered framework 199–200
precautionary principle 84
pre-reflective consciousness 310
prima facie duties: machine ethics 240–1, 243, 245, 271; top-down engineering 258
principle abstinence 282
principle eliminativism 282
principles of safety integration 477
prioritization, rule 259
Prisoner's Dilemma 61
privacy 12; care robots 130; childcare robots 178; and childcare robots 159; home robots 496–8; nursing by machines 486; privacy-enhancing robots 494–5; research priorities 21; roboethics 83; social robots 498–502; solitude and robots 503n9, 504n10; values in computer systems 65, 66
privacy, accuracy, intellectual property, and access (PAPA) 499
Privileges and Immunities Clause 442–3
"Probablistic Love Assembly" 220
Probably Approximately Correct (PAC) learning 378–9
Probo robot 164, 174
procreation 47, 219–20
product development and testing 408, 411
product liability 476–8
professional codes of conduct *see* codes of ethics
professional ethics and research priorities 21
programming: auditing robot compliance 40; autonomous vehicles 507–9; care robots 132; machine ethics 236, 243, 244–7; Natural Deduction Language 293–4; *see also* developers and development
projection and relational artifacts 108, 114
proofs 293–4, 296

property: artificial intelligence as 460–3; moral status and appearances 122; roboethics 83
proto-ethics 525
proto-morality 310
psyche and superintelligence 72
psychological abuse 176
psychological issues: associative learning 60–1; childcare robots 167–74, 179; love and sex robots 216–7, 222
psychopathy and emotions 118
public reason requirement for legal personhood 446–7
punishment 60–1, 429–32

Qatar 20, 234, 481
QRIO robot 164

Rawls, John 245
Reactive Attachment Disorder 173–4
realism 392–3
reality and appearances 102
reasoning: by analogy 269; case-based reasoning 243, 283–7; defeasible reasoning 278; ethical reasoning 89, 267–73; inductive reasoning 243, 245, 378–9; moral competence in social robots 229; moral judgments 304; moral patiency 517; moral reasoning 118, 283–8; phenomenal consciousness 311; practical reason 149, 150; *see also* decision making; logic
reciprocity: capabilities approach and care robots 153; caregiving 147; patient lifting robot design 206; relational artifacts 108; virtue ethics and care robots 143–4
reflective equilibrium 245
Regan, Tom 397
regulations *see* European regulations; law and policy
reinforcement learning 25
Reiter, R. 278
relational artifacts: attachment theory 171; children's relationship to 110–3, 162–3; definition 107; depression 130; nurturance 108; privacy 501–2
relativism: machine ethics 241; Method of Abstraction 323
remorse 101, 102, 104
Renaldo, Ken 494
Rendezvous with Rama (Clarke) 45

replacement robots 201, 204
representation and moral norms 226
reproduction, human 219–20
reproductive technology 219–20, 364
research: alternative laws of robotics 408, 411; European regulations 474–5, 484; focus of 5–6; machine ethics 241–3; priorities for robust and beneficial AI 17–8, 19–26; research roadmap, robotics 82–4
responsibility: agency, culpability, and liability 90–1; alternative laws of robotics 410; aresponsible morality 332–7; artificial agents 343–4; care, elements of 196, 197; childcare robots 177–8; computer ethics 339; consciousness 303; European regulations on robotics 476–81; and learning robots 377–81; machine ethics 241; patient lifting robots 205–6, 207; real *vs.* artificial intelligence trustees 437–9; robot ethics 90–3; trustees, artificial intelligence as 428–32; values in computer systems 65
responsibility objection 334–7
responsiveness: alternative laws of robotics 408–9; care, elements of 196, 197; patient lifting robot design 206
restraint and childcare robots 160, 177, 178
revised laws of robotics 44–5
reward 60–1
Rey, Georges 455
RIBA robot 138, 187, 204
Ridge, M. 282
rights 12; artificial agents 342; and childcare robots 159; consciousness 302–3; expanding moral circle 512; information ethics 520–4; intelligence and 96; legal personhood 424n36; machine ethics 240; moral agency 512–6; moral agents and moral patients 318; moral paciency 516–20; moral philosophy viewpoints 527–8; 'other minds' problem 524–7; *see also* constitutional personhood
Riken Institute 204
RIKEN-TRI Collaboration Center 138
RI-MAN robot 187, 204
risk and risk management: laws of robotics, implications of 50–1; superintelligence 74
Robeyns, I. 127
Robins, B. 131
Robodoc® system 485
Roboethics Atelier Project 82

"robo-paths" 92
robot ethics 8; aim of 87–93; bottom-up interdisciplinary discourse 79–84; categories 139–40; emergence of 2; logicist methodology for engineering 291–7; moral realism *vs.* social relationism 385–8; three dimensions of 187–8; *see also* laws of robotics
"robotic divide" 482, 487, 489
robotics, laws of *see* laws of robotics
robotics, origins of 33–6
robotics research roadmap 82–4
robots: definition 476; emotions 118–9; as ethical agents 88; European regulations 476–81; feminist theory 366–70; form of 35–6; impact of 36; key elements 35; robotics culture 108–9; therapeutic robots 109; views on 81
Robots and Empire (Asimov) 45
The Robots of Dawn (Asimov) 45
Robovie 131
robustness and research priorities 19–26
Romanian orphanages 172–3
Roomba 407
Rorschach effect 110–4
Ross, W. D. 240, 247
Roy, D. 255
RUBI project 165–6
rules: ethical reasoning 269, 270; LIDA model of cognition 307, 308; particularism and generalism 283–7; top-down engineering 253, 256–7, 259; *see also* deontology; laws of robotics
rules of war engagement 93
R.U.R. (Čapek) 35
Rushton, A. 174
Russel, J.A. 216
Russell, Stuart 7, 22, 24
Rutherford, Ernest 23–4
Rzepka, Rafal 242

safety: alternative laws of robotics 408; childcare robots 158–61, 175, 177; European regulations 476–8; healthcare 194; implicit ethical agents 234; laws of robotics 406
Saffo, Paul 499
Sanders, J.W.: artificial agents 344; morality threshold 337; "On the Morality of Artificial Agents" 349, 353–7

Savage, L. J. 269
Scalia, Anton 501
Scassellati, B. 263
Schräder, P. 485
science and technology studies 350–1, 356–7
science fiction: androids 213; human-robot relationships 35, 80; themes 3–4; unethical behavior by machines 238; *see also* Asimov, Isaac; laws of robotics
scientific neutralism 391
Scott, G. E. 514
Scuola di Robotica 82
search and seizure law 495–6
Searle, John 236, 420–1, 451–3, 465, 466–7
Second Restatement of Trusts 427, 433
secure attachment 168, 171–2
security: childcare robots 160–1; long-term research priorities 25; short-term research priorities 21, 23
select groups of humans, benefit to 74
self awareness 97, 99, 422, 450, 455
self-destruction, robot 36, 38, 407
self-driving cars *see* autonomous vehicles
self-expansion model of love 218
self-interest and moral competence 228
self-regulation 228
Selman, Bart 13, 14
semantics 37, 43, 46, 293; *see also* language; terminology
Sen, Amartya 126, 127
senses 45–6, 164–5
sensorimotor condition and consciousness 400
sentience: consciousness 303; expanding moral circle 389; relational artifacts 111
sentimentality 166
service robots 201, 496
sex *see* love and sex robots
shallow care 129
shame and morality 62
Shanahan, Murray 304
Sharkey, A. 138, 187
Sharkey, Noel 138, 187, 493
Shaw, J. C. 294
shopping assistant robots 494
short-term ethical concerns 81, 82–3
short-term research priorities 20–3
Silverman v. United States 497
Simon, Herbert 233–4, 294
simple recurrent network *285,* 285–6, 290

Simroid robot 517
simulation, culture of 108–9
Singer, Irving 219
Singer, M. G. 55
Singer, Peter 388–9, 512
Singularity Institute 405
SIROCCO (System for Intelligent Retrieval of Operationalized Cases and Codes) 243
Skyrms, B. 61
slavery and slaves 122, 460–3
Sloman, Aaron 454
small world/grand world problem 269
smart grids 501
SmartPaint 330–1
Smit, I. 300, 480
Snell, Joel 217
social biofeedback 172
social change 65
social construction of technology 351
social good 150
social impact of technology 190
social interaction 392
social intuitionist model 304
social issues in machine learning 381
social justice 92
social meaning 492, 502
social mechanisms and machine ethics 262–4, 301
social organizations and censureship 342
social policy *see* law and policy
social referencing 172
social relationism: defined 385; expanding moral circle 388–90; moral landscape of conscious well-being 390–1; 'Turing Dream' 391–3; *vs.* moral realism 385–8
social relativism: sociality and consciousness 395–7; synthetic-biological humanoids 398–9
social robots 162, 225–9, 498–502
social utility: ethical reasoning 269; laws of robotics 405
society (terminology) 384
sociobiological model of morality 61
sociomateriality: dysmmetries 366; frames and accountability 371–2; human-machine interface 370–1; mutual constitutions 362–6; robots 366–70
sociotechnical systems: responsibility and agency 90–3; robotics as 88

software: computer ethics 339–40, 341; level of abstraction 329–30; verification of 22
soldiers 131
solipsism 394
solitude and robots 499–500, 503n9, 504n10
soul 102–3, 446–7
South Korea 176
Sparrow, L. 129, 133, 145
Sparrow, R. 129, 133, 145, 166
Special Law for Family Violence Criminal Prohibition (South Korea) 176
speciecism 121
species, robots as new 81
specificity problem 276–7
speculation and science fiction 3–4
speech, freedom of 441–2
speed: decision making 235; methodology for engineering robot ethics 296, 297
spy robots *see* surveillance
stabilization 25–6
stakeholders 48
standard deontic logic 294, 296
Stanford University 21, 26
state and state-transitions 324–9, 331–2, 337, 343, 357, 363
stoicism 251–2
Stone, Christopher 417, 423
strong artificial intelligence 96
subjectivity, imitation of 120–1
subordinate and superordinate robots 47
suffering: constitutional personhood 453, 455; moral patiency 516–9; moral realism *vs.* social relationism 389–90; moral status 99–100, 303; 'Turing Dream' 392; Turing Triage Test 101–2, 104
suicide 278–9
Sullins, John 356
Super, D. E. 196–7
superintelligence 7, 14, 26, 69–74
superogatory action 277
suprarational faculties 250–1, 263, 300–1
surgical robots 186, 201, 485
surveillance: direct surveillance 491–2; privacy 503n4; spy robots 493–6; "Teddycam" 488
"survival of the most moral" 255
symmetry principle 272
sympathy 101, 102, 104
Symposium (Plato) 219
syntax 293

synthetic biology 397–9
System for Intelligent Retrieval of Operationalized Cases and Codes (SIROCCO) 243

Tamagotchi fad 213
Tanaka, F. 164, 168
Tapus, A. 131
taxonomy of robotics 84
Taylor, Thomas 512, 519
technological determinism 356–7
technological imperative 49
"technological unemployment" 13
technology: addiction to 83; limits of 405; social impact 190
technoscience: dissymmetries 366; feminist theories 363–72
"Teddycam" 488
teleological objection 333
teleological theories 240
tele-presence and European regulations 482–4, 488, 489
telepresence robots 201
terminology: agents 320–1; care 192; care robots 187; human and humanity 44, 45; intelligence 80, 96–7; love 217–9; Method of Abstraction terms 322–3; moral realism and social relationism 384–5; responsibility 334–5; robotics 35; robots 81, 476; society 384; superintelligence 70; values 191–2; *see also* language; semantics
testing 242, 246–7, 277
themes of science fiction 3–4
therapeutic robots 138
thermostats, futuristic 330, 337–8
third-party doctrine 497
Thirteenth Amendment 450
Thompson, Charis 364
Thompson, E. 394–5
Thomson, Judith 284
three laws of robotics *see* laws of robotics
threshold of morality 337–8
Tikka, Heidi 370–1
Tmsuk robot 158–9
tolerance and the laws of robotics 51
top-down approaches: machine ethics 252–3, 256–9, 263; machine ethics engineering 252–3
Torrance, Steve 303, 311, 517

touch: childcare robots 164–5; feminist robotics 367–8; in healthcare 194
tracking children 158–9
traditional transitional objects 108
transhumanism 5
transition system *see* state and state-transitions
transitional objects 108
tree searching techniques 58
trespass violations 495
trial and error in bottom-up engineering 253
Trigg, Randall 372
Tronto, Joan: care values 195–6; moral elements 201, 205; value-sensitive design 189
trust: bottom-up engineering 256; care centered value sensitive design 203; care robots 131; healthcare practices 195; machine ethics 252; robot ethics 292; social robots 500
trustees, artificial intelligence as: judgment issues 432–7; legal question 427–8; real *vs.* artificial intelligence trustees 437–9; responsibility objection 428–32; scenario 424–6
truth 223
Truth-Teller program 243
Turing, Alan 56, 95, 96, 321, 419–20
'Turing Dream' 391–3
Turing Test: consciousness 400; and constitutional personhood 464; critics of 97; imitation game 419–20; intelligence 321; trustees, artificial intelligence as 434, 437; *see also* comparative Moral Turing Test; Moral Turing Test
Turing Triage Test 95, 96, 98–9, 100, 104
Turkle, Sherry 162–3, 217, 222
Two Treatises of Government (Locke) 460–1
typed variable 322

"uncanny valley" hypothesis 131
undesired consequences 26
Unimate 34
United Kingdom 175, 176–7, 493
United Nations: Convention for Certain Conventional Weapons 13; Convention on the Rights of the Child 159, 175; human rights law 481; privacy rights 486
United States child neglect legislation 175–6
United States v. Miller 497
United States v. Place 495

United States v. Taborda 495, 503n4
United States v. United States District Court 494
universality: ethical considerations 527; Kantian machine ethics 276; maxims 244
unmanned aerial vehicles *see* drones
unstructured decision making 50
Upheavals of Thought (Nussbaum) 118
users: moral agency 354; responsibility of 479
utilitarianism: artificial moral agents 58; capabilities approach and 126, 127; deontology compared to 275; deterrence theory of punishment 432; emotions 118; ethical reasoning 269–70; machine ethics 240–4; methodology for engineering robot ethics 294–5; moral theory 54–5; top-down engineering 257–8; and virtue ethics 260

validity 21, 22–3, 24–5
Vallor, S. 187
valuation of individual humans 38
values: care robots, design of 125–6; computer systems 65–7; defining 191–2; healthcare and care ethics 193–8; machine ethics 251; patient lifting robot design 207–8; phenomenal-valuational holism 390; superintelligence 73; top-down engineering 257
value-sensitive design: care centered value-sensitive design 198, 202–8; care robots 188–9; value, definition of 192
variables 322
variance 221
Velmans, Max 514
Verbeek, P. 190, 191
verification 21, 22, 24
A Vindication of the Rights of Brutes (Taylor) 512, 519
A Vindication of the Rights of Men (Wollstonecraft) 512
A Vindication of the Rights of Woman (Wollstonecraft) 512, 519–20
virtual pornography 488
virtue ethics 60; capabilities approach and care robots 153; care robots 143–6; implicit ethical machines 234; machine ethics 241; machine ethics engineering 260–1
voice recognition technology 494
volitional decision making 306–9

wages 13
wait mode and the laws of robotics 39–40
Wallach, Wendell: APACHE system 131; behavioral standards 23; care robots 134; fitness functions and consciousness 310; hybrid development approach 300; LIDA 301–2; moral judgments and moral decisions 307; "Why Machine Ethics?" 480
weak artificial intelligence 96
weapons 21, 92–3, 486–7
web-based knowledge discovery systems 242
webbots 330, 338, 479
Weizenbaum, Joseph 108, 494
well-being 390
Westin, Alan 500
'whip lashing' 366
"Why You Can't Make a Computer That Feels Pain" (Dennett) 518–9, 527
Wiener, Norbert 406

Wilson, Melanie 194
Winch, Peter 100n11, 103
Wittgenstein, L. 100, 393
Wolf, Marty 354
Wollstonecraft, Mary 512, 519–20
Wolmark, Jenny 367
Woolley, Richard 24
working animals 122
World Health Organization 173–4, 193
World Medical Association 483
WowWee RobotSapien V2 498
WowWee Rovio 496, 498
wrongdoing and autonomous moral agents 90

xenophobia 482

Yujin Robotics 165

zeroth law 44, 45, 47, 259

Printed and bound by CPI Group (UK) Ltd, Croydon, CR0 4YY
22/09/2024
01036666-0003